D0889966

Computer Structures: Readings and Examples

McGraw-Hill computer science series

RICHARD W. HAMMING
Bell Telephone Laboratories

EDWARD A. FEIGENBAUM
Stanford University

Computer Structures: Readings and Examples

C. Gordon Bell
Professor of Computer Science and Electrical Engineering
Carnegie-Mellon University

Allen Newell
University Professor
Carnegie-Mellon University

McGraw-Hill Book Company
New York   St. Louis   San Francisco   Düsseldorf
London   Mexico   Panama   Rio de Janeiro
Singapore   Sydney   Toronto

To Brigham, Laura,
Paul

Computer Structures: Readings and Examples

Copyright © 1971 by McGraw-Hill, Inc. All rights reserved. Printed in the United States of America. No part of this publication may be reproduced, stored in a retrieval system, or transmitted, in any form or by any means, electronic, mechanical, photocopying, recording, or otherwise, without the prior written permission of the publisher.

Library of Congress Catalog Card Number 75-109245

ISBN 07-004357-4

567890 HDBP 75432

This book was set in News Gothic by Graphic Services, Inc., printed on permanent paper by Halliday Lithograph Corporation, and bound by The Book Press, Inc. The designer was Elliot Epstein; the drawings were done by John Cordes, J. & R. Technical Services, Inc. The editors were Richard Dojny and J. W. Maisel. William P. Weiss supervised production.

Preface

The structures that we call computer systems continue to grow in complexity, in size, and in diversity. This book is linked firmly to the nature of this growth. The book is about the upper levels of computer structure: about instruction sets, which define a computer system at the programming level; and about organizations of processors, memories, switches, input-output devices, controllers, and communication links, which provide the ultimate functioning system. These levels are just emerging into well-defined systems levels—with developed symbolic techniques of analysis and synthesis and accumulated engineering know-how, all expressed in a crystallized representation. These aspects of computer systems have always existed, of course, but only in rudimentary form. The classical four-box picture of a computer (arithmetic unit, memory, input-output, and control) is certainly an effective organization of components to process information. But multiple-processors hierarchies of memories and remote communications force the top level of organization into a distinct level, requiring analysis and rational design. Similarly, the 25 instructions of the IBM 701 computer (developed around 1953) is certainly an instruction set—indeed one worthy of study. But processors with dozens of registers and almost unlimited logical circuitry, again force the instruction set to become a topic of rational analysis and design.

This book is tied to the emergence of these upper levels of organization: eight years ago (a computer engineer's half dozen) would have been too early to write this book; eight years hence would be too late. Eight years ago the diversity and complexity of computer structures was not sufficient to justify the attention this book provides. This book would have been too thin. Eight years hence textbooks will exist that treat these levels systematically. This book will then appear too descriptive.

But right now, as these aspects of computer structure are emerging, and with systematic treatment still precluded, there is a need to make available material on these levels for systematic reference and study. Our choice has been to present a large set of examples, which illustrate the various design options and structural possibilities, both in instruction sets and in overall configurations. These examples are descriptions of actual computer systems, taken from the technical literature or from technical reports and manuals. Descriptions of actual systems are to be much preferred over idealized abstractions. The latter can reflect the real issues only after successful systematization.

Not only are the chapters about actual computers, they present much detail. The complexity of computers resides in part in their size and the multiplicity of their parts—e.g., to their having 200 instructions rather than 20, or having to service 50 Teletypes rather than 2. It seems essential to describe computer systems in their entirety, rather than via simplified vignettes. Again, this view stems from the existing state of the art. Eight years hence, it will not necessarily hold.

We fall from grace on all the above principles, providing occasionally descriptions of paper machines and partial descriptions of partial systems. But our feeling that detail and reality is important remains. This is why this book is so large; and fit for study rather than for reading.

The book presents a large number of examples. Variation needs to be presented along all the major dimensions that instruction sets and system configurations currently exhibit. Thus, as a glance at the table of contents will show, the examples in the book are hardly picked at random. The variation is empirical. It exists in the population of computers that have actually been built. This characteristic of the book stems, again, from our assessment that the upper levels of computer structure are still in an essentially descriptive and empirical state of development. However, as the book documents, ample variation occurs in existing computer systems. The evidence presented here should finally lay to rest the remarks—once echoed almost universally and still heard occasionally—that nothing has happened in computer structure since the von Neumann machine.

Dimensions of variations imply a framework, for dimensions do not by themselves arise from a population of systems. They require the aid, witting or not, of a conceptual framework. As the first three chapters of the book testify, we have most wittingly created a framework, and have had no hesitation in imposing it throughout the book. However, in keeping with our view already expressed, this framework is primarily descriptive. It has come inductively from the common lore, from our own experiences as designers, and from the effort of putting this book together. This attempt at systematization has given rise to two notations: one for instruction sets (ISP) and the other for configurations of major components (PMS). But, again, these notations are primarily descriptive.

So much for what the book actually tries to provide. What are our goals for it? The first is educational. There are three distinct populations of professionals whose education is to be served by this book: the computer engineer, who will design physical computer systems; the computer scientist, who is concerned primarily with the programming level and with various abstract views of information processing; and the electrical engineer, who sees computer systems simply as one part of a larger technology.

For all of these, we see no sense in talking of elementary versus advanced treatments of computer structure. There is surely ''less'' versus ''more,'' but consistent with our view of the current art, no vertical stratification of education is possible in instruction sets and device configurations. It is sufficient, in the present day, for these aspects of computer systems to become accepted as worthy of study in their own right.

This book will hardly make easy fare for undergraduate students, who do not have an instructor somewhat skilled in the art that is being taught. However, this book is meant for study. A good instructor can, we feel, develop an excellent course (or part thereof) in computer structures, taking this book as the basic material. In addition to the three introductory chapters, Chapter 5 (on the DEC PDP-8), by providing a complete example of a computer system with descriptions at all systems levels, helps to tie the aspects of computer structure discussed in this book to the view students will pick up from a traditional course in logical design.

It goes without saying that for the computer engineer and designer, the material of this book should be fully assimilated. In designing a new computer system, or subsystem thereof, he should be familiar with all that this book has to offer—the design choices, the structural variations possible, the experiments of the past and the design needs they attempted to satisfy. Given that systematic analysis does not yet exist, there is no substitute for extensive, critical understanding of the existing examples of designed systems. We assume the student of computer engineering comes to this book with a working knowledge of logical design. He should find it possible to realize many of the systems described in this book at the next lower levels of logic structure.

For the computer scientist, the levels of computer structure discussed in this book constitute a substantial part of what he should know about the physical devices that underlie his science. As we pass downward from these levels to lower ones—to register-transfer systems, sequential logic circuits, combinatory circuits, continuous circuits and on down—the relevance of each level gradually fades. The levels of this book, along with the register-transfer level constitute the main aspects of computer structure that the computer scientist must understand. It does not matter that they are, as yet, basically empirical and descriptive. The computer scientist undoubtedly will not be able to carry through the design of the systems described in this book in terms of the lower logic levels, but this is not necessary for an appropriate grasp of these upper levels of computer structure. Indeed, this is what it means for distinct systems levels to exist.

For the electrical engineer, this book undoubtedly presents more examples than he cares to know (or needs to). But an appropriate sampling, plus the overview presented in the first three chapters, is appropriate to give him some insight into the elaborate growth that has occurred on top of the basic digital technology created within electrical engineering.

The student of systems engineering may also find the material presented here useful, as an example of a class of complex systems which has evolved several distinct levels of representation. Again, the book undoubtedly presents too massive a dose of detail for him, but the overview in the first chapters, plus a sampling throughout the space of computer systems, should prove highly instructive.

We have goals for the book in addition to the educational ones. We think the book can serve as a useful reference for the practicing computer engineer. The time is past when every computer engineer knows about all computer systems because he has lived through all of computer history. That position is now reserved for those of us who are past forty (and still active). For the rest, a source book that provides the cumulated design experience of the field is a useful substitute, especially so if it contains enough detail so that a designer can reasonably evaluate the actual computer systems that embody a particular design alternative.

Behind the goal of the book as a guide for the practicing computer designer lies the feeling that the field of computer engineering needs to develop a sense of history and of looking to the past for guidance. The fantastic advance in basic logic technology—in speed, cost, and reliability— makes each day seem an absolutely new one. But, of course, it is not. Many alternative designs have been tried out in past systems, in ways relevant to current design. Thus, we have the goal of saving some of the past in a form accessible to the future needs of computer design. This goal is mixed with a certain archival feeling. Many of the systems in this book have never been documented, other than in manuals and various elementary how-to programming books.

A final goal comes from our feelings as computer scientists that the variety of computer systems is a phenomena worthy of study in its own right. This book carries, therefore, an invitation to taxonomy—to asking how to classify the diversity of forms of computer systems that are coming into existence. Taxonomic endeavors usually take place in a field of natural systems, particularly biological systems. It may seem strange that a domain of artificial systems calls for taxonomic activity. But the demand for empirical classification exists whenever there is a population of significant size and rich structure. Rudimentary classification efforts have occurred for many populations of artifacts—for ships, for aircraft, for houses. This book should amply confirm that computer systems are complex and diverse enough—and undergoing enough continual proliferation and evolution—to command significant taxonomic endeavor.

Enough is said in the first two chapters about the new notations introduced in the book, so that nothing substantive need be added here. We apologize for inflicting new notation on the reader. We feel that good notations are really quite important for the aspects of computer structure described in this book. Much would be gained by the whole field of computers—by users, programmers, engineers, planners, buyers, sellers, manufacturers, students, and scientists—if relatively uniform notations came into common use. Although we have no illusions about the perfection of the notations we have introduced, we would be most happy if they cause a rise in concern for standard notations and nomenclature.

A large number of distinct systems are described in substantial detail. We have redescribed many of the systems in the common notation introduced in the book. The accuracy of all these descriptions is a major problem. Even where the papers are reproduced from the literature, this problem of accuracy remains—although then it is not ours alone. Even though we have taken pains to obtain accurate information on the systems and to portray them faithfully in our various descriptions and figures, there is no way we can be responsible for their ultimate accuracy. The PMS and ISP figures, in particular, cannot be guaranteed to be accurate representations of the systems they purport to describe. Ultimately, one would like to have simulation languages for such notations and to verify (up to the usual criteria of a debugged program) that a system given by, say, an ISP description, simulates the behavior of the target machine. But that day is still far off.

Our most fundamental acknowledgment is to the contributors to this volume, not only for the articles they have written, but for the computers they have designed and built, thereby creating a population of fascinating artifacts worthy of study. An additional reason for reprinting their articles rather than simply describing their computer systems is the importance of having available the views of the designers themselves about the nature of their systems.

The research on the basic ideas underlying the notations was supported by Advanced Research Projects Agency of the Office of the Secretary of Defense (F 44620-67-C-0058) and is monitored by the Air Force Office of Scientific Research.

We would like to extend an acknowledgment to the organizations that have produced all of these computers, oftentimes it would seem in defiance of the laws of economics. Perhaps, as the old saw has it, a computer manufacturer is simply a computer's way of breeding another computer. This might account for the tenacity shown by computer manufacturers in spawning the vast numbers of computer systems that provide our field of study. Within this general acknowledgment, we would like to extend a very specific one to all the people in these organizations who helped make information available to us—the manuals, photographs, dates, etc., that this book has demanded in such great quantity.

We are indebted to the students who have read and criticized the various PMS and ISP figures: Richard Dove, Wayne Kohl, Michael Knudsen, Paul Mobus, and Charles Pfferkorn. Ken Fitzgerald and Anita Jones of IBM were kind enough to read the introduction to the IBM System/360.

Professor David L. Parnas initially reviewed the text and contents, thus providing many helpful suggestions. Our other colleagues, especially Professors Angel Jordan, Alan Perlis, Herbert Simon and Everard M. Williams deserve a special thanks for their patience and encouragement.

Finally, we would like to thank those who were a part of the machine that assembled the book: the editors of McGraw-Hill; Mrs. Mary Ross who assembled the bibliography, figures, and contributor articles; Mrs. Mildred Sisko who typed the PMS and ISP Appendix; and especially Mrs. Dorothy Josephson who not only typed nearly all drafts of the book, but also the final PMS figures, and ISP Appendices.

C. Gordon Bell
Allen Newell

Acknowledgments

R. H. Allmark and J. R. Lucking: Design of an Arithmetic Unit Incorporating a Nesting Store, *Proceedings of the International Federation of Information Processing Congress 1962,* pp. 694–698, North Holland Publishing Co., Amsterdam, Holland, by permission from American Federation of Information Processing Societies (AFIPS), Spartan Books, Washington, D.C.

R. L. Alonso, H. Blair-Smith, and A. L. Hopkins: Some Aspects of the Logical Design of a Control Computer, A Case Study, *Transactions on Electronic Computers,* vol. EC-12, no. 6, pp. 687–697, December, 1963, by permission of the authors and the Institute of Electrical and Electronics Engineers (IEEE).

James P. Anderson, Samuel A. Hoffman, Joseph Shifman, and Robert J. Williams: D825—A Multiple Computer System for Command and Control, *Proceedings of the AFIPS Fall Joint Computer Conference,* vol. 22, pp. 86–96, 1962, by permission from AFIPS, Spartan Books, Washington, D.C. The authors acknowledge:

> The authors wish to acknowledge the outstanding efforts of their many colleagues at Burroughs Laboratories who have contributed so well and in so many ways to all stages of D825 design, development, fabrication, and programming. It would be impossible to cite all of these efforts. The authors also wish to acknowledge the contributions of Mr. William R. Slack and Mr. William W. Carver, also of Burroughs Laboratories. Mr. Slack has been closely associated with the D825 from its original conception to its implementation in hardware and software. Mr. Carver made important contributions to the writing and editing of this paper.

George H. Barnes, Richard M. Brown, Maso Kato, David J. Kuck, Daniel L. Slotnick, and Richard A. Stokes: The ILLIAC IV Computer, *Transactions on Computers,* vol. C-17, no. 8, pp. 746–757, August 1968, by permission of the authors and the IEEE. The authors acknowledge:

> This work was supported in part by the Department of Computer Science, University of Illinois, Urbana, Illinois, and in part by the Advanced Research Projects Agency as administered by the Rome Air Development Center, Griffiss Air Force Base, Rome, New York, under Contract USAF 30 (602)4144.
>
> The authors are pleased to acknowledge their indebtedness to the group at the Westinghouse Electric Corporation that initiated the parallel computer effort. The work of W. C. Borck, A. B. Carroll, J. R. Hudson, W. H. Leonard, R. C. McReynolds, and G. Shapiro formed the basis for the subsequent efforts. Of particular importance is the work of J. G. Gregory in tuning the conceptual design to the real world of technology.

Theodore R. Bashkow, Azra Sasson, and Arnold Kronfeld: System Design of a FORTRAN Machine, *Transactions on Electronic Computers,* vol. EC-16, no. 4, pp. 485–499, August 1967, by permission of the authors and the IEEE. The authors acknowledge:

> This research is supported by the Air Force Office of Scientific Research Contract AF19(628)—2798.

G. A. Blaauw and F. P. Brooks, Jr.: The Structure of System/360, Part I—Outline of the Logical Structure, *IBM Systems Journal,* vol. 3, no. 2, pp. 119–135, 1964, by permission from the *IBM Systems Journal.*

Erich Bloch: The Engineering Design of the Stretch Computer, *Proceedings of the Eastern Joint Computer Conference, 1959,* pp. 48–58, by permission of the author and the Institute of Electrical and Electronics Engineers. The author acknowledges:

> The efforts and contributions of many people have gone into the engineering design of the Stretch computer. To mention all would be impossible. However, the following individuals and their groups were responsible for the units indicated; Mr. R. T. Blosk for the Instruction Unit, Mr. J. F. Dirac for the Look-ahead Units, Messrs. J. A. Hipp and O. L. MacSorley for the Arithmetic Units, and Mr. L. O. Ulfsparre for the Memory Bus. The Systems Development was under the guidance of Messrs. S. W. Dunwell and R. E. Merwin.

Arthur W. Burks, Herman H. Goldstine, and John von Neumann: Preliminary Discussion of the Logical Design of an Electronic Computing Instrument, "Collected Works of John von Neumann," vol. V, pp. 34–79, General Editor: A. H. Taub, Macmillan Company, by permission from Pergamon Press, New York, 1963. The authors acknowledge:

> This report has been prepared in accordance with the terms of Contract W-36-034-0RD-7481 between the Research and Development Service, Ordnance Department, U.S. Army and the Institute for Advanced Study.
>
> The authors wish to express their thanks to Dr. John Tukey, of Princeton University, for many valuable discussions and suggestions.

John W. Carr III: UNIVAC Scientific (1103A) Instruction Logic, pp. 77–83; IBM 650 Instruction Logic, pp. 93–98; Instruction Logic of the Soviet

Strela (Arrow), pp. 111–115; Instruction Logic of the MIDAC, pp. 115–121, chap. 2, Programming and Coding, "Handbook of Automation, Computation, and Control," vol. 2, edited by Eugene M. Grabbe, Simon Ramo, and Dean Wooldridge, Copyright © 1959 John Wiley & Sons, Inc., New York, reprinted by permission.

J. Presper Eckert, Jr., James R. Weiner, H. Frazer Welsh, and Herbert F. Mitchell: The UNIVAC System, *American Institute of Electrical Engineers-Institute of Radio Engineers Conference*, pp. 6–16, December, 1951, by permission of the authors and the IEEE. The authors acknowledge:

The UNIVAC System has been an over-all company project and hundreds of people have participated. It is, therefore, difficult to acknowledge the contributions of individuals. However, special mention must be made of the contributions of Mr. H. Lukoff, Mr. E. I. Blumenthal, Mr. L. D. Wilson, and Mr. J. D. Chapline, Jr. To the Census Bureau a great debt of gratitude is owed for their continuous support of the project.

W. S. Elliott, C. E. Owen, C. H. Devonald, and B. G. Maudsley: The Design Philosophy of Pegasus, A Quantity-production Computer, *Proceedings of the Institution of Electrical Engineers*, London, Pt. B, vol. 103, Supplement 2, pp. 188–196, 1956, by permission of the Institution of Electrical Engineers. The authors acknowledge:

The authors would like to acknowledge the contributions that Mr. C. Strachey and Dr. D. B. Gillies, of the National Research Development Corporation, and Dr. J. M. Bennett and Mr. T. G. H. Braunholtz, of Ferranti, Ltd., made to the logical design of Pegasus: particular thanks are due to Mr. C. Strachey for originating the order code.

They also thank Ferranti, Ltd., and the National Research Development Corporation for permission to publish the paper.

R. R. Everett: The Whirlwind I Computer, *Review of Electronic Digital Computers, Joint Computers American Institute of Electrical Engineers-Institute of Radio Engineers Conference*, pp. 70–74, February, 1952, by permission of the author and the IEEE.

Thomas W. Kampe: The Design of a General-purpose Microprogram-controlled Computer with Elementary Structure, Institute of Radio Engineers, *Transactions on Electronic Computers*, vol. EC-9, no. 2, pp. 208–213, June, 1960, by permission of the author and the IEEE. The author acknowledges:

The author wishes to thank his co-designers, R. Compton and T. Hayata, for their assistance during the design of the SD-2 computer and for their suggestions on this paper.

T. Kilburn, D. B. G. Edwards, M. J. Lanigan, and F. H. Sumner: One-level Storage System, *Institute of Radio Engineers Transactions*, vol. EC-11,

no. 2, pp. 223–235, April, 1962, by permission of the authors and the IEEE. The authors acknowledge:

The authors gratefully acknowledge the contributions made to this work by all members of the Atlas computer team at both Manchester University and Ferranti Ltd.

B. W. Lampson, W. W. Lichtenberger, and M. W. Pirtle: A User Machine in a Time-sharing System, *Proceedings of the Institute of Electrical and Electronics Engineers*, vol. 54, no. 12, pp. 1766–1774, December, 1966, by permission of the authors and the IEEE. The authors acknowledge:

The work for this paper was supported in part by the Advanced Research Projects Agency, Department of Defense, Contract SD-185.

The software portion of the system was designed and written in part by L. P. Deutsch, who is entitled to equal credit with the authors for the ideas in this paper. L. Barnes also contributed significantly to the final result.

M. Lehman: A Survey of Problems and Preliminary Results Concerning Parallel Processing and Parallel Processors, *Proceedings of the Institute of Electrical and Electronics Engineers*, vol. 54, no. 12, pp. 1889–1901, December, 1966, by permission of the author and the IEEE. The author acknowledges:

This paper reports on a group activity in which each individual member had his own specific assignments and in addition participated in regular discussions on all aspects of the project. Credit is therefore due to all members of the group which, during the period covered by the contents of this paper, included G. C. Driscoll, J. M. Lee, A. P. Mullery, J. L. Rosenfeld, H. P. Schlaeppi, and M. Weitzman. I should also like to express my sincere thanks to Dr. H. A. Ernst for the constructive criticism, advice, and encouragement offered during preparation of this paper. My sincere thanks are also due to members of the Graphics and Design Department at the Thomas J. Watson Research Center, and in particular to G. Massi and Mrs. M. J. LaMarre for their preparation of the charts and figures. Last, my thanks to Mrs. J. Galto for her infinite patience in the repeated retypings of the manuscript.

A. L. Leiner, W. A. Notz, J. L. Smith, and A. Weinberger: PILOT, The NBS Multicomputer System, *Proceedings of the Eastern Joint Computer Conference*, 1958, pp. 71–75, by permission of the authors and the IEEE. The authors acknowledge:

The authors wish to acknowledge the valuable contributions of their colleagues H. Loberman and W. Youden, who helped to develop the logical design and programming procedures for this system.

William Lonergan and Paul King: Design of the B 5000 System, *Datamation*, vol. 7, no. 5, pp. 28–32, May, 1961, by permission of, published and Copyrighted © 1961 by F. D. Thompson Publications, Inc., Greenwich, Conn.

Richard E. Monnier, Thomas E. Osborne, and David S. Cochran: The HP Model 9100A Computing Calculator. This chapter is a compilation of three articles: A New Electronic Calculator with Computerlike Capabilities, by Richard E. Monnier, pp. 3–9; Hardware Design of the Model 9100A Calculator, by Thomas E. Osborne, pp. 10–13; and Internal Programming of the 9100A Calculator, by David S. Cochran, pp. 14–16, which appeared in the *Hewlett-Packard Journal*, volume 20, no. 1, September, 1968, by permission of the *Hewlett-Packard Journal*.

R. E. Porter: The RW-400—A New Polymorphic Data System, *Datamation*, vol. 6, no. 1, pp. 8–14, January/February, 1960, by permission of, published and Copyrighted © 1960 by F. D. Thompson Publications, Inc., Greenwich, Conn.

J. C. Shaw, A. Newell, H. A. Simon, and T. O. Ellis: A Command Structure for Complex Information Processing, Western Joint Computer Conference 1958, by permission of the authors and the IEEE.

W. Y. Stevens: The Structure of System/360, Part II—System Implementations, *IBM Systems Journal*, vol. 3, no. 2, pp. 136–143, 1964, by permission from the *IBM Systems Journal*.

James E. Thornton: Parallel Operation in the Control Data 6600, *Proceedings of the AFIPS Fall Joint Computer Conference*, Pt. II, vol. 26, pp. 33–40, 1964, by permission from AFIPS, Spartan Books, Washington, D.C.

W. L. van der Poel: ZEBRA, A Simple Binary Computer, *Proceedings of an International Conference on Information Processing*, Paris, UNESCO House, June, 1959, pp. 361–365, by permission from AFIPS, Spartan Books, Washington, D.C.

Helmut Weber: A Microprogrammed Implementation of EULER on IBM System/360 Model 30, *Communications of the Association for Computing Machinery*, vol. 10, no. 9, pp. 549–558, September, 1967, Copyright © 1967 Association for Computing Machinery, Inc., by permission of the author and the Association for Computing Machinery, Inc. The author acknowledges:

> I wish to thank Jack Carman, who wrote the I/O Control Program and the Operating System linkage for the EULER system and Miss Sheila Morrison who helped prepare the figures. I am also grateful for the valuable criticism offered by the referee, W. C. McGee, as well as by Professor N. Wirth and E. Satterthwaite.

J. H. Wilkinson: The Pilot ACE, by permission from Automatic Digital Computation, pp. 5–14, National Physical Laboratory, Teddington, England, March 25–28, 1953.

M. V. Wilkes and J. B. Stringer: Micro-programming and the Design of the Control Circuits in an Electronic Digital Computer, *Proceedings of the Cambridge Philosophical Society*, Pt. 2, vol. 49, pp. 230–238, April, 1953, by permission of the authors and the Cambridge Philosophical Society, Cambridge, England. The authors acknowledge:

> The authors wish to express their thanks to Mr. A. L. Freedman and Mr. W. Renwick for assisting them in clarifying a number of points, and to Professor D. R. Hartree, F.R.S., for his generous help with the preparation of the paper.

Joseph E. Wirsching: NOVA: A List-oriented Computer, *Datamation*, vol. 12, no. 12, pp. 41–43, December, 1966, by permission of, published and Copyrighted © 1966 by F. D. Thompson Publications, Inc., Greenwich, Conn. The author acknowledges:

> This work was performed under the auspices of the U.S. Atomic Energy Commission.

Several organizations have contributed to the writing and production of this book by giving us permission to use material from their publications. In many cases they have also supplied us with original copies. We have credited their text, tables, pictures, and diagrams when they are used. This cooperation has been invaluable. The specific organizations are:

Adams's Associates: *Computer Characteristics Quarterly.* (Adams, 1966–1968)

Computers and Automation magazine

Control Data Corporation, 8100 34th Avenue South, Minneapolis, Minnesota

Datamation magazine

Digital Equipment Corporation, 146 Main Street, Maynard, Massachusetts

Hewlett-Packard Company, 1501 Page Mill Road, Palo, California

International Business Machines Corporation, White Plains and Poughkeepsie, New York

Massachusetts Institute of Technology, Cambridge, Massachusetts

National Science Foundation

Olivetti Underwood Corporation, 1 Park Avenue, New York, New York

Scientific Data Systems, 1649 Seventeenth Street, Santa Monica, California

Contributors

R. H. Allmark
R. L. Alonso
James P. Anderson
Theodore R. Bashkow
George H. Barnes
G. A. Blaauw
H. Blair-Smith
Erich Bloch
F. P. Brooks, Jr.
Richard M. Brown
Arthur W. Burks
John W. Carr III
David S. Cochran
C. H. Devonald
D. B. G. Edwards
J. Presper Eckert, Jr.

W. S. Elliott
T. O. Ellis
R. R. Everett
Herman H. Goldstine
Samuel A. Hoffman
A. L. Hopkins
Thomas W. Kampe
Maso Kato
T. Kilburn
Paul King
David J. Kuck
Arnold Kronfeld
B. W. Lampson
M. J. Lanigan
A. L. Leiner
M. Lehman

W. W. Lichtenberger
William Lonergan
J. R. Lucking
B. G. Maudsley
Herbert F. Mitchell
Richard E. Monnier
W. A. Notz
Thomas E. Osborne
C. E. Owen
M. W. Pirtle
R. E. Porter
Azra Sasson
J. C. Shaw
Joseph Shifman
H. A. Simon
Daniel L. Slotnick

J. L. Smith
W. Y. Stevens
Richard A. Stokes
J. B. Stringer
F. H. Sumner
James E. Thornton
W. L. van der Poel
John von Neumann
Helmut Weber
A. Weinberger
James R. Weiner
H. Frazer Welsh
M. V. Wilkes
J. H. Wilkinson
Robert J. Williams
Joseph E. Wirsching

Contents[1]

---

[1]This is a "virtual" contents, which means that because many of the computers are relevant to more than one part and section, we have used italic type to indicate a nonsequential mapping for computers placed out of "physical" order. The reader might read (reference) the book according to the virtual order.

Part 3 The Instruction-set Processor Level: Variations in the Processor

Part 1

The structure of computers

Chapter 1

This book presents many examples of computer systems. It presents them in enough detail so that meaningful engineering study and analysis are possible. Most of these examples are presented by using the original descriptions of them in the technical literature. Others have been redescribed by us, especially where the original descriptions existed only in technical manuals. In both cases there are considerable discussion and analysis of the computer structures: what problems they were intended to solve, what solutions were adopted, and how these solutions have fared. Yet the emphasis has remained on detailed descriptions precise enough so that the systems themselves are available for independent study.

Why should one want to produce such a book? Collections of reprintings from the technical literature are common in many science and engineering fields, e.g., "Programming Systems and Languages" [Rosen, 1967]. We have departed from this traditional exercise in two ways, both of which seem important to us. First, we have presented substantial amounts of detail: in effect, block diagrams of computer structures and the equivalents of programming manuals. These constitute neither good reading nor a way of communicating the "essential ideas" in the field. Second, we have introduced a system of notation and have used it not only in the parts we ourselves have written but also to provide additional (sometimes redundant) descriptions of computer systems in the reprinted articles. Why should there be a book like this? The reasons are several and require some background discussion.

Computer systems

Computer systems are one example of man's more complex artificial systems.[1] They have existed as successful engineering products long enough to undergo radical evolution and to give rise to a number of basic, unique technologies. They are sufficiently complex that they have given rise to a science, that is, to a continuing, institutionalized endeavor to understand what sort of beast has been brought forth.[2] Our fundamental interest is in the development of this science and technology of computers (one of us also likes to build computers). To understand why this particular book seems to us to be the right way to push this development at this particular time requires characterizing the current state of computer-systems technology.

A computer system is complex in several ways. Figure 1 shows the most important. There are at least four levels of system description, possibly five, that can be used for a computer. These are not alternative descriptions in the sense that anything said one way can be said another. On the contrary, each level arises from abstraction of the levels below it. Each does a job that the lower levels could not perform because of the unnecessary detail they would be forced to carry around.

A system (at any level) is characterized by a set of components, of which certain properties are posited, and a set of ways of combining components to produce systems. When formalized appropriately, the behavior of the systems is determined by the behavior of its components and the specific modes of combination used.

---

[1] We need not argue that they are his most complex system. That view is myopic. Setting aside quasi-natural systems, such as cities and economies, it is still the case that a modern aircraft carrier is more complex than a modern computer by any reasonable measure.
[2] Here uniqueness can be claimed, perhaps, since few other artifactual systems (again, excluding the quasi-natural ones) provide new phenomena that require sustained scientific investigation to understand them. There certainly is no science of aircraft carriers. But there is a computer science.

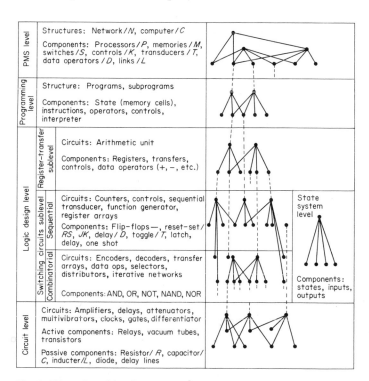

Fig. 1. Hierarchy of levels: computer structure.

Elementary circuit theory is an almost prototypic example. The components are R's, L's, C's, and voltage sources. The mode of combination is to run wires between the terminals of components, which corresponds to an identification of current and voltage at these terminals. The algebraic and differential equations of circuit theory provide the means whereby the behavior of a circuit can be computed from the properties of its components and the way the circuit is constructed.

There is a recursive feature to most system descriptions. A system, composed of components structured in a given way, may be considered a component in the construction of yet other systems. There are, of course, some primitive components whose properties are not explicable as the resultant of a system of the same type. For example, a resistor is not to be explained by a subcircuit but is taken as a primitive. Sometimes there are no absolute primitives, it being a matter of convention what basis is taken. For example, one can build logical design systems from many different primitive sets of logical operations (AND and NOT, NAND, OR and NOT, etc.).

A system level, as we have used the term in Fig. 1, is characterized by a distinct language for representing the system (that is, the components, modes of combination, and laws of behavior). These distinct languages reflect special properties of the types of components and of the way they combine. Otherwise, there would be no point in adopting a special representation. Nevertheless, these levels exist in the system analyst's way of describing the same physically existing system. The fact that the languages are highly distinct makes it possible to be confident about the existence of different system levels. Where we are fuzzy, as in the existence of an additional intermediate level, it is because new representations have not yet congealed into distinct formal languages. As we noted, within each level there exists a whole hierarchy of systems and subsystems. However, as long as these are all described in the same language, e.g., a subroutine hierarchy, all given in machine-assembly language, they do not constitute separate system levels.

With this general view, let us work through the levels of computer systems, starting at the bottom. Each level in Fig. 1 actually has two languages or representations associated with it: an algebraic one and a graphical one. These are isomorphic to each other, the same entities, properties, and relations being given in both.

The lowest level in Fig. 1 is the *circuit level*. Here the components are R's, L's, C's, voltage sources, and nonlinear devices. The behavior of the system is measured in terms of voltage, current, and magnetic flux. These are continuously varying quantities associated with various components, and so there is continuous behavior through time. The components have a discrete number of terminals, whereby they can be connected to other components. Figure 2 shows both an algebraic and graphical description of an inverter circuit, as well as an algebraic and graphical description of its behavior. We note that its structure is specified first as a circuit (a directed graph), with symbols for the arcs and nodes. The particular circuit still is an abstraction because the transistor Q1, the resistor R, and the stray capacitors C_s are given only token values. The structure can be described symbolically by first writing the relationship describing each of the components (i.e., Ohm's law, Faraday's law, etc.) and then the equation which describes the interconnection of the components (i.e., Kirchhoff's laws). We observe the behavior of the circuit (probably using an oscilloscope) by applying an input $e_i(t)$ and observing an output $e_o(t)$. Alternatively, if we solve the equations which specify the structure, we obtain expressions which describe the behavior explicitly.

The circuit level is not in fact the lowest level that might be used in describing a computer system. The devices themselves require a different language, either that of electromagnetic theory or of quantum mechanics (for the solid-state devices). It is usually an exercise in a course on Maxwell's equations to show that circuit theory can be derived as a specialization under appropriately restricted boundary conditions. Actually, even at its level of abstraction, circuit theory is not quite adequate to describe computer technology since there are a number of mechanical devices which must be represented. Magnetic tapes and drums are most likely

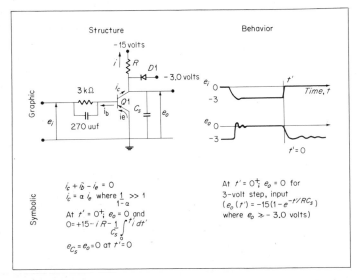

Fig. 2. Electronic-circuit level: inverter circuit.

to come to mind first, but card readers, card punches, and Teletype terminals are other examples. These devices obey laws of motion and are analyzed in units of mass, length, and time.

The next level is the *logic level*. It is unique to digital technology, whereas the circuit level (and below) is what digital technology shares with the rest of electrical engineering. The behavior of a system is now described by discrete variables which take on only two values, called 0 and 1 (or + and −, true and false, high and low). The components perform logical functions: AND, OR, NOT, NAND, etc. Systems are constructed in the same way as at the circuit level, by connecting the terminals of components, which thereby identify their behavioral values. The laws of boolean algebra are used to compute the behavior of a system from the behavior and properties of its components.

The previous paragraph described *combinatorial circuits* whose outputs are directly related to the inputs at any instant of time. If the circuit has the ability to hold values over time (store information), we get *sequential circuits*. The problem that the combinatorial-level analysis solves is the production of a set of outputs at time t as a function of a number of inputs at the same time t. As described in textbooks, the analysis abstracts from any transport delays between input and output; however, in engineering practice the analysis of delays is usually considered to be still part of the combinatorial level. In Fig. 3 we show a combinatorial network formed from combinatorial elements which realize three boolean output expressions, O_1, O_2, and O_3, as a function of the input boolean variables A and B. Note that in the symbolic representation of the structure we can write an expression that reflects the structure of the combinatorial network, but, on reduction, the boolean equations no longer reflect the actual structure of the combinatorial circuit but become a model to predict its behavior.

The representation of a sequential switching circuit is basically the same as that of a combinatorial switching circuit, although one needs to add memory components, such as a delay element (which produces as output at time t the input at time t − τ). Thus the equations that specify structure must be difference equations involving time. Again, there is a distinction (even in representation) between *synchronous* circuits and *asynchronous* circuits, namely, whether behavior can be represented by a sequence of values at integral time points (t = 1, 2, 3, . . .) or must deal in continuous time. But this is a minor variation. Figure 4 gives a sequential logic circuit in both an algebraic and a graphical form and shows also the representation of the behavior of the system.

Now it is clear that logic circuits are simply a subspecies of general circuits. Indeed, to design the logic components one constructs circuit-level descriptions of them. For instance, Fig. 5

shows a circuit for a NAND (or NOR) gate plus a table of its behavior. It is evident that its behavior corresponds to that of the NAND gate only if certain restrictions hold; namely, that one does not look at the voltage (which is identified as the behavior variable in the logic circuit) during certain periods when it is transient ("settling down," to use the common phrase). Thus the logic level is an instance of the circuit level only in the same sense that the circuit level is an instance of Maxwell's equations—as a limiting case in which certain features are deliberately ignored.

One buys a great deal from the specialization to logic circuits, since one can compute the behavior of circuits at the logic level that are extremely complex at the circuit level. The techniques for doing so use an entirely different mathematical apparatus. In general, we cross into another level when the representation at the previous level provides information that is no longer relevant. A lower level is concerned with explaining the behavior of a certain structure, whereas the next highest level takes the lower level as given (a primitive). The higher level is concerned not about internal behavior but only how primitives are combined.

A glance at Fig. 1 shows that we have described only the lower part of the logic level. There is another part, called the *register-transfer level* (or RT level). This is still an uncertain level, a matter

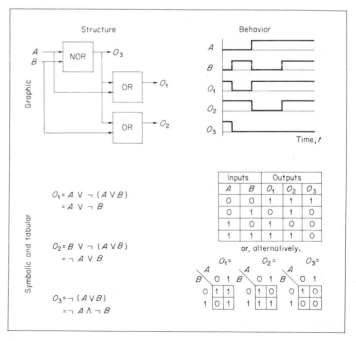

Fig. 3. **Combinatorial-switching-circuit sublevel of the logic level: realization of three logic expressions.**

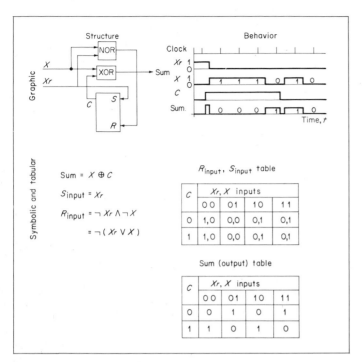

Fig. 4. Sequential-switching-circuit sublevel of the logic level: computation of x + 1 from serial input string x.

we will discuss after we have finished describing it. The components of an RT system are registers and functional transfers between registers. A register is a device that holds a set of bits.[1] The behavior of the system is given by the time course of values of these registers, i.e., their bit sets.

The system undergoes discrete operations, whereby the values of various registers are combined according to some rule and then are stored in another register (thus "transferred"). The law of combination may be almost anything, from the simple unmodified transfer $(A \leftarrow B)$ to logical combination $(A \leftarrow B \wedge C)$ to arithmetic $(A \leftarrow B + C)$. Thus a specification of the behavior, equivalent to the boolean equations of sequential circuits or the differential equations of the circuit level, is a set of expressions (often called productions) which give the conditions under which such transfers will be made. In Fig. 6 we give a picture of an RT system to compute the sum of integers. The figure includes the specification

[1]This assumes that the elementary state variable of the system holds a bit (i.e., one of two values, such as 0 or 1). This need not be; sometimes the elementary variable holds a decimal digit (one of 10 values) or a character (one of, say, 48 values). For present purposes we can talk in terms of bits, without losing anything thereby.

of its behavior and a table that shows the resulting behavior over time. Here the graphical structure of the system includes registers (N, I, S), transfers $(S \leftarrow S + 1)$, data operators $(S + 1, I > N, \text{etc.})$. The flowchart shows the behavior of the control with time.

The register-transfer level is still uncertain because there is substantial agreement neither on the exact language to be used for the level nor on the techniques of analysis and synthesis that go with it. As we will note below, for both the circuit level and the logic-circuit level there exist well-defined representations, guaranteed, so to speak, by standard textbooks and college courses that teach these levels. Standard texts on digital computers make only informal use of the RT level.

We have indeed a systems level in emergence here. If one restricts the transfer operations to boolean operations and thinks of a register as simply a set of 1-bit memories, one can write a set of logic equations for any register-transfer system. Furthermore, if one considers the role of logic design in digital computers, this has encompassed both sequential circuits and the register-transfer

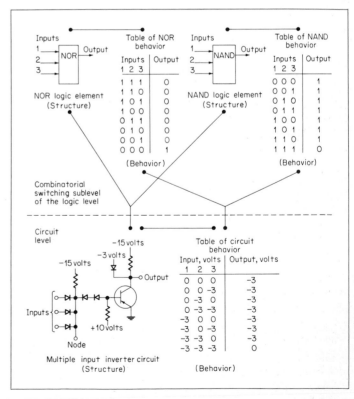

Fig. 5. Change of representation at the circuit level combinatorial-switching sublevel boundary.

level. The practicing logic designer (by now an institutionalized position, on a par with that of circuit designer) has sequential and combinatorial circuits as his basic analytic tools, and he attempts to design systems on the register-transfer level (e.g., central processors) with these as tools. The register-transfer level has emerged from the informal attempts to create a notation closer to the job to be done.

Recently there have been a number of efforts to construct formalized register-transfers systems. Most of them are built around the construction of a programming system or language that permits computer simulation of systems on the RT level. Although there is agreement on the basic components and types of operations, there is much less agreement on the representation of the laws of the system (corresponding to the production system in Fig.

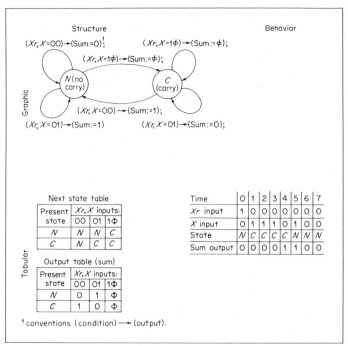

Fig. 7. State-system representation of the logic level: computation of x + 1 from serial input string x.

6) or on the way to represent the dynamic behavior (corresponding to the behavior table in the figure).

There is another representation used at the logic level, the *state-system* representation, but it has been put at one side in Fig. 1. The state system is the most general representation of a discrete system available.[1] A system is represented as capable of being in one of N abstract states at any instant of time. (For digital systems, N is finite or enumerable.) Its behavior is specified by a transition function that takes as arguments the current state and the current input and determines the next state (and the concomitant output). A digital computer is, in principle, representable as a state system, but the number of states is far too large to make it useful to do so. Instead, the state system becomes a useful representation in dealing with various subparts of the total machine, such as the sequential circuit that controls a magnetic tape. Here the number of states is small enough to be tractable. Thus, we have placed state systems at one side as an auxiliary to the logic level. In Fig. 7 we give the common representations of the state system. Co-

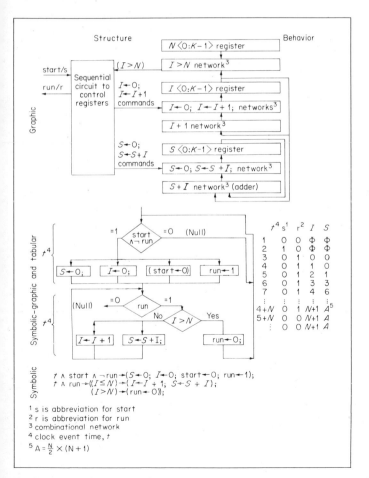

Fig. 6. Register-transfer sublevel of the logic level: computation of the sum of integers.

[1] There have been energetic attempts to apply the state-system approach to control systems of a more general nature [Zadeh and Desoer, 1963], although they do not concern us here.

incidently, we use the representations of Fig. 7 for the sequential switching circuit of Fig. 4. That is, Fig. 7 may be viewed as an abstraction of the physical system in Fig. 4. To the logic designer the state system is a useful abstraction of a logic design. A design usually passes through the following problem representations:

1   The problem exists in a natural language.

2   The problem is converted to a state diagram (output as a function of state, and input).

3   The state diagram is represented as a state table and output table.

4   States are assigned (physical memory elements are used).

5   The excitation table and output-tables are formed.

6   The excitation and output logic equations are written (constrained by the actual logic elements).

7   The sequential circuit is drawn.

Let us go to the next higher level, the *program level*. This not only is a unique level of description for digital technology (as was the logic level) but is uniquely associated with computers, namely, with those digital devices that have a central component that interprets a programming language. There are many uses of digital technology, especially in instrumentation and digital controls, which do not require such an interpretation device and hence have a logic level but no program level.

The components of the program level are a set of memories and a set of operations. The memories hold data structures which represent things both inside and outside the memory, e.g., numbers, payrolls, molecules, other data structures, etc. The operations take various data structures as inputs and produce new data structures, which again reside in memories. Thus the behavior of the system is the time pattern of data structures held in its memories. The unique feature of the program level is the representation it provides for combining components, that is, for specifying what operations are to be executed on what data structures. This is the program, which consists of a sequence of instructions. Each instruction specifies that a given operation (or operations) be executed on specified data structures. Superimposed on this is a control structure that specifies which instruction is to be interpreted next. Normally this is done in the order in which the instructions are given, with jumps out of sequence specified by branch instructions. Again, Fig. 8 shows a simple program, the data structures, and the behavior.

Two things separate the logic level from the program level. First, computer systems at the logic level are parallel devices, with all components active simultaneously. At the program level, computers are represented essentially as serial devices. Second, the program level, but not the logic level, is essentially linguistic in nature. At the program level things can be named, abbreviations can be used, decisions can be made, instructions are interpreted — all concepts that are strikingly absent from physical systems. Of course, they are not "really" absent since one can give a full description of the operation of a program at the logic level. But one does so by carrying in mind the set of physical behaviors discovered for computers that make them show the appropriate linguistic behavior at the program level. Thus, one does not "go to ALPHA if accumulator is negative'; one has a logic circuit that transfers the contents of the address field of the instruction register to the program counter, ANDing that transfer with the sign of the accumulator, so that it does not take place if the accumulator is not negative. Such a translation reveals how distinct is the system boundary between the register-transfer level and the program level. The size of the gap is also revealed in the ability of people to become expert programmers without knowing anything about any representations below the programming level.

The program level constitutes an entire technology in its own right, and one that carries within it most of the emergent characteristics of computer systems that make them worthy of a science. Among the programming languages alone, there are levels of language which are so distinct from each other as to constitute system levels fully as important as the ones exhibited in Fig. 1. Nevertheless, from the viewpoint of someone basically concerned with hardware systems, these can all be accounted a single level, at least for the present. The one aspect of programming systems that should be of most concern, that of operating systems, is still in such a fragmented state that it does not even begin to be a distinct system level.

One peculiarity of the program level is that there exists no universal representation for it, as there does for the circuit or logic-circuit level (and, it is to be hoped, soon for the register-transfer level). Each machine has its own machine language (and its own assemblers and command languages built on those machine languages). Each of these languages forms a complete system at the program level, applicable only to the machine in question. There is no universal machine language, although there is much in common at a conceptual level between all existing machine languages. There has existed a long-standing attempt within the programming field to develop an UNCOL (for Universal Computer Oriented Language) [Steel, 1961] that would play this role, but it has never been successful. The reasons are not far to seek. The role of the machine language is to be interpreted by the machine in order to produce behavior. It is not free to have arbitrarily desirable properties from our human viewpoint, since its details affect the efficient operation of the computer too much — how much space is devoted to the program, how much time is saved by a special order oriented to matrix multiply, etc. UNCOL was also attempting to fill the same role as machine languages, being one from which to compile a machine code for an arbitrary machine. Another reason why there has been no universal programming representation is that each particular machine language is a *language,* and so a universal description would seem to be a description of a class of languages. This is by no means impossible, as the wide use of notations such as Backus Normal Form (BNF) show.[1] Nevertheless, it has contributed to the lack of any universal notation.

We now move to the fourth and last level. In Fig. 1 it is called the Processor-Memory-Switch level, or PMS level for short. The name is not recognized, nor is any other, since the level exists only informally. Nevertheless, its existence is hardly in doubt. It is the view one takes of a computer system when one considers only its most aggregate behavior. It then consists of central processors, core memories, tapes, disks, input/output processors, communication lines, printers, tape controllers, busses, Teletypes, scopes, etc. The system is viewed as processing a medium, information, which can be measured in bits (or digits, characters, words, etc.). Thus the components have capacities and flow rates as their operating characteristics. All details of the program are suppressed, although many gross distinctions of encoding and infor-

[1]We will propose a notation later. See also the work by F. Haney in his Generalized Instruction System (GIS) [Haney, 1968].

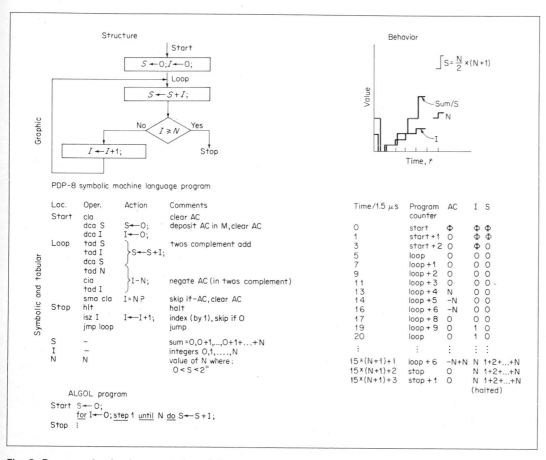

Fig. 8. Programming level: computation of the sum of integers.

mation type remain, depending on the analysis. Thus one may distinguish program from data, or file space from resident monitor. One may remain concerned with the fact that input data are in alphameric and must be converted into binary, or are bit-serial and must be converted to bit-parallel.

We might characterize this level as the "chemical engineering view of a digital computer," which likens it more to a continuous-process petroleum-distilling plant than to a place where complex FORTRAN programs are applied to matrices of data. Indeed, this system level is more nearly an abstraction from the logic level than from the program level, since it returns to a simultaneously operating flow system.

One might question whether there is a distinct systems level here. In the early days of computers almost all computer systems could be represented as in the diagram in M.I.T.'s Whirlwind computer programming manual in Fig. 9: with classic boxes of memory (storage), control, arithmetic, and input/output. Actually, this view of the computer in 1953 was considerably advanced; few texts on the logic design of computers in the 1960s have such a detailed model. This model has secondary memory (magnetic tape and drums in the Whirlwind's case). The most interesting aspect of the model, which text writers omit, is any kind of switching (the bus of Fig. 9). The bus provides a communication path to link the other components. Certainly the pushbuttons (actually the console) is novel for such a model. Compare this with the diagram of a modern computer system in Fig. 10, which shows a two-processor UNIVAC 1108, the level of abstraction being the same as in Fig. 9. The arithmetic element of Fig. 9 has disap-

peared and is replaced by a processor (a combined control and arithmetic element) in Fig. 10. The central control of Fig. 9 is now distributed throughout the remaining components. The control in Fig. 10 is a combined unit for transforming a serial character-information stream into words. It also manages the transmission of a word vector between the primary memory and a terminal or a secondary memory. The Resource Allocation Diagram is introduced in Fig. 10 to describe the allocation (use), hence behavior, of the PMS components as a function of time. Chapter 2 describes these figures more fully.

Another indication of the emergence of the PMS level lies in the models used in most operations-research types of studies on computer systems. Again, in the early 1960s these were practically nonexistent. Now, with the advent of multiprogramming, multiprocessing, and time sharing, and the imminent arrival of computer networks, there are substantial numbers of such studies. The level of abstraction is always one that considers only flows and stocks of information, measured in bits (or an equivalent), perhaps divided into several subtypes. The concerns are bottle-necks, capacities, total flow rates, queuing problems, buffer sizes, and the like. All this indicates a system level above both the logic level and the program level.

There is no uniform language for representation at this level and even, as we noted, no standard name. We have used the term PMS in analogy to the use of RT for the register-transfer level. Processors, memories, and switches are the main kinds of components out of which systems at this level are built. If one names a number of components at the PMS level, as we did previously, one finds few switches in the list. "Busses" in our list would be one, although many would think first of their data transfer characteristics. But, as this book amply shows, what makes the PMS level both interesting and complex is the existence of switches which govern the pattern of information flow through the system. One reason why they seem buried is their association with other components as addressing systems. There are other components besides processors, memories, and switches, namely, links, transducers, and controls. But the first three, P, M, and S, seem appropriate to characterize the level.

It is not known whether there will be yet other systems levels, say one above the PMS level, as networks come into existence. The simplicity of the top level argues against it, but that may only show our narrow vision. It is important to realize that these levels are not sacrosanct. They depend strongly on physical technology. Thus, as we move toward integrated circuitry, there may emerge representations other than register-transfer diagrams, and the latter may never develop into a clear systems level. One could even

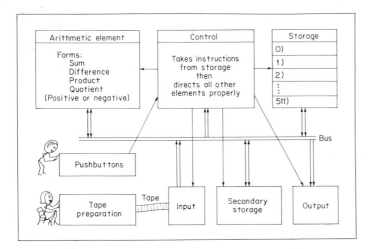

Fig. 9. Automatic digital computation. (*From the Whirlwind Computer Manual, M.I.T. By permission of the publishers.*)

imagine something happening to the circuit level, as continuous distributions became more important (although the use of equivalent circuits is well embedded in the engineering culture). We are not concerned with predicting any particular changes. We wish only to emphasize that the system-levels diagram of Fig. 1 is a reflection both of current technology and of our ways of analyzing given physical systems. As such, these levels have a certain impermanency about them.

What is the problem?

The systems levels we have just described correspond to the technologies that are available for the analysis and synthesis of computer systems. Each of these levels exists, in fact, precisely to the extent that a technology has become well developed. Thus both the circuit level and the lower half of the logic level (combinatorial and sequential circuits) are highly polished technologies. They are what one learns today, if one wants to become a computer engineer. Textbooks exist, courses are taught, and there is a flourishing, cumulative technical literature. As we progress up the systems levels, matters become progressively worse. The register-transfer level is not yet well established, although there is considerable

current activity in the area, and the next few years may see its universal establishment. Although programming is certainly well defined, each machine is a king in his own court, with no common technology of the program level that is relevant to the design of computer systems. The latter phrase must be added since we are taking a very specialized viewpoint here. We do not consider the world of programming research at all, it being entirely divorced from computer-systems design.[1] Finally, at the top, there is practically no consensus on the nature of the systems level.

There is nothing very surprising about this state of affairs. It reflects accurately the fundamental fact that only in the past few years have computer systems become complex enough for the higher levels to emerge as distinct systems levels. When most computers could be described in the diagram of Fig. 9—and such a diagram was reprinted innumerable times in the first decade—there was no need to have a technology at the PMS level. When registers were so expensive that one could count the registers of a processor on the fingers of one hand (no thumbs allowed), one did not need a register-transfer language in order to describe the

[1] This is not entirely true. Each level must provide coupling with adjacent levels. A major issue in computer-design is the trade-off between hardware and software.

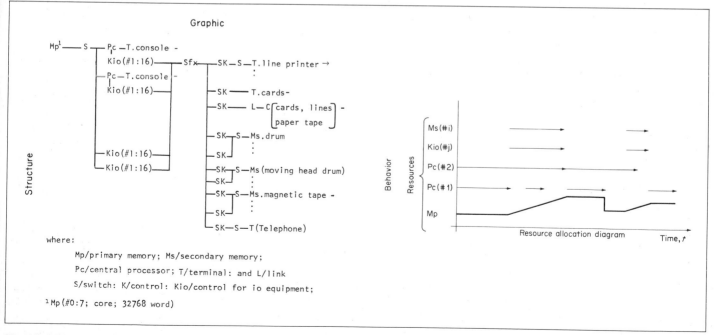

Fig. 10. PMS level: UNIVAC 1108.

flows. In both cases, an informal block diagram conveyed all the information adequately.

The question of the programming level is somewhat different, since this level has existed as a formal language from the very start. Here the key aspect, it seems to us, is that, since well-defined languages existed, there was little pressure to find a better one. The fact that such languages were completely idiosyncratic to the machine, since they emerged as a product of the design itself, simply did not worry anyone overly much. Each language provided a design framework one could work into, and this seemed to suffice. It led, it is true, to the game of "We have another bit left in the mode field of the instruction—got another mode you'd like?" But this has only made computer designers feel that creating an order code was something of an art.

Thus we feel that the increased complexity of computer systems is making these higher system levels of increasing importance. Since this is only the second decade of the serious development of computer systems, these upper levels are not in very good shape. For instance, textbooks devote very little attention to the area. Textbooks (especially good ones) tend to be technique-oriented, giving most attention to what is known. (When we were students we always used to wonder why there were no mathematics texts which told you about the problems that were not solvable in closed form.) Thus the present need for some material at these higher levels constitutes a major motivation for this book.

There is a second feature of the current scene that enters into our motivation for this book. Around 1,000 different computer systems have been built. This represents a substantial amount of pragmatic experimentation. This is especially true at the programming level and PMS level, and also to some extent at the register-transfer level. Many things have been tried, many found worthwhile, and many found wanting. A good deal of reinvention goes on. Thus we are concerned that this history of experimentation not be lost. It is true that, if the underlying technology changes enough, the experience may become largely irrelevant, but this does not appear to us to be an imminent development.

We will admit also to a third concern, which does not stem from our role as computer engineers concerned with design, but from our role as computer scientists, fascinated with the phenomena of computers. The variety of about 1,000 computers represents the beginning of a proliferation of a species. It is not under biological control but rather under economic and intellectual control. Nevertheless, it is in every sense of the word an evolutionary population. We find ourselves feeling a little like naturalists must have felt when confronted with the proliferation of the organic world. We were at one time tempted to call this book "Computer Botany" and at another "Computer Taxonomy." We feel that the attempt to gather, document, and classify these existing computers is a worthy endeavor in its own right. One might think that all this material is easily available. But the record fades rapidly, especially when much of it exists only as manufacturers' manuals and papers in assorted proceedings.

The main reasons for producing this book and for its particular character are by now evident. There is a need for material on the upper levels of computer systems, both for teaching new students of computer science and engineering and for making the past record available for professional designers. Since the technologies are not well developed for the upper levels, it is not possible to write a textbook, making use only of well-accepted techniques, notations, and results. Instead, one settles for making available a collection of examples of systems, so that they can be studied and analyzed directly.

Notations

It remains to say a word about two notations we have introduced, both about our motivations for doing so and about their character. Some, but not all, of this is already implicit in the foregoing account.

We started simply to produce a set of readings in computer systems, motivated by the lack of detailed examples we could use in a course one of us (GB) was giving on computer design. As noted, we felt the need to expose the students to real examples of complex computer structures. As we gathered material we became impressed (depressed is actually a better term) with the diversity of ways of describing these higher levels. Even more, the amount of clumsy description—downright verbosity—even in purely technical manuals acted as a further depressant. The thought of putting such a congeries of descriptions between hard covers for one person to peruse and study was almost too much to contemplate. Gradually, we began to rewrite and condense many of the descriptions. As we did so, a set of common notations developed. Becoming aware of what was happening, we devoted a substantial amount of attention and effort to creating notational systems that have some consistency and, we hope, some chance of doing the job required. These are the PMS descriptive system for the PMS level (sic) and the ISP (Instruction-set processor) descriptive system for the program level. Each of these requires some comment on its nature and the role we think it should play.

The PMS descriptive system is meant to provide a notation for the top level of computer systems. Figure 10 is given in this notation. On the surface it is largely self-explanatory, given the mnemonics of P for processor, M for memory, S for switch, T for transducer (hence also terminal), and K for control (since C is for computer). There is also L for link, but in most computer structures it is unnecessary to distinguish a separate link component, except to show connectivity. (It does become appropriate if communication delays exist.)

There is an issue about whether this small set of components is an appropriate set of primitives, but the issue is not of major proportions. The real issues in the development of the notation come from the stress of two opposite forces. On the one hand, one wants extremely compact notations for expressing computer systems. The systems are large in any event, and if there is much extra notational freight in the way of fixed formats, forced writing of what is already known and assumed, etc., then the notation will be neither useful nor used. On the other hand, there is a tremendous variety and quantity of information that potentially must be capable of being written into a description: word size, capacity, flow, operation rate, data-types, variations of operation rate for different classes of instructions, parity checking, technology, and on and on. Thus one needs a notation that responds to both these demands—and without being hopelessly complex and difficult to learn. Our attempt at a solution involves a basically simple language with comprehensive (and we think natural) ways of systematic abbreviation and abstraction.

The ISP descriptive system is meant to provide a uniform way of describing instruction sets, that is, of giving the information contained in a programming manual. It must provide the instruction format, the registers referenced by the instructions, the rules of interpretation of the instruction, and the semantics of each instruction in the processor's repertoire. It must be able to do this for any existing computer, plus the expected extensions into the future. Its homeliest virtue is to make it possible to read the descriptions of the forty-odd computer systems described in this book, without having to fight a new notation for each system, and still to know in detail what the instructions really do.

Our attempt at a solution turns out not to be a generalized sort of instruction. Rather, it is very similar in flavor to a register-transfer scheme. The differences lie in being able to suppress all timing information and all detail that is not essential to understanding the instructions. ISP is not a variety of UNCOL, in which one can program; rather it is a language in which one can describe what any particular instruction set does. We thus avoid many of the pitfalls of the UNCOL-like efforts.

There is a price to be paid for introducing new notations, for they must be learned. We feel that the two systems we have introduced here are natural enough to require almost no learning

for superficial use (e.g., looking at Fig. 10) and only modest amounts for full exploitation. They seem to us vastly preferable to the array of ad hoc notations that we were faced with initially (and with which we almost faced the reader). Still we are aware of the price.

A word should be said about antecedents. The PMS descriptive system is close to the way computer scientists talk informally about the top level of computer systems; no one effort in the environment stands out as a predecessor. Some notations, such as CPU (for central processing units), have become widespread. We clearly have assimilated them. Our modifications, such as Pc instead of CPU, are dictated entirely by the attempt to build a consistent notation over the whole range of computer systems. With respect to ISP, we have been heavily influenced by the work on register-transfer languages.[1] The one that we used most as a kernel from which to grow ISP was the work of Darringer and Parnas [Darringer, 1969]. In particular, their decision to work within the framework of ALGOL suited our own sensibilities, even though the final version of ISP departs from a sequential algorithmic language in a number of respects.

Finally, a word should be said about innocence and aspirations. We are putting PMS and ISP forward as two notations. They are that. But they also imply a particular view of digital processing. Thus they are not entirely innocent. It would be appropriate to explore fully this view and to justify the particular decompositions and definitions used. This is not to say that these views are peculiarly ours. They are implicit in the informal use of similar descriptive systems. However, the attempt to formalize a notation makes them more accessible. We accept the obligation to perform such an exploration. But this volume is not the place to do so, for that would turn it into something between a treatise and a textbook. For this book, it is appropriate to take these notations at face value. We have a companion volume in preparation that attempts the other job. This is an aspiration.

We have other aspirations as well. Notations in the computer world should turn into working tools. There are many tasks, such as the communicative one of this book, where the notation by itself is useful. Others are easy to imagine: writing specifications for new machines; being sure what the computer salesmen are selling; standardization of programming manuals, so that learning about a new machine is easier; etc. But there are other tasks where the

---

[1] We have not been influenced in a direct way by the work of Iverson [Falkoff, Iverson, and Sussenguth, 1964] in the sense of patterning our notation after his. Nevertheless, his creation of a full description of the IBM System/360 in APL stands as an important milestone in moving toward formal descriptions of machines.

notations must become formal programming languages, so that analysis and synthesis procedures can be carried on automatically in their terms. As we have noted, the development of ISP and PMS germinated from purely notational issues. We have not let our aspirations to turn them into simulation languages delay our use of them for purely descriptive purposes. Thus we accept the obligation also to develop them as operational tools. That is also an aspiration and cannot be dealt with anywhere within this book.

Plan of the book

We now have enough background to explain the structure of the book. Two other chapters complete the introductory part. Chapter 2 provides an exposition of the PMS and ISP descriptive systems. As we have just noted, this does not attempt to explore seriously the view of digital processing implicit in these notations, although it does provide a small amount of motivation. A summary of the language conventions and parameter values is given at the end of the book in the appendix.

Chapter 3 provides a description of the space of computer systems. One can view all computer systems as occupying a space whose dimensions are the various important systems features. Many features of the actual systems are relatively locked together. For example, word size and number of instructions in the repertoire covary; no 12-bit machine has 200 instructions but several with over 32 bits do. Thus the number of significant dimensions of variation is much less than the total number of features of computer systems. Such a space provides a basic frame in which to choose representative computer systems for inclusion in the book. We hope Chap. 3 will also justify our feeling that there is a diversity and proliferation of computer systems that is worthy of serious study.

The remainder of the book is divided into five parts (2 to 6, with the introduction constituting Part 1), and each part into sections. Each chapter gives a description of a computer system that is an instance of the part and section. Usually a chapter describes only one computer or computer system, although there are a few exceptions in Part 6 on computer families.

A word needs to be said about the "Virtual" Table of Contents. Many of the example computers are relevant to more than one part and section. Physically, they have to be located at one place. But we have permitted multiple entries in the Contents, so that, for instance, Chap. 33 on the IBM 1800 appears in Sec. 1 of Part 2 as an example of a one-address ISP, in Sec. 1 of Part 4 as a terminal control, and finally in Sec. 2 of Part 5 as an example of a PMS with one central processor and multiple input/output processors (1 Pc, multi-Pio); physically it is located in the latter section. By using different type faces we hope the reader will not become confused between virtual and actual.

There is little point in outlining the content of the various parts and sections here. This is better done at the end of Chap. 3 after the computer space has been laid out.

References

Brackets are used to enclose author(s) and year of publication, e.g., [Darringer, 1969] or [Falkoff, Iverson, and Sussenguth, 1964]. A list of all the references in a chapter is given in code at the end of the chapter. The code refers to the bibliography at the end of the book. This 7- or 8-character code is as follows:

Characters 1:4	First four characters of the last name of author (or first author)
Character 5	First initial of author (or first author)
Characters 6:7	Year of publication—1900
Character 8	(Optional) a, b, c, . . . , used to denote multiple referenced publications of author in a year.

References

DarrJ69; FalkA64; HaneF68; RoseS67; SteeT61; ZadeL63.

Chapter 2

The PMS and ISP descriptive systems

The task of this chapter is to provide an introduction to the PMS descriptive system for the top computer-system level and to the ISP descriptive system for the program level. We take the view that informal notations exist and are in use. PMS and ISP are an attempt to tidy up these notations—to make them consistent and more powerful. Thus we depend on the reader already to understand implicitly much of the notation and how it is to be used. In consequence, there is no attempt in this chapter to provide a formal treatment of the whole system. The appendix 1, at the end of the book contains a complete summary of the notation rules, including the component attributes and values, and their abbreviations (i.e., the main technical vocabulary). We will provide a brief discussion of the conceptual view underlying the two systems, since it is an appropriate way to make the notation understandable. But this is informal and heuristic.

The two descriptive systems are not independent. There is a common set of notational conventions for abbreviating, for giving parameter values, and so on. (The Appendix separates them.) Likewise, there exists, in effect, an ISP description for every PMS component, or, conversely, ISP statements imply particular PMS component structures. A natural way is to present PMS first, which will also serve to introduce the main notational devices. Then we will give ISP. Finally, we will add more comments on the relationship between PMS and ISP.

PMS level of description

Digital systems can be characterized most generally as systems that at any time exist in one of a discrete set of states and that undergo discrete changes of state with time. This is a highly abstract view. Nothing is said about what physical state corresponds to a system state; nothing is said about what laws of physics transform the system from one state to another. The states are given abstract labels: S_1, S_2, ... The transitions are provided by a state-transition table with many entries of the form: If the system is in state S_i and the input is I_j, then the system is transformed to state S_k and evokes output O_l. (Alternatively, a state diagram has the same information.) The virtue of this "state-system" view is that it truly seems to capture what we mean by a discrete (or digital) system. Its disadvantage lies in this same comprehensiveness, which makes it impossible to deal with large

systems because of their immense number of states (of the order of $10^{10^{10}}$ states for a big computer).[1]

Existing digital computers can be viewed as discrete state systems that are specialized in three ways. These three specializations make possible a much more compact and useful description of these systems, the one that we call the PMS description.

First, the state is realized by a medium, called information, which is stored in memories. Thus, a core store of N words each of 32 bits is a digital device that can exist in one of 2^{32N} states. Similarly, all the states of a processor are made explicit in a set of registers: an accumulator, an address register, an instruction register, status register, etc. Each holds a specified number of bits. No permanent information is kept in digital devices except as encoded in bits in a memory. There are two qualifications to this blanket statement. First, the basic unit of information need not be the bit; it could be any base: One can have ternary machines, decimal machines, etc. Second, the sequential logic circuits that carry out operations in the system have intermediate states. But this is a strictly temporary affair while the operation is occurring, for example, the intermediate, inaccessible, partial results during a multiply operation. At the end—when the smoke has cleared, so to speak—all information carried over to the next operation has been encoded into bits in memories somewhere. At the PMS level we care only about the end result of such operations.

The second specialization of the general state-system view is that current digital computer systems consist of a small number of discrete subsystems linked together by flows of information. There is a distinct component called the memory, another called the central processor, another called the card reader, etc. This is analogous to the lumped-parameter specialization at the circuit level. Thus the natural representation of a digital computer system is as a graph which has component systems at the nodes and information flows as branches. Now, in fact, the discrete character of digital encoding in bits prevents there being any truly continuous digital devices (in analogy to the continuously distributed parameter circuits). But one can have distributed networks with very small components. Such iterated arrays are a topic of much

[1] As we noted in Fig. 1 of Chap. 1, we actually describe some parts of the control mechanisms of computers by state-system diagrams; however, these are exceedingly small pieces. An example may be seen in Fig. 7 on page 7.

current investigation, as the possibility of manufacturing them by integrated-circuit techniques has emerged. These distributed networks look very different from the computer systems of today, although they are still digital systems. Thus, the representation as a flow network with functionally specialized nodes is a real specialization.

The third specialization of the general state-system viewpoint is that associated with each component in a digital system is a small number of discrete operations for changing its own state or the state of neighboring components. All transitions must occur through the application of these few operations, which are evoked as a function of the current state of the component. The total behavior of the system is built up from the repeated execution of the operations as the conditions for their execution become realized by the results of prior operations. The general state-system view is more general. The state-transition table for a system may exhibit an arbitrary pattern of immediate state transitions, without regard to how such transition would be physically realized.

To summarize, within this specialized view one wants a way of describing a system of an interconnected set of *components,* which are individual devices that have associated with them a set of *operations* that work on a medium of *information,* measured in bits (or some other base).

The major complication in this picture is the amount of detail involved in describing actual computers. It takes a whole manual, for instance, to describe the operations of a major computer, such as the IBM 7090. Thus the descriptive system must permit very compressed descriptions. It must also permit description of only those aspects of the components that are of interest, ignoring the rest. And what is of interest at the PMS level? Besides a description of the gross structure of a computer system, it is primarily the analysis of the amounts of information held in various components, the flows of information between components, and the distribution of the control that accomplishes these flows.

Thus a PMS-level description is analogous to the chemical engineer's diagram of a refinery in which he is interested in various kinds of liquid and gas flow. He has to account for matter and energy loss with the system at various stages involving the transduction of materials from one form to another. A specific chemical plant's external performance is measured in terms of its production flow rate for a given cost. With computers, external performance is concerned with the economical accomplishment of discrete tasks, but at the PMS level this translates into operation rates and cost of operations.

For the PMS level we ignore all the fine structure of information processing and consider a system consisting of components that work on a homogeneous medium called information. Information comes in packets, called *i-units* (for information units), and is measured in bits (or equivalent units, such as characters). I-units have the sort of hierarchical structure indicated by the phrase: A record consists of 300 words; a word consists of 4 bytes; a byte consists of 8 bits. A record, then, contains $300 \times 4 \times 8 = 9,600$ bits. Each of these numbers—300, 4, 8—is called a *length,* since one often thinks of an i-unit as a spatial sequence of the next lower i-units of which it is composed. For example, one speaks of "word length" and of a record being "300 words long."

Other than being decomposable into a hierarchy of factors, i-units have no other structure at the PMS level. They do have a *referent,* that is, a *meaning.* Thus it is possible to say of an i-unit that it refers to an employer's payroll, to the pressure of a boiler, or to a prime number satisfying certain conditions. To do so, of course, the i-units *encode* the information necessary to make the reference. At the PMS level we are not concerned with what is referred to, but only with the fact that certain components transform i-units but do not modify their meaning. In fact, these meaning-preserving operations are the most basic information-processing operations of all, and they provide the basic classification of computer components.

PMS primitives

In PMS there are seven basic component types, each distinguished by the kinds of operations it performs:

Memory, M. A component that holds or stores information (i.e., i-units) over time. Its operations are reading i-units out of the memory and writing i-units into the memory. Each memory that holds more than a single i-unit has associated with it an *addressing system* by means of which particular i-units can be designated or selected. A memory can also be considered as a switch to a number of submemories. The i-units are not changed in any way by being stored in a memory.

Link, L. A component that transfers information (i.e., i-units) from one place to another in a computer system. It has fixed ports. The operation is that of transmitting an i-unit (or a sequence of them) from the component at one port to the component at the other. Again, except for the change in spatial position, there is no change of any sort in the i-units.

Control, K. A component that evokes the operations of other components in the system. All other components are taken to consist of a set of discrete operations, each of which, when evoked, accomplishes some discrete transformation of state.

With the exception of a processor, P, all other components are essentially passive and require some other active agent (a K) to set them into small episodes of activity.

Switch, S. A component that constructs a link between other components. Each switch has associated with it a set of possible links, and its operations consist of setting some of these links and breaking others.

Transducer, T. A component that changes the i-unit used to encode a given meaning (i.e., a given referent). The change may involve the medium used to encode the basic bits (e.g., voltage levels to magnetic flux, or voltage levels to holes in a paper card), or it may involve the structure of the i-unit (e.g., bit-serial to bit-parallel). Note that T's are meaning-preserving but not necessarily information-preserving (in number of bits), since the encodings of the (invariant) meaning need not be equally optimal.

Data-operation, D. A component that produces i-units with new meanings. It is this component that accomplishes all the data-operations, e.g., arithmetic, logic, shifting, etc.

Processor, P. A component that is capable of interpreting a program in order to execute a sequence of operations. It consists of a set of operations of the types already mentioned—M, L, K, S, T, and D—plus the control necessary to obtain instructions from a memory and interpret them as operations to be carried out.

Throughout PMS (and ISP, too) an operation is taken to mean a transformation of bits from one specific memory to another. For instance, it is an operation to transmit a word of information from memory M to memory M′; it is a different operation to transmit a word from memory M′ to M″. Similarly, it is an operation to add the contents of memory M to that of M′ and a different operation to add the contents of M′ to M″.

The reason for emphasizing this point is that one often talks as if addition were an operation, ignoring the specific locus of the operands. In a discussion of computer systems, an operation must include specification of the locus of its operands. The reason is that the physical devices that realize operations are always localized in space. If, for instance, we wish to have a physical device that corresponds to addition on operands anywhere in some memory, we must couple the physical device that adds with other devices that either transmit information to and from the memory to the adder or (more exotic) that modify the adder to have different cells of memory as its terminals. Thus the symbol + is to be taken as an incomplete specification of an operation.

Computer model (*in PMS*)

Components of the seven types can be connected to make *stored-program digital computers,* abbreviated by C. For instance, the classical configuration for a computer is

$$C := Mp—Pc—T—X$$

Here Pc indicates a *central processor* and Mp a *primary memory,* namely, one which is directly accessible from a P and holds the program for it. T is a transducer connected to the external environment, represented by X. (The colon-equals (:=) indicates that C is the name of what follows to the right.) Thus a computer is a central processor connected to its primary memory on the one hand and to a transducer on the other, which is what an input/output device is.

Actually the classic diagram had four components, since it decomposed the Pc into a control (K) and an arithmetic unit or data-operation (D):

$$Mp—K—T\,|\,Ms^1—X \qquad \text{or} \qquad Mp—D—T\,|\,Ms—X$$
$$\quad\ |\qquad\qquad\qquad\qquad\quad\diagdown\!\diagup$$
$$\quad\ D\qquad\qquad\qquad\qquad\quad\ K$$

where the solid information-carrying lines are for instructions and their data, and the dotted lines signify control.

Often logic operations were lumped with control, instead of with data operations, but this no longer seems to be the appropriate way to decompose the system functionally.

If we associate local control of each component with the appropriate component, we get

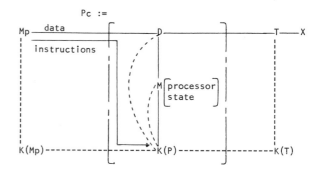

where the solid lines carry the information in which we are interested, and the dotted lines carry information about when to evoke operations on the respective components. The solid information-

---

[1] The " | " expresses mutually exclusive alternatives. Here, a T or Ms exists at the periphery.

carrying lines between K and Mp are instructions. Now, suppressing the K's, then lumping the processor state memory, the data operators, and the control of the data-operations, and processor state memory to form a central processor, we again get

Mp—Pc—T—X

Computer systems can be described in PMS at varying levels of detail. For instance, in the diagrams above we did not write in the links (L's) as separate components. These would be of interest only if the delays in transmission were significant to the discussion at hand or if the i-units transmitted by the L were different from those available at its terminals. Since this is not usually the case in current computers, one indicates simply that two components (e.g., an Mp and a Pc) are connected together. Similarly, often the encoding of information into i-units is unimportant; then there is no reason to show the T's. The same statement holds for K's. Sometimes one wants to show the locus of control, say when there is one control for many components, as in a tape controller, but often this is not of interest. Then there is no reason to show K's in a PMS diagram.

As a somewhat different case, D's never occur in PMS diagrams of computers, since in the present design technology D's occur only as subcomponents of P's. If we were to make PMS-type diagrams of analog computers, D's would show extensively as multipliers, summers, integrators, etc. There would be few memories and variable switches. The rather large patchboard would be represented as a very elaborate manually fixed switch.

Components are often decomposable into arrangements of other components. Thus, most memories are composed of a switch—the addressing switch—and a number of submemories. Thus a memory is recursively defined. The decomposition stops with the unit memory, which is one that stores only a single i-unit and hence requires no addressing. Likewise, a switch is often composed of a cascade of one-way to n-way switches. For example, the switch that addresses a word on a multiple-headed disk might look like

— S(random) — S(random) — S(linear) — S(cyclic) — M(word)
    ⋱   ⋱   ⋱   ⋱
    ⋮   ⋮   ⋮   ⋮

The first S(random) selects a specific Ms.disk_drive_unit; the second S (random) is a switch with random addressing that selects the head (hence the platter and side); S(linear) is a switch with linear accessing that selects the track; and S(cyclic) is a switch with cyclic addressing that finally selects the M(word) along the circular track. Note that the switches are realized by differing technologies. The first two S(random)'s are generally electronic (AND-OR gates) with selection times of $10 \sim 100$ microseconds or perhaps electromechanical (relay). The S(linear) is the electromechanical action of a stepping motor or a pneumatic-driven, servomechanism-controlled arm which holds the read-write heads; the selection time for a new track is $50 \sim 500$ milliseconds. Finally, the S(cyclic) is determined by the rotation time of the disk and requires from $16 \sim 60$ milliseconds, depending on the speed $(3,600 \sim 1,000$ rpm).

We can write such decompositions of a component into subcomponents either when we actually know the structure of the component or even when we know only the behavior. For example, we could write a memory as random access (M.random) even if it was, in fact, cyclic, as long as its behavior as far as the larger system was concerned took no account of its cyclic character, accepting the average access time as the random-access time.

When people speak of the control element of a computer, they often refer mainly to the processors—not to the control of a disk or magnetic tape, which, however, can often be more complex. When we suppress detail, the control often disappears from a PMS diagram. Similarly, when we agglomerate primitive components (as we did above when combining Mp and K(Mp) to be just Mp) into the physically distinct subparts of a computer system, a separate control, K, often occurs. The functionally and physically separate control[1] has evolved in the past decade. These controls, often as big as a Pc, can be computers with stored control programs. When we decompose a compound control, we find data-operations (D) for calculating addresses or for error detection and error correction data; transducers (T) for changing logic signal levels and information flow widths; memory (M) as it is used in D, T, K, and for buffering; and finally a large control (K) which coordinates the activities of all the other primitives.

It should be clear from the above discussion that components are named according to the function they perform and that they can be composed of many different types of components. Thus, a control (K) may have memory (M) as a subcomponent, and a memory M may have a transducer (T) as well as a switch (S) as subcomponents. All these subcomponents exist to accomplish the total function of the component and do not make the component also some other type. For instance, the M that does a transduction (T) from voltages on its input wires to magnetism in its cores and a second transduction from magnetism to voltages on its output wires does not thereby become a transducer as far as the total

---

[1] A variety of names for K's are used: controller, adapter, channel, buffer, interface, etc.

system functioning is concerned. To the rest of the system all the M can do is to remember i-units, accepting and delivering them in the same form (voltages). In the Appendix at the end of this book we define for each type both a simple component and a compound component, reflecting in part this fact that complex subsystems can be put together to perform a single function from the viewpoint of the total system. For example, a typewriter may have 4∼6 simple information transduction channels.

PMS notation

In the above discussions we used various notations to designate additional specifications for a component, for example, Mp for a functional classification, and S(cyclic) for a type of access function. There are many other additional specifications one wants to give—so many that it makes no sense to enumerate them all in advance. A fixed position notation, such as standard function notation, $F(x,y,z)$, where the first, second, and third argument places have fixed interpretation, is not suitable. Instead we agree on a single general way of providing additional specifications. If X is a component, we can write

$$X(a_1:v_1;a_2:v_2; \ldots)$$

to indicate that X is further specified by attribute a_1 having value v_1, attribute a_2 having value v_2, etc. Each *parameter* (as we call the pair a:v) is well defined independently of whatever other parameters are given; hence there is no significance to the order in which they are written or the number which have to be written.

According to this notation we should have written M(function: primary) or S(access-function:random) rather than Mp or S(random). This shows immediately the price paid for the general convention: It requires an excessive amount of writing (which would be even more apparent if a large number of parameters were given), and the extra information seems to be redundant in some cases. We compensate for these disadvantages by several conventions for abbreviating and abstracting parameters. All these conventions are listed in the Appendix. Let us illustrate them by showing some alternative ways of writing Mp:

M(function:primary)	Complete specification.
M(primary)	Drop the attribute "function," since it can be inferred from the value.
M.primary	Use the value outside the parentheses, concatenated with a dot.
M.p	Use an explicitly given abbreviation, namely, primary/p (only if it is not ambiguous).
Mp	Drop the concatenation marker (the dot), if it is not needed to recover the two parts (all components are given by a single capital letter—here M).

Each of these rules corresponds to a natural tendency to abbreviate when redundant information is given; each has as its condition that recovery must be possible.

In the full description in the appendix each component is defined and given a large number of parameters, i.e., attributes with their domain of values. Throughout, we use the slash (/) to introduce abbreviations or aliases as we go.[1] Thus p is introduced as an abbreviation for "primary" by writing primary/p when "primary" is given as one of the values of the attribute "function" of a memory with respect to processors (see page 607). The list of parameters in the Appendix does not exhaust those aspects of a component that one might want to talk about. For instance, there are many distinct dimensions for any component in addition to the information dimension: packaging, physical size, physical location, energy use, cost, weight, style and color, reliability, maintainability, etc. Furthermore, each of these dimensions includes an entire set of parameters, just as the information dimension breaks out into the set of parameters we have given in the Appendix. Thus the descriptive system is an open one, and new parameters are definable at any occasion.

The very large number of parameters provides one of the major challenges to creating a viable scheme to describe computer systems. We have responded to this in part by providing automatic ways in which one can compress the descriptions by appropriate abbreviation while still avoiding a highly cryptic encoding of each separate aspect. Abstraction is another major area in which some conventions can help to handle the large numbers of parameters. It often happens that one has only imperfect information about an attribute, or one wishes to give its value only approximately or partially. For instance, one attribute of a processor is the time taken by its operations. This attribute can be defined with a complex value:

Pc(operation-times: add:4 μs, store:4 μs, load:4 μs,
multiply:16 μs, . . .)

That is, the value is a list of times for each separate operation. However, one might wish to give only the range of these numbers;

---

[1] There is no difficulty in distinguishing this use from the use of the slash as a division sign; the latter takes priority, since it is the more specific use of the slash.

this is done without introducing a new attribute (i.e., operation-time-range) simply by indicating that the value is a range:

Pc(operation-time: 4 ~16 μs)

Similarly, one could have given typical times or average times (under some assumed frequency mix of instructions):

Pc(operation-time: 4 μs)
Pc(operation-time: average: 8.1 μs)

The primary advantage of this notational convention, which permits descriptions of values to be used in place of actual values whenever desired, is that it keeps the number of attributes that have to be defined much smaller than otherwise.

A PMS example using the DEC PDP-8

Let us now describe the PMS structure of an actual, though small, general-purpose computer, the DEC LINC-8, which is a PDP-8 with a LINC processor. Figure 1 gives the detailed PMS diagram. In explaining it, we will concentrate on making the notation clear rather than on discussing substantive features of the system (which are described in Chap. 5). A simplified PMS diagram of the system shows its essential structure:

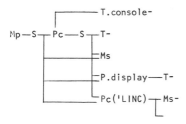

This shows the basic Mp-Pc-T-X structure of a C with the addition of a secondary memory (Ms) and two processors, one of which, Pc('LINC), has its own Ms. Two switches are used: the I/O Bus which permits access to all the devices, and the Data Break to Mp via Pc for high-data-rate devices. There are many other switches in the actual system, as one can see from Fig. 1; for example, Mp is really one to eight separate modules connected by a switch S to Pc. Also there are many T's connected to the input/output switch, Sio, which we collapsed as a single T, and similarly for S(' Data Break).

Consider the Mp module. The specifications assert that it is made with core technology, that its word size is 13 bits (12 data bits plus one other with a different function); that its size is 4,096

words; and that its operation time is 1.5 μs. We could have written the same information as

M(function:primary; technology:core; operation-time: 1.5 μs; size: 4096 w; word: (12 + 1) b)

In Fig. 1 we wrote only the values, suppressing the attributes, since moderate familiarity with memories permits an immediate inference about what attributes are involved. For example, it is common knowledge that computer memories store information in words; therefore 4096 w must be the number of words in the memory. As another example, we did not specify the function of the additional bit in the word when we wrote (12 + 1) b. An informed reader will assume this to be a parity bit, since this is the common reason for having an extra bit in a word. If the extra bit had some unusual function, we would have needed to define it. That is, in the absence of additional information, the most common interpretation is to be assumed.

In fact, we could have been even more cryptic and still communicated with most readers:

M.core(1.5 μs/w; 4 kw; 12 b)

This corresponds to the phrase "A 12-bit, 1.5-μs, 4k core store," which is intelligible to any computer engineer. The 4 kw stands for $4 \times 1,024 = 4,096$, which again is known to computer engineers; however, if someone less informed took it to be $4 \times 1,000 = 4,000$, no real harm would be done.

Consider the magnetic tapes for Pc. Since there are eight possible tapes that make use of the same controller, K, through a switch S, we label them #0 through #7. Actually, # is an abbreviation for index, which is an attribute like any other, whose values are integers. Since the attribute is a unique character, we do not have to write #:3 (although we could). The additional parameters give information about the physical attributes of the encoding. These are alternative values, and any tape has only one of them. We use a vertical bar (|) to indicate this (as in BNF notation for grammars). Thus, 75 | 112 in/s says that one can have a tape with a speed of 75 inches per second or one with 112 inches per second, but not a tape which can be switched dynamically to run at either speed.

For many of the components no further information is given. Thus, knowing that M.magnetic_tape is connected to a control and from there to the Pc tells generally what that K does. It is a "tape controller" which evokes all the actions of the tape, such as read, write, rewind; therefore these actions do not have to be done by Pc. The fact that there is only one K for many Ms's implies that only one tape can be accessed at a time. Other infor-

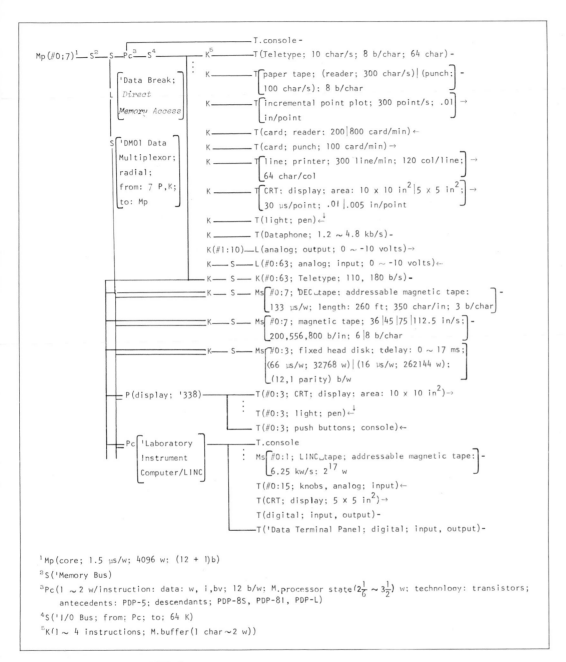

^{1}Mp(core; 1.5 μs/w; 4096 w: (12 + 1)b)

^{2}S('Memory Bus)

^{3}Pc(1 ~2 w/instruction: data: w, i,bv; 12 b/w: M.processor state ($2\frac{1}{6}$ ~ $3\frac{1}{2}$) w; technology: transistors; antecedents: PDP-5; descendants; PDP-8S, PDP-8I, PDP-L)

^{4}S('I/O Bus; from; Pc; to; 64 K)

^{5}K(1 ~ 4 instructions; M.buffer(1 char ~2 w))

Fig. 1. DEC LINC-8-PDP-8 PMS diagram.

mation could be given, although that just provided is all that is usual in specifying a controller in an overall description of a system. (The next level of detail goes to the structure of the actual operations and instructions and belongs to the ISP level, not the PMS level.)

We have used several different ways of saying the same thing in Fig. 1 in order to show the range of descriptive notations. Thus the 64 Teletypes are shown by describing a single connection through a switch and putting the number of links in the switch above the connecting line.

Consider, finally, the Pc in Fig. 1. We have given a few parameters: the data-types, the processor state, the descendants, etc. These few parameters hardly define a processor. Several other important parameters are easily inferred from the Mp. The basic operation time in a processor is a small multiple of the read time of its Mp. Thus it is predictable that Pc stores and reads information in 2×1.5 μs (one for instruction fetch, one for data fetch). Again, where this is not the case (as in the CDC 6600) it is necessary to say so. Similarly, the word size in the Pc is the same as the word size of the Mp: 12 data bits. More generally, the Pc must have instructions that take care of evoking all the components of the PMS structure. These instructions do not see the switches and controls as distinct entities; rather, they speak directly to the operation of the M's and T's connected via these switches and controls.

Other summary parameters could have been given for the Pc. None of them would come close to specifying its behavior uniquely, although to those knowledgeable in computers still more can be inferred from the parameters given. For instance, knowing both the data-types available in a Pc and the number of instructions, one can come very close to predicting exactly what the instructions are. Nevertheless, the way to describe a Pc in full detail is not to add larger and larger numbers of summary parameters. It is more direct and more revealing to develop a description at the level of instructions, which is the ISP description.

Let us end this introduction to the PMS descriptive system by returning to a critical item in its design philosophy. A descriptive scheme for systems as complex and detailed as digital computers must have the ability to range from extremely complete to highly simplified descriptions. It must permit highly compressed descriptions as well as extensive ones and must permit the selective suppression or amplification of whatever aspects of the computer system are of interest to the user. PMS attempts to fulfill these criteria by providing simple conventions for detailed description with additional conventions that permit abbreviation and abstractions, almost without limit. The result is a notation that may seem somewhat fluid, especially on first contact in such a brief intro-

duction as this. But once assimilated, PMS seems to allow some of the flexibility of natural language within enough notational controls to enhance communication considerably.

ISP level of description

The behavior of a processor is completely determined by the nature and sequence of its operations. This sequence is completely determined by a set of bits in Mp, called the program, and a set of interpretation rules that specify how particular bit configurations evoke the operations. Thus, if we specify the nature of the operations and the rules of interpretation, the actual behavior of the processor depends solely on the particular program in Mp (and also on the initial state of data). This is the level at which the programmer wants the processor described—and which the programming manual provides—since he himself wishes to determine the program. Thus the ISP (Instruction-set processor) description must provide a scheme for specifying any set of operations and any rules of interpretation.

Actually, the ISP descriptive scheme need only be general enough to cover some broad range of possibilities adequate for past and current generations of machines along with their likely descendants. As we saw earlier when discussing the PMS level, there are certain restrictions that can be placed on the nature of a computer system, specializing it from the more general concept of a discrete state system. It processes a medium, called information; it is a system of discrete components linked together by information transfers; and each component is characterized by a small set of operations. These assumptions are built into the PMS descriptive scheme in an integral way. Similarly, for the ISP level we can add two more such restrictions, which will in turn provide the shape of its descriptive scheme.

The first specialization is that a program can be conceived as a distinct set of *instructions*. Operationally, this means that some set of bits is read from the program in Mp to a memory within P, called the instruction register, M.instruction/M.i. This set of bits then determines the immediately following sequence of operations. Only a single operation may be determined, as in setting a bit in the internal state of the P; or a substantial number of operations may be determined, as in a "repeat" instruction that evokes a search through Mp. In a typical one- or two-address machine the number of operations per instruction ranges from two to five. In any event, after this sequence of operations has occurred, the next instruction to be fetched from Mp is determined and obtained. Then the entire cycle repeats itself.

The cycle of activity we have just described is called the *interpretation cycle*, and the part of the P that performs it is called the *interpreter*. The effect of each instruction can be expressed entirely in terms of the information held in memories at the end of the cycle (plus any changes made to the outside world). During execution, operations may have internal states of their own as sequential circuits which are not represented as bits in memories. But by the end of the interpretation cycle, whatever effect is to be carried on to a later time has been staticized in bits in some memory.[1]

The second additional specialization is on the data-operations. A processor's total set of operations can be divided into two parts. One part contains those necessary to operate other components given in the PMS diagram: links, switches, memories, transducers, etc. The operations associated with these components and the extent to which they can be indirectly controlled from P are highly restrained by the basic nature of the components and their controls. The second part contains those operators associated with a processor's D component. So far we have said nothing at all about them, except to exclude them completely from all PMS components except P. These are the operations that produce bit patterns with new meaning—that do all the "real" processing or changing of information.[2] If it were not for data-operations, the system would merely transmit information. As we noted in our original definitions (page 17) a P (including a D) is the only component capable of directly changing information. A P can create, modify, and destroy information in a single operation. As we noted earlier, D's are like the primitive components in an analog computer. Later, when we express instruction sets as simple arithmetic expressions, the D's are the primitive operators, for example,

$+, -, \times, /, \times 2^n, \wedge, \vee, \oplus$, concatenation, etc., which are evoked by the instruction-set-interpreter part of a processor.

The specialization is that all the data-operations can be characterized as working on various *data-types*. For example, there is a data-type called the signed integer, and there are data-operations that add two signed integers, subtract them, multiply them, take their absolute value, test for which of the two is greater, etc. A data-type is a compound of two things: the referent of the bit pattern (e.g., that this set of bits refers to an integer in a certain range) and the representation in the bit pattern (e.g., that bit 31 is the sign, and bits 30 to 0 are the coefficients of successive powers of 2 in the binary representation of the integer). Thus a processor may have several data-types for representing numbers: unsigned integers, signed integers, single precision floating point, double precision floating point, etc. Each of these is a distinct data-type, because it requires distinct operations to process it. On occasion, operations for several data-types may all be encoded into a single instruction with a data-type subfield that selects whether the data are fixed or floating point. The operations are still separate, no matter how packaged, and so their data-types remain distinct.

With these two additional specializations—instructions and data-types—we can define an ISP description of a processor. A processor is completely described at the ISP level by giving its *instruction set* and its *interpreter* in terms of its *operations, data-types,* and *memories*.

Let us concentrate first on the instruction set, leaving the interpreter until later. The effect of each instruction is described by an *instruction-expression*, which has the form

condition → action-sequence

The *condition* describes when the instruction will be evoked, and the *action-sequence* describes what transformations of data take place between what memories. The right arrow (→) is the control action (of a K) of evoking an *operation*.

Recall that all operations in a computer system result in modifications of bits in memories. Thus each action in a sequence ultimately has the form

memory-expression ← data-expression

The left arrow (←) is the transmit operation of a link and corresponds to the ALGOL assign operation. The left side must describe the memory location that is affected; the right side must describe the information pattern that is to be placed in that memory location. The details of data expressions and memory expressions are patterned on standard mathematical notation and are communi-

---

[1] This description holds true for a P with a single active control (the interpreter). Some P's (e.g., the CDC 6600) have several active controls and get involved in "overlapping" several instructions and in reordering operations according to the data and devices available. With these, a more complex statement is required to express the same general restriction we have been stating for simple P's: that the program can be decomposed into a sequence of bit sets (the instructions), each of which has local control over the behavior of the P for a limited period of time, with all interinstruction effects being staticized as bits in M's.

[2] In principle, this view that only D components do "real" processing is false. It can be shown that a universal Turing machine can be built from M, S, L, and K components. The key operation is the write operation into M, which suffices to construct arbitrary bit patterns under suitably controlled switches. Hence arbitrary data operations can be built up. The stated view is correct in practice in that the data-operations provided in a P are highly efficient for their bit transformations. Only the foolish add integers in a modern computer by table look-up.

cated most easily by examples. The same is true of the condition, which is a standard expression involving boolean values and relations among memory contents.

Before we get to the examples, let us note two features of the action sequence. The first is that each action in the sequence may itself be conditional, i.e., of the form, "condition → action-sequence." The second is that some actions are sequentially dependent on each other, because the result of one is used as an input to the other; on other occasions a set of actions are independent and can occur in parallel. The normal situation is the parallel one. Thus, in the action sequence

$$Y_1 \leftarrow X_1; \ Y_2 \leftarrow X_2; \ Y_3 \leftarrow X_3; \ Y_4 \leftarrow X_4$$

all the transfers of information may be considered simultaneous. In particular, all the X's have their values defined by the situation before the transfer. For example, if A and B are two registers, then

$$(A \leftarrow B; \ B \leftarrow A)$$

exchanges the contents of A and B. When sequence is required, the term "next" is used; thus

$$(A \leftarrow B; \ next \ B \leftarrow A)$$

transfers the contents of B to A and then transfers it back to B, leaving both A and B holding the original contents of B (and so this contrived example is essentially just $A \leftarrow B$).

An ISP example using the DEC PDP-8

The memories, operations, instructions, and data-types all need to be declared for a processor. Again these are most easily explained by example, although full definitions are given in the Appendix at the end of the book. Consequently, let us examine the ISP description of the Pc of the PDP-8, given in Fig. 2 (the PDP-8 is explained fully in Chap. 5). Throughout the book the ISP descriptions of computers follow a more highly structured format than the ISP notation requires, in order to help the reader see the similarities among the computers.

Processor state. We first need to specify the memories of the Pc in detail, providing names for the various bits. Thus,

AC⟨0:11⟩   *the accumulator*

is a memory called AC, with 12 bits, labeled at 0 and 11 from the left. Comments are given in italics[1]—in this case that AC is

---

[1]There are a few features of the notation, such as the use of italics, which are not easily carried over into current computer character sets. Thus, the ISP of Fig. 2 is a publication language.

called the accumulator (by the designers of the PDP-8). AC corresponds to an actual register in the Pc. However, the ISP does not imply any particular implementation, and names may be assigned to various sets of bits purely for descriptive convenience. The colon is used to denote a range or list of values. Alternatively, we could have listed each bit, separating the bit names by commas, as

AC⟨0,1,2,3,4,5,6,7,8,9,10,11⟩

Having defined a second memory, L (which has only a single bit), one could define a combined register, LAC, in terms of L and AC as

LAC⟨L,0:11⟩: = L□AC

The colon-equal (: =) is used for definition, and the middle square box (□) denotes concatenation. Note that the bit named L of register LAC merely happens to correspond to the 1-bit L register.

Primary memory state. In dealing with addressed memory, either Mp or various forms of working memory within the processor, we need to indicate multidimensional arrays. Thus

Mp[0:7777$_8$]⟨0:11⟩

gives primary memory as consisting of 10000_8 (i.e., base 8) words of 12 bits each, being addressed as indicated. Such an address does not necessarily reflect the switching structure through which the address occurs, though it often will. (Needless to say, it reflects only addressing space, and not how much actual M is available in a PMS structure.) In general, only memory within the processor will occur as operands of the processor's operators. The one exception is primary memory (Mp), which was defined as a memory external to a P but directly accessible from it.

In writing memories it is natural to use base 10 for all numbers and to consider the basic i-unit of the memory to be a bit. This is always assumed unless otherwise indicated. Since we used base 8 numbers above for specifying the addressing range, we indicated the change of number base by a subscript, in standard fashion. If a unit of information other than the bit were to be used, we would subscript the angle brackets. Thus

Mp[0:7777$_8$]⟨0:1⟩$_{64}$

reflects the same memory. The choice carries with it, of course, some presumption of organization in terms of base 64 characters, but this would show up in the specification of the operators (and is not true, in fact, of the PDP-8). We can also have multidimensional memories (i.e., arrays), though no examples occur in

Fig. 2. These add the extra dimensions with an extra pair of brackets, for example,

M[a:b][c:d]$\cdots$[g:h]⟨x:y⟩

The PDP-8 memory might better be described as:

Mp[0:7][0:31][0:127]⟨0:11⟩

representing 8 memory fields with 32 pages per field, 128 words per page, and 12 bits per word.

Instruction format. It is possible to have several names for the same set of bits; e.g., having defined instruction⟨0:11⟩ we define the format of the instruction as follows:

op⟨0:2⟩ := instruction⟨0:2⟩
indirect_bit/ib := instruction⟨3⟩
page_0_bit/p := instruction⟨4⟩
page_address⟨0:6⟩ := instruction⟨5:11⟩

The colon-equal (:=) is used to allow us to assign names to various parts of the instruction. In effect, we are making a definition which is equivalent to the conventional diagram for the instruction:

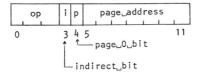

Notice that in page_address the names of all the bits have been shifted, e.g., page_address⟨4⟩ := instruction⟨9⟩.

The Appendix gives the permissible alphabet of symbols for ISP. In general, a "name" can be any combination of uppercase and lowercase letters and numerals, not including names which would be considered numbers (integers, mixed numbers, fractions, etc.). A compound name can be sequences of names separated by spaces ( ). In order to make certain compound names more readable, a space symbol (_) may optionally be used to signify the non-printing character. Periods (.) and hyphens (-) are also used.

The instruction set. With all the registers defined, we can give the instructions. These are shown on the second page of Fig. 2 (there are some unexplained parts left on the bottom of the first page, to which we will return). The second page is actually a single expression, named Instruction_execution, which consists of a list of instructions. They are listed vertically down the page for ease of reading. Each instruction consists of a condition and an action

sequence, separated by the condition arrow ($\rightarrow$). In this case the condition is an expression of the form (op = octal digit). Recall that op is instruction⟨0:2⟩, and so this expresses the condition that the operation code of the machine have a particular value. Each condition has been given a name in passing; e.g., "and" is the name of (op = 0). This provides the correspondence between the operation code and the mnemonic name of the operation code. If this correspondence had been established elsewhere, or if we did not care what numerical operation code the "and" instruction is, we could have written

and $\rightarrow$ (AC $\leftarrow$ AC $\wedge$ M[z])

We would not have known what condition the name "and" stood for but could have surmised (with little difficulty) that it was simply an equality test on the operation code. We will do this on a number of the ISP descriptions later in the book. Most generally the form of an instruction is written as

two's complement add/tad(:= op = 1) $\rightarrow$
(L$\square$AC $\leftarrow$ L$\square$AC + M[z])

Here, we simultaneously define the action of the tad instruction, its name, an abbreviation for the name, and the conditions for tad's execution. The parentheses are, in effect, a remark to allow an inline definition. For example, the above single ISP statement is equivalent to

two's complement add/tad $\rightarrow$ (L$\square$AC $\leftarrow$ L$\square$AC + M[z])

followed by

tad := (op = 1)

All the instructions in the list constitute the total instruction repertoire of the Pc. Since all the conditions are disjoint, one and only one condition will be satisfied when a given instruction is interpreted; hence one and only one action sequence will occur. Actually, all operation codes might not be present, and so there would be some illegal op codes that would evoke no action sequence. The act of selection is usually called operation decoding. Again, ISP implies no particular mechanism by which this is carried out. Normally a logic circuit works directly on the op part of the instruction register, and the way op codes are assigned is significant for the complexity of this decoding circuit. Thus, sometimes one exhibits the instructions in a two-dimensional decoding diagram that makes it evident what these bit patterns are (see Fig. 2 in Chap. 5), rather than in a linear list.

It might be wondered why we do not in general introduce some

Pc State

   AC<0:11>   *Accumulator*

   L   *Link bit/AC extension for overflow and carry*

   PC<0:11>   *Program Counter*

   Run   *1 when Pc is interpreting instructions or "running"*

   Interrupt_state   *1 when Pc can be interrupted; under programmed control*

   IO_pulse_1; IO_pulse_2; IO_pulse_4   *IO pulses to IO devices*

Mp State
Extended memory is not included.

   M[0:7777$_8$]<0:11>

      Page_0[0:177$_8$]<0:11> := M[0:177$_8$]<0:11>   *special array of directly addressed memory registers*

      Auto_index[0:7]<0:11> := Page_0[10$_8$:17$_8$]<0:11>   *special array when addressed indirectly, is incremented by 1*

Pc Console State
Keys for start, stop, continue, examine (load from memory), and deposit (store in memory) are not included.

   Data switches<0:11>   *data entered via console*

Instruction Format

   instruction/i<0:11>

      op<0:2>   := i<0:2>   *op code*

      indirect_bit/ib   := i<3>   *0, direct; 1 indirect memory reference*

      page_0_bit/p   := i<4>   *0 selects page 0; 1 selects this page*

      page_address<0:6> := i<5:11>

      this_page<0:4>   := PC'<0:4>

      PC'<0:11>   := (PC<0:11> -1)

      IO_select<0:5>   := i<3:8>   *selects a T or Ms device*

      io_p1_bit   := i<11>   *these 3 bits control the selective generation of -3 volts,*

      io_p2_bit   := i<10>   *0.4 μs pulses to I/O devices*

      io_p4_bit   := i<9>

      sma   := i<5>   μ *bit for skip on minus AC, operate 2 group*

      sza   := i<6>   μ *bit for skip on zero AC*

      snl   := i<7>   μ *bit for skip on non zero Link*

Effective Address Calculation Process

   z<0:11> := (*effective*

    ¬ib → z'';

    ib ∧ (10$_8$ ≤ z'' ≤ 17$_8$) → (M[z''] ← M[z''] + 1; next);   *auto indexing*

    ib → M[z''])

   z'<0:11> := (¬ ib → z''; ib → M[z''])

   z''<0:11> := (page_0_bit → this_page□page_address;   *direct address*

               ¬page_0_bit → 0□page_address)

μ *microcoded instruction or instruction bit(s) within an instruction*

Fig. 2. DEC PDP-8 ISP description.

Instruction Interpretation Process

Run ∧ ¬ (Interrupt_request ∧ Interrupt_state) → (*no interrupt interpreter*
    instruction ← M[PC]; PC ← PC + 1; next   *fetch*
    instruction_execution);   *execute*
Run ∧ Interrupt_request ∧ Interrupt_state → (*interrupt interpreter*
    M[0] ← PC; Interrupt_state ← 0; PC ← 1)

Instruction Set and Instruction Execution Process

Instruction_execution := (
    and (:= op = 0) → (AC ← AC ∧ M[z]);   *logical and*
    tad (:= op = 1) → (L☐AC ← L☐AC + M[z]);   *two's complement add*
    isz (:= op = 2) → (M[z'] ← M[z] + 1; next   *index and skip if zero*
                       (M[z'] = 0) → (PC ← PC + 1));
    dca (:= op = 3) → (M[z] ← AC; AC ← 0);   *deposit and clear AC*
    jms (:= op = 4) → (M[z] ← PC; next PC ← z + 1);   *jump to subroutine*
    jmp (:= op = 5) → (PC ← z);   *jump*
    iot (:= op = 6) → (μ *in out transfer, microprogrammed to generate up to 3 pulses*
        io_p1_bit → IO_pulse_1 ← 1; next   *to an io device addressed by IO_select*
        io_p2_bit → IO_pulse_2 ← 1; next
        io_p4_bit → IO_pulse_4 ← 1);
    opr (:= op = 7) → Operate_execution   *the operate instruction is defined below*
)   *end Instruction execution*

Operate Instruction Set
The microprogrammed operate instructions: operate group 1, operate group 2, and extended arithmetic are defined as a separate
instruction set.

    Operate_execution := (
        cla (:= i<4> = 1) → (AC ← 0);   *clear AC. Common to all operate instructions.*
        opr_1 (:= i<3> = 0) → (*operate group 1*
            cll (:= i<5> = 1) → (L ← 0); next   μ *clear link*
            cma (:= i<6> = 1) → (AC ← ¬ AC);   μ *complement AC*
            cml (:= i<7> = 1) → (L ← ¬ L); next   μ *complement L*
            iac (:= i<11> = 1) → (L☐AC ← L☐AC + 1); next   μ *increment AC*
            ral (:= i<8:10> = 2) → (L☐AC ← L☐AC × 2 {rotate});   μ *rotate left*
            rtl (:= i<8:10> = 3) → (L☐AC ← L☐AC × 2² {rotate});   μ *rotate twice left*
            rar (:= i<8:10> = 4) → (L☐AC ← L☐AC / 2 {rotate});   μ *rotate right*
            rtr (:= i<8:10> = 5) → (L☐AC ← L☐AC / 2² {rotate})); μ *rotate twice right*
        opr_2 (:= i<3,11> = 10) → (*operate group 2*
            skip condition ⊕ (i<8> = 1) → (PC ← PC + 1); next   μ *AC,L skip test*
                skip condition := ((sma ∧ (AC < 0)) ∨ (sza ∧ (AC = 0)) ∨ (snl ∧ L))
            osr (:= i<9> = 1) → (AC ← AC ∨ Data switches);   μ *"or" switches*
            hlt (:= i<10> = 1) → (Run ← 0));   μ *halt or stop*

        EAE (:= i<3,11> = 11) → EAE_instruction_execution)   *optional EAE description*

additional conventions into the language, e.g., list the instructions in a table with their mnemonic names in a special column, rather than write the whole affair as an expression. (In fact, if you examine the first page of Fig. 2, you will note that the entire description of the PDP-8 Pc is a single expression.) The reason is that although many processors fit such a format very well, not all do so, e.g., microprogrammed machines. By making the ISP description a general expression for evoking action-sequences, we obtain the generality we need to cover all the variations. We will have two examples with the PDP-8 itself: the microprogrammed feature and the fact that the interpretive cycle simply becomes part of the total expression for the behavior of the processor.

Let us now consider the action-sequence. We use standard mathematical infix notation. Thus we write

$$AC \leftarrow AC \wedge M[z]$$

This indicates that the word in Mp at address z is ANDed with the accumulator and the result left in the accumulator. It is assumed that the operation designated by $\wedge$ is well understood. (The $\leftarrow$, of course, is the transmit operation.) Each processor will have a basic set of operations that work on data-types of the machine. Here the data-type is simply the 12-bit word viewed as an array of bits.

Operators need not involve memories actually within the Pc (the processor state). Thus,

$$Mp[z] \leftarrow Mp[z] + 1$$

expresses a change in a word in Mp directly. That this must be mechanized in the PDP-8 by means of some register in Pc is irrelevant to the ISP description.

We also use functional notation; for example,

$$AC \leftarrow abs(AC)$$

replaces the contents of the AC with its absolute value. When an action has an unspecified function or operation we generally write

$$A \leftarrow f(A,B, \ldots) \quad \text{or} \quad A \leftarrow u\,B \quad \text{or} \quad A \leftarrow B\,b\,C$$

for function, unary operation, and binary operation, respectively.

Effective-address calculation process. In the examples just given we used z as the address in Mp. This is the effective address and

is defined as a conditional expression (in the manner of ALGOL or LISP):

$$
\begin{aligned}
z\langle 0{:}11\rangle := (\\
\neg\ ib \rightarrow z'';\\
ib \wedge (10_8 \leqslant z'' \leqslant 17_8) \rightarrow (M[z''] \leftarrow M[z''] + 1);\ \text{next}\\
ib \rightarrow M[z''])
\end{aligned}
$$

The right arrow ($\rightarrow$) is analogous to the conditional sign used in the main instruction, equivalent to the "if . . . then . . ." of ALGOL. The parentheses are used to indicate grouping in the usual fashion. However, we arrange expressions on the page to make reading easier.

As the expression for z shows, we permit conditionals within conditionals and also the nesting of definitions (z is defined in terms of z''). Again, we should emphasize that the structure of such definitions may reflect the underlying hardware organization, but it need not. When describing existing processors, as in this book, the ISP description often reflects the hardware. But if one were designing a processor, the ISP expressions would be stated as design objectives for the RT structure, and the latter might differ considerably.

Special note should be taken of the opr instruction (op = 7) in Fig. 2, since it provides a microprogramming feature. There are two separate options depending on instruction$\langle 3\rangle$ being 0 or 1. But common to both is the operation of clearing the AC (or not), associated with instruction$\langle 4\rangle$. Then, within one option (instruction$\langle 3\rangle$ = 0) there are a series of independently executable actions (following the clearing of L); within the other (instruction$\langle 3\rangle$ = 1), there are three independently settable control actions. The nested conditionals and the use of "next" to force sequential behavior make it easy to see exactly what is going on (in fact a good deal easier than describing it in natural language, as we have been doing).

The instruction interpreter. We now have all the instructions defined for the PDP-8, including the effective-address computation (z). It remains to define the interpreter. From a hardware point of view, an interpreter consists of the mechanisms for fetching a new instruction, for decoding that instruction and executing the operations so designated, and for determining the next instruction. A substantial amount of this total job has already been taken care of in the part of the ISP that we have just explained. Each instruction carries with it a condition that amounts to one fragment of the decoding operation. Likewise, any further decoding of the instruction that might be done in common by the interpreter (rather than by the individual operation circuits) is implied in the expressions for each instruction, and by the expression for the effective address. The only thing that is left is to fetch the next instruction and to execute it.

In a standard machine, there is a basic principle that defines operationally what is meant by the "next instruction." Normally the current instruction address is incremented by 1, but other principles are used (e.g., on a processor with a cyclic Mp). In addition, several specific operations exist in the repertoire that can affect what program is in control. The basic principle acts like a default condition: If nothing specific happens to determine program control, the normal "next" instruction is taken. Thus, in the PDP-8 we get an interpretation process that is essentially the classic fetch-execute cycle (ignoring interrupts):

$$\text{Run} \to (\text{instruction} \leftarrow M[PC];\ PC \leftarrow PC + 1;\ \text{next } \textit{fetch}$$
$$\text{Instruction\_execution}) \qquad\qquad \textit{execute}$$

The sequence is evoked so long as Run is true (i.e., its bit value is 1). The processor will simply cycle through the sequence, fetching and then executing the instruction. In the PDP-8 there exists a halt operation that sets Run to be 0, and the console keys can, of course, stop the computer. It should be noted that the ISP descriptions in this book do not, generally, include console behavior.

A state diagram (Fig. 3) is useful to represent the behavior of the instruction-interpretation process. As an instruction is interpreted, the system moves from state to state. Any of the states can be null, in which case a simple transition is to be made to the successor of the null state. The K(instruction interpreter) con-

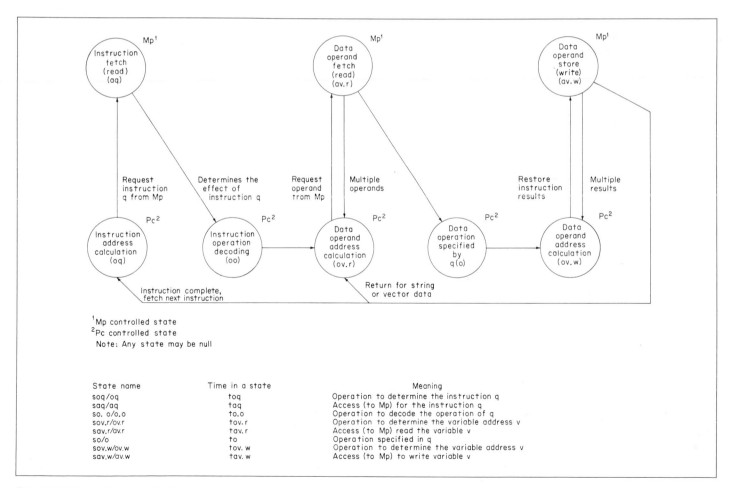

Fig. 3. ISP interpretation state diagram.

trols these movements according to the information in the instruction. Which states are null and which of multiple alternative transitions occur depend on the instruction being interpreted.

Within each state, various operations are carried out, under the control of subordinate K's. Note that the upper states in Fig. 3 are controlled by the Mp whereas the lower ones are controlled by the Pc. We have tried to use a simple mnemonic scheme to label these states: o for operation, q for instruction, a for access, r for read, and w for write. Similarly, we prefix the state with t to indicate the time duration of the state, and we may prefix the state by s.

Figure 3 is somewhat more detailed than is usual. We will use it in Chap. 3 to describe a number of different processors. However, the figure simplifies the familiar fetch-execute cycle:

$$\text{Fetch:} \quad \{oq, aq\}$$
$$\text{t.fetch} = toq + taq$$
$$\text{Execute:} \{oo, ov.r, av.r, o, ov.w, av.w\}$$
$$\text{t.execute} = too + tov.r + tav.r + \cdots + tov.r$$
$$+ tav.r + \cdots + to + tov.w + tav.w$$

Consider, by way of example, the tad instruction of the PDP-8, using the general state diagram of Fig. 3. From the ISP, the net effect is

$$\text{Run} \rightarrow (\text{instruction} \leftarrow M[PC]; \ PC \leftarrow PC + 1; \ \text{next}$$
$$\text{tad} \ (:= op = 1) \rightarrow (L\square AC \leftarrow L\square AC + M[z]))$$

where

$$z\langle 0:11 \rangle := (\textit{specifies the effective-address calculation process})$$

The state diagram has more detail to explain the computer's behavior with respect to timing and its temporary registers. (Note a complete state diagram for the physical PDP-8 is given in Fig. 11 of Chap. 5.) The actual state table appears on page 31.

Notice again that the ISP description does not determine the way the processor is to be organized to achieve this sequencing or to take advantage of the fact that many instructions lead to similar sequences. All it does is specify unambiguously what operations must be carried out for a program in Mp. The ISP description does specify the actual format of the instruction and how it enters into the total operation, although sometimes indirectly. For example, in the case of the and instruction ($op = 0$), the definition of AC shows that the AC does not depend on the instruction, and the definition of z shows that z depends on other fields of the

instruction (indirect_bit, page_0_bit, page_address). Likewise, the form of the ISP expression shows that AC and PC both enter into the instruction implicitly. That is, in the ISP description all dependence on memory is explicit.[1]

Data-types and data-operations

This completes the description of the ISP for the PDP-8. For more complex machines the number of data-types and the operations on them are much more extensive. Then the data-types may be declared independently of the instruction set, in the same manner as we declared memory.

In fact, the one major piece of organization in the structure of processors at the ISP level that has not appeared in our example involves the data-types. Each data-type has a set of operations that are proper to it. Add, subtract, multiply, and divide are all proper to any numerical data-type, as well as absolute value and negation. Not all of these need exist in a computer just because it has the data-type, since there are several alternative bases, as well as some levels of completeness. For instance, notice that the PDP-8 first of all does not have multiply and divide (unless one has its special option), thus having a relatively minimal level of arithmetic operations, and second, it does not have a subtract operation, using a two's complement add, which permits negation ($-AC$) to be accomplished by complementation ($\neg AC$) followed by add 1. Still, the options are rather few, provided one has decided to include a given data-type in the repertoire. In the Appendix at the end of the book are given with each of the data-types (or classes thereof) the sets of operations that are proper to that data-type.

The PDP-8, for example, does not have several data representations for what is, externally considered, the same entity. An operator that does a floating add and one that does an integer add are not the same. However, we will denote both by the same symbol (in this case, $+$), indicating the difference parenthetically after the expression. Alternatively, the specification of the data type can be attached to the data. Thus, in the IBM 7094 we have the instructions

---

[1] This is not correct, actually. In physically realizing an ISP description, additional memories may be utilized (they may even be necessary). It can be said that in the ISP description these memories are implicit. However, a consistent and complete description of an ISP can be made without use of these additional memories whereas with, say, a single-address machine it does not seem possible to describe each instruction without some reference to the implicit memories—as we see in the effective-address calculation procedures where definitions look much like registers.

States	Time	ISP effect	Operational description
soq ↑	toq	MA ← PC; PC ← PC + 1	Calculate the address of the instruction, q, and calculate the address of the next instruction, q + 1. The address is stored in the address register, MA, used to control the access.
S.fetch saq ↓	taq	MB ← M[MA]	Fetch the data from memory location, M[MA] (i.e., essentially M[PC]), and place the result in a buffer (temporary) register.
soo ↑	too	IR ← MB⟨0:2⟩	Calculate and decode the instruction.
sov.r S.execute	tov.r	MA ← f(MB,IR)	Calculate the address of the data.
sav.r	tav.r	MB ← M[MA]	Fetch the data from Mp.
so ↓	to	L □ AC ← L □ AC + MB	Do the operation specified by the instruction.

Add → (AC ← AC + M[e]);
Add and carry logical word/ACL → (
    AC ← AC + M[e] {unsigned.integer});
Floating add/FAD → (AC ← AC + M[e] {sf});
Unnormalized floating add/UFA → (AC ← AC + M[e] {suf});
Double-precision floating add/DFAD → (
    ACMQ ← ACMQ + M[e]□M[e + 1] {df});
Double-precision unnormalized floating add/DUFA → (
    ACMQ ← ACMQ + M[e] □ M[e + 1] {duf})

The first one, without a special indicator of data-type, is taken to be integer addition; the next, unsigned integer; the next, single precision floating point; the next, unnormalized single precision floating point; the next, double precision floating point; and the last, unnormalized double precision floating point. Although there are often clues that could be used to infer which form of addition is being defined (e.g., double precision takes two words) we label all but the integer operation.

We use braces { } to differentiate which operation is being performed in the above examples. Thus, above, the data-type is enclosed in braces and refers to all the memory elements (operands) of the expression. Alternatively, we use braces as a modifier on any memory to signify the information meaning. For example, a fixed point to floating point data-conversion operation would be given as

AC{floating} ← AC{fixed}

We also use braces as a modifier for the operation-type. For example, shifting (left or right) can be a multiplication or division by a base, but it is not always an arithmetic operation. In the PDP-8, for instance, we have

L □ AC ← L □ AC × 2 {rotate}

where the end bits L and AC⟨11⟩ are connected when a shift occurs (the operator is also referred to as a circular shift).

In general, the nature of the operations used in processors are sufficiently familiar to the computer professional that no definitions are required, and they can all be taken as primitive. It is necessary only to have agreed upon conventions for the different data representations used. The Appendix provides the basic abbreviations. In essence, a data-type is made up recursively of a concatenation of subparts, which themselves are data-types. This concatenation may be an iteration of a data-type to form an array. Fig. 4 shows the structure of various data-types and how each is built from more primitive data-types.

If required, an operation can be defined in terms of other (presumably more primitive) operations. It is necessary first to define the data format explicitly (including perhaps some additional memory). Variables for the operands are permitted in the natural way. For example, binary single-precision floating-point multiplication on a 36-bit machine could be defined in terms of the data fields as follows:

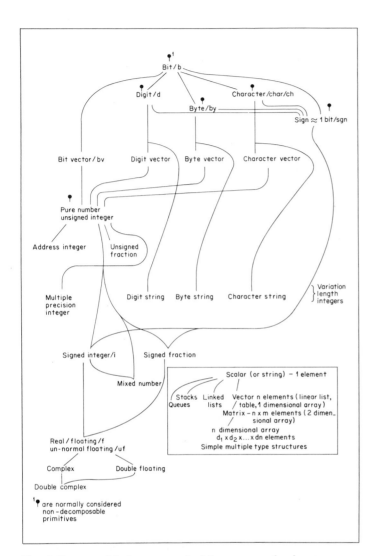

Fig. 4. Common data-types recognized by processor hardware.

$$\text{sf mantissa/mantissa} := \langle 0{:}27\rangle$$
$$\text{sf sign/sign} \qquad\quad := \langle 0\rangle$$
$$\text{sf exponent/exponent} := \langle 28{:}35\rangle$$
$$\text{sf exponent\_sign} \quad := \langle 28\rangle$$
$$x1 := x2 \times x3\{sf\} := ($$
$$\quad x1 \text{ mantissa} := x2 \text{ mantissa} \times x3 \text{ mantissa};$$
$$\quad x1 \text{ exponent} := x2 \text{ exponent} + x3 \text{ exponent};$$
$$\quad \text{next } x1 := \text{normalize }(x1)\ \{sf\})$$

where normalize is

$$x1 := \text{normalize}(x2)\ \{sf\} := ($$
$$\quad (x1 \text{ mantissa} = 0) \rightarrow (x1 \text{ exponent} := 0);$$
$$\quad ((x2 \text{ mantissa} \neq 0) \wedge (x2\langle 0\rangle = x2\langle 1\rangle)) \rightarrow ($$
$$\qquad x1 \text{ mantissa} := x2 \text{ mantissa} \times 2;$$
$$\qquad x1 \text{ exponent} := x2 \text{ exponent} - 1; \text{ next}$$
$$\qquad x1 := \text{normalize}(x2)\ \{sf\}))$$

Three additional aspects need to be noted with respect to data-types: two substantive and one notational. First, not everything one does with an item of data makes use of all the properties of its data-type. For example, numbers have to be moved from place to place. This operation is not a numerical operation and does not depend on the item being a number. In fact, for the purpose of data transmission, the item is only a word (assuming it fits into a single word) and can be treated as such. Second, one can often embed one kind of operation in another, so as to coalesce data-types. We saw this to a small extent in the example above of the PDP-8 arithmetic operations. A more pervasive example is encoding the Mp addresses into the same integer data-type as is used for regular arithmetic. Then there need be no separate data-type for addresses.[1] The upshot of both these aspects can be seen below where we present an outline structure of data-types that shows how one data-type can be embedded in another for various purposes.

Data-types embedded in other data-types for common operations

word
    integer
        fraction
        mixed
        unsigned integer
        address integer
    boolean vector
        boolean (single bit)
        integer sign (divide or multiply by two operations)
        field
    single precision floating
        single precision unnormalized floating
double word
    double precision integer
        fraction
        mixed
    double precision floating point
        double precision unnormalized floating point
character string
    digit string

---

[1]However logical such a course may seem, it is not always done this way. For example, the IBM 7090 (and other members of that family) have a 15-bit address data-type and a 36-bit integer data-type, with separate operations for each.

The notational aspect is our use in ISP of a mnemonic abbreviation scheme for data-types. We have already used sf for single precision floating point. More generally, as Table 1 shows, an abbreviation is made up of a letter giving the precision, a letter giving the name, and a letter giving the length. A full treatment can be found in the Appendix.

The simple naming convention does not take into account all that is known about a data-type. The information carrier for the data is only partially included in the length characteristic. Thus the carrier should also include the data base and the sign convention for representing negative numbers. The common sign conventions are sign magnitude, true complement (i.e., two's complement for base 2), and radix-1 complement (i.e., one's complement for base 2).

For each of the data-types the processor must have the implied operators. In fact, being able to represent a particular entity is useful only if particular transformations can be carried out on the entity. The most primitive operation is data movement (i.e., transmission). Data movement can be thought of as a complex operation consisting of accessing (locating), reading, and writing. Data-types which represent numbers require the ability to perform the arithmetic operations $+$, $-$, $\times$, $/$, abs ( ), sqrt, max, min, etc. The address integer is a special case of an arithmetic quantity, and often only additive arithmetic operations ($+$ and $-$) are available for it. Boolean scalars (or vectors) require some subset of the 16 logical operations (sufficient subsets are $\neg$, $\wedge$ or $\neg$, $\vee$). When character strings are represented, the concatenation, deletion, and transmission operations are required. Alternatively, we can look to string processing languages like SNOBOL or COMIT to see the operations they require. If the strings also represent numeric quantities, then the arithmetic operations are necessary. Almost all arithmetic and symbolic data require relational operations between two quantities, yielding a boolean result (true or false). These relational operators are $=$ and $\neq$, but for arithmetic quantities includes $>$, $\geq$, $<$, $\leq$. The more complex structured data-types (e.g., vectors and arrays) also have a range of certain primitive operations such as scalar accessing and transmission. Typical operations of vectors are search and element-by-element compare operations.

Relationship between PMS and ISP

In the introduction to this chapter we discussed briefly the relationship between PMS and ISP. With the two described, we can now be more precise. There are really two questions here. First, where do these two descriptive systems fit in with respect to the general hierarchical view of computer structures discussed in

Table 1   Abbreviations used to name data-types

Precision	Data-type-name	Length-type
fractional/f	boolean/b	°scalar
quarter/q	sign	vector/v
half/h	decimal digit/digit/d	matrix
°single/s	octal digit/octal/o	array
double/d	character/char/ch/c	string/st
triple/t	byte/by	
quadruple/q	syllable	
multiple/m	word/w	
+ integer (eq. 10)	signed integer/i	
	unsigned integer/ui	
	fraction/fr	
	fixed/mixed/mx	
	floating/real/f	
	unnormalized-floating/uf	
	complex real/complex/cx	

Examples:

w	word
bv	boolean vector
i	integer
sfr	single precision fraction
mx	mixed
di	double integer
10d	10 decimal digit (scalar)
3.ch	3 character (scalar)
ch.st	character string
sf	single precision floating
suf	single precision unnormalized floating
df	double precision floating
duf	double precision unnormalized floating

* May be optionally omitted from name

Chap. 1. Second, what is the relationship between a PMS diagram of a processor and the ISP of that same processor. The questions are related, but each is best answered separately.

With respect to the first question, the PMS system describes the topmost system level (recall Fig. 1 of Chap. 1), above the programming, logic, and circuit levels. It lacks a characteristic that all these other levels share, namely, that of providing a complete description of the computer's performance. The programming manual (with timing) tells everything that is significant about the performance of the computer (if it runs error-free). The same is true of the full description at the register-transfer level, the logic-circuit level, and on down to the electrical circuit level. But the PMS level is only an approximate description, from which only certain aspects of the system's performance can be calculated.

The ISP does not constitute a distinct system level. Rather, it describes the interface between two levels, the register-transfer level and the programming level. It is used to define the components of the programming level—instructions, operations, and sequences of instructions—in terms of the next lower level. In principle, and usually in fact, the language of the lower level is used to describe the components and modes of connections, one level up. In many ways ISP is a register-transfer language (in symbolic rather than graphical form—but as we noted in Chap. 1, there appear always to be two such isomorphic notations at each system level). However, ISP has been extended by allowing the instruction-expression to be a general linguistic expression for a computation, just as if ISP were FORTRAN or ALGOL. This is what permits us to talk of ISP as not necessarily determining the exact set of physical registers and transfer paths. The instruction-expressions describe the functions to be performed without entirely committing to the RT structure.

If the ISP is the interface language between the RT and programming levels, what is its relationship to PMS, which is one level above? Every PMS component has associated with it a set of operations and a control structure for getting those operations executed in connection with the arrival of various external signals. As we noted earlier in the chapter, there is an ISP description for each operation in its context of control. That is, ISP is the interface language for describing *all* PMS components in terms of the register-transfer level, not just P. It happens that only one of these PMS components, the processor, carries with it an entire new systems level—the programming level. All the other components have no analog of the programming level and interface directly to the register-transfer level (or even in simple cases to the logic-circuit level). Precisely because of the simplicity, we have not bothered to develop ISP descriptions of other components of components other than processors.

The second question, namely, the relation between the ISP and PMS descriptions of the same processor, arises from the ability to represent PMS components recursively as PMS structures made up from more elementary PMS components. Thus, Mp(32 kw, 16 b) can be considered as compounded of 32k memories, M(1 w, 16 b), with an addressing switch, S.random. Indeed, if one carries this to the limit, where the M's are single bit memories (flip-flops), the S's are one bit gates, a couple of specific K's are defined for AND and OR, etc., then it is possible to draw a PMS diagram isomorphic to any logic circuit. Thus, a processor (P) can be represented as a PMS involving M's, K's, D's, S's, etc., and at varying levels of detail. Since we also have a description of this same P in ISP, it is appropriate to consider the correspondence.

First of all, every memory in the ISP description corresponds to a memory in the PMS description. The data operations in ISP imply corresponding D's in PMS and every occurrence of transmit (←) implies a corresponding link between the M's and D's on the right hand side and the M on the left, being written into. That the instructions of the ISP are evoked only under certain conditions implies that a control (K.operation-decode) exist in the PMS structure. Similarly, the simple, two-state stored-program model (instruction-fetch, instruction-execute) for the interpreter implies an interpreter control (K.interpreter). The action-sequence of each instruction, if it contains any semi-colons or next's, requires additional K and possibly additional M (if the structure involves embedded operations such as $(A + B) \times (C + D)$). Thus for every ISP component there is an implied component in the PMS structure of the processor.

The PMS diagram model for a computer shown initially on page 17 has the "natural units" implied by the ISP description (with the exception of the instruction format part) as suggested on page 24. The data-operations D are therefore implied each time an operation is written. Each process implies a control which we lump into the single K of the figure. The model also shows both the arrival of instructions and the flow of data between the processor (P) and memory (Mp).

There are several memories within Pc which are not explicitly shown on page 17. These include temporary memory within D and the K for carrying out complex arithmetic operations. The interpreter control has temporary memory, of course. Finally, other kinds of memories have been omitted to simplify the model. In multiprogrammed computers a mapping control and memory would be used, and in pipeline or highly parallel processors there would be temporary memory for various buffering (e.g., instructions and data). The Appendix lists the various memories of the processor.

K(P), the control for the processor above, controls data movement among the Mp and M.processor_state and evokes the data-operations of D. Functionally, K(P) can be broken into several parts, each of which is responsible for a part of the overall instruction interpretation and execution process, and each corresponds to a part of the ISP description. This decomposition is allowed in PMS, and if we did so, each component would contain an independent control for its own domain, e.g., a K(D), K(Mp), K(Instruction-set interpreter). More elaborate processor structures imply having controls for functions like multiprogram mapping. The K(Instruction-set interpreter) is the supervisory component which causes other processor K's to be utilized in a complex processor. In an ISP description of a C, the interpreter usually selects only the next instruction and then after decoding (or examining it) proceeds to have the instruction executed by K(instruction execution).

Resource Allocation. At the PMS level the concept of resources, their uses and allocation, becomes a major focus of analysis. This is obvious by now in multiprogramming and multiprocessing systems where many programs share the same Mp and hence must be allocated space. But this holds equally well at all levels of detail.

By giving a resource allocation diagram along with the state diagram (Fig. 5) we show the relationship of resources, their function, and time for the instruction-interpretation process. In Fig. 5 the add instruction for a simple 1 accumulator computer consisting of 1Pc-2Mp is given. The interpretation for Fig. 5 in ISP is as follows:

1 Calculates the address of instruction q in state soq. $t_1 - t_0 = toq$.
   PC ← PC + 1; next *advance the program counter*

2 The instruction is fetched (accessed) from Mp in state saq. $t_2 - t_1 = taq$.
   M.instruction ← Mp[PC]; next

3 The operation o to be performed and the address part, v, for the data in M.instruction to be added to A are obtained in state soo + sov.r. $t_3 - t_2 = too + tov.r$
   M.address ← M.instruction ⟨v⟩; next

4 The data Mp[v] are fetched in state sav.r. $t_4 - t_3 = tav.r$
   M.temporary ← Mp[M.address]; next

5 The operation part o of the instruction is carried out on A; that is, the actual addition is performed on the data previously accessed in the state so. $t_5 - t_4 = to$.
   A ← M.temporary + A; next

In the state diagram, each state represents the time spent for a given activity. The two states at the top of the state diagram (Fig. 5) are waiting for primary memory accesses, and the three lower states represent processor activity waits. If we were to specialize the state diagram for the conventional 1 address/instruction computer, we would need one additional state, representing operand storage, sav.w, and this would occur after state, so. Note that we have ignored the operation decoding state, so.o. Of course, conditional state transformation paths have to be added to describe all instructions (e.g., a complement-the-accumulator instruction has only states soq, saq, and so). Similarly, we could

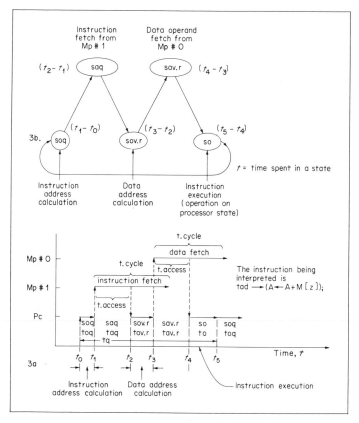

Fig. 5. State and resource allocation diagram for a 1Pc-2Mp add instruction-interpretation process.

make a more general state diagram to handle the different processors (e.g., multiple addresses/instruction, stack, and general registers), as shown in Fig. 4. At the PMS level, a derivative of the state diagram, the resource allocation diagram is more useful because it relates to the physical structure.

A resource allocation diagram expresses the above instruction activity in terms of the time each unit is occupied with a particular activity. In this diagram a slightly more complex computer structure with two primary memories has been assumed. In the case of the add instruction, the long memory-cycle time suggests that two memories can be used so that an operand be fetched while the instruction memory restoration occurs. These diagrams show the time various resources are utilized; thus performance and utilization can be measured.

Resource allocation diagrams can express other time scales. Interest in operating-system software analysis is often in the activities on a longer time scale of the resources utilization as a function of various programs and subprograms. They may show Mp memory occupancy in a multiprogrammed environment. Some other time scales of particular interest are the instruction(s), short instruction sequences or subprograms, and the program times. The first two time scales are influenced predominantly by the hardware, and the latter time scale is influenced by software and the external environment.

The resource allocation diagrams also can describe the utilization of the C's resources over time (e.g., throughout the instruction-interpretation process) and provide a basis for more detailed analysis and design.

The design problem at the PMS-ISP interface is mainly one of resources scheduling.

1 A fixed set of operations have to be performed on the jobs (here, a job is an instruction).

2 Each instruction may create a few other small but definitive subjobs.

3 There can be a fixed set of operators which handle various parts of the operations.

4 Jobs (or instructions) enter P sequentially.

We may ask:

1 How many operators of each type do we have?

2 What is the scheduling policy for assigning instructions to the operators?

3 How many instructions can be in P at one time, and in what order must the processing be performed? How are the jobs interlocked?

We do not attempt to answer the above questions but intend only to show the relationship of the various parts which define the problem. ISP implies a certain structure (conversely, PMS behavior is specified in terms of the ISP language). A particular ISP structure and a program denote a certain path through a state space as specified by a state diagram. Finally, the physical resources (in PMS) are constrained to operate according to the state diagram as expressed by using a resources allocation diagram. The resource allocation diagram can then be used to evaluate the structure's performance (in PMS) at a higher level (e.g., the number of instructions/second it executes).

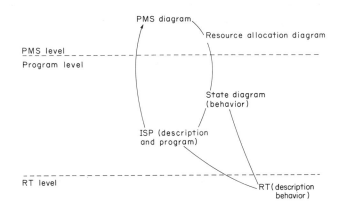

Summary

The ISP descriptions of computers are usually given as an appendix to a chapter. We organize the description into the following units:

Memory Declaration	P State Mp State P Console State
Formats and Operators	Instruction Format Data-type Formats and Special Data Operation Definitions Effective-address Calculation Process
Interpreter and the Instruction-set Execution	Instruction Interpretation Process Instruction-set and Instruction Execution Process

The above description format conveys a rather narrow-minded view of the ISP structure of computer systems. However, almost all present computers fit easily into such a format. We do not presume to say whether it will suffice for future ISPs.

With the introduction given here and with the definitions and example in the Appendix at the end of the book, it should be possible to understand all the PMS diagrams and ISP descriptions used throughout the book.

Chapter 3

The computer space

Introduction

The preceding two chapters have provided a view of a computer system as an organized hierarchy of many levels: physical devices, electronic circuits, logic circuits, register-transfer systems, programs, and PMS systems. We must remember that these are *levels of description* for what, after all, remains the same physical system. Each higher level describes more of the total system, but with a loss of detail. As this is an engineered system, great care is taken that each level represent adequately all the behavior necessary to determine the performance of the system. In natural systems too there are often many levels of description (e.g., in biological systems, from the molecule to the organelle to the cell to the tissue to the organ to the organism).

However, in natural systems we usually depend on statistics to eliminate the details of lower levels and permit aggregation, and they always do so imperfectly. In computer systems, on the other hand, the aggregation is intended to be perfect. It fails, of course, and so both error detection and error correction exist as fundamental activities in computer systems. But these imperfections are ascribed to the system itself and not to our description of it, which is just the opposite from how we treat natural systems. Only the PMS level of description is natural, in the sense of not being the intended result of the design. This is because performance is defined ultimately at the programming level. The aggregations and simplifications that go into a PMS description (e.g., measuring power by bits per second) are approximations, just as they are for any natural system (e.g., measuring the productivity of the economy by gross national product).

We have provided descriptive systems for the top levels of the hierarchy: the PMS level and the ISP level, the latter defining the basic components of the programming level in terms of the RT level just below. These are the two descriptions that are of most concern in the overall design of a computer system. We did not define the lower levels, because they go beyond the focus of this book. Neither did we define the program level, partly because there exists no uniform description (no common programming language) and partly because the computer designer works mostly at the interface, defining the instruction set. This latter is what the ISP provides.[1]

PMS and ISP permit the description of an indefinite number of computer systems—indeed, all that come within the scope of the current design art. (They might even be taken as a definition of what that current art is.) Some $10^4 \sim 10^5$ individual computer systems have in fact come into existence, each of which can be described in PMS and ISP. They are not all radically individual. There are about 10^3 types of computer systems represented, if we define two systems with the same Pc to be of the same type. (By exercising various options, a single computer type could take on 10^5 different forms.)

Of these thousand-odd types, we present in this book just 40.[2] What sort of total population do we have here? What does our miniscule sample look like when compared with the whole? More fundamentally, what are the significant aspects of the computer systems that should be used in a comparison or classification? These are the questions we will try to deal with in this chapter. We can be neither comprehensive nor elegant. There has simply not yet been done the necessary study on which to base an adequate taxonomy of computer systems. But we can present a rough picture based on the common lore of the field, filled in with our own predilections.

For any system, either an entire computer, C, or a component, such as P, M, or S, it is convenient to distinguish its function, its performance, and its structure. The system is designed to operate in some task environment; to accomplish such tasks is its *function*. How well it does these tasks is its *performance*. Evaluation of performance is normally restricted to these tasks. Although it is always noteworthy when a system can perform adequately outside its specified domain (e.g., when a business computer is also a good control computer), it is rarely worth noting when a system cannot perform those tasks it was not built to perform. Thus, function denotes scope, and performance denotes an evaluation within that scope.

Structure denotes those aspects of the system that allow it to perform. This includes descriptions of its subcomponents and how they are organized. Performance of subcomponents often may be considered structure as far as the whole system is concerned, especially if the performance can be taken as given. For example, early digital transmission-oriented telephone lines came in two capacities, ~ 200 bits/sec and $\sim 2,000$ bits/sec. From the viewpoint of the telephone system, these are performance measures;

---

[1] An increasingly popular view is that the program and RT levels (with ISP in between) are one, thus erasing the difference between hardware and software. The boundary appears to us not quite so invisible. We take the important task to be drawing the boundary in the right place for any specific design.

[2] Counting each of the families in Part 6 as one computer. The IBM System/360 is actually a series.

from the viewpoint of a computer system with remote terminals, these are structural parameters.

Typically, design proceeds in a context in which the function of the to-be-developed system is taken as given and certain structures are available; the problem is to construct a structure that achieves adequate performance.

These terms apply to any designed system. For example, consider automotive vehicles. Function is a classification by use: cars to carry people, trucks to carry goods, racers to win competitions, antiques to satisfy nostalgia and collectors' pride. Performance is those aspects of behavior relevant to function: maximum speed, power-to-weight ratio, cargo capacity, run versus not run for an antique, and so on. Structure is such things as number of wheels, shape of the vehicle, stroke volume, and gear ratios. Structure determines performance, although from the standpoint of design, of course, causality runs the other way: from function to performance to structure.

There are, then, three main ways to classify or describe a computer system: according to its function, its performance, or its structure. Each consists in turn of a number of dimensions. It is useful to think of all these dimensions as making up a large space in which any computer system can be located as a point. In such a space all the thousand computer types built to date constitute a sparse scatter, clustering (it is to be hoped) in various regions that make sense functionally and economically. The 40 computer types in this book sample this larger scatter in some way, to give a picture both of the entire space and of the part already explored.

How many dimensions are there in this computer space? Indefinitely many, if one wants to locate a computer with ultimate precision. In fact, if one wants to go all the way, one might as well give the PMS and ISP descriptions (and down through the RT, logic, circuit, and device levels). The virtue of thinking of such a space is to abstract to a small number of dimensions, and to select those that are most relevant. Of the functions, one wants those that most influence the design; of the performance, one wants those that make the largest difference; of structure those that not only affect performance but represent possible design choices by the computer engineer. In addition, one wants dimensions along which there is significant variation. Those aspects of computer systems which are common to all, such as the use of binary devices, though of supreme interest are not part of the computer space.

What are the dimensions of the computer space? As we remarked earlier, there is no sufficiently comprehensive theory of computer systems to tell us. Considerable lore has grown up from experience to date in designing machines. But at some point one must simply propose a set of dimensions and let them justify themselves after the fact. Table 1 gives our set for function and structure. Table 3 (page 52) gives our set for performance. Table 1 gives only a single dimension for computer system function and 19 for computer structure; Table 3 gives 8 for performance. However, the dimensions are not all independent. Many of the structure dimensions are highly (though not perfectly) correlated. Thus, in Table 1 we have put the structure dimensions in seven horizontal groups, with the one at the left-hand side being the most relevant. (In the first structure group, we have also added two temporal dimensions, since a strong correlation with time exists.) For performance, the dimensions form a tree structure, where the higher dimensions are essentially aggregate summaries of the lower ones. Finally, there is a general correlation between overall performance and the various structure dimensions, in Table 1, with increasing performance as one moves down the dimensions. We have left off two important dimensions because we do not have values; these are reliability (mean time between failures per operation) and physical size density (e.g., bits/ft^3), both of which increase with generation.

With each dimension we have indicated the range of possible values. For some (Pc.speed, for example) this is a numerical quantity. However, for most, the range is a discrete set of design choices, which may or may not have a simple ordering. Clearly, these discrete values are selections from a meaningful subspace of design choices, but mostly we do not know how to construct that subspace. The values given are those that have arisen in practice, and they serve to classify the computers in the book. Obtaining a more rational subspace is a task for future research.

The body of the chapter will be taken up with a discussion of each of these dimensions, where we will discuss further their definition, the basis for their selection, and the reasons behind the arrangements of Tables 1 and 3. We give the entire set of dimensions here at the beginning, both for later reference and to emphasize the view of a single computer space in which computer systems can be located. We will refer to Tables 1 and 3 from now on simply as the computer space or, more narrowly, as the computer structure space, the computer performance space, etc.

History

Like all systems subject to variation and selection, computers have evolved through time. So striking and rapid has been this evolution that the concept of "generation" has become firmly embedded in the computer engineering culture (to say nothing of the marketing culture and the view of the lay public). It is at best an ambiguous term, having none of the sharpness of its root term in biological evolution, where it is possible to draw a strict genealogical tree.

Nevertheless, the term is useful in stressing that the history of computer systems is not just a story of particular men discovering or building particular things, but of a somewhat more impersonal and widespread series of advances that have changed computer systems radically.

The generations are best defined solely in terms of logic technology: The first generation is that of vacuum tubes (1945 ~ 1958), the second generation is that of transistors (1958 ~ 1966), and the third generation is that of integrated circuits (1966~). In fact, current usage describes hybrid logic technology machines, such as the IBM System/360, as third generation, and so this extension must be included. What will be called fourth generation is yet to emerge; most likely it will be medium and large scale integrated circuits with possibly integrated circuit primary memory.

It is a measure of American industry's generally ahistorical view of things that the title of "first" generation has been allowed to be attached to a collection of machines which were some generations removed from the beginnings by any reasonable accounting. Mechanical and electromechanical computers existed prior to electronic ones. Furthermore, they were the functional equivalents of electronic computers and were realized to be such. They were also separated by a wide gap in performance and structure, both from each other and from vacuum tube machines. Thus, by reasonable reckoning, we are currently in the fifth generation of computers, not the third. But usage is now too well established to change.

Actually, it was not always viewed thus. Figure 1 reproduces a genealogical tree of the early computers prepared by the Na-

Fig. 1. The "family tree" of computer design. The remarkable growth of electronic computing systems in the Western world began primarily through government support of research and development in the universities. The need for data-processing facilities of increased capacity inspired further support for their development in both educational institutions and private industry. The current generation of computers is predominantly the result of development by private industry. The tree lists many of the machines developed in these ways. At the roots are the contributions of many existing technologies to the rapid growth from electromechanical to electronic systems. Some of the milestones are ENIAC (Electronic Numerical Integrator and Computer), the first electronic computer; EDVAC (Electronic Discrete Variable Automatic Computer), the first internally stored-program computer and first acoustic delay-line storage; MADM (Manchester Automatic Digital Machine), the first index registers (B lines) and first cathode-ray-tube electrostatic storage; MTC (Memory Test Computer), the first core-storage computer. (*Courtesy of National Science Foundation.*)

Table 1   The computer-space dimensions

Computer function
Scientific
Business
Control
Communications
(switching|store and forward)
File control
Terminal
Time sharing

Logic technology	Generation	Historical date	Pc.speed (sec)	Cost/operation ($/bit/s)
Mechanical				
Electromechanical		1930	10^{-1}	1000
(Fluidics)		(1970)	10^{-2}	
Vacuum tube	first	1945	10^{-3}	10
Transistor	second	1958	10^{-5}	-1
Hybrid		1964	10^{-6}	
Integrated/IC	third	1966	10^{-7}	0.1
Medium to large-scale integrated/ MSI $\sim$ LSI	fourth?	197?	10^{-8}	0.01

Word size	Base	Data-types
8 b	binary	word
12 b	decimal	integer|address (integer)
16 b		bit|bit vector
24 b		instruction
32 b		floating point
48 b		character
64 b		character string
	character (6b)	word vector
	character (8b)	vector
		matrix
		array
		lists, stacks

Addresses/instruction	M.processor state (excluding program counter)
0 address (stack)	stack
1 address	1 Accumulator
1 + x (index) address	accumulator and index registers
1 + g (general register) address	general registers array
2 address	
3 address	no explicit state
n + 1 address	
Language determined	
Compound	
Microprogrammed	

PMS structure	Switching	Processor function
1Pc	1:n (duplex)	P.microprogram
1Pc(interrupt)		Pc
1Pc-nPio	n:m (time-multiple x)	Pc (no io)
1 Pc-nPio-P(display)		Pio
2C (duplex)	2:n (dual-duplex)	P.display
nPc(multiprocessing)	n:m (cross-point)	
nPc-P(array|special algorithm)		P.array
nPc(parallel processing)		P.vector move
C (network)		P.algorithm
Network	n/2:n/2 (non-hierarchy)	P.language

Accessing algorithm	Mp.size	Ms.size	Mp.speed (b/s)	Ms.speed (b/s)
Linear (stack)				
Linear (queue)				
Bilinear		tape (large)		$>10^5$
Cyclic-random		disk (medium) |magnetic card (large)|		
Cyclic	drum (large)	drum (small) |photostore (large)	$>10^6$	
Random	core (medium)	core (smaller)	$>10^7$	$>10^7$
Content	film (small)		$>10^8$	
Associative	integrated circuit		$>10^9$	

Mp concurrency	Interprocess communication
1 program	subroutines and traps
1 program with interrupts	interrupts
1 program with multiple concurrent	interprocessor interrupts
subprograms (for example, 1Pc-nPio)	
Monitor or fixed program(M) + 1 program	extracodes (programmed operators for
m + n swapped programs	monitor calls)
m + n programs (multiprogramming)	
No relocation	
1 segment	
2 segments (pure, impure)	
$>$2 segments	
Pages	
m + n segments with shared programs	intersegment communication
Fixed length, paged segments	
Multiple-length paged segments	
Variable-length segments	
Named segments	

Processor concurrency

Serial by bit
Parallel by word
Multiple instruction streams, 1Pc
Multiple data streams (arrays)
1 instruction buffer
n instruction buffer
Look-aside memories
Pipeline processing

tional Science Foundation in 1959. Notice that the Harvard Mark machines, which were constructed from relays (hence electro-mechanical) are accorded the place of honor as first generation (but Babbage is nowhere to be seen).

It is not appropriate to provide here an adequate history of computer technology. The early story has often been told, starting with Babbage and early mechanical calculators, through Hollerith punched cards, on to the relay calculators at Bell Laboratories and Harvard, up to the birth of electronic machines with ENIAC, and finally to the stored-program concept with the von Neumann machine at the Institute for Advanced Studies (IAS), EDSAC at Cambridge University, and EDVAC at the University of Pennsylvania (with the contemporary developments by ZUSE in Germany often left out). And there have been a few scattered attempts to tell some of the story of the last three generations. But to date no really satisfactory historical account has been given. This is due in part to recency and in part to the difficulties of evaluating and sorting out the significant developments of a very complex technology undergoing rapid growth.

What is appropriate here is to view the evolution of computer systems as measured by the dimensions of computer space and to localize the examples of this book in relation to calendar time and other computers. The concept of generation has led others to attempt the same thing by constructing a family tree, Fig. 1 being but one example. But the relationships between computers is not nearly as simple as such a tree implies. We prefer to plot a straightforward time chart,[1] as shown in Fig. 2, in which we group the machines by manufacturer and within each group, by acknowledged family relationship (for example, 701–704–709–etc.). There is clearly relatively closer kinship within a company than between companies. One advantage of such a time chart is its depiction of the life history of a single system, showing how long it takes for computer systems to go from paper through prototype to production.

Not all computer types are shown on the chart, there being about 250 out of the estimated 1,000 types. Lack of space (and of perseverance) accounts for the omissions. The major United States manufacturers, as well as some minor ones, and all machines of substantial historical interest are represented. All the machines discussed in this book are gathered together on a separate line (though they also occur elsewhere, if appropriate). Foreign machines are omitted, unless they are described in this book. In addition, the machines of many early minor manufacturers are missing (ALWAC, ELECOM, etc.).

The second part of the time chart arranges many computers by word size, to give the reader our classification. Unfortunately, only a few samples are given, owing to space limitations. Thus, the density on the graph does not indicate the true density of existing machines. Many small computers, which are dedicated to a particular task, are beginning to be built and a comparatively small number of very large computers have been built. On the bottom fine line we place the machines in this book.

The third part of the time chart deals with technology by listing events along various dimensions that have been significant in the evolution of computers. Besides the dimensions in the computer space we have also added some dimensions describing software systems. Although we have not been able to deal with the programming level in this book (except for the ISP interface), its development is clearly as important as that of the hardware, and there exists strong mutual interaction between the two.

The fourth (and final) part of the time chart gives selected technological events leading up to the development of the computer. It includes the early work of Babbage, desk calculators, and the Bell Labs and Harvard calculators.

Many stories can be read from the chart. For example, note that the early Bell Telephone Laboratories relay calculator was used remotely at Dartmouth in 1940, about 20 years prior to remote use of time-shared computers. Note also that successful manufacturers tend to have a small number of computer families, but add members as the technology dictates. (We omit the exodus of computer companies.) We hope the reader gets as much enjoyment from browsing the chart as we have (even after we put it together!).

The computer space in Table 1 and the time chart in Fig. 2 provide an overall framework. We are now ready to consider each of the dimensions individually, starting with those of system function, then the performance, and finally structure.

---

[1] Whereas we have checked the Time Chart numerous times for accuracy, we make no claim about the number of errors it still has. We have relied on the following source data: (1) Original papers. These are mostly shown on the chart as "p". Normally the reader can infer that the work presented in a paper occurs prior to the actual publication. There are notable exceptions (e.g., the core memory, and Atlas papers) which were first published to lay claims to certain ideas. (2) Historical reviews. Primary historical papers include: Rosen [1969] and Serrell [1962]. Secondary historical review papers include: Bowden [1953], Campbell [1952], Chase [1952], Nisenoff [1966], and Samuel [1957]. (3) Encyclopedia. (4) Computer surveys. Two sources have been used: The Adams Associates *Computer Characteristics Quarterly,* published since 1960 [Adams, 1960; Adams Assoc., 1966, 1967, and 1968]; and Martin H. Weik's four *Surveys of Domestic Electronic Digital Computer Systems* [Weik, 1955; Weik, 1961 (third); and Weik, 1964 (fourth)]. The Adams' Charts give the date of first delivery, and the Weik Survey gives the date the computer was first operating. (5) Manufacturer, organization or person supplied dates. In a few cases we have asked directly for specific operational and delivery information.

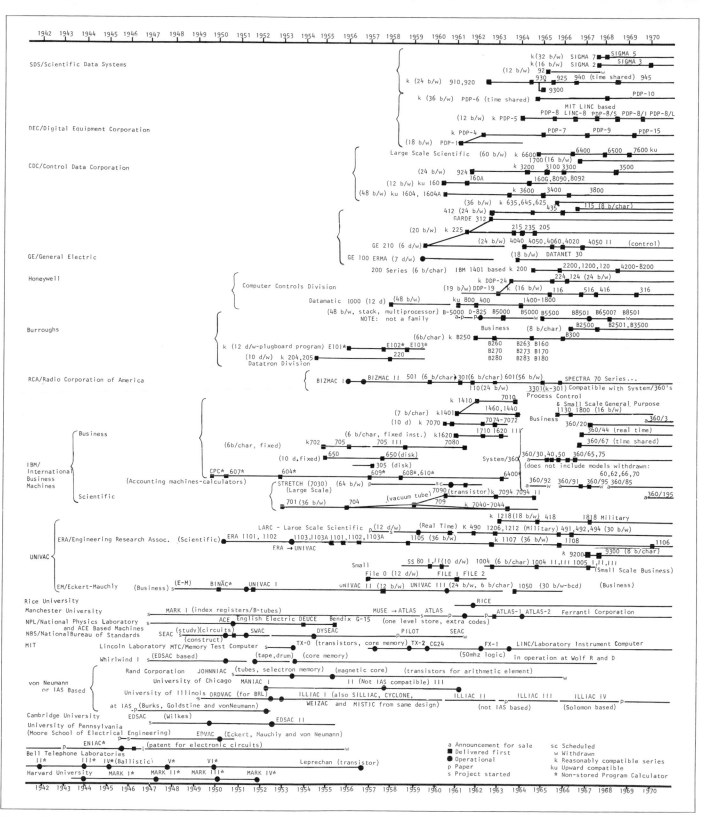

Fig. 2a. Time chart: computers by originator.

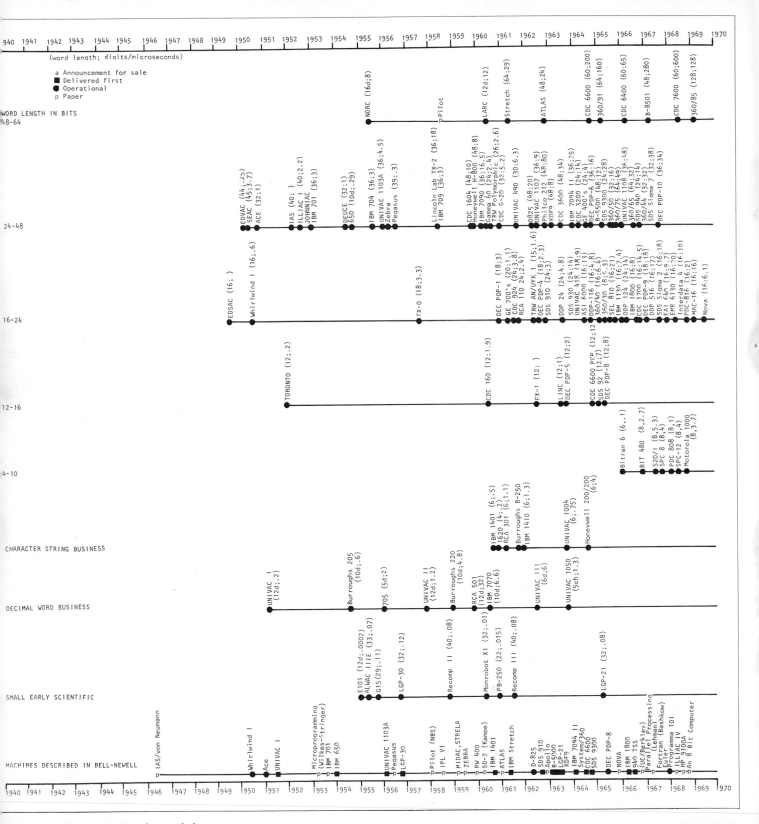

g. 2b. Time chart: computers by word size.

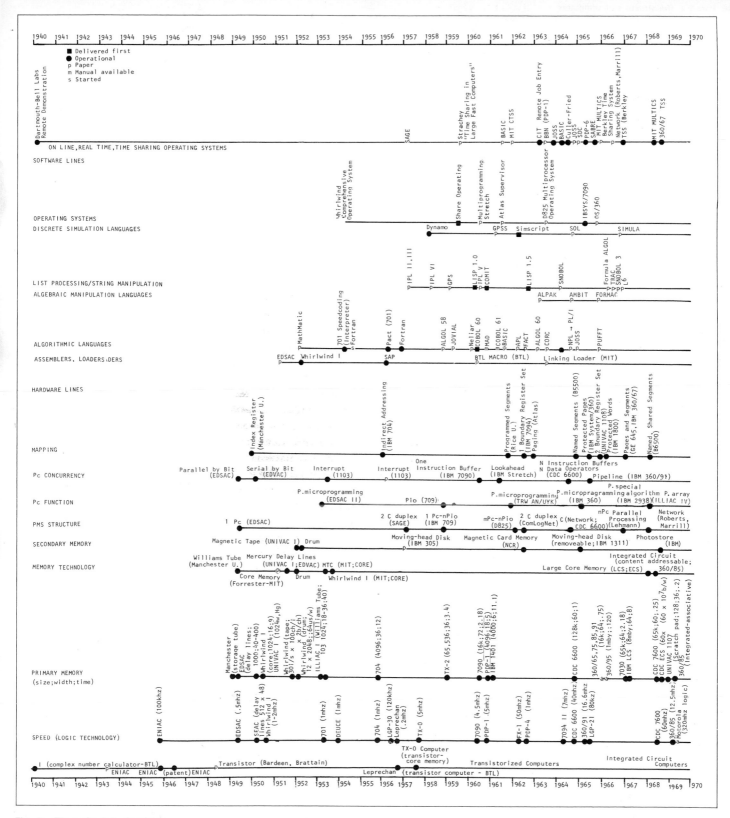

Fig. 2c. Time chart: technology.

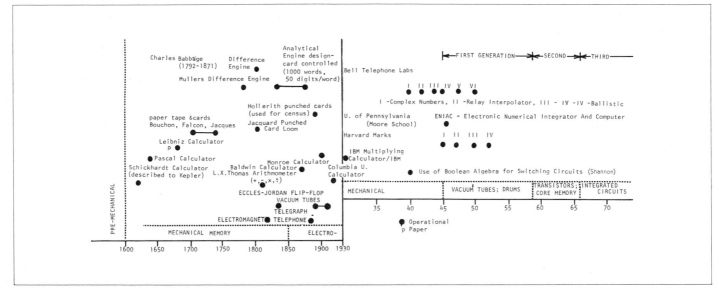

Fig. 2d. Time chart: pre-computer technology.

Function

The most striking fact about function is the existence of only a single dimension, and with only a few values. Perhaps we have taken a simplistic view of the functions that computers perform, but we think our computer space represents reality: To wit, there is remarkably little shaping of computer structure to fit the function to be performed.

At the root of this lies the general-purpose nature of computers, in which all the functional specialization occurs at the time of programming and not at the time of design. However, it might seem that specialized environments would not require all the generality, so that functional adaptation would still be possible. But this appears not to be so for two reasons. First, the level of operations of the Pc (as defined in the ISP) is too basic to reflect the kind of specialization offered by the environment (think of information-transfer or conditional-transfer operations). Second, all environments ultimately require a variety of tasks in addition to the main specialized task. These include at least language compilation or assembly, readable formatted output, debugging aids, and other utility routines. By the time these have been added, a substantial requirement for generality has been generated.

However, this is not the whole story. A second part is the difference between the computer type and the specific configuration assembled for a task. The latter is often carefully specialized to the function to be performed. But this is mostly the amount of Mp, the amount of types of Ms, and the number and types of T's. Within limits, these are all items that can be attached to any type of computer (i.e., to any Pc) and are handled in an environment-independent way. Thus there is little specialization of computer types, but great specialization of particular configurations. That this should be the case indicates something about the nature of the functional specialization—that it can be expressed adequately in gross PMS terms, as more bits of storage and more data rate.

There is still more to the story. Some functional specialization exists, as indicated in the dimension. This depends primarily on two kinds of things beyond the reach of the configurational adaptation described above. The first consists of demands for reliability, ruggedness, small size, etc. These have strong effects on design, but *below* the ISP and PMS levels. The second consists of demands for large amounts of processing power. One response to this again affects design at the lower levels of logic, devices, and circuitry and has little impact on design at the ISP and PMS level. But response is also possible in terms of the data-types that are built into the ISP. Large machines have data-types that are appropriate to their tasks (with operations to match), and these affect the design. In fact, this effect is the substance of the functional specialization shown in the computer-space dimension.

Finally, there is one last part of the story, and it is the most interesting of all. Various groups of computer engineers have felt strongly from time to time that functional specialization should exist, and they have set out to create such machines. These efforts have often produced machines that were different from the existing main line of computers, i.e., were appropriately specialized. But the net effect of almost all such attempts has been that the new idea was seen to be good in general for all computers and was taken back into the main line of computers. Thus, what started out to be a functional separation turned out to be simply a way to produce rapid development of a more universally applicable computer. A classic example is the expansion of input/output facilities in creating a functionally specialized business machine, which simply led to better I/O facilities for all computers. We will have more to say about such examples as we discuss the values along the dimension.

Computer-system function

Scientific. The first machines were clearly designed for scientific calculations. In fact, Aberdeen Proving Grounds funded the early work on the ENIAC for the computation of ballistic firing tables. And the image used frequently by the early computer designers was the computer as a statistical clerk, the arithmetic unit being the desk calculator, the memory the work sheet, and the program the instructions that the mathematician gave to the clerk.

From a design standpoint, scientific computation has posed two striking requirements. The first is the great accuracy of the numbers, which has led to word lengths of 36 to 60 bits (11 to 18 decimal digits of significance) and arises from the propagation of roundoff error during repeated arithmetic operations. The second is the emphasis on fast arithmetic operations, i.e., for arithmetic power. In the early machines the standard rule for estimating computation times was to count the number of multiplications in a program; all else could be neglected. The arithmetic unit has developed to where the floating point multiply is hardly more expensive than floating point add. This requirement on fast arithmetic, however, has really been directed at the logical design level, not at the ISP or PMS level. Thus, the main effect at the ISP is the adoption of long word lengths, floating point data-types (in addition to integers), and an extensive repertoire of arithmetic operations in the ISP. The main PMS effect is the emphasis on the classic "statistical clerk" PMS design.

The press for increased arithmetic processing has led in recent times to the development of various forms of Pc concurrency, as in the look-ahead of Stretch (Chap. 34) and the n-instruction buffer of the CDC 6600 (Chap. 39). This might be considered a unique functional specialization for scientific computation. It is too early to tell, but it is our impression that, although the needs for scientific computation initiated the exploration of concurrency and parallelism, we will eventually see them in all computers above a certain power, whatever the task domain. Physical limits on component speed and signal propagation will make these techniques universally attractive.

A better case for permanent specialization can be made in the special algorithm computers, which compute the fast Fourier transform or do vector operations. Here we finally have systems whose whole design is responsive to a narrow class of problems. This may extend to the very special kinds of Pc parallelism exhibited by the ILLIAC IV (Chap. 27), although there is substantial generality in such systems.

Business. In the early days of electronic computing it was felt by many that there was a major functional separation between business computing and scientific computing.[1] Scientific problems were "large computing–small input/output"; business problems were "small computing–large input/output." Certainly most of the existing computers, designed for scientific computation, had poor input/output facilities. The IBM 701, for example, used the Pc to control everything dynamically, actually catching the bits from running tapes on the fly (by executing well-timed small loops). These design efforts for business computers resulted in the IBM 702 (and subsequently the IBM 705, 708, and 7080). This machine had two major innovations for IBM: It used characters, and it had a PMS structure that permitted more flexible and voluminous input/output. The latter feature was immediately incorporated into scientific computers, e.g., into the 709, and then into all large scientific computers as separate input/output control (either Kio or Pio), for it was realized that there were also demands on input/output for scientific calculation. Thus the bifurcation was temporarily halted.

The specialization to characters as a basic type (as opposed to long words) was already present in the IBM 702 but did not have its effect until 5 years later with the development of the IBM 1401 (Chap. 18). The latter machine was adapted to business, both in being character-based and in being small enough so that small businesses could afford it. It was extremely successful (many thousands were produced) and certainly represents a successful func-

---

[1] Such feelings are still extant, but we are concerned here not with the validity of the feelings but with what they led to at a particular period of computer development.

tional specialization for business. However, it is interesting that the specialization has not been maintained, for the IBM System/360 (Chaps. 43 and 44) is again a single machine, although it has in essence two internal ISP's, one centered around characters and the other around floating point data-types, that is, a business and a scientific specialization residing side by side.[1]

Control. The third functional value is a computer used for control in real time. Examples are process-control computers, aerospace computers, and laboratory instrument-control computers. The role of the computer is to act as a sophisticated control (K) in some larger physical process, and thus it plays a subordinate role. Their relatively late arrival was due to the high cost and unreliability of early computers, as well as to the lack of necessary interface equipment.

The functional specialization is seen most strongly in the word size, which reflects the appropriate numerical data-type. The numbers used in control processes are generated by physical devices and are rarely better than 0.1 percent accurate. Since elaborate arithmetic calculations are not called for, the numbers, and hence the word size, can be around 12 bits. Most control computers have been 12 to 18 bits/word. A second specialization, again reflecting appropriate data-types, is that all control computers are binary and have boolean operations. This arises because many of the external conditions to be sensed and effected are binary in nature.

About the only other functional specialization of control computers is the interrupt[2] capability to allow them to respond to many potentially simultaneous external conditions in real time. This provides apparent parallelism, though still using a sequential processor. This is another possible example of functional specialization leading to reunification rather than divergence, for it has again been widely accepted that all general-purpose computers must have good interrupt capabilities. However, in actuality, interrupts, though not existing in early computers, were developed to obtain good input/output facilities, not for control computers.

Chapters 7 and 29 give examples of aerospace computers, and Chap. 33 describes the IBM 1800, which is specifically designed for process control. As these examples show, a complex ISP is not

---

[1] The story above has been told exclusively in terms of IBM machines. Although this does not distort the picture too strongly in terms of total movements of the field, since IBM dominated the market, concurrent developments were taking place throughout the field. UNIVAC I was the first computer built by a manufacturer and did not have the idiosyncrasies we ascribe to IBM; on the other hand, the marketing effort for it was nil.
[2] Apparently introduced in the UNIVAC 1103.

necessarily required. This in part reflects the fact that control computers may retain their programs over their whole lifetime, so that programming and reprogramming is less important. (It is not absent, however, and so this is not a very strong functional adaptation.)

Communication. The functional specialization of communication could be taken as a subfunction of a control computer. The function is mainly to behave as a switch. In a message-switching application the computer transfers messages from terminals (and links) into primary (and sometimes secondary) memories and then transfers them to other terminals (and links). In message switching, messages are first stored and then forwarded. The computer in a telephone exchange functions as a very sophisticated switch control. Here the computer reads the off-the-hook signal, detects the dialed numbers, rings the dialed parties, and finally sets the switches to connect the telephones together. In some instances, when it answers information inquiries about new telephone numbers or reroutes calls to other phones, it functions as a memory. Thus a communications computer is functionally a switch or a control for a switch.

The main distinction between control computers and communications computers is that the task environment of the latter, since it consists of digitally encoded messages (even in the case of the voice telephone exchange), can be handled directly by the communications computer. That is, the communications computer can do the work of transshipment and storage as well as control.

There are no pure examples of communications computers in this book. However, the Pio's serve essentially the same function within a single computer (Part 4, Sec. 1), and they can profitably be examined from this viewpoint.

File Control. We list this as a separate specialization only because a number of computers have been built to do exactly this task. The specialization is easily described: It is a communication computer with the messages being characters (since they are built for business), and with the large memory (the file) being considered to be part of the system. There are no examples of file-control computers in this book, but the early IBM 305 and UNIVAC file computers serve this function. An IBM 1800 is used as the control for a 10^{12}-bit photo-optical memory, for example.

Terminal. Since it is possible to obtain a separate computer system whose only function is to run a display, we have listed this as a separate functional specialization. In fact, it is better viewed (and almost always occurs) as a component of a larger computer system, i.e., as a special Pio. The DEC 338 is such a P.display and is described both later in this chapter and in detail in Chap. 25.

Time-sharing. The requirement to have a large number of users in simultaneous conversational interaction with a single large machine has bred a new specialization, that of the time-sharing computer. All the computers described above can be time-shared (even if they do not have interrupts or inherent multiprogramming). However, the emphasis on this mode of operation with the particular timing and flexibility requirements of human users doing general computing at consoles in multiple software systems has led to a number of innovations in design. The most important is the virtual-memory techniques for achieving multiprogramming (described in Part 3, Sec. 6). There is also substantially increased complexity of PMS structure to handle the integration of large files, swapping memories, and the huge software systems that seem to be endemic to time-sharing systems. It is still too early to tell whether any of the design responses will produce permanent specialization or will again simply be the first instigation of design features that will become universally used.

In summary, we see that there is functional specialization and that it translates mostly into total size of the machine and into the data-types available. Many of the other design aspects created in response to functional specialization have instead become the common property of all machines.

Performance

For a device that does a complex job, it is meaningless to ask for a single precise index of performance. It is like asking for the average speed of a given model of car over its lifetime without specifying who will own it, where he will drive it, and what sort of terrain he will encounter along the way. Notice that the difficulty is as much in the complexity of the task environment as in the complexity of the internal workings of the machine. Specify everything about the environment, and the performance can often be given in a single figure. It may be hard to determine, but at least it is well defined. If you know the terrain and road conditions perfectly and how the car was driven, then from the structure of the car it is possible to figure out the instantaneous velocity and from this to construct the average speed.

To put this in terms of computers, given a particular configuration for a computer system, given a particular program, and given a particular set of input data, it is possible to determine all aspects of the performance: how long it took, how much space was used, whether it was correct, and so on. But we are not interested in such specifics. We want to know how well the computer system performs, given some vague notion of the kind of task—programs and data—that will be used with it. Although we know that we cannot have adequate measures, we believe that there is something that can be said about the performance—that tells us that a CDC 6600 is many times more powerful in actual performance than a PDP-8.

An interesting way to look at the problem of specifying performance is to play a simple game: We will give you a number, say 4. You are to give the best description of computer systems involving only that many parameters (equivalently, dimensions or attributes). That is, what is the best description of a computer that can be stated in four numbers? The game is easier to play if we speak of the dimensions, rather than the information content of the description (in bits, say).[1] We have still not defined "best," of course. It can be taken to mean the best prediction of the relative ordering of the computer system; better on the index means better on the same task.[2]

To start at the beginning, what single number would you give to characterize a computer's power? Such a question makes most people uncomfortable, since strong feelings exist for at least two kinds of numbers, dealing with speed and memory, respectively. If forced, we would probably settle for something related to processing speed. The cycle time of the primary memory is a possibility because for simple machines it determines (limits) the operation rate. It is a structural parameter, but that is no reason to avoid it as a performance index. The average number of instructions per second, or operations per second, is a better indicator. Since the latter does not take into account the size of the word being processed, perhaps average bits processed per second is the best single number. (We measure this number at the processor, and it may include both the instruction and data streams.)

To take an average we must adopt some weightings. The simplest scheme is simply to add all the instruction (or operation) times and divide by their number. This is equivalent to weighting them equally, the rare ones and the common ones. If we want to do better than that we need some data. Several sets of relative frequencies, of instruction types, called "mixes," have been used in the literature. Table 2 gives four examples. The Gibson mix is

[1] It is not fair, of course, to invent tricks to encode many conceptually independent dimensions into a single one, just to beat the limit. On the other hand, composite dimensions, such as average operation time, are perfectly acceptable.
[2] Definitional precision is not appropriate, since we are not attempting to deal seriously with the technical questions of indices, only to illustrate the issues.

Table 2 Instruction-mix weights for evaluating computer power

	Arbuckle [1966]	Gibson[1]	Knight (scientific)	Knight (commercial)
Fixed + / −	. . .	6	$10(25)^2$	$25(45)^2$
×	. . .	3	6	1
÷	. . .	1	2	
Floating + / −	9.5		10	
Floating ×	5.6			
Floating ÷	2.0			
Load/store	28.5	25 (move)		
Indexing	22.5			
Conditional branch	13.2	20		
Compare	. . .	24		
Branch on character	. . .	10		
Edit	. . .	4		
I/O initiate	. . .	7		
Other	18.7	. . .	72	74

[1] Published reference unknown.

[2] Extra weight for either indirect addressing or index registers.

probably the best known. The best source for such data comes from instruction counts of running programs.

Knight takes the view (Fig. 3) that a single number can be used to indicate power, and his formula has been evaluated for some 300 computers [Knight, 1966]. His formula is the product of three factors: processing time, memory size (in words), and word length. The formula was derived (roughly) to measure power so that technological change could be modeled. Applying the formula is like measuring automotive-vehicle power as a product of speed, weight, and the number of wheels. (Such an indicator is roughly proportional to a car's momentum.) Thus, although it is a reasonable single-number indication for power, a computer buyer could not use it directly.

Taking averages, as in the case of mixes, suggests a more sophisticated approach. A collection of programs, called a "bench mark," is developed that does a variety of different tasks. Then the one number is the time it takes to do this collection. Such a bench mark generates its own frequencies of occurrence of the primitive instructions. It brings in a number of additional dimensions that affect performance: the instruction code, the size of Mp, programming skill, input/output devices, etc. It also carries with it an implicit frequency of different kinds of task demands (how much of the set involves compiling, how much number crunching, how much I/O, etc.).

There are severe practical problems in carrying out such measurements on many computers, since the problems must be coded and run on all the systems. It is somewhat easier if the task set is restricted to programs coded in a procedure-oriented language, such as FORTRAN, where all computers accept FORTRAN. Nevertheless, although it has often been done to compare two systems, only occasionally has it been done for even a modest number. We feel that for a general-purpose computer the compiler-derived bench mark is a reasonable single-performance number. Much actual use will be with the compiler, and good compilers produce code to rival hand coding, so that special features of the machine are utilized. Cox [1968] compares several, using hand coding and compilers for several tasks.

There is a difficulty with the bench-mark scheme that is inherent in its strongest advantage, that of doing a total problem and thus integrating all features of the computer. The number obtained depends not only on the type of computer, for example, an IBM 704, but on the exact configuration, for example, 16 kwords of Mp versus 32 kwords, and even on the operating system and the software (which version of FORTRAN). Thus, although the number perhaps comes closest to an adequate single-performance figure, it becomes much less of a parameter characterizing the structure of the computer than one characterizing a contingent total system.

Let us underscore again the distinction between the computer type and the particular configuration (possibly including basic software) assembled in a particular installation. Computer systems are designed with certain forms of variability. To specify a CDC 1604 is to specify many things, such as the ISP of the Pc, the cycle time of Mp, the K's used to control secondary memories (Ms), and interfaces to the external world. But it leaves open many other

$$P = \frac{10^{12} \; [32{,}000 \; (36\text{-}7)]^i}{t_0 + t_{I/o}} \; \frac{[(L\text{-}7) \; (T) \; (WF)]^i}{}$$

$$t_c = 10^4[C_1 A_{FI} + C_2 A_{FL} + C_3 M + C_4 D + C_5 L]$$

$$
\begin{aligned}
t_{I/O} = &\; P \times OL_1 \, [10^6 \, (W_{I1} \times B \times 1/K_{I1}) + (W_{O1} \times B \times 1/K_{O1}) \\
&+ N(S_1 + H_1)] \, R_1 \\
&+ (1\text{-}P) \, OL_2 \, [10^6 \, (W_{I2} \times B \times 1/K_{I2}) + (W_{O2} \times B \times 1/K_{O2}) \\
&+ N(S_2 + H_2)]
\end{aligned}
$$

Variables—attributes of each computing system

P = the computing power of the nth computing system
L = the word lengths (in bits)
T = the total number of words in memory
t_c = the time for the Central Processing Unit to perform 1 million operations
$t_{I/O}$ = the time the Central Processing Unit stands idle waiting for I/O to take place
A_{FI} = the time for the Central Processing Unit to perform 1 fixed point addition
A_{FL} = the time for the Central Processing Unit to perform 1 floating point addition
M = the time for the Central Processing Unit to perform 1 multiply
D = the time for the Central Processing Unit to perform 1 divide
L = the time for the Central Processing Unit to perform 1 logic operation
B = the number of characters of I/O in each word
K_{I1} = the Input transfer rate (characters per second) of the primary I/O system
K_{O1} = the Output transfer rate (characters per second) of the primary I/O system
K_{I2} = the Input transfer rate (characters per second) of the secondary I/O system
K_{O2} = the Output transfer rate (characters per second) of the secondary I/O system
S_1 = the start time of the primary I/O system not overlapped with compute
H_1 = the stop time of the primary I/O system not overlapped with compute
S_2 = the start time of the secondary I/O system not overlapped with compute
H_2 = the stop time of the secondary I/O system not overlapped with compute
R_1 = 1 + the fraction of the useful primary I/O time that is required for non-overlap rewind time

Semi-constant factors		Values	
Symbol	Description	Scientific computation	Commercial computation
WF	the word factor		
	a. fixed word length memory	1	1
	b. variable word length memory	2	2
C_1	weighting factor representing the percentage of the fixed add operations		
	a. computers without index registers or indirect addressing	10	25
	b. computers with index registers or indirect addressing	25	45

Fig. 3. Knight's functional model algorithm to calculate P for any computer system. (*Courtesy of Datamation, vol. 12, no. 9, September, 1966, page 42.*)

		Scientific	Commercial
C_2	weighting factor that indicates the percentage of floating additions	10	0
C_3	weighting factor that indicates the percentage of multiply operations	6	1
C_4	weighting factor that indicates the percentage of divide operations	2	0
C_5	weighting factor that indicates the percentage of logic operations	72	74
P	percentage of the I/O that uses the primary I/O system		
	a. systems with only a primary I/O system	1.0	1.0
	b. systems with a primary and secondary I/O system	variable	variable
W_{I1}	number of input words per million internal operations using the primary I/O system		
	a. magnetic tape I/O system	20,000	100,000
	b. other I/O systems	2,000	10,000
W_{O1}	number of output words per million internal operations using the primary I/O system	the values are the same as those given above for W_{I1}	
W_{I2} W_{O2}	number of input/output words per million internal operations using the secondary I/O system	the values are the same as those given above for W_{I1}	
N	number of times separate data is read into or out of the computer per million operations	4	20
OL_1	overlap factor 1—the fraction of the primary I/O system's time not overlapped with compute		
	a. no overlap—no buffer	1	1
	b. read or write with compute—single buffer	.85	.85
	c. read, write and compute—single buffer	.7	.7
	d. multiple read, write and compute—several buffers	.60	.60
	e. multiple read, write and compute with program interrupt—several buffers	.25	.55
OL_2	overlap factor 2—the fraction of the secondary I/O system's time not overlapped with compute	values are the same as those given above for OL_1, a through e	
i	the exponential memory weighting factor	.5	.333

things, e.g., the types and sizes of Ms and the size of Mp. On some computers it can even leave open part of the ISP (e.g., the multiply/divide options on many small machines), or the speed of the Pc and Mp (e.g., in the IBM System/360).

When we ask questions about computer systems, we should be clear whether we are talking about a computer "type," such as CDC 1604, or whether we are talking about a particular installation, with all the variability specified. It is possible to describe either with PMS and ISP, provided we recognize that the diagrams for the types represent maximal possibilities for assembling particular systems. This is how almost all the PMS and ISP diagrams in this book were prepared. From the point of view of our "number game," if we are talking about computer types, we might prefer numbers that do not depend on the particular configuration.

If two numbers were available for describing performance, what would they be? Clearly there are several directions to go. One could fractionate the bench mark, so that one has a bench mark for arithmetic-rich tasks and a bench mark for others (a composite of compiling and data processing). One could decompose the processing rate into, say, operations per second and word size (from which bits per second can be recaptured approximately). Alternatively, one could retain only a single number for processing rate and add a measure of the memory available, e.g., size of Mp (in bits). Of the three we would choose the latter, especially if we were talking about a particular installation rather than computer types, for which Mp size remains variable.

We can continue this game through several numbers. Table 3 shows some of our choices. Various parameters drop out or change only when they are decomposed into other parameters from which they can be recovered. Thus, initially Mp must be measured into bits, but when the word size is given, Mp is more reasonably measured in words. One of the reasons for exposing such a list is to emphasize its judgmental and approximate character. There is as yet no way to validate such proposals for brief descriptions.

If we had bench marks, which are themselves only approximations at measuring performance, we might look at how well the parameters in Table 3 predict the bench marks. But there remain the difficulties of how to take into account the additional aspects of the total system (e.g., compiler efficiency) that are implied in the bench mark. Alternatively, one might want to construct a mixed description of bench-mark numbers and measurements of the kind in Table 3. Then the relationship between bench marks and these other measurements would become an indirect measure of the efficiency of the rest of the system.

We have discussed performance in a crude and cavalier way, but this accurately reflects the state of the art. There are no precise measures for performance. There are precise structure and performance measures of individual components (e.g., memory size, and speed and word length, and processor instruction times). When designers (and users) are faced with obtaining a certain total performance for a given cost, the only method is that of the bench mark, because the task is such a significant variable. If performance is to be increased, unless the task is sufficiently trivial, it is difficult to predict what effect changing even the most direct structural variables will have (e.g., memory speed).

Structure

We now turn from function and performance, which provide design constraints and objectives, to the dimensions of structure, which provide the space in which the design is actually cast. A structural dimension is one in which the designer can attain any of the values along the dimension by relatively direct means. Thus a machine is completely specified by listing all its values along the structural dimensions. From this, the system's function and its performance within that function can be determined.

What dimensions should be selected for structure? The viewpoint is distinctly different from that of performance, where one

**Table 3 Performance parameters specification
(as a function of an allowable number of parameters)**

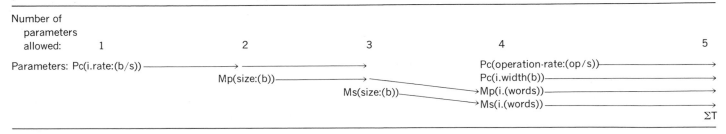

Number of parameters allowed:	1	2	3	4	5
Parameters:	Pc(i.rate:(b/s)) ———————————→	———————————→		Pc(operation-rate:(op/s)) ———————→	
		Mp(size:(b))————————→		Pc(i.width(b)) ————————→	
			Ms(size:(b))———————→	Mp(i.(words))————————→	
				Ms(i.(words)) ———————→	
					ΣT

averages and combines many features to summarize effective output. This tends to obscure structure. For structure, one wants maximally independent aspects which are easily obtained if selected as a design choice. For example, if the computer designer had only a single dimension to describe a computer, he would undoubtedly select the logic technology used in the Pc and K's. This tells him a good deal about many aspects of the computer's structure. In fact, the technology and the average bits processed per second by the Pc are correlated, and so each can be used to predict the other, though only imperfectly. If one is interested in performance, effective bits per second is preferred; if one is interested in design, technology is preferred.

The computer space in Table 1 presents our choice of the major structure dimensions. There is even less means to validate the choice of dimensions here than there is for performance. Nevertheless, there are a few hallmarks. Perhaps the most important is redundancy (the opposite side of the coin from independence, mentioned above). Several dimensions of structure may covary, so that giving any one of them is tantamount to giving the others. This covariation need not come from physical dependence; it may arise from the nature of an appropriate design and good engineering practice. Such a cluster of covarying dimensions is likely to indicate an important dimension (which one among the correlates is to be used is a secondary matter). Table 1 is organized in terms of such clusters, with one of each selected as the main representative and placed at the left.

A second hallmark derives from the hierarchical nature of computer systems. Generally a description of a system consists of the union of the description of its parts, plus a description of the interconnections. This is the basic style of PMS, for example. But there are a few features that affect the total system, i.e., affect many components. These are usually rather important. Technology is a prime example.

Yet a third clue is that the dimensions discriminate the actual population of computers. If all machines had single-address instructions, for instance, there would be no sense in using number of addresses per instruction as a dimension. Any computer engineer who had studied machines at all would know this to be true of all computers. Thus one looks for dimensions that spread the machines out evenly into a substantial number of categories.

If the dimensions of the space are known, a computer is supposed to be defined by a single point. For most existing computers this is actually the case. However, if a computer system were complicated enough, say consisting of several processors, each built with different technologies and having a different number of addresses per instruction, then such a representation would not be

possible. For instance, the Rice University computer uses vacuum tubes, transistors, and integrated-circuit logic. But such complexities are rare; time and good engineering practice work against it. If it were necessary to consider such cases, then additional dimensions (e.g., for secondary and tertiary logic) could be added, or several points in the space for a given computer could be used.

The computer-structure space is thus our choice of the seven most important dimensions. It is our response, so to speak, to playing the number game, given only seven descriptors. They are arranged in order of importance, although clearly no simple way exists to validate such an order. But, if we were to have only three attributes to describe the structure of a computer system, we would pick logic technology, word size, and PMS structure (i.e., what processors exist with what functions).

At this point we are ready to proceed through the space, describing the various dimensions and discussing how the computer systems in this book illustrate various points along them. We take up each major dimension separately. A few of the correlated dimensions are accorded separate sections, but most are discussed along with the main dimension.

Technology

Computers are constrained by the physical technology from which they are constructed. It is not just that new technologies provide greater speed, size, and reliability at less cost, although of course they do that. But technologies dictate the kinds of structures that can be considered and thus come to shape our whole view of what a computer is. For instance, the emergence of the PMS system level is due to advances in technology. Prior to transistor technology, it did not make sense to think of elaborate PMS structures. The costs of the various parts were too high and the reliabilities were too low. When, occasionally, such a machine was in fact designed, it invariably proved too far ahead of its time to succeed. An example in this book might be the RW-40, described in 1960 (Chap. 38). A more classic example is the Analytic Engine of Babbage, which he designed in 1844 and was never able to complete.[1] The technology of the time was entirely mechanical, and its crude state accounts for a large share of the failure. Thus the technology is by all odds the most important single attribute to know about the computer system.

Many technologies go into making up a computer. Each type of component typically uses a different one. In current (so-called

---

[1] Thus, the first real digital computer established the precedent of failing by a large margin to meet the expected dates of completion and full operation.

third-generation) machines the Pc may use hybrid- and integrated-circuit technology for its logic, thin-film technology for the Pc generalized registers, core technology for the Mp, electromechanical technology for tapes and disks (with integrated circuits for logic), mechanical technology for card punches and typewriters, and even manual technology for mounting tapes and disk packs. The existence of all these technologies poses major issues of systems balance, issues which are only imperfectly resolved. For example, it remains true in the current generation that input/output is not in balance with the internal structures. This is due to the crude state of terminal technology, so that it appears to cost too much to provide an appropriate solution.[1]

The heterogeneity of technologies is not a consequence of cost/benefit analysis; rather, each represents the forefront technology for the type of device shown. (There is, of course, cost/performance exchange for any component, but this is usually within a technology.) Thus there is a sense in which the leading technology can be used to represent them all. This is the technology used for the logic level and is the one listed in the computer space. If it is known that transistor logic is used in the Pc of a computer, it is a safe prediction that Ms is electromechanical, Mp is core, Tio is electromechanical printers and punches, etc. This reflects the fact that technology develops and hence becomes locked with calendar time. Thus a prediction is from logic technology to date and then to all other things known to be current at that date.

This correlation of date with technology is given in the computer space along with the generation. It can also be seen in the time chart. The correspondences must be taken as very rough only. The technologies are listed in increasing power (and decreasing cost). The dates run in exactly the same order. The one exception is fluidics, which has been introduced very recently and is a special technology for ruggedness, reliability, and direct external coupling in certain control systems. (Small fluidic computers are at the early prototype stage.)

Alongside the technology dimension we list the dimensions: Pc speed (operations per second), and cost (dollars per million operations), all of which vary directly (or inversely) with logic technology. In general, costs are extremely difficult to determine, espe-

cially when technological costs are of interest rather than market costs (which reflect numerous other factors). Nevertheless the effect of technology on costs has been so striking (while simultaneously pushing up performance along all other dimensions) that it seemed necessary to give a measure of cost in Table 1, no matter how crude.

We have indicated only a few of the dimensions that are correlated with technology. In fact, the only dimensions in Table 1 that are independent of technology are the word length and the Pc addresses/instruction. All the rest show dependence on technology. For some, such as memory speed and size, there is a direct correlation. For others, such as PMS structure and Pc concurrency, the development of more complex versions—the leading edge, so to speak—depends on technology, but there is free use of all versions that are in existence at any given time. There are still other dimensions of importance, not shown in Table 1, that have also changed with technology, e.g., electric-power consumption.

One way to see both what varies and what is independent of technology is to compare selected machines. For instance, Whirlwind (Chap. 6), a first-generation system, and the IBM 1800 (Chap. 33), a third-generation system, have reasonably similar ISP descriptions, if one ignores index registers, which were not invented at the time of Whirlwind's design. However, they have very different PMS structures. In Whirlwind, the early system, transferred information between Tio's and Ms was under program control of the Pc. The existing Pc registers and transfer gates were used because it was too expensive to have separate ones. In the 1800, which uses hybrid circuits, it is economical to have additional subsystems devoted to special functions; hence there are many Pio's operating independently of the main Pc. It was not cost alone that limited the complexity of first-generation vacuum-tube systems. The large physical size of tubes introduced substantial transmission delays; their large power consumption added dependency on a cooling system; and their limited life and deteriorating nature constrained the number of tubes that could be used in a system requiring high reliability.

The IBM 700 scientific series (701, 704, 709, 7090, 7040, 7044, 7094 I and II) offers another comparison, where there is an evolving structure over time, hence across technologies, but where for reasons of compatibility the ISP's have remained almost constant (except for the 701). Again we see radical increases both in performance (Pc speed increases by a factor of 5 from the 701 to the 704 and another 10 to the 7094 II) and PMS complexity. But various other features, though not affecting compatibility, were locked in with the ISP and remained fairly constant. For example, Mp size went to 32 kw (kilowords) early in the series with the 704; and

---

[1] Although beside the point of the current discussion, one reason why these imbalances appear to be "permanent" is that the time constant for change in the technology is of the same order as the time constant for human beings (i.e., systems analysts, programmers, and users) to understand the imbalance. Before system imbalance is diagnosed and solved, the terms of the problem change, inducing new imbalances.

it took a jerry-rigged modification to get 64 kw on a 7094 toward the end of the lifetime of the series (see Chap. 41, page 517).

Throughout this section we have referred to technology as the dominant factor in the computer. Does this mean that computer development waits upon new fundamental windfalls? We have been lucky in getting the transistor and, to a lesser degree, the integrated circuit from external efforts. However, core memories were invented for the computer and resulted because of need. Read-only memories have also resulted both from development at the circuit level and from pressure above, requiring the memories to be developed. All the electromechanical secondary memories (i.e., magnetic tape, drums, disks, and photostores) have resulted from the computer's needs. Thus, although technology is dominant, the computer often forces the development.

The Pc operation rate is strongly correlated with logic technology, as we have indicated in the computer space. Our discussion about technology and generations is also about operation rate. The principal reason for the higher operation rate is because of faster logic technology. Technology also has a secondary effect on increasing speed. More reliable devices allow large computers to be built. Smaller devices allow higher device densities, thus decreasing stray capacitance and inductance and shortening transmission delays. Smaller components also allow increased interconnection density.

Operation rate is also relatively highly correlated with total performance. If we hold the structure and concurrency constant, the simplest way to increase performance is by increasing the clock rate. The increase in the performance/cost ratio over the past two decades of computer evolution has made their primary gains through higher operation rates. The two 16-bit computers already mentioned, Whirlwind (Chap. 6) and the IBM 1800 (Chap. 33), provide a nice comparison of the evolution. With a difference of 10 years and two generations, their cost ratio is ~10:1 whereas performance is ~1:5 and the internal clock rates are also ~1:5.[1]

Information structure: word length, information base, and data-types

All computers structure their information in a hierarchy of units, which we defined as an i-unit in Chap. 2. For example, the IBM System/360 starts with the bit; then the byte, which is 8 bits; then the word, which is 4 bytes; then the record, which is a variable number of words. In between, playing minor roles, are decimal digits (4 bits), the halfword, and the double word. A number of features of the design are related to this hierarchical organization of data. Before we consider them, we need to characterize the organization itself. One characteristic of this organization, the word length (in bits), gives most of the information, the rest of the hierarchy adding only a little.

Let us see why this is so. At the bottom there is the bit, encoded in two-state devices. Although other numbers of states are possible, and ternary (three-state) machines have been proposed occasionally, digital technology has developed exclusively to handle binary information. There are several reasons for this. The first is the requirement for high reliability and high signal-to-noise ratios in the basic devices. Generally a basic n-state device (that is, one not built up from other k-state devices) is realized by breaking a continuous physical dimension, such as voltage, current, or magnetic flux, into n discrete levels or regions. Reliability and signal-to-noise ratio then depend on keeping adequate separation. This is easiest to do with two states (e.g., in the limit they become on-off devices) and becomes progressively more difficult as n increases. The second reason is the simplicity of the logical design for binary representations. A basic device for combining two ternary digits must deal with $3 \times 3 = 9$ configurations, rather than $2 \times 2 = 4$ configurations for the binary case. This also gets worse as n increases.

A final reason—the *coup de grace*, so to speak—is that no one has ever found striking advantages for the resulting processing structure in having more than two states. Thus there are no compelling reasons to suffer the first two disadvantages. In short, what might have been an important dimension on which to distinguish computers, namely, the number of states in the basic encoding, turns out instead to be one of the great uniformities in digital technology.

Information base. That the physical devices deal ultimately in bits does not imply that the information processing must be organized in terms of bits. It is possible to select an arbitrary *base* (one with any number of states) and construct the entire ISP in its terms. A base unit is represented physically, of course, as a set of bits. If one wanted a base 13 machine, for example, one would have to use at least 4 bits (with 16 states) to encode it. But no operations at the ISP level would refer to anything but base units and data structures built up from sets of base units, and there would be no way to manipulate directly the bits that represented the base. Thus, using a base other than binary obtains whatever advantages might accrue to n-state units, without any of the disadvantages at the device level.

---

[1] However, it is not as dramatic an example as we could find. By picking a better third-generation example we might get a cost ratio of ~100:1 and a performance ratio of ~1:10.

Computers have been built with a variety of different bases, the main ones being binary, decimal, and character. The character has shifted between a 6-bit character and an 8-bit character (byte).[1] The arguments for bases other than binary (which represents the natural base of the computer) all hinge on the alphabets used externally by human beings and the desire to avoid conversions into a different representation inside the computer. With universal acceptance of higher languages, such as FORTRAN and ALGOL, this argument has also lost much of its force. In fact, all third-generation machines are binary. Nevertheless, in the fifties there was much controversy over which base to use, and the machines presented in this book exhibit all three bases.

There is little difference between binary and decimal computers in their ISP organization. However, there is a great difference between these two and character machines. The latter are designed for handling text and are constructed to deal with variable-length strings of characters. Correspondingly, they deemphasize numerical computation. Both these decisions affect the ISP considerably. Thus, in the computer space we indicate the base dimension along with the word-length dimension. The two together make up a single dimension.

Word length. Let us now examine the role of word length. The word is the first major information unit above the base. It is defined as n bits for a binary computer or n digits for a decimal computer (character machines being excluded as not having a fixed word length). Sometimes there are intermediate units, but they always play a minor role and we can disregard them at this stage. As we noted earlier, the main determinant of word length has been the function of the total system: large word lengths for arithmetic systems, small word lengths for control systems (and character strings for business). Thus, only within narrow limits is the word length a free design choice.

However, the interesting thing about word length is not so much its determinant as the way it affects other aspects of the total system design. This starts with a design decision that the unit of information transfer between components will be a word. As soon as this becomes the case, then registers in various components must hold a word, since that is what arrives or is to be transmitted. Thus the word becomes the information unit of the Mp, and most of the registers of the Pc hold one word. The instruction is designed to fit into one word, since that is the number of bits that is obtained "at once" and hence can be used to effect the next time increment of processing.

Once these basic features are set, others follow. An integer number of any smaller units, such as the character, should fit into a word, since otherwise a set of words will not provide a homogeneous sequence of subunits. (That is, only five 6-bit characters fit into 32 bits, so that a set of 32-bit words filled with 6-bit characters has a number of 2-bit holes in it. This can complicate algorithms that deal with long character strings.) The constraint of compatibility is not so strong with Ms, since speeds are slow enough to permit conversion algorithms (either hardware or software). Still, the system is simpler (and therefore usually will work better) if incommensurabilities of information units do not exist. Thus, to pick an example, the number of parallel tracks on magnetic tapes tends to divide evenly into the word length. IBM tapes for the 700 series of 36-bit machines have six data tracks; for the System/360, which has a 32-bit word, the tapes have eight data tracks.

There is an interesting correlation between the word length of a computer and the number of data-types that it makes available. As we saw in Chap. 2, the operations in a computer can be classified according to the type of data they operate upon. Each data type tends to have a certain set of operations appropriate to it (for example, $+$, $-$, $\times$, and $/$ for numbers) and the decision to include a data-type carries with it the decision to include its operations. Thus the number of operations tends to grow with the number of data-types. The total amount of hardware in a computer grows as the word size (because data paths are word-parallel[2]) and also as the number of operations. Thus machines with large word size tend to be large machines and have many data-types and many operations. ("Large" as an adjective for machines invariably means big and expensive, hence—given economics—capable of doing large amounts of processing.)

There are two additional, somewhat independent, features that support the relationship between word size, number of data-types, and size of computer. First, with a large system there will already be available many of the pieces necessary to add additional operations. That is, the marginal cost of a new operation goes down as the system grows. Therefore, given a large system, there is a tendency to add more operations. The number of operations per data-type is not easy to increase; rather, one adds new data-types. Second, with small word lengths, one cannot define many worthwhile data-types that will fit into a word, and multiple-word data-types are left to the programmer to define with software. With large word lengths there are many different worthwhile data-types that fit into the word, for instance, decompositions of the word into partial words, or into character strings. Each of these requires

---

[1] Seven bits have been proposed for communication purposes but have never been made the basis of a machine, as far as we know.

[2] The issue of bit-serial versus bit-parallel is discussed subsequently.

additional operations, since the initial data-types involve the entire word or some large part of it (i.e., the word, address, and integer operations).

In sum, the word length stands as an indicator of many aspects of the machine. It not only tells something about the basic organization of many components but indicates how big the computer is, both in number of data-types and number of operations. Figure 2 shows time lines of well-known computers with their word length, with a special time line for the ones in this book. Five groups are suggested in the figure which classify these computers.[1] The classes overlap, and to separate a computer into one of two classes requires more knowledge (e.g., the number of data-types). For example, the 24-bit SDS 9300 and CDC 3200 appear in the same class with the 36-bit IBM 7090 just because both machines have floating point hardware and, in fact, perform comparably for arithmetic tasks.

The one design choice that makes word length have few of the consequences just described is making a computer bit-serial rather than bit-parallel. In many machines information transfers are conducted on a single bit stream (especially Pc-Mp transfers). Coincident with this is the construction of operations on a bit-by-bit basis. This works well for arithmetic and logical operations. Time is traded for hardware. The cost of the system becomes independent of word length, but the processing rates go down correspondingly. This design decision was an extremely important one when logic was expensive and unreliable. It has become less so in the current era, where processors and transfer paths are relatively few in number while both the cost and the reliability of components have improved. However, as large parallel processors are considered ($\sim 10^3$ P's), bit-serial processors again become a serious design alternative. (See the serial computers of Part 3, Sec. 2.)

In summary, word length is an important dimension, and we find many characteristics either proportional to or inversely proportional to it. To be sure, these relations hold only for current design practice, as we have seen with the bit-serial designs. The main-line computers in Part 2 are ordered according to increasing word length.

Data-types. We have presented the number of data-types as being correlated with word length and also with computer size through the effect on number of operations. Although far from perfect, there is a rough order in which specific data-types are included in a computer. We have listed the main types in such an order in the data-type dimension of the computer space. (See Chap. 2

for their definitions.) To be located at a point on this dimension (say at floating point) means to have all the data types below it on the dimension, (i.e., word, address, integer, boolean.) Occasionally machines which violate this have arisen. Decimal machines do not generally have boolean data-types, and there has been some attempt at machines with only floating point, i.e., without a separate integer type (e.g., the CDC G20[2]).

The reason behind this cumulation of data-types in a fixed order is that certain general tasks must be performed by any computer. It must transmit data between the Pc and Mp, and this transmission has nothing to do with the meaning or content of the data; thus there is always the "unit of transmission," which is the word (except on character machines). Next, all computers manipulate addresses to achieve generality (e.g., to compile), providing for a second data-type. Next come integers, since almost all algorithms make use of arithmetic (this could conceivably be absent in some communications computers), and on up to floating point numbers, multiple precision, and vector and string operations. At each stage the uses are more specialized so that lower ones cannot be eliminated, except for a few cases such as handling addresses as regular integers.

Addresses per instruction and processor state

The number of addresses in an instruction has been a traditional way of describing processors (i.e., their ISP's) and hence the computer systems containing these processors.[3] We use it in Parts 2 and 3 to separate the different processors.

Originally the dimension was simple: one-, two-, three-, and four-address machines were constructed. It has become somewhat more complex. A "one plus one" machine has one address for data and one for determining the next instruction, and is to be distinguished from a two-address machine, which uses both addresses for data. Index registers and so-called general registers provide instruction schemes which lie somewhere between one- and two-address organizations. When processors admit several instruction formats or variable-length instructions, matters become even more complicated.

A correlated dimension in the computer space is the amount of processor state, that is, the number of bits that exist in the processor, as described in the ISP. This is the amount of information that can be held at the end of one instruction to provide the processing context for the next instruction. It consists of a number of status and mode bits (in modern machines packaged into regis-

---

[1] The class number is essentially $[\log_2(\text{Mp word length}) - 2]$.

[2] Originally the Bendix G-20.

[3] Although used mostly to describe Pc's, the description applies to any processor.

ters, but in earlier machines simply scattered around in the processor), the next instruction address, the accumulator and other arithmetic registers, the index registers, and other general registers making up a "scratch-pad" memory. It is a simpler descriptor of the ISP than addresses per instruction, since it is independent of the number and variety of instruction formats. It is easy to define processor state generally for any ISP, but difficult to define addresses per instruction.

The processor state is not the total number of bits in the processor, since there may be registers in the physical system that are used within the interpretation of one instruction but which carry no information between instructions. Address registers for obtaining operands from Mp are the most common such "underground" or "temporary" registers, but there can be others. We implied this distinction by defining processor state in terms of the ISP rather than the physical processor.

The correlation between the processor state and the number of addresses per instruction is not simple, since it rests on two separate issues. For the first, note that larger programs perform transformations on the state of Mp (or even Ms or Tio's) and are not concerned with the state of the processor. Processor state enters only because, in decomposing the total algorithm into a series of small steps, it is not possible (or efficient) to make each step a transformation from Mp to Mp. Basically, this happens because the instruction does not hold enough information to specify the Mp-to-Mp transformations. For example, if one wants to add two numbers, two operands are required, and an instruction must contain at least two addresses; if it does not, then an intermediate state (i.e., processor state) must be created to hold the information while the additional instructions are fetched. Thus, one-address organizations require the most processor state, with less for two- and three-address organizations. This consideration stops at three (two operands and a result) because only a few elementary operations are more than binary. The processor state cannot be eliminated entirely, however, since there must be at least an instruction address (a program register) to maintain continuity of the program.

The second source of correlation between processor state and instructions per address comes from differential access time to processor registers and to Mp. As long as there is an appreciable differential, substantial gain, processing power can be obtained from increasing processor state. This derives, again, from the structure of algorithms which generate intermediate results that are used almost immediately afterward and then are of no further interest. Rapid temporary storage and retrieval are beneficial under these conditions. Thus, working against higher address

organization is the extra time to store in Mp results that need only temporary storage. Thus, also, index registers and general registers almost always imply increased processor state, although they need not do so logically (that is, the registers could exist in Mp and still have their effect on the instruction format).

With interrupts and multiprogramming the processor state gains additional significance, since it is the amount of information that has to be saved and restored when switching programs. For example, in the Honeywell H-800, an early three-address computer, the processor state per program consisted only of the program counter and index registers, and when io-halts occurred during processing, the Pc was switched immediately to another program. Eight programs could run concurrently (by having a total processor state of 64 program registers). In present computers with general-register state, often $25 \sim 100$ words must be stored, which implies an appreciable time for switching contexts.

We can now consider briefly the different organizations according to addresses per instruction. To show the common similarities, we give in Fig. 4 a state diagram that can be used for all processors. In common is the basic idea of the stored program: Fetch an instruction, determine what the instruction is to do, then execute it (the fetch-execute cycle). Other than this, only a part of the state diagram will be applicable to a given processor type.

As shown in the computer space, the addresses-per-instruction dimension starts with zero addresses, then one address, then one plus indexing, one plus general registers, and on up to two, three, and variable addresses. However, from an expository viewpoint one should follow a different course, starting with single-address machines, then indexing, then two- and three-address machines, then general registers, and finally the zero-address and variable-address organizations. This not only puts the more common organizations first but makes it easy to relate the organizations to each other.

P(1 address) and P(1 + index address). These Pc's constitute most first-, second-, and simple third-generation computers. The earliest outline of the structure was the IAS computer (Chap. 4), which has come to be known as the von Neumann computer. Although fundamentally like the IAS computer, EDSAC's adaptation appears to be the closest prototype to this class. Although EDSAC is not described, it influenced M.I.T.'s Whirlwind I significantly (Chap. 6).

A significant change to the IAS machine was the addition of the index register (called B-tubes) in the Manchester University machine in the early 1950s. The evolution can be seen by comparing the first and third generations using Whirlwind (Chap. 6) and

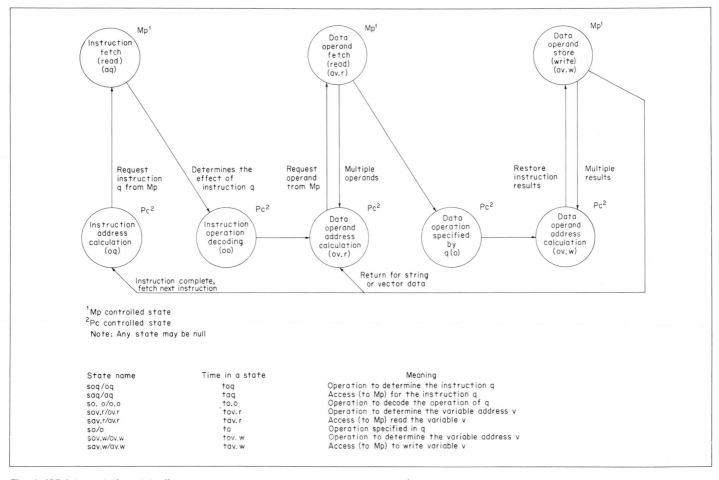

Fig. 4. ISP interpretation state diagram.

the IBM 1800 (Chap. 33) or looking at the IBM 701–7094 evolution in Part 6, Sec. 1. Index registers are motivated by the frequent occurrence, in 1 address systems, of circuitous address calculations that involve first computing the address (e.g., the index of an array in Mp) and then planting it just ahead in the instruction stream in order to make use of it as an address. Providing a set of index registers introduces a second address into the instruction, even though of extremely limited function. Thus we classify processors with indexing as having $(1 + x)$ addresses per instruction.[1] An alternative view of index registers suggests that they double the number of data-types by allowing operations on vector data elements rather than just scalars.

[1] Indirect addressing, on the other hand, does not add to the addresses per instruction; rather, it introduces a second operation per instruction.

For the 1 address processor, the processor state (Mps) typically consists of the program counter (instruction location counter), an Accumulator/AC, a Multiplier-Quotient register/MQ (the extension of AC), and one or more Index registers/X/XR.

With only one address in the instruction, the one arithmetic register, A, must be used for temporary results. Thus an effective-address integer (z) is computed as a function of the address part (v part) of the instruction (q) and the index registers. This process is typically

$$z := v + X[j]$$

where X[j] is the jth index registers as specified in the instruction.

There are several forms for the transmission operators between A and Mp.

$A \leftarrow z$	*load immediate*
$A \leftarrow Mp[x]$	*load direct*
$A \leftarrow Mp[Mp[x]]$	*load indirect*
$M[x] \leftarrow A$	*store direct*
$Mp[Mp[z]] \leftarrow A$	*store indirect*

In indirect operations a convention may be required to determine what address in Mp[z] is to be used.

Similarly, the binary operations ($+$, $-$, $\times$, $/$, $\wedge$, $\vee$, $\oplus$, concatenation, etc.) are generally of the form[1]

$$A \leftarrow A \; b \; Mp[z]$$

Rarely do we find the symmetrical operation form

$$Mp[z] \leftarrow A \; b \; Mp[z]$$

For unary operations ($\neg$, $-$, abs, sin, cos, etc.) the most common forms are

$$A \leftarrow u \; A$$
$$A \leftarrow u \; Mp[z]$$

Rarely do we find

$$Mp[z] \leftarrow u \; Mp[z]$$
$$Mp[z] \leftarrow u \; A$$

In both the above cases, exclusion of the operations that place results in Mp[z] stems from the added cost of including the symmetrical function and the marginal utility of such a function, which stems from the result of applying u not being available for further processing.

The transmission, unary, and binary operators account for almost all operations in these computers. If we allow A to stand for any part of the Mps, rather than just the accumulator, then the instructions not included above are input/output data transmission, e.g.,

$$Mp \leftarrow T \quad \text{and} \quad T \leftarrow Mp$$

and conditional execution

$$(\text{branch if zero AC}) \rightarrow ((AC = 0) \rightarrow (P \leftarrow z))$$

Having index registers requires operations to process them. At a minimum they must be loaded and stored (usually from and to Mp), i.e.,

$Mp[z] \leftarrow X$	*store index*
$X \leftarrow Mp[z]$	*load index register*

[1] Any of the addressing modes suggested above can be used for an operand: that is, z *immediate,* Mp[z] *direct,* and Mp[MP[z]] *indirect.*

But simple operations on an X are also desirable; for example,

$$X \leftarrow X + 1$$

Here X is used to point to (access) the next element in a vector. More complex operations can be carried out by placing X in the A register, via the program steps:

$A \leftarrow X$	*load A with X*
$A \leftarrow f(A)$	*manipulate A*
$X \leftarrow A$	*load X with A*

An operation to add k to X would then be

$A \leftarrow X$;	next
$A \leftarrow A + k$;	next
$X \leftarrow A$	

instead of

$Mp[z] \leftarrow X$;	next
$A \leftarrow Mp[z]$;	next
$A \leftarrow A + k$;	next
$Mp[z] \leftarrow A$;	next
$X \leftarrow Mp[z]$	

which assumes no transmission paths between X and A. Ideally we would like to perform any operation directly on X as simply

$$X \leftarrow X + k$$

From this begins the idea that X should look like the main arithmetic register, A. This is, no doubt, one evolutionary path to general-register processors.

Part 2, Sec. 1 is devoted entirely to 1 address computers in the first three generations. They were the "main line" of computer development.

P(2 address) and P(3 address). The computers in Part 3, Sec. 1 have instructions which contain multiple addresses per instruction. The addresses (v) specify operands in Mp (Fig. 4). The Mps decreases as the number of addresses per instruction increases, since the operands need not be held temporarily between instructions (i.e., each instruction performs a complete operation).

The instruction form for the 3 address computer is

$$Mp[v_3] \leftarrow Mp[v_1] \; b \; Mp[v_2]$$

where b is a binary operator, and v_1, v_2, and v_3 are the addresses specifying the operands. In the case of unary operations, u, v_2 is usually blank. In the case of a binary operation and a three-address computer, the states are oq, aq, oo, ov.r, av.r, ov.r, av.r, o, ov.w, av.w (Fig. 4). MIDAC (Chap. 14) and Strela (Chap. 15) are typical three-address computers.

A 2 address computer does not necessarily require more processor state than a 3 address computer, since the operations can correspond to

$$Mp[v_2] \leftarrow Mp[v_2] \; b \; Mp[v_1]$$

and

$$Mp[v_2] \leftarrow u \; Mp[v_1]$$

However, sometimes extra Mps is usual. The RW-400 (Chap. 38) has an accumulator, and operations generally terminate with results both in primary memory, $Mp[v_2]$, and in the accumulator. The branch on accumulator instructions allows results to be checked directly without referring to Mp. An especially nice instruction in 2 address computers is the transmission instruction (a special-case unary operation): $Mp[v_2] \leftarrow Mp[v_1]$.

The IBM 1401 (Chap. 18) has two registers, A_address and B_address, which hold v_1 and v_2 and can be loaded by the v_1 and v_2 parts of the instruction. These registers point to (address) operands and do not contain data. The remaining processor state is the Instruction_address. The 1401 has instructions with no address parts, and these instructions take as operand addresses the values of A_address and B_address as of the previous instruction. The 1401 instruction-interpreter state diagram is given in Chap. 18 (Fig. 3). The state-diagram specialization (Fig. 4) is roughly:

oq, aq, oo $\{ov.r_1, av.r_1, ov.r_2, av.r_2, o, ov.w_2, av.w_2\} \cdots$
$$\{ov.r_1, av.r_1, ov.r_2, av.r_2, o, ov.w_2, av.w_2\}$$

where the sequence delimited by the $\{\cdots\}$ is the operation on a character; because the 1401 operates on variable-length strings, it is repeated until the end of the string.

P(n + 1 address). Processors with n + 1 addresses deviate only slightly from the n-address processors above. The final, or +1, address explicitly specifies the address of the next instruction. As such, it can be used with any instruction set. There are two reasons why +1 addressing is used. First, freedom is provided in the placement of each instruction within the program address space. Second, the next instruction address can be calculated in parallel with the execution of the current instruction.

For computers with cyclic memories (Part 3, Sec. 2), the +1 address allows both data and the next instruction to be specified independently, providing the opportunity to arrange the program and data in an optimum fashion. Since each instruction completion time depends on the location of data, it is desirable that the next instruction location be variable rather than the implicit next address used for most processors. This is almost universal practice in computers with Mp.cyclic (see LGP-30 in Chap. 16 for an exception).

Microprogrammed processors may use the +1 address to locate the next instruction, and there may be several such next addresses. Microprogram subroutines tend to be short (intrinsic to interpreting an instruction set), and there are many jump addresses. The increased speed from not having to compute the next instruction address is worth the added space cost. The IBM System/360 Model 30 (Chap. 32) shows the use of multiple (+1) addresses and if classified according to our scheme would be at least a P(microprogram; 3 + 1 address).

P(general register). The general register processor has a small array of registers that can be used for multiple functions. These have fast access compared with the Mp, so that it pays to do as much processing as possible within them. Since the general register array is small, it requires only a small address (3 to 8 bits). Thus the instruction format contains fields for one (or more) general registers. There must still exist addressing for Mp, though this never exceeds a single address. Thus we classify general registers machines as (1 + g) addresses per instruction.

The organization of a (1 + g) system can vary from something very close to a (1 + x) organization, in which essentially every instruction involves some Mp information, to an organization in which the only Mp instructions are transfers between Mp and Mps (the processor state holding the general registers), and there is a two- or three-address instruction set involving only Mps (see the CDC 6600 in Chap. 39). That is, from a data point of view the Mps acts like a directly addressable Mp.

The processor state of a general register processor is invariably held entirely within the general register array (rather than having additional independent registers). This is due in part to an already available mechanism (the array) and in part to the need for program switching, which is somewhat simplified by having all the Mps held in a single homogeneous memory.

The general registers typically perform a variety of functions:

1. Arithmetic registers (accumulator and the accumulator extension for the multiplier-quotient).

2. Index registers.

3. A second index register or base register; if the program addresses (v) are short, a base register is needed to address any area of Mp.

4. Subroutine linkage registers.

5 Program flag (sense) registers for boolean variables.

6 Stack pointer (P may have multiple simultaneously active stacks).

7 Address pointers to data arrays and lists.

8 Temporary data storage for intermediate results.

9 Temporary program storage for short program loops.

The power of a general register processor is obtained because the registers can serve many functions. Thus the operations on these registers can be extensive, because the operations need not be duplicated in other parts of the structure. For example, special operations for index registers are not necessary because the operations for integers apply universally to both the accumulator and index registers. Of course, such generality requires compromises. The stack computer is faster for problems which can utilize stacks, whereas the general register Pc must utilize Mp for the stack(s) and does not have the encoding efficiency of a pure stack processor (see below). In addition, the assignment (and reassignment) of general registers is most crucial, since they are a scarce resource with many uses. A general register organization allows processors with a high degree of parallelism to be constructed, since several instruction subsequences can be executed concurrently.

The actual number of registers is rather critical and depends not only on the algorithms of tasks coded but also on the technology. In multiprogramming and interrupt computers, the program switching time increases with the number of registers. Thus the upper bound on the number of registers is both cost and program switching time.

We would expect to find instructions which produced the following affects.

Format	Addresses/instruction
$G[g] \leftarrow u\ G[g]$	1g
$G[g_1] \leftarrow u\ G[g_2]$	2g
$Mp[v] \leftarrow u\ Mp[v]$	1
$Mp[v_1] \leftarrow u\ Mp[v_2]$	2
$G[g] \leftarrow u\ Mp[v]$	1 + g
$Mp[v] \leftarrow u\ G[g]$	1 + g
$G[g] \leftarrow G[g]\ b\ Mp[v]$	1 + g
$G[g_1] \leftarrow G[g_1]\ b\ G[g_2]$	2g
$G[g_1] \leftarrow G[g_2]\ b\ G[g_3]$	3g
$Mp[v] \leftarrow G[g]\ b\ Mp[v]$	1 + g
$Mp[v_1] \leftarrow Mp[v_2]\ b\ Mp[v_3]$	3

where

u are unary operators $(\neg\ |\ -\ |\ \text{abs}(\ )\ |\ -\text{abs}(\ )\ |\ \text{etc.})$
b are binary operators $(+\ |\ -\ |\ /\ |\ \times\ |\ \wedge\ |\ \vee\ |\ \oplus\ |\ \text{etc.})$
G is the general-register array
g, g_1, g_2, g_3 are instruction parts specifying a general register, G
v, v_1, v_2, v_3 are Mp addresses specified as a function of instruction and general registers (for example, $v := (\text{address} + G[g])$ or $v := (\text{address} + G[g_1] + G[g_2])$) in the IBM System/360.

General registers can be thought of as an outgrowth (generalization) of the $1 + x$ processors, as we have already suggested. Alternatively, they can be thought of as evolving from a 2 or 3 address structure. The UNIVAC 1103A, a 2 address processor (Chap. 13), was no doubt a forerunner of the general register UNIVAC 1107 and 1108. Pegasus (Chap. 9) is, we think, about the earliest computer to use general registers (1956). In Part 2, Sec. 2 we discuss four general registers computers.

P.stack (0 *addresses per instruction*). From a PMS viewpoint the P.stack is built around having a first-in–last-out memory (M.stack) as part of the processor state. Conceptually, it is built around the fact that computations can often be sequenced so that no explicit names (i.e., addresses) are required for temporary results. All operations are performed on the top of the stack. As each partial result is computed, it is pushed down in the stack and appears again to participate as an operand at exactly the appropriate point in later calculation. Thus the stack operates as an implicit memory for all intermediate products and not only are transfers between P and Mp avoided but space in the instruction for Mp addresses is eliminated.

Instructions in such a system consist only of operations, since all their operands are in the stack. Thus the instruction format is that of zero addresses per instruction. There must, of course, be some addressing of Mp (just as in a general-register organization). However, the addresses for Mp themselves sit in the stack so that the instruction contains only the transfer (load or store) operation, not the address. There still must exist some way of getting fresh data in the stack, and all P.stacks have at least one operation that loads an address written in the program stream onto the top of the stack.

Why there should be this happy correspondence between calculations and memory to be performed and stack memories requires a little explication. It rests fundamentally on the phrase structuring of calculation in which each partial result is required at one and only one point, so that each subcomputation can be nested in the program (and hence its result nested in the stack) in the same order as it will occur as operand to the one operation that uses it.

There are several arguments against a P.stack. Multiple stacks are often required. Part of the power of a P.stack is derived from having higher-speed Mps for the stack. Yet only the top few (2 ∼ 8) registers of the stack can be in Mps. When M.stack overflows into Mp, the speed of operations can become much worse than not having a stack at all. A simpler implementation, for example, P.general_registers, is as fast and perhaps more general. Another difficulty with the stack is the inability to access other than the top. If full addressing is provided, then the organization has become almost general register. Yet another difficulty arises from inhomogeneity of data-types, especially if several of them are packed into a single word (the width of the stack). Thus, for instance, in one stack machine (the Burroughs B 5000 in Chap. 22) there is a completely separate nonstack ISP for string manipulation.

A simple numerical computation is given in Table 4 as a comparison of the P.stack, P.1 address, and P.general_registers. Here, the P.stack is probably shown at its best as there are no array-indices calculations or program-flow manipulations involving testing, etc. The criteria we measure are the algorithm encoding space and the problem running time.

The kinds of instructions interpreted by a P.stack are typically:

Operation	Interpreter state sequence	Example
Load	oq, aq, oo, ov.r, av.r	M.stack-top ← Mp[v]
Store	oq, aq, oo, ov.w, av.w	Mp[v] ← M.stack-top
Unary operation	oq, aq, oo, o(u)	M.stack-top ← u M.stack-top
Binary operation	oq, aq, oo, o(b)	M.stack-top ← M.stack-top b M. stack-top − 1

Variable numbers of addresses per instruction. Although there are a few operations that require the specification of three or more addresses, these are of such low frequency that no machine has ever been built (or seriously proposed, for that matter) that has more than three data addresses and one next-instruction address. (Some of the microprogrammed processors have more than one next-instruction address, and they often do several operations in parallel in one instruction.)

However, there have been developed processors that can have a variable number of operands. Most of these involve the use of an instruction that is larger than a single Mp word. Thus, bringing in the first word of an instruction, which contains the operation code, determines how many additional operands are needed and

hence how many additional words to obtain from Mp. (In a character-based system this may require several reads per operand; in a word-based system this may be one or two operands per read.) The gain in such a system is the higher average density of operations per instruction, bought at the price of extra Mp accesses.

Most such variable-address processors have a mixture of one, two, and three addresses per instruction—simply a mix of the types already considered. The fundamental limit to such variability is the processor state (plus the additional within-instruction temporary state). This, of physical necessity, must be finite, and the number of addresses must yield an amount of information that is less than this total state. Otherwise the processor cannot hold onto it to process it.[1] Thus the various processors which claim to operate from a higher language (see the P.languages of Part 4, Sec. 4) must in fact either translate into another simpler programming language, as does the FORTRAN machine (Chap. 31), or become an interpreter which processes a small amount of a language statement before the rest.

PMS structure

The idea that there is significant higher organization to computers is relatively new. Texts on logical design of computers develop a model based on an arithmetic section, input/output devices, a memory for holding instructions and data, and a single control to force the other components to interact. A PMS diagram of an early model is given in Fig. 5 (X represents an external agent, usually a man). The Whirlwind I manual-model figure (page 10) used in Chap. 1 was rather highly developed because it had a secondary memory and switching. Figure 6 is a PMS diagram which reflects this more accurate model. Often computer designers lump the devices at the periphery and call them all input/output; these devices are both input/output terminals (T) and secondary memories (Ms).

[1] If it processes a large amount of information, but in pieces (i.e., sequentially in real time), it is not really executing a single instruction based on all the addresses but has decomposed the total computation, just as a single address organization has.

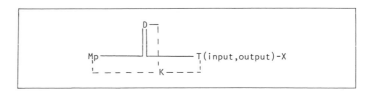

Fig. 5. Early model of a stored program digital computer PMS diagram.

Table 4   Comparison of stack, general registers, and accumulator Pc for evaluating the expression: f = (a − b)/(c − d × e)

	Pc.stack [stack contents]	*Pc.general register*	*Pc.1address*
	Push a [a]	Load G[1], a	Load d
	Push b [a, b]	Subtract G[1], b	Multiply e
	Subtract [a − b]	Load G[2], d	Inverse subtract c[1]
	Push c [a − b, c]	Multiply G[2], e	Store temporary
	Push d [a − b, c, d]	Inverse subtract G[2], c[1]	Load a
	Push e [a − b, c, d, e]	Divide G[1], G[2]	Subtract b
	Multiply [a − b, c, d × e]	Store G[1], f	Divide temporary
	Subtract [a − b, c − d × e]		Store f
	Divide [(a − b)/(c − d × e)]		
	Pop f [ ] − stores stack at location, f		
Program size:			
Address integer/ai	6 ai	6 ai + 8 ai(gr)	8 ai
Operation parts/o	4 o	7 o	8 o
Number of Mp references for data:			
Program size for hypothetical example machines:	$6 \times (18 + 1)$ $\underline{4 \times 6}$ 138	$6 \times (18 + 6 + 4^2)$ $\underline{1 \times (6 + 2 \times 4^2)}$ 182	$8 \times (18 + 6)$ 192
Program size in bits among specific C's:	B8501[3]:168	IBM System/360:208(above[1]) :224(actual) + base register overhead (0 ∼ 192)[4]	IBM 7090:288(above[1]) 360(actual)

[1] Not an instruction in the specific-example machines.

[2] Assume 16 general registers.

[3] The Burroughs Corporation B8501 Pc.stack (discontinued).

[4] Not completely true, since System/360 has only a 12-bit address and uses base registers. Some overhead should be assumed. Worst case (but not unreasonable) is 6 × 32 or 192-bit overhead.

If we separate each component according to its function, assign control (K) to each element, and finally introduce the processor (P), we get the structure of Fig. 7. Of course, a large part of P is a data operator (D). The processor has the behavioral properties attributed to the structure of Fig. 5. If we include the control within each component, we get Fig. 8 from Fig. 7.

To consider larger structures, consisting of several Mp's, P's, Ms's, and T's, one might think to expand the system as shown in Fig. 9, in which we connect everything through a single switch. If the central S has sufficient power for multiple conversations, this indeed provides maximum generality. However, although

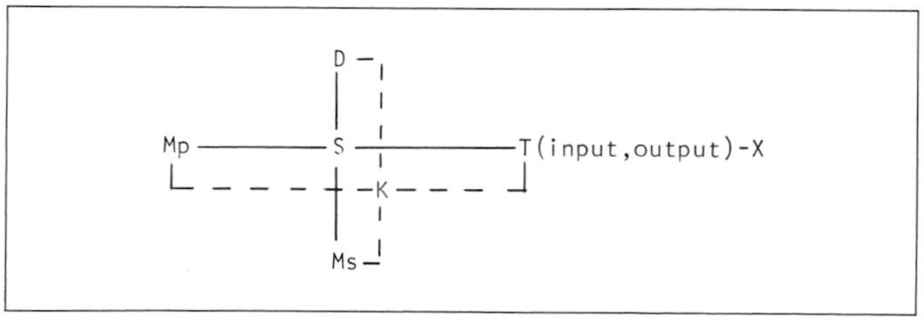

Fig. 6. Early computer model (with Ms and S) PMS diagram.

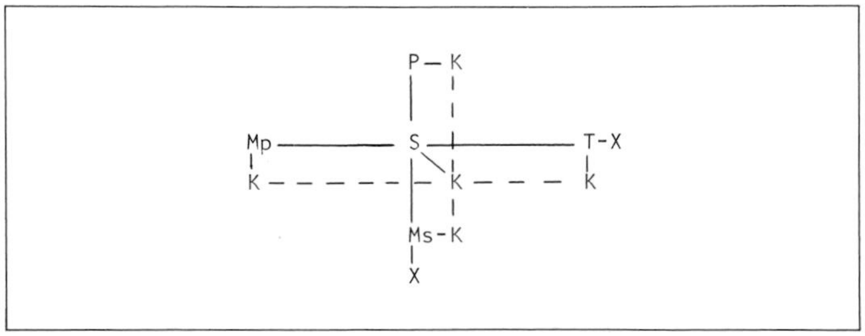

Fig. 7. General computer model (with distributed control) PMS diagram.

designs have been proposed for such a system, technology and economics have so far prohibited their actual realization. Instead, there has developed the general latticelike structure shown in Fig. 10. Each switch in this structure connects components on one side with components on the opposite side (the S interconnecting the P's being the exception).

The lattice structure of Fig. 10 is hierarchical in the sense that the Mp's form the inner core and one travels out toward the periphery in moving from left to right. With this movement there is a general decrease in data rate, being highest through the Mp-P switch and lower as one moves to the right.

The model has five switches (S). One switch connects the computer's peripheral devices with the external environment (human beings, other processes, etc.). Three switches appear alike in the way they interconnect Mp-P, P-K, and K-(T|Ms), respectively. However, they are usually quite different. We would expect any P to connect with any Mp. We probably would expect to have only one or two Pio's connected to a given set of K's. Most certainly one or two K's would manage a given set of Ms's or T's. Thus the structure nearest the periphery becomes more like a tree, rather than a lattice (examples are provided in Figs. 11 and 12). The last switch in Fig. 10, unlike the above four, provides intercommunication among the processors. In any multiprocessor structure (even 1Pc-nPio) there must be communication among the processors. A switch of this type is organized as a nonhierarchy and appears like a conventional telephone exchange, since any P can call another. On the other hand, the amount of communication (measured in bits) is rather low.

The P's and (usually) Mp's have their controls associated with them, and we have not bothered to show such K's in the diagram. The K's that are shown provide control for the T's and Ms's. These are separated in the figure because they are separated in current computer systems and made into identifiable physical components. Under current technology they are expensive devices, so that one K per T or Ms is not economical. Therefore, each K needs to be

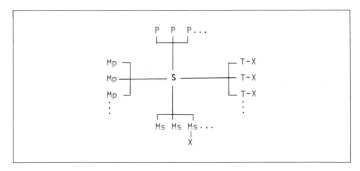

Fig. 9. General computer model (with multiple components) PMS diagram.

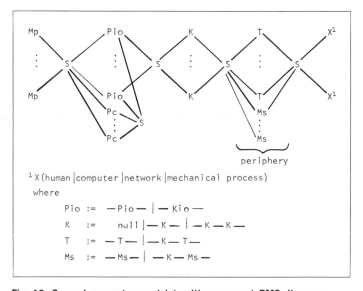

Fig. 10. General computer model (multiprocessors) PMS diagram.

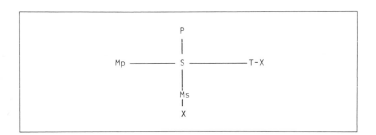

Fig. 8. General computer model (without K) PMS diagram.

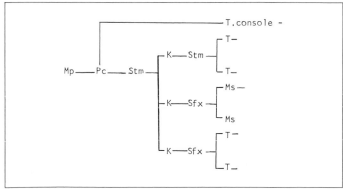

Fig. 11. Tree-structured computer (1Pc) PMS diagram.

shared among a set of T's and Ms's. (That is, one purchases a single magnetic-tape controller for, say, four magnetic tapes.) The shared K also explains why only one of a given class of devices (e.g., magnetic tapes) can operate at a time. As technology changes (especially costs), these separate K's may disappear.

Nearly all the computers discussed in this book fit the lattice model of Fig. 10. However, it is not unlikely that structures will be or have been built that do not conveniently fit it. For example, NOVA (Chap. 26) does not fit the model nicely, although the more complex ILLIAC IV arithmetic-computer portion (Chap. 27) does.

The values along the PMS structure dimension of the computer space have been generated from the general model and laid out in the order of their evolution. This evolution is strictly from less complex to more. The seemingly more complex network structures, such as the duplexed computers, are not necessarily as complex as a single multiprocessor computer. Duplex computers have been used for some time. The slow evolution to the parallel processor structure is due primarily to limitations in technology. A structured computer with a distributed control is more expensive than a tightly integrated design with shared function. In addition, multiprogramming—a question of software—must be present to allow multiprocessing.

The PMS structure plays only a minor role in obtaining multiprocessing and parallel processing. The classical debate about building large computers has always been resolved by building a single large processor (e.g., the CDC 6600 and Stretch, Chaps. 39 and 34). Proponents of multiprocessors say that one can always add several large processors to a structure and increase the per-

formance of a one-processor structure. In Part 6, Sec. 3, when we discuss the IBM System/360, we advocate multiprocessing.

Today there is no parallel processing in the form suggested in Chap. 37. We include a discussion of parallel processing on the bet that it will come in the future. Part 5 is dedicated to moving along the PMS structure dimension.

The simple 1 Pc structure shown in Fig. 11 is a tree. Although there are no values on the information rates, the nature of the fixed[1] and time-multiplexed switches indicates that perhaps the top two T's, one Ms, and one of the bottom T's can *all* be active at a given time. In Fig. 12 a 1 Pc, 2 Pio computer is given. Here we note that the control of one secondary memory is by a Kio rather than the Pio. (The Kio cannot fetch its next instruction from Mp and must rely on Pc for control.) Note that there is necessarily a lattice connection between the 2 Mp and the Pc, 2 Pio, and Kio. The special cases of P.displays multiprocessors, P(array | wired algorithm), and parallel processing are all realized from the general model of Fig. 10.

Switching

A principal issue of a computer design at the PMS level is switching (as we indicated in the preface). Unfortunately, we do not illuminate switching problems in this book except to provide examples. The switching dimension of the computer space is correlated with PMS structure, as we have just seen. To have a more complex structure, more complex intercommunication (switching) is required. Figure 13 shows the various logical switches, together with some of the more common implementations. The switch parameters are also given in the Appendix of this book. Each of the switching issues will be discussed in turn as they apply to various parts of the structural model (Fig. 10). The reader should note that Fig. 13 has relatively primitive switches. More complex switches can be formed by cascading (connecting) the primitives together. (A noncomputer example is the manner in which telephone exchanges are constructed and interconnected together.)

Processor-memory switching. Only recently, with the advent of multiple processors, has memory-processor switching become an important problem. But the Mp-P switch makes multiprocessing possible, and it is a determining factor in both performance and reliability.

The structure of the processor-memory switch for computers which have multiple memories and multiple processors is a lattice if simultaneous memory/processor dialogues are allowed. A cross-

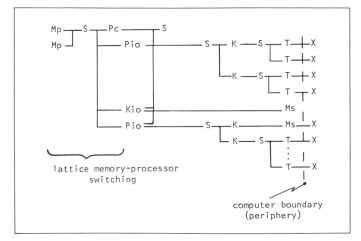

Fig. 12. Tree-structured computer (1Pc-2Pio and lattice Mp-P switch) PMS diagram.

[1] A relative value for the attribute that denotes the time a switch is closed. Fixed usually denotes a time duration such that more than 1 i-unit is transmitted.

Group I. Hierarchical switches for connecting a_m components to b_n components for 2-way conversations. The logical structures are first given, followed by common physical realizations. For the physical realizations links are required between pairs of components. Not all physical realizations are given; it is assumed the roles of the a's and b's can be interchanged.

a_1— S —b_1
.1 S(gate; 1 a; 1 b)

a_1— L — S — b_1
.1a S(gate; *switching at b*)

a_1— S — L —b_1
.1b S(gate; *switching at a*)

a_1— S — L — S —b_1
.1c S(gate; *switching at a,b*)

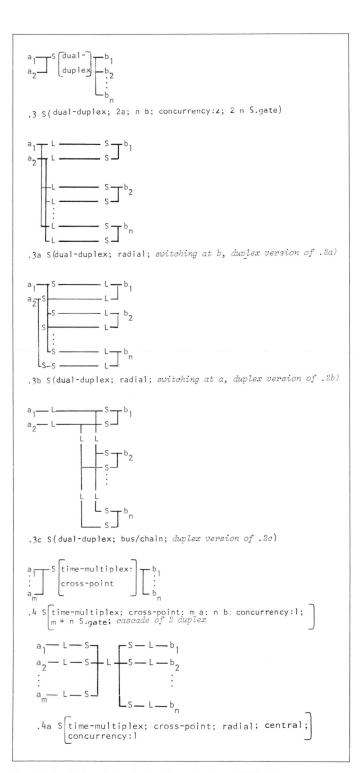

a_1— S(duplex) — b_1, b_2, ..., b_n
.2 (duplex 1 a: n b; concurrency:1; n S.gate)

.2a S(duplex; radial; *switching at b*)

.2b S(duplex; radial; *switching at a*)

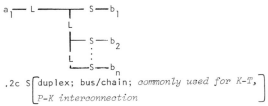

.2c S[duplex; bus/chain; *commonly used for K-T, P-K interconnection*]

.3 S(dual-duplex; 2a; n b; concurrency:2; 2 n S.gate)

.3a S(dual-duplex; radial; *switching at b, duplex version of .2a*)

.3b S(dual-duplex; radial; *switching at a, duplex version of .2b*)

.3c S(dual-duplex; bus/chain; *duplex version of .2c*)

.4 S[time-multiplex; cross-point: m a: n b: concurrency:1; m + n S.gate; *cascade of 2 duplex*]

.4a S[time-multiplex; cross-point; radial; central; concurrency:1]

Fig. 13. Logical and physical switch structures PMS diagrams.

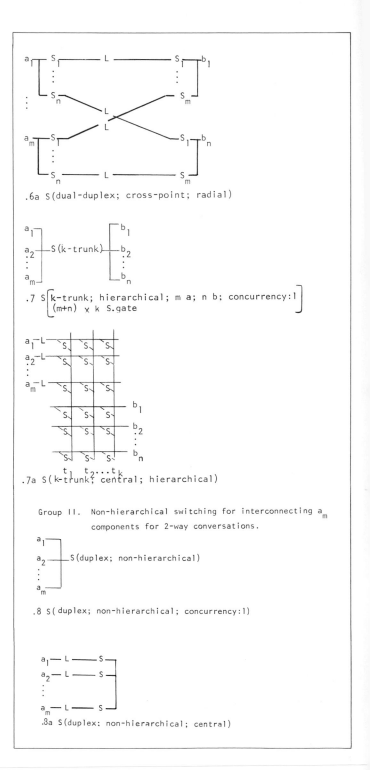

.4b S(time-multiplex; cross-point: bus/chain)

.5 $\begin{bmatrix} \text{cross-point; m a; n b: concurrency:min(m,n)} \\ \text{m} \times \text{n S.gate} \end{bmatrix}$

.5a S(cross-point: radial; *Links to a or b may be null*)

.5b S(cross-point: bus/chain: *used for Mp–P interconnection*)

.6 S $\begin{bmatrix} \text{dual-duplex cross-point; m a; n b; concurrency:} \\ \text{min(m,n); 2} \times \text{m} \times \text{n S.gate} \end{bmatrix}$

.6a S(dual-duplex; cross-point; radial)

.7 S $\begin{bmatrix} \text{k-trunk; hierarchical; m a; n b; concurrency:1} \\ \text{(m+n)} \times \text{k S.gate} \end{bmatrix}$

.7a S(k-trunk; central; hierarchical)

Group II.   Non-hierarchical switching for interconnecting a_m components for 2-way conversations.

.8 S(duplex; non-hierarchical; concurrency:1)

.8a S(duplex: non-hierarchical; central)

Fig. 13. (Continued)

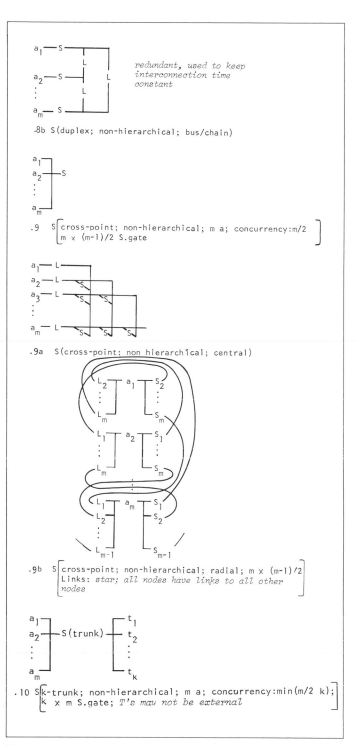

.8b S(duplex; non-hierarchical; bus/chain)

redundant, used to keep interconnection time constant

.9 S$\begin{bmatrix}\text{cross-point; non-hierarchical; m a; concurrency:m/2}\\\text{m x (m-1)/2 S.gate}\end{bmatrix}$

.9a S(cross-point; non hierarchical; central)

.9b S$\begin{bmatrix}\text{cross-point; non-hierarchical; radial; m x (m-1)/2}\\\textit{Links: star; all nodes have links to all other}\\\textit{nodes}\end{bmatrix}$

.10 S$\begin{bmatrix}\text{k-trunk; non-hierarchical; m a; concurrency:min(m/2 k);}\\\text{k x m S.gate; } \textit{T's may not be external}\end{bmatrix}$

Fig. 13. (Continued)

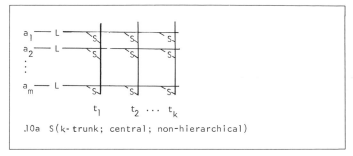

.10a S(k-trunk; central; non-hierarchical)

Fig. 13. (Continued)

point switch provides redundancy and is used to form the lattice structure. To vary from the full-duplex/duplex switch (for m-memories and one processor, or p-processors and one memory) requires more components to be devoted to the switching, to buffering, and to arbitration control. Hence duplex switches are used on most multiprocessor computers. The processor-memory switching possibilities can be seen nicely in Fig. 13. The important switch parameters are the number of memories, the number of processors, and the number of simultaneous processor-memory dialogues. In current designs P always originates the dialogue, which is generally taken to mean the reading or writing of a given word in Mp. The range of complexity is roughly

S(null; 1M; 1P; concurrency: 1)|

S(simplex[1] | half-duplex[2] | full-duplex[3]; $\qquad$ (mM; 1P)|(1M; pP);
 concurrency:1)|

S(time-multiplex cross-point; mM; pP; concurrency:1)|

S(cross-point; mM; pP; concurrency: min(m,p))

An S.duplex can be used to increase the number of processors which can be connected to the memory system while not having to provide additional switch points on each memory. For example, in the CDC 3600 [Casale, 1962] a basic S(8M; 4P; concurrency: 4) is expanded by placing another S(1M; 6P; concurrency: 1) in series to give a possible overall S(8M; 24P; concurrency: 4). This scheme was used to provide multiple processor accesses to the memories.

Processor-control switching. The first switching problem developed with the need to communicate with several input/output devices. This switching is hierarchical in nature; one (or two) processors

[1] A switch which allows communication in one direction between two ports.

[2] A switch which allows communication in either direction but only one direction at a time.

[3] A switch which allows concurrent communication between two ports.

maintain control of many K's by giving a K a single instruction task. At the completion of the task the K signals the processor that the task has been completed.

The switch provides a link between processor and controls for the secondary memory or the terminals and is parameterized by the number of processors, the number of controls, the number of simultaneous conversations, and who originates the dialogue. In these switches the control of information transmission is always by the processor. The evolution has been approximately as follows:

1   S(null; 1P; 1K; concurrency: 1; initiator: P)
    P and K are connected during data transfers.

2   S(simplex|half-duplex|full-duplex/duplex; 1P; 1K; concurrency: 1; initiator: P, K)
    Each K operates independently because it can return or request communication with P when control task is completed.

3   S(dual-duplex; 2P; 1K; concurrency: 2; initiator: P, K)
    Duplex paths from dual P's to each K for reliability.

4   S(cross-point; pP; kK; concurrency: min (p,k) initiator: P,K)
    General case of multiple P's and K's with communication among the components.

The early machines used the first structure, and concurrent operation of controls was possible only by starting several controls and by very carefully programming the timing for the data transfers. Two conditions occurred to cause this: The buffering for a T or an Ms was associated with the processor, and the control could not signal the processor. Although rather trivial to implement, the idea (item 2 above) of allowing a K to signal the processor did not occur until after the idea of arithmetic processor traps were incorporated into processors. The interrupt was used as the method by which a K communicated its desire to converse with a P. The early IBM 709 provided a separate, independent processor for handling the communication with input/output equipment. Simultaneous processor-to-input/output or secondary-memory dialogues could take place (provided the devices were connected to the right processor). In most of the early computers, part of the control function (data buffering) was associated with the Pc, and, as such, only one device could operate at a time. This stemmed from the comparatively high cost of registers, so that links were established for a fixed period of time during a complete block transfer of data.

In some of the military computers a duplicate set of K's is provided for reliability. The more elaborate switching structures (types 3 or 4 above) are rarely used between Pio's and K's; thus to work on a peripheral requires the use of the rest of the computer. The S. dual-duplex is becoming more common; it provides a method of off-line operation for maintaining better component utilization and a more reliable structure.

Control-terminal and control-secondary-memory switching. The switches which link a control with a particular terminal or secondary memory are generally fairly straightforward. Normally, a fixed duplex switch is used. However, a dual-duplex switch is used if multiple access paths to the component are required. The switch links a secondary memory to a control during the transmission of relatively long information units (e.g., records). A typical example of such a switch is the bus structure used when magnetic tape units connect to a common control. Only one of the units operates at a time (although all can be rewinding simultaneously). The switches are far less interesting than those above. Because they are nearer the periphery, failure in them does not imply a failure in the complete system.

Processor function

The emergence of complex PMS structures is coincident with the development of functionally specialized processors. In the simple computers of Figs. 5 to 9 there is place only for Pc. In the general lattice there can be a Pc specialized to perform no input/output operations; one or more Pio's specialized to communicate with the T's and Ms's and even to organize information in Mp for transshipment; additional Pio's specialized to handle graphic displays (hence P.display); and even P's specialized to work on specific data-types (for example, P.array) or specific algorithms (e.g., the fast Fourier transform). In addition, any of these processors may be realized by microprogramming, which is to say, by having its ISP interpreted by a specialized P.microprogram.

Although the existence of various functionally specialized processors is coupled most closely with the PMS structure dimension, the processors themselves are defined primarily by the data-types they can process. In this they agree entirely with the computer-system-function dimension. Possibly the processor-function dimension should be considered simply an extension of the computer-system-function dimension. On the other hand, the inclusion of microprogrammed processors really extends the PMS structure dimension to where a P can be seen as a cascade of two P's.

The processor-function dimension in the computer space is laid out in an evolutionary way, so that its correspondence with PMS structure is clear. P.microprogram is put at the beginning of the dimension ahead of Pc, not because it occurs earlier in evolutionary development, but because it extends the PMS dimension down into the processor. Any of the P's along the dimension can be attained by a P.microprogram.

As an actual dimension characterizing a total computer it must be viewed cumulatively (similarly to the data-type dimension). Thus, if a computer has a Pio, it also has a Pc, and if it has a P.array it also has the prior ones. There are numerous exceptions to this, such as small Pc's with P.displays (hence with no Pio's). This evolutionary ordering does not correspond to complexity or number of data-types in the P. Pc and P.array are the most complex; Pio and P.vector_move are least.

We will make a few brief comments on each functional type, taking them in the order of the dimension.

Microprogram processor (P.microprogram). The term microprogramming was introduced initially in "The Best Way to Design an Automatic Calculating Machine" (Wilkes, 1951a). We use "microprogrammed" to mean that an ISP is defined by an interpreter program residing in an internal Mp, processed by an internal processor (the P.microprogram). Thus the structure is really an external processor (ISP) being defined by the computer formed as

P := Mp(internal; read-only)—P.microprogram

The operations that microprogram processors perform are primitive in comparison with other processors. The task of the microprocessor is to interpret the instructions of the ISP it is realizing. This involves mostly data transfers among the registers of the processor state (Mps) plus simple boolean tests. Although it must handle all the data-types of the larger ISP, it does so only as bit fields to be extracted and transferred from one register to another. The complex data operations (e.g., multiplication) are carried out by other units (D's). In fact, if a complex instruction set were to be used for the P.microprogram, the external processor might as well be implemented directly in hardware. In very minimal P's, for example, C(PDP-8) in Chap. 5, the ISP is essentially already at the level of a microprogram ISP, as shown by the inclusion of instruction that can be microcoded.

The long lag between the idea of microprogramming and its more widespread adoption is due to several reasons. Early ISP's were comparatively straightforward, so that a microprogram approach was not economically justified. The interpretation overhead time is higher than with the hardwired approach, and unless complex functions are realized this time becomes objectionable. In addition, suitable read-only memories were not developed until the mid 1960s (though it is unclear whether this is cause or effect). An additional feature of using a P.microprogram is the ability to realize several ISP's within a single physical processor. IBM has exploited this feature extensively in the System/360 (Part 6, Sec. 3), which is by far the most ambitious use of microprogramming. One can argue that without the additional payoff, which was used to ease the transition to a new incompatible computer system by providing emulation of the old system, the microprogramming would be marginal.

Several P.microprogram design approaches have emerged: Kampe (Chap. 29) presents a design based on a short word; the internal processor is very much like a conventional processor. At the other extreme, the IBM System/360 (Chap. 32) is based on a long word which allows multiple operations to be coded in parallel. (The parallel operations are necessary to gain an acceptable performance level.) Thompson Ramo Wooldridge called their AN/UYK a "stored logic" computer, and it provided the ability to use primary memory for defining the ISP. The IBM System/360 Model 25 (page 567) also uses this approach. The Hewlett-Packard desk calculator (Chap. 20) shows the use of microprogramming on a relatively circumscribed, but complex, task.

Central processors (Pc). These processors interpret an instruction set for manipulating arithmetic, logical, and symbolic data-types. In all simple systems it is the only processor and thus does all tasks. The growth of processor specialization can be described in terms of relieving the Pc of simpler functions that require substantial processing time but do not make full use of the devices within the Pc, such as the arithmetic units. Crucial to this issue is the time it takes the Pc to switch from one task to another (recall the discussion on Mps, the processor state), since many of the jobs that are extracted to specialized processors are demand jobs, such as input/output.

With the removal of tasks from the Pc, it becomes more specialized. A very pure example of this is the Pc of the CDC 6600 (Chap. 39), which has no input/output instructions of any kind in the Pc. That is, not only has the control and management of communication and transmission with the T's and Ms's been removed from the Pc, but the act of initiation has been removed as well and placed in the Pio's. Thus, the 6600 Pc is just an engine for working on the arithmetic, logical, and symbolic (address) data-types.

The mixture of operations to be performed in most complex algorithms prevents specialization of the Pc from going very far, e.g., from there being a P.arithmetic, for with every switch between capabilities distributed in distinct P's there must be intercommunication of the components, which introduces an overhead cost in processing time.

Input/output processors (Pio). The Pio specializes in the management of peripherals (secondary memories and terminals). They are also called peripheral processors, data channels, and channels.[1] The tasks a Pio and its subordinate peripherals perform are the transmission of information between Ms and Mp; the transmission of information between some extra computer real-time system (e.g., human); and the transmission of information outside the C, via a T to some other information media (e.g., a card reader, card punch, line printer, etc.). All the above tasks are similar and often are considered the same, though in principle they can be quite different. A task in this environment is the management of some quanta of information, whether it be one bit or character, a voice message, or a record or file from magnetic disk or magnetic tape. Thus a Pio does not usually change any information; it is merely an interpreter for moving information. There are three exceptions: Computation is required for error and correction and/or detection; computation is required if recoding and reformatting are done; and computation is required when search operations are carried out on Ms without Pc intervention.

To accomplish the above tasks requires a fairly simple instruction set. Typically it contains jump (branch); data transmission within Mp to initialize process variables; simple counting ability, e.g., to control error retries; subroutine calling; interrupt process handling; initializing KMs or KT; testing the state of KMs or KT; and sometimes code conversion (data in one code format is converted to another code). Thus substantial arithmetic and logic facility is not needed. Part 4, Sec. 1 provides a detailed discussion of Pio's.

Display processors (P.display). The P.display is a complex Pio that processes information for display terminals. The data-type is a representation of a complex graphic object, e.g., lines, points, curves, and spatially localized text. The representations vary considerably from system to system, using various list pointers and vector encodings. The operations on the data-types include the maintenance of the display (due to the short-term persistence of the CRT); the selective modification of the representation under commands from the T.display or the Pc, such as adding or deleting a line, inserting text, etc.; the control of T.inputs such as keyboards, light pens, joysticks; and the performance of more complex spatial transformations, such as translation, rotation, scale change, and determination of hidden lines.

[1]These terms are usually used without distinguishing between a Pio and a Kio, that is, whether the device interprets a sequential program (and thus is capable of sustained independent activity) or only decodes a single instruction.

The P.display is a good example of a highly complex but specialized data-type for which there are substantial local operations to perform, that is, where no interaction is needed with a complex algorithm (that requires the Pc). Users of displays wish to correct, modify, and transform the display in geometrically simple ways (in effect, edit and view) between processing of the graphic information by complex algorithms. Thus the graphic display is a prime candidate for the development of a specialized processor.

The DEC 338 (Chap. 25) is typical of these processors, being neither the simplest nor the most complex (e.g., it does not have rotation or hidden line elimination instructions).

Array processors (P.array). The array processor might be considered a more general Pc. It has been proposed or discussed in the literature for some time. (See bibliography for Chap. 27, page 329.) The information unit processed is an array of one (vector) or two (matrix) dimensions. Instructions are provided to operate on these data. The specification of algorithms for a P.array is based on the assumption that an operation can be carried out in parallel for array elements. Actually, both serial (sequential) and parallel (concurrent) execution can be implemented. Both structures have the same logical characteristics, from an ISP viewpoint, and may differ only in execution rate. The three array processors, ILLIAC IV (Chap. 27), NOVA (Chap. 26), and the IBM 2938 (page 577), are discussed in Part 4, Sec. 2 (page 315).

Vector-move processors. The vector-move processor is a special-case P.array. It is capable only of moving a word vector at some location in Mp to some other location within Mp. Because of its limited instruction set, such a P is found only in computers which require constant Mp shuffling. This condition arises either because of a hierarchy of Mp speeds or because the programs must have a particular structure before they can be interpreted by the processor. A time-shared computer might require such a processor for multiprogram memory management. It is therefore common to find block (vector) transmission instructions in a Pc. The IBM System/360 has Pio(Storage channel) for this function (page 577).

Special algorithm processors (P.algorithm). Only a small number of special algorithm processors have been specified and/or implemented. High performance is almost guaranteed by hardwiring and through specialization. The time to fetch the algorithm (instruction fetch time) and many of the references to Mp for temporary data are eliminated by hardwiring. A hardwired algorithm can easily outperform a stored program by a factor of $10 \sim 100$. The lack of these processors in systems stems mainly from lack of market demand.

It is not clear that the special algorithm processors meet our criteria for being a processor, because of the rather limited functions they perform. In fact, some so-called processors are just K's, or D's since they have no instruction location counter and interpret only a single instruction at a time, requesting each new instruction from a superior component.

Algorithms which have been hardwired (or proposed) include the fast Fourier transform using the Cooley-Tukey algorithm; cross-correlation, autocorrelation, and convolution processing; polynomial and power-series evaluation; floating-point array processing; and neural network simulation.[1]

Language processors (P.language). Language P's interpret a language that has been designed to some external criteria, such as a procedure-oriented language (ALGOL or FORTRAN) or a list language (IPL-VI). Thus complexity takes the form of a complex data-type for the "instruction," rather than a complex data-type for processing (e.g., floating complex numbers). If such processors were extended to do all the things a Pc also does, then they would become more complex than a Pc. However, to date, most of them are experimental and focus exclusively on language interpretation.

In Part 4, Sec. 4, several examples are presented. It is worthy of note that of the three P.languages only EULER (chap. 32) has been implemented in hardware using a P.microprogram.

Memory access

The most useful classification of memories is according to their accessing algorithm.[2] These are queue (i.e., access according to first-in–first-out discipline); stack (i.e., access according to first-in–last-out discipline); linear (e.g., a tape with forward read and rewind); bilinear (e.g., a tape with forward and backward read); cyclic (e.g., a drum); random (e.g., core); and content and associative. All these memories are explicitly addressed except the stack and queue, which deliver an implicitly specified i-unit on each read.

Memory size and basic operation times (i.e., the time constants in the access algorithm) are important too, of course. But once a distinction is made between Mp and Ms, then for any given technological era there have existed characteristic sizes and speeds

for memories of a specified access algorithm. Where there has been variation, either it has been linear with size (e.g., buying two boxes of magnetic core Mp versus buying one) or there has been a narrow range of cost/performance tradeoff (as in data rate for magnetic tapes, in which modest increases in density and tape speed can be bought for substantially increased dollars). Table 5 shows the relative price, size, and performance of various memories. The memory-size versus information-rate plot (Fig. 14) shows the clustering of memories and their suitability for a particular function.

From a technology standpoint, Mp's have been constrained to either cyclic- or random-access memories (although one can easily construct any type from random-access memories). In Part 2, Sec. 1 we have not separated the machines according to whether they used cyclic- or random-access memories. The early first-generation computers used cyclic-access memories. Part 3, Sec. 2 presents only the cyclic-access memories.

Similarly, Ms's have been constrained to be cyclic or linear, although quasi-random access has been achieved with some disks and magnetic-card memories (random by block and linear or cyclic within a block). Any Ms's can be part of almost any computer structure. Thus there is no large effect of Ms structure on the main design features of computer systems, and they are not discussed to any extent in the remainder of the book. Our discussion of memory type below deals exclusively with Mp and Mps.

Stack and queue memories (M.stack, M.queue). Data elements in a stack and queue are not accessed explicitly, as we noted above. The stack has some rather unique properties that aid in the compilation and evaluation of nested arithmetic expressions. Although there are no machines employing stacks exclusively for primary memory, there are stacks in some arithmetic processors. Part 3, Sec. 5 is devoted to processors with stack memories (i.e., with stacks in the processor state).

The IPL-VI machine (Chap. 30) is the only computer in the book to have its entire memory organized as a list of stacks. Although no hardware exists that *inherently* behaves as a stack or queue,[3] it can be simulated by a random-access memory. A shift register capable of shifting in either of two directions is a stack.

Cyclic-access memories (Mp.cyclic). Nearly all the first-generation (vacuum tube) computers had Mp.cyclic. The Mp.cyclic acoustic, magnetostrictive delay line, and magnetic drum provided an in-

---

[1] Chasm: A Macromodular Computer for Analog Neuron Models [Molnar, 1967].

[2] Access for writing should be distinguished from access for reading. Memories are conceivable with arbitrarily different read and write access algorithms (e.g., random read and cyclic write). However, in general, the two access algorithms are tightly coupled, and normally only the read access algorithm is given.

[3] Small (10 ~ 1,000 word) queue- and stack-accessed memories are especially easy to build with large-scale integrated-circuit technology.

Table 5 Memory characteristics

Memory module	Function	Access method	Memory size		Memory performance		Cost/bit($)[1]
			Module size (bits)	Modules/ computer	Access time sec	Data rate (bits/sec)	
Punched paper card	permanent, archival	random + linear	(500 ~ 1,000)/ card; ~ 1,000 card/unit	1 ~ 2	$10^0 ~ 10^3$	10^4	$2 × 10^{-6} + 2 × 10^{-1}$
Magnetic card	secondary, archival	linear + constant + cyclic	$3 × 10^9$	1 ~ 4	$10^{-1} ~ 10^0$	$0.4 × 10^6$	$1.5 × 10^{-8} + 5 × 10^{-5}$
Magnetic tape	secondary, archival	linear	$2 × 10^8$	1 ~ 16	$10^0 ~ 10^2$	$0.4 ~ 4 × 10^6$	$2 × 10^{-7} + 10^{-4}$
Moving-head disk pack	secondary, files swapping	linear + cyclic	$2 × 10^8$	1 ~ 16	$10^{-1} ~ 10^0$	$2.5 × 10^6$	$3 × 10^{-6} + 10^{-4}$
Fixed-head disk	secondary, files swapping	cyclic	$5 × 10^7$	1 ~ 40	$~10^{-2}$	$10^6 ~ 10^7$	10^{-3}
Drum	secondary, swapping	cyclic	$(1 ~ 5) × 10^7$	1 ~ 10	$(5 ~ 30) × 10^{-3}$	$10^6 ~ 10^7$	10^{-3}
Bulk core memory	primary and/or secondary, swapping	random	10^7	1 ~ 8	$(2 ~ 10) × 10^{-6}$	$10^6 ~ 10^8$	$0.02 ~ 0.05$
High-speed core or thin-film memory	primary	random	$10^5 ~ 10^6$	1 ~ 16	$(0.2 ~ 2) × 10^{-6}$	$10^7 ~ 10^8$	$0.05 ~ 0.25$
Integrated circuit (scratch-pad memory)	primary, processor state	random	$10^3 ~ 10^5$	1	$~10^{-7}$	10^9	$0.25 ~ 1.0$
Integrated circuit (content addressable)	primary, cache	content, random	$2 × 10^5$	1 ~ 2	$~10^{-7}$	10^9	$1 ~ 3$
Read only (capacitor, inductor)	processor instruction-set definition	random	$(1 ~ 5) × 10^5$	1	$10^{-6} ~ 10^{-7}$	$10^8 ~ 10^9$	$10^{-3} ~ 10^{-2}$

[1] The first component is the memory media (e.g., a disk pack), and the second component is the transducer (e.g., a disk drive).

expensive, simple, producible memory. By the second generation the cost of Mp.random (though still more expensive than an Mp.cyclic) was about equal to the processor logic. The incremental cost for an Mp.random in a large system was then small, whereas the performance gain could be a factor of up to 3,000 (access time of 10 microseconds versus 30 ~ 30,000 microseconds). Some of the first-generation machines were reimplemented using transistors (the LGP-30 became the LGP-21). Only a few new cyclic access machines were introduced in the second generation. Most notable was the low-cost Packard-Bell PB-250 using transistor logic and magnetostrictive delay lines (a derivative of the Bendix G-15 and NPL ACE).

Nearly all these computers use some form of n + 1 addressing.

The memory is organized on a digit-by-digit serial basis for a word (e.g., ZEBRA with binary and IBM 650 with decimal). Hence, the arithmetic or logic function hardware is implemented for only a single digit. An operation is done for the entire word by iterating over all digits in time; thus the cost of a serial computer is nearly independent of its word length.

Because of the cyclic and synchronous nature of these Mp's, it is difficult to synchronize them with secondary memories and terminals (which are also synchronous). The very early machines had no large secondary memories. In some cases, where magnetic tape was used, it was added at very low performance (low density, low speed, and, therefore, low data rates) so that synchronization was not a problem. In other cases a small random-access core

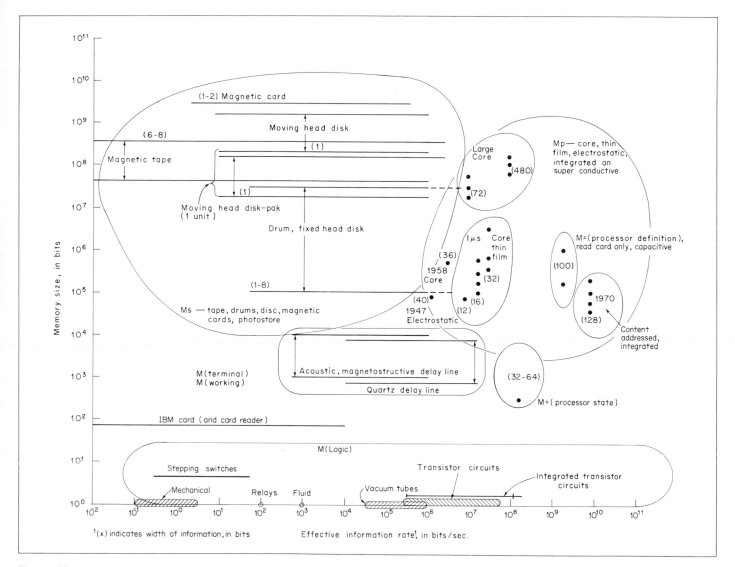

Fig. 14. Memory size versus effective information rate.

memory was added to provide synchronization between the two memories (for example, IBM 650).

Random-access memories (Mp.random). Random-access memories were used late in the first generation, and they have remained the predominant memory during the second and third generations. It is unlikely that their popularity will decline unless content-addressable memories can be constructed sufficiently cheaply (if then). The earliest first-generation random-access memories were

electrostatic and depended on maintaining a charge on plates of an array of capacitors. The most common was the Williams tube (invented by F. H. Williams at the University of Manchester) which works in essence like a CRT, with the beam used to charge a capacitor array at the tube face [Williams and Kilburn, 1949]. Other schemes included an array of capacitors which were selected by digital logic (Pilot, Chap. 35).

Late in the first generation Forrester [1951] invented the core memory, which rapidly became the predominant primary-memory component. It is unlikely that it will be replaced in the near future; the most likely candidate is large-scale integrated-circuit arrays of flip-flops.

The random-access memory seems nearly perfect for the Mp's of present computers. Of course, enthusiasm for this memory may be based on not knowing how computers would have developed if we had not had them. However, with little or no effort an M.random can be a stack, a queue, a linear, a cyclic, and even (within limits) a content or associative memory. It is an organization which is very hard to beat.

Content-addressable and associative memories. It is possible to conceive of many exotic accessing capabilities, and numerous proposals have been made involving either theoretical structures or experimental prototypes. Since no particular varieties have become widespread, terminology is still variable. Content-addressable memories are usually taken to mean a collection of cells of predetermined size (i.e., a fixed i-unit) such that if one presents as "address" the contents of a predetermined part of the cell (the tag or content address) then the contents of the entire cell will be retrieved. An associative memory is usually taken to mean a system such that, when presented with an item of information, it delivers one or more "associated" items of information. The principle of association is variable, yielding different kinds of associative memories. Content-addressable memories provide a form of association, as do all memories, in fact. Thus the term "associative memory" tends to denote forms of association different from familiar ones—forms that presumably have less sharp constraints imposed by the structure of memory (as opposed to the structure of the information in the memory).

No examples exist of a computer with a content-addressable memory as its primary-memory structure. However, both the IBM 360 Models 67 (page 571) and Model 85 (page 574) use 8 and ~1,000-word content-addressable memories, respectively, to increase performance (in both cases they are transparent to the program). The CDC 6600 instruction buffer is in effect a small content-addressable memory. In the above three cases, the content-addressable memories vary in size and position in the structure; however, the pattern of use is common. There is a large but slower Mp.random behind the content-addressable memory. The purpose of the fast small content-addressable memory is to hold local, current data so that an access will not have to be made to the random-access memory.

Small prototype associative addressable M's have been constructed, but they are normally based on random-access memories under the control of special hardware. There are immediate uses for content-addressable memories with a large information-content address. For example, the read-only memories for microprogram processors use long words principally because content-addressable memories are not available. Ideally a microprogrammed processor would like to look at a fairly large processor state to determine what action is to be taken in the microprogram. It is interesting to speculate about the evolution of computers if a content-addressable memory had been developed in place of the random-access memory.

Mp concurrency

Multiprogramming is the simultaneous existence of multiple, independent programs within Mp being processed sequentially or in parallel by one or more processors. Multiprogramming provides each user program with a memory space independent of other users. It may provide, in addition, the sharing by several users (for independent use, not for communication) of a block of Mp, which thus does not have to be duplicated. For example, operating systems software, including compilers, assemblers, loaders, and editors, can be usefully shared.

The ability to have multiple programs gives rise to a corresponding problem of communication between programs. We have defined this as a correlated dimension in the computer space (interprogram communication) and will discuss it in the next section. The issues it raises are just the opposite from those raised by the requirement for multiple programs, which are discussed in this section. Here we are concerned with protecting one program from another—with assuring that no unjustified communication will occur—and with obtaining appropriate space in Mp so that multiple programs can run.

The requirement for protection is obvious. If two independent programs are to be resident in Mp at the same time, they must not have access to each other's space. Not only would such access (especially for writing) have disastrous consequences when the programs are running, but they would be entirely unpredictable and undebuggable from the viewpoint of the programmer of each individual program. Thus this requirement is absolute; i.e., it must be highly reliable. This implies a hardware solution, although purely software schemes are possible in special cases.

The requirement for appropriate space is somewhat more subtle. Certainly there must be enough space in Mp for all the programs that are to be resident simultaneously. It must be possible to find that space, assign it to a new program, and make it available again when that program is finished. But what kind of space will do? Must it be a single interval of Mp, large enough for the total program with data? And if the program is assembled or compiled in Mp and is removed temporarily to make room for another program, must it be brought back into the exact same addresses into which it was originally assembled?

The key issue resides in the kind of intercommunications that hold within a program and its data, for these determine how and in what way a program is interconnected and depends on the specific Mp addresses that it occupies. These connections are of two kinds: explicit addresses present in the program and data and implicit relations between addresses due to addressing algorithms (e.g., that programs are laid sequentially in Mp, or that the elements of an array are to be accessed by indexing and hence must occupy consecutive addresses). Again, although some purely software solutions to the space issue exist, hardware is involved in a fundamental way.

Thus, the two main questions of program concurrency[1]—protection and space assignment—imply basic design features of a computer system. It might seem that they imply separate features and should be separate dimensions in the computer space. In fact, each proposal for how to solve the space-assignment problem also contains a particular proposal for the protection problem. Thus we treat them as a single dimension.

Virtual-address space and Mp mapping. Before considering various solutions to Mp concurrency (i.e., the values along the dimension), let us introduce two concepts in terms of which all current solutions can be understood. Consider a particular program, PROGRAM-1, one of many that might wish to reside in the Mp. PROGRAM-1 assumes a set of addresses, some explicitly and some implicitly, in the addressing algorithm it uses. PROGRAM-1 requires a memory space that has addresses that satisfy all these requirements, the implicit and explicit ones. Other than that it does not care how these addresses are realized. Let us call this address space required by PROGRAM-1 its *virtual memory*, Mv. Thus, each program has its own virtual memory. (You might think of this as having its own Mp, except, as we shall see, this Mp may be many times bigger than any actual Mp and still be entirely feasible.)

Actually to run PROGRAM-1 requires that it be placed in the real Mp in such a way that the real addresses of Mp containing it satisfy all the requirements, that is, that it be a faithful image of the virtual memory. Thus there must be some *memory mapping* that maps the actual addresses into the actual memory. Once PROGRAM-1 is placed in Mp there must be some process that takes each virtual address (as it occurs to be processed in an

instruction) and finds the actual address in Mp, so that the correct contents can be obtained.

This might seem simply a complicated and abstract way to view matters, but it becomes essential as soon as we realize that the computer can have hardware memory mappings other than the familiar direct-addressing structure of Mp. Furthermore, if this mapping is given the right properties, it may solve some of the space-assignment and protection problems for Mp concurrency. What we have really done is to divorce the addressing required by the programs from that provided by the physical computer, so that we can redesign it (via the memory mapping) to meet new design requirements that were not apparent when the original random-addressing schemes were created.

Let us make the notion of memory mapping more precise. The program contains virtual addresses, z (that is, symbols in the program that denote addresses are taken to denote addresses in Mv). During the execution of the program, whenever there is a reference to an address z (either explicitly via an address calculation or implicitly via, say, getting the next instruction), a computation occurs on z to obtain the actual address in Mp. This computation is part of the Pc, just as is an automatic indexing or indirect-addressing calculation. It takes as input not just the virtual address z but information on where the program is located in Mp. The latter information is called the *map*, and a program's map information is determined when it is placed into Mp on a given run. Thus, using our ISP notation, and calling the address calculation f, we get

$$Mv[z] := Mp[f(z,map)]$$

That is, the information in virtual memory at virtual address z is the same as the information in actual memory at address f(z,map).

This whole scheme is built to permit programs to be placed in Mp's in various ways, e.g., relocated or scattered around, and still make it possible to run the program. Any such scheme brings a solution to the protection problem, namely, that for some values of z the above calculation cannot take place or is invalid (i.e., there is no mapping for z). This can correspond to a violation of protection, which can then be prevented. All calculations may even be permissible, but f is so arranged that it never produces an address in anyone else's part of Mp.

The memory map is part of each user's program. With many users, it must reside in Mp, since there will not be enough space in Mps to hold a large amount of mapping information. However, when a program is being executed, some part of the mapping information becomes part of the Mps (i.e., at least the Mp address

<hr>

[1] See also Randell and Kuehner [1968].

of the rest of the map). In addition, the map may contain special access control information, such as whether a part may be read, read as data, written, or read as program. The map can also collect statistical information concerning whether a part of the program has been used or has been changed (written).

Random-access memories for Mp constrain the mapping by requiring linear addresses of the form Mp[0:p], since the mapping calculation must be economical (as it is performed with very high frequency). We would not consider a map structure which provides every word in Mv to be mapped into an arbitrary word in Mp, for this would require a map exactly the same size as Mv. With many programs in Mp, there would be little room for anything but maps. Similarly, the amount of processing in f, the calculation, must be very minimal. These two aspects constrain the mapping scheme strongly.

The constraint to linear addresses appears to force the structure of virtual memory to consist of a multidimensional array. This can be one-dimensional, Mv[0:n], or two-dimensional, Mv[0:s][0:m]. It could be of higher dimension, but the need seems not to have been felt (since within any single dimension one can have multidimensional arrays as one normally does in a regular Mp). However, the two-dimensional array, which also is called segmented addresses, since it can be taken as a discrete collection of s + 1 segments each of m + 1 linear addresses, has advantages in terms of the mappings; namely, segments can be placed disjointly in Mp without fear that virtual-address calculations will cross from one segment to another.

With this introduction to the problems of multiprogramming we will look at some of the hardware schemes. Table 6 provides a summarization of them, including a brief description of how each scheme operates.

No special mapping hardware. If no hardware exists in the Pc to accomplish a memory address mapping, then when the address

Table 6   Memory-allocation methods

Hardware designation (arranged in order of increasing hardware complexity)	Method of memory allocation among multiple users	Limits of particular method (example of use)
No relocation Mv ≤ Mp: Conventional computer—no memory-allocation hardware	No special hardware. Completely done by interpretive programming.	Completely interpretive programming required. Very high cost in time is paid for generality. (JOHNNIAC interpreting JOSS).
1 + 1 users. Protection bit for each memory cell	A protection bit is added to each memory cell. The bit specifies whether the cell can be written or accessed.	Only 1 special user + 1 other user is allowed. User programs must be written at special locations or with special conventions, or loaded or assembled into place. The time to change bits if a user job is changed makes the method nearly useless. No memory allocation by hardware. (IBM 1800)
1 + 1 users. Protection bit for each memory page.	A protection bit is added for each page. (See above scheme.)	No memory allocation by hardware. (SDS Sigma 2)
Page-locked memory	Each block of memory has a user number which must coincide with the currently active user number.	Not general. Expensive. Memory relocation must be done by conventions or by relocation software. A fixed, small number of users are permitted by the hardware. No memory allocation by hardware. A program cannot be moved until it is run to completion. (IBM System/360)

Relocation and protection: Mv ≤ Mp:

One protection count and one field register (addresses formed and checked by logical operations)	All programs are written as though their origin were location 0. The count register determines the number of high-order bits to be examined. The field register is then compared for identity with the requested address.	Memory allocation blocks must be in power of 2. Unless blocks are the same size, the memory utilization can be poor. Although faster than the following scheme (which requires a hardware adder), the inflexibility of location and size makes it restrictive. (IBM 7040)
One set of protection and relocation registers (base address and limit registers). Also called boundary registers.	All programs written as though their origin were location 0. The relocation register specifies the actual location of the user, and the protection register specifies the number of words allowed.	As users enter and leave, primary-memory holes form, requiring the moving of users. Pure procedures can be implemented only by moving impure part adjacent to pure part. (CDC 6600, PDP-6)
Two sets of protection and relocation registers. Two segments.	Similar to above. Two discontiguous physical areas of memory can be mapped into a homogeneous virtual memory.	Similar to above. Simple, pure procedures with one data array area can be implemented. (UNIVAC 1108, PDP-10)
$n \geq 3$ sets of protection and relocation registers.	Similar to above. More similar to page mapping.	Has not been used in any conventional computer.

Mapping, Mv ≥ Mp:

Memory page mapping	For each page (2^6 to 2^{12} words) in a user's virtual memory, corresponding information is kept concerning the actual physical location in primary or secondary memory. If the map is in primary memory, it may be desirable to have "associative registers" at the processor-memory interface to remember previous reference to virtual pages, and their actual locations. Alternatively, a hardware map may be placed between the processor and memory to transform processor virtual addresses into physical addresses.	Relatively expensive. Not as general as following method for implementing pure procedures. (Atlas, CDC-3500, SDS-940)
Memory page/segmentation mapping	Additional address space is provided beyond a virtual memory above by providing a segment number. This segment number addresses or selects the page tables. This allows a user an almost unlimited set of addresses. Both segmentation and page map look-up is provided in hardware. May be thought of as two-dimensional addressing.	Expensive. Little experience to judge effectiveness. (GE 645, IBM 360/67)
Indirect references through a descriptor table to segments.	All data are considered part of a descriptor array which is referred to by a number. A descriptor table indexed by the descriptor number is used to locate the array in Mp and give its size.	An indirect reference must be made to the description table in Mp. (B 5500)

z is encountered in the program, the information at Mp[z] will be obtained. There are still, however, two different ways to obtain the effect of a virtual memory.

First, one can operate interpretively, with a software system taking the place of hardware. That is, the programs of all the users are in a nonmachine language (e.g., a higher procedure-oriented language), and each access in the language is processed by the software interpreter before an access is made to Mp. It is clear that all the logical power of a memory mapping is available with this scheme. The only drawback is the loss of efficiency from the interpretation, which may range from a factor of 5 to 100. Consequently this scheme is used only in special circumstances, such as multiuser time-shared conversational algebraic languages.

The second scheme is to modify the code at the time it is placed in the Mp for a given run, so that all addresses in the code correspond to the actual Mp addresses used. That is, an assembly or translation operation is performed each time the program is placed in Mp. The advantage of this scheme is that no further address calculations are necessary. There are three disadvantages. Assembly operations are expensive so that, although the scheme is tolerable if the program is brought in once and run to completion, it is not tolerable if programs are continually being swapped in and out of Mp. In addition, the program must be laid into continuous intervals of Mp corresponding to predetermined segments of the program, for assembly occurs on a static representation of the program and cannot unravel the potential effect of address algorithms. Finally, the size of Mv (i.e., the addresses used externally) must be not greater than Mp.

Relative to these software schemes—one interpretive and very expensive and one involving assembly (i.e., compilation) and loading—the hardware schemes to be described appear as address interpreters, where the cost of continuous interpretation has been made tolerable.

Protection for words or pages hardware. There are three schemes in Table 6 that provide a means of protecting one part of Mp against references from other programs. The rationale for these designs is that there will be only two users (or user classes), one user being superior and assumed perfect (its program debugged). References to Mp via the imperfect program to a perfected and superior part of Mp are forbidden. These schemes provide no method of hardware mapping, and physical addresses are the same as virtual addresses. In the simplest scheme, as in the IBM 1800 (Chap. 33), a protect bit is added to every word in Mp, that is,

$$Mp[0: 2^{16} - 1]\langle 0: (w - 1), \text{protect\_bit}\rangle$$

Every reference Mv[z] takes place as

$$Mv[z] : = (\neg Mp[z]\langle \text{protect\_bit}\rangle \rightarrow Mp[z];$$
$$Mp[z]\langle \text{protect\_bit}\rangle \rightarrow \text{protection violation} \leftarrow 1)$$

That is, any reference to a word with a protect bit causes an error. The other two schemes protect on the basis of blocks of words.

Protection and relocating register(s) hardware. A protection and relocation register mechanism is used in four schemes of Table 6. These provide either one concatenated, one additive, two additive, or n additive register pairs for mapping a single program into one, one, two, or n nonadjacent blocks in Mp. The authors know of no schemes where more than three registers are used; this would really be akin to using a more general page map. Generally, these schemes restrict Mv ≤ Mp.

An additive protection and relocation register pair is shown in Fig. 15 in which four users are occupying a Mp[0:7999]. Each user program is written to occupy a continuous address space in a virtual Mv. Thus in ISP, when Pc is running programs for user-j, which address Mv[z], with z varying from 0 to $v_j - 1$ the mapping uses actual memory. The action is

$$Mv[z] : = ((z < \text{Protection}) \rightarrow Mp[z + \text{Relocation}];$$
$$z \geq \text{Protection} \rightarrow (\text{Protection violation} \leftarrow 1))$$

Protection and Relocation are the two registers that specify mapping. The implementation of this scheme generally takes the form of adding the contents of the relocation register after all address calculations have taken place. Thus, in PMS we might think of the structure

Mp—K(address translation)—Pc.
M('Protection,Relocation)

Page-map hardware. Figure 16 shows the memory allocation using a page map. Note that, of the 4,096 words it is possible to define by the map, the range 1,024 to 2,047 is actually undefined. Along with the map containing the addresses to words in actual Mp, it is desirable to have accessor protection control information. Such information might specify:

1 No restrictions (any form of reading or writing can take place).
2 Read only as data.
3 Read only as a program.
4 Writing.
5 Undefined.

6 Defined but located in Ms.

7 This page has been written in (to know whether a copy in Ms has to be updated).

8 This page has been accessed.

This scheme is essentially a generalization of n protect/relocate registers but includes more control bits, suggested above, and restricts each block to be the same size. Note that Mv can be greater than Mp. In addition, parts of the virtual memory may remain unused.

There are two ways the above scheme is usually implemented:

1 A complete map is first considered as a conventional, explicitly addressed M whose addresses correspond to the virtual-address pages. At a given page-memory address the contents of the map specifies the address in Mp. The map is similar to an indirect reference. However, the map is usually about 10 times faster and about 1/1,000 the size, since it keeps track only of pages, not words. The PMS structure is

Mp—M.map—Pc

2 The map is retained in Mp and referenced by a protection and relocation register which are set for the particular active user. In order to avoid making references to Mp for each word reference to Mv by a Pc, a small, fast M(content address) is placed between Pc and Mp. The PMS structure is

$$Mp \left\{ \begin{array}{c} \overline{\hspace{1em} L(\text{data}) \hspace{1em}} \\ \\ \leftarrow K(\text{address translation}) \leftarrow L(\text{addresses}) \leftarrow \end{array} \right\} Pc$$
$$M(\text{content address}; \ 8 \sim 16 \ \text{words})$$

Memory-segmentation hardware. Figure 9 (page 574) in the introduction to the IBM System/360 shows the logical mapping process for a segmented memory. There is provision for a very large two-dimensional virtual-address space. This scheme is discussed extensively in the literature [Arden et al., 1966; Dennis, 1965; Gibson, 1966]. The physical implementation is similar to that of paging. Note that two levels of mapping are provided: the segment map and the page maps. The two levels facilitate the sharing of a single segment by two jobs.

The Burroughs B 5000 (Chap. 22) and the later B 6500 have a mapping that is more closely integrated into the Pc because they

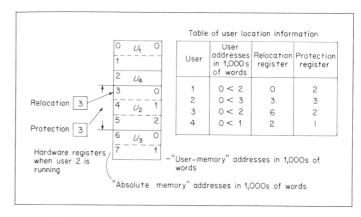

Fig. 15. Memory allocation using a boundary (relocation and protection) register.

provide a variable-sized address space (not paged) within a segment. The segments are named, and a large number of segments exist.

Interprogram communication

The dimension of interprogram communication is completely correlated with the multiprogramming dimension as we have previously noted. To have a problem of intercommunication, there must be a structure of components that require communication. At the simplest level the dimension is represented by a single program, and there is no need for intercommunication. Variables of the

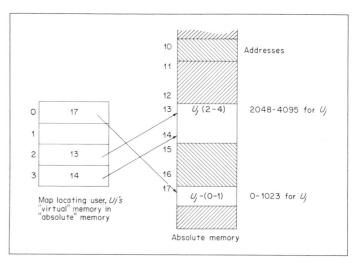

Fig. 16. Memory allocation using a page allocation map.

program are completely accessible to the whole program, and the address space is essentially uniform.

The second value of the dimension, subroutine calling, produces a hierarchy of communication contexts. There is not a fixed number of levels to the hierarchy, since each subroutine may call others *ad nauseum*. When subroutines are present, address names and values within the subroutine become addresses which are local to that part of the subprogram. Such a structuring is apparent when looking at the higher-level languages such as FORTRAN, ALGOL, and PL/I, where there are explicit statements for controlling the names (addresses) that are available to each of the parts of the program. The concept of subroutine structure has been with us almost from the first programs.

The next value of the dimension relates to signaling within a single process. It is akin to subroutines embedded in hardware. These are called extracodes and were perhaps first suggested for the Atlas (Chap. 23). Each extracode can be looked at as just a call to a specific subroutine. The variables of the user (caller's) program are made available to the called (extracode defined) program. The calling usually is accompanied by a context shift, in which a completely different program (one that is used by any number of calling programs) takes command to interpret the instruction. This scheme is used in systems which are controlled by a special software monitor. When a function such as the input or output of a file is required, the main program issues a call to the monitor to make the transfer. (In theory, the monitor knows about conditions in the system and has the capability to perform the complex function.) A central monitor control can then begin to run another program if the request is one which would normally halt the computer. This form of communication is useful to supply extra facilities to users and to have a method of knowing what the users are doing (e.g., so that equipment will be better utilized).

As more complex program structures are directly represented by the hardware, the intercommunication complexity also increases beyond the simple subroutine call. If a segmented-memory scheme is used, the problem of communicating between the segments can be solved in a range of ways. The value of the range would be somewhere between ignoring the problem with the hardware and providing methods for naming of addresses between the communicating segments.

In the above cases, the communication among the various programs or parts of programs is done explicitly by one program to another program. The instruction trap does not fit this view so nicely. Here, conditions occurring within a single process which are not explicitly called cause another part of the program to be called. Typical conditions which cause traps are arithmetic results outside expected range or erroneous program conditions (e.g., trying to call someone else's program). The trap causes a change in context that is synchronized with the process causing it. Trapping is a form of program interruption; a trap is an intraprocess interrupt as distinct from interprocess interrupts.

Intercommunication between two independent processes (being carried out by two independent components) is usually accomplished by using the program interrupt. The interrupting process requests that a program interrupt occur in a component (interruptee). The interrupter's request is acknowledged by the interruptee, and a change of process state occurs in the interruptee; a new process is then run in the interruptee on behalf of the interrupter. The program interrupt is used among processors in a multiprocessor system and between 1Pc and nPio's. A control K may also use the program-interrupt request to communicate with its superior Pio or Pc. For example, a Pio does not usually have the logical capability to execute an algorithm which would decide that action is to be taken for various error conditions.

Usually the interruptee is equipped with certain logic which is capable of arranging priorities of requesting interrupters. The typical kinds of interrupt requests are component faults (e.g., parity error), a timer has counted down, and various task completions (e.g., a program has completed, a tape unit has rewound, a disk arm has stopped moving, a certain record has been found on tape, a buffer is full).

State diagrams would show how each of the communication methods above are similar to one another. A typical interrupt state diagram is shown in Fig. 17. There are four states: normal process interpretation, process state saving, interrupt process interpretation, and process state restoration. The sequence is as follows:

1   Normal instruction interpretation is occurring in the interruptee.

2   The interrupter requests an interrupt.

3   After some delay, t.acknowledgment, a state is reached in which part of the interruptee's process state is saved.

4   After t.acknowledgment + t.save, a program is running in the interruptee in response to the interrupter.

5   The interrupt program is run for t.interrupt.

6   At the completion of the interrupt program, the original process state is restored in the interrupter.

7   After t.restore, normal processing resumes in the interrupter.

The significant attributes of the system are the various times required to move from state to state. These times are directly related to the amount of process state which must be saved (and restored) when switching context.

The intercommunication problem is probably the least understood dimension in the computer space. It is rather intimately related to the ISP, in that the various calling methods (implicitly and explicitly) depend on the ISP. Also, the amount of processor state (a function of the ISP) affects the response time for making context transitions. Most interrupt systems allow several independent classes and/or sources of interrupters. The classes are arranged in priority so that lower-level interrupters are ignored until higher-level interrupt programs are run to completion (see Chap. 42 on the SDS 910-9300 series). The design problems associated with intercommunication are not those of implementation but of knowing what should be implemented. The PMS structure part and the corresponding register-transfer implementations for intercommunication are, by comparison, straightforward.

Processor concurrency

Concurrency (parallelism) in the processor is the number of events or logical operations that are happening at a given time. If the basic logic technology is held constant, decreasing the processing time (increasing the power) requires increasing the number of parallel operations. An exact measure of parallelism can be made in terms of the number of n-bit operations made per clock pulse. The parallelism in a structure is also a measure of its complexity; to have a highly parallel structure implies control structure together with multiple data paths (and operations) which can be concurrently evoked.

Processor parallelism is also necessary to overcome Mp speed technological boundaries. Thus it is difficult to isolate completely the processor from the memory.

Flynn [1966] categorized high-speed processors by whether there are single or multiple instruction streams and whether each stream has single or multiple data streams. The CDC 6600 and IBM Stretch are examples of a single instruction stream and a single data stream. An ILLIAC IV processor has a single instruction stream with multiple data streams. Thus, the single instruction stream and multiple data stream are a form of array processing in which an instruction performs an operation on multiple data elements.

The CDC 6600 main processor has multiple instructions of a single stream in the fetch, buffering, and decoding process at a given time. In addition, instructions are being executed in parallel

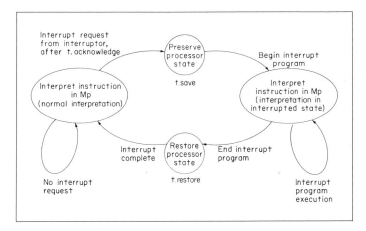

Fig. 17. State diagram for the interrupt process.

by the 10 parallel data-operations. The 6600 has functionally different data operators, although a system could exist in which these operators are the same, or, if the operator were much faster, a single unit could be used sequentially. Depending on the utilization of the 10 data units, there could be a computer with several processors which share a common set of data-operations. The 6600's peripheral processors are implemented in a mode whereby several instructions streams are processed in parallel by a single processor. The simplicity of the shared processor for multiprocessing or parallel processing thereby provides still another form of parallelism. The following subsections discuss particular forms of parallelism. At one end of the dimension there is the most primitive structure, a serial processor, and at the other end there are pipeline processors.

Serial processors. At the most elementary level only one bit of an n-bit word is operated on at a given time. There is no concurrency, and even the most trivial operations on n bits requires a time of n. The bit-serial processor was used in the first generation because the cyclic primary memories to which it connected were fundamentally bit-serial (see page 73). Although the processor memory could be made to operate on a parallel basis where words were available in one unit of time, such a tradeoff was not worthwhile because of the relatively long access time to Mp. The word lengths for serial processors tended to be relatively long, because the cost is independent of word length (see page 216).

Parallel-by-word processors. The simple parallel-by-word processor is the most common processor of the first to third generation. This occurred in part because Mp became parallel by word. Within the processor we assume that almost every internal register-transfer operation requires one or more clock times. (A simple multiply operation usually takes between n/2 and 2n clock times.) We do not mean to rule out multiple simultaneous internal operations within the processor, but they are exceptions. With only a view of a processor's registers, it is easy to tell if multiple operations are possible. Most of these processors do only one operation at a time. As a rule, the simple processor is locked to the primary-memory cycle time (usually core). Approximately $2 \sim 10$ events (clock times) are available within the processor. For example, the PDP-8 (Chap. 5) has four events, and the IBM 7090 (Chap. 41) has 10 events. A precise measure of parallelism would count the number of operations per clock time for given program conditions.

Multiple instruction streams, 1 Pc. The only example of this structure in the book is the CDC 6600. Opportunities for such a structure are possible with the parallel computer suggested by Lehmann (Chapter 37).

Multiple data streams. The most obvious implementation of multiple data streams with one or more instruction streams is the array processor. Part 4, Section 2 is devoted to these structures.

1-Instruction buffer. The 1-instruction buffer is a form of looking ahead in the instruction-interpretation cycle and is about the simplest form of parallelism in a parallel-by-word processor. A single register is assigned the role of holding the next instruction to be interpreted. The IBM 7094 Instruction Backup Register (Chap. 41) is typical of this case. In the 7094 two instructions are fetched at a time. More generally the next instruction would be fetched during the execution of the current instruction.

n-Instruction buffering. Multiple instruction buffering is a generalization of the 1-instruction buffer above. It can take several forms depending on the algorithms used to fetch the next instruction (i.e., the look-ahead) and the organization of the memory holding the instructions. Stretch (Chap. 34) and the CDC 6600 (Chap. 39) use instruction buffers. A small, restricted content-addressable memory holds a block of instructions. In the simplest case of these computers a block of memory, relative to the instruction counter, is kept in the local instruction buffer memory.

Look-aside buffering (slave) memories. Look-aside is a more general form of instruction buffering because both instructions and commonly accessed data tend to migrate to the faster look-aside memory. This scheme is discussed for the IBM System/360 Model 85 (page 574). The look-aside memory suggested by Wilkes [1965] is a content-addressable memory for retaining the active (most recently used) memory words.

Pipeline processing. Pipeline (assembly-line) concurrency is the name given to a system of multiple functional units, each of which is responsible for partial interpretation and execution of the instruction stream. A pipeline processor has several partially completed instructions in process at one time. Each processor stage operates on a specific part of the instruction, e.g., instruction fetch, effective-address calculation, operand fetching, execution of operation specified by the instruction, and results storing. A PMS diagram for a pipeline processor is given in Fig. 19. Thus there is a separate functional unit for each state suggested by the state diagram of Fig. 4. There must be interlocks so that sequence is preserved, i.e., so that results are not used until they are available. Figure 18 shows a time/function diagram of a pipeline processor. There are at least three instructions being interpreted simultaneously. Although we have not extended Fig. 18, we would expect the processor in the sketch to operate on about eight instructions

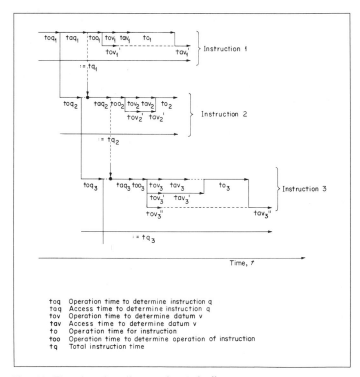

toq	Operation time to determine instruction q
taq	Access time to determine instruction q
tov	Operation time to determine datum v
tav	Access time to determine datum v
to	Operation time for instruction
too	Operation time to determine operation of instruction
tq	Total instruction time

Fig. 18. Time-function diagram for a pipeline processor.

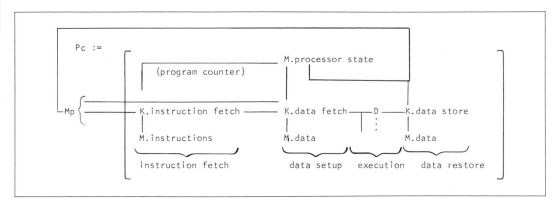

```
Pc :=
                                         M.processor state
        (program counter)

 Mp {        K.instruction fetch        K.data fetch    D    K.data store
                                                        ⋮
        M.instructions                  M.data               M.data

            instruction fetch           data setup   execution   data restore
```

Fig. 19. Example of processor parallelism by spatially independent control function (pipeline processing) PMS diagram.

at one time. Note that the processor sometimes completes later instructions first. In this model there is only one instruction fetching, one operand fetching, and one operand storing unit, while there are multiple data operation units. The particular number of each type of unit is obviously not fixed for all structures but depends heavily on the memory system, the number of instruction streams, and the ISP.

A processor may require many data-operation units in order to avoid bottlenecks. Each unit is independent and may be functionally capable of carrying out only selected tasks. Multiple data-operations are normally desirable in a pipeline processor so that several operations can be carried out at a time, since most of the processing time within the processor is spent on the operations (e.g., multiplication, division, shifting, etc.)

in a multidimensional space. The previous discussion has enumerated the values of one dimension, while (in effect) holding the values of other dimensions constant. The dimensions are highly correlated, especially with cost and evolutionary time. We have been brief in presenting the dimensions because the book is primarily about computer examples. However, one should be able to recognize the dimensions and values when they are encountered within the context of a particular computer.

The remainder of the book is organized around these dimensions. The examples lose the identity of dimensions because they are descriptions of points in the space (computers). Furthermore, the descriptions themselves are not especially organized around these dimensions but are based on the designer's own view of his machine.

Conclusions

You now have our view of the important aspects of the stored-program computer. We have tried to organize the parameters as dimensions so that a computer can be viewed as a point (or points)

References

AdamA60,66,67,68; AdamC60; ArbuR66; ArdeB66; BowdB53; CampR50; CasaC62; ChasG52; CoxJ68; DennJ65; FlynM66; ForrJ51; GibsC66; KnigK66; MolnC67; NiseN66; RandB68; RoseS69; SamuA57; SerrR62; WeikM55,61,64; WilkM51a,65; WillF49.

The Instruction-set Processor: main-line computers

To have a "main line" of computers is to have a family that predominates through the generations. Predominance can probably best be measured by the percentage of distinct computers produced within the family, as opposed to outside it. Members of the family need not all be identical; especially evolution over time can be tolerated. But it must be the case that there is at any moment a "standard" design which is seen as emerging from the just prior "standard" design.

Within these definitions there indeed has been a main line in computer systems. It is based on the Burks, Goldstine, and von Neumann memorandum, reprinted as Chap. 4. The most striking characteristic is the evolution from 1 address organization (1), through index-register (1 + x) to general-register (1 + g) organization. Left outside the main line have been multiple-address organizations, character machines, and stack machines. This seems to be an appropriate description, even though a character machine (variable-length character string), the IBM 1401, probably holds the record for number of machines produced (when each model of the IBM System/360 is counted as a separate computer).

A second characteristic feature has been the PMS structure, which has evolved from a single P to a Pc-nPio structure. This has not been uniform within the family, since it applies only to the larger members; the small machines, such as the PDP-8 (Chap. 5), have no separate Pio's. It might seem that all computer systems, both within and without the family, have evolved in this same way. But this disregards the history of computer development. For a while, in the early fifties, there were seen to be two main lines of potential development: scientific computers, featuring large computation and small input/output, and business computers, featuring small computation and large input/output. The latter started to develop into the Pc-nPio structure (with the IBM 702) but, instead of a separate line developing, scientific computers (with the IBM 704 and UNIVAC computers) adopted the more powerful input/output structure. Again, despite its success, the 1401 has not bred a new generation of computer systems in its image, either within IBM (where one might argue that the overriding consideration was to have a uniform series) or by IBM's competitors.

A third characteristic of the main line is the use of binary as opposed to decimal as the basic radix of the machine. This affects both the arithmetic and whether logical processing (on bit vectors) can be done. The issue seems almost settled in the third generation, with smaller machines being binary and larger machines having multiple data-types. The last serious venture into a large pure decimal machine was the UNIVAC LARC, delivered in 1960. In retrospect, the difference in organizations between binary and decimal machines seems small enough so that we have included them all in the same section.

There are a number of striking features that are characteristic of the main line but do not differentiate it from any of the alternatives that have actually been produced. These features include the stored-program concept; the use of sequential instructions of the operator-operand variety; the use of the word as an information unit, within the range of 12 to 64 bits; and a processor state of less than 100 words. Alternative organizations are conceivable, though they have clearly not seemed practical to computer designers. For instance, in the early fifties there was an attempt to construct an electronic plugboard machine, after the fashion of the ENIAC and the IBM CPC (Card Programmed Calculator). And we see in the new programmed desk calculators (Part 3, Sec. 4) yet another organization that is rather far from the main line (but because of low cost may yet be a part of the future main line). These desk calculators, by the way, are decimal, rather than binary.

Section 1

Processors with one address per instruction

This section is principally concerned with the ISP. It is the largest section in the book, reflecting the dominance of the one-address organization during the first two generations. Machines with index registers are included, but not machines with general registers, which are discussed in Sec. 2. Some processors store two single-address instructions per word, following the pattern of the IAS[1] (von Neumann) machine (Chap. 4). In machines with short word lengths, one single-address instruction is stored in one or two words, for example, in the 16-bit IBM 1800 (Chap. 33) and in the 12-bit PDP-8 (Chap. 5). The evolution of these machines can be seen by comparing first- and third-generation machines (e.g., Whirlwind and the IBM 1800). In general, the section is arranged by increasing word length, alternatively complexity and performance.

Preliminary discussion of the logical design of an electronic computing instrument

This article (Chap. 4) is important for historical as well as technical reasons. It is one of a series[2] written in 1946 prior to building the first fully stored-program computer. Although its authors were not engineers, it is written with the caution of those responsible for the implementation of a rather significant development task. The major problems for the computer are identified, the alternatives analyzed, and a rationale for each decision is given. If computer designers were all required to analyze and describe their machines in such a fashion prior to building them, there would be fewer, but better, computers. Some of the especially enjoyable aspects of the discussion include:

1 Selection of word length and number base.

2 Discussion of the instructions needed.

3 Concern for the input/output structure and the idea of displays (now almost a reality).

4 Rationale for not including floating-point arithmetic (caution about the technology).

5 The lack of necessity for the rather trivial binary-decimal conversion hardware and the idea of cost effectiveness.

6 Analysis of the addition, multiplication, and division hardware implementation. (This description includes a nice, one-page discussion of the average carry length for addition.)

It is difficult to say which machines have been influenced by this memorandum since the idea of data and instructions stored together in a homogeneous primary memory is so basic to all computers. The idea of the single-address instruction set and format is at the heart of all the machines discussed in this section. However, it did not have index registers. Many of the machines with long word length, like IAS, use the two-instructions-per-word format.

Subsequent machines built with only minor variations include ORDVAC; ILLIAC I at the University of Illinois with a 40-bit electrostatic memory and vacuum-tube logic; AVIDAC, ORACLE, MANIAC I, WEIZAC, SILLIAC, BESK, DASK, CSIRAC, and JOHNNIAC at the RAND Corporation with a 40-bit core memory and transistor logic [Gruenberger, 1968]. Other similar computers include the IBM 701 with a 36-bit word, electrostatic memory and vacuum-tube logic; and the CDC 1604, with a 48-bit word, core memory, and transistor logic (possibly influenced by MANIAC II).

The DEC PDP-8

The PDP-8 is included as Chap. 5 to illustrate the effects of a 12-bit word length. It is given in detail using a "top-down" approach in order that the student may thoroughly understand it by simulating it, interpreting it, writing microprograms that

---

[1] Institute for Advanced Study, Princeton University, Princeton, N.J.

[2] The articles in the series were:
1. On the Principles of Large Scale Computing Machines (1946) [Goldstine and von Neumann, 1963a].
2. Preliminary Discussion of the Logical Design of an Electronic Computing Instrument, pt. I, vol. 1 (1946) [Burks, Goldstine, and von Neumann, 1963].
3. Planning and Coding of Problems for an Electronic Computing Instrument, pt. II, vols. 1, 2, 3 (1947–1948) [Goldstine and von Neumann, 1963b, 1963c, 1963d].

emulate it, making incremental modifications to it, and completely redesigning it.[1]

The PDP-8, although not the first 12-bit computer, achieved a status that made it the first standard for small, low cost dedicated computers. There is an active market now for computers in this size and price range to which the marketing culture has responded with the names microcomputer, minicomputer and midicomputer for 8- to 12-, 12- to 16- and 16- to 24-bit word-length computers, respectively.[2]

The PDP-8 has a nearly minimal processor state because the address and ISP integers are 12 bits. Twelve bits is just large enough to represent data from external physical process environments (analog signals) and also just right to address a 4096 word memory. System software (editors, assemblers, compilers, etc.) can surprisingly all fit into this sized memory.[3] The processor state is only 26 bits, and the predecessor PDP-5 had a hardwired state of only 14 bits.

The PDP-8 is also discussed in Part 5, Sec. 2, page 396.

The Whirlwind I computer

Whirlwind I is based on Wilkes' EDSAC at Manchester University. Chapter 6 describes the computer and gives a brief description of vacuum-tube logic and electrostatic storage-tube technology. The PMS structure of Whirlwind I with core memory is given in Fig. 1.

The Memory Test Computer (MTC) of M.I.T.'s Lincoln Laboratory was the first computer to use a core memory. MTC was built to test the memory which Whirlwind I received in August, 1953. Subsequent modifications included the addition of another 2,048-word magnetic-core memory in September, 1953.

The machine's construction and technology are outstanding. It has effective marginal checking and preventive-maintenance test facilities. At the time the machine was dismembered and moved from M.I.T., it had a use time availability of greater than 95 percent. Although Whirlwind I left M.I.T. in 1960, the machine was reassembled and was operational as late as 1966.

The machine's PMS structure is a simple 1 Pc. The K to Mp block transfers are via the Pc on a one-at-a-time, programmed basis. A single data transfer can be initiated to a particular device, thus providing some opportunity for input/output and processing concurrency. The simple structure is due to the high

[1] Perhaps also because of one of the author's (GB) obvious attachment.

[2] See the computers in this size range Chapter 3, Figure 2, page 43.

[3] Conceivably a corollary to Parkinson's law: Programs expand to fill every word in the primary memory of a computer.

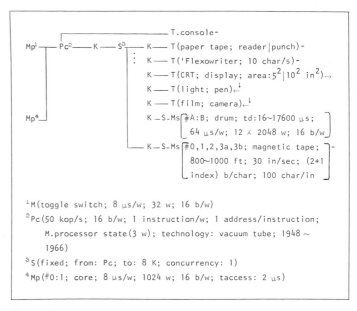

Fig. 1. Whirlwind I PMS diagram.

register costs of the vacuum-tube technology; thus only a single central processor register is provided to hold (or buffer) data during a K transmission to a T or Ms. Appendix 1 of Chap. 6, which is from the programming manual, gives its instruction set.

The IBM 1800

The IBM 1800 (Chap. 33) is a third-generation, 16-bit computer. It is discussed in Part 5, Sec. 2, page 396.

Some aspects of the logical design of a control computer: a case study

Chapter 7 presents the aerospace computer Apollo designed by M.I.T.'s Instrumentation Laboratory. It is presented in contrast to the general-purpose 16-bit computers, Whirlwind (Chap. 6) and the IBM 1800 (Chap. 33). The Apollo computer uses a M(read only) because it is obviously a problem to reload programs. Kampe's SD-2 (Chap. 29) and Apollo (Chap. 7) are both controllers and have other similar design constraints. The IBM 1800 is also used for control purposes. In fact, the computers in this section up to and including the 24-bit SDS 910-9300 series are all designed for control environments. However, all the latter machines have a goal of generality not present in the Apollo.

The SDS 910-9300 series

The SDS 910-9300 computers are illustrative of typical, second-generation, 24-bit computers. The computers are discussed in Part 6, Sec. 2, page 542. Chapter 42 also attempts to show how implementation affects performance for the series.

The LGP-30 and LGP-21

The LGP-30 and later LGP-21 is presented in Chap. 16 and discussed in Part 3, Sec. 2, page 216.

IBM 650 instruction logic

The IBM 650 (Chap. 17) is a one plus one address computer. Its attributes as a cyclic-memory computer, though hardly apparent at the ISP level, are discussed in Part 3, Sec. 2, page 216.

The IBM 7094 I, II

Part 6, Sec. 1 shows the evolution of the IBM 36-bit scientific computers. The IBM 7094 II (Chap. 41) is presented for many reasons (page 517). Among them are its effect on the later IBM System/360 and its position as the standard large scientific computer of the late fifties and early sixties.

The UNIVAC system

The UNIVAC system, first delivered in March, 1951, was later known as UNIVAC I. UNIVAC (UNIVersal Automatic Computers) was the second computer[1] to be manufactured by the Eckert-Mauchly Computer Corporation, subsequently a division of Remington-Rand.[2]

[1] The Eckert-Mauchly BINAC was apparently the first computer to be manufactured by a corporation.

[2] Eckert-Mauchly Computer Corporation was initially independent of Remington-Rand.

UNIVAC is a single-address, decimal computer with 12 digits/word. Two instructions are stored per word. In effect, UNIVAC is a decimal version of the IAS computer. The Mp consists of 1,000 words, made up of 10 words/delay line. Each delay line requires 404 microseconds to recirculate.

UNIVAC is significant because it was the most important computer during the early 1950s. Its performance record is discussed in Chap. 8. The UNIVSERVO magnetic-tape system was rather advanced for 1950, considering performance, error checking, and buffering. Particularly nice is the ability to partition the input/output system for off-line printing and key punching.

One-level storage system

The 48-bit Atlas was developed at Manchester University and subsequently manufactured by Ferranti Corp. (now part of International Computers and Tabulators). The development began about 1960, and the paper was written in 1962. The importance of Atlas with respect to current and future machines is discussed in Part 3, Sec. 6, page 274.

The engineering design of the Stretch computer

The IBM Stretch (also called the IBM Model 7030) single-address computer (Chap. 34) is one of the earliest computers built to provide maximum computing power subject to no apparent cost, size, and producibility constraints. A discussion of its importance is given in Part 5, Sec. 2, page 396.

Chapter 4

Preliminary discussion of the logical design of an electronic computing instrument[1]

Arthur W. Burks / Herman H. Goldstine /
John von Neumann

PART I

1. Principal components of the machine

1.1.   Inasmuch as the completed device will be a general-purpose computing machine it should contain certain main organs relating to arithmetic, memory-storage, control and connection with the human operator. It is intended that the machine be fully automatic in character, i.e. independent of the human operator after the computation starts. A fuller discussion of the implications of this remark will be given in Sec. 3 below.

1.2.   It is evident that the machine must be capable of storing in some manner not only the digital information needed in a given computation such as boundary values, tables of functions (such as the equation of state of a fluid) and also the intermediate results of the computation (which may be wanted for varying lengths of time), but also the instructions which govern the actual routine to be performed on the numerical data. In a special-purpose machine these instructions are an integral part of the device and constitute a part of its design structure. For an all-purpose machine it must be possible to instruct the device to carry out any computation that can be formulated in numerical terms. Hence there must be some organ capable of storing these program orders. There must, moreover, be a unit which can understand these instructions and order their execution.

1.3.   Conceptually we have discussed above two different forms of memory: storage of numbers and storage of orders. If, however, the orders to the machine are reduced to a numerical code and if the machine can in some fashion distinguish a number from an order, the memory organ can be used to store both num-

bers and orders. The coding of orders into numeric form is discussed in 6.3 below.

1.4.   If the memory for orders is merely a storage organ there must exist an organ which can automatically execute the orders stored in the memory. We shall call this organ the *Control*.

1.5.   Inasmuch as the device is to be a computing machine there must be an arithmetic organ in it which can perform certain of the elementary arithmetic operations. There will be, therefore, a unit capable of adding, subtracting, multiplying and dividing. It will be seen in 6.6 below that it can also perform additional operations that occur quite frequently.

The operations that the machine will view as elementary are clearly those which are wired into the machine. To illustrate, the operation of multiplication could be eliminated from the device as an elementary process if one were willing to view it as a properly ordered series of additions. Similar remarks apply to division. In general, the inner economy of the arithmetic unit is determined by a compromise between the desire for speed of operation—a non-elementary operation will generally take a long time to perform since it is constituted of a series of orders given by the control—and the desire for simplicity, or cheapness, of the machine.

1.6.   Lastly there must exist devices, the input and output organ, whereby the human operator and the machine can communicate with each other. This organ will be seen below in 4.5, where it is discussed, to constitute a secondary form of automatic memory.

2. First remarks on the memory

2.1.   It is clear that the size of the memory is a critical consideration in the design of a satisfactory general-purpose computing

[1]From A. H. Taub (ed.), "Collected Works of John von Neumann," vol. 5, pp. 34–79, The Macmillan Company, New York, 1963. Taken from report to U. S. Army Ordnance Department, 1946. See also Bibliography Burks, Goldstine and von Neumann, 1962*a*, 1962*b*, 1963; and Goldstine and von Neumann 1963*a*, 1963*b*, 1963*c*, 1963*d*.

machine. We proceed to discuss what quantities the memory should store for various types of computations.

2.2. In the solution of partial differential equations the storage requirements are likely to be quite extensive. In general, one must remember not only the initial and boundary conditions and any arbitrary functions that enter the problem but also an extensive number of intermediate results.

a For equations of parabolic or hyperbolic type in two independent variables the integration process is essentially a double induction. To find the values of the dependent variables at time $t + \Delta t$ one integrates with respect to x from one boundary to the other by utilizing the data at time t as if they were coefficients which contribute to defining the problem of this integration.

Not only must the memory have sufficient room to store these intermediate data but there must be provisions whereby these data can later be removed, i.e. at the end of the $(t + \Delta t)$ cycle, and replaced by the corresponding data for the $(t + 2\Delta t)$ cycle. This process of removing data from the memory and of replacing them with new information must, of course, be done quite automatically under the direction of the control.

b For total differential equations the memory requirements are clearly similar to, but smaller than, those discussed in (a) above.

c Problems that are solved by iterative procedures such as systems of linear equations or elliptic partial differential equations, treated by relaxation techniques, may be expected to require quite extensive memory capacity. The memory requirement for such problems is apparently much greater than for those problems in (a) above in which one needs only to store information corresponding to the instantaneous value of one variable [t in (a) above], while now entire solutions (covering all values of all variables) must be stored. This apparent discrepancy in magnitudes can, however, be somewhat overcome by the use of techniques which permit the use of much coarser integration meshes in this case, than in the cases under (a).

2.3. It is reasonable at this time to build a machine that can conveniently handle problems several orders of magnitude more complex than are now handled by existing machines, electronic or electro-mechanical. We consequently plan on a fully automatic electronic storage facility of about 4,000 numbers of 40 binary digits each. This corresponds to a precision of $2^{-40} \sim 0.9 \times 10^{-12}$, i.e. of about 12 decimals. We believe that this memory capacity exceeds the capacities required for most problems that one deals with at present by a factor of about 10. The precision is also safely higher than what is required for the great majority of present day problems. In addition, we propose that we have a subsidiary memory of much larger capacity, which is also fully automatic, on some medium such as magnetic wire or tape.

3. First remarks on the control and code

3.1. It is easy to see by formal-logical methods that there exist codes that are *in abstracto* adequate to control and cause the execution of any sequence of operations which are individually available in the machine and which are, in their entirety, conceivable by the problem planner. The really decisive considerations from the present point of view, in selecting a code, are more of a practical nature: simplicity of the equipment demanded by the code, and the clarity of its application to the actually important problems together with the speed of its handling of those problems. It would take us much too far afield to discuss these questions at all generally or from first principles. We will therefore restrict ourselves to analyzing only the type of code which we now envisage for our machine.

3.2. There must certainly be instructions for performing the fundamental arithmetic operations. The specifications for these orders will not be completely given until the arithmetic unit is described in a little more detail.

3.3. It must be possible to transfer data from the memory to the arithmetic organ and back again. In transferring information from the arithmetic organ back into the memory there are two types we must distinguish: Transfers of numbers as such and transfers of numbers which are parts of orders. The first case is quite obvious and needs no further explication. The second case is more subtle and serves to illustrate the generality and simplicity of the system. Consider, by way of illustration, the problem of interpolation in the system. Let us suppose that we have formulated the necessary instructions for performing an interpolation of order n in a sequence of data. The exact location in the memory of the $(n + 1)$ quantities that bracket the desired functional value is, of course, a function of the argument. This argument probably is found as the result of a computation in the machine. We thus need an order which can substitute a number into a given order—in the case of interpolation the location of the argument or the group of arguments that is nearest in our table to the desired value. By means of such an order the results of a computation can be introduced into the instructions governing that or a different computation. This makes it possible for a sequence of instructions to be used with different sets of numbers located in different parts of the memory.

To summarize, transfers into the memory will be of two sorts: *Total substitutions*, whereby the quantity previously stored is cleared out and replaced by a new number. *Partial substitutions* in which that part of an order containing a *memory location-number*—we assume the various positions in the memory are enumerated serially by memory location-numbers—is replaced by a new memory location-number.

3.4.   It is clear that one must be able to get numbers from any part of the memory at any time. The treatment in the case of orders can, however, be more methodical since one can at least partially arrange the control instructions in a linear sequence. Consequently the control will be so constructed that it will normally proceed from place n in the memory to place $(n + 1)$ for its next instruction.

3.5.   The utility of an automatic computer lies in the possibility of using a given sequence of instructions repeatedly, the number of times it is iterated being either preassigned or dependent upon the results of the computation. When the iteration is completed a different sequence of orders is to be followed, so we must, in most cases, give two parallel trains of orders preceded by an instruction as to which routine is to be followed. This choice can be made to depend upon the sign of a number (zero being reckoned as plus for machine purposes). Consequently, we introduce an order (*the conditional transfer order*) which will, depending on the sign of a given number, cause the proper one of two routines to be executed.

Frequently two parallel trains of orders terminate in a common routine. It is desirable, therefore, to order the control in either case to proceed to the beginning point of the common routine. This *unconditional transfer* can be achieved either by the artificial use of a conditional transfer or by the introduction of an explicit order for such a transfer.

3.6.   Finally we need orders which will integrate the input-output devices with the machine. These are discussed briefly in 6.8.

3.7.   We proceed now to a more detailed discussion of the machine. Inasmuch as our experience has shown that the moment one chooses a given component as the elementary memory unit, one has also more or less determined upon much of the balance of the machine, we start by a consideration of the memory organ. In attempting an exposition of a highly integrated device like a computing machine we do not find it possible, however, to give an exhaustive discussion of each organ before completing its description. It is only in the final block diagrams that anything approaching a complete unit can be achieved.

The time units to be used in what follows will be:

$$1 \ \mu\text{sec} = 1 \text{ microsecond} = 10^{-6} \text{ seconds}$$
$$1 \text{ msec} = 1 \text{ millisecond} = 10^{-3} \text{ seconds}$$

4.   The memory organ

4.1.   Ideally one would desire an indefinitely large memory capacity such that any particular aggregate of 40 binary digits, or *word* (cf. 2.3), would be immediately available—i.e. in a time which is somewhat or considerably shorter than the operation time of a fast electronic multiplier. This may be assumed to be practical at the level of about 100 μsec. Hence the availability time for a word in the memory should be 5 to 50 μsec. It is equally desirable that words may be replaced with new words at about the same rate. It does not seem possible physically to achieve such a capacity. We are therefore forced to recognize the possibility of constructing a hierarchy of memories, each of which has greater capacity than the preceding but which is less quickly accessible.

The most common forms of storage in electrical circuits are the flip-flop or trigger circuit, the gas tube, and the electro-mechanical relay. To achieve a memory of n words would, of course, require about $40n$ such elements, exclusive of the switching elements. We saw earlier (cf. 2.2) that a fast memory of several thousand words is not at all unreasonable for an all-purpose instrument. Hence, about 10^5 flip-flops or analogous elements would be required! This would, of course, be entirely impractical.

We must therefore seek out some more fundamental method of storing electrical information than has been suggested above. One criterion for such a storage medium is that the individual storage organs, which accommodate only one binary digit each, should not be macroscopic components, but rather microscopic elements of some suitable organ. They would then, of course, not be identified and switched to by the usual macroscopic wire connections, but by some functional procedure in manipulating that organ.

One device which displays this property to a marked degree is the iconoscope tube. In its conventional form it possesses a linear resolution of about one part in 500. This would correspond to a (two-dimensional) memory capacity of $500 \times 500 = 2.5 \times 10^5$. One is accordingly led to consider the possibility of storing electrical charges on a dielectric plate inside a cathode-ray tube. Effectively such a tube is nothing more than a myriad of electrical capacitors which can be connected into the circuit by means of an electron beam.

Actually the above mentioned high resolution and concomitant memory capacity are only realistic under the conditions of television-image storage, which are much less exigent in respect to the reliability of individual markings than what one can accept in the storage for a computer. In this latter case resolutions of one part in 20 to 100, i.e. memory capacities of 400 to 10,000, would seem to be more reasonable in terms of equipment built essentially along familiar lines.

At the present time the Princeton Laboratories of the Radio Corporation of America are engaged in the development of a storage tube, the *Selectron*, of the type we have mentioned above. This tube is also planned to have a non-amplitude-sensitive switching system whereby the electron beam can be directed to a given spot on the plate within a quite small fraction of a millisecond. Inasmuch as the storage tube is the key component of the machine envisaged in this report we are extremely fortunate in having secured the cooperation of the RCA group in this as well as in various other developments.

An alternate form of rapid memory organ is the acoustic feedback delay line described in various reports on the EDVAC. (This is an electronic computing machine being developed for the Ordnance Department, U.S. Army, by the University of Pennsylvania, Moore School of Electrical Engineering.) Inasmuch as that device has been so clearly reported in those papers we give no further discussion. There are still other physical and chemical properties of matter in the presence of electrons or photons that might be considered, but since none is yet beyond the early discussion stage we shall not make further mention of them.

4.2.   We shall accordingly assume throughout the balance of this report that the Selectron is the modus for storage of words at electronic speeds. As now planned, this tube will have a capacity of $2^{12} = 4,096 \approx 4,000$ binary digits. To achieve a total electronic storage of about 4,000 words we propose to use 40 Selectrons, thereby achieving a memory of 2^{12} words of 40 binary digits each. (Cf. again 2.3.)

4.3.   There are two possible means for storing a particular word in the Selectron memory—or, in fact, in either a delay line memory or in a storage tube with amplitude-sensitive deflection. One method is to store the entire word in a given tube and then to get the word out by picking out its respective digits in a serial fashion. The other method is to store in corresponding places in each of the 40 tubes one digit of the word. To get a word from the memory in this scheme requires, then, one switching mechanism to which all 40 tubes are connected in parallel. Such a switching scheme seems to us to be simpler than the technique needed in the serial system and is, of course, 40 times faster. We accordingly adopt the parallel procedure and thus are led to consider a so-called *parallel machine*, as contrasted with the serial principles being considered for the EDVAC. (In the EDVAC the

peculiar characteristics of the acoustic delay line, as well as various other considerations, seem to justify a serial procedure. For more details, cf. the reports referred to in 4.1.) The essential difference between these two systems lies in the method of performing an addition; in a parallel machine all corresponding pairs of digits are added simultaneously, whereas in a serial one these pairs are added serially in time.

4.4.   To summarize, we assume that the fast electronic memory consists of 40 Selectrons which are switched in parallel by a common switching arrangement. The inputs of the switch are controlled by the control.

4.5.   Inasmuch as a great many highly important classes of problems require a far greater total memory than 2^{12} words, we now consider the next stage in our storage hierarchy. Although the solution of partial differential equations frequently involves the manipulation of many thousands of words, these data are generally required only in blocks which are well within the 2^{12} capacity of the electronic memory. Our second form of storage must therefore be a medium which feeds these blocks of words to the electronic memory. It should be controlled by the control of the computer and is thus an integral part of the system, not requiring human intervention.

There are evidently two distinct problems raised above. One can choose a given medium for storage such as teletype tapes, magnetic wire or tapes, movie film or similar media. There still remains the problem of automatic integration of this storage medium with the machine. This integration is achieved logically by introducing appropriate orders into the code which can instruct the machine to read or write on the medium, or to move it by a given amount or to a place with given characteristics. We discuss this question a little more fully in 6.8.

Let us return now to the question of what properties the secondary storage medium should have. It clearly should be able to store information for periods of time long enough so that only a few per cent of the total computing time is spent in re-registering information that is "fading off." It is certainly desirable, although not imperative, that information can be erased and replaced by new data. The medium should be such that it can be controlled, i.e. moved forward and backward, automatically. This consideration makes certain media, such as punched cards, undesirable. While cards can, of course, be printed or read by appropriate orders from some machine, they are not well adapted to problems in which the output data are fed directly back into the machine, and are required in a sequence which is non-monotone with respect to the order of the cards. The medium should be capable of remembering very large numbers of data at a much smaller price than electronic devices. It must be fast enough so that, even when it has to be used frequently in a problem, a large percentage of the total solution time is not spent in getting data into and out of this medium and achieving the desired positioning on it. If this condition is not reasonably well met, the advantages of the high electronic speeds of the machine will be largely lost.

Both light- or electron-sensitive film and magnetic wires or tapes, whose motions are controlled by servo-mechanisms integrated with the control, would seem to fulfil our needs reasonably well. We have tentatively decided to use magnetic wires since we have achieved reliable performance with them at pulse rates of the order of 25,000/sec and beyond.

4.6.   Lastly our memory hierarchy requires a vast quantity of dead storage, i.e. storage not integrated with the machine. This storage requirement may be satisfied by a library of wires that can be introduced into the machine when desired and at that time become automatically controlled. Thus our dead storage is really nothing but an extension of our secondary storage medium. It differs from the latter only in its availability to the machine.

4.7.   We impose one additional requirement on our secondary memory. It must be possible for a human to put words on to the wire or other substance used and to read the words put on by the machine. In this manner the human can control the machine's functions. It is now clear that the secondary storage medium is really nothing other than a part of our input-output system, cf. 6.8.4 for a description of a mechanism for achieving this.

4.8.   There is another highly important part of the input-output which we merely mention at this time, namely, some mechanism for viewing graphically the results of a given computation. This can, of course, be achieved by a Selectron-like tube which causes its screen to fluoresce when data are put on it by an electron beam.

4.9.   For definiteness in the subsequent discussions we assume that associated with the output of each Selectron is a flip-flop. This assemblage of 40 flip-flops we term the *Selectron Register*.

5.   The arithmetic organ

5.1.   In this section we discuss the features we now consider desirable for the arithmetic part of our machine. We give our tentative conclusions as to which of the arithmetic operations should be built into the machine and which should be programmed. Finally, a schematic of the arithmetic unit is described.

5.2.   In a discussion of the arithmetical organs of a computing machine one is naturally led to a consideration of the number system to be adopted. In spite of the longstanding tradition of building digital machines in the decimal system, we feel strongly in favor of the binary system for our device. Our fundamental unit of memory is naturally adapted to the binary system since we do not attempt to measure gradations of charge at a particular point in the Selectron but are content to distinguish two states. The flip-flop again is truly a binary device. On magnetic wires or tapes and in acoustic delay line memories one is also content to recognize the presence or absence of a pulse or (if a carrier frequency is used) of a pulse train, or of the sign of a pulse. (We will not discuss here the ternary possibilities of a positive-or-negative-or-no-pulse system and their relationship to questions of reliability and checking, nor the very interesting possibilities of carrier frequency modulation.) Hence if one contemplates using a decimal system with either the iconoscope or delay-line memory one is forced into a binary coding of the decimal system—each decimal digit being represented by at least a tetrad of binary digits. Thus an accuracy of ten decimal digits requires at least 40 binary digits. In a true binary representation of numbers, however, about 33 digits suffice to achieve a precision of 10^{10}. The use of the binary system is therefore somewhat more economical of equipment than is the decimal.

The main virtue of the binary system as against the decimal is, however, the greater simplicity and speed with which the elementary operations can be performed. To illustrate, consider multiplication by repeated addition. In binary multiplication the product of a particular digit of the multiplier by the multiplicand is either the multiplicand or null according as the multiplier digit is 1 or 0. In the decimal system, however, this product has ten possible values between null and nine times the multiplicand, inclusive. Of course, a decimal number has only $\log_{10}2 \sim 0.3$ times as many digits as a binary number of the same accuracy, but even so multiplication in the decimal system is considerably longer than in the binary system. One can accelerate decimal multiplication by complicating the circuits, but this fact is irrelevant to the point just made since binary multiplication can likewise be accelerated by adding to the equipment. Similar remarks may be made about the other operations.

An additional point that deserves emphasis is this: An important part of the machine is not arithmetical, but logical in nature. Now logics, being a yes-no system, is fundamentally binary. Therefore a binary arrangement of the arithmetical organs contributes very significantly towards producing a more homogeneous machine, which can be better integrated and is more efficient.

The one disadvantage of the binary system from the human point of view is the conversion problem. Since, however, it is completely known how to convert numbers from one base to another and since this conversion can be effected solely by the use of the usual arithmetic processes there is no reason why the computer itself cannot carry out this conversion. It might be argued that this is a time consuming operation. This, however, is not the case. (Cf. 9.6 and 9.7 of Part II. Part II is a report issued under the title *Planning and Coding of Problems for an Electronic Computing Instrument*.[1]) Indeed a general-purpose computer, used as a scientific research tool, is called upon to do a very great number of multiplications upon a relatively small amount of input data, and hence the time consumed in the decimal to binary conversion is only a trivial percentage of the total computing time. A similar remark is applicable to the output data.

In the preceding discussion we have tacitly assumed the desirability of introducing and withdrawing data in the decimal system. We feel, however, that the base 10 may not even be a permanent feature in a scientific instrument and consequently will probably attempt to train ourselves to use numbers base 2 or 8 or 16. The reason for the bases 8 or 16 is this: Since 8 and 16 are powers of 2 the conversion to binary is trivial; since both are about the size of 10, they violate many of our habits less badly than base 2. (Cf. Part II, 9.4.)

5.3.   Several of the digital computers being built or planned in this country and England are to contain a so-called "floating decimal point". This is a mechanism for expressing each word as a characteristic and a mantissa—e.g. 123.45 would be carried in the machine as (0.12345,03), where the 3 is the exponent of 10 associated with the number. There appear to be two major purposes in a "floating" decimal point system both of which arise from the fact that the number of digits in a word is a constant, fixed by design considerations for each particular machine. The first of these purposes is to retain in a sum or product as many significant digits as possible and the second of these is to free the human operator from the burden of estimating and inserting into a problem "scale factors"—multiplicative constants which serve to keep numbers within the limits of the machine.

There is, of course, no denying the fact that human time is consumed in arranging for the introduction of suitable scale factors. We only argue that the time so consumed is a very small percentage of the total time we will spend in preparing an interesting problem for our machine. The first advantage of the floating point is, we feel, somewhat illusory. In order to have such a floating point one must waste memory capacity which could otherwise be used for carrying more digits per word. It would therefore seem

to us not at all clear whether the modest advantages of a floating binary point offset the loss of memory capacity and the increased complexity of the arithmetic and control circuits.

There are certainly some problems within the scope of our device which really require more than 2^{-40} precision. To handle such problems we wish to plan in terms of words whose lengths are some fixed integral multiple of 40, and program the machine in such a manner as to give the corresponding aggregates of 40 digit words the proper treatment. We must then consider an addition or multiplication as a complex operation programmed from a number of primitive additions or multiplications (cf. §9, Part II). There would seem to be considerable extra difficulties in the way of such a procedure in an instrument with a floating binary point.

The reader may remark upon our alternate spells of radicalism and conservatism in deciding upon various possible features for our mechanism. We hope, however, that he will agree, on closer inspection, that we are guided by a consistent and sound principle in judging the merits of any idea. We wish to incorporate into the machine—in the form of circuits—only such logical concepts as are either necessary to have a complete system or highly convenient because of the frequency with which they occur and the influence they exert in the relevant mathematical situations.

5.4.   On the basis of this criterion we definitely wish to build into the machine circuits which will enable it to form the binary sum of two 40 digit numbers. We make this decision not because addition is a logically basic notion but rather because it would slow the mechanism as well as the operator down enormously if each addition were programmed out of the more simple operations of "and", "or", and "not". The same is true for the subtraction. Similarly we reject the desire to form products by programming them out of additions, the detailed motivation being very much the same as in the case of addition and subtraction. The cases for division and square-rooting are much less clear.

It is well known that the reciprocal of a number a can be formed to any desired accuracy by iterative schemes. One such scheme consists of improving an estimate X by forming $X' = 2X - aX^2$. Thus the new error $1 - aX'$ is $(1 - aX)^2$, which is the square of the error in the preceding estimate. We notice that in the formation of X', there are two bona fide multiplications—we do not consider multiplication by 2 as a true product since we will have a facility for shifting right or left in one or two pulse times. If then we somehow could guess $1/a$ to a precision of 2^{-5}, 6 multiplications—3 iterations—would suffice to give a final result good to 2^{-40}. Accordingly a small table of 2^4 entries could be used to get the initial estimate of $1/a$. In this way a reciprocal $1/a$

---

[1] See Bibliography [Goldstine and von Neumann, 1963b, 1963c, 1963d]. References in this chapter are all to this report.

could be formed in 6 multiplication times, and hence a quotient b/a in 7 multiplication times. Accordingly we see that the question of building a divider is really a function of how fast it can be made to operate compared to the iterative method sketched above: In order to justify its existence, a divider must perform a division in a good deal less than 7 multiplication times. We have, however, conceived a divider which is much faster than these 7 multiplication times and therefore feel justified in building it, especially since the amount of equipment needed above the requirements of the multiplier is not important.

It is, of course, also possible to handle square roots by iterative techniques. In fact, if X is our estimate of $a^{1/2}$, then $X' = \frac{1}{2}(X + a/X)$ is a better estimate. We see that this scheme involves one division per iteration. As will be seen below in our more detailed examination of the arithmetic organ we do not include a square-rooter in our plans because such a device would involve more equipment than we feel is desirable in a first model. (Concerning the iterative method of square-rooting, cf. 8.10 in Part II.)

5.5. The first part of our arithmetic organ requires little discussion at this point. It should be a parallel storage organ which can receive a number and add it to the one already in it, which is also able to clear its contents and which can transmit what it contains. We will call such an organ an *Accumulator*. It is quite conventional in principle in past and present computing machines of the most varied types, e.g. desk multipliers, standard IBM counters, more modern relay machines, the ENIAC. There are of, course, numerous ways to build such a binary accumulator. We distinguish two broad types of such devices: static, and dynamic or pulse-type accumulators. These will be discussed in 5.11, but it is first necessary to make a few remarks concerning the arithmetic of binary addition. In a parallel accumulator, the first step in an addition is to add each digit of the addend to the corresponding digit of the augend. The second step is to perform the carries, and this must be done in sequence since a carry may produce a carry. In the worst case, 39 carries will occur. Clearly it is inefficient to allow 39 times as much time for the second step (performing the carries) as for the first step (adding the digits). Hence either the carries must be accelerated, or use must be made of the average number of carries or both.

5.6. We shall show that for a sum of binary words, each of length n, the length of the largest carry sequence is on the average not in excess of $^{2}\log n$. Let $p_n(v)$ designate the probability that a carry sequence is of length v or greater in the sum of two binary words of length n. Then clearly $p_n(v) - p_n(v + 1)$ is the probability that the largest carry sequence is of length exactly v and the weighted average

$$a_n = \sum_{v=0}^{n} v[p_n(v) - p_n(v + 1)]$$

is the average length of such carry. Note that

$$\sum_{v=0}^{n} [p_n(v) - p_n(v + 1)] = 1$$

since $p_n(v) = 0$ if $v > n$. From these it is easily inferred that

$$a_n = \sum_{v=1}^{n} p_n(v)$$

We now proceed to show that $p_n(v) \leq \min[1, (n - v + 1)/2^{v+1}]$. Observe first that

$$p_n(v) = p_{n-1}(v) + \frac{1 - p_{n-v}(v)}{2^{v+1}} \qquad \text{if} \qquad v \leq n$$

Indeed, $p_n(v)$ is the probability that the sum of two n-digit numbers contains a carry sequence of length $\geq v$. This probability obtains by adding the probabilities of two mutually exclusive alternatives: *First:* Either the $n - 1$ first digits of the two numbers by themselves contain a carry sequence of length $\geq v$. This has the probability $p_{n-1}(v)$. *Second:* The $n - 1$ first digits of the two numbers by themselves do not contain a carry sequence of length $\geq v$. In this case any carry sequence of length $\geq v$ in the total numbers (of length n) must end with the last digits of the total sequence. Hence these must form the combination 1, 1. The next $v - 1$ digits must propagate the carry, hence each of these must form the combination 1, 0 or 0, 1. (The combinations 1, 1 and 0, 0 do not propagate a carry.) The probability of the combination 1, 1 is $\frac{1}{4}$, that one of the alternative combinations 1, 0 or 0, 1 is $\frac{1}{2}$. The total probability of this sequence is therefore $\frac{1}{4}(\frac{1}{2})^{v-1} = (\frac{1}{2})^{v+1}$. The remaining $n - v$ digits must not contain a carry sequence of length $\geq v$. This has the probability $1 - p_{n-v}(v)$. Thus the probability of the second case is $[1 - p_{n-v}(v)]/2^{v+1}$. Combining these two cases, the desired relation

$$p_n(v) = p_{n-1}(v) + \frac{1 - p_{n-v}(v)}{2^{v+1}}$$

obtains. The observation that $p_n(v) = 0$ if $v > n$ is trivial.

We see with the help of the formulas proved above that $p_n(v) - p_{n-1}(v)$ is always $\leq 1/2^{v+1}$, and hence that the sum

$$\sum_{i=v}^{n} [p_i(v) - p_{i-1}(v)] = p_n(v)$$

is not in excess of $(n - v + 1)/2^{v+1}$ since there are $n - v + 1$ terms in the sum; since, moreover, each $p_n(v)$ is a probability, it is not greater than 1. Hence we have

$$p_n(v) \leqq \min\left[1, \frac{n - v + 1}{2^{v+1}}\right]$$

Finally we turn to the question of getting an upper bound on $a_n = \sum_{v=1}^{n} p_n(v)$. Choose K so that $2^K \leqq n \leqq 2^{K+1}$. Then

$$a_n = \sum_{v=1}^{K-1} p_n(v) + \sum_{v=K}^{n} p_n(v) \leqq \sum_{v=1}^{K-1} 1 + \sum_{v=K}^{n} \frac{n}{2^{v+1}} = K - 1 + \frac{n}{2^K}$$

This last expression is clearly linear in n in the interval $2^K \leqq n \leqq 2^{K+1}$, and it is $= K$ for $n = 2^K$ and $= K + 1$ for $n = 2^{K+1}$, i.e. it is $=^2\log n$ at both ends of this interval. Since the function $^2\log n$ is everywhere concave from below, it follows that our expression is $\leqq^2\log n$ throughout this interval. Thus $a_n \leqq {}^2\log n$. This holds for all K, i.e. for all n, and it is the inequality which we wanted to prove.

For our case $n = 40$ we have $a_n \leqq \log_2 40 \sim 5.3$, i.e. an average length of about 5 for the longest carry sequence. (The actual value of a_{40} is 4.62.)

5.7. Having discussed the addition, we can now go on to the subtraction. It is convenient to discuss at this point our treatment of negative numbers, and in order to do that right, it is desirable to make some observations about the treatment of numbers in general.

Our numbers are 40 digit aggregates, the left-most digit being the sign digit, and the other digits genuine binary digits, with positional values $2^{-1}, 2^{-2}, \ldots, 2^{-39}$ (going from left to right). Our accumulator will, however, treat the sign digit, too, as a binary digit with the positional value 2^0—at least when it functions as an adder. For numbers between 0 and 1 this is clearly all right: The left-most digit will then be 0, and if 0 at this place is taken to represent a $+$ sign, then the number is correctly expressed with its sign and 39 binary digits.

Let us now consider one or more unrestricted 40 binary digit numbers. The accumulator will add them, with the digit-adding and the carrying mechanisms functioning normally and identically in all 40 positions. There is one reservation, however: If a carry originates in the left-most position, then it has nowhere to go from there (there being no further positions to the left) and is "lost". This means, of course, that the addend and the augend, both numbers between 0 and 2, produced a sum exceeding 2, and the accumulator, being unable to express a digit with a positional value 2^1, which would now be necessary, omitted 2. That is, the

sum was formed correctly, excepting a possible error 2. If several such additions are performed in succession, then the ultimate error may be any integer multiple of 2. That is, the accumulator is an adder which allows errors that are integer multiples of 2—it is an adder *modulo* 2.

It should be noted that our convention of placing the binary point immediately to the right of the left-most digit has nothing to do with the structure of the adder. In order to make this point clearer we proceed to discuss the possibilities of positioning the binary point in somewhat more detail.

We begin by enumerating the 40 digits of our numbers (words) from left to right. In doing this we use an index $h = 1, \ldots, 40$. Now we might have placed the binary point just as well between digits j and $j + 1$, $j = 0, \ldots, 40$. Note, that $j = .0$ corresponds to the position at the extreme left (there is no digit $h = j = 0$); $j = 40$ corresponds to the position at the extreme right (there is no position $h = j + 1 = 41$); and $j = 1$ corresponds to our above choice. Whatever our choice of j, it does not affect the correctness of the accumulator's addition. (This is equally true for subtraction, cf. below, but not for multiplication and division, cf. 5.8.) Indeed, we have merely multiplied all numbers by 2^{j-1} (as against our previous convention), and such a "change of scale" has no effect on addition (and subtraction). However, now the accumulator is an adder which allows errors that are integer multiples of 2^j it is an adder modulo 2^j. We mention this because it is occasionally convenient to think in terms of a convention which places the binary point at the right end of the digital aggregate. Then $j = 40$, our numbers are integers, and the accumulator is an adder modulo 2^{40}. We must emphasize, however, that all of this, i.e. all attributions of values to j, are purely convention—i.e. it is solely the mathematician's interpretation of the functioning of the machine and not a physical feature of the machine. This convention will necessitate measures that have to be made effective by actual physical features of the machine—i.e. the convention will become a physical and engineering reality only when we come to the organs of multiplication.

We will use the convention $j = 1$, i.e. our numbers lie in 0 and 2 and the accumulator adds modulo 2.

This being so, these numbers between 0 and 2 can be used to represent all numbers modulo 2. Any real number x agrees modulo 2 with one and only one number $\bar{x}$ between 0 and 2—or, to be quite precise: $0 \leqq \bar{x} < 2$. Since our addition functions modulo 2, we see that the accumulator may be used to represent and to add numbers modulo 2.

This determines the representation of negative numbers: If $x < 0$, then we have to find the unique integer multiple of 2, $2s$

$(s = 1, 2, \ldots)$ such that $0 \leqq \bar{x} < 2$ for $\bar{x} = x + 2s$ (i.e. $-2s \leqq x < 2(1 - s)$), and represent x by the digitalization of $\bar{x}$.

In this way, however, the sign digit character of the left-most digit is lost: It can be 0 or 1 for both $x \geqq 0$ and $x < 0$, hence 0 in the left-most position can no longer be associated with the $+$ sign of x. This may seem a bad deficiency of the system, but it is easy to remedy—at least to an extent which suffices for our purposes. This is done as follows:

We usually work with numbers x between -1 and 1—or, to be quite precise: $-1 \leqq x < 1$. Now the $\bar{x}$ with $0 \leqq \bar{x} < 2$, which differs from x by an integer multiple of 2, behaves as follows: If $x \geqq 0$, then $0 \leqq x < 1$, hence $\bar{x} = x$, and so $0 \leqq \bar{x} < 1$, the left-most digit of $\bar{x}$ is 0. If $x < 0$, then $-1 \leqq x < 0$, hence $\bar{x} = x + 2$, and so $1 \leqq \bar{x} < 2$, the left-most digit of $\bar{x}$ is 1. Thus the left-most digit (of $\bar{x}$) is now a precise equivalent of the sign (of x): 0 corresponds to $+$ and 1 to $-$.

Summing up:

The accumulator may be taken to represent all real numbers modulo 2, and it adds them modulo 2. If x lies between -1 and 1 (precisely: $-1 \leqq x < 1$)—as it will in almost all of our uses of the machine—then the left-most digit represents the sign: 0 is $+$ and 1 is $-$.

Consider now a negative number x with $-1 \leqq x < 0$. Put $x = -y$, $0 < y \leqq 1$. Then we digitalize x by representing it as $x + 2 = 2 - y = 1 + (1 - y)$. That is, the left-most (sign) digit of $x = -y$ is, as it should be, 1; and the remaining 39 digits are those of the *complement* of $y = -x = |x|$, i.e. those of $1 - y$. Thus we have been led to the familiar representation of negative numbers by *complementation*.

The connection between the digits of x and those of $-x$ is now easily formulated, for any $x \gtreqless 0$. Indeed, $-x$ is equivalent to

$$2 - x = \{(2^1 - 2^{-39}) - x\} + 2^{-39} = \left(\sum_{i=0}^{39} 2^{-i} - x\right) + 2^{-39}$$

(This digit index $i = 1, \ldots, 39$ is related to our previous digit index $h = 1, \ldots, 40$ by $i = h - 1$. Actually it is best to treat i as if its domain included the additional value $i = 0$—indeed $i = 0$ then corresponds to $h = 1$, i.e. to the sign digit. In any case i expresses the positional value of the digit to which it refers more simply than h does: This positional value is $2^{-i} = 2^{-(h-1)}$. Note that if we had positioned the binary point more generally between j and $j + 1$, as discussed further above, this positional value would have been $2^{-(h-j)}$. We now have, as pointed out previously, $j = 1$.) Hence its digits obtain by subtracting every digit of x from 1—by complementing each digit, i.e. by replacing 0 by 1 and 1 by

0—and then adding 1 in the right-most position (and effecting all the carries that this may cause). (Note how the left-most digit, interpreted as a sign digit, gets inverted by this procedure as it should be.)

A subtraction $x - y$ is therefore performed by the accumulator, Ac, as follows: Form $x + y'$, where y' has a digit 0 or 1 where y has a digit 1 or 0, respectively, and then add 1 in the right-most position. The last operation can be performed by injecting a carry into the right-most stage of Ac—since this stage can never receive a carry from any other source (there being no further positions to the right).

5.8.   In the light of 5.7 multiplication requires special care, because here the entire modulo 2 procedure breaks down. Indeed, assume that we want to compute a product xy, and that we had to change one of the factors, say x, by an integer multiple of 2, say by 2. Then the product $(x + 2)y$ obtains, and this differs from the desired xy by $2y$. $2y$, however, will not in general be an integer multiple of 2, since y is not in general an integer.

We will therefore begin our discussion of the multiplication by eliminating all such difficulties, and assume that both factors x, y lie between 0 and 1. Or, to be quite precise: $0 \leqq x < 1$, $0 \leqq y < 1$.

To effect such a multiplication we first send the multiplier x into a register AR, the *Arithmetic Register*, which is essentially just a set of 40 flip-flops whose characteristics will be discussed below. We place the multiplicand y in the *Selectron Register*, SR (cf. 4.9) and use the accumulator, Ac, to form and store the partial products. We propose to multiply the entire multiplicand by the successive digits of the multiplier in a serial fashion. There are, of course, two possible ways this can be done: We can either start with the digit in the lowest position—position 2^{-39}—or in the highest position—position 2^{-1}—and proceed successively to the left or right, respectively. There are a few advantages from our point of view in starting with the right-most digit of the multiplier. We therefore describe that scheme.

The multiplication takes place in 39 steps, which correspond to the 39 (non-sign) digits of the multiplier $x = 0, \xi_1, \xi_2, \ldots, \xi_{39} = (0.\xi_1\xi_2, \ldots, \xi_{39})$, enumerated backwards: $\xi_{39}, \ldots, \xi_2, \xi_1$. Assume that the $k - 1$ first steps ($k = 1, \ldots, 39$) have already taken place, involving multiplication of the multiplicand y with the $k - 1$ last digits of the multiplier: $\xi_{39}, \ldots, \xi_{41-k}$; and that we are now at the kth step, involving multiplication with the kth last digit: ξ_{40-k}. Assume furthermore, that Ac now contains the quantity p_{k-1}, the result of the $k - 1$ first steps. [This is the $(k - 1)$st partial product. For $k = 1$ clearly $p_0 = 0$.] We now form $2p_k = p_{k-1} + \xi_{40-k}y$, i.e.

$$2p_k = p_{k-1} + y_k, \qquad y_k \begin{cases} = 0 & \text{for} & \xi_{40-k} = 0 \\ = y & \text{for} & \xi_{40-k} = 1 \end{cases} \tag{1}$$

That is, we do nothing or add y, according to whether $\xi_{40-k} = 0$ or 1. We can then form p_k by halving $2p_k$.

Note that the addition of (1) produces no carry beyond the 2^0 position, i.e. the sign digit: $0 \leq p_h < 1$ is true for $h = 0$, and if it is true for $h = k - 1$, then (1) extends it to $h = k$ also, since $0 \leq y_k < 1$. Hence the sum in (1) is ≥ 0 and < 2, and no carries beyond the 2^0 position arise.

Hence p_k obtains from $2p_k$ by a simple right shift, which is combined with filling in the sign digit (that is freed by this shift) with a 0. This right shift is effected by an electronic shifter that is part of Ac.

Now

$$p_{39} = 2^{-1}\llbracket 2^{-1}[2^{-1}\{ \cdots (2^{-1}\xi_{39}y + \xi_{38}y) \cdots \} + \xi_2 y] + \xi_1 y \rrbracket$$
$$= \sum_{i=1}^{39} 2^{-i} \xi_i y = xy$$

Thus this process produces the product xy, as desired. Note that this xy is the exact product of x and y.

Since x and y are 39 digit binaries, their exact product xy is a 78 digit binary (we disregard the sign digit throughout). However, Ac will only hold 39 of these. These are clearly the left 39 digits of xy. The right 39 digits of xy are dropped from Ac one by one in the course of the 39 steps, or to be more specific, of the 39 right shifts. We will see later that these right 39 digits of xy should and will also be conserved (cf. the end of this section and the end of 5.12, as well as 6.6.3). The left 39 digits, which remain in Ac, should also be rounded off, but we will not discuss this matter here (cf. *loc. cit.* above and 9.9, Part II).

To complete the general picture of our multiplication technique we must consider how we sense the respective digits of our multiplier. There are two schemes which come to one's mind in this connection. One is to have a gate tube associated with each flip-flop of AR in such a fashion that this gate is open if a digit is 1 and closed if it is null. We would then need a 39-stage counter to act as a switch which would successively stimulate these gate tubes to react. A more efficient scheme is to build into AR a shifter circuit which enables AR to be shifted one stage to the right each time Ac is shifted and to sense the value of the digit in the rightmost flip-flop of AR. The shifter itself requires one gate tube per stage. We need in addition a counter to count out the 39 steps of the multiplication, but this can be achieved by a six stage binary counter. Thus the latter is more economical of tubes and has one additional virtue from our point of view which we discuss in the next paragraph.

The choice of 40 digits to a word (including the sign) is probably adequate for most computational problems but situations certainly might arise when we desire higher precision, i.e. words of greater length. A trivial illustration of this would be the computation of π to more places than are now known (about 700 decimals, i.e. about 2,300 binaries). More important instances are the solutions of N linear equations in N variables for large values of N. The extra precision becomes probably necessary when N exceeds a limit somewhere between 20 and 40. A justification of this estimate has to be based on a detailed theory of numerical matrix inversion which will be given in a subsequent report. It is therefore desirable to be able to handle numbers of $39k$ digits and signs by means of program instructions. One way to achieve this end is to use k words to represent a $39k$ digit number with signs. (In this way 39 digits in each 40 digit word are used, but all sign digits excepting the first one, are apparently wasted; cf. however the treatment of double precision numbers in Chapter 9, Part II.) It is, of course, necessary in this case to instruct the machine to perform the elementary operations of arithmetic in a manner that conforms with this interpretation of k-word complexes as single numbers. (Cf. 9.8–9.10, Part II.) In order to be able to treat numbers in this manner, it is desirable to keep not 39 digits in a product, but 78; this is discussed in more detail in 6.6.3 below. To accomplish this end (conserving 78 product digits) we connect, via our shifter circuit, the right-most digit of Ac with the left-most non-sign digit of AR. Thus, when in the process of multiplication a shift is ordered, the last digit of Ac is transferred into the place in AR made vacant when the multiplier was shifted.

5.9.   To conclude our discussion of the multiplication of positive numbers, we note this:

As described thus far, the multiplier forms the 78 digit product, xy, for a 39 digit multipler x and a 39 digit multiplicand y. We assumed $x \geq 0$, $y \geq 0$ and therefore had $xy \geq 0$, and we will only depart from these assumptions in 5.10. In addition to these, however, we also assumed $x < 1$, $y < 1$, i.e. the x, y have their binary points both immediately right of the sign digit, which implied the same for xy. One might question the necessity of these additional assumptions.

Prima facie they may seem mere conventions, which affect only the mathematician's interpretation of the functioning of the machine, and not a physical feature of the machine. (Cf. the corresponding situation in addition and subtraction, in 5.7.) Indeed, if x had its binary point between digits j and $j + 1$ from the left (cf. the discussion of 5.7 dealing with this j; it also applies to k below), and y between k and $k + 1$, then our above method of multiplication would still give the correct result xy, provided that the position of the binary point in xy is appropriately assigned. Specifically: Let the binary point of xy be between digits l and $l + 1$. x has the binary point between digits j and $j + 1$, and its sign digit is 0, hence its range is $0 \leqq x < 2^{j-1}$. Similarly y has the range $0 \leqq y < 2^{k-1}$, and xy has the range $0 \leqq xy < 2^{l-1}$. Now the ranges of x and y imply that the range of xy is necessarily $0 \leqq xy < 2^{j-1} 2^{k-1} = 2^{j+k-2}$. Hence $l = j + k - 1$. Thus it might seem that our actual positioning of the binary point—immediately right of the sign digit, i.e. $j = k = 1$—is still a mere convention.

It is therefore important to realize that this is not so: The choices of j and k actually correspond to very real, physical, engineering decisions. The reason for this is as follows: It is desirable to base the running of the machine on a sole, consistent mathematical interpretation. It is therefore desirable that all arithmetical operations be performed with an identically conceived positioning of the binary point in Ac. Applying this principle to x and y gives $j = k$. Hence the position of the binary point for xy is given by $j + k - 1 = 2j - 1$. If this is to be the same as for x, and y, then $2j - 1 = j$, i.e. $j = 1$ ensues—that is, our above positioning of the binary point immediately right of the sign digit.

There is one possible escape: To place into Ac not the left 39 digits of xy (not counting the sign digit 0), but the digits j to $j + 38$ from the left. Indeed, in this way the position of the binary point of xy will be $(2j - 1) - (j - 1) = j$, the same as for x and y.

This procedure means that we drop the left $j - 1$ and right $40 + j$ digits of xy and hold the middle 39 in Ac. Note that positioning of the binary point means that $x < 2^{j-1}$, $y < 2^{j-1}$ and xy can only be used if $xy < 2^{j-1}$. Now the assumptions secure only $xy < 2^{2j-2}$. Hence xy must be 2^{j-1} times smaller than it might be. This is just the thing which would be secured by the vanishing of the left $j - 1$ digits that we had to drop from Ac, as shown above.

If we wanted to use such a procedure, with those dropped left $j - 1$ digits really existing, i.e. with $j \neq 1$, then we would have to make physical arrangements for their conservation elsewhere. Also the general mathematical planning for the machine would be definitely complicated, due to the physical fact that Ac now holds a rather arbitrarily picked middle stretch of 39 digits from among the 78 digits of xy. Alternatively, we might fail to make such arrangements, but this would necessitate to see to it in the mathematical planning of each problem, that all products turn out to be 2^{j-1} times smaller than their *a priori* maxima. Such an observance is not at all impossible; indeed similar things are unavoidable for the other operations. [For example, with a factor 2 in addition (of positives) or subtraction (of opposite sign quantities). Cf. also the remarks in the first part of 5.12, dealing with

keeping "within range".] However, it involves a loss of significant digits, and the choice $j = 1$ makes it unnecessary in multiplication.

We will therefore make our choice $j = 1$, i.e. the positioning of the binary point immediately right of the sign digit, binding for all that follows.

5.10. We now pass to the case where the multiplier x and the multiplicand y may have either sign $+$ or $-$, i.e. any combination of these signs.

It would not do simply to extend the method of 5.8 to include the sign digits of x and y also. Indeed, we assume $-1 \leqq x < 1$, $-1 \leqq y < 1$, and the multiplication procedure in question is definitely based on the $\geqq 0$ interpretations of x and y. Hence if $x < 0$, then it is really using $x + 2$, and if $y < 0$, then it is really using $y + 2$. Hence for $x < 0$, $y \geqq 0$ it forms

$$(x + 2)y = xy + 2y$$

for $x \geqq 0$, $y < 0$ it forms

$$x(y + 2) = xy + 2x$$

for $x < 0$, $x < 0$, it forms

$$(x + 2)(y + 2) = xy + 2x + 2y + 4$$

or since things may be taken modulo 2, $xy + 2x + 2y$. Hence correction terms $-2y$, $-2x$ would be needed for $x < 0$, $y < 0$, respectively (either or both).

This would be a possible procedure, but there is one difficulty: As xy is formed, the 39 digits of the multiplier x are gradually lost from AR, to be replaced by the right 39 digits of xy. (Cf. the discussion at the end of 5.8.) Unless we are willing to build an additional 40 stage register to hold x, therefore, x will not be available at the end of the multiplication. Hence we cannot use it in the correction $2x$ of xy, which becomes necessary for $y < 0$.

Thus the case $x < 0$ can be handled along the above lines, but not the case $y < 0$.

It is nevertheless possible to develop an adequate procedure, and we now proceed to do this. Throughout this procedure we will maintain the assumptions $-1 \leqq x < 1$, $-1 \leqq y < 1$. We proceed in several successive steps.

First: Assume that the corrections necessitated by the possibility of $y < 0$ have been taken care of. We permit therefore $y \gtreqless 0$. We will consider the corrections necessitated by the possibility of $x < 0$.

Let us disregard the sign digit of x, which is 1, i.e. replace it by 0. Then x goes over into $x' = x - 1$ and as $-1 \leqq x < 0$, this x' will actually behave like $(x - 1) + 2 = x + 1$. Hence our multiplication procedure will produce $x'y = (x + 1)y = xy + y$, and therefore a correction $-y$ is needed at the end. (Note that we did not use the sign digit of x in the conventional way. Had we done so, then a correction $-2y$ would have been necessary, as seen above.)

We see therefore: Consider $x \gtreqless 0$. Perform first all necessary steps for forming $x'y(y \gtreqless 0)$, without yet reaching the sign digit of x (i.e. treating x as if it were ≥ 0). When the time arrives at which the digit ξ_0 of x has to become effective—i.e. immediately after ξ_1 became effective, after 39 shifts (cf. the discussion near the end of 5.8)—at which time Ac contains, say, $\bar{p}$ (this corresponds to the p_{39} of 5.8), then form

$$\bar{\bar{p}} \begin{cases} = \bar{p} & \text{if} \quad \xi_0 = 0 \\ = \bar{p} - y & \text{if} \quad \xi_0 = 1 \end{cases}$$

This $\bar{\bar{p}}$ is xy. (Note the difference between this last step, forming $\bar{\bar{p}}$, and the 39 preceding steps in 5.8, forming $p_1, p_2, \ldots, p_{39}$.)

Second: Having disposed of the possibility $x < 0$, we may now assume $x \geq 0$. With this assumption we have to treat all $y \gtreqless 0$. Since $y \geq 0$ brings us back entirely to the familiar case of 5.8, we need to consider the case $y < 0$ only.

Let y' be the number that obtains by disregarding the sign digit of y' which is 1, i.e. by replacing it by 0. Again y' acts not like $y - 1$, but like $(y - 1) + 2 = y + 1$. Hence the multiplication procedure of 5.8 will produce $xy' = x(y + 1) = xy + x$, and therefore a correction x is needed. (Note that, quite similarly to what we saw in the first case above, the suppression of the sign digit of y replaced the previously recognized correction $-2x$ by the present one $-x$.) As we observed earlier, this correction $-x$ cannot be applied at the end to the completed xy' since at that time x is no longer available. Hence we must apply the correction $-x$ digitwise, subtracting every digit at the time when it is last found in AR, and in a way that makes it effective with the proper positional value.

Third: Consider then $x = 0, \xi_1, \xi_2, \ldots, \xi_{39} = (\xi_1, \xi_2 \ldots \xi_{39})$. The 39 digits $\xi_1 \ldots \xi_{39}$ of x are lost in the course of the 39 shifts of the multiplication procedure of 5.8, going from right to left. Thus the operation No. $k + 1$ ($k = 0, 1, \ldots, 38$, cf. 5.8) finds ξ_{39-k} in the right-most stage of AR, uses it, and then loses it through its concluding right shift (of both Ac and AR). After this step $39 - (k + 1) = 38 - k$ further steps, i.e. shifts follow, hence before its own concluding shift there are still $39 - k$ shifts to come. Hence the positional values are 2^{39-k} times higher than they will be at the end. ξ_{39-k} should appear at the end, in the correcting term $-x$, with the sign $-$ and the positional value $2^{-(39-k)}$. Hence we may inject it during the step $k + 1$ (before its shift) with the

sign $-$ and the positional value 1. That is to say, $-\xi_{39-k}$ in the sign digit.

This, however, is inadmissible. Indeed, ξ_{39-k} might cause carries (if $\xi_{39-k} = 1$), which would have nowhere to go from the sign digit (there being no further positions to the left). This error is at its origin an integer multiple of 2, but the $39 - k$ subsequent shifts reduce its positional value 2^{39-k} times. Hence it might contribute to the end result any integer multiple of $2^{-(38-k)}$—and this is a genuine error.

Let us therefore add $1 - \xi_{39-k}$ to the sign digit, i.e. 0 or 1 if ξ_{39-k} is 1 or 0, respectively. We will show further below, that with this procedure there arise no carries of the inadmissible kind. Taking this momentarily for granted, let us see what the total effect is. We are correcting not by $-x$ but by $\sum_{i=1}^{39} 2^{-i} - x = 1 - 2^{-39} - x$. Hence a final correction by $-1 + 2^{-39}$ is needed. Since this is done at the end (after all shifts), it may be taken modulo 2. That is to say, we must add $1 + 2^{-39}$, i.e. 1 in each of the two extreme positions. Adding 1 in the right-most position has the same effect as in the discussion at the end of 5.7 (dealing with the subtraction). It is equivalent to injecting a carry into the right-most stage of Ac. Adding 1 in the left-most position, i.e. to the sign digit, produces a 1, since that digit was necessarily 0. (Indeed, the last operation ended in a shift, thus freeing the sign digit, cf. below.)

Fourth: Let us now consider the question of the carries that may arise in the 39 steps of the process described above. In order to do this, let us describe the kth step ($k = 1, \ldots, 39$), which is a variant of the kth step described for a positive multiplication in 5.8, in the same way in which we described the original kth step *loc. cit.* That is to say, let us see what the formula (1) of 5.8 has become. It is clearly $2p_k = p_{k-1} + (1 - \xi_{40-k}) + \xi_{40-k}y'$, i.e.

$$2p_k = p_{k-1} + y'_{k'} \qquad y'_k \begin{cases} = 1 & \text{for} \quad \xi_{40-k} = 0 \\ = y' & \text{for} \quad \xi_{40-k} = 1 \end{cases} \tag{2}$$

That is, we add 1 (y's sign digit) or y' (y without its sign digit), according to whether $\xi_{40-k} = 0$ or 1. Then p_k should obtain from $2p_k$ again by halving.

Now the addition of (2) produces no carries beyond the 2^0 position, as we asserted earlier, for the same reason as the addition of (1) in 5.8. We can argue in the same way as there: $0 \leq p_h < 1$ is true for $h = 0$, and if it is true for $h = k - 1$, then (1) extends it to $h = k$ also, since $0 \leq y'_k \leq 1$. Hence the sum in (2) is ≥ 0 and < 2, and no carries beyond the 2^0 position arise.

Fifth: In the three last observations we assumed $y < 0$. Let us now restore the full generality of $y \gtreqless 0$. We can then describe the equations (1) of 5.8 (valid for $y \geqq 0$) and (2) above (valid for $y < 0$) by a single formula,

$$2p_k = p_{k-1} + y_k''$$

$$y_k'' \begin{cases} = y\text{'s sign digit} & \text{for} & \xi_{40-k} = 0 \\ = y \text{ without its sign digit} & \text{for} & \xi_{40-k} = 1 \end{cases} \qquad (3)$$

Thus our verbal formulation of (2) applies here, too: We add y's sign digit or y without its sign, according to whether $\xi_{40-k} = 0$ or 1. All p_k are $\geqq 0$ and < 1, and the addition of (3) never originates a carry beyond the 2^0 position. p_k obtains from $2p_k$ by a right shift, filling the sign digit with a 0. (Cf. however, Part II, Table 2 for another sort of right shift that is desirable in explicit form, i.e. as an order.)

For $y \geqq 0$, xy is p_{39}, for $y < 0$, xy obtains from p_{39} by injecting a carry into the right-most stage of Ac and by placing a 1 into the sign digit in Ac.

Sixth: This procedure applies for $x \geqq 0$. For $x < 0$ it should also be applied, since it makes use of x's non-sign digits only, but at the end y must be subtracted from the result.

This method of binary multiplication will be illustrated in some examples in 5.15.

5.11.   To complete our discussion of the multiplicative organs of our machine we must return to a consideration of the types of accumulators mentioned in 5.5. The static accumulator operates as an adder by simultaneously applying static voltages to its two inputs—one for each of the two numbers being added. When steady-state operation is reached the total sum is formed complete with all carries. For such an accumulator the above discussion is substantially complete, except that it should be remarked that such a circuit requires at most 39 rise times to complete a carry. Actually it is possible that the duration of these successive rises is proportional to a lower power of 39 than the first one.

Each stage of a dynamic accumulator consists of a binary counter for registering the digit and a flip-flop for temporary storage of the carry. The counter receives a pulse if a 1 is to be added in at that place; if this causes the counter to go from 1 to 0 a carry has occurred and hence the carry flip-flop will be set. It then remains to perform the carries. Each flip-flop has associated with it a gate, the output of which is connected to the next binary counter to the left. The carry is begun by pulsing all carry gates. Now a carry may produce a carry, so that the process needs to be repeated until all carry flip-flops register 0. This can be detected by means of a circuit involving a sensing tube connected to each carry flip-flop. It was shown in 5.6 that, on the average, five pulse times (flip-flop reaction times) are required for the complete carry. An alternative scheme is to connect a gate

tube to each binary counter which will detect whether an incoming carry pulse would produce a carry and will, under this circumstance, pass the incoming carry pulse directly to the next stage. This circuit would require at most 39 rise times for the completion of the carry. (Actually less, cf. above.)

At the present time the development of a static accumulator is being concluded. From preliminary tests it seems that it will add two numbers in about 5 μsec and will shift right or left in about 1 μsec.

We return now to the multiplication operation. In a static accumulator we order simultaneously an addition of the multiplicand with sign deleted or the sign of the multiplicand (cf. 5.10) and a complete carry and then a shift for each of the 39 steps. In a dynamic accumulator of the second kind just described we order in succession an addition of the multiplicand with sign deleted or the sign of the multiplicand, a complete carry, and a shift for each of the 39 steps. In a dynamic accumulator of the first kind we can avoid losing the time required for completing the carry (in this case an average of 5 pulse times, cf. above) at each of the 39 steps. We order an addition by the multiplicand with sign deleted or the sign of the multiplicand, then order one pulsing of the carry gates, and finally shift the contents of both the digit counters and the carry flip-flops. This process is repeated 39 times. A simple arithmetical analysis which may be carried out in a later report, shows that at each one of these intermediate stages a single carry is adequate, and that a complete set of carries is needed at the end only. We then carry out the complement corrections, still without ever ordering a complete set of carry operations. When all these corrections are completed and after round-off, described below, we then order the complete carry mentioned above.

5.12.   It is desirable at this point in the discussion to consider rules for rounding-off to n-digits. In order to assess the characteristics of alternative possibilities for such properly, and in particular the role of the concept of "unbiasedness", it is necessary to visualize the conditions under which rounding-off is needed.

Every number x that appears in the computing machine is an approximation of another number x', which would have appeared if the calculation had been performed absolutely rigorously. The approximations to which we refer here are not those that are caused by the explicitly introduced approximations of the numerical-mathematical set-up, e.g. the replacement of a (continuous) differential equation by a (discrete) difference equation. The effect of such approximations should be evaluated mathematically by the person who plans the problem for the machine, and should not be a direct concern of the machine. Indeed, it has to be handled by a mathematician and cannot be handled by the machine, since its nature, complexity, and difficulty may be of any kind, depending upon the problem under consideration. The approximations which concern us here are these: Even the elementary operations of arithmetic, to which the mathematical approximation-formulation for the machine has to reduce the true (possibly transcendental) problem, are not rigorously executed by the machine. The machine deals with numbers of n digits, where n, no matter how large, has to be a fixed quantity. (We assumed for our machine 40 digits, including the sign, i.e. $n = 39$.) Now the sum and difference of two n-digit numbers are again n-digit numbers, but their product and quotient (in general) are not. (They have, in general, $2n$ or ∞-digits, respectively.) Consequently, multiplication and division must unavoidably be replaced by the machine by two different operations which must produce n-digits under all conditions, and which, subject to this limitation, should lie as close as possible to the results of the true multiplication and division. One might call them pseudo-multiplication and pseudo-division; however, the accepted nomenclature terms them as multiplication and division with round-off. (We are now creating the impression that addition and subtraction are entirely free of such shortcomings. This is only true inasmuch as they do not create new digits to the right, as multiplication and division do. However, they can create new digits to the left, i.e. cause the numbers to "grow out of range". This complication, which is, of course, well known, is normally met by the planner, by mathematical arrangements and estimates to keep the numbers "within range". Since we propose to have our machine deal with numbers between -1 and 1, multiplication can never cause them to "grow out of range". Division, of course, might cause this complication, too. The planner must therefore see to it that in every division the absolute value of the divisor exceeds that of the dividend.)

Thus the round-off is intended to produce satisfactory n-digit approximations for the product xy and the quotient x/y of two n-digit numbers. Two things are wanted of the round-off: (1) The approximation should be good, i.e. its variance from the "true" xy or x/y should be as small as practicable; (2) The approximation should be unbiased, i.e. its mean should be equal to the "true" xy or x/y.

These desiderata must, however, be considered in conjunction with some further comments. Specifically: (a) x and y themselves are likely to be the results of similar round-offs, directly or indirectly inherent, i.e. x and y themselves should be viewed as unbiased n-digit approximations of "true" x' and y' values; (b) by talking of "variances" and "means" we are introducing statistical concepts. Now the approximations which we are here considering are not really of a statistical nature, but are due to the peculiarities (from our point of view, inadequacies) of arithmetic and of digital representation, and are therefore actually rigorously and uniquely determined. It seems, however, in the present state of mathematical science, rather hopeless to try to deal with these matters rigorously. Furthermore, a certain statistical approach, while not truly justified, has always given adequate practical results. This consists of treating those digits which one does not wish to use individually in subsequent calculations as random variables with equiprobable digital values, and of treating any two such digits as statistically independent (unless this is patently false).

These things being understood, we can now undertake to discuss round-off procedures, realizing that we will have to apply them to the multiplication and to the division.

Let $x = (.\xi_1 \ldots \xi_n)$ and $y = (.\eta_1 \ldots \eta_n)$ be unbiased approximations of x' and y'. Then the "true" $xy = (.\xi_1 \ldots \xi_n\xi_{n+1} \ldots \xi_{2n})$ and the "true" $x/y = (.\omega_1 \ldots \omega_n\omega_{n+1}\omega_{n+2} \ldots)$ (this goes on *ad infinitum*!) are approximations of $x'y'$ and x'/y'. Before we discuss how to round them off, we must know whether the "true" xy and x/y are themselves unbiased approximations of $x'y'$ and x'/y'. xy is indeed an unbiased approximation of $x'y'$, i.e. the mean of xy is the mean of $x(= x')$ times the mean of $y(= y')$, owing to the independence assumption which we made above. However, if x and y are closely correlated, e.g. for $x = y$, i.e. for squaring, there is a bias. It is of the order of the mean square of $x - x'$, i.e. of the variance of x. Since x has n digits, this variance is about $1/2^{2n}$ (If the digits of x', beyond n are entirely unknown, then our original assumptions give the variance $1/12.2^{2n}$.) Next, x/y can be written as $x.y^{-1}$, and since we have already discussed the bias of the product, it suffices now to consider the reciprocal y^{-1}. Now if y is an unbiased estimate of y', then y^{-1} is not an unbiased estimate of y'^{-1}, i.e. the mean of y's reciprocal is not the reciprocal of y's mean. The difference is $\sim y^{-3}$ times the variance of y, i.e. it is of essentially the same order as the bias found above in the case of squaring.

It follows from all this that it is futile to attempt to avoid biases of the order of magnitude $1/2^{2n}$ or less. (The factor $1/12$ above may seem to be changing the order of magnitude in question. However, it is really the square root of the variance which matters and $\sqrt{(1/12)} \sim 0.3$ is a moderate factor.) Since we propose to use $n = 39$, therefore $1/2^{78}(\sim 3 \times 10^{-24})$ is the critical case. Note that this possible bias level is $1/2^{39}(\sim 2 \times 10^{-12})$ times our last significant digit. Hence we will look for round-off rules to n digits for the "true" $xy = (.\xi_1 \ldots \xi_n\xi_{n+1} \ldots \xi_{2n})$ and $x/y = (.\omega_1 \ldots \omega_n\omega_{n+1}\omega_{n+2} \ldots)$. The desideratum (1) which we formulated previously, that the variance should be small, is still valid. The desideratum (2), however, that the bias should be zero, need, according to the above, only be enforced up to terms of the order $1/2^{2n}$.

The round-off procedures, which we can use in this connection, fall into two broad classes. The first class is characterized by its ignoring all digits beyond the nth, and even the nth digit itself, which it replaces by a 1. The second class is characterized by the procedure of adding one unit in the $(n + 1)$st digit, performing the carries which this may induce, and then keeping only the n first digits.

When applied to a number of the form $(.\nu_1 \ldots \nu_n\nu_{n+1}\nu_{n+2} \ldots)$ (*ad infinitum*!), the effects of either procedure are easily estimated. In the first case we may say we are dealing with $(.\nu_1, \ldots, \nu_{n-1})$ plus a random number of the form $(.0 \ldots, 0\nu_n\nu_{n+1}\nu_{n+2} \ldots)$, i.e. random in the interval $0, 1/2^{n-1}$. Comparing with the rounded off $(.\nu_1\nu_2 \ldots \nu_{n-1}1)$, we therefore have a difference random in the interval $-1/2^n, 1/2^n$. Hence its mean is 0 and its variance $\frac{1}{3} \cdot 2^{2n}$. In the second case we are dealing with $(.\nu_1 \ldots \nu_n)$ plus a random number of the form $(.0 \ldots 00\nu_{n+1}\nu_{n+2} \ldots)$, i.e. random in the interval $0, 1/2^n$. The "rounded-off" value will be $(.\nu_1 \ldots \nu_n)$ increased by 0 or by $1/2^n$, according to whether the random number in question lies in the interval $0, 1/2^{n+1}$, or in the interval $1/2^{n+1}$, $1/2^n$. Hence comparing with the "rounded-off" value, we have a difference random in the intervals $0, 1/2^{n+1}$, and $0, -1/2^{n+1}$, i.e. in the interval $-1/2^{n+1}, 1/2^{n+1}$. Hence its mean is 0 and its variance $(\frac{1}{12})2^{2n}$.

If the number to be rounded-off has the form $(.\nu_1 \ldots \nu_n\nu_{n+1}\nu_{n+2} \ldots \nu_{n+p})$ (p finite), then these results are somewhat affected. The order of magnitude of the variance remains the same; indeed for large p even its relative change is negligible. The mean difference may deviate from 0 by amounts which are easily estimated to be of the order $1/2^n \cdot 1/2^p = 1/2^{n+p}$.

In division we have the first situation, $x/y = (.\omega_1 \ldots \omega_n\omega_{n+1}\omega_{n+2} \ldots)$, i.e. p is infinite. In multiplication we have the second one, $xy = (.\xi_1 \ldots \xi_n\xi_{n+1} \ldots \xi_{2n})$, i.e. $p = n$. Hence for the division both methods are applicable without modification. In multiplication a bias of the order of $1/2^{2n}$ may be introduced. We have seen that it is pointless to insist on removing biases of this size. We will therefore use the unmodified methods in this case, too.

It should be noted that the bias in the case of multiplication can be removed in various ways. However, for the reasons set forth above, we shall not complicate the machine by introducing such corrections.

Thus we have two standard "round-off" methods, both unbiased to the extent to which we need this, and with the variances

$1/3 \cdot 2^{2n}$, and $(\frac{1}{12})2^{2n}$, that is, with the dispersions $(1/\sqrt{3})(1/2^n)$ $= 0.58$ times the last digit and $(1/2\sqrt{3})(1/2^n) = 0.29$ times the last digit. The first one requires no carry facilities, the second one requires them.

Inasmuch as we propose to form the product $x'y'$ in the accumulator, which has carry facilities, there is no reason why we should not adopt the rounding scheme described above which has the smaller dispersion, i.e. the one which may induce carries. In the case, however, of division we wish to avoid schemes leading to carries since we expect to form the quotient in the arithmetic register, which does not permit of carry operations. The scheme which we accordingly adopt is the one in which ω_n is replaced by 1. This method has the decided advantage that it enables us to write down the approximate quotient as soon as we know its first $(n - 1)$ digits. It will be seen in 5.14 and 6.6.4 below that our procedure for forming the quotient of two numbers will always lead to a result that is correctly rounded in accordance with the decisions just made. We do not consider as serious the fact that our rounding scheme in the case of division has a dispersion twice as large as that in multiplication since division is a far less frequent operation.

A final remark should be made in connection with the possible, occasional need of carrying more than $n = 39$ digits. Our logical control is sufficiently flexible to permit treating k ($=2, 3, \ldots$) words as one number, and thus effecting $n = 39k$. In this case the round-off has to be handled differently, cf. Chapter 9, Part II. The multiplier produces all 78 digits of the basic 39 by 39 digit multiplication: The first 39 in the Ac, the last 39 in the AR. These must then be manipulated in an appropriate manner. (For details, cf. 6.6.3 and 9.9–9.10, Part II.) The divider works for 39 digits only: In forming x/y, it is necessary, even if x and y are available to $39k$ digits, to use only 39 digits of each, and a 39 digit result will appear. It seems most convenient to use this result as the first step of a series of successive approximations. The successive improvements can then be obtained by various means. One way consists of using the well known iteration formula (cf. 5.4). For $k = 2$ one such step will be needed, for $k = 3, 4$, two steps, for $k = 5, 6, 7, 8$ three steps, etc. An alternative procedure is this: Calculate the remainder, using the approximate, 39 digit, quotient and the complete, $39k$ digit, divisor and dividend. Divide this again by the approximate, 39 digit, divisor, thus obtaining essentially the next 39 digits of the quotient. Repeat this procedure until the full $39k$ desired digits of the quotient have been obtained.

5.13. We might mention at this time a complication which arises when a floating binary point is introduced into the machine. The operation of addition which usually takes at most $\frac{1}{10}$ of a multiplication time becomes much longer in a machine with floating binary since one must perform shifts and round-offs as well as additions. It would seem reasonable in this case to place the time of an addition as about $\frac{1}{3}$ to $\frac{1}{2}$ of a multiplication. At this rate it is clear that the number of additions in a problem is as important a factor in the total solution time as are the number of multiplications. (For further details concerning the floating binary point, cf. 6.6.7.)

5.14. We conclude our discussion of the arithmetic unit with a description of our method for handling the division operation. To perform a division we wish to store the dividend in SR, the partial remainder in Ac and the partial quotient in AR. Before proceeding further let us consider the so-called *restoring* and *non-restoring* methods of division. In order to be able to make certain comparisons, we will do this for a general base $m = 2, 3, \ldots$.

Assume for the moment that divisor and dividend are both positive. The ordinary process of division consists of subtracting from the partial remainder (at the very beginning of the process this is, of course, the dividend) the divisor, repeating this until the former becomes smaller than the latter. For any fixed positional value in the quotient in a well-conducted division this need be done at most $m - 1$ times. If, after precisely $k = 0, 1, \ldots, m - 1$ repetitions of this step, the partial remainder has indeed become less than the divisor, then the digit k is put in the quotient (at the position under consideration), the partial remainder is shifted one place to the left, and the whole process is repeated for the next position, etc. Note that the above comparison of sizes is only needed at $k = 0, 1, \ldots, m - 2$, i.e. before step 1 and after steps $1, \ldots, m - 2$. If the value $k = m - 1$, i.e. the point after step $m - 1$, is at all reached in a well-conducted division, then it may be taken for granted without any test, that the partial remainder has become smaller than the divisor, and the operations on the position under consideration can therefore be concluded. (In the binary system, $m = 2$, there is thus only one step, and only one comparison of sizes, before this step.) In this way this scheme, known as the *restoring* scheme, requires a maximum of $m - 1$ comparisons and utilizes the digits $0, 1, \ldots, m - 1$ in each place in the quotient. The difficulty of this scheme for machine purposes is that usually the only economical method for comparing two numbers as to size is to subtract one from the other. If the partial remainder r_n were less than the dividend d, one would then have to add d back into $r_n - d$ in order to restore the remainder. Thus at every stage an unnecessary operation would be performed. A more symmetrical scheme is obtained by *not restoring*. In this method (from here on we need not assume the positivity of divisor and dividend)

one compares the signs of r_n and d; if they are of the same sign, the dividend is repeatedly subtracted from the remainder until the signs become opposite; if they are opposite, the dividend is repeatedly added to the remainder until the signs again become like. In this scheme the digits that may occur in a given place in the quotient are evidently $\pm 1, \pm 2, \ldots, \pm (m - 1)$, the positive digits corresponding to subtractions and the negative ones to additions of the dividend to the remainder.

Thus we have $2(m - 1)$ digits instead of the usual m digits. In the decimal system this would mean 18 digits instead of 10. This is a redundant notation. The standard form of the quotient must therefore be restored by subtracting from the aggregate of its positive digits the aggregate of its negative digits. This requires carry facilities in the place where the quotient is stored.

We propose to store the quotient in AR, which has no carry facilities. Hence we could not use this scheme if we were to operate in the decimal system.

The same objection applies to any base m for which the digital representation in question is redundant—i.e. when $2(m - 1) > m$. Now $2(m - 1) > m$ whenever $m > 2$, but $2(m - 1) = m$ for $m = 2$. Hence, with the use of a register which we have so far contemplated, this division scheme is certainly excluded from the start unless the binary system is used.

Let us now investigate the situation in the binary system. We inquire if it is possible to obtain a quasi-quotient by using the non-restoring scheme and by using the digits 1, 0 instead of 1, -1. Or rather we have to ask this question: Does this quasi-quotient bear a simple relationship to the true quotient?

Let us momentarily assume this question can be answered affirmatively and describe the division procedure. We store the divisor initially in Ac, the dividend in SR and wish to form the quotient in AR. We now either add or subtract the contents of SR into Ac, according to whether the signs in Ac and SR are opposite or the same, and insert correspondingly a 0 or 1 in the right-hand place of AR. We then shift both Ac and AR one place left, with electronic shifters that are parts of these two aggregates.

At this point we interrupt the discussion to note this: multiplication required an ability to shift right in both Ac and AR (cf. 5.8). We have now found that division similarly requires an ability to shift left in both Ac and AR. Hence both organs must be able to shift both ways electronically. Since these abilities have to be present for the implicit needs of multiplication and division, it is just as well to make use of them explicitly in the form of explicit orders. These are the orders 20, 21 of Table 1, and of Table 2, Part II. It will, however, turn out to be convenient to arrange some details in the shifts, when they occur explicitly under the control of those orders, differently from when they occur implicitly under the control of a multiplication or a division. (For these things, cf. the discussion of the shifts near the end of 5.8 and in the third remark below on one hand, and in the third remark in 7.2, Part II, on the other hand.)

Let us now resume the discussion of the division. The process described above will have to be repeated as many times as the number of quotient digits that we consider appropriate to produce in this way. This is likely to be 39 or 40; we will determine the exact number further below.

In this process we formed digits $\xi'_i = 0$ or 1 for the quotient, when the digit should actually have been $\xi_i = -1$ or 1, with $\xi'_i = 2\xi'_i - 1$. Thus we have a difference between the true quotient z (based on the digits ξ_i) and the quasi-quotient z' (based on the digits ξ'_i), but at the same time a one-to-one connection. It would be easy to establish the algebraical expression for this connection between z' and z directly, but it seems better to do this as part of a discussion which clarifies all other questions connected with the process of division at the same time.

We first make some general remarks:

First: Let x be the dividend and y the divisor. We assume, of course, $-1 \leq x < 1$, $-1 \leq y < 1$. It will be found that our present process of division is entirely unaffected by the signs of x and y, hence no further restrictions on that score are required.

On the other hand, the quotient $z = x/y$ must also fulfil $-1 \leq z < 1$. It seems somewhat simpler although this is by no means necessary, to exclude for the purposes of this discussion $z = -1$, and to demand $|z| < 1$. This means in terms of the dividend x and the divisor y that we exclude $x = -y$ and assume $|x| < y$.

Second: The division takes place in n steps, which correspond to the n digits $\xi'_1, \ldots, \xi'_n$ of the pseudo-quotient z', n being yet to be determined (presumably 39 or 40). Assume that the $k - 1$ first steps ($k = 1, \ldots, n$) have already taken place, having produced the $k - 1$ first digits: $\xi'_1, \ldots, \xi'_{k-1}$; and that we are now at the kth step, involving production of the kth digit: ξ'_k. Assume furthermore, that Ac now contains the quantity r_{k-1}, the result of the $k - 1$ first steps. (This is the $(k - 1)$st partial remainder. For $k = 1$ clearly $r_0 = x$.) We then form $r_k = 2r_{k-1} \mp y$, according to whether the signs of r_{k-1} and y do or do not agree, i.e.

$$r_k = 2r_{k-1} \boxplus y$$

$$\boxplus \begin{cases} \text{is } - \text{ if the signs of } r_{k-1} \text{ and } y \text{ do agree} \\ \text{is } + \text{ if the signs of } r_{k-1} \text{ and } y \text{ do not agree} \end{cases}$$

Let us now see what carries may originate in this procedure. We can argue as follows: $|r_h| < |y|$ is true for $h = 0 (|r_0| =$

$|x| < |y|)$, and if it is true for $h = k - 1$, then (4) extends it to $h = k$ also, since r_{k-1} and $\boxplus y$ have opposite signs. The last point may be elaborated a little further: because of the opposite signs

$$|r_k| = 2|r_{k-1}| - |y| < 2|y| - |y| = |y|$$

Hence we have always $|r_k| < |y|$, and therefore *a fortiori* $|r_k| < 1$, i.e. $-1 < r_k < 1$.

Consequently in equation (4) one summand is necessarily > -2, < 2, the other is ≥ 1, < 1, and the sum is > -1, < 1. Hence we may carry out the operations of (4) modulo 2, disregarding any possibilities of carries beyond the 2^0 position, and the resulting r_k will be automatically correct (in the range > -1, < 1).

Third: Note however that the sign of r_{k-1}, which plays an important role in (4) above, is only then correctly determinable from the sign digit, if the number from which it is derived is ≥ -1, < 1. (Cf. the discussion in 5.7.) This requirement however is met, as we saw above, by r_{k-1}, but not necessarily by $2r_{k-1}$. Hence the sign of r_{k-1} (i.e. its sign digit) as required by (4), must be sensed before r_{k-1} is doubled.

This being understood, the doubling of r_{k-1} may be performed as a simple left shift, in which the left-most digit (the sign digit) is allowed to be lost—this corresponds to the disregarding of carries beyond the 2^0 position, which we recognized above as being permissible in (4). (Cf. however, Part II, Table 2, for another sort of left shift that is desirable in explicit form, i.e. as an order.)

Fourth: Consider now the precise implication of (4) above. $\xi'_k = 1$ or 0 corresponds to $\boxplus = -$ or $+$, respectively. Hence (4) may be written

$$r_k = 2r_{k-1} + (1 - 2\xi'_k)y$$

i.e.

$$2^{-k}r_k = 2^{-(k-1)}r_{k-1} + (2^{-k} - 2^{-(k-1)}\xi'_k)y$$

Summing over $k = 1, \ldots, n$ gives

$$2^{-n}r_n = x + \left\{ (1 - 2^{-n}) - \sum_{k=1}^{n} 2^{-(k-1)}\xi'_k \right\} y$$

i.e.

$$x = \left(-1 + \sum_{k=1}^{n} 2^{-(k-1)}\xi'_k + 2^{-n} \right) y + 2^{-n}r_n$$

This makes it clear, that $\bar{z} = -1 + \sum_{k=1}^{n} 2^{-(k-1)}\xi'_k + 2^{-n}$ corresponds to true quotient $z = x/y$ and $2^{-n}r_n$, with an absolute value $< 2^{-n}|y| \leq 2^{-n}$, to the remainder. Hence, if we disregard the term -1 for a moment $\xi'_1, \xi'_2, \ldots, \xi'_n, 1$ are the $n + 1$ first digits of what may be used as a true quotient, the sign digit being part of this sequence.

Fifth: If we do not wish to get involved in more complicated round-off procedures which exceed the immediate capacity of the only available adder Ac, then the above result suggests that we should put $n + 1 = 40, n = 39$. The $\xi'_1, \ldots, \xi'_{39}$ are then 39 digits of the quotient, including the sign digit, but not including the right-most digit.

The right-most digit is taken care of by placing a 1 into the right-most stage of Ac.

At this point an additional argument in favor of the procedure that we have adopted here becomes apparent. The procedure coincides (without a need for any further corrections) with the second round-off procedure that we discussed in 5.12.

There remains the term -1. Since this applies to the final result, and no right shifts are to follow, carries which might go beyond the 2^0 position may be disregarded. Hence this amounts simply to changing the sign digit of the quotient $\bar{z}$: replacing 0 or 1 by 1 or 0, respectively.

This concludes our discussion of the division scheme. We wish, however, to re-emphasize two very distinctive features which it possesses:

First: This division scheme applies equally for any combinations of signs of divisor and dividend. This is a characteristic of the non-restoring division schemes, but it is not the case for any simple known multiplication scheme. It will be remembered, in particular, that our multiplication procedure of 5.9 had to contain special correcting steps for the cases where either or both factors are negative.

Second: This division scheme is practicable in the binary system only; it has no analog for any other base.

This method of binary division will be illustrated on some examples in 5.15.

5.15. We give below some illustrative examples of the operations of binary arithmetic which were discussed in the preceding sections.

Although it presented no difficulties or ambiguities, it seems best to begin with an example of addition.

Binary notation		Decimal notation (fractional form)
Augend	0.010110011	179/512
Addend	0.011010111	215/512
Sum	0.110001010	394/512
(Carries)	1111 111	

In what follows we will not show the carries any more.

We form the negative of a number (cf. 5.7):

Binary notation		Decimal notation (fractional form)
	0.101110100	372/512
Complement:	1.010001011	
	1	
	1.010001100	-1 $+140/512$

A subtraction (cf. 5.7):

Binary notation		Decimal notation (fractional form)
Subtrahend	0.011010111	215/512
Minuend	0.110001010	394/512
Complement of subtrahend	1.100101000	
	1	-1 $+297/512$
Difference	0.010110011	179/512

Some multiplications (cf. 5.8 and 5.9):

Binary notation		Decimal notation (fractional form)
Multiplicand............................	0.101	5/8
Multiplier................................	0.011	3/8
	0101	
	0101	
	0	
Product	0.001111	15/64

Binary notation		Decimal notation (fractional form)
Multiplicand............................	1.101	−3/8
Multiplier................................	1.011	−5/8
	0101	
	0101	
	1	
	.101111	
Correction 1†	1 1	
	1.110111	
Correction 2‡ (Complement of the multiplicand).	0.010	
	1	
	0.001111	15/64

A division (cf. 5.14):

Binary notation		Q.D.§	Decimal notation (fractional form)
Divisor....................................	1.011000		−5/8
Dividend	0.001111		15/64
	0.011110	0	
	1.011000		
	1.110110		
	1.101100	1	
	0.100111		
	1		
	0.010100		
	0.101000	0	
	1.011000		
	0.000000		
	0.000000	0	
	1.011000		
	1.011000		
	0.110000	1	
	0.100111		
	1		
	1.011000		
	. . . . . .	1	
Quotient (uncorrected).....................	0.10011		
" (corrected)......................	1.100111		−1 + 39/64 = −25/64

† For the sign of the multiplicand.   ‡ For the sign of the multiplier.   § Quotient digit.

Note that this deviates by $\frac{1}{64}$, i.e. by one unit of the right-most position, from the correct result $-\frac{3}{8}$. This is a consequence of our round-off rule, which forces the right-most digit to be 1 under all conditions. This occasionally produces results with unfamiliar and even annoying aspects (e.g. when quotients like $0:y$ or $y:y$ are formed), but it is nevertheless unobjectionable and self-consistent on the basis of our general principles.

6. The control

6.1. It has already been stated that the computer will contain an organ, called the control, which can automatically execute the orders stored in the Selectrons. Actually, for a reason stated in 6.3, the orders for this computer are less than half as long as a forty binary digit number, and hence the orders are stored in the Selectron memory in pairs.

Let us consider the routine that the control performs in directing a computation. The control must know the location in the Selectron memory of the pair of orders to be executed. It must direct the Selectrons to transmit this pair of orders to the Selectron register and then to itself. It must then direct the execution of the operation specified in the first of the two orders. Among these orders we can immediately describe two major types: An order of the first type begins by causing the transfer of the number, which is stored at a specified memory location, from the Selectrons to the Selectron register. Next, it causes the arithmetical unit to perform some arithmetical operations on this number (usually in conjunction with another number which is already in the arithmetical unit), and to retain the resulting number in the arithmetical unit. The second type order causes the transfer of the number, which is held in the arithmetical unit, into the Selectron register, and from there to a specified memory location in the Selectrons. (It may also be that this latter operation will permit a direct transfer from the arithmetical unit into the Selectrons.) An additional type of order consists of the transfer orders of 3.5. Further orders control the inputs and the outputs of the machine. The process described at the beginning of this paragraph must then be repeated with the second order of the order pair. This entire routine is repeated until the end of the problem.

6.2. It is clear from what has just been stated that the control must have a means of switching to a specified location in the Selectron memory, for withdrawing both numbers for the computation and pairs of orders. Since the Selectron memory (as tentatively planned) will hold $2^{12} = 4,096$ forty-digit words (a word is either a number or a pair of orders), a twelve-digit binary number suffices to identify a memory location. Hence a switching mecha-nism is required which will, on receiving a twelve-digit binary number, select the corresponding memory location.

The type of circuit we propose to use for this purpose is known as a decoding or many-one function table. It has been developed in various forms independently by J. Rajchman [Rajchman, 1943] and P. Crawford [Crawford, 19??]. It consists of n flip-flops which register an n-digit binary number. It also has a maximum of 2^n output wires. The flip-flops activate a matrix in which the interconnections between input and output wires are made in such a way that one and only one of 2^n output wires is selected (i.e. has a positive voltage applied to it). These interconnections may be established by means of resistors or by means of non-linear elements (such as diodes or rectifiers); all these various methods are under investigation. The Selectron is so designed that four such function table switches are required, each with a three digit entry and eight (2^3) outputs. Four sets of eight wires each are brought out of the Selectron for switching purposes, and a particular location is selected by making one wire positive with respect to the remainder. Since all forty Selectrons are switched in parallel, these four sets of wires may be connected directly to the four function table outputs.

6.3. Since most computer operations involve at least one number located in the Selectron memory, it is reasonable to adopt a code in which twelve binary digits of every order are assigned to the specification of a Selectron location. In those orders which do not require a number to be taken out of or into the Selectrons these digit positions will not be used.

Though it has not been definitely decided how many operations will be built into the computer (i.e. how many different orders the control must be able to understand), it will be seen presently that there will probably be more than 2^5 but certainly less than 2^6. For this reason it is feasible to assign 6 binary digits for the order code. It thus turns out that each order must contain eighteen binary digits, the first twelve identifying a memory location and the remaining six specifying an operation. It can now be explained why orders are stored in the memory in pairs. Since the same memory organ is to be used in this computer for both orders and numbers, it is efficient to make the length of each about equivalent. But numbers of eighteen binary digits would not be sufficiently accurate for problems which this machine will solve. Rather, an accuracy of at least 10^{-10} or 2^{-33} is required. Hence it is preferable to make the numbers long enough to accommodate two orders.

As we pointed out in 2.3, and used in 4.2 *et seq.* and 5.7 *et seq.*, our numbers will actually have 40 binary digits each. This allows 20 binary digits for each order, i.e. the 12 digits that specify a memory location, and 8 more digits specifying the nature of the operation (instead of the minimum of 6 referred to above). It is convenient, as will be seen in 6.8.2. and Chapter 9, Part II, to group these binary digits into *tetrads,* groups of 4 binary digits. Hence a whole word consists of 10 tetrads, a half word or order of 5 tetrads, and of these 3 specify a memory location and the remaining 2 specify the nature of the operation. Outside the machine each tetrad can be expressed by a base 16 digit. (The base 16 digits are best designated by symbols of the 10 decimal digits 0 to 9, and 6 additional symbols, e.g. the letters a to f. Cf. Chapter 9, Part II.) These 16 characters should appear in the typing for and the printing from the machine. (For further details of these arrangements, cf. *loc. cit.* above.)

The specification of the nature of the operation that is involved in an order occurs in binary form, so that another many-one or decoding function is required to decode the order. This function table will have six input flip-flops (the two remaining digits of the order are not needed). Since there will not be 64 different orders, not all 64 outputs need be provided. However, it is perhaps worthwhile to connect the outputs corresponding to unused order possibilities to a checking circuit which will give an indication whenever a code word unintelligible to the control is received in the input flip-flops.

The function table just described energizes a different output wire for each different code operation. As will be shown later, many of the steps involved in executing different orders overlap. (For example, addition, multiplication, division, and going from the Selectrons to the register all include transferring a number from the Selectrons to the Selectron register.) For this reason it is perhaps desirable to have an additional set of control wires, each of which is activated by any particular combination of different code digits. These may be obtained by taking the output wires of the many-one function table and using them to operate tubes which will in turn operate a one-many (or coding) function table. Such a function table consists of a matrix as before, but in this case only one of the input wires are activated. This particular table may be referred to as the recoding function table.

The twelve flip-flops operating the four function tables used in selecting a Selectron position, and the six flip-flops operating the function table used for decoding the order, are referred to as the *Function Table Register,* FR.

6.4.   Let us consider next the process of transferring a pair of orders from the Selectrons to the control. These orders first go into SR. The order which is to be used next may be transferred directly into FR. The second order of the pair must be removed from SR (since SR may be used when the first order is executed), but cannot as yet be placed in FR. Hence a temporary storage

is provided for it. The storage means is called the *Control Register,* CR, and consists of 20 (or possibly 18) flip-flops, capable of receiving a number from SR and transmitting a number to FR.

As already stated (6.1), the control must know the location of the pair of orders it is to get from the Selectron memory. Normally this location will be the one following the location of the two orders just executed. That is, until it receives an order to do otherwise, the control will take its orders from the Selectrons in sequence. Hence the order location may be remembered in a twelve stage binary counter (one capable of counting 2^{12}) to which one unit is added whenever a pair of orders is executed. This counter is called the *Control Counter,* CC.

The details of the process of obtaining a pair of orders from the Selectron are thus as follows: The contents of CC are copied into FR, the proper Selectron location is selected, and the contents of the Selectrons are transferred to SR. FR is then cleared, and the contents of SR are transferred to it and CR. CC is advanced by one unit so the control will be prepared to select the next pair of orders from the memory. (There is, however, an exception from this last rule for the so-called transfer orders, cf. 3.5. This may feed CC in a different manner, cf. the next paragraph below.) First the order in FR is executed and then the order in CR is transferred to FR and executed. It should be noted that all these operations are directed by the control itself—not only the operations specified in the control words sent to FR, but also the automatic operations required to get the correct orders there.

Since the method by means of which the control takes order pairs in sequence from the memory has been described, it only remains to consider how the control shifts itself from one sequence of control orders to another in accordance with the operations described in 3.5. The execution of these operations is relatively simple. An order calling for one of these operations contains the twelve digit specification of the position to which the control is to be switched, and these digits will appear in the left-hand twelve flip-flops of FR. All that is required to shift the control is to transfer the contents of these flip-flops to CC. When the control goes to the Selectrons for the next pair of orders it will then go to the location specified by the number so transferred. In the case of the unconditional transfer, the transfer is made automatically; in the case of the conditional transfer it is made only if the sign counter of the Accumulator registers zero.

6.5.   In this report we will discuss only the general method by means of which the control will execute specific orders, leaving the details until later. It has already been explained (5.5) that when a circuit is to be designed to accomplish a particular elementary operation (such as addition), a choice must be made between a static type and a dynamic type circuit. When the design of the control is considered, this same choice arises. The function of the control is to direct a sequence of operations which take place in the various circuits of the computer (including the circuits of the control itself). Consider what is involved in directing an operation. The control must signal for the operation to begin, it must supply whatever signals are required to specify that particular operation, and it must in some way know when the operation has been completed so that it may start the succeeding operation. Hence the control circuits must be capable of timing the operations. It should be noted that timing is required whether the circuit performing the operation is static or dynamic. In the case of a static type circuit the control must supply static control signals for a period of time sufficient to allow the output voltages to reach the steady-state condition. In the case of a dynamic type circuit the control must send various pulses at proper intervals to this circuit.

If all circuits of a computer are static in character, the control timing circuits may likewise be static, and no pulses are needed in the system. However, though some of the circuits of the computer we are planning will be static, they will probably not all be so, and hence pulses as well as static signals must be supplied by the control to the rest of the computer. There are many advantages in deriving these pulses from a central source, called the *clock*. The timing may then be done either by means of counters counting clock pulses or by means of electrical delay lines (an RC circuit is here regarded as a simple delay line). Since the timing of the entire computer is governed by a single pulse source, the computer circuits will be said to operate as a synchronized system.

The clock plays an important role both in detecting and in localizing the errors made by the computer. One method of checking which is under consideration is that of having two identical computers which operate in parallel and automatically compare each other's results. Both machines would be controlled by the same clock, so they would operate in absolute synchronism. It is not necessary to compare every flip-flop of one machine with the corresponding flip-flop of the other. Since all numbers and control words pass through either the Selectron register or the accumulator soon before or soon after they are used, it suffices to check the flip-flops of the Selectron register and the flip-flops of the accumulator which hold the number registered there; in fact, it seems possible to check the accumulator only (cf. the end of 6.6.2). The checking circuit would stop the clock whenever a difference appeared, or stop the machine in a more direct manner if an asynchronous system is used. Every flip-flop of each computer will be located at a convenient place. In fact, all neons will be located on one panel, the corresponding neons of the two machines being placed in parallel rows so that one can tell at a glance (after the machine has been stopped) where the discrepancies are.

The merits of any checking system must be weighed against its cost. Building two machines may appear to be expensive, but since most of the cost of a scientific computer lies in development rather than production, this consideration is not so important as it might seem. Experience may show that for most problems the two machines need not be operated in parallel. Indeed, in most cases purely mathematical, external checks are possible: Smoothness of the results, behavior of differences of various types, validity of suitable identities, redundant calculations, etc. All of these methods are usually adequate to disclose the presence or absence of error *in toto*; their drawback is only that they may not allow the detailed diagnosing and locating of errors at all or with ease. When a problem is run for the first time, so that it requires special care, or when an error is known to be present, and has to be located—only then will it be necessary as a rule, to use both machines in parallel. Thus they can be used as separate machines most of the time. The essential feature of such a method of checking lies in the fact that it checks the computation at every point (and hence detects transient errors as well as steady-state ones) and stops the machine when the error occurs so that the process of localizing the fault is greatly simplified. These advantages are only partially gained by duplicating the arithmetic part of the computer, or by following one operation with the complement operation (multiplication by division, etc.), since this fails to check either the memory or the control (which is the most complicated, though not the largest, part of the machine).

The method of localizing errors, either with or without a duplicate machine, needs further discussion. It is planned to design all the circuits (including those of the control) of the computer so that if the clock is stopped between pulses the computer will retain all its information in flip-flops so that the computation may proceed unaltered when the clock is started again. This principle has already demonstrated its usefulness in the ENIAC. This makes it possible for the machine to compute with the clock operating at any speed below a certain maximum, as long as the clock gives out pulses of constant shape regardless of the spacing between pulses. In particular, the spacing between pulses may be made indefinitely large. The clock will be provided with a mode of operation in which it will emit a single pulse whenever instructed to do so by the operator. By means of this, the operator can cause the machine to go through an operation step by step, checking the results by means of the indicating-lamps connected to the flip-flops. It will be noted that this design principle does not exclude the use of delay lines to obtain delays as long as these are only used to time the constituent operations of a single step, and have no part in determining the machine's operating repetition rate. Timing coincidences by means of delay lines is excluded since this requires a constant pulse rate.

6.6.   The orders which the control understands may be divided into two groups: Those that specify operations which are performed within the computer and those that specify operations involved in getting data into and out of the computer. At the present time the internal operations are more completely planned than the input and output operations, and hence they will be discussed more in detail than the latter (which are treated briefly in 6.8). The internal operations which have been tentatively adopted are listed in Table 1. It has already been pointed out that not all of these operations are logically basic, but that many can be programmed by means of others. In the case of some of these operations the reasons for building them into the control have already been given. In this section we will give reasons for building the other operations into the control and will explain in the case of each operation what the control must do in order to execute it.

In order to have the precise mathematical meaning of the symbols which are introduced in what follows clearly in mind, the reader should consult the table at the end of the report for each new symbol, in addition to the explanations given in the text.

6.6.1.   Throughout what follows $S(x)$ will denote the memory location No. x in the Selectron. Accordingly the x which appears in $S(x)$ is a 12-digit binary, in the sense of 6.2. The eight addition operations $[S(x) \rightarrow Ac+$, $S(x) \rightarrow Ac-$, $S(x) \rightarrow Ah+$, $S(x) \rightarrow Ah-$, $S(x) \rightarrow Ac + M$, $S(x) \rightarrow Ac - M$, $S(x) \rightarrow Ah + M$, $S(x) \rightarrow Ah - M]$ involves the following possible four steps:

First: Clear SR and transfer into it the number at $S(x)$.

Second: Clear Ac if the order contains the symbol c; do not clear Ac if the order contains the symbol h.

Third: Add the number in SR or its negative (i.e. in our present system its complement with respect to 2^1) into Ac. If the order does not contain the symbol M, use the number in SR or its negative according to whether the order contains the symbol $+$ or $-$. If the order contains the symbol M, use the number in SR or its negative according to whether the sign of the number in SR and the symbol $+$ or $-$ in the order do or do not agree.

Fourth: Perform a complete carry. Building the last four addition operations (those containing the symbol M) into the control is fairly simple: It calls only for one extra comparison (of the sign in SR and the $+$ or $-$ in the order, cf. the third step above), and it requires, therefore, only a few tubes more than required for the first four addition operations (those not containing the symbol M).

These facts would seem of themselves to justify adding the operations in question: plus and minus the absolute value. But it should be noted that these operations can be programmed out of the other operations of Table 1 with correspondingly few orders (three for absolute value and five for minus absolute value), so that some further justification for building them in is required. The absolute value order is frequently in connection with the orders L and R (see 6.6.7), while the minus absolute value order makes the detection of a zero very simple by merely detecting the sign of $-|N|$. (If $-|N| \geqq 0$, then $N = 0$.)

6.6.2.   The operation of $S(x) \rightarrow R$ involves the following two steps:

First: Clear SR, and transfer $S(x)$ to it.

Second: Clear AR and add the number in the Selectron register into it. The operation of $R \rightarrow Ac$ merits more detailed discussion, since there are alternative ways of removing numbers from AR. Such numbers could be taken directly to the Selectrons as well as into Ac, and they could be transferred to Ac in parallel, in sequence, or in sequence parallel. It should be recalled that while most of the numbers that go into AR have come from the Selectrons and thus need not be returned to them, the result of a division and the right-hand 39 digits of a product appear in AR. Hence while an operation for withdrawing a number from AR is required, it is relatively infrequent and therefore need not be particularly fast. We are therefore considering the possibility of transferring at least partially in sequence and of using the shifting properties of Ac and of AR for this. Transferring the number to the Selectron via the accumulator is also desirable if the dual machine method of checking is employed, for it means that even if numbers are only checked in their transit through the accumulator, nevertheless every number going into the Selectron is checked before being placed there.

6.6.3.   The operation $S(x) \times R \rightarrow Ac$ involves the following six steps:

First: Clear SR and transfer $S(x)$ (the multiplicand) into it.

Second: Thirty-nine steps, each of which consist of the two following parts: (a) Add (or rather shift) the sign digit of SR into the partial product in Ac, or add all but the sign digit of SR into the partial product in Ac—depending upon whether the right-most digit in AR is 0 or 1—and effect the appropriate carries. (b) Shift Ac and AR to the right, fill the sign digit of Ac with a 0 and the digit of AR immediately right of the sign digit (positional value 2^{-1}) with the previously right-most digit of Ac. (There are ways to save time by merging these two operations when the right-most digit in Ar is 0, but we will not discuss them here more fully.)

Third: If the sign digit in SR is 1 (i.e. $-$), then inject a carry into the right-most stage of Ac and place a 1 into the sign digit of Ac.

Fourth: If the original sign digit of AR is 1 (i.e. −), then subtract the contents of SR from Ac.

Fifth: If a partial carry system was employed in the main process, then a complete carry is necessary at the end.

Sixth: The appropriate round-off must be effected. (Cf. Chapter 9, Part II, for details, where it is also explained how the sign digit of the Arithmetic register is treated as part of the round-off process.)

It will be noted that since any number held in Ac at the beginning of the process is gradually shifted into AR, it is impossible to accumulate sums of products in Ac without storing the various products temporarily in the Selectrons. While this is undoubtedly a disadvantage, it cannot be eliminated without constructing an extra register, and this does not at this moment seem worthwhile.

On the other hand, saving the right-hand 39 digits of the answer is accomplished with very little extra equipment, since it means connecting the 2^{-39} stage of Ac to the 2^{-1} stage of AR during the shift operation. The advantage of saving these digits is that it simplifies the handling of numbers of any number of digits in the computer (cf. the last part of 5.12). Any number of $39k$ binary digits (where k is an integer) and sign can be divided into k parts, each part being placed in a separate Selectron position. Addition and subtraction of such numbers may be programmed out of a series of additions or subtractions of the 39-digit parts, the carry-over being programmed by means of $Cc \rightarrow S(x)$ and $Cc' \rightarrow S(x)$ operations. (If the 2^0 stage of Ac registers negative after the addition of two 39-digit parts, a carry-over has taken place and hence 2^{-39} must be added to the sum of the next parts.) A similar procedure may be followed in multiplication if all 78 digits of the product of the two 39-digit parts are kept, as is planned. (For the details, cf. Chapter 9, Part II.) Since it would greatly complicate the computer to make provision for holding and using a 78 digit dividend, it is planned to program $39k$ digit division in one of the ways described at the end of 5.12.

6.6.4.   The operation of division $Ac \div S(x) \rightarrow R$ involves the following four steps:

First: Clear SR and transfer $S(x)$ (the divisor) into it.

Second: Clear AR.

Third: Thirty-nine steps, each of which consists of the following three parts: (a) Sense the signs of the contents of Ac (the partial remainder) and of SR, and sense whether they agree or not. (b) Shift Ac and AR left. In this process the previous sign digit of Ac is lost. Fill the right-most digit of Ac (after the shift) with a 0, and the right-most digit of AR (before the shift) with 0 or 1,

depending on whether there was disagreement or agreement in (a). (c) Add or subtract the contents of SR into Ac, depending on the same alternative as above.

Fourth: Fill the right-most digit of AR with a 1, and change its sign digit.

For the purpose of timing the 39 steps involved in division a six-stage counter (capable of counting to $2^6 = 64$) will be built into the control. This same counter will also be used for timing the 39 steps of multiplication, and possibly for controlling Ac when a number is being transferred between it and a tape in either direction (see 6.8.).

6.6.5.   The three substitution operations $[At \rightarrow S(x), Ap \rightarrow S(x),$ and $Ap' \rightarrow S(x)]$ involve transferring all or part of the number held in Ac into the Selectrons. This will be done by means of gate tubes connected to the registering flip-flops of Ac. Forty such tubes are needed for the total substitutions, $At \rightarrow S(x)$. The partial substitution $Ap \rightarrow S(x)$ and $Ap' \rightarrow S(x)$ requires that the left-hand twelve digits of the number held in Ac be substituted in the proper places in the left-hand and right-hand orders, respectively. This may be done by means of extra gate tubes, or by shifting the number in Ac and using the gate tubes required for $At \rightarrow S(x)$. (This scheme needs some additional elaboration, when the order directing and the order suffering the substitution are the two successive halves of the same word; i.e. when the latter is already in FR at the time when the former becomes operative in CR, so that the substitution effected in the Selectrons comes too late to alter the order which has already reached CR, to become operative at the next step in FR. There are various ways to take care of this complication, either by some additional equipment or by appropriate prescriptions in coding. We will not discuss them here in more detail, since the decisions in this respect are still open.)

The importance of the partial substitution operations can hardly be overestimated. It has already been pointed out (3.3) that they allow the computer to perform operations it could not otherwise conveniently perform, such as making use of a function table stored in the Selectron memory. Furthermore, these operations remove a very sizeable burden from the person coding problems, for they make possible the coding of classes of problems in contrast to coding each individual problem separately. Because $Ap \rightarrow S(x)$ and $Ap' \rightarrow S(x)$ are available, any program sequence may be stated in general form (that is, without Selectron location designations for the numbers being operated on) and the Selectron locations of the numbers to be operated on substituted whenever that sequence is used. As an example, consider a general code for nth order integration of m total differential equations for p steps of independent variable t, formulated in advance. Whenever a problem requiring this rule is coded for the computer, the general integration sequence can be inserted into the statement of the problem along with coded instructions for telling the sequence where it will be located in the memory [so that the proper $S(x)$ designations will be inserted into such orders as $Cu \rightarrow S(x)$, etc.]. Whenever this sequence is to be used by the computer it will automatically substitute the correct values of m, n, p and Δt, as well as the locations of the boundary conditions and the descriptions of the differential equations, into the general sequence. (For the details of this particular procedure, cf. Chapter 13, Part II.) A library of such general sequences will be built up, and facilities provided for convenient insertion of any of these into the coded statement of a problem (cf. 6.8.4). When such a scheme is used, only the distinctive features of a problem need be coded.

6.6.6. The manner in which the control shift operations $[Cu \rightarrow S(x), Cu' \rightarrow S(x), Cc \rightarrow S(x),$ and $Cc' \rightarrow S(x)]$ are realized has been discussed in 6.4 and needs no further comment.

6.6.7. One basic question which must be decided before a computer is built is whether the machine is to have a so-called floating binary (or decimal) point. While a floating binary point is undoubtedly very convenient in coding problems, building it into the computer adds greatly to its complexity and hence a choice in this matter should receive very careful attention. However, it should first be noted that the alternatives ordinarily considered (building a machine with a floating binary point vs. doing all computation with a fixed binary point) are not exhaustive and hence that the arguments generally advanced for the floating binary point are only of limited validity. Such arguments overlook the fact that the choice with respect to any particular operation (except for certain basic ones) is not between building it into the computer and not using it at all, but rather between building it into the computer and programming it out of operations built into the computer. (One short reference to the floating binary point was made in 5.13.)

Building a floating binary point into the computer will not only complicate the control but will also increase the length of a number and hence increase the size of the memory and the arithmetic unit. Every number is effectively increased in size, even though the floating binary point is not needed in many instances. Furthermore, there is considerable redundancy in a floating binary point type of notation, for each number carries with it a scale factor, while generally speaking a single scale factor will suffice for a possibly extensive set of numbers. By means of the operations already described in the report a floating binary point can be programmed. While additional memory capacity is needed for this, it is probably less than that required by a built-in floating binary

point since a different scale factor does not need to be remembered for each number.

To program a floating binary point involves detecting where the first zero occurs in a number in Ac. Since Ac has shifting facilities this can best be done by means of them. In terms of the operations previously described this would require taking the given number out of Ac and performing a suitable arithmetical operation on it: For a (multiple) right shift a multiplication, for a (multiple) left shift either one division, or as many doublings (i.e. additions) as the shift has stages. However, these operations are inconvenient and time-consuming, so we propose to introduce two operations (L and R) in order that this (i.e. the single left and right shift) can be accomplished directly. These operations make use of facilities already present in Ac and hence add very little equipment to the computer. It should be noted that in many instances a single use of L and possibly of R will suffice in programming a floating binary point. For if the two factors in a multiplication have no superfluous zeros, the product will have at most one superfluous zero (if $\frac{1}{2} \leqq X < 1$ and $\frac{1}{2} \leqq Y < 1$, then $\frac{1}{4} \leqq XY < 1$). This is similarly true in division (if $\frac{1}{4} \leqq X < \frac{1}{2}$ and $\frac{1}{2} \leqq Y < 1$, then $\frac{1}{4} < X/Y < 1$). In addition and subtraction any numbers growing out of range can be treated similarly. Numbers which decrease in these cases, i.e. develop a sequence of zeros at the beginning, are really (mathematically) losing precision. Hence it is perfectly proper to omit formal readjustments in this event. (Indeed, such a true loss of precision cannot be obviated by any formal procedure, but, if at all, only by a different mathematical formulation of the problem.)

6.7. Table 1 shows that many of the operations which the control is to execute have common elements. Thus addition, subtraction, multiplication and division all involve transferring a number from the Selectrons to SR. Hence the control may be simplified by breaking some of the operations up into more basic ones. A timing circuit will be provided for each basic operation, and one or more such circuits will be involved in the execution of an order. The exact choice of basic operations will depend upon how the arithmetic unit is built.

In addition to the timing circuits needed for executing the orders of Table 1, two such circuits are needed for the automatic operations of transferring orders from the Selectron register to CR and FR, and for transferring an order from CR to FR. In normal computer operation these two circuits are used alternately, so a binary counter is needed to remember which is to be used next. In the operations $Cu' \rightarrow S(x)$ and $Cc \rightarrow S(x)$ the first order of a pair is ignored, so the binary counter must be altered accordingly.

The execution of a sequence of orders involves using the various

Table 1

	Symbolization		
	Complete	Abbreviated	Operation
1	$S(x) \rightarrow Ac+$	x	Clear accumulator and add number located at position x in the Selectrons into it.
2	$S(x) \rightarrow Ac-$	$x-$	Clear accumulator and subtract number located at position x in the Selectrons into it.
3	$S(x) \rightarrow AcM$	xM	Clear accumulator and add absolute value of number located at position x in the Selectrons into it.
4	$S(x) \rightarrow Ac - M$	$x - M$	Clear accumulator and subtract absolute value of number located at position x in the Selectrons into it.
5	$S(x) \rightarrow Ah+$	xh	Add number located at position x in the Selectrons into the accumulator.
6	$S(x) \rightarrow Ah-$	$xh-$	Subtract number located at position x in the Selectrons into the accumulator.
7	$S(x) \rightarrow AhM$	xhM	Add absolute value of number located at position x in the Selectrons into the accumulator.
8	$S(x) \rightarrow Ah - M$	$x - hM$	Subtract absolute value of number located at position x in the Selectrons into the accumulator.
9	$S(x) \rightarrow R$	xR	Clear register† and add number located at position x in the Selectrons into it.
10	$R \rightarrow A$	A	Clear accumulator and shift number held in register into it.
11	$S(x) \times R \rightarrow A$	xX	Clear accumulator and multiply the number located at position x in the Selectrons by the number in the register, placing the left-hand 39 digits of the answer in the accumulator and the right-hand 39 digits of the answer in the register.
12	$A \div S(x) \rightarrow R$	$x \div$	Clear register and divide the number in the accumulator by the number located in position x of the Selectrons, leaving the remainder in the accumulator and placing the quotient in the register.
13	$Cu \rightarrow S(x)$	xC	Shift the control to the left-hand order of the order pair located at position x in the Selectrons.
14	$Cu' \rightarrow S(x)$	xC'	Shift the control to the right-hand order of the order pair located at position x in the Selectrons.
15	$Cc \rightarrow S(x)$	xCc	If the number in the accumulator is ≥ 0, shift the control as in $Cu \rightarrow S(x)$.
16	$Cc' \rightarrow S(x)$	xCc'	If the number in the accumulator is ≥ 0, shift the control as in $Cu' \rightarrow S(x)$.
17	$At \rightarrow S(x)$	xS	Transfer the number in the accumulator to position x in the Selectrons.
18	$Ap \rightarrow S(x)$	xSp	Replace the left-hand 12 digits of the left-hand order located at position x in the Selectrons by the left-hand 12 digits in the accumulator.
19	$Ap' \rightarrow S(x)$	xSp'	Replace the left-hand 12 digits of the right-hand order located at position x in the Selectrons by the left-hand 12 digits in the accumulator.
20	L	L	Multiply the number in the accumulator by 2, leaving it there.
21	R	R	Divide the number in the accumulator by 2, leaving it there.

† Register means arithmetic register.

timing circuits in sequence. When a given timing circuit has completed its operation, it emits a pulse which should go to the timing circuit to be used next. Since this depends upon the particular operation being executed, these pulses are routed according to the signals received from the decoding and recoding function tables activated by the six binary digits specifying an order.

6.8. In this section we will consider what must be added to the control so that it can direct the mechanisms for getting data into and out of the computer and also describe the mechanisms themselves. Three different kinds of input-output mechanisms are planned.

First: Several magnetic wire storage units operated by servo-mechanisms controlled by the computer.

Second: Some viewing tubes for graphical portrayal of results.

Third: A typewriter for feeding data directly into the computer, not to be confused with the equipment used for preparing and printing from magnetic wires. As presently planned the latter will consist of modified Teletypewriter equipment, cf. 6.8.2 and 6.8.4.

6.8.1. Since there already exists a way of transferring numbers between the Selectrons and Ac, therefore Ac may be used for transferring numbers from and to a wire. The latter transfer will be done serially and will make use of the shifting facilities of Ac. Using Ac for this purpose eliminates the possibility of computing and reading from or writing on the wires simultaneously. However, simultaneous operation of the computer and the input–output organ requires additional temporary storage and introduces a synchronizing problem, and hence it is not being considered for the first model.

Since, at the beginning of the problem, the computer is empty, facilities must be built into the control for reading a set of numbers from a wire when the operator presses a manual switch. As each number is read from a wire into Ac, the control must transfer it to its proper location in the Selectrons. The CC may be used to count off these positions in sequence, since it is capable of transmitting its contents to FR. A detection circuit on CC will stop the process when the specified number of numbers has been placed in the memory, and the control will then be shifted to the orders located in the first position of the Selectron memory.

It has already been stated that the entire memory facilities of the wires should be available to the computer without human intervention. This means that the control must be able to select the proper set of numbers from those going by. Hence additional orders are required for the code. Here, as before, we are faced with two alternatives. We can make the control capable of executing an order of the form: Take numbers from positions p to $p + s$ on wire No. k and place them in Selectron locations v to $v + s$. Or we can make the control capable of executing some less complicated operations which, together with the already given control orders, are sufficient for programming the transfer operation of the first alternative. Since the latter scheme is simpler we adopt it tentatively.

The computer must have some way of finding a particular number on a wire. One method of arranging for this is to have each number carry with it its own location designation. A method more economical of wire memory capacity is to use the Selectron memory facilities to remember the position of each wire. For example, the computer would hold the number t_1 specifying which number on the wire is in position to be read. If the control is instructed to read the number at position p_1 on this wire, it will compare p_1 with t_1; and if they differ, cause the wire to move in the proper direction. As each number on the wire passes by, one unit is added or subtracted to t_1 and the comparison repeated. When $p_1 = t_1$ numbers will be transferred from the wire to the accumulator and then to the proper location in the memory. Then both t_1 and p_1 will be increased by 1, and the transfer from the wire to accumulator to memory repeated. This will be iterated, until $t_1 + s$ and $p_1 + s$ are reached, at which time the control will direct the wire to stop.

Under this system the control must be able to execute the following orders with regard to each wire: Start the wire forward, start the wire in reverse, stop the wire, transfer from wire to Ac,

and transfer from Ac to wire. In addition, the wire must signal the control as each digit is read and when the end of a number has been reached. Conversely, when recording is done the control must have a means of timing the signals sent from Ac to the wire, and of counting off the digits. The 2^6 counter used for multiplication and division may be used for the latter purpose, but other timing circuits will be required for the former.

If the method of checking by means of two computers operating simultaneously is adopted, and each machine is built so that it can operate independently of the other, then each will have a separate input-output mechanism. The process of making wires for the computer must then be duplicated, and in this way the work of the person making a wire can be checked. Since the wire servomechanisms cannot be synchronized by the central clock, a problem of synchronizing the two computers when the wires are being used arises. It is probably not practical to synchronize the wire feeds to within a given digit, but this is unnecessary since the numbers coming into the two organs Ac need not be checked as the individual digits arrive, but only prior to being deposited in the Selectron memory.

6.8.2.   Since the computer operates in the binary system, some means of decimal-binary and binary-decimal conversions is highly desirable. Various alternative ways of handling this problem have been considered. In general we recognize two broad classes of solutions to this problem.

First: The conversion problems can be regarded as simple arithmetic processes and programmed as sub-routines out of the orders already incorporated in the machine. The details of these programs together with a more complete discussion are given fully in Chapter 9, Part II, where it is shown, among other things, that the conversion of a word takes about 5 msec. Thus the conversion time is comparable to the reading or withdrawing time for a word—about 2 msec—and is trivial as compared to the solution time for problems to be handled by the computer. It should be noted that the treatment proposed there presupposes only that the decimal data presented to or received from the computer are in tetrads, each tetrad being the binary coding of a decimal digit—the information (precision) represented by a decimal digit being actually equivalent to that represented by 3.3 binary digits. The coding of decimal digits into tetrads of binary digits and the printing of decimal digits from such tetrads can be accomplished quite simply and automatically by slightly modified Teletype equipment, cf. 6.8.4 below.

Second: The conversion problems can be regarded as unique problems and handled by separate conversion equipment incorporated either in the computer proper or associated with the mechanisms for preparing and printing from magnetic wires. Such converters are really nothing other than special purpose digital computers. They would seem to be justified only for those computers which are primarily intended for solving problems in which the computation time is small compared to the input-output time, to which class our computer does not belong.

6.8.3.   It is possible to use various types of cathode ray tubes, and in particular Selectrons for the viewing tubes, in which case programming the viewing operation is quite simple. The viewing Selectrons can be switched by the same function tables that switch the memory Selectrons. By means of the substitution operation $Ap \rightarrow S(x)$ and $Ap' \rightarrow S(x)$, six-digit numbers specifying the abscissa and ordinate of the point (six binary digits represent a precision of one part in $2^6 = 64$, i.e. of about 1.5 per cent which seems reasonable in such a component) can be substituted in this order, which will specify that a particular one of the viewing Selectrons is to be activated.

6.8.4.   As was mentioned above, the mechanisms used for preparing and printing from wire for the first model, at least, will be modified Teletype equipment. We are quite fortunate in having secured the full cooperation of the Ordnance Development Division of the National Bureau of Standards in making these modifications and in designing and building some associated equipment.

By means of this modified Teletype equipment an operator first prepares a checked paper tape and then directs the equipment to transfer the information from the paper tape to the magnetic wire. Similarly a magnetic wire can transfer its contents to a paper tape which can be used to operate a teletypewriter. (Studies are being undertaken to design equipment that will eliminate the necessity for using paper tapes.)

As was shown in 6.6.5, the statement of a new problem on a wire involves data unique to that problem interspersed with data found on previously prepared paper tapes or magnetic wires. The equipment discussed in the previous paragraph makes it possible for the operator to combine conveniently these data on to a single magnetic wire ready for insertion into the computer.

It is frequently very convenient to introduce data into a computation without producing a new wire. Hence it is planned to build one simple typewriter as an integral part of the computer. By means of this typewriter the operator can stop the computation, type in a memory location (which will go to the FR), type in a number (which will go to Ac and then be placed in the first mentioned location), and start the computation again.

6.8.5.   There is one further order that the control needs to execute. There should be some means by which the computer can signal to the operator when a computation has been concluded, or when the computation has reached a previously determined point. Hence an order is needed which will tell the computer to stop and to flash a light or ring a bell.

References

BurkA62a, BurkA62b; CrawP??; GoldH63a, b, c, d; RajcJ43.

Chapter 5

The DEC PDP-8

Introduction[1]

The PDP-8 is a single-address, 12-bit-word computer of the second generation. It is designed for task environments with minimum arithmetic computing and small Mp requirements. For example, it can be used to control laboratory devices, such as gas chromotographs or sampling oscilloscopes. Together with special T's, it is programmed to be a laboratory instrument, such as a pulse height analyzer or a spectrum analyzer. These applications are typical of the laboratory and process control requirements for which the machine was designed. As another example, it can serve as a message concentrator by controlling telephone lines to which typewriters and Teletypes are attached. The computer occasionally stands alone as a small-scale general-purpose computer. Most recently it was introduced as a small-scale general-purpose time-sharing system, based on work at Carnegie-Mellon University and DEC. It is used as a KT(display) when it has a P(display; '338); this C is discussed in Chap. 25. The PDP-8 has achieved a production status formerly reserved for IBM computers; about 5,000 have been constructed.

PDP-8 differs from the character-oriented 8-bit computer in Chap. 10; it is not unlike the 16-bit computers, such as the IBM 1800 in Chap. 33. The PDP-8 is typical of several 12-bit computers: the early CDC-160 series (1960), CDC-6600 Peripheral and Control Processor (Chap. 39), the SDS-92, M.I.T. Lincoln Laboratory's Laboratory Instrument Computer LINC (1963), Washington University's Programmed Console (1967), and the SCC 650 (1966).

The PDP-5 (transistor, 1963), PDP-8 (1965), PDP-8/S (serial, 1966) and PDP-8/I (integrated circuit, 1968), PDP-8/L (integrated circuit, 1968) constitute a series of computers based on evolving technology. All of these have identical ISP's. Their PMS structures are nearly identical, and all components other than Pc and Mp are compatible throughout the series. The LINC-8-338 PMS structure is presented in Fig. 1. A cost performance tradeoff took place in the PDP-8 (parallel-by-word arithmetic) and PDP-8/S (serial-by-bit arithmetic) implementations. A PDP-8/S is one-fifteenth of a PDP-8 at one-half the cost. The performance factors can be attributed to 8/1.5 or 5.3 for Mp speed and a factor of about 3 for logical organization, even though the same 2-megahertz logic clock is used in both cases. The PDP-8 is about 6.7 times a PDP-5.

[1]The initials in the title stand for Digital Equipment Corporation Programmed Data Processor.

The ISP of the PDP-8 Pc is about the most trivial in the book. It has only a few data operators, namely, ←, +, − (negate), ¬, ∧, / 2, × 2, (optional) ×, /, and normalize. It operates on words, integers, and boolean vectors. However, there are microcoded instructions, which allow compound instructions to be formed in a single instruction.

The computer is straightforward and illustrates the levels discussed in Chap. 1. We can easily look at it from the "top down." The C in PMS notation is

```
C('PDP-8; technology:transistors; 12 b/w;
    descendants:'PDP-8/S, 'PDP-8/I, 'PDP-8/L;
    antecedents: 'PDP-5;
    Mp(core; #0:7; 4096 w; tc:1.5 μs/w);
    Pc(Mps(2 ~ 4 w);
        instruction length:1 | 2 w
        address/instruction:1;
        operations on data/od:(←, +, ¬, ∧, −(negate), × 2,
        / 2, +1)
        optional operations:(×, /, normalize);
        data-types:word, integer, boolean vector;
        operations for data access:4);
    P(display; '338);
    P(c; 'LINC);
    S('I/O Bus; 1 Pc; 64 K);
    Ms(disk, 'DECtape, magnetic tape);
    T(paper tape, card, analog, cathode-ray tube))
```

ISP

The ISP is presented in Appendix 1 of this chapter (including the optional Extended Arithmetic Element/EAE). The 2^{12}-word Mp is divided into 32 fixed-length pages of 128 words each. Address calculation is based on references to the first page, Page_0, or to the current page of the Program Counter/PC. The effective-address calculation procedure provides for both direct and indirect reference to either the current page or the first page. This scheme allows a 7-bit address to specify local page addresses.

A 2^{15}-word Mp is available on the PDP-8, but addressing greater than 2^{12} words is comparatively inefficient. In the extended range, two 3-bit registers, the Program Field and Data Field Registers, select which of the eight 2^{12}-word blocks are being actively addressed as program and data.

There is an array of eight registers, called the Auto_index registers, which resides in Page_0. This array (Auto_index[0: 11]⟨0:7⟩:= M[10_8:17_8]⟨0:11⟩) possesses the useful property that whenever an indirect reference is made to it, a 1 is first added

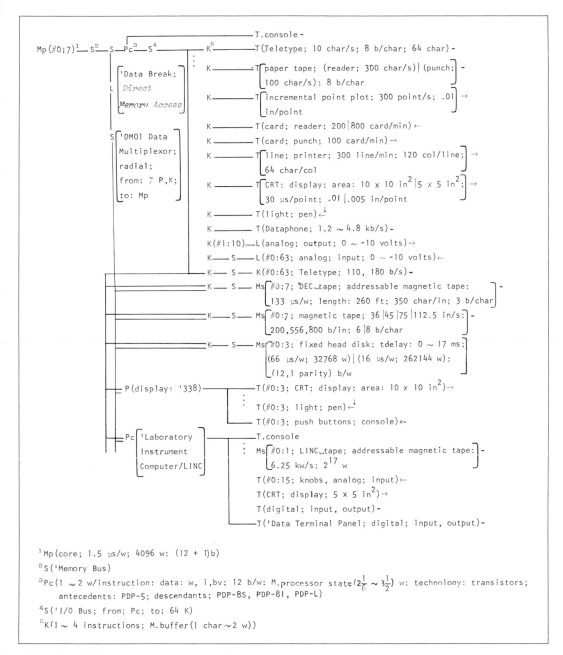

Fig. 1. DEC LINC-8-338 PMS diagram.

to its contents. (That is, there is a side effect to referencing.) Thus, address integers in the register can select the next member of a vector or string for accessing.

The instruction-set-execution definition can also be presented as a decoding diagram or tree (Fig. 2). Here, each block represents an encoding of bits in the instruction word. A decoding diagram allows one more descriptive dimension than the conventional, linear ISP description, revealing the assignment of bits to the instruction. Figure 2 still requires ISP descriptions for Mp, Mps, the instruction execution, the effective-address calculation, and the interpreter. Diagrams such as Fig. 2 are useful in the ISP

design to determine which instruction numbers are to be assigned to names and operations and instructions which are free to be assigned (or encoded).

There are eight basic instructions encoded by 3 bits, that is $op\langle 0:2\rangle := i\langle 0:2\rangle$, where instruction/$i\langle 0:11\rangle$. Each of the first six instructions (where $0 \leq op < 6$) have the 4 address operand determination modes (thus yielding essentially 24 instructions). The first six instructions are:

data transmission: deposit and clear-accumulator/dca
two's complement add to the accumulator/tad

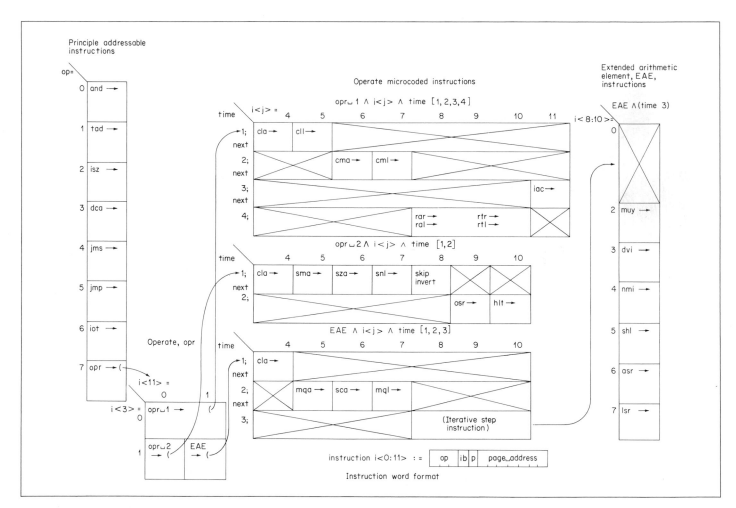

Fig. 2. DEC PDP-8 instruction-decoding diagram.

binary arithmetic: two's complement add to the accumulator/tad

binary boolean: and to the accumulator/and

program control: jump/set program counter/jmp
jump to subroutine/jms
index memory and skip if results are zero/isz

Note that the add instruction, tad, is used for both data transmission and arithmetic.

The subroutine-calling instruction, jms, provides a method for transferring a link to the beginning (or head) of the subroutine. In this way arguments can be accessed indirectly, and a return is executed by a jump indirect instruction to the location storing the returned address. This straightforward subroutine-call mechanism, although inexpensive to implement, requires reentrant and recursive subroutine calls to be interpreted by software, rather than by hardware. A stack, as in the DEC 338 (Chap. 25), would be nicer.

The input_output instruction/iot $(:= \text{op} = 6)$ uses the remaining 9 bits of the instruction to specify instructions to input/output devices. The 6 io_select bits select 1 of 64 devices. The 3 bits, io_p1_bit, io_p2_bit, io_p4_bit, command the selected device by conditionally providing three pulses in sequence. The instructions to a typical io device are:

io_p1_bit $\to$ (IO_skip_flag[io select] $\to$ (PC $\leftarrow$ PC + 1))
testing a condition of an IO device output to a device input from a device

io_p4_bit $\to$ (Output_data[io select] $\leftarrow$ AC)

io_p2_bit $\to$ (AC $\leftarrow$ Input_data[io select])

There are three microcoded instruction groups selected by op = 7. The instruction decoding diagram (Fig. 2) and the ISP description (Appendix 1 of this chapter) show the microinstructions which can be combined in a single instruction. These instructions are: operate group 1 $(:= (\text{op} = 7) \wedge \neg\, i\langle 3\rangle)$ for operating on the processor state; operate group 2 $(:= (\text{op} = 7) \wedge (i\langle 3,11\rangle = 10_2))$ for testing the processor state; and the extended arithmetic element group $(:= ((\text{op} = 7) \wedge (i\langle 3,11\rangle = 11_2)))$ for multiply, divide, etc. Within each instruction the remaining bits, $\langle 4{:}10\rangle$ or $\langle 4{:}11\rangle$, are extended instruction (or opcode) bits; that is, the bits are microcoded to select instructions. In this way an instruction is actually programmed (or microcoded). For example, the instruction set_link $\to$L $\leftarrow$1 is formed by coding the two microinstructions, clear link, next, complement link.

$$\text{opr\_}1 \to (i\langle 5\rangle \to L \leftarrow 0;\ \text{next}$$
$$i\langle 7\rangle \to L \leftarrow \neg L)$$

Thus, in operate group 1, the instructions clear link, complement link, and set link are formed by coding instruction$\langle 5,7\rangle$ = 10, 01, and 11, respectively. The operate group 2 instruction is used for testing the condition of the Pc state. This instruction uses bits 5, 6, and 8 to code tests for the accumulator. The AC skip conditions are coded $(0 \sim 7)$ as never, always, $=0$, $\neq 0$, <0, ≥ 0, ≤ 0, and >0. If all the nonredundant and useful variations in the two operate groups were available as separate instructions in the manner of the first seven (dca, tad, etc.), there would be approximately $7 + 12(\text{opr\_}1) + 10(\text{opr\_}2) + 6(\text{EAE}) = 35$ instructions in the PDP-8.

The optional Extended Arithmetic Element/EAE includes additional Multiplier Quotient/MQ and Shift Counter/SC registers and provides the hardwired operations multiply, divide, logical shift left, arithmetic shift, and normalize. The EAE is defined on the last page of Appendix 1.

The interrupt scheme

External conditions in the input/output devices can request that Pc be interrupted. Interrupts are allowed if (Interrupt_state = 1). A request to interrupt clears Interrupt_state (Interrupt_state $\leftarrow$ 0), and Pc behaves as though a jump to subroutine 0 instruction, jms 0, had been given. A special iot instruction (instruction = 6001_8) followed by a jump to subroutine indirect to 0 instruction (instruction = 5200_8) returns Pc to the interruptable state with Interrupt_state = 1. The program time to save M(processor state/ps) is 6 Mp accesses (9 microseconds), and the time to restore Mps is 9 Mp accesses (13.5 microseconds).

Only one interrupt level is provided in the hardware. If multiple priority levels are desired, programmed polling is required. Most io devices have to interrupt because they do not have a program-controlled enable switch for the interrupt. For multiple devices approximately 3 cycles (4.5 μs) are required to poll each interrupter.

PMS structure

The PMS structure of the LINC-8-338 consisting of a Pc('LINC), Pc('PDP-8), and P.display('338) is shown in Fig. 1. The PDP-8 is just a single Pc. The Pc('LINC) is a very capable Pc with more instructions than the main Pc. It is available in the structure to interpret programs written for the C('LINC), a computer developed by M.I.T.'s Lincoln Laboratory as a laboratory instrument computer for biomedical and laboratory applications. Because of the rather limited ISP in Pc, one would hardly expect to find all the components present in Fig. 1 in an actual configuration.

The S between the Mp and the Pc allows eight Mp's. This S is actually S('Memory Bus; 8 Mp; 1 Pc; (P requests); time-multiplexed; 1.5 μs/w). Thus the switch makes Mp logically equivalent to a single Mp(32768 w). There are two other L's which are connected to the Pc, excluding the T.console. They are L('I/O Bus) and L('Data Break; *Direct Memory Access*). These links become switches when we consider the physical structure. Associated with each device is a switch, and the bus links all the devices; the L('I/O Bus) is really an S('I/O Bus). Each time a K connects to it, the S is included in the K. A simplified PMS diagram (Fig. 3) shows the structure and the logical-physical transformation. Thus, the I/O Bus is

S('I/O Bus; duplex; bus; time-multiplexed, 1 Pc; 64 K; *Pc controlled, K requests;* t:4.5 μs/w)

The S('I/O Bus) is the same for the PDP-5, 8, 8/S, 8/I, and 8/L. Hence, any K can be used on any of the above C's. The I/O Bus is the link to the K's for Pc-controlled data transfers. Each word transferred is designated by a Pc instruction. However, the I/O Bus allows a K to request Pc's attention via the interrupt request signal. The Pc polls the K's to find the requesting K if multiple interrupt requests occur. A detailed structure of the Pc-Mp (Fig. 4) shows these L('I/O Bus, 'Data Break) connections to the registers and control in the notation used by DEC. This diagram is essentially a functional block diagram.

The S('I/O Bus) in Fig. 1 is only an abstract representation of the structure. Since it is a bus structure, the S can be expanded into L's and simple S's as shown in Fig. 3. The termination of the L in Pc is given in Fig. 3. The corresponding logic at a K is given in Fig. 5 in terms of logic design elements (AND's and OR's). (Fig. 5 also shows the S('I/O Bus) structure of Figs. 1 and 3). The operation of S('I/O Bus) shown in Fig. 5 starts when Pc sends a signal to select (or address) a particular K, using the IO_select ⟨0:5⟩ signals to form a 6-bit code to which K responds. Each K is hardwired to respond to a unique code. The local control, K[j], select signal is then used to form three local commands when ANDed with the three iot command lines from Pc, io_p1_bit, io_p2_bit, and io_p4_bit. Twelve data bits are transmitted either to or from Pc, indirectly under K's control. This is accomplished by using the AND-OR gates in K for data input to Pc, and the AND gate for data input to K. The data lines are connected to AC as shown in Fig. 4. A single *skip* input is used so that Pc can test a status bit in K. A K communicates to Pc via the interrupt request line. Any K wanting attention simply ORs its request signal into the interrupt request signal. Program polling in Pc then selects the specific interrupter. Normally, the K signal causing an interrupt is also connected to the skip input.

The L('Data Break; *Direct Memory Access*) provides a direct access path for a P or K to Mp via Pc. The number of access ports to memory can be expanded to eight by using the S('DM01 Data Multiplexer). The S is requested from a P or K. The P or K supplies an Mp address, a read or write access request, and then either accepts or supplies data for the Mp accessed word. In the configuration (Fig. 1), P('LINC) and P('338) are connected to S('DM01) and make requests to Mp for both their instructions and data in the same way as the Pc. The global control of these processor programs is via the S('I/O Bus). The Pc issues start and stop commands, initializes their state, and examines their final state when a program in the other P halts or requires assistance.

When a K is connected to L('Data Break) or to S('DM01 Data Multiplexer), the K only accesses Mp for data. The most complex function these K's carry out is the transfer of a complete block of data between the Mp and an Ms or a T, for example, K('DECtape, disk). A special mode, the three-cycle data break, is controlled by Pc so that a K may request the next word from a queue in Mp. In this mode the next word is taken from the queue (block) in Mp, and a counter is reduced each time K makes a request. With this scheme, a word transfer takes three Mp cycles: one to add one to the block count, one to add one to the address pointer, and one to transmit the word.

The DECtape was derived from M.I.T.'s Lincoln Laboratory LINCtape unit. Data are explicitly addressed by blocks (variable

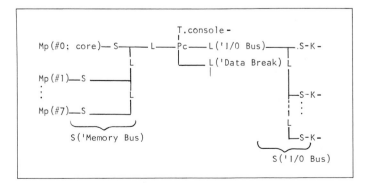

Fig. 3. DEC PDP-8 PMS diagram (simplified).

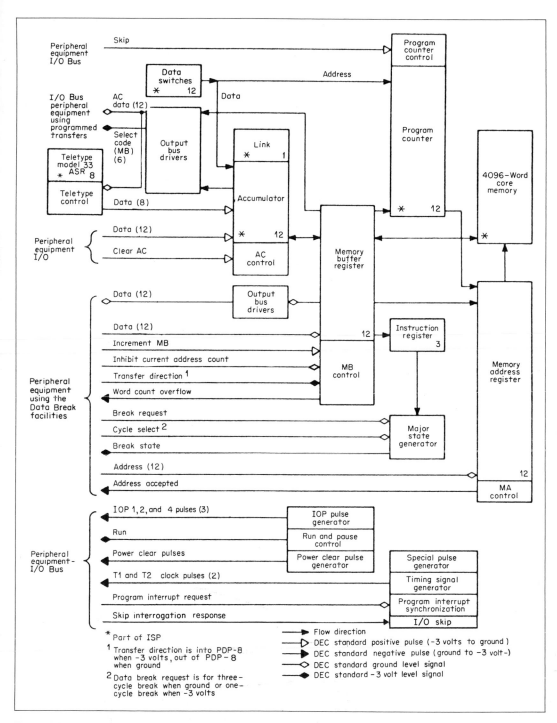

Fig. 4. DEC PDP-8 timing and control-element block diagram.
(*Courtesy of Digital Equipment Corporation.*)

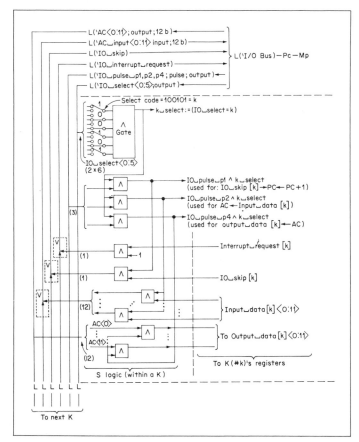

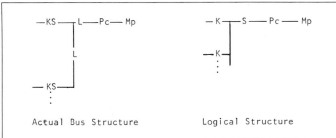

Fig. 5. DEC PDP-8 S('I/O Bus) logic and PMS diagrams.

tion serially by bit, there are special input/output instructions in the Pc to sample the line and to convert the sampled bits to coded characters. There are 11 bits transmitted per character (although other codings use 7, 7.42, 7.5, and 10 bits per character). Of the 11 bits, there are 3 control, 1 parity, and 7 information bits. The action of the Pc instruction, which is issued 5×11 (55) times for every character, is to control the line by forming the 7-bit characters. The instruction is a good example of tradeoff in the hardware/software domain toward almost pure software; the only hardware state associated with a telephone line is a 1-bit register to hold the state of the outgoing line, and a single AND gate to sample the incoming line state. This sampling process requires about 0.3 per cent of Pc-Mp capacity per active line (each of $10 \sim 15$ char/s). In general, the PDP-8 hardware controls are minimal—in turn fairly elaborate control programs must be used as part of them.

Computer levels

In this section we describe all the systems levels in the PDP-8 computer from the top down. The reader should already have a sketchy knowledge of the PDP-8 because the registers and ISP have been exposed. Here, we wish to clarify how it operates. A map of the hierarchy is given in Fig. 6, starting from PMS to ISP and down through logic design to circuit electronics. These description levels are subdivided to provide more organizational detail. For example, the register-transfer level has the more detailed registers, data operators, functional units, and macro logic of the processor, whereas the next logic level below has sequential and combinational networks, and the sequential and combinatorial elements.

It should be apparent that the relationship of the various description levels constitutes a tree structure where the organizationally complex computer is the top node and each descending description level represents increasing detail (or smaller component size), until the final circuit element level is reached. For simplicity, only a few of the many possible paths through the structural description tree are illustrated. For example, the path showing mechanical parts is missing. The path shown proceeds from the PDP-8 computer to the processor and from there to the arithmetic unit or, more specifically, to the AC register of the arithmetic unit. Next, the macro logic implementing the register-transfer operations and functions for the jth bit of the AC is given; the flip-flops and gates needed for this particular implementation are shown. Finally, on the last segment of the path, come the electronic circuits and components of which flip-flops and NAND gates are constructed.

but by convention 128 w). Thus information in a block can be replaced or rewritten at random. This operation is unlike queue-accessed tape (conventional IBM format magnetic tape) in which data can be appended only to the end of a file.

The control for the T(telephone) links 64 Teletypes or typewriters to the Pc. The final K which connects to a line is on a bit-serial basis. Since a telephone line sends and receives informa-

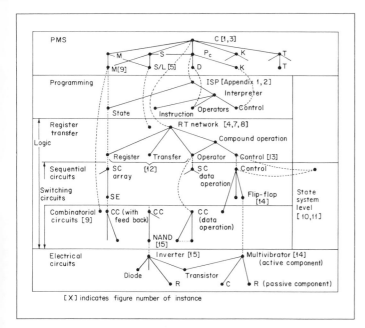

Fig. 6. DEC PDP-8 hierarchy of descriptions.

Abstract representations

Figure 6 also lists some of the methods used to represent the physical computer abstractly at the different description levels. As mentioned previously, only a small part of the PDP-8 description tree is represented here. The many documents, schematics, diagrams, etc., which constitute the complete representation of even this small computer include logic diagrams, wiring lists, circuit schematics and printed-circuit board layout masks, production description diagrams, production parts lists, testing specifications, programs for testing and diagnosing faults, and manuals for modification, production, maintenance, and use. As the discussion continues down the abstract description tree, the reader will observe that the tree conveniently represents the constituent objects of each level and their interconnection at the next highest level. Each level in the abstract-description tree will be described in order.

The PMS level

The simplified PMS structure in Fig. 3 has been reduced from Fig. 1. The computer is small enough so that the physical delineation of the PMS components, such as K's and S's, is less pronounced than in larger systems. In fact, in the case of the S('Memory Bus, 'I/O Bus), the S's are actually within the K and

Mp, as shown in Fig. 5. The implementation of these switches within the K and Mp was shown in Fig. 5. In Fig. 7 we present a more conventional functional diagram and the equivalent PMS diagram of the computer, with Pc decomposed into K, processor state (Mps), and D. The functional diagram has the same components of the characteristic elementary computer model, namely, K, D, M, and T(input, output). These figures give a somewhat general idea of what processes can occur in the computer, and how information flows, but it is apparent that at least another level is needed to describe the internal structure and behavior of the Mp and Pc. We should look at these primitives (although still together as a C) at the register-transfer level.

Programming level (ISP)

The ISP interpretation is given in Appendix 1 of this chapter and is the specification of the programming machine. In addition, it constrains the physical machine's behavior to have a particular ISP. The ISP has been discussed earlier in the chapter.

Register-transfer level

The C can also be represented at the register-transfer level by using PMS. Figure 4 (by DEC) shows the register-transfer level;

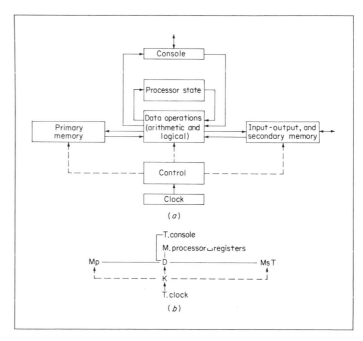

Fig. 7. DEC PDP-8 function block and PMS diagrams. (*a*) **Processor functional block diagram.** (*b*) **Pc PMS diagram.**

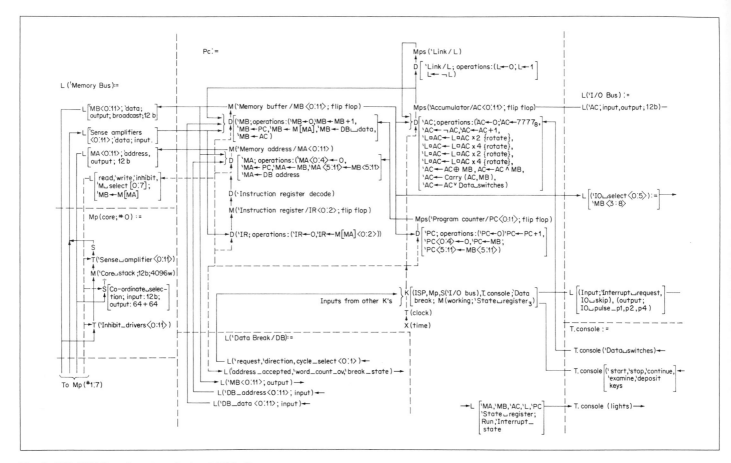

Fig. 8. DEC PDP-8 register-transfer-level PMS diagram.

only registers, operations, and L's are important at this level. We still lack information about the conditions under which operations are evoked. Figure 8 is a PMS diagram of Pc-Mp registers. Here we show considerably more detail (although we do not bother with electrical pulse voltages and polarities) than in Fig. 4. We declare the Pc state (including the temporary register) within Pc. The figure also gives the permissible data operations, D, which are permitted on the registers. It should be clear from this that the logical design level for the registers and the operators can easily be reached. The K logic design cannot be reached until we use the programming level constraints (ISP), thus defining the conditions for evoking the data operators.

The core memory. The Mp structure is given in Fig. 8. A more detailed block diagram which shows the core stack with its twelve

64×64 1-bit core planes is needed. Such a diagram, though still a functional block diagram, takes on some of the aspects of a circuit diagram because a core memory is largely circuit-level details. The Mp (Fig. 9) consists of the component units: the two address decoders (which select 1 each of 64 outputs in the X and Y axis directions of the coincident current memory); selection switches (which transform a coincident logic address into a high-current path to switch the magnetic cores); the 12 inhibit drivers (which switch a high current or no current into a plane when either a 0 or 1 is rewritten); 12 sense amplifiers (which take the induced low sense voltage from a selected core from a plane being switched or not switched and transform it into a 1 or 0); and the core stack, an array $M[0:7777_8]\langle 0:11\rangle$. Since this is the only time the Mp is mentioned, Fig. 9 also includes the associated circuit-level hardware needed in the core-memory operation, such as power supplies, timing, and logic signal level conversion amplifiers. The timing signals are generated within Pc(K) and are shown together with Pc's clock in Fig. 10.

The process of reading a word from memory is:

1 A 12-bit selection address is established on the MA⟨0:11⟩ address lines, which is 1 of 10000_8 (or 4096_{10}) unique numbers. The upper 6 bits, ⟨0:5⟩, select 1 of 64 groups of Y addresses and the lower 6 bits, ⟨6:11⟩, select 1 of 64 groups of X addresses.

2 The read logic signal is made a 1.

3 A high-current path flows via the X and Y selection switches. In each of the X and Y directions 64 × 12 cores

have selection current. Only one core in each plane is selected since Ix = Iy = Iswitching/2, and the current at the selected intersection = Ix + Iy = Iswitching.

4 If a core is switched to 0 (by having Iswitching amperes through it), then a 1 was present and is read at the output of the plane (bit) sense amplifiers. A sense amplifier receives an input from a winding that threads every core of every bit within a core plane [0:7777_8]. All 12 cores of the selected word are reset to 0. The sense time at which the sense amplifier is observed is tms (memory strobe), and the strobe in effect creates MB ← M[MA].

5 The read current is turned off.

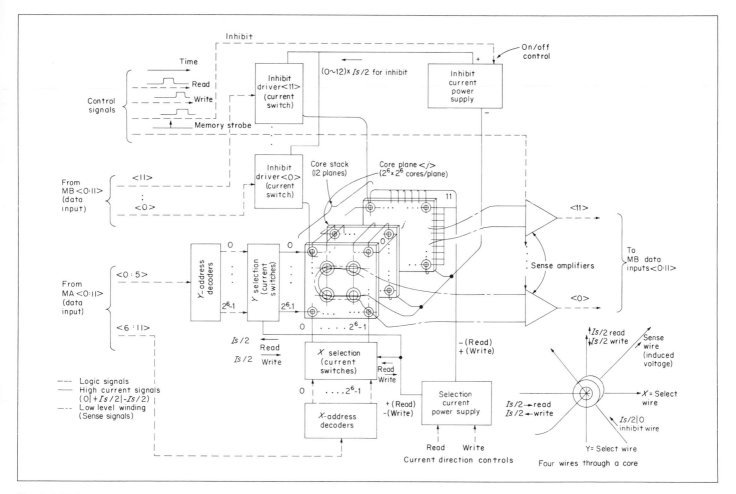

Fig. 9. DEC PDP-8 four-wire coincident current (three dimensions) core-memory-logic block diagram.

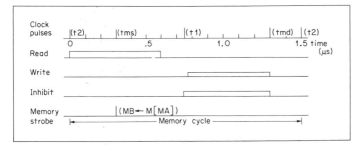

Fig. 10. DEC PDP-8 clock and memory timing diagram.

6   The write and inhibit logic signals are turned on. The bit inhibit signal is present or not, depending on whether a 0 or 1, respectively, is written into a bit.

7   A high-current path flows via the X and Y selection switches, but in an opposite direction to the read case (2 above). If a 1 is written, no inhibit current is present, and the net current in the selected core is $-I$switching. If a 0 is written, the current is $-I$switching $+(I$switching$/2)$ and the core remains reset.

8   The inhibit and write logic signals are turned off, and the memory cycle is completed.

Registers and operations. As Fig. 8 shows, the registers in the Pc cannot be uniquely assigned to a single function. In a minimal machine such as the PDP-8, functional separation is not economical. Thus there are not completely distinct registers and transfer paths for memory, arithmetic, and program and instruction flow. (This sharing complicates understanding of the machine.) However, Fig. 8 clarifies the structure considerably by defining all the registers in Pc (including temporaries). For example, the Memory Buffer/MB is used to hold the word being read from or written to Mp. MB also holds one of the operands for binary operations (for example, $AC \leftarrow AC \wedge MB$). MB is also used as an extension of the Instruction Register/IR during the instruction interpretation. The additional registers, not in the ISP, are:

Memory Buffer/MB⟨0:11⟩	holds memory data, instruction, and operands
Memory Address/MA⟨0:11⟩	holds address of word in Mp being accessed
Instruction Register/IR⟨0:2⟩	holds the value of current instruction being performed

State_register₃	a ternary state register holding the major state of memory cycle being performed
Fetch/F := (State_register = 0)	memory cycle to fetch instruction
Defer/D/Indirect := (State_register = 1)	memory cycle to get address of operand
Execute/E := (State_register = 2)	memory cycle to fetch (store) operand and execute the instruction

Figure 8 has been concerned with the static definition (or declaration) of the information paths, the operations, and state. The ISP interpretation (Appendix 1) is the specification for the physical machine's behavior. As the temporary hardware registers are added, a more detailed ISP definition could be given in terms of time and temporary registers. Instead, we give a state diagram (Fig. 11) to define the actual Pc which is constrained by both the ISP registers, the temporary registers implied by the implementation, and time. The relationship among the state diagram, the ISP description, and the logic is shown in the hierarchy of Fig. 6. In the relationships of the figures, we observe that the ISP definition does not have all the necessary detail for fully defining a physical Pc. The physical Pc is constrained by actual hardware logic and lower-level details even at the circuit level. For example, a core memory is read by a destructive process and requires a temporary register (MB) to hold the value being rewritten. This is not representable within a single ISP language statement since we define only the nondestructive transfer ←, but it can be considered as the two parallel operations MB ← M[MA]; M[MA] ← 0. The problem of explaining rewriting of core using ISP is also difficult, because explicit time is not in the ISP language (although we can define clock events, or at least relative time).

The state diagram (Fig. 11) describes the implementation behavior using the registers and register operations (Fig. 8) and the temporary registers declared above.

The implementation is fundamentally Mp-timing-based, as we see from both the state diagram and the times when the four clock signals are generated (Fig. 10). Thus there are three (State_register = 0,1,2) × 4 (clock), that is, 12 major states, in the implementation. We use the IR to obtain two more states, F2b and F3b,

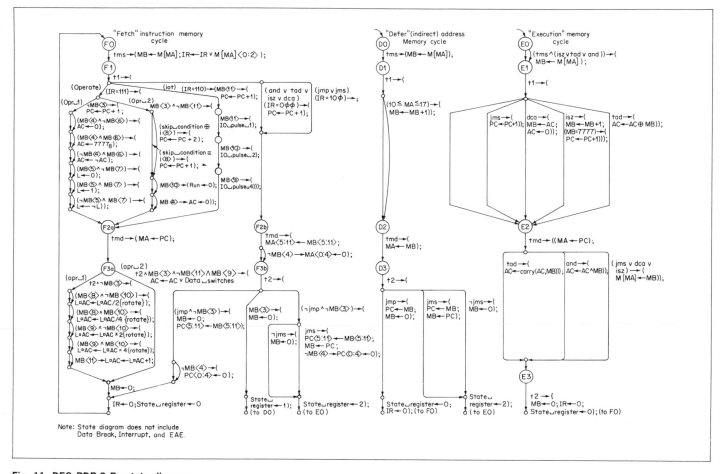

Fig. 11. DEC PDP-8 Pc state diagram.

for the description. The State_register values 0, 1, and 2 correspond to fetching, deferring (indirect addressing, i.e., fetching an operand address), and executing (fetching or storing data, then executing) the instruction. The state diagram does not describe the Extended Arithmetic Element/EAE operation, the interrupt state, and the data break states (these add 12 more states). The initialization procedure, including the T.console state diagram, is also not given. One should observe that when t2 occurs at the beginning of the memory cycle, a new State_register value is selected. The State_register value is always held for the remainder of the cycle; i.e., only the sequences (F0 → F1 → F2 → F3 or D0 → D1 → D2 → D3 or E0 → E1 → E2 → E3) are permitted.

Figure 8 alludes to Pc(K), that is, the sequential network used for controlling Pc. The inputs and the present state (including clocks) determine the operations to be issued on the registers.

Logic design level (registers and data operations)

Proceeding from the register-transfer and ISP descriptions, the next level of detail is the logic module. Typical of the level is the 1-bit logic module for an accumulator bit, AC⟨j⟩, illustrated in Fig. 12. The horizontal data inputs in the figure are to the logic module from AC⟨j⟩, MB⟨j⟩, IO Bus⟨j⟩, and Data_switch⟨j⟩. The vertical control signal inputs command the register operations (i.e., the transfers); they are labeled by their respective ISP operations (for example, $AC \leftarrow MB \wedge AC$, $AC \leftarrow AC \times 2$ {rotate}). The sequential network Pc(K) (Fig. 8) generates these control signal inputs.

Logic design level (Pc control, Pc(K) sequential network)

The output signals from the Pc(K) (Fig. 8) can be generated in a straightforward fashion by formulating the boolean expressions

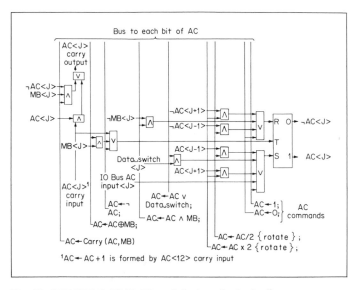

Fig. 12. DEC PDP-8 AC⟨J⟩ bit register-transfer logic diagram.

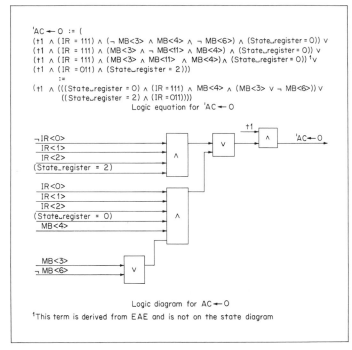

Fig. 13. DEC PDP-8 Pc(K) 'AC ← O signal-logic equations and diagram.

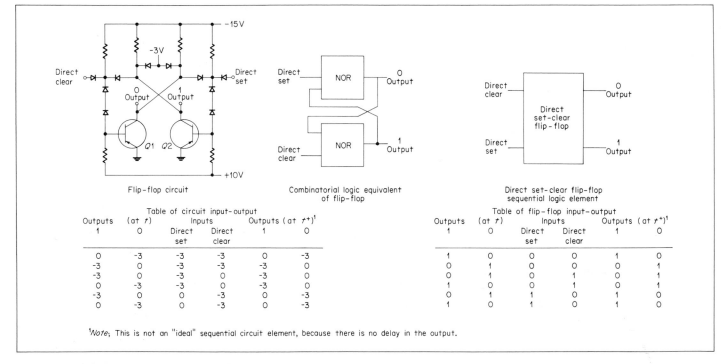

Fig. 14. DEC PDP-8 sequential-element circuit and logic diagrams.

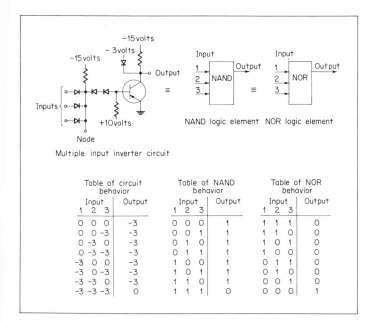

Fig. 15. DEC PDP-8 combinational element circuit and logic diagrams.

Table of circuit behavior:

Input 1 2 3	Output
0 0 0	-3
0 0 -3	-3
0 -3 0	-3
0 -3 -3	-3
-3 0 0	-3
-3 0 -3	-3
-3 -3 0	-3
-3 -3 -3	0

Table of NAND behavior:

Input 1 2 3	Output
0 0 0	1
0 0 1	1
0 1 0	1
0 1 1	1
1 0 0	1
1 0 1	1
1 1 0	1
1 1 1	0

Table of NOR behavior:

Input 1 2 3	Output
1 1 1	0
1 1 0	0
1 0 1	0
1 0 0	0
0 1 1	0
0 1 0	0
0 0 1	0
0 0 0	1

directly from the state diagram in Fig. 11. For example, the AC ← 0 control signal is expressed algebraically and with a combinatorial network in Fig. 13. Obviously these boolean output control signals are functions which include the clock, the State_register, and the states of the arithmetic registers (for example, A = 0, L = 0, etc.). The expressions should be factored and minimized so as to reduce the hardware cost of the control for the interpreter. Although we are rather cavalier about Pc(K), it constitutes about one-half the logic within Pc.

Circuit level

The final level of description is the circuits which form the logic functions of storage (flip-flops) and gating (NAND gates). Figures 14 and 15 illustrate some of these logic devices in detail.

In Fig. 14 a direct set and direct clear flip-flop, a sequential-logic element, is described in terms of circuit implementation, combinational logic equivalent, a table of its behavior, and its algebraic behavior. Note that this is not an ideal element, because it has no delay and responds directly and immediately to an input. Some idealized sequential logic elements are used in the PDP-8 (but not illustrated), including the RS (Reset-Set), T(Trigger), JK, and D(Delay). A delay in the flip-flops makes them behave in the same way as the ideal primitives in sequential-circuit theory. The outputs require a series delay, Δt, such that, if the inputs change at time t, the outputs will not change until $t + \Delta t$. In fact, the PDP-8 uses capacitor-diode gates at the flip-flop inputs to delay the inputs.

Figure 15 illustrates the combinatorial logic elements used in the PDP-8. The circuit selection is limited to the inverter circuit with single or multiple inputs. These are more familiarly called NAND gates or NOR gates, depending on whether one uses positive and/or negative logic-level definitions.

Conclusion

We could continue to discuss the behavior of the transistor as it is used in these switching-circuit primitives but will leave that to books on semiconductor electronics and physics. It is hoped that the student has gained a grasp of how to think about the hierarchical decomposition of computers into particular levels of analysis (and synthesis).

APPENDIX 1   DEC PDP-8 ISP DESCRIPTION

Appendix 1

DEC PDP-8 ISP Description

Pc State

AC<0:11>	*Accumulator*
L	*Link bit/AC extension for overflow and carry*
PC<0:11>	*Program Counter*
Run	*1 when Pc is interpreting instructions or "running"*
Interrupt_state	*1 when Pc can be interrupted; under programmed control*
IO_pulse_1; IO_pulse_2; IO_pulse_4	*IO pulses to IO devices*

Mp State
Extended memory is not included.

M[0:7777$_8$]<0:11>

Page_0[0:177$_8$]<0:11> := M[0:177$_8$]<0:11>   *special array of directly addressed memory registers*

Auto_index[0:7]<0:11> := Page_0[10$_8$:17$_8$]<0:11>   *special array when addressed indirectly, is incremented by 1*

Pc Console State
Keys for start, stop, continue, examine (load from memory), and deposit (store in memory) are not included.

Data switches<0:11>   *data entered via console*

Instruction Format

instruction/i<0:11>

op<0:2>	:= i<0:2>	*op code*
indirect_bit/ib	:= i<3>	*0, direct; 1 indirect memory reference*
page_0_bit/p	:= i<4>	*0 selects page 0; 1 selects this page*
page_address<0:6>	:= i<5:11>	
this_page<0:4>	:= PC'<0:4>	
PC'<0:11>	:= (PC<0:11> -1)	
IO_select<0:5>	:= i<3:8>	*selects a T or Ms device*
io_p1_bit	:= i<11>	*these 3 bits control the selective generation of -3 volts,*
io_p2_bit	:= i<10>	*0.4 µs pulses to I/O devices*
io_p4_bit	:= i<9>	
sma	:= i<5>	µ *bit for skip on minus AC, operate 2 group*
sza	:= i<6>	µ *bit for skip on zero AC*
snl	:= i<7>	µ *bit for skip on non zero Link*

Effective Address Calculation Process

z<0:11> := (*effective*

¬ib → z'';

ib ∧ (10$_8$ ≤ z'' ≤ 17$_8$) → (M[z''] ←M[z''] + 1; next);   *auto indexing*

ib → M[z''])

z'<0:11> := (¬ ib → z''; ib → M[z''])

z''<0:11> := (page_0_bit → this_page□page_address;   *direct address*

¬page_0_bit → 0□page_address)

µ *microcoded instruction or instruction bit(s) within an instruction*

APPENDIX 1   DEC PDP-8 ISP DESCRIPTION (Continued)

Instruction Interpretation Process

   Run ∧ ¬ (Interrupt⌣request ∧ Interrupt⌣state) → (*no interrupt interpreter*

      instruction ← M[PC]; PC ← PC + 1; next   *fetch*

      instruction⌣execution);   *execute*

   Run ∧ Interrupt⌣request ∧ Interrupt⌣state → (*interrupt interpreter*

     M[0] ← PC; Interrupt⌣state ← 0; PC ← 1)

Instruction Set and Instruction Execution Process

   Instruction⌣execution := (

     and (:= op = 0) → (AC ← AC ∧ M[z]);   *logical and*

     tad (:= op = 1) → (L□AC ← L□AC + M[z]);   *two's complement add*

     isz (:= op = 2) → (M[z'] ← M[z] + 1; next   *index and skip if zero*

                 (M[z'] = 0) → (PC ← PC + 1));

     dca (:= op = 3) → (M[z] ← AC; AC ← 0);   *deposit and clear AC*

     jms (:= op = 4) → (M[z] ← PC; next PC ← z + 1);   *jump to subroutine*

     jmp (:= op = 5) → (PC ← z);   *jump*

     iot (:= op = 6) → (μ *in out transfer, microprogrammed to generate up to 3 pulses*

        io⌣p1⌣bit → IO⌣pulse⌣1 ← 1; next   *to an io device addressed by IO⌣select*

        io⌣p2⌣bit → IO⌣pulse⌣2 ← 1; next

        io⌣p4⌣bit → IO⌣pulse⌣4 ← 1);

     opr (:= op = 7) → Operate⌣execution   *the operate instruction is defined below*

)   *end Instruction execution*

Operate Instruction Set

The microprogrammed operate instructions: operate group 1, operate group 2, and extended arithmetic are defined as a separate instruction set.

   Operate⌣execution := (

     cla (:= i$<$4$>$ = 1) → (AC ← 0);   *clear AC. Common to all operate instructions.*

     opr⌣1 (:= i$<$3$>$ = 0) → (*operate group 1*

       cll (:= i$<$5$>$ = 1) → (L ← 0); next   μ *clear link*

       cma (:= i$<$6$>$ = 1) → (AC ← ¬ AC);   μ *complement AC*

       cml (:= i$<$7$>$ = 1) → (L ← ¬ L); next   μ *complement L*

       iac (:= i$<$11$>$ = 1) → (L□AC ← L□AC + 1); next   μ *increment AC*

       ral (:= i$<$8:10$>$ = 2) → (L□AC ← L□AC × 2 {rotate});   μ *rotate left*

       rtl (:= i$<$8:10$>$ = 3) → (L□AC ← L□AC × 2^2 {rotate}); μ *rotate twice left*

       rar (:= i$<$8:10$>$ = 4) → (L□AC ← L□AC / 2 {rotate});   μ *rotate right*

       rtr (:= i$<$8:10$>$ = 5) → (L□AC ← L□AC / 2^2 {rotate})); μ *rotate twice right*

     opr⌣2 (:= i$<$3,11$>$ = 10) → (*operate group 2*

       skip condition ⊕ (i$<$8$>$ = 1) → (PC ← PC + 1); next μ *AC,L skip test*

         skip condition := ((sma ∧ (AC $<$ 0)) ∨ (sza ∧ (AC = 0)) ∨ (snl ∧ L))

       osr (:= i$<$9$>$ = 1) → (AC ← AC ∨ Data switches);   μ *"or" switches*

       hlt (:= i$<$10$>$ = 1) → (Run ← 0));   μ *halt or stop*

     FAE (:= i$<$3,11$>$ = 11) → EAF⌣instruction⌣execution)   *optional EAE description*

APPENDIX 1   DEC PDP-8 ISP DESCRIPTION (Continued)

KT and KMs State
  Each K may have any or all of the following registers.   There can be up to 64 optional K's.

  Input_data[0:77$_8$]<0:11>   *64 input buffers*

  Output_data[0:77$_8$]<0:11>   *64 output buffers*

  IO_skip_flag[0:77$_8$]   *64 test conditions*

  IO_interrupt_request[0:77$_8$]   *1 signifies a request.   If interrupt_state = 1, then an*
                                                      interrupt occurs.

    Interrupt_request := (*"or" of all requests from each IO device*
      max(IO_interrupt_request[0:77$_8$]))

Extended Arithmetic Element, EAE (optional)
  Provides additional arithmetic instructions (or operators) including $\times$*, /, normalize, logical shift and arithmetic shift.*
  EAE State

    MQ<0:11>   *Multiplier Quotient*

    SC<0:4>   *Shift Counter*

  Instruction Format and Data

    mds<0:11>   *multiplier divisor shift data*

    s<0:4> := mds<7:11>   *shift count parameter*

Instruction Set for EAE
    EAE_instruction_execution := (next

      mqa (:= i<5>) → (AC ← AC ∨ MQ);   *MQ into AC*

      sca (:= i<6>) → (AC ← AC ∨ SC):   *SC into AC*

      mql (:= i<7>) → (MQ ← AC; AC ← 0); next   *AC into MQ, clear AC*

    Note only one of nmi, shl, asr, lsr, muy, or dvi can be given at a time.

      i<8:10> = 00₆ → ;   *IO operation*

      ¬ nmi →(mds ← M[PC]; PC ← PC + 1); next

      muy (:= i<8:10> = 2) → (L□AC□MQ ← MQ × mds; SC ← 0)   *multiply*

      dvi (:= i<8:10> = 3) → (MQ ← L□AC□MQ/mds;   *divide*

                    L□AC ← L□AC□MQ mod mds; SC ← 0):

      nmi (:= i<8:10> = 4) → (AC□MQ ← normalize(AC□MQ);   *normalize(AC,MQ) into SC*

                    SC ← normalize_exponent(AC□MQ));

      shl (:= i<8:10> = 5) → (L□AC□MQ ← L□AC□MQ × 2^{s+1}: SC ← 0);   *shift left*

      asr (:= i<8:10> = 6) → (L□AC□MQ ← L□AC□MQ / 2^{s+1}: SC ← 0):   *shift right*

      lsr (:= i<8:10> = 7) → (L□AC□MQ ← L□AC□MQ / 2^{s+1}{logical};   *logical shift*

                    SC ← 0)
)   *end EAE instruction execution*

Chapter 6

The Whirlwind I computer[1]

R. R. Everett

Project Whirlwind is a high-speed computer activity sponsored at the Digital Computer Laboratory, formerly a part of the Servomechanisms Laboratory, of the Massachusetts Institute of Technology (M.I.T.) by the Office of Naval Research (O.N.R.) and the United States Air Force. The project began in 1945 with the assignment of building a high-quality real-time aircraft simulator. Historically, the project has always been primarily interested in the fields of real-time simulation and control; but since about the beginning of 1947 most of its efforts have been devoted to the design and construction of the digital computer known as Whirlwind I (WWI). This computer has been in operation for about 1 year and an increasing proportion of project effort now is going into application studies.

Applications for digital computers are found in many branches of science, engineering, and business. Although any modern general-purpose digital computer can be applied to all these fields, a machine is generally designed to be most suited to some particular area. Whirlwind I was designed for use in control and simulation work such as air traffic control, industrial process control, and aircraft simulation. This does not mean that Whirlwind will not be used on applications other than control. About one-half the available computing time for the next year will be assigned to engineering and scientific calculation including research in such uses supported by the O.N.R. through the M.I.T. Committee on Machine Methods for Computation.

These control and simulation problems result in a specialized emphasis on computer design.

Short register length

WWI has 16 binary digits and the control problems are usually very simple mathematically. Furthermore, the computer is almost always part of a feedback rather than an open-ended system. Consequently, roundoff errors are seldom troublesome and the register length can be shortened to something comparable to the sensitivity of the physical quantities involved, perhaps five decimal places or less.

WWI has a register length of 16 binary digits including sign or about four and one-half decimals. The register length was

[1]*AIEE-IRE Conf.*, 70–74 (1951)

chosen as the minimum that would provide a usable single-address order, in this case five binary digits for instruction and 11 binary digits for address. In a future machine we would probably increase this register length to 20 or 24 binary digits to get additional order flexibility; the increased numerical precision is less important.

For scientific and engineering calculation, greater than 16-digit precision is often required. There is available a set of multiple-length and floating point subroutines which make the use of greater precision very easy. It is true that these subroutines are slow, bringing effective machine speed down to about that obtained by acoustic memory machines. It is much more efficient occasionally to waste computing time this way than continuously to waste a large part of the storage and computing equipment of the machine by providing an unnecessarily long register.

High operating speed

WWI performs 20,000 single-address operations per second. Control and simulation problems require very high speeds. The necessary calculations must be carried out in real time; the more complex the controlled system is, the faster the computer must be. There is no practical upper limit to the computing speed that could be used if available.

Where the problems are large enough, and these problems are, one high-speed machine is much better than two simpler machines of half the speed. Communication between machines presents many of the same problems that communication between human beings presents.

Great effort was put into WWI to obtain high speed. The target speed was 50,000 single-address operations per second, and all parts of the machine except storage meet this requirement. The actual WWI present operating speed of 20,000 single-address operations per second is on the lower edge of the desired speed range.

Large internal storage

WWI now has 1,280 registers. A large amount of high-speed internal storage is needed since it is not in general possible to use slow auxiliary storage because of the time factor. In many cases a magnetic drum can be useful since its access time is short compared to the response times of real systems. Even with a drum there is considerable loss of computing and programming efficiency due to shuffling information back and forth between drum and computer.

WWI is designed for 2,048 registers of storage. Until recently there has been available only about 300 registers. This number, while small, has been adequate for much useful work. Very recently a second bank of new-model storage tubes has been added. These new tubes operate at 1,024 spots per tube bringing the total WWI storage to 1,280 registers. These tubes have been in the computer and under test for 2 months and in active use for about 2 weeks. In the next few months the tubes in the first bank will be replaced by new-model storage tubes bringing the total storage to 2,048. This number is on the lower end of what the project considers desirable. What the computer business needs, has needed, and will probably always need is a bigger, better, and faster storage device.

Extreme reliability

In a system where much valuable property and perhaps many human lives are dependent on the proper operation of the computing equipment, failures must be very rare. Furthermore, checking alone, however complete, is inadequate. It is not enough merely to know that the equipment has made an error. It is very unlikely that a man, presumably not too well suited to the work during normal conditions, can handle the situation in an emergency. Multiple machines with majority rule seem to be the best answer. Self-correcting machines are a possibility but appear to be too complicated to compete, especially as they provide no standby protection.

The characteristics of the Whirlwind I computer may be recapitulated as follows:

Register length	16 binary digits, parallel
Speed	20,000 single-address operations per second
Storage capacity	Originally 256 registers Recently 320 registers Presently 1,280 registers Target 2,048 registers
Order type	Single-address, one order per word
Numbers	Fixed point, 9's complement
Basic pulse	1 megacycle
repetition frequency	2 megacycles (arithmetic element only)

Tube count	5,000, mostly single pentodes
Crystal count	11,000

There are 32 possible operations, of which about 27 are assigned. They are of the usual types: addition, subtraction, multiplication, division, shifting by an arbitrary number of columns, transfer of all or parts of words, subprogram, and conditional subprogram. There are terminal equipment control orders and there are some special orders for facilitating double-length and floating-point operations.

One way to increase the effective speed of a machine is to provide built-in facilities for operations that occur frequently in the problems of interest. An example is an automatic co-ordinate transformation order. The addition of such facilities does not affect the general-purpose nature of the machine. The machine retains its old flexibility but becomes faster and more suited to a certain class of problems.

From March 14, 1951, at which time we began to keep detailed records, until November 22, 1951 a total of 950 hours of computer time were scheduled for applications use. The machine has been running on two shifts or a total of about 3,000 hours during this interval. The two-thirds time not used for applications has been used for machine improvement, adding equipment, and preventive maintenance.

Of the 950 hours available, 500 have been used by the scientific and engineering calculation group, the rest for control studies. The limited storage available until recently has been admittedly a serious handicap to the scientific and engineering applications people. There has not been room in storage for the lengthy subroutines necessary for convenient use of the machine. The largest part of their time has been spent in training, in setting up procedures, and in preparing a library of subroutines.

A partial list of the actual problems carried out by the group includes:

1 An industrial production problem for the Harvard Economics School

2 Magnetic flux density study for our magnetic storage work

3 Oil reservoir depletion studies

4 Ultra-high frequency television channel allocation investigation for Dumont

5 Optical constants of thin metal films

6 Computation of autocorrelation coefficients

7 Tape generation for a digitally-controlled milling machine

The scientific and engineering applications time on Whirlwind I has been organized in a manner patterned after that originated by Dr. Wilkes at EDSAC. The group of programmers and mathematicians assigned to WWI assist users in setting up their own problems. Small problems requiring only a few seconds or minutes of computer time are encouraged. Applications time is assigned in 1-hour pieces two or three times a day. No program debugging is allowed on the machine. Program errors are deduced by the programmer from printed lists of results, storage contents, or order sequences as previously requested from the machine operator. The programmer then corrects his program which is rerun for him within a day or perhaps within a few hours.

Every effort is made to reduce the time-consuming job of printing tabulated results. In many cases a user desires large amounts of tabulated data only because he doesn't really know what answers he wants and so asks for everything. Such users are encouraged to ask only for pertinent results in the form of numbers or curves plotted by the machine on a cathode-ray tube and automatically photographed. If these results prove inadequate or the user gets a better idea of his needs, he is allowed to rerun his program, again asking only for what appear to be significant results. Figure 1 shows a sample curve plotted by the computing machine showing calibrated axes and decimal intercepts.

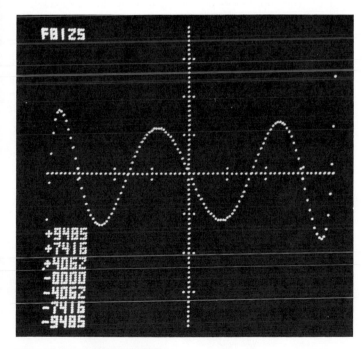

Fig. 1. Sample computer output.

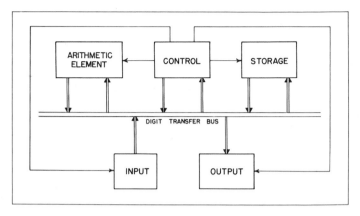

Fig. 2. Simplified computer block diagram.

WWI system layout

Figure 2 shows the major parts of any computer such as WWI. The major elements of the computer communicate with each other via a central bus system.

WWI is basically a simple, straightforward, standard machine of the all-parallel type. Unfortunately, the simple concept often becomes complicated in execution, and this is true here. WW's control has been complicated by the decision to keep it completely flexible, the arithmetic element by the need for high speed, the storage by the use of electrostatic storage tubes, the terminal equipment by the diversity of input and output media needed.

Control

The WW control is divided into several parts, as shown in Fig. 3.

Central control

The central control of the machine is the master source of control pulses. When necessary the central control allows one of the other controls to function. In general there is no overlapping of control operation; except for terminal equipment control, only one of the controls is in operation at any one time.

Storage control

Storage control generates the sequence of pulses and gates that operate the storage tubes. Central control instructs the storage control either to read or to write.

Arithmetic control

Arithmetic control carries out the details of the more complex arithmetic operations such as multiplication and division. The

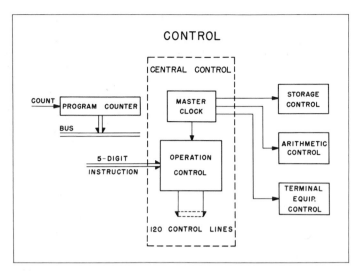

Fig. 3. Control.

setup of these operations plus the complete controlling of the simpler operations such as addition are carried out by central control.

Terminal equipment control

Terminal equipment control generates the necessary control pulses, delay times, and interlocks for the various terminal equipment units.

Program counter

The program counter which keeps track of the address of the next order to be carried out is considered as part of control. This is an 11-binary counter with provision for reading to the bus.

Most of the functions of these subsidiary controls could be combined with the central control. The major reason they are not is that they were designed at different times. The arithmetic element and its control came first, followed by central control. At the time central control was designed, the necessary characteristics of storage control were unknown. In fact, the machine was designed so that any parallel high-speed storage could be used. The form of terminal equipment control was also unknown at this time. Since flexibility was a prime specification, it was felt preferable to build separate flexible controls for the various parts of the computer than to try to combine all the needed flexibility in one central control.

In a new machine we would attempt to combine control functions where possible, hoping to have enough prior knowledge

about component needs to eliminate subsidiary controls completely. We would still insist on a large degree of control flexibility.

Master clock

The master clock consists of an oscillator, pulse shaper and divider that generate 1- and 2-megacycle clock pulses, and a clock pulse control that distributes these clock pulses to the various controls in the machine. It is this unit that determines which of the subsidiary controls actually is controlling the machine. This unit also stops and starts the machine and provides for push-button operation.

Operation control

The operation control, see Fig. 4, was designed for maximum flexibility and minimum number of operation digits, and, consequently, minimum register length. It is of the completely decoding type.

The operation switch is a 32-position crystal matrix switch that receives the 5-bit instruction from the bus and in turn selects one of 32 output lines corresponding to the 32 built-in operations.

There are 120 gate tubes on the output of the operation control. Pulses on the 120 output lines go to the gate drivers, pulse drivers, and control flip-flops all over the machine; 120 is a generous number. The suppressors of these gate tubes are connected to vertical wires that cross the 32 output lines from the operation switch. Crystals are inserted at the desired junctions to turn on those gate tubes that are to be used for any operation.

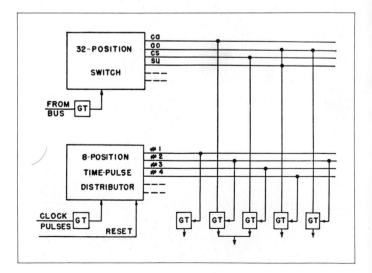

Fig. 4. Operation control.

The time pulse distributor consists of an 8-position switch driven from a three binary-digit counter. Clock pulses at the input are distributed in sequence on the eight output lines. The control grids of the output gate tubes are connected to these timing lines. The output of the operation control is thus 120 control lines on each of which can appear a sequence of pulses for any combination of orders at any combination of times.

Central control

The Central Control of the machine is shown in Fig. 5. The control switch is in the foreground with the operation matrix to the right.

Electrostatic storage

The electrostatic storage shown in Fig. 6 consists of two banks of 16 storage tubes each. There is a pair of 32-position decoders

Fig. 6. View of electrostatic storage.

Fig. 5. View of central control.

set up by address digits read in from the bus. There is a storage control that generates the sequence of pulses needed to operate the gate generators, et cetera. A radio frequency pulser generates a high power 10-megacycle pulse for readout.

Each digit column contains, besides the storage tubes, write plus and write minus gate generators and a signal plate gate generator for each tube. Ten-megacycle grid pulses are used for readout in order to get the required discrimination between the fractional volt readout pulses and the 100-volt signal plate gates. For each storage tube there is a 10-megacycle amplifier, phase-sensitive detector and gate tube, feeding into the program register. The program register is used for communicating with the storage tubes. Information read out of the tubes appears in the program register. Information to be written into the tubes must be placed in the program register.

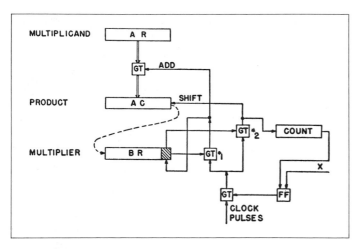

Fig. 7. Arithmetic element.

1   Power supply control and meters.

2   Neon indicators for all flip-flops in the machine.

3   Switches for setting up special conditions.

4   Manual intervention switches.

5   Oscilloscopes for viewing wave forms. A probe and amplifier system allows viewing any wave form in the computer on one scope at test control.

6   Test equipment to provide synchronizing, stop, or delay pulses at any step of any order of a program, allowing viewing wave forms on the fly anywhere in the machine.

An important part of the test facilities is the test storage, a group of 32 toggle-switch registers plus five flip-flop registers that can be inserted in place of any five of the toggle-switch registers. This storage has proved invaluable not only for testing control and

Arithmetic element

The arithmetic element, see Fig. 7, consists of three registers, a counter, and a control.

The first register is an accumulator (AC) which actually consists of a partial-sum or adding register and a carry register. The accumulator holds the product during multiplication.

The second or A-register holds the multiplicand during multiplication. All numbers entering the arithmetic element do so through AR.

The third or B-register holds the multiplier during multiplication. The accumulator and B-register shift right or left. A high-speed carry is provided for addition. Subtraction is by 9's complement and end-around-carry. Multiplication is by successive additions, division by successive subtractions, and shift orders provide for shifting right or left by an arbitrary number of steps, with or without roundoff.

The arithmetic element is straightforward except for a few special orders and the high speed at which it operates. Addition takes 3 microseconds complete with carry; multiplication, 16 microseconds average including sign correction.

In Fig. 8 are shown several digits of the arithmetic element. The large panels are accumulator digits. Above the accumulator is the B-register, below it the A-register.

Test control

Test control, shown in Fig. 9, is used at present both for operating and for trouble shooting the computer. The control includes:

Fig. 8. View of arithmetic element.

Fig. 9. View of test control.

arithmetic element before electrostatic storage was available but also for testing electrostatic storage itself. When not in use for test purposes test storage earns its keep as part of the terminal equipment system. The toggle-switches hold a standard read-in program; the flip-flop registers are used as in-out registers for special purposes.

Checking

Logical checking facilities built into WWI are rather inconsistent. A complete bus transfer checking system has been provided, duplicate checking of some terminal equipment is permitted, but little else is thoroughly checked. We felt that it was worthwhile to thoroughly check some substantial portion of the machine. This portion would then serve as a prototype for studying the tube circuitry used throughout the machine. We did not feel it was worthwhile to check all the machine, a procedure that requires a great deal of added equipment and logical complexity plus a substantial loss in computing speed.

Operating experience has shown us that it is not worthwhile to provide detailed logical checking of a machine. In a new machine we would leave out the transfer checking. The amount of information and security given by the detailed checking system is not enough to warrant the expense of building and maintaining it.

This decision is based on the expectation that a computing machine should operate 95 per cent of total time or better and that the average time between random failures should be of the order of 5 to 10 hours or approximately 10^9 operations.

In our opinion the way to achieve the extremely high reliability needed in some real-time control problems is to provide three or more identical but distinct machines, thus obtaining error correction as well as detection, plus such features as standby, safety, and damage control. Even so the failure probability of each machine must be kept low by proper design, marginal checking, and preventive maintenance.

Extremely high reliability means a reliability far beyond that achieved in existing machines and not conveniently represented as a per cent. Consider a system consisting of three machines, each operable 98 per cent of the time and each averaging 10 hours between random errors.

One machine will be out of operation $\frac{1}{2}$ hour per day.

Two machines will be out of operation $\frac{1}{4}$ hour per month.

All three machines will be out of operation 4 minutes per year. Furthermore undetected random errors might occur on the average of once a year. Such reliability is needed in some systems.

Our decision to omit detailed checking does not extend to checking devices intended to detect programming errors. Devices to check for overflow from the arithmetic element or for non-existent order configurations are necessary. Programmers make many mistakes. Techniques for dealing with programming errors are very important and need future development.

Terminal equipment

At the present time, Whirlwind is using the following terminal equipment:

1 A photoelectric paper tape reader

2 Mechanical paper tape readers and punches

3 Mechanical typewriters

4 Oscilloscope displays 5 to 16 inches in diameter with phosphors of various persistencies including a computer-controlled scope camera

5 Inputs from various analogue equipments needed for control studies

6 Outputs to analogue equipment

To be added during the next year:

1 Magnetic Tape (units by Raytheon). One such unit is now being integrated with machine.

2 Magnetic drums (units by Engineering Research Associates, Inc.).

3 Many more analogue inputs and outputs.

This great complexity of terminal equipment requires a flexible switching system. There is a single in-out register (IOR) through which most of the data passes.

There is a switch which is set up by an order to select the desired piece of terminal equipment. Other orders put data into IOR or remove data from IOR. The in-out control provides the necessary control pulses to go with each type of equipment. In general the computer continues to run during terminal equipment wait times; suitable interlocks are provided to prevent trouble. This complete equipment has not yet been fully installed.

References

Whirlwind: EverR51; SerrR62; TaylN51.
EdSAC: SamuA57; WilkM56.

APPENDIX 1 WHIRLWIND I INSTRUCTION CODE[1]

The complete WWI instruction code is given below in tabular form.   The notations used in the table, together with their meanings, are as follows.

```
---- ≡ remains unchanged
IOR = In-Out Register
PC  = Program Counter
AC  = Accumulator, AC(i) ≡ digit i of AC, 0 ≤ i ≤ 15
BR  = B-Register, BR(i) ≡ digit i of BR, 0 ≤ i ≤ 15
AR  = A-Register, AR(i) ≡ digit i of AR, 0 ≤ i ≤ 15
ρ   ≡ round-off from BR; if BR(0)=1, ρ = 2^{-15}
                         if BR(0)=0, ρ =0
CM  = Core memory
```

```
x  = address of a storage register 0 ≤ x ≤ 2047
n  = positive integer 0 ≤ n ≤ 511 mod 32
SAM = Special Add Memory
C( ) = original contents of register ( )
C_i( ) = original contents of digit i of register ( )
F{ } = fractional part of the quantity in { }
I{ } = integral part of the quantity in { }
--→ = becomes
```

```
x(i) = digit i of register x, 0 ≤ i ≤ 15
x(k-ℓ) = digits k thru ℓ of register x, 0 ≤ k < ℓ ≤ 15
AC+BR = the composite 32 digit register (including
        sign)composed of the AC and BR taken in
        that order
(AC+BR)(i) = digit i of (AC+BR), 0 ≤ i ≤ 31
⊕ · Boolean "exclusive or" operator
⊗ · Boolean "and" operator
```

Instruction	Function	Binary Code	Dec. Equiv.	AC	BR*	AR	SAM	x	Comments	Time
si pqr	select in-out unit or stop the computer	00000	0	----	----	----	----	----	The unit selected is designated by the digits pqr, and is started. si 0 will stop the computer. si 1 is a "conditional" stop Program alarm possible if si selects Magnetic or Photo Electric Tape Reader without necessary rd's	31 μsec.
		00001	1		----	----	----	----	Illegal instruction.	
bi x	block transfer in (n words to CM)	00010	2	x + n	----	x	----	first word of block	±n·2^{-15} must be stored in AC at the time the computer executes the bi. If n=0 one word will be read but not transferred. It is simply discarded.	For drum 8 msec. average and 16 msec. max. for first word. 32 μsec. for ea. additional word.
rd x	read	00011	3	C(IOR)	----	C(IOR)	----	----	After word is transferred from IOR to AC, the IOR is cleared. The address of rd has no significance.	1 μsec.
bo x	block transfer out (n words from CM)	00100	4	x + n	----	x	----	----	±n·2^{-15} must be stored in AC. If n=0, no recording will take place.	same as for bi.
rc x	record	00101	5	----	----	----	----	----	If rc is used as a display instruction, the IOR is cleared.	24 μsec.
sd x	sum digits	00110	6	$C_i(AC) \oplus C_i(x)$ --→ AC(i) i = 0,..., 15	----	C(x)	clear	----	Adds digits without carry ⊕ | 0 1 | 0 | 0 1 | 1 | 1 0	24 μsec.
ts x	transfer to storage	01000	8	----	----	----	----	C(AC)		24 μsec.
td x	transfer digits	01001	9	----	----	----	----	x(0-4) unchanged x(5-15)≡ AC(5-15)		32 μsec.
ta x	transfer address	01010	10	----	----	----	----	x(0-4) unchanged x(5-15)≡ AR(5-15)	ta normally follows an sp, cp, sf, or ao	32 μsec.
ck x	check	01011	11	----	----	----	----	----	Computer stops on "check-register alarm" if C(AC)≠C(x). (Note that +0 ≠ -0).	24 μsec.
ab x	add BR	01100	12	C(BR)+C(x)	----	C(x)	clear	C(BR)+C(x)	possible Arith. Overflow Alarm if |C(x)+C(BR)| ≥ 1 dm x or clh 16 puts C(AC) into BR	32 μsec.
ex x	exchange	01101	13	C(x)	----	C(x)	----	C(AC)	ex 0 will clear AC without clearing BR	32 μsec.
cp x	conditional transfer control (conditional program)	01110	14	----	----	y + 1 (digits 5-16)	----	----	If C(AC) ≥ +0 proceed to next instruction. If C(AC) ≤ -0, execute sp x. y is location of cp x	16 μsec.
sp x	transfer control (subprogram)	01111	15	----	----	y+1 (digits 5-16)	----	----	Take next instruction from register x. PC --→ x y is location of sp x	16 μsec.
ca x	clear and add	10000	16	C(x)+C(SAM)2^{-15}	clear	C(x)	clear	----	possible Arith. Overflow Alarm if |C(x)+C(SAM)2^{-15}| ≥ 1	24 μsec.
cs x	clear and subtract	10001	17	-C(x)+C(SAM)2^{-15}	clear	C(x)	clear	----	possible Arith. Overflow Alarm if |-C(x)+C(SAM)2^{-15}| ≥ 1	24 μsec.
ad x	add	10010	18	C(AC)+C(x)	----	C(x)	clear	----	possible Arith. Overflow Alarm if |C(AC)+C(x)| ≥ 1	24 μsec.
su x	subtract	10011	19	C(AC)-C(x)	----	C(x)	clear	----	possible Arith. Overflow Alarm if |C(AC)-C(x)| ≥ 1	24 μsec.
cm x	clear and add magnitude	10100	20	|C(x)|+C(SAM)2^{-15}	clear	|C(x)|	clear	----	possible Arith. Overflow Alarm if |C(x)+C(SAM)2^{-15}| ≥ 1	24 μsec.
sa x	special add	10101	21	F{C(AC)+C(x)}	----	C(x)	I{C(AC)+ C(x)}; ±1 or 0		Sign of SAM determined by sign of overflow. Previous contents of SAM cleared without alarm.	24 μsec.
ao x	add one	10110	22	C(x)+(1x2^{-15})	----	C(x)	clear	C(x) + (1x2^{-15})	possible Arith. Overflow Alarm if C(x)+(1x2^{-15})≠1	32 μsec.
dm x	difference of magnitudes	10111	23	|C(AC)| - |C(x)|	C(AC)	|C(x)|	clear	----	if |C(AC)| = |C(x)| result is -0	24 μsec.
mr x	multiply and round-off	11000	24	C(AC)·C(x) + ρ	----	|C(x)|	clear	----	Sign of AC is determined by sign of product.	36-43 μsec.
mh x	multiply and hold	11001	25	C(AC)· C(x) (digits 1-15)	|C(AC)·C(x)| (digits 16-30) BR(0)=BR(14) BR 15=0	|C(x)|	clear	----	Sign obtained same as for mr. Result in (AC+BR) is a double register product.	Same as for mr
dv x	divide	11010	26	± 0 (sign of quotient)	|C(AC)| |C(x)| (16 digits)	|C(x)|	clear	----	Divide Error Alarm if |C(AC)| > |C(x)|. Arith. Overflow Alarm if |C(AC)| = |C(x)| ≠ 0 and dv x by slr 15. If C(AC)=C(x)=0, the quotient is 0.	73 μsec.
slr n	shift left and round-off (n places)	11011 0	27	F{C(AC+BR)2^n} + ρ (n taken mod 32)	clear	----	clear	----	possible Arith. Overflow Alarm if I{ C(AC+BR)2^n } + ρ ≠ 0. The sign digit is not shifted. Digits shifted out of AC 1 are lost. Negative numbers are complemented in AC before shifting and after rounding off. Digit 6 of slr n must be zero.	16-41 μsec.
slh n	shift left and hold (n places)	11011 1	27	F{C(AC+BR)2^n} (n taken mod 32)	F{|C(AC+BR)2^n|} (digits 16-(31-n)) BR(j) --→ 0 j=16-n,...,15 n>0	----	clear	----	The sign digit is not shifted. Negative numbers are complemented in AC before and after the shift. Digit 6 of slh n must be a one.	Same as for slr
srr n	shift right and round-off	11100 0	28	C(AC)2^{-n} + ρ (n taken mod 32)	clear	----	clear	----	possible Arith. Overflow Alarm on srr 0 (this instruction simply causes roundoff and clears BR). The sign digit is not shifted. Negative numbers are complemented in AC before shifting and after rounding off. Digit 6 of srr n must be a zero.	Same as for slr
srh n	shift right and hold	11100 1	28	C(AC)2^{-n} (n taken mod 32)	|C(AC+BR)2^{-n}| (digits 16-31)	----	clear	----	The sign digit is not shifted. Negative numbers are complemented in AC before and after the shift. Digit 6 of srh n must be a one.	Same as for slr
sf x	scale factor	11101	29	C(AC+BR)2^n (digits 1-15)	C(AC+BR)2^n (digits 16-(31-n)) BR(j) --→ 0 j=16-n,...,15 n>0	n·2^{-15}	clear	n·2^{-15} x(5-15)	X(0-4) unaffected, n is such that ½ ≤ |C(AC+BR) 2^n| < 1, if C(AC+BR) = 0, then n = 33. Negative numbers are complemented in AC before and after the multiplication.	33-81 μsec.
clc n	cycle left and clear	11110 0	30	(AC+BR)(n+i)_{32} --→ AC(i) i=0,...,15	clear	----	----	----	Sign digit is shifted with all other digits. Digits shifted left out of AC 0 are carried around into BR 15. No roundoff. No complementing of AC either before or after the shift. Digit 6 of clc n must be a zero. Instruction clc 0 clears AC without affecting AC.	Same as for slr
clh n	cycle left and hold	11110 1	30	(AC+BR)(n+i)_{32} --→ AC(i) i=0,...,15	(AC+BR)(n+i+16)_{32} --→ BR(i) i=0,...,15	----	----	----	Sign digit is shifted with all other digits. Digits shifted left out of AC 0 are carried around into BR 15. No complementing of AC either before or after the shift. Digit 6 of clh n must be a one. Instruction clh 0 does nothing.	Same as for slr
md x	multiply digits	11111	31	$C_i(AC) \otimes C_i(x)$ --→ AC(i) i = 0,..., 15	----	-(Final AC)	----	----	Multiplies digits with no carry ⊗ | 0 1 | 0 | 0 0 | 1 | 0 1	24 μsec.

Note: In operations mr, mh, dv, slr, srr, srh, sf, the C(BR) is assumed to be the magnitude of the least significant part of AC + BR. For the ab and dm operations, the BR is treated just as any storage register.

[1] Whirlwind I Instruction Code came from "Comprehensive System Manual, A System of Automatic Coding for the Whirlwind Computer," published by Massachusetts Institute of Technology, Digital Computer Laboratory, Cambridge, Mass.

Chapter 7

Some aspects of the logical design of a control computer: a case study[1]

R. L. Alonso / H. Blair-Smith / A. L. Hopkins

Summary   Some logical aspects of a digital computer for a space vehicle are described, and the evolution of its logical design is traced. The intended application and the characteristics of the computer's ancestry form a framework for the design, which is filled in by accumulation of the many decisions made by its designers. This paper deals with the choice of word length, number system, instruction set, memory addressing, and problems of multiple precision arithmetic.

The computer is a parallel, single address machine with more than 10,000 words of 16 bits. Such a short word length yields advantages of efficient storage and speed, but at a cost of logical complexity in connection with addressing, instruction selection, and multiple-precision arithmetic.

1.   Introduction

In this paper we attempt to record the reasoning that led us to certain choices in the logical design of the Apollo Guidance Computer (AGC). The AGC is an onboard computer for one of the forthcoming manned space projects, a fact which is relevant primarily because it puts a high premium on economy and modularity of equipment, and results in much specialized input and output circuitry. The AGC, however, was designed in the tradition of parallel, single-address general-purpose computers, and thus has many properties familiar to computer designers [Richards, 1955], [Beckman et al., 1961]. We will describe some of the problems of designing a short word length computer, and the way in which the word length influenced some of its characteristics. These characteristics are number system, addressing system, order code, and multiple precision arithmetic.

A secondary purpose for this paper is to indicate the role of evolution in the AGC's design. Several smaller computers with about the same structure had been designed previously. One of these, MOD 3C, was to have been the Apollo Guidance Computer, but a decision to change the means of electrical implementation (from core-transistors to integrated circuits) afforded the logical designers an unusual second chance.

It is our belief, as practitioners of logical design, that designers, computers and their applications evolve in time; that a frequent reason for a given choice is that it is the same as, or the logical next step to, a choice that was made once before.

A recent conference on airborne computers [*Proc. Conf. Space-borne Computer Eng.*, Anaheim, Calif., Oct. 30–31, 1962] affords a view of how other designers treated two specific problems: word length and number system. All of these computers have word lengths of the order of 22 to 28 bits, and use a two's complement system. The AGC stands in contrast in these two respects, and our reasons for choosing as we did may therefore be of interest as a minority view.

2.   Description of the AGC

The AGC has three principal sections. The first is a *memory*, the fixed (read only) portion of which has 24,576 words, and the erasable portion of which has 1024 words. The next section may be called the *central section;* it includes, besides an adder and a parity computing register, an instruction decoder (SQ), a memory address decoder (S), and a number of addressable registers with either special features or special use. The third section is the *sequence generator* which includes a portion for generating various microprograms and a portion for processing various interrupting requests.

The backbone of the AGC is the set of 16 write busses; these are the means for transferring information between the various registers shown in Fig. 1. The arrowheads to and from the various registers show the possible directions of information flow.

In Fig. 1, the data paths are shown as solid lines; the control paths are shown as broken lines.

Memory: fixed and erasable

The Fixed Memory is made of wired-in "ropes" [Alonso and Laning, 1960], which are compact and reliable devices. The number of bits so wired is about 4×10^5. The cycle time is 12 μsec.

The erasable memory is a coincident current system with the same cycle time as the fixed memory. Instructions can address registers in either memory, and can be stored in either memory.

[1]*IEEE Trans.*, EC-12 (6), 687–697 (December, 1963)

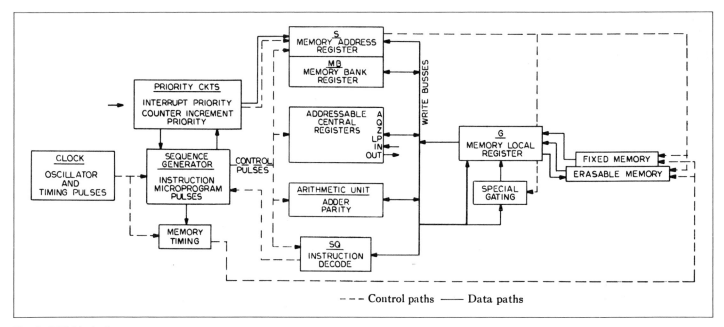

Fig. 1. AGC block diagram.

The only logical difference between the two memories is the inability to change the contents of the fixed part by program steps.

Each word in memory is 16 bits long (15 data bits and an odd parity bit). Data words are stored as signed 14 bit words using a one's complement convention. Instruction words consist of 3 order code bits and 12 address code bits.

The contents of the address register S uniquely determine the address of the memory word only if the address lies between octal 0000 and octal 5777, inclusive. If the address lies between octal 6000 and octal 7777, inclusive, the address in S is modified by the contents of the memory bank register MB. The modification consists in adding some integral multiples of octal 2000 to the address in S before it is interpreted by the decoding circuitry. The memory bank register MB is itself addressable; its address, however, is *not* modified by its own contents.

Transfers in and out of memory are made by way of a memory local register G. For certain specific addresses, the word being transferred into G is not sent directly, but is modified by a special gating network. The transformations on the word sent to G are right shift, left shift, right cycle, and left cycle.

Central section

The middle part of Fig. 1 shows the central section in block form. It consists of the address register S and the memory bank register

MB both of which were mentioned above. There is also a block of addressable registers called "central and special registers," which will be discussed later, an arithmetic unit, and an instruction decoder register SQ.

The arithmetic unit has a parity generating register and an adder. These two registers are not explicitly addressable.

The SQ register bears the same relation to instructions as the S register bears to memory locations; neither S nor SQ are explicitly addressable.

The central and special registers are A, Q, Z, LP, and a set of input and output registers. Their properties are shown in Table 1.

Sequence generator

The sequence generator provides the basic memory timing, the sequences of control pulses (microprograms) which constitute an instruction, the priority interrupt circuitry, and a number of scaling networks which provide various pulse frequencies used by the computer and the rest of the navigation system.

Instructions are arranged so as to last an integral number of memory cycles. The list of 11 instructions is treated in detail in Sec. 6. In addition to these there are a number of "involuntary" sequences, not under normal program control, which may break into the normal sequence of instructions; these are triggered either by external events, or by certain overflows within the AGC, and

Table 1 Special and central registers

Register(s)	Octal address	Purpose and/or properties
A	0000	Central accumulator. Most instructions refer to A.
Q	0001	If a transfer of control (TC) occurred at L, $(Q) = L + 1$.
Z	0002	Program counter. Contains $L + 1$, where L is the address of the instruction presently being executed.
LP	0003	Low product register. This register modifies words written into it by shifting them in a special way.
IN	. . .	Several registers which are used for sampling either external lines, or internal computer conditions such as time or alarms.
OUT	. . .	Several output registers whose bits control switches, networks, and displays.

may be divided into two categories: counter incrementing and program interruption.

Counter incrementing may take place between any two memory cycles. External requests for incrementing a counter are stored in a counter priority circuit. At the end of every memory cycle a test is made to see if any incrementing requests exist. If not, the next normal memory cycle is executed directly, with no time between cycles. If a request is present, an incrementing memory cycle is executed. Each "counter" is a specific location in erasable memory. The incrementing cycle consists of reading out the word stored in the counter register, incrementing it (positively or negatively), or shifting it, and storing the results back in the register of origin. All outstanding counter incrementing requests are processed before proceeding to the next normal memory cycle. This type of interrupt provides for asynchronous incremental or serial entry of information into the working erasable memory. The program steps may refer directly to a "counter" to obtain the desired information and do not have to refer to input buffers. Overflows from one counter may be used as the input to another. A further property of this system is that the time available for normal program steps is reduced linearly by the amount of counter activity present at any given time.

Program interruption occurs between normal program steps

rather than between memory cycles. An interruption consists of storing the contents of the program counter and transferring control to a fixed location. Each interrupt line has a different location associated with it. Interrupting programs may not be interrupted, but interrupt requests are not lost, and are processed as soon as the earlier interrupted program is resumed. Calling the resume sequence, which restores the program counter, is initiated by referencing a special address.

3. Word length

In an airborne computer, granted the initial choice of parallel transfer of words within it, it is highly desirable to minimize the word length. This is because memory sense amplifiers, being high-gain class A amplifiers, are considerably harder to operate with wide margins (of temperature, voltages, input signal) than, say, the circuits made up of NOR gates. It is best to have as few of these as possible. Furthermore, the number of ferrite-plane inhibit drivers equals the number of bits in a word in this case. Similarly, the time required for a carry to propagate in a parallel adder is proportional to the word length, and in the present case, this factor could be expected to affect the microprogramming of instructions. The initial intent, then, was to have as short a word length as possible.

Another initial choice is that the AGC should be a "common storage" machine, which means that instructions may be executed from erasable memory as well as from fixed memory, and that data (obviously constants, in the case of fixed memory) may be stored in either memory. This in turn means that the word sizes of both types of memory must be compatible in some sense; for the AGC, the easiest form of compatibility is to have equal word lengths. So-called "separate storage" solutions which allow different word lengths for instructions and data can be made to work [Walendziewicz, 1962] but they have a drawback in that three memories are then required: a data memory (erasable), and two fixed memories, one for instructions and one for constants. In addition, we have found that separate storage machines are more awkward to program, and use memory less efficiently, than common storage machines.

There are three principal factors in the choice of word length. These are:

1 Precision desired in the representation of navigational variables.

2 Range of the input variables which are entered serially and counted.

3   Instruction word format. Division of instruction words into two fields, one for operation code and one for address.

As a start, the choice of word length (15 bits) for two previous machines in this series was kept in mind as a satisfactory word length from the point of view of mechanization; *i.e.*, the number of sense amplifiers, inhibit drivers, the carry propagation time, etc., were all considered satisfactory. The act of "choosing" word length really meant whether or not to alter the word length, at the time of change from MOD 3C to the AGC, and in particular whether to increase it. The influence of the three principal factors will be taken up in turn.

Precision of data words

The data words used in the AGC may be divided roughly into two classes: data words used in elaborate navigational computations, and data words used in the control of various appliances in the system. Initial estimates of the precision required by the first class ranged from 27 to 32 bits, $0(10^{8\pm1})$. The second class of variables could almost always be represented with 15 bits. The fact that navigational variables require about twice the desired 15-bit word length means that there is not much advantage to word sizes between 15 and 28 bits, as far as precision of representation of variables is concerned, because double-precision numbers must be used in any event. Because of the doubly signed number representation for double-precision words, the equivalent word length is 29 bits (including sign), rather than 30, for a basic word length of 15 bits.

The initial estimates for the proportion of 15-bit vs 29-bit quantities to be stored in both fixed and erasable memories indicated the overwhelming preponderance of the former. It was also estimated that a significant portion of the computing had to do with control, telemetry and display activities, all of which can be handled more economically with short words. A short word length allows faster and more efficient use of erasable storage because it reduces fractional word operations, such as packing and editing; it also means a more efficient encoding of small integers.

Range of input variables

As a control computer, the AGC must make analog-to-digital conversions, many of which are of shaft angles. Two principal forms of conversion exist: one renders a whole number, the other produces a train of pulses which must be counted to yield the desired number. The latter type of conversion is employed by the AGC, using the counter incrementing feature.

When the number of bits of precision required is greater than the computer's word length, the effective length of the counter must be extended into a second register, either by programmed scanning of the counter register, or by using a second counter register to receive the overflows of the first. Whether programmed scanning is feasible depends largely on how frequently this scanning must be done. The cost of using an extra counter register is directly measured in terms of the priority circuit associated with it.

In the AGC, the equipment saved by reducing the word length below 15 bits would probably not match the additional expense incurred in double-precision extension of many input variables. The question is academic, however, since a lower bound on the word length is effectively placed by the format of the instruction word.

Instruction word format

An initial decision was made that instructions would consist of an operation code and a single address. The straightforward choices of packing one or two such instructions per word were the only ones seriously considered, although other schemes, such as packing one and a half instructions per word, are possible [England, 1962]. The previous computers MOD 3S and MOD 3C had a 3-bit field for operation codes and a 12-bit field for addresses, to accommodate their 8 instruction order codes and 4096 words of memory. In the initial core-transistor version of the AGC (*i.e.*, MOD 3C), the 8 instruction order codes were in reality augmented by the various special registers provided, such as shift right, cycle left, edit, so that a transfer in and out of one of these registers would accomplish actions normally specified by the order code (see Sec. 6). These registers were considered to be more economical than the corresponding instruction decoding and control pulse sequence generation. Hence the 3 bits assigned to the order code were considered adequate, albeit not generous. Furthermore, as will be seen, it is possible to use an indexing instruction so as to increase to eleven the number of explicit order codes provided for.

The address field of 12 bits presented a different problem. At the time of the design of MOD 3C we estimated that 4000 words would satisfy the storage requirements. By the time of redesign it was clear that the requirement was for 10^5 words, or more, and the question then became whether the proposed extension of the address field by a bank register (see Sec. 7) was more economical than the addition of 2 bits to the word length. For reasons of modularity of equipment, adding 2 more bits to the word length would result in adding 2 more bits to all the central and special registers, which amounts to increasing the size of the nonmemory portion of the AGC by 10 per cent.

In summary, the 15-bit word length seemed practical enough so that the additional cost of extra bits in terms of size, weight, and reliability did not seem warranted. A 14-bit word length was thought impractical because of the problems with certain input variables, and it would further restrict the already somewhat cramped instruction word format. Word lengths of 17 or 18 bits would result in certain conceptual simplicities in the decoding of instructions and addresses, but would not help in the representation of navigational variables. These require 28 bits, and so they must be represented to double precision in any event.

4. Number representation

Signed numbers

In the absence of the need to represent numbers of both signs, the discussion of number representation would not extend beyond the fact that numbers in AGC are expressed to base two. But the accommodation of both positive and negative numbers requires that the logical designer choose among at least three possible forms of binary arithmetic. These three principal alternatives are: (1) one's complement, (2) two's complement, and (3) sign and magnitude [Richards, 1955].

In one's complement arithmetic, the sign of a number is reversed by complementing every digit, and "end around carry" is required in addition of two numbers.

In two's complement arithmetic, sign reversal is effected by complementing each bit and adding a low order *one*, or some equivalent operation.

Sign and magnitude representation is typically used where direct human interrogation of memory is desired, as in "post-mortem" memory dumps, for example. The addition of numbers of opposite sign requires either one's or two's complementation or comparison of magnitude, and sometimes may use both. No advantage is offered in efficiency with the possible exception of sign changing, which only requires changing the sign bit. A disadvantage is engendered in magnetic core logic machines by the extra equipment needed for subtraction or conditional recomplementation.

The one's complement notation has the advantage of having easy sign reversal, which is equivalent to Boolean complementation; hence a single machine instruction performs both functions. Zero is ambiguously represented by all *zero's* and by all *one's*, so that the number of numerical states in an n-bit word is $2^n - 1$.

Two's complement arithmetic is advantageous where end around carry is difficult to mechanize, as is particularly true in serial computers. An n-bit word has 2^n states, which is desirable

for input conversions from such devices as pattern generators, geared encoders, or binary scalers. Sign reversal is awkward, however, since a full addition is required in the process.

The choice in the case of the AGC was to use one's complement arithmetic in general processing, and two's complements for certain input angle conversions. Since the only arithmetic done in the latter case is the addition of plus or minus *one*, the two's complement facility is provided simply by suppressing end around carry and using the proper representation of minus *one*. The latter is stored as a fixed constant, so that no sign reversal is required.

Modified one's complement system

In a standard one's complement adder, overflow is detected by examining carries into and out of the sign position. These overflow indications must be "caught on the fly" and stored separately if they are to be acted upon later. The number system adopted in the AGC has the advantage of being a one's complement system with the additional feature of having a static indication of overflow. The implementation of the method depends on the AGC's not using a parity bit in most central registers. Because of certain modular advantages, 16, rather than 15, columns are available in all of the central registers, including the adder. Where the parity bit is not required, the extra bit position is used as an extra column. The virtue of the 16-bit adder is that the overflow of a 15-bit sum is readily detectable upon examination of the two high order bits of the sum (see Fig. 2). If both of these bits are the same, there is no overflow. If they are different, overflow has occurred with the sign of the highest order bit.

The interface between the 16-bit adder and the 15-bit memory is arranged so that the sign bit of a word coming from memory enters both of the two high order adder columns. These are denoted S_2 and S_1 since they both have the significance of sign bits. When a word is transferred from the accumulator A to memory, only one of these two signs can be stored. Our choice was to store the S_2 bit, which is the standard one's complement sign except in the event of overflow, in which case it is the sign of the two operands. This preservation of sign on overflow is an important asset in dealing with carries between component words of multiple-precision numbers (see Sec. 5).

In a standard one's complement system, a series of additions may result in subtotals which overflow, yet still produce a valid sum so long as the total does not exceed the capacity of one word. In a modified one's complement system, however, where sign is preserved on overflow, this is no longer true; and the total may depend on the order in which the numbers are added; this is not a serious drawback, but it must be accounted for in all phases of logical design and programming.

	STANDARD					MODIFIED						
	S_1	4	3	2	1	S_2	S_1	4	3	2	1	
EXAMPLE 1: Both operands positive; Sum positive, no overflow. Identical results in both systems.	0	0	0	0	1	0	0	0	0	0	1	
	0	0	0	1	1	0	0	0	0	1	1	
	0	0	1	0	0	0	0	0	1	0	0	
EXAMPLE 2: Both operands positive; positive overflow. Standard result is negative; Modified result is positive using S_2 as sign of the answer. Positive overflow indicated by $S_1 \cdot \bar{S}_2$.	0	1	0	0	1	0	0	1	0	0	1	
	0	1	0	1	1	0	0	1	0	1	1	
	1	0	1	0	0	0	1	0	1	0	0	
EXAMPLE 3: Both operands negative; Sum negative, no overflow. End around carry occurs. Identical results in both systems using either S_1 or S_2 as the sign of the answer.	1	1	1	1	0	1	1	1	1	1	0	
	1	1	1	0	0	1	1	1	1	0	0	
	1	1	0	1	0	1	1	1	0	1	0	
					1 carry						1 carry	
	1	1	0	1	1	1	1	1	0	1	1	
EXAMPLE 4: Both operands negative; negative overflow. Standard result is positive; modified result is negative using S_2 as the sign of the answer. Negative overflow indicated by $\bar{S}_1 \cdot S_2$.	1	0	1	1	0	1	1	0	1	1	0	
	1	0	1	0	0	1	1	0	1	0	0	
	0	1	0	1	0	1	0	1	0	1	0	
					1 carry						1 carry	
	0	1	0	1	1	1	0	1	0	1	1	
EXAMPLE 5: Operands have opposite sign; Sum positive. Identical results in both systems.	1	1	1	1	0	1	1	1	1	1	0	
	0	0	0	1	1	0	0	0	0	1	1	
	0	0	0	0	1	0	0	0	0	0	1	
					1 carry						1 carry	
	0	0	0	1	0	0	0	0	0	1	0	
EXAMPLE 6: Operands have opposite sign; sum negative. Identical results in both systems.	1	1	1	0	0	1	1	1	1	0	0	
	0	0	0	0	1	0	0	0	0	0	1	
	1	1	1	0	1	1	1	1	1	0	1	

Fig. 2. Illustrative example of properties of modified one's complement system.

5. Multiple precision arithmetic

A short word computer can be effective only if the multiple-precision routines are efficient corresponding to their share of the computer's word load. In the AGC's application there is enough use for multiple-precision arithmetic to warrant consideration in the choice of number system and in the organization of the instruction set. Although the limited number of order codes prohibits multiple-precision instructions, special features are associated with the conventional instructions to expedite multiple-precision operations.

Independent sign representation

A variety of formats for multiple-precision representation are possible; probably the most common of these is the identical sign representation in which the sign bits of all component words agree. The method used in the AGC allows the signs of the components to be different.

Independent signs arise naturally in multiple-precision addition and subtraction, and the identical sign representation is costly because sign reconciliation is required after every operation. For example, $(+6, +4) + (-4, -6) = (+2, -2)$, a mixed sign representation of $(+1, +8)$. Since addition and subtraction are the most frequent operations, it is economical to store the result as it occurs and reconcile signs only when necessary. When overflow occurs in the addition of two components, a *one* with the sign of the overflow is carried to the addition of the next higher components. The sum that overflowed retains the sign of its operands. This overflow is termed an *interflow* to distinguish it from an overflow that arises when the maximum multiple-precision number is exceeded.

The independent sign method has a pitfall arising from the fact that every number has two representations, either one of which may occur as a sum. There are some numbers for which one of the representations exceeds the capacity of the most significant component. The overflow is false in the sense that the double-precision capacity is not exceeded, only the single word capacity of the upper component. Sign reconciliation can be used in this case to yield an acceptable representation. This problem can be avoided if all numbers are scaled so that none are large enough to produce false overflows. Such a restriction is not necessary, however, since the false overflow condition arises infrequently and can be detected at no expense in time. The net cost of reconciliation is therefore very low.

Multiplication and division

For triple and higher orders of precision, multiplication and division become excessively complex, unlike addition and subtraction where the complexity is only linear with the order of precision.

The algorithm for double-precision multiplication is directly applicable to numbers in the independent sign notation. False overflow does not arise, and the treatment of interflow is simplified by an automatic counter register which is incremented when overflow occurs during an add instruction. The sign of the counter increment is the same as the sign of the overflow; and the increment takes place while one of the product components of next higher order is stored in that counter.

Double-precision division is exceptional in that the independent sign notation may not be used; both operands must be made positive in identical sign form, and the divisor normalized so that the left-most nonsign bit is *one*.

Triple precision

A few triple-precision quantities are used in the AGC. These are added and subtracted using independent sign notation with interflow and overflow features the same as those used for double-precision arithmetic.

6.   Instruction set

Basic design criteria

The implicit requirements for any von Neumann-type machine demand that facilities exist for:

1   Fetching from memory

2   Storing in memory

3   Negating (complementing)

4   Combining two operands (*e.g.*, addition)

5   Address modification (more generally, executing as an instruction the result of arithmetic processing)

6   Normal sequencing (to each location from which an instruction can be executed there corresponds one location whose contents are the next instruction)

7   Conditional sequence changing, or transfer of control

8   Input

9   Output

An instruction can, of course, provide several of these facilities. For instance, some computers have an instruction that subtracts the contents of a memory location from an accumulator and leaves the result in that memory location and in the accumulator; this instruction fulfills all of requirements 1–4 above. Requirement 5 is met in a somewhat primitive manner if instructions can be executed from erasable memory, and is met elegantly by the use of index registers. Still another scheme, somewhat similar to one used in the Bendix G-20, is employed in the AGC. Requirement 6 is usually fulfilled by having an instruction location counter which contains the address of the next instruction to be executed, and is incremented by *one* when an instruction is fetched. Alternatively, each instruction may include the address of the next instruction, as is often done in machines having drum memories. In the AGC, as in most short-word computers, the former method, with one single-address instruction per word, is clearly the simplest and cheapest. Requirement 7 is generally met by examining a condition such as the sign of an accumulator and, if the condition is satisfied, either incrementing the instruction location counter (skipping), or using an address included in the instruction as that of the next instruction (conditional transfer of control). An unconditional transfer of control is usual but not necessary, since any desired condition can be forced. Most machines have special input-output instructions to satisfy requirements 8 and 9. In the AGC, however, since input and output is through addressable registers, input is subsumed under fetching from memory, and output under storing in memory. Counter incrementing and program interruption aid these functions also.

Further criteria

The major goals in the AGC were efficient use of memory, reasonable speed of computing, potential for elegant programming, efficient multiple precision arithmetic, efficient processing of input and output, and reasonable simplicity of the sequence generator. The constraints affecting the order code as a whole were the word length, one's complement notation, parallel data transfer, and the characteristics of the editing registers. The ground rules governing the choice of instructions arose from these goals and constraints.

a   Three bits of an instruction word are devoted to operation code.

b   Address modification must be convenient and efficient.

c   There should be a multiply instruction yielding a double length product.

d   Treatment of overflow on addition must be flexible.

e   A Boolean combinatorial operation should be available.

f   No instruction need be devoted to input, output, or shifting.

This list is by no means complete, but gives a good indication of what kind of computer the AGC has to be. In the following paragraphs the ways in which the instructions fulfill the above requirements are described.

Details of the instruction set

In the listing that follows, L denotes the location of the instruction; K denotes the data address contained in the instruction. Parentheses mean "content of," and the leftward arrow means that the register named at the arrowhead is set to the quantity named to the right.

$L: TC\ K;\ Transfer\ Control$

$Q \leftarrow L + 1;$ go to K.

This is the primary method of transferring control to any stated location, and thus meets part of requirement 7. The setting of the return address register Q renders complex subroutines feasible. TC Q may be used to return from a subroutine (with no other TC's) because the binary number "$L + 1$" is the same as the binary word "TC $L + 1$," by virtue of the TC code being all zeros. TC A behaves like an "execute" instruction, executing whatever instruction is in A, because Q follows A in the address pattern, see Table 1.

$L: CCS\ K;\ Count,\ Compare,\ and\ Skip$

If $(K) > +0, A \leftarrow (K) - 1$, no skip; if $(K) = +0, A \leftarrow +0$, skip to $L + 2$; if $(K) < -0, A \leftarrow 1 - (K)$, skip to $L + 3$; if $(K) = -0, A \leftarrow +0$, skip to $L + 4$.

This instruction fulfills the remainder of requirement 7 and provides several features. It is clear that in a machine with a 3-bit

operation code there should be only one code devoted entirely to branching, if at all possible. It is inefficient to program a zero test using only a sign-testing code; it is even more inefficient to program a sign test using only a zero-testing code. This instruction was therefore designed to test both types of conditions simultaneously. It has to be a four-way branch, and since there is only one address per instruction, it follows that CCS must be a skipping-type branch.

The function of (K) delivered to A is the diminished absolute value (DABS). It serves two primary purposes: to do most of the work in generating an absolute value, and to apply a negative increment to the contents of a loop-counting register, so that CCS has some of the properties of TIX in the IBM 704.

$L: INDEX\ K;\ Index\ using\ K$

Use $(L + 1) + (K)$ as the next instruction.

In a short-word machine where there is no room in the instruction word to specify indexing or indirect addressing, this code meets requirement 5 in a way far superior to forming an instruction and placing it in A or in erasable memory for execution. INDEX operates on whole words, so that the operation code as well as the address may be modified. It may be used recursively (consider the implications of several INDEX's in succession, assuming that no operation codes are modified). Finally, it permits more than 8 operation codes to be specified in 3 bits, since overflow of the indexing addition is detectable.

$L: XCH\ K;\ Exchange$

$(A) \longleftrightarrow (K)$.

This instruction meets requirements 1, 2, and 8. When K is in fixed memory, it is simply a data-fetching (clear and add) code. Its use with erasable memory aids efficiency by reducing the need for temporary storage. XCH is also an important input instruction in a machine where addressable counters, incremented in response to external events, are an input medium, because a counter can be read out and reset (to zero or any desired value) by XCH with no chance of missing a count.

$L: CS\ K;\ Clear\ and\ Subtract$

$A \leftarrow -(K)$.

CS is the primary means of sign-changing and logical negation, and so fulfills requirements 1 and 3. Since there is no clear and add instruction, it is the usual operation for nondestructive readout of erasable memory in simple data transfers, that is, when no addition or other arithmetic is required. Usually the programming can be arranged so that complementing during transfer is acceptable; otherwise the CS can be followed by CS A before storing.

$L: TS\ K;\ Transfer\ to\ Storage$

$K \leftarrow (A);$ if (A) includes $\pm$ overflow, $A \leftarrow \pm 1$, skip to $L + 2$.

This instruction is the primary means of transfers to memory and output, satisfying requirements 2 and 9. It is also the most convenient method of testing for overflow. Since A and the other central registers have two sign positions, overflow indication is retained in a central register. TS always stores (A) and tests whether overflow is present. If K is in erasable memory and is not a central register, the lower-order sign bit S_1 is not transmitted; this is the process or overflow correction. If positive overflow indication is present in A, TS skips over the next instruction and sets $A \leftarrow +1$ ($+1$ denotes octal 000001); if negative overflow is present, TS skips over the next instruction and sets $A \leftarrow -1$ (-1 denotes octal 177776); otherwise (A) are unchanged. The sequence

TS   K
XCH   ZERO   (ZERO in fixed memory)

suffices to store in K an overflow-corrected word of a multiple-precision sum and leave in A the interflow to the next higher-order part. TS A skips if either type of overflow is present, but leaves all 16 bits of (A) unchanged.

Finally, a computed transfer of control may be achieved by TS Z because Z is the program counter; only the low-order 12 bits of (A) are significant, being the address of the instruction to which control is transferred. Overflow in (A) in this case does not affect the transfer but sets $A \leftarrow \pm 1$.

L: AD K; Add

$A \leftarrow (A) + (K)$; if the final (A) includes $\pm$ overflow, OVCTR $\leftarrow$ (OVCTR) ± 1.

Addition is the most frequently used combinatorial operation (requirement 4). The property of OVCTR is used chiefly in developing double-precision products and quotients, partly because the additions in these processes are less susceptible to false overflow than are multiple-precision additions.

L: MASK K; Mask

$A \leftarrow (A) \cap (K)$.

This is the only combinatorial Boolean instruction, and may be used with CS to generate any Boolean function.

Extracodes

The AGC instruction set was carried over in large part from its ancestor, MOD 3C [Alonso et al., 1961]. All instructions of MOD 3C were retained in the AGC, modifications and additions being adopted where a substantial increase in computing power could be obtained at small cost. The MOD 3C instruction set was like the one described above for the AGC with two major exceptions: first, instead of a mask instruction, MOD 3C had a multiply instruction. Second, the transfer to storage instruction did not include the property of skipping on overflow, although it did have properties which aided masking.

After the design of MOD 3C was completed, it was discovered that the INDEX instruction could be used to expand the instruction set beyond eight instructions by producing overflow in the instruction word following the INDEX. For example, the addition of octal 47777 to the instruction word "CS K" in the course of an INDEX instruction will cause negative overflow, producing MP K, a multiply instruction with operand address K.

In order to implement the extracodes in the AGC, it was necessary to provide a path from the high-order 4 bits of the adder to the unaddressable sequence selection register SQ. Part of this path is the unaddressable buffer register B; these requirements helped to suggest the benefits of retaining two sign bit positions in all the central registers.

In principle, eight additional instruction codes can be obtained by causing overflow, but we did not feel obliged to use them all. Because every extracode must be indexed, the instructions chosen for this class had two properties to some degree: they are normally indexed, or they take long enough so that the cost of indexing without address modification is small. All the extracodes are combinatorial, and therefore relate to requirement 4.

L: MP K; Multiply

$A \leftarrow$ upper part, $LP \leftarrow$ lower part, of $(A) \cdot (K)$; the two words of the product agree in sign, which is determined strictly by the sign bits of the operands.

Experience with MOD 3C showed that it was worthwhile making a completely algebraic, self-contained multiply instruction, especially in doing double-precision multiplication whose operands have independent signs. The AGC multiply is much faster than that of MOD 3C, being limited by adder carry propagation time rather than core-switching time.

L: DV K; Divide

$A \leftarrow$ quotient, $Q \leftarrow - |$remainder$|$, of $(A)/(K)$; $LP \leftarrow$ nonzero number with the sign of the quotient.

Many facets of AGC design originally adopted for other reasons combined to make a divide instruction inexpensive. The foremost of these is the nature of the editing registers, which are in the standard erasable memory and have no special wiring. The special properties of these registers are supplied by a shift or cycle of the word being written into the memory local register G, when the address of an editing register is selected. The central loop of DV selects such an address and inhibits memory operations, so that all the left shifts required in division are accomplished in the G register while the editing register itself remains unchanged. The microprogrammed nature of order construction makes a restoring algorithm more efficient than a nonrestoring one. The quotient delivered to A has a sign determined according to normal algebraic rules by the signs of (A) and (K); the same sign is available in LP to aid in determining the correct sign of the remainder from those of the divisor and quotient in case the quotient has been absorbed by subsequent processing. DV is not usually indexed, but it pays such large benefits in space and time, especially in double-precision division, that the cost of extracode indexing is negligible. If the divisor is less in magnitude than the dividend, or is zero, the quotient has correct sign and, in general, maximum magnitude. No infinite loop results in any case.

L: SU K; Subtract

$A \leftarrow (A) - (K)$; if the final (A) includes $\pm$ overflow, $OVCTR \leftarrow (OVCTR) \pm 1$.

The primary justification for this instruction is that it allows multiple-precision addition subroutines to be changed into multiple-precision subtract subroutines merely by changing the indexing quantity. There are occasions in the middle of involved calculations where it is clumsy to construct a subtraction out of complementations and additions, especially when the sign of an overflow is of interest. Since SU differs from AD only in that the operand from memory is read out of the complement side of the buffer register B rather than the direct side, its cost is virtually zero. This last is not necessarily true when using core-transistor logic, or two's complement notation.

7. Expansion of memory addressing

The AGC's 12-bit address field is insufficient for specifying directly all the registers in its memory. This predicament seems increasingly to afflict most computers, either because indirect addressing is assumed as a necessary evil from the start or, as was our case, because our earliest estimates of memory requirements were wrong by a factor of two or three. The method of indirect addressing we arrived at uses a bank register MB, but with an important modification: the 5-bit number stored in MB has no effect unless the address is in the range (octal) 6000 to 7777. The MB register contents are not interpreted as higher-order bits of the address; they are interpreted as integers which specify which bank of 1024 words is meant in the event of the address part of the instruction being in the ambiguous range. The over-all map of memory is shown in Table 2. The unambiguous, fixed memory addresses domain has come to be known as "fixed-fixed."

It is interesting that this method of extending the addressing capability was not the result of trying to improve upon more conventional methods, but was almost a consequence of the phys-

Table 2   Address part of an instruction word

(Decimal)	
0–3071	Fixed and erasable memory; unambiguous addresses.
3072–4095	Fixed memory, ambiguous address. Contents of MB used to resolve the ambiguity. Up to 32 such banks are possible.

ical difference between fixed and erasable memory. Since all data other than constants are concentrated in the erasable memory, these had to be exempt from modification by the MB register. An alternative arrangement, whereby only the addresses of instructions (as opposed to the addresses within an instruction word) are modified, would be deficient in that it would allow only instructions to be stored in banks; there would be no way to refer to constants stored in banks, or to use bank addresses to store arguments of arithmetic operations. The possibility of using two bank registers is worthy of serious consideration [Casale, 1962], but it did not occur to us.

In addition to the addresses in erasable, it is necessary to exempt the addresses of interrupting programs (*i.e.*, the addresses to which a program interrupt transfers control) from the influence of the MB register. It was clear that it would be valuable to have a large body of unambiguous addresses for use in executive and dispatcher programs.

The most frequent and critical applications of bank changing are in the AGC's interpretive mode. Most of the programs relevant to navigation are written in a parenthesis-free pseudocode notation for economy of storage. An interpretive program executes these pseudocode programs by performing the indicated data accesses and subroutine linkages.

The format of the notation permits two macrooperators (*e.g.*, "double-precision vector dot product") or one data address to be stored in one AGC word. Thus data addresses appear as full 15-bit words, potentially capable of addressing up to 32,768 registers. Each such address is examined in the interpreter and the contents of the bank register are changed if necessary; preparation is also made for subsequent return if a subroutine call is being made.

The structure of the interpretive program, and its relationship to the computer characteristics discussed in this paper will not be taken up here except to point out that parenthesis-free notation is particularly valuable in a short-word computer such as the AGC. It permits a very substantial expansion of the address and pseudo-operation fields without sacrificing efficiency in program storage [Muntz, 1962].

The conversion of a 15-bit address into a bank number and an ambiguous 12-bit address is as follows: the top 5 bits correspond directly to the desired bank number. The remaining lower-order 10 bits, logically added to octal 6000, form the proper ambiguous address. If the 15-bit address is less than octal 6000, however, the address is in erasable or fixed-fixed memory. In this case the logical addition of octal 6000 is suppressed.

It is possible to have a program in one bank call a closed subroutine in another bank, and then have control returned to the proper place in the bank of origin. This is done by means of a short bank switching routine which is in fixed-fixed memory.

One potential awkwardness about this method of extending memory addresses is the possible requirement for a routine in one bank to have access to large amounts of data stored in another. There are many programming solutions to this problem, obviously at a cost in operating speed; a better solution would be to have two bank registers. No problems of this nature have yet materialized, however.

References

AlonR63; AlonR60; AlonR61; AlonR62; BeckF61; CasaC62; EnglW62; HopkA63; MuntC62; RichR55; WaleW62; *Proc. Conf. Spaceborne Computer Eng.*; Anaheim, Calif., Oct. 30–31, 1962.

APPENDIX 1   BACKGROUND FOR AGC DESIGN

Name, date completed	Memory size (F = fixed E = erasable)	Number of bits	Number of instructions	Purpose of design	Features incorporated at this stage
MOD 1, 1960	F:448 E: 64	11 and parity	4 plus involuntary	Feasibility Prototype	Counter increments, Interrupts, Core-Transistor Logic, Pulse rate outputs, Editing registers, Wired-in fixed memory, Interpretive programs.
MOD 2, not built	about 4000 total	23 and parity	16 plus indirect	Unmanned Space Probe	"Extended Operation" subroutine linkages (only instance).
MOD 3S, 1962	F: 3584 E: 512	15 and parity	8	Earth Satellite	Modified one's complement, Parallel adder, Addressable central registers.
MOD 3C, 1962	F: greater than 10^4 E: greater than 10^3	15 and parity	8 and involuntary	Apollo Guidance	CCS, INDEX, MULTIPLY instructions, Overflow counter, Bank switching.
AGC, 1963	F: greater than 10^4 E: greater than 10^3	15 and parity	11 and involuntary	Apollo Guidance	DV, SU, MSK instructions, Editing memory buffer, All transistor NOR logic instead of core-transistor logic, Extracodes, Parenthesis-free interpreter.

Chapter 8

The UNIVAC system[1]

J. Presper Eckert, Jr. / James R. Weiner
H. Frazer Welsh / Herbert F. Mitchell

Organization of the UNIVAC system

In March 1951, the first UNIVAC[2] system formally passed its acceptance tests and was put promptly into operation by the Bureau of the Census. Since the UNIVAC is the first computer which can handle both alphabetic and numerical data to reach full-scale operation so far, its operating record and a review of the types of problems to which it has been applied provide an interesting milestone in the ever-widening field of electronic digital computers.

The organization of the UNIVAC is such that those functions which do not directly require the main computer are performed by separate auxiliary units each having its own power supply. Thus the keyboard to magnetic tape, punched card to magnetic tape and tape to typewritten copy operations are delegated to auxiliary components.

The main computer assembly includes all of those units which are directly concerned with the main or central computer operations. A block diagram of this arrangement is shown in Fig. 1. All of the elements shown are contained within the central computer casework except the supervisory control desk (SC) and the Uniservos,[2] to which the lines in the upper right section of the diagram connect.

The supervisory control, in addition to all the necessary control switches and indicator lights, contains an input keyboard. Also cabled to the supervisory control is a typewriter which is operable by the main computer. By means of these two units, limited amounts of information can be inserted or removed either at the will of the operator or by the programmed instructions.

The input-output circuits operate on all data entering or leaving the computer. The input and output synchronizers properly time the incoming or outgoing data for either the Uniservos (tape devices) or the supervisory control devices. The input and output registers (I and O) are each 60 word (720 characters) temporary storage registers which are intermediate between the main computer and the input-output devices.

The high-speed bus amplifier is a switching central through which all data must pass during transfer between any arithmetic register and the main memory or between the memory and the input-output registers. The arithmetic registers are shown along the bottom of diagram each connected to the high speed bus system.

The L-, F-, X-, and A-registers are each of one word or 12-character capacity and are directly concerned with the arithmetic operations. The V- and Y-registers are of 2- and 10-word capacity, respectively. They are used solely for multiple word transfers within the main memory. Associated with the arithmetic registers are the algebraic adder (AA), the comparator (CP), and the multiplier-quotient counter (MQC).

Addition-subtraction instructions

The addition-subtraction operations are performed in conjunction with the comparator since all numerical quantities are absolute magnitudes with an algebraic sign attached. Before either an addition or subtraction is performed, the two quantities, one already in the A-register and the other either from the memory or from the X-register, depending upon the particular instruction, are compared for magnitude and sign. The adder inputs can then be switched so as always to produce a noncomplemented result for any operation. The choice of adder input arrangement is therefore under the control of the comparator. The comparator also determines the proper sign for the result according to the usual algebraic rules.

One additional function performed by the comparator for addition and subtraction is to control the complementer. This determination is based upon which operation (+, or −) is indicated, and, whether the signs are like or unlike. For a subtract instruction, the sign of the subtrahend is reversed before entering the comparator. The comparator then compares the signs of the quantities in order to determine whether the two quantities are subtracted or added.

Multiplication instruction

The multiplication process requires the services of the adder, the comparator, the multiplier-quotient counter and the four arithmetic registers. During the first step of multiplication the X-reg-

[1]*AIEE-IRE Conf.*, 6–16, December, 1951.
[2]Registered trade mark.

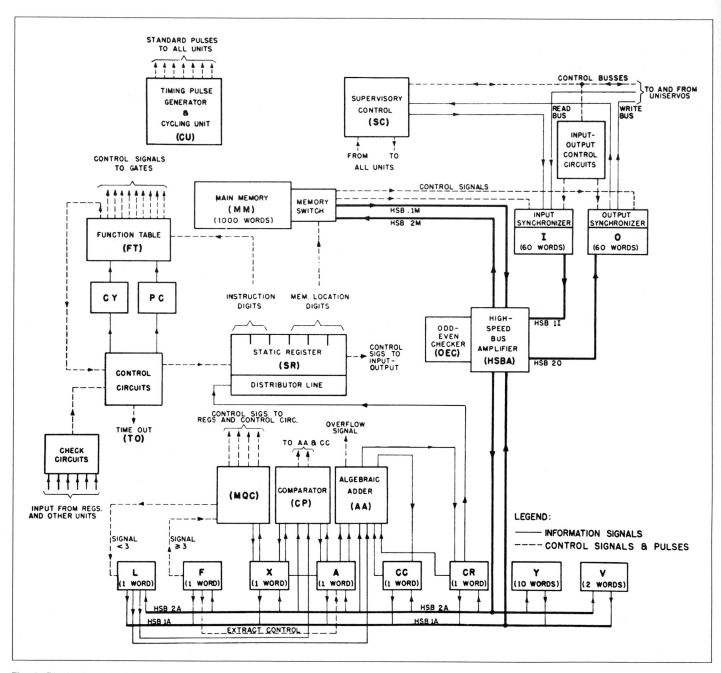

Fig. 1. Block diagram of UNIVAC.

ister receives the multiplier from the memory and the comparator determines the sign of the final product by comparing the signs of the multiplier and multiplicand. During the next three steps the multiplicand, which has been stored in the L-register by some previous instruction, is transferred three times to the A-register through the algebraic adder. The result, three times the multiplicand, is then stored in the F-register. During the next 11 steps of multiplication, the successive multiplier digits, beginning with the least significant, are transferred from the X-register to the multiplier-quotient counter. The multiplier-quotient counter then determines whether each particular multiplier digit is less than three, or greater than or equal to three.

If the former, the L-register releases the multiplicand to the A-register via the adder, and the multiplier-quotient counter is stepped downward one unit. If the multiplier digit is equal to or greater than three, the multiplier-quotient counter sends a signal to the F-register which releases three times the multiplicand to the A-register and the multiplier-quotient counter is stepped three times. Thus a multiplier digit of seven would be processed as two transfers from the F-register to the A-register and one transfer from the L-register to the A-register, or a total of three transfers.

When the multiplier-quotient counter reaches zero, the next multiplier digit is brought in from the X-register, while the A-register, containing the first partial product, is shifted one position to the right.

During the final step of multiplication, the sign is attached to the product which has been built up in the A-register. One of the several available multiplication instructions causes the least significant digits, as they are shifted beyond the limits of the A-register, to be transferred to the X-register where they replace the multiplier digits as they are moved to the multiplier-quotient counter. Thus 22 place products can be obtained as well as 11 place.

Division instruction

The division operation is performed by a nonrestoring method. The divisor is stored in the L-register by some previous instruction and the dividend is brought from the memory and put in the A-register during the first step of the division instruction. As in multiplication, the signs of the two operands are compared in the comparator at this time and the sign of the quotient is then stored in the comparator pending completion of the division operation. The principal stages of division consist of transferring the divisor from the L-register to the A-register through the complementer and adder as many times as required to produce a quantity less than zero in the A-register, the dividend having been first shifted one

position to the left. The multiplier-quotient counter counts each transfer, thereby building up the first quotient digit. As soon as the quantity in the A-register, (neglecting its original sign) goes negative, the digit in the multiplier-quotient counter, not counting the transfer which causes the remainder to go negative, is transferred to the X-register and the remainder in the A-register is shifted one place to the left. The divisor is then added to the A-register until the quantity becomes positive. This time the multiplier-quotient counter must give the complement of the number of transfers for the real quotient digit. Special complementing read-out gates provide this method of interpreting the multiplier-quotient counter.

The X-register therefore collects the quotient, digit by digit, from the multiplier-quotient counter until the full 11 digits have been obtained. The quotient is then transferred to the A-register and the sign from the comparator (CP) is affixed during the final stage of the divide instruction.

The other internal operations of the UNIVAC include many transfer instructions by which words may be moved among the registers and memory with and without clearing, the extraction instruction by which certain digits of a word may be extracted into another word according to the parity of the corresponding digits of an extractor word; shift instructions; and special control instructions such as breakpoint, transfer of control, (explained in subsequent paragraphs) and stop.

Basic operating cycle

The basic operating cycle of the UNIVAC is founded upon single address instructions which specify the memory location of one word. In the case of the arithmetic instructions which require two operands, one of the operands must be moved into the proper register by some previous instruction. In order to control the sequence of instructions, a special counter, called the control counter (CC), retains the memory location from which the succeeding instruction word is to be obtained. Each time a new instruction word is received from the memory, the quantity in the control counter is passed through the adder where a unit is added to it. Therefore the normal sequence is to refer to successive memory locations for successive instruction words. Initially the control counter is cleared to zero and the first group of instructions must, therefore, be placed in memory locations from zero upward. A transfer of control instruction enables the programmer to change the control counter reading whenever desired and thus shift from one sequence to another. After a transfer of control takes place, the new number in the control counter is increased by unity each time a new instruction word is obtained from the memory.

Transfer of control instructions

The transfer of control instructions are of three types, the unconditional transfer which changes the control counter reading without question, and two conditional instructions which require that either equality or a specific inequality exists between the words in the A-register and the L-register. In the former case the quantities must be identical for transfer of control to occur and in the latter the quantity in the A-register must be greater than the quantity in the L-register for the control counter reading to be changed.

Since the UNIVAC can handle alphabetic as well as numerical data, these conditional transfer instructions are as useful for alphabetizing as they are to determine if a certain iterative arithmetic process has been performed often enough to come within specified numerical tolerances.

Control register

Since six characters (intermixed alphabetic and numerical) are sufficient to specify an instruction and there are 12 characters per word, each instruction word can represent two independent instructions. A 1-word register, called the control register (CR), has been provided which stores each instruction word as it comes from the memory. Thus one memory referral is sufficient for a pair of instructions and the control register stores both halves so that the second instruction is available as soon as the first has been completed.

The general term control circuits includes all those elements which work together to process the instruction routine. As each instruction word reaches the control register, the first half of it is passed immediately into the static register (SR). The static register drives the main function table and memory switch. The instruction digits are translated by the function table into the appropriate control signals for the instruction called for. The memory switch selects the location called for by the memory location digits and opens the proper memory channel to the high-speed bus system at the proper time. Since the memory is constructed of 100 channels, each holding ten words, the memory switch is a combination of spatial and temporal selection.

Cycle counter

Implicit within each instruction, as translated by the function table, is an ending signal which causes the computer to move on to the next instruction. The key to this sequence is the cycle counter (CY), which is advanced by the ending pulse. The cycle counter is a 2-stage 4-position counter, which is connected into

the function table. By virtue of this relation, CY develops signals in addition to those developed by the instruction, which, for example, can cause the control register to transfer the second half of the instruction word into the static register when the first half has been completed. Similarly, after the second half instruction is finished the cycle counter causes the reading of the control counter to pass into the memory location section of the static register and thus cause the next instruction word to be transferred from the memory to the control register. When the word reaches the control register, the cycle counter also causes the control counter reading to be increased by unity. The four cycles are designated by the first four Greek letters α (transfer CC to SR), β (transfer memory to CR), γ (perform first instruction), and δ (perform second instruction).

Program counter

The multistage instructions, such as multiplication, are guided through their various steps by the program counter (PC). The program counter has four stages or 16 positions. All multistage instructions can be performed within this number of steps.

Checking circuits

The checking circuits of the UNIVAC are of two main types, odd-even checkers and duplicated equipment with comparison circuits. The odd-even checker depends upon the design of the pulse code used within the computer. This code provides seven pulse positions for every character. Six of the seven positions are significant as the actual code while the seventh is the odd-even channel. If the number of pulses or ones within the first six channels of any character is even, a one is placed in the seventh channel to make the total odd. Thus, the total number of ones across the seven channels is always odd. By means of a binary counter and a few gates, an odd-even checker has been constructed which examines every seven pulse group which passes through the high speed bus amplifier. In this connection, mention must be made of the periodic memory check which interrupts operation every five seconds to pass the entire contents of the memory over the high speed bus system and, consequently, through the odd-even checker. Any discrepancy is immediately signalled to the supervisory control and further operation ceases.

The duplicated equipment type of checking consists of duplicating the most essential part of the arithmetic circuits and their controls and producing simultaneously independent results, which can then be compared for equality. For this type of checking, the A-, F-, X-, and L-registers, algebraic adder, comparator, multiplier-quotient counter, and the high speed bus amplifier are duplicated.

The memory is not duplicated, but is checked by the periodic memory check mentioned previously. Various sections of the control circuits are duplicated such as the program counter and cycle counter.

Timing pulse generator and cycling unit

The timing pulse generator and cycling unit (CU) are the source of the basic timing signals throughout the computer. The timing pulses occur at 2.25 megacycles per second. The cycling unit subdivides this rate into the character rate and word rate. The character rate is one seventh of the basic pulse rate since there are seven pulses for each character. There are 12 characters per word but space for a 13th character is included in a word time and is called the space between words. This time is used for switching purposes.

The cycling unit, therefore, develops the word signals at $\frac{1}{7} \times \frac{1}{13}$ or $\frac{1}{91}$ of the basic pulse rate. Within the cycling unit (CU) are numerous duplications and comparisons to ensure complete reliability.

Input-output circuits

The operation of the input-output system is dovetailed as efficiently as possible with the operation of the arithmetic circuits. Whenever possible, parallel operations are allowed to proceed so as to minimize the time lost on internal operation while the slower input-output operations are taking place.

The principal input-output instructions are handled in a manner identical to that for the internal operations, except that now the function table develops signals which bring the input-output control circuits into operation. The information supplied to the input-output control circuits by the function table includes the following:

1 Which of the ten possible Uniservos is being called on

2 Whether it is a read or write, that is, an input or output operation

3 If it is "read," the direction in which the tape is to move

The input-output control circuits, therefore, begin by testing whether or not the Uniservo indicated now is in use or not. If it is already in use, everything else waits until that Uniservo is free. Next, the input-output control circuits test to determine whether the Uniservo selected last moved backward or forward.

If the previous direction does not agree with the new direction called for, the input-output control circuits generate the proper signals to prepare the Uniservo to move in the opposite direction. If the instruction is to rewind a Uniservo, the input-output control circuits then direct the center drive of the selected Uniservo to rewind the tape to the beginning and stop.

As soon as the instruction has proceeded to the point where the input-output control circuits need no further information from the function table, the instruction ending signal is generated and the internal circuits proceed to the next instruction, even while the reading, writing or rewinding continues. The UNIVAC can process an input, an output and several rewind operations while simultaneously carrying on internal computation.

So far the method by which the words are transferred from the I-register to the memory has not been mentioned. This operation is combined with certain read instructions in a manner not immediately obvious. There are two instructions which read from the tape to the I-register, one causing the tape to move forward, the other causing it to move backward. There are two other input instructions similar to those just mentioned, but they have the additional operation of first reading from the I-register to the memory and then reading a new group of 60 words from tape into the I-register. Thus the first type of input instruction reads from tape to the I-register only. It must be followed by the second type of instruction in order first to clear the I-register and then read in the second block of 60 words.

The output instructions do not operate in this way but instead read directly from memory to the O-register and then to the tape as one instruction.

A third type of checking circuit occurs in the input-output control circuits which counts the number of characters transferred from the tape in each block. Since there must always be 720 characters per block, the 720 checker signals any discrepancy to the supervisory control.

One other phase of the input-output operation concerns the two supervisory control input-output instructions. One of them permits a single word to be typed in from the input keyboard and the other causes a single word to be typed out automatically.

Auxiliary equipment

The two principal auxiliary devices mentioned earlier were the Unityper,[1] which converts keyboard operations to tape recording, and the Uniprinter,[1] which converts magnetic recording to typewritten copy.

[1] Registered trade mark.

Unityper. A simple block diagram of the Unityper is shown in Fig. 2. Each keyboard operation pulses the input to an encoding function table which, in turn, drives the appropriate heads for recording the particular combination on the tape. Simultaneously, the same pulse triggers a motor delay flop which operates the tape motor for an interval sufficient to move the tape across the head for the distance required to record one character. However, there is a punched paper loop system associated with the Unityper for the purpose of providing the typist with various guideposts individually set up for each problem. The loop control system serves three distinct control functions. First, it allows the programmer to set up various numbers of characters for the individual items being entered for a given problem. If the typist ever enters other than the specified number of characters, the loop control signals an error. Although the basic word length is 12 characters, the programmer may subdivide or group the words to suit any length of item. The loop can then be punched with what are called "force check" punches. Whenever the typist completes a correctly entered item, she must operate a release key before entering the next item. If the forced check is released too early an error is created, or if an additional character is typed after the forced check should have been released, an error is similarly indicated.

The second function of the loop is to control the erase operation. The erase operation is the only way in which an error can be recalled. When the erase key is operated, the loop and tape

are both stepped backward until a stop punch (usually associated with each forced check) is encountered. Thus the entire erroneous item is erased, and at a much higher rate than that at which the backspace key can be operated. The backspace, incidentally, cannot cancel an error indication, but it can be used to correct a wrongly typed character if the typist recognizes it.

The third function of the loop system is to enter, automatically, various fill-in characters. Under one such system of operation, the loop control records the characters only at the behest of the operator. This function is useful where individual entries, such as personal names, do not fill out all of the space allotted. The other operation is fully automatic in which the loop assumes full control to record, for example, a group of fill-in characters later to be replaced by computed data within the central computer.

The block diagram therefore shows the loop motor connected to the same delay flop that steps the tape motor. The same signal which moves the two motors also sets a second delay flop ($DF2$) which produces a delayed probing pulse. The probing pulse examines the paper loop photoelectrically for the new combination. A third delay flop ($DF3$) produces another probing pulse after the relays associated with the loop photocells have had time to set up. If any automatic function is indicated by the photocells, the probing pulse passes through the interpreting relays, enters the encoding function table to generate the fill-in characters, and thus starts the cycle over again. All automatic functions take place at about 22 characters per second.

Numerous odd-even checks are introduced in the Unityper to provide checks on tape and loop motion and on the recorded code combination.

Uniprinter. The Uniprinter is shown in simplified block diagram in Fig. 3. Its operation is a simple cycle which is initiated by a start button. The start button triggers the motor flip-flop (MFF). The motor pulls the tape across the reading head until a combination is detected. The presence of pulses on any of the seven lines between the reading head and the relay decoding function table is sufficient to restore the motor flip-flop (MFF) and stop the tape motion. Simultaneously a print delay flop ($DF1$) is triggered. During the delay flop interval, the decoding relays are given time to set up. When the delay flop recovers, a pulse is sent through the relay table which reappears at one of the typewriter magnetic actuators. As the typebar reaches the platen, a printer action switch (PAS) is operated which pulses the motor flip-flop and starts a new search for the next character on the tape. The odd-even properties of the UNIVAC pulse code are utilized for checking purposes.

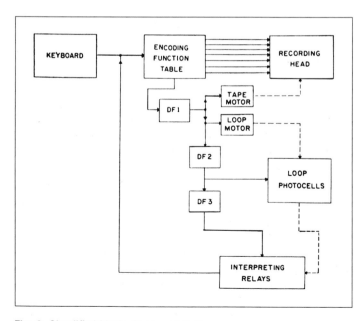

Fig. 2. Simplified block diagram of Unityper.

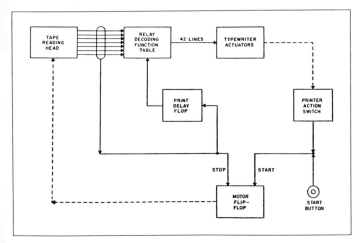

Fig. 3. Simplified block diagram of Uniprinter.

Engineering aspects

The entire UNIVAC system is constructed of circuits which are as conservative as is consistent with the desired reliability and speeds of operation. The circuits have been designed as building blocks and the entire computer is constructed around these blocks.

One of the most important of these blocks is the pulse reshaping circuit which consists of a timing pulse gate and a fast acting flip-flop which generates the pulse envelope equivalent of the gated timing pulses. Two polarities of timing pulse are used, the one being capable of tripping the flip-flop into one state, the other polarity of tripping it to the other state. As a deteriorated pulse envelope is applied to the timing pulse gate input, either one or the other polarity of pulse is always gated. The flip-flop therefore produces a sharpened and correctly timed output waveform.

The gating and switching circuits in the central computer are constructed of germanium crystal diodes, which include the main and subordinate function tables.

The registers are all circulating delay type using a mercury tank of one, two, or ten word-times of delay, except the static register. The latter is composed of 27 flip-flops which are required to maintain the static signals applied to the function tables, for at least an entire word-time.

The switching time allowed by the seven pulse-times of the space between words is, in general, not sufficient for a new function table excitation to stabilize. Therefore the time-out system used successfully in the BINAC, also is employed in the UNIVAC. Whenever an ending pulse is generated, or any other pulse which indicates that a new set of control signals are required from the function table, an interval of one word-time is introduced to allow the function table signals to reach equilibrium. The time-out interval is controlled by a single fast-acting flip-flop. All gates attached to the function table signals which are critical as to opening and closing can be inhibited by the time-out flip-flop during time out. Regardless of the presence of the function table signals, the gate does not operate until the time-out flip-flop releases it. Thus, the burden of speed imposed by the short space between words has been shifted to a single flip-flop which can accommodate the needs of the entire computer.

The UNIVAC uses the excess-three pulse code system which requires a second binary adder after the main binary adder in order to provide the excess-three correction after each addition. On the other side of the ledger, the complementing operation for subtraction and division is very much simplified, since the substitution of ones for zeros and vice versa is sufficient to form a complement. The excess-three part of the pulse code occupies the four least significant digit positions. The next two positions beyond the excess-three digits are used as zone indicators. When these digits are both zero, the last four positions are interpreted as a numerical quantity; when nonzero, an alphabetic or punctuation symbol is indicated. The seventh channel is the check pulse channel.

The adder is provided with an alphabetic bypass circuit which allows an alphabetic letter to enter one input and emerge unscathed provided a numeral enters the other input. Thus additive numerical constants can be combined with instruction words to adjust the memory location part of an instruction without affecting the alphabetic instruction symbols.

The power supply for the computer is separately housed. It can be placed any reasonable distance from the central computer. Almost all rectification is done by dry disc rectifiers. The power supply provides all a-c and d-c potentials to the central computer, supervisory control, directly-connected printer, and the Uniservos.

A complete fusing system has been included which serves both as protection and as a short-circuit isolating means. Each section, of which there are 39, is locally fused, enabling the engineer to locate a short within only 12 chassis, rather than the total of 468.

An automatic voltage monitoring system may be used to test every d-c voltage at the rate of one per second. A meter movement relay signals any discrepancy from standard. Similarly, overheat thermostats detect any unfavorable temperature condition in the bays or mercury tanks.

Cooling for the power supply and central computer is provided by three blowers. Local cooling in the Uniservos is provided by small fans in each unit. The operating statistics of the UNIVAC are as follows:

Tape reading and recording:
  Pulse density: 120 per inch
  Tape speed: 108 inches per second
  Input block size: 60 words: 720 characters
  Tape width: $\frac{1}{2}$ inch: 8 channels

Internal operations:
  Memory capacity: 1,000 words; 12,000 characters
  Memory construction: 100 mercury channels; 10 words/channel

Access time:
  Average: 202 microseconds
  Maximum: 404 microseconds

Word length:
  12 characters
  9 pulses
  (include space between words = 7 pulses)

Basic pulse rate:
  2.25 megacycles
  Addition: 525 microseconds
  Subtraction: 525 microseconds
  Multiplication: 2,150 microseconds
  Division: 3,890 microseconds
  (All times shown include time for obtaining instructions and operands from memory)

Applications of UNIVAC

Types of problems for which UNIVAC is applicable

True to its name, Universal Automatic Computer, the UNIVAC system is capable of handling data processing or calculation in virtually all fields of human endeavor. It is particularly well suited to applications requiring large volumes of input or output data, or both.

For convenience and classification, applications of the UNIVAC will be treated under four headings: scientific, statistical, logistical, and commercial. The scientific problem usually, though not always, has relatively small amounts of input and output data, with emphasis on computation. The statistical problem has relatively large volumes of input data with a small volume of output data and simple processing procedures. The commercial and logistical problems both have relatively large amounts of input and output data with processing requirements varying from slight to relatively great. A number of problems in each of these four fields have been studied and found suited for solution on the UNIVAC system. Several in each field have actually been processed on the computer.

Scientific problems

A general-purpose matrix algebra routine designed to add, subtract, multiply, and reciprocate matrices of orders up to 300 has been prepared and applied to a number of matrices. Inverses have been calculated for three different matrices of orders 40, 50, and 44. The error matrices for the first two of these inverses also were calculated. In both, the largest error term was of the order of 10^{-8}. A triple product matrix was formed from component matrices ranging from 5 by 40 to 40 by 40. A check product was obtained by reversing the sequence of multiplications, verifying the original product to within 2 units in the 11th place. The computer time required for these calculations was 1 hour and 15 minutes to calculate the inverse of order 50, 45 minutes to determine its error matrix. The other calculations were proportionately shorter. In all of this work, magnetic tapes were used as temporary storage for the bulk of the matrix elements involved. The high speed of the tape reading units more than kept up with the computer's need for data. No mathematical checks, other than the over-all check mentioned, were included in the computation, the self-checking features of the system making these completely unnecessary.

A second computation—that of obtaining six different specific solutions to a system of 385 simultaneous equations—was completed in 27 minutes on the computer. The system of equations arose from a second order nonlinear differential equation of gas flow through a turbine. The error terms resulting from the substitution of the computed unknowns into the basic equation were of the order of 10^{-11}.

The third example is that of a 2-dimensional Poisson equation, using a 22 by 22 mesh. Each iteration required 13 seconds and produced a maximum separation of successive surfaces of the order of 10^{-8} after approximately 300 iterations.

Statistical problems

In the second major field of statistical computation, the Census problem has been a prime example. The Census problem produces a part of the Second Series Population on Tables for the 1950 Decennial Census.

The Second Series contains 30 types of tables covering the statistics of our population—age, sex, race, country of birth, education, occupation, employment, and income. These tables are to be compiled for every county, and for every city, rural farm, and rural nonfarm area within a county.

The preparation of these tables by the UNIVAC system requires three major steps:

1   Tabulation of each individual's characteristics by groups of about 7,000

2 Arranging these groups by cities, counties

3 Assembling from the tabulations the data required for each table

The raw data were prepared in the form of a punched card for each individual in the United States. The data from these enumeration cards are then transcribed onto magnetic tape. From these tapes, the computer processes the data sequentially through the three steps, producing output tapes from which the tables are printed on Uniprinters. The only manual operations encountered in this entire procedure are the handling of the original punched cards, mounting and demounting tape reel (the equivalent of 9,700 cards), and the removal of the printed tables from the Uniprinters.

The most important feature of the present procedure is the elimination of handling and sorting tremendous quantities of punched cards. Each handling of the card stacks is a source of potential error and delay. The UNIVAC memory permits the simultaneous accumulation of the 580 tallies which describe our population for each local area being studied by the UNIVAC system.

Commercial problems

In the commercial field, the UNIVAC system has handled premium billing for a life insurance company. This program produces premium notices, dividends, and commissions. In a particular example worked out, approximately 1,000,000 bills, 340,000 dividends, and 100,000 commissions have to be produced monthly. The necessary information for processing a particular policy is contained in 240 digits, or, in special cases, 480. This compactness is made possible by a logical system of 40 symbols, comprising both alphabetic and numeric characters, which denote over 90 definitions. The UNIVAC processes the policies as directed by the symbols, policy dates, and policy numbers.

The problem includes inserting over 250,000 changes each month before further handling is done. After this step, the policies to be processed are selected from a file of 1,500,000 items. Next, a list is produced of the cases which have symbols indicating that special notices must be sent to the policyholders. Following the calculation of dividends and commissions, additional lists are produced: one group contains information pertaining to commissions and agents; another contains information regarding dividends; and finally, there is a listing of option changes for later insertion into the policy files. Policies requiring premium notices are then edited and the notices are automatically printed from the data contained on magnetic tapes.

The UNIVAC time needed for a program of this proportion is about 135 hours a month. The average computer time per policy processed is less than 0.5 second. The average time for all change insertions, printing, calculations, and unityping is 9 seconds per item.

Logistical problems

In the field of logistics, five major studies have been conducted, four of these resulting in actual problems executed on the computer.

The first is the type of computation in which the basic purpose is to determine quantitively whether a given operational or mobilization plan can be logistically supported. The ultimate desired is to find, by calculation, the optimum program for carrying out such plans. At the time of writing, only a small model has been actually run on UNIVAC, but full size models will be run within the next few weeks. Two computations have been executed, one a set of three tables of thousands of lines each, giving a detailed breakdown of machine deployment, fuel requirements, and overhaul requirements. The other problem was a computation of the amounts of critical raw materials required to construct a given number of each type of equipment, these requirements being phased by quarters over a 2-year period. The fourth problem, which was actually computed, was a sample of a similar calculation in which every pound of critical raw material required each month for the ultimate construction of a complete building program was computed.

The UNIVAC program which was prepared is capable of accommodating every type of equipment, individually tailored construction schedules, detailed bills of materials running into the millions of items and of determining the actual amounts of alloy elements based on thousands of tables of percentages for the many alloys employed. The demonstration showed that this computation for 400 pieces of equipment of a given type could be executed in three hours of computer time. The last problem in this field has not yet been run, but the study has shown that the entire gamut of stock control for a large supply office can be covered by the computer in approximately 3 weeks time.

This program involves the maintenance of stock balances of hundreds of thousands of stock items for many service points and provides for the preparation of stock transfer orders, purchase requisitions, critical lists and summary reports.

Performance record of the UNIVAC

Acceptance tests

The Acceptance Tests, prepared jointly by the Bureau of Standards and Bureau of Census, are fully discussed in the following paper by Dr. Alexander and Mr. McPherson.[1] However, a few comments

[1] Paper not included in this book. See McPherson and Alexander [1951].

concerning them from the engineering point of view are appropriate.

The Census computer was given two tests; the first, a test of its computational ability; the second, a test of its input-output system which particularly stressed the tape reading and recording abilities.

The Central Computer Acceptance Test A consisted of two parts. During Part 1, every available internal operation, except input-output operations, was performed. Among these operations were addition, subtraction, comparisons, division, and three different types of multiplication operations. Each of the arithmetic operations handled a pair of 11-decimal digit quantities. Altogether there were about 2,500 operations in the routine, yet the entire routine required only 1.26 seconds to do. The routine was performed 808 times in 17 minutes making a total of about 2,000,000 operations in all.

The second part of Test A included the solution of a heat distribution equation, a short routine involving the input-output device and a sorting routine. The sorting routine arranged ten numerical quantities each containing 12 decimal digits in correct numerical order in about 0.2 second. All three routines took a total of $1\frac{1}{2}$ minutes to perform. They were performed twice for each test and when added to Part 1 made a total of 20 minutes for unit test A.

The Acceptance Test B examined the input-output tape devices (Uniservos). During the first part of Test B, 2,000 blocks or about 1.4 million digits, which included every available character (numeric and alphabetic) were recorded on a tape and then read back into the computer with the tape moving backward. The information read back was then compared with the original data read out. The recording operation required about 4 minutes while reading back and comparison required about 8 minutes. The second part of Test B consisted of recording and reading over one spot of tape for 700 passes in order to determine the readability of tape as it wears. This test required 13 minutes and when combined with Part 1, made a total of approximately 25 minutes for Test B. This test was repeated 19 times.

The first test run passed in 6.6 hours (minimum theoretical time: 6.0 hours) and the second test was passed in 9.47 hours (minimum theoretical time: 7.45 hours). Of the 2.02 hours down time, 1.45 hours were accumulated at one time with the remaining 0.58 hours spread over the rest of the test.

The Uniprinter test required that a block of information (60 words) be printed 200 times in tabular form. The minimum time for printing was five hours. The test was passed in 6.16 hours.

The card-to-tape test required that ten good reels of tape be produced in 12 hours. There were certain restrictions as to reading accuracy and other criteria of reproducing ability which defined "good" reels. In 10 hours, the converter had prepared over 15 reels, 14 reels had been tested, 11 of the 14 were found satisfactory and the converter was accepted for payment.

Although the test was run on only one of two converters, the Bureau of Census put both card-to-tape machines into operation and after six months of use, the acceptance test was run on the second card-to-tape converter. This test differed to some extent from the first test in that the Census Bureau was satisfied with the reading ability of the machines and did not require a digit-by-digit verification of the information. However, a new stipulation was added that, after the engineers had checked the converter out preparatory to running the test, the converter was to be used in actual operation for eight hours before doing the remainder of the test with no engineering intervention between the two portions of the test. The first part was run on Friday, October 5, 1951; the device remained idle Saturday and Sunday and was turned on Monday morning to complete the test. It passed with flying colors, preparing ten acceptable reels (out of ten reels) plus two decks of check cards in slightly less than 7 hours. Both card-to-tape converters now are in Washington and the remainder of the system is in operation by the Bureau of the Census on the Eckert-Mauchly premises in Philadelphia.

Reliability and factors affecting performance

The first UNIVAC system now has been operating for approximately 8 months. In that time, much has been learned about how UNIVACs should be operated and maintained. The situation has been somewhat complicated by having to shake down the equipment while in the customer's possession; that is, there were certain faults in the system from both engineering and production standpoints which could only become apparent in the course of time and under actual operation conditions. For example, weak tubes or faulty solder joints did not reveal their presence at the time of installation. Another type of difficulty only became apparent under certain duty cycle conditions imposed by various types of problems. Because only certain problems present this particular duty cycle, these troubles remained in the machine causing intermittent stoppages until they could be tracked down.

Patient isolation and elimination of such problems, most of which have occurred only with conditions of operation infrequently encountered, is a powerful, though sometimes painful proving ground for the engineering group charged with such responsibility. The experience and depth of judgment acquired by such a group in the course of performing such work have become unmistakably apparent in the already noted improved performance of following UNIVACs and generally advanced ability to predict and realize performance in any large scale and complex apparatus of the same character.

Some of the troubles encountered are interesting to study in detail. On a rather complicated routine requiring the use of a number of Uniservos, all ran smoothly for 15 minutes. At that time, one of the Uniservos executing a backward read somewhere in the middle of the reel, did not stop at the end of the block but continued to run until it ran off the end of the tape. After much work, it was shown that a cycling unit signal was being overloaded because it was being used both by a multiplication instruction and the backward read which were occurring simultaneously. The input precessor loop was cleared as a result and the count of the pulses coming off the tape was thereby lost. Once the trouble was found, it was simple to remedy.

Another rather interesting case occurred intermittently over an extended period. Normally when reading out of the memory, the contents should not be cleared. Occasionally, however, reading from the memory also caused the contents to be cleared. As the trouble only remained for a period of seconds or, at most, a few minutes, it was somewhat difficult to localize. Of course, parasitic oscillations of some sort were suspected and, in fact, the trouble was traced to the actual source on a logical basis; but the source, a high power cathode follower, showed no evidence of oscillation. Before the problem was remedied, various combinations of parasitic suppressors were tried; the trouble would vanish for perhaps a week and then return. The oscillation finally cropped up during a maintenance shift, was found to be in the suspect tube at 100 megacycles and was eliminated rather easily.

Other types of troubles that have occurred include intermittent parasitic oscillations in other circuits, bounce in Uniservo relay circuits, various mechanical problems in Uniservos, time constants not consistent with the longest duty cycle signals, and various types of noise in the input circuits. The tubes, which initially were bothersome, have now stabilized to the point where two tubes per week (on the average) stop the computer during computation.

All of the above troubles and others not discussed here have contributed to lost computing time on the UNIVAC. However, they cannot influence future operation because the reasons for them have been found and eliminated. The fact that these troubles will not occur in future UNIVACs cannot be emphasized too strongly.

Under a contract with the Bureau of Census, Eckert-Mauchly Computer Corporation maintains the Census installation. This system is operated 24 hours a day, seven days a week, except for four 8-hour preventive maintenance shifts each week. This allows approximately 32 hours for regular maintenance and 136 hours for operation or 21 and 79 per cent respectively. Table 1 shows the engineering time spent on the computer system during typical weeks of operation. The figures are given both in hours and percentages. Both nonscheduled engineering time as well as preventive maintenance time are shown. The sum of the two gives the total engineering time spent on the computer per week. It should be noted that this is actual engineering time and does not include time that the computer may have been shut down while waiting for an engineer to report. According to our maintenance contract, this must be within a half hour during regular working hours and within two hours at all other times. Attention should be given to the fact that the preventive maintenance time does not total exactly 32 hours each week. This is due in part to a half-hour period each morning devoted to checking and cleaning the mechanical portions of Uniservos. It is expected that this work will be taken over by the UNIVAC operators since the procedures and the techniques involved are quite simple.

In addition, one extra shift was required the week ending June 3 and three extra shifts the week ending October 7, 1951. These shifts were required to incorporate engineering changes which had been developed over a period of time and could not be incorporated in the equipment during the normal preventive main-

Table 1

Week ending 1951	Nonscheduled engineering Hours	Per Cent	Preventive maintenance Hours	Per Cent	Total engineering time Hours	Per Cent	Percentage of nonscheduled engineering
June 3	18.9	11.3	40	23.8	58.9	35.1	14.8
26	20.5	12.2	34	20.2	54.5	32	15.3
July 14	14.7	8.8	33	19.6	47.7	28	10.9
21	19.4	11.6	34.5	20.5	53.9	32	14.5
28	39.2	23.3	34.5	20.5	73.7	43.8	29.4
Aug. 4	26.2	15.6	33	19.6	59.2	35.2	19.4
Sept. 2	28.8	17.1	34.5	20.5	63.3	37.7	21.6
9	16.1	9.6	34.5	20.5	50.6	30	12.1
16	22.6	13.5	33	19.6	55.6	33	16.7
23	42.3	25.2	34.5	20.5	76.8	45.7	31.7
30	21.8	13.0	34.5	20.5	56.3	33.5	16.3
Oct. 7	15.9	9.5	56	33.3	71.9	42.8	14.2
14	14.0	8.3	34.5	20.5	48.5	28.9	10.5
21	10.4	6.2	34.5	20.5	44.9	26.7	7.8
28	20.8	12.4	33	19.6	53.8	32	15.4
Nov. 4	40.4	24.0	34.5	20.5	74.9	44.6	30.3
11	10.1	6.0	34.5	20.5	44.6	26.5	7.6
18	30.5	18.2	34.5	20.5	65	38.7	22
25	13.7	8.2	34.5	20.5	48	28.6	10
Dec. 2	14.8	8.7	34.5	20.5	49.3	29.3	12.6
9	19.6	11.7	34.5	20.5	54.1	32.2	14.7

tenance time. The nonscheduled engineering time has varied from as little as 10.1 hours or 6 per cent to 42.3 hours or 25 per cent. The last column in the Table shows the amount of nonscheduled engineering time as compared to the allowable operating time (total time less preventive maintenance time). Here there is a variation of from 7.6 to 31.7 per cent and an average for the weeks shown of 16.9 per cent. It is believed that these figures, while good for the first months of operation of a new piece of equipment, will show definite improvement over the next year.

Although the opportunity to prove or disprove the following theory of operation has not presented itself, it is believed logical that optimum use of the UNIVAC equipment might be obtained by means of scheduling preventive maintenance only at such times as it is indicated in the judgment of competent operators. In other words, there are many occasions preceding a scheduled maintenance shift when the system is performing very well. At such times, it is extremely inefficient to shut down the operation in order to provide maintenance. For many reasons, however, it has been impossible to operate and maintain the first system in this way. It is hoped that such operation will be possible in following installations.

It should be realized that the UNIVAC system requires a supervisor of the same caliber as the one required for a large punched card installation. However, the large group of operating personnel would be replaced by a small group of well-trained extremely competent people thoroughly familiar with the details of the computer and associated equipment. The time spent in providing a high degree of training for these people is more than repaid in increased operating efficiency and consequently higher work output. For example, situations arise in the course of running a problem where a correct operational decision can save hours of elapsed computation. Also, a competent operator will recognize malfunctions sufficiently early to prevent serious delays. He is capable of deciding whether to continue with machine operation or to stop to diagnose. The second UNIVAC system which is ready for installation in Washington, will be operated by a group of engineers who have been trained in operation and maintenance. This procedure, it is believed, will result in the UNIVAC system being of maximum benefit to the Air Comptroller's Office.

Evaluation of UNIVAC design

Checking features

Maintenance of the UNIVAC has been vastly simplified by use of duplicate arithmetic and control equipment and other checking methods. Many factors which would have led to undetected errors

have, by virtue of duplication, immediately stopped the computer. Although checking by means of inverse operations can provide operational checks on the arithmetic circuits, there is some question as to whether it provides as good a check as duplication. However, in connection with odd-even codes, it may conceivably be comparable. It should be remembered, however, that this is from an operational standpoint and not a maintenance standpoint. When the control equipment is considered it is difficult to visualize a check that is as good as duplicated equipment. Other checks that are utilized in UNIVAC include the periodic memory check, intermediate line function table checker, function table output checker, memory switch checker, and 720 checker.

As explained earlier in the paper, the periodic memory check is accomplished by reading out of all memory channels sequentially and performing an odd-even check on each digit as it passes through the high speed bus amplifier. The period at which the check is repeated may be varied over a large interval. At present, it is set at 5 seconds, the check taking 52 milliseconds or about 1 per cent of the computing time.

The function table has a check at the very input by bringing in the check pulse in each character so that if an odd-even error occurs between the control register and the static register, no order will be set up and the computer will grind to a halt! If the input sets up properly but an error occurs farther on in the table, but not ahead of the intermediate lines (the linear set into which the input combinations are decoded), the error is caught at this point. The intermediate lines are broken into groups in such a way that an error is indicated when more than one line is set up in one group or the entire set. There is an exception to this in some groups where no error is indicated by this checker if more than one line is set up within the group.

This has been allowed only in those cases where it has been shown that setting up two or more lines will cause some other checker or checkers to indicate the trouble.

If the error occurs beyond the intermediate lines, the output checker then comes into play. This checker makes an odd-even count on the number of gates used on each instruction: dummy lines having been added so that the count is normally always odd.

The memory switch or tank selector checker ensures that one and only one memory channel is selected on any instruction. It checks each of the two digit positions separately indicating which if either, is in error.

The 720 checker counts the digits coming off the tape and if there are either more or less than 720 in one block, the computer stops; by examining the indicators on the supervisory control console, the operator can determine the number of digits actually ead. By means of some rather simple manipulations, the operator can then reread the block without losing his place in the routine; and if the information is then read correctly, he may again start the computer on the routine. The same procedure may be followed if an odd-even error is made in reading from the tape.

Many checks other than those mentioned before have been built into the UNIVAC. On the basis of operating experience, the engineers cannot recommend too strongly the use of built-in checking facilities. All in all, the faith that can be put into results obtained from an unchecked computer comparable in size to UNIVAC is in the writers' opinion exceedingly low.

More than this, however, the methods by which the UNIVAC is checked have been of extreme usefulness in trouble shooting. The duplication of circuits has amply repaid the increase of space and the number of components required by this checking system.

General comments

After evaluating UNIVAC performance over a period of eight months, the over-all picture of the UNIVAC design, in the minds of its designers, is extremely good. Certain phases of its design exceeded expectations, while of course, other phases were somewhat disappointing. The first eight months of actual operation have taught more than years of experimentation with laboratory models. Many improvements have already been conceived of this experience and are continuing daily to increase reliability.

The other major factor influencing computer design, cost, has been duly considered in the UNIVAC design; and it is being met with plans for a continuing full-scale production of UNIVAC systems. As the production techniques are developed concurrently with the engineering design details, the UNIVAC becomes the realization of a hope which has long been in the minds of its designers: An economical, completely reliable commercial computer for performing the routine mental work of the world much as automatic machinery has taken over the routine mechanical work of the manufacturer.

References

McPhJ51.

Section 2

Processors with a general register state

The processors described in this section all have a processor state consisting of registers which are used for multiple (i.e., general) purposes. Perhaps a better name might be processors with a state consisting of a register array(s). The following machines are fairly similar in their ISP structure: Pegasus (Chap. 9), the DEC PDP-6,10, the SDS Sigma 5 and 7, and the UNIVAC 1107 and 1108. However, other computers including an 8-bit character computer (Chap. 10) and the CDC 6600 (Chap. 39) also use arrays of registers.

The general register organization appears as a compromise between the 1 and 2 address organizations. It avoids some of the extra instructions for shuffling data, inherent in a 1 address system, but avoids taking the space for a full additional address. The index register organization is also such a compromise, but one that is specialized to address calculations. The general register organization moves further toward a full 2 address organization without much additional cost. This assumes a small relative cost for a small amount of memory that is significantly faster than the larger Mp.

The design philosophy of Pegasus, a quantity-production computer

Chapter 9 describes Pegasus's logical organization and the technology from which it was implemented. The technology includes vacuum tubes, a cyclic memory, and dynamic logic based on delay lines. Pegasus has the nicest ISP processor structure discussed in this section—perhaps in the book. It is included because it is probably the first machine to use an array of general registers as accumulators, multiplier-quotient registers, index registers, etc. This ISP organization should be compared with the IBM System/360 (Chap. 43). Note that the multiple-register organization is independent of Mp.cyclic. This organization improves performance by generality.

The structure of System/360
Part I—outline of the logical structure

The IBM System/360 is described in Part 6, Sec. 3, and is included mainly because of the very large number of such systems that have been built.

An 8-bit-character computer

This computer (Chap. 10) has been invented by the authors to show the composite features of a small character/word-oriented computer. In reality, 8-bit machines turn out to look either like 16-bit machines, because the Mp size accessed is usually $>2^8$ words, or like character-string processors. Because of the primitive nature of this machine, it is a possible alternative to the larger more complex microprogrammed processors for defining more complex ISP's.

Parallel operation in the Control Data 6600

The CDC 6600, described in Chap. 39, has three arrays of eight registers each. Two of the arrays are used rather generally, and the third array is used to access words in Mp. The design of the CDC 6600 is a classic because of the computing power it provides. It is also worth studying as an example of a Pc assigned exclusively to data operation, with all concern with the larger PMS structure located in Pio's. A discussion of it is given in Part 5, Sec. 4, page 470.

Chapter 9

The design philosophy of Pegasus, a quantity-production computer[1]

W. S. Elliott / C. E. Owen / C. H. Devonald
B. G. Maudsley

Summary The paper gives an historical account of the development of the packaged method of construction of computers, and the advantages of this method are discussed. The packages used in the computer Pegasus are described from both an electronic and a mechanical point of view. The specification of the machine is given and the arguments which led to this specification are discussed. The detailed logical design procedure leading from the specification to the wiring lists is described. The method of maintenance and some reliability figures are given.

Introduction

The development of standard plug-in unit circuits ('packages') for digital computers began in this country [England] in 1947, and some of the advantages of the method have been discussed in earlier papers [Elliott, 1951; Johnston, 1952; Elliott et al., 1952; Elliott et al., 1953]. The advantages start in the design stage of a new computer project and follow through production and commissioning to maintenance.

In the design stage, what is known as 'logical' design is separated from engineering design. Once the packages have been designed by electronic engineers and the rules for their interconnection have been laid down, the 'logical designers' (usually, but not necessarily, mathematicians) can begin organizing the packages into various computers to carry out different functional requirements. The electronic and mechanical design work invested in the packages is thus drawn on for more than one computer design, and each computer can be assembled from stock parts without further engineering effort. Design time and cost are therefore much reduced.

In production, whether we consider one design of computer or several designs using the same packages, costs and time are also much reduced. Quantity production lines for the relatively few types of standard package are set up, and are common to different computer designs, thus reducing inspection and planning costs. Standard cabinet work has been designed for Pegasus, and this

too can be taken from stock or established production lines to make other computers.

In commissioning a computer, because all the packages have been pretested, when power is first applied to the complete machine it is known that a large part is already fault-free. It remains to detect a few errors which may have been made in the interconnections.

Perhaps an even more important consideration is ease and speed of maintenance. Test programmes will usually indicate the part of the machine in which a fault is occurring. Several monitor sockets are located on the front of each package, and by inspection the faulty package is speedily found and replaced.

The package method has been criticized on the grounds of the cost and questionable reliability of plugs and sockets, and some redundancy of components.

The authors believe that the many advantages far outweigh the cost of plugs and sockets. The present trend is to use copper-etched printed circuits, and these fall naturally into the plug-in unit idea, the plug contacts being part of the printed wiring; there has been no trouble in Pegasus from plugs and sockets. Component redundancy in Pegasus is about 10% of the diodes and a few resistors, the cost of redundant components being about £150.

Electrical design of the packages

Circuits used for arithmetic and switching operations

Historical. A previous data-processing machine [Elliott et al., 1952; Elliott et al., 1956b] used 330 kc/s serial-digital circuits; they had originally been designed for 1 Mc/s operation, but 330 kc/s was chosen to suit an anticipation-pulse cathode-ray-tube store. This frequency has been retained to the present time because it suits the magnetostriction delay-line store [Fairclough, 1956] and the magnetic-drum store [Merry and Maudsley, 1956]. Experience with the data processor led to work (commenced in 1951) on a new set of circuits [Elliott et al., 1952], particular emphasis being

[1]*Proc. IEE*, pt. B, vol. 103, supp. 2, pp. 188–196, 1956.

laid on flexibility of use and ability to work without error in high electrical interference fields. These circuits form the basis of those in Pegasus.

Operations to be carried out. The following well-known operations are used to build up the logical structure of the computer:

a   'And.' This operation, which may be carried out between two or more input serial trains of pulses, produces an output train in which pulses occur only when pulses are present at the same time on all inputs.

b   'Or.' This operation produces an output train in which pulses occur at all times when a pulse is present on any of a number of inputs.

c   'Not.' 1's are changed into 0's and 0's into 1's; this is achieved by inverting the pulse train.

d   *Digit Delay.* The passing of a pulse train through a digit delay produces a pulse train similar to the input, but each pulse is one pulse position later in timing and restandardized in shape.

All operations in the computer, including addition, subtraction, and staticizing, are carried out by combinations of these elements. There is no circuit specifically for addition, and there are, in general, no flip-flops such as are often used for staticizing or storing a single digit. A similar philosophy was arrived at independently by the designers of SEAC and DYSEAC [Elbourne and Witt, 1953], but the detailed working out is considerably different.

Digit waveforms. The timing of digit pulses throughout the machine is controlled by a common 'clock' waveform—a 3 microsec square wave (Fig. 1a) in which the positive-going portions define digit positions.

The digit pulses, which are routed about the machine and applied to logical circuits, are generally of the form shown in Fig. 1b; as generated, they have their leading edges well in advance of the clock pulse and are of a greater amplitude. This means that considerable distortion of the pulse is tolerable, since only the portion which coincides with positive clock pulse is of consequence. Digit pulse trains are 'clocked' ('and' operation with clock) only at their entry into a storage system or into a digit-delay circuit.

Inverted pulses are also employed: as an illustration, consider the operation 'A and not B'. Pulses A and B (Fig. 1) are on two lines and are of the same nominal timing, and we wish to form A . $\bar{B}$ (symbolic representation of 'A and not B'). To do this pulse

B is inverted (forming $\bar{B}$, or 'not B') and is used to gate pulse A and prevent its passage. The inverted pulse $\bar{B}$ will be a little late on B, which also may have been later than A, as shown in Fig. 1c; thus when A and $\bar{B}$ are 'anded' together a spike may be produced, as shown in Fig. 1e. This spike, however, lies between clock pulses and so will be rejected on clocking.

The pulse system used allows several logical operations to be performed in cascade without any loss in nominal timing, so easing the problem of logical design (particularly by permitting afterthoughts). The maximum number of logical operations performed

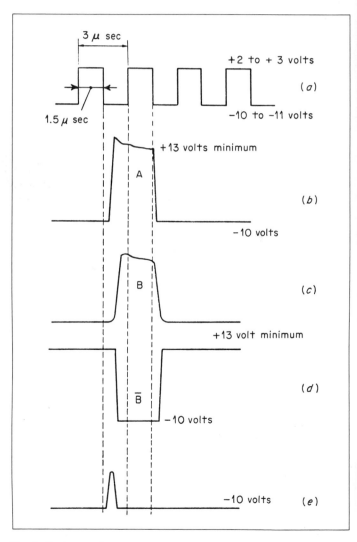

Fig. 1. Basic waveforms.

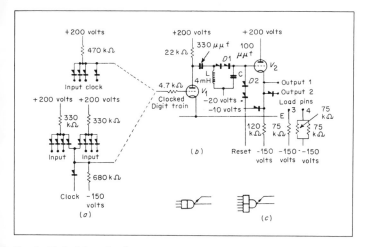

Fig. 2. Digit-delay circuit.

in cascade in Pegasus is five, though up to 12 could be performed in special circumstances.

The logical circuits. Each of the logical packages has more than one circuit unit. A circuit unit is defined as that part of a package which has input and output pins, and no connections to other parts of the package other than supplies. We may make the following generalizations:

a Each unit has an 'and' gate at its input.

b Each unit has a cathode-follower output (half a 12AT7 valve).

c Each unit has an additional output via a germanium diode for making 'or' gate connections.

[Note: There are exceptions to (*a*) and (*c*) on one package type.]

There are three possibilities for the part of the circuit unit between the input 'and' gate and the output cathode-follower, namely a digit delay (half a 12AT7 valve), an inverter (half a 12AT7 valve), and a direct connection. Space does not permit a description of all the circuits, so it is proposed to deal only with the digit delay.

The circuit is shown in Fig. 2, and some typical waveforms are shown in Fig. 3. The input circuit can be of two forms, namely a 3-input 'and' gate and two such gates with their outputs 'or-ed' together. In both cases there is a further gating with a clock pulse. The clocked digits from the gate input circuit are applied to the grid of V_1, the anode voltage of which falls, so building up a

current in L. When V_1 is cut off at the end of the digit, this current flows through diodes D_1 and charges up a storage condenser, C, which is discharged at the end of the next clock pulse by a 'reset' pulse applied through D_2. The reset pulse supply is a common computer supply whose amplitude and phasing relative to the clock pulse is shown in Fig. 3.

It will be noted that the reset pulse is also present at a time, just after V_1 is cut off, when the current in the inductor is about to charge the storage condenser. This merely has the effect of deferring the charging of C until the end of the reset pulse, the

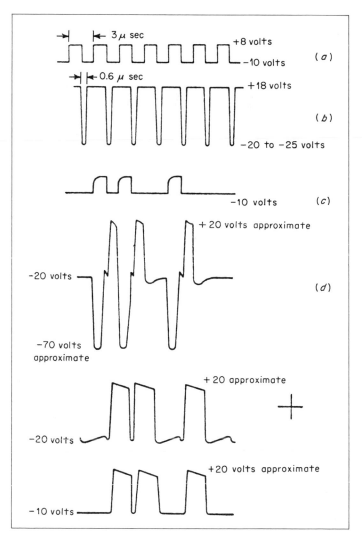

Fig. 3. Digit-delay waveforms.

current in the meantime continuing to flow through the diodes with little loss in the stored energy of L, since the voltage across L is low at this time.

The output cathode-follower V_2 is caught at -10 volts in the negative direction by a diode; this safeguards the crystal-diode circuits driven by it in the event of failure of the h.t. supply or V_2, and it removes residual ripple on the bottom of the input waveform, and thus reduces the back voltage and hence leakage in diodes of gates driven by the output.

The second output through a diode can be used in conjunction with similar outputs from other circuits and a resistor (pins 3 and 4) to make an 'or' (up to about 16-way).

In general, each output circuit has two available load resistors, disposed between direct and 'or' outputs according to a set of rules which are applied for each case. The number of units which can be driven by an output can vary between three and 16 according to circumstances; where more have to be driven than the rules allow, use is made of 'booster' cathode-followers available on one of the packages.

Some examples of the use of the logical circuits

Two examples will be given, the first being a simple arrangement—the staticizor—which is used frequently, and the second being a complicated arrangement—the adder/subtracter—which is used infrequently. The symbols used to indicate the circuit units are shown in Figs. 2c and 5b.

The staticizor. The function of a staticizor is to remember the fact that a digit occurred at a particular time, for an indefinite period, the method generally used in Pegasus being shown in Fig. 4. A digit delay with a twin 'and' gate input has its output connected to one of its inputs. It is turned on by gate 1, which causes a digit to circulate as long as the inputs to gate 2 remain positive.

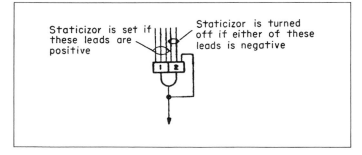

Fig. 4. The staticizor.

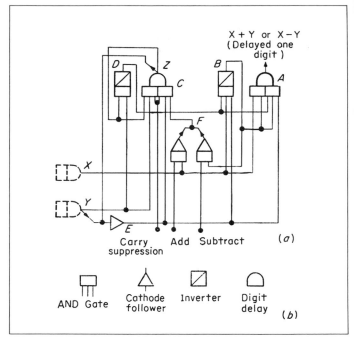

Fig. 5. The adder/subtracter.

It is normally turned off by an inverted pulse (a '0' following a series of 1's) on one of the gate 2 inputs.

The adder/subtracter. Figure 5 shows an adder/subtracter unit with inputs X and Y and an output $X + Y$ for the sum or $X - Y$ for the difference. There are two further input control leads marked 'add' and 'subtract'. If the 'add' lead is held positive while the 'subtract' lead is held negative, the unit acts as an adder. If the 'subtract' lead is held positive and the 'add' lead negative, the unit acts as a subtracter. Carry suppression is controlled by the lead marked 'carry suppression'. Carries are allowed to propagate when this lead is held positive, so that a negative signal on this lead will suppress carry.

Table 1 gives the digits appearing at the outputs of logical elements in the adder/subtracter unit for all combinations of input and carry digits when the unit is operating as an adder.

Arrangement of circuits based on packages

It was required to base the logical circuits on a standard size of package which could also be used for other circuits, e.g. a nickel-line 1-word store [Fairclough, 1956]. A unit which could accommodate three valves and had a 32-way plug was decided on; the

Table 1   Digits at various internal points of the adder/subtracter unit when set to add, for all combinations of the input and carry digits

Inputs digits		Present carry digit	Digits at internal points					
			A (Sum)	B •	C (Next carry)	D	E	F
X	Y	Z						
0	0	0	0	1	0	1	0	0
0	0	1	1	1	0	1	1	0
0	1	0	1	1	0	1	1	0
0	1	1	0	1	1	0	1	0
1	0	0	1	1	0	1	0	1
1	0	1	0	0	1	1	1	1
1	1	0	0	0	1	1	1	1
1	1	1	1	1	1	0	1	1

Note.—A and C are at the grids of the digit delay units.

problem then was to arrange the various circuits in such a way as to enable a computer to be designed using a minimum total number of packages without too many types. Five types were arrived at and these are shown in Fig. 6.

As an example of the factors involved, consider package types 1 and 2. The circuit units based on package type 1 can perform all the functions of those on type 2. However, there are many uses for a digit-delay circuit with a single 'and' gate input (package type 2), and since three units of this kind (instead of two for a 2- 'and'-gate input delay) can be based on one package, a saving can be effected. In Pegasus this saving amounts to 32 packages, which is considered to be well worth an extra package type.

In addition to the five logical packages, a further 16 types (three of which are peculiar to each computer) are required. The numbers used for the various functions are given below:

		Number
Logical types	Type 1	113
	Type 2	64
	Type 3	55
	Type 4	45
	Type 8	37
Nickel-line 1-word store		61
Drum-store packages (8 types)		38
Input/output packages (3 types)		17
Clock and reset waveforms (3 types)		14
Total		444

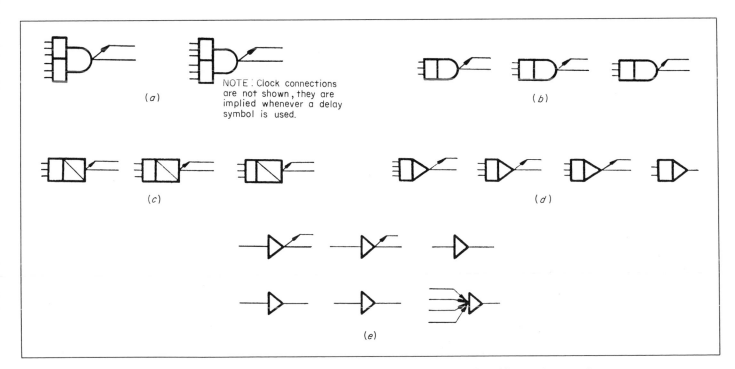

(a)

NOTE : Clock connections are not shown, they are implied whenever a delay symbol is used.

(b)

(c)

(d)

(e)

Fig. 6. Contents of logical packages. The arrowhead on an output lead denotes the presence of an OR crystal connection.

The magnetic-drum store and the circuit packages used with it are described in another paper [Merry and Maudsley, 1956], as is the nickel-line store [Fairclough, 1956].

The mechanical design of the packages

General form

Each standard package consists of three main parts, namely the valve panel, the component panel and the plug.

The valve panel is an aluminium pressing, there being three types—a 3-valve type, a 2-valve type and a blank. The package type number is marked on the panel by two dots according to the standard resistor colour code.

The component panel houses up to 100 components, including small transformers, chokes and coils, the panel and the handle being made in one piece from sheet insulating material. This design provides a minimum resistance to airflow over the valves and gives ample protection to the valves against accidental damage.

The plugs and sockets are used in multiples of eight connections. Most of the packages have four plugs providing 32 connections, but up to 64 are possible in each package. The plug contacts are made of brass and are heavily silver-plated. The socket uses a proprietary valve-holder contact, which can readily be replaced if damaged.

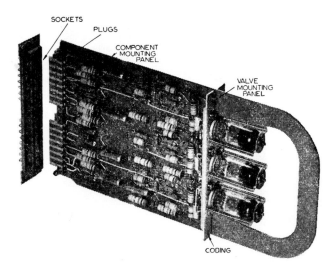

Fig. 7. Standard package.

This combination of plug and socket has a consistently low contact resistance (0.003 ohm at 1 amp); the insertion and withdrawal force is about 4 oz per contact.

The wiring of the packages

At present packages are wired and soldered by hand. The wiring is point-to-point, and within the limitations of layout for efficient performance, wire lengths are standardized for mass production on automatic wire-cutting and stripping machines. The symmetry of the eyelet positions makes it possible to use components which are preformed to a standard pitch and would allow for automatic preforming and insertion of components.

Experimental packages have been produced by photo-etched wiring and dip soldering.

Specification of the computer Pegasus

Summary specification

A detailed specification would cover the ground of the programming manual [Pegasus Programming Manual, Ferranti Ltd., London] and would be out of place here.

Pegasus is a binary serial-digital computer. The word length is 42 binary digits, of which 39 digits are used for a number and its sign (negative numbers are represented by their complements with respect to two), one digit is used for a parity check and the other two are gap digits. The length of an order is 19 binary digits, so that one word may consist of two orders, the remaining digit being a 'stop-go' digit. If the 'stop-go' digit is a '0', the computer will stop before obeying the orders in the word, but will proceed unhindered if the digit is a '1'.

There is a 2-level store, a magnetic drum holding 5120 words and an immediate-access or computing store of 55 single-word magnetostriction delay lines.

An order is made up of seven N-digits, three X-digits, six F-digits and three M-digits, the N-digits being the most significant and the M-digits the least significant. The N-digits allow 128 addresses in the immediate-access store (of which only 63 are used). The registers in this store are shown in Fig. 8. The X-digits refer to one of the accumulators, the registers corresponding to N-addresses 0–7. Thus the order code is a 2-address code with one address referring to only a limited part of the store. The F-digits indicate the function of the order. A list of functions and their corresponding F values are given in the appendix of this chapter. The M-digits indicate a modifier for the order: they select one of the accumulators, and the modification process is to add certain parts of the contents of the selected accumulator to the order before it is

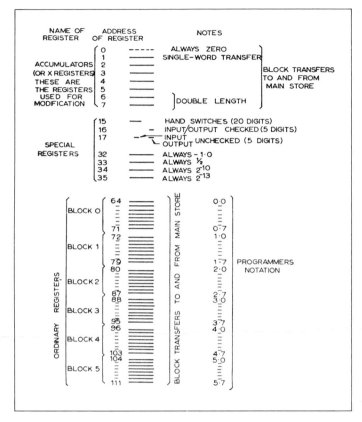

Fig. 8. Allocation of addresses in store.

obeyed, the part chosen depending on the function of the order to be modified. Figure 9 gives a schematic representation of the modification process. The effect of modifying an order depends on the function of the order and can be to make the effective order length 22 digits. This extension is necessary when specifying an address in the main store.

Transfers of information can take place between the computing store and the main store, and vice versa, either in single words or in blocks of eight words. For single-word transfers, only the register with address 1 in the computing store is involved. For block transfers the address on the drum of the first word of the block must be divisible by eight, and the registers in the computing store that are involved will be one of the discrete blocks indicated in Fig. 8.

Input and output is by means of punched paper tape. An 'external conditioning' order is included in the code to enable a choice of input and output equipment to be made. In the standard machine, two tape readers are used.

All stored information is checked (when read) by means of a parity digit, which is such that the total number of 1's in any correctly stored word is odd. The input and output of decimal characters on tape can be checked by a similar process.

The considerations which led to the specification and the logical design

The main features of the design are

a The use of a computing store from which all orders and numbers are taken while computing

b The provision of multiple accumulators

c The provision of special orders and facilities for dealing easily with 'red tape'[1]

The computing store. The use of a fast-access store from which all numbers and orders are taken increases the speed of the machine and eliminates the need for optimum programming. It is this computing store which makes it possible to use an inexpensive magnetic drum (with a relatively long access time) as the main store, and yet have a machine which is fast and relatively simple to programme. On the other hand, programmes have more 'red tape' and are not as simple as with single-level storage.

Transfer between levels is in blocks of eight words; this is a simplification and saves time. One block holds a reasonable amount of programme and other blocks hold data. Four blocks in all (32 words) would be just sufficient, and Pegasus was originally designed with this number. The design was subsequently modified to six blocks, which is quite adequate, in conjunction with the seven accumulators. Any further increase in the size of the computing store would be achieved by increasing the size, not the number, of blocks. As it is there is an economic balance between the usefulness and the cost of the computing store.

[1] 'Red tape' is an expression for the non-arithmetic orders in a programme.

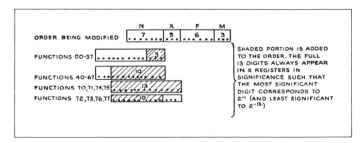

Fig. 9. Order-modification process.

The provision of several accumulators. This is the most novel feature of the logical design of Pegasus. It is generally agreed that the simplest order code from the user's aspect is the 3-address code with orders of the form, $A + B \rightarrow C$. An examination of this form of code, however, shows that in many cases two of the addresses are the same, so that the order takes the 2-address form, $A + B \rightarrow A$. A further examination shows that in a large proportion of cases the address A is confined to a very few addresses. This leads to the suggestion of a code of the form $N + X \rightarrow X$, where X covers only a small part of the store while N covers the whole store. This will have the advantage of yielding a reasonably short order. In Pegasus two such orders are incorporated in one word, leaving sufficient digits to specify a modification register (a Mancunian B-line) in each order.

The extreme case of this code is, of course, the single-address code, where X is confined to one address, the accumulator. However, experience had convinced the programmers collaborating in the design of Pegasus that, with single-address codes, a large number of orders are concerned solely with transfers of numbers from one register to another; the single accumulator is a restriction through which all numbers must pass and in which all operations have to be performed.

In the Manchester University computer the B-lines serve two very valuable but distinct purposes: they allow order modification and rudimentary arithmetic (such as counting) to be done without disturbing the accumulator. It was felt that fuller arithmetic and logical facilities on these B-lines would have been extremely valuable. The seven accumulators in Pegasus, used for modification and arithmetic, are a development of the B-line concept.

Special facilities for dealing with 'red tape'. The difficulties associated with the 2-level storage system have been greatly reduced by having an order-modification procedure which depends on the function of the order (Fig. 9). This method of modifying orders, used in conjunction with order 66 of the code (the unit-modify order), enables the counting through blocks of information to be done with relative ease.

The use of the group-4 orders of the code enables counters to be set conveniently and a constant (up to 127) to be placed in an accumulator, the constant being the value of the N-digits of the order. Order 67 (the unit-count order) enables the counting of cycles of operations to be dealt with in a simple way. A jump to another part of the programme can be programmed to take place automatically when the required number of cycles has been performed.

Having a large number of jump instructions greatly helps in organizing a programme. In particular, one order enables a jump to be made depending on the condition of an accumulator (being zero, for example), and another order on the complementary condition (being not zero). When only one of these orders is available it is necessary to think ahead to see whether or not the correct condition will be satisfied. Although the eight jump instructions included in the code were felt initially to be enough, it is now suggested by programmers that even more such orders would be helpful.

The logical shift orders, 52 and 53, are also included to simplify 'red tape'. In particular, they are used for packing and unpacking words holding several items of information.

As a result of including these various orders, the order code of Pegasus is quite large. It is worth remarking, however, that by a sensible grouping of the orders in the code the remembering of the code is a very simple task. A sensible arrangement of the code tends to reduce the amount of equipment needed to engineer it. For example, when the equipment for dealing with group 0 of the code has been allocated, groups 1 and 4 require the addition of only three gates.

Facilities for checking programmes. The features mentioned above make the computer easier to programme, and there are other facilities in Pegasus that make it easier to check out and develop new programmes. These include causing the machine to stop obeying orders, either under programme control or when the programme is in error. In particular, the machine stops if an order for writing in the main store is reached and an overflow indicator is set. A further aid when testing new programmes is the automatic punching out of all main-store addresses appearing in block-transfer orders. When this information is examined an indication of the course of a programme is readily obtained. The punching can be inhibited by a switch when a return to full-speed running is needed.

Machine rhythm

The logical design of Pegasus is built around a nucleus that deals with the simple arithmetic orders, groups 0, 1 and 4, of the code. This nucleus contains the control section, i.e. the order register and order decoding equipment, and the mill in which these orders are executed. The design of this nucleus could not begin until a basic rhythm for dealing with the extraction from the computing store and the execution of such a pair was determined. When the outline of this nucleus was clear, the equipment for dealing with the remaining orders in the code was designed to fit it.

The following arguments led to the basic rhythm. Since the orders of groups 0, 1 and 4 are similar in many respects, for definiteness, it will be sufficient to consider a particular order, 11 of the code, say. This is an order which takes two numbers from the computing store and replaces one of them by their sum. It would take a prohibitive amount of equipment to extract these numbers, add them together and have the least significant digit of the sum available for replacing in the store in the same digit time as the least significant digits of the two components taken out of the store. In practice, some four digit times at least would be needed for this sequence of operations. Thus, it would be impossible to return the sum to the store in the same word as the operands are extracted without having an entry point to each register which is in a different timing from the normal circulation entry. To produce two such entry points to each register would mean more equipment associated with each register, which was considered an uneconomical use of extra equipment. Instead, it was decided to delay the sum so that it could enter the register in the computing store in the next word time in standard timing. This involves one common delaying circuit instead of one for every register. Such an order therefore takes two word times to execute. It may be argued that this second word time could be made to overlap with the first word time for the next order. Two reasons oppose this: the new contents of the register being changed might be required by the next order; and two different sets of equipment for selecting a storage register would be needed if numbers were to be extracted from one and replaced in another register in the same word time.

Thus, the execution of a pair of orders taken from the computing store requires four word times. The reasons for opposing the overlapping of the execution of two orders also oppose the extraction of an order pair while the previous pair is being dealt with. Five word times are therefore needed for the process of extracting and obeying a pair of simple arithmetic orders. More time may be needed for some of the other orders in the code.

The basic 3-beat rhythm is thus established:

a Extract the order pair from the computing store.

b Obey the first order of the pair.

c Obey the second order.

The duration of beat (*a*) is one word time; beats (*b*) and (*c*) are each two word times long for orders in groups 0, 1, 4 and 6 of the code, but may be longer for other orders.

Times for typical operations

The times for the various arithmetic operations are:

	millisec
Addition and subtraction	0.3
Multiplication	2.0
Division	5.4

These times include an allowance for the time to extract the orders.

Some times for standard subroutines are:

	millisec
Exponential function	29
Sine function	24
Logarithmic function	34

Finally, to give some indication of the time for a typical problem, a set of 50 simultaneous equations (with a single right-hand side) takes about $10\frac{3}{4}$ min. Of this time, 3 min 8 sec is for input, 7 min 17 sec is for calculation and 18 sec is for output.

Realizing the specification

The detailed logical design

It would take too long to describe fully the detailed logical design. One aspect is worth mentioning, however, namely the avoidance of all 'exceptions' in the results of orders. As an example of an exception consider the overflow indicators, which should be set whenever the final result of an order is outside the permissible range of numbers. In multiplication this can occur only when both the multiplier and the multiplicand are −1, and this is likely to occur very infrequently. Rather than provide equipment to sense this infrequent case, it is easier to put a footnote in the programming manual, where the overflow indicator is described, pointing out the exception. It was felt, however, that such exceptions should be avoided even at the expense of extra equipment or extra complication. For this and other reasons concerned with facilitating machine use, the logic of Pegasus is quite complicated.

The end-product of the detailed logical design is a series of diagrams with symbols corresponding to the circuit units of the packages, as shown, for example, in Fig. 5. The inputs and outputs of the units on these diagrams correspond to the pins of the sockets into which the packages plug. Thus, the wiring lists of connections of these pins can be produced from these logical diagrams. The first step in the production of these lists is to allocate a position in the cabinets to each logical circuit in such a way as to reduce the amount of wire needed. When the layout has been completed, the last stage of producing the wire lists can proceed.

General construction of machine

The main units are shown in Fig. 10.

The package frame. This unit is a simple light-alloy frame supporting diecast light-alloy frame racks to which the back socket panels are fixed. The packages slide into grooves in the rack and plug into sockets at the back, a polarizing feature preventing the insertion of a package upside down. If electrical or magnetic

screening is necessary between any packages, a special metal plate is inserted in slots in the cast rack and is fixed by a single screw in the back panel. Coded aluminium strips containing coloured plastic studs which identify the position of each package are fixed to the front of each casting.

Arrangement of the packages. There are 200 packages per cabinet, arranged in ten horizontal rows of 20 units per row. The metal valve panels are placed so that the edges almost touch. The component panel of each unit is in register with the unit in the corresponding position in each of the other rows, thereby providing vertical chimneys for cooling the components secured to these

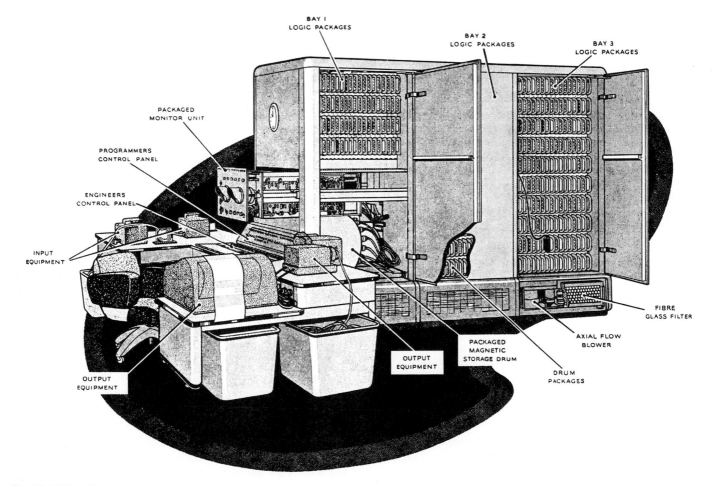

Fig. 10. Main units.

panels. Warm air from the main source of heat, the valves, is prevented by the valve panels from reaching the more temperature-sensitive components, such as diodes, secured to the component panel.

The back panel wiring. For locating long signal wires between sockets a system of plastic strips is used, which hold the wires at definite positions given by the instructions on the wiring lists. The exact route of every wire is predetermined, thus making wiring and inspection more reliable and fault finding and maintenance easier.

Final assembly. The completely wired frame is assembled in its cabinet, which has already been fitted with the control and auxiliary supply circuit unit, heater transformers, fuses, cooling assembly and cableforms. The work of connecting the cableforms, heaters and earths can be done by relatively unskilled labour working to clearly written instructions and diagrams.

The cooling system. Each cabinet has its own cooling system as an integral part of the construction; there is therefore no difficulty in cooling cabinets added to existing computers. Two axial-flow turbo blowers are mounted in the base beneath an airtight pressure chamber, each providing 300 ft³/min of air at a total pressure head of 1 in (water gauge). The maximum temperature rise is 10° C.

The power supply. A separate cubicle houses metal rectifiers, shunt stabilizing valves and control circuits. The power is obtained from the mains through a motor-alternator set, the output of which is stabilized to 2%, the main purpose of this set being to act as a buffer against switching surges and other mains voltage variations. The valve heaters in the computer are energized from the stabilized alternator output, which is expected to extend the valve life.

Maintenance

General

All digital computers so far have a fault rate which cannot be ignored. When the best has been done in the choice of components, circuits and mechanical construction, attention must be paid to the following points to get the best out of a machine:

a   Rapid fault location

b   Getting the machine working again as soon as possible after locating a fault

c   Preventive maintenance

Fault location

There are parity-checking circuits on both the main and the high-speed stores. Errors of a single digit in the stores stop the machine. The fault can then be quickly located by examination of the monitors.

For other faults the general method is to run a test programme (assuming the fault is not in the main control) which will indicate the area of the fault. Detailed examination can then be carried out with the monitors.

All outputs of circuit units are readily accessible at monitoring sockets on the front of each package, and in addition about 80 points can be directly selected by switches from the monitoring position: these include all store lines and a number of key waveforms. Fault-finding is normally a matter of tracing 0's and 1's through the machine with reference to logical diagrams rather than electronic circuit diagrams.

A variety of triggers can be selected for the monitor time-bases, these including

a   Trigger at any word position within a drum revolution (128 different times selectable by switches)

b   Trigger at any word time of any selected order

These triggers and some other monitoring facilities are produced by 19 standard packages and are found to be well worth the extra equipment.

Fault repair

Once a faulty package has been located, the machine can be got working again immediately by replacement of the package with a spare; repair of the faulty package can be done at leisure with the aid of a package tester. With this equipment a package can quickly be given a series of standard tests; each is selected by switches, and the performance is measured either by observation of meters or a built-in oscillograph.

During commissioning not one case was found of the first machine doing other than what one would expect from the logical diagram (except for a very few cases of incorrect wiring).

Preventive maintenance

The machine h.t. supplies are reduced while the test programmes are being run. This marginal testing shows up incipient faults such as deterioration in valves, crystal diodes or resistors. The machine is at present kept in good running order down to 10% margins (the supplies are normally controlled to about 1% of nominal), although correct running at about 20% reduction has been observed.

Conclusions

The first machine has been computing regularly for only a few months and has been on regular preventive maintenance (about 1 hour per day) for a few weeks. Error-free runs of over 30 hours are common, and at the time of writing there has been no error for $55\frac{3}{4}$ hours' running. The majority of package replacements are done during routine maintenance.

The packaged method of construction of computers has proved to have great advantages in design, construction and operation.

References

ElliW56a; ElboR53; ElliW51, 52, 53, 56b; FairJ56; JohnD52; MerrI56; Pegasus Programming Manual, Ferranti Ltd., London; Pegasus Maintenance Manuals, Ferranti Ltd., London.

APPENDIX

The Pegasus Order Code

00 $x' = n$
01 $x' = x + n$
02 $x' = -n$
03 $x' = x - n$
04 $x' = n - x$
05 $x' = x \,\&\, n$
06 $x' = x \not\equiv n$
07 Not allocated

10 $n' = x$
11 $n' = n + x$
12 $n' = -x$
13 $n' = n - x$
14 $n' = x - n$
15 $n' = n \,\&\, x$
16 $n' = n \not\equiv x$
17 Not allocated

20 $(pq)' = n \cdot x$
21 $(pq)' = n \cdot x + 2^{-39}$
22 $(pq)' = p + 2^{-38}q + nx$

23 $(nq)' = n + 2^{-38}q$ $\left\{\begin{array}{l}\text{this order assumes that any}\\\text{overflow is due to opera-}\\\text{tions in 7. Clears overflow}\\\text{unless } n' \text{ overflows}\end{array}\right.$

$\left.\begin{array}{l}24\\25\end{array}\right\} q' + 2^{-38}\left(\dfrac{p'}{n}\right) = \dfrac{x + 2^{-38}q}{n}$ $\left\{\begin{array}{l}0 \le p'/n < 1 \quad \text{(unrounded}\\\qquad\qquad\qquad \text{division)}\\-\frac{1}{2} \le p'/n < \frac{1}{2} \quad \text{(rounded}\\\qquad\qquad\qquad \text{division)}\end{array}\right.$

26 $q' + 2^{-38}\left(\dfrac{p'}{n}\right) = \dfrac{x}{n}; \; -\frac{1}{2} \le p'/n < \frac{1}{2}$ (rounded single-length division

27 Not allocated

$\left.\begin{array}{l}30\\31\\32\\33\\34\\35\\36\\37\end{array}\right\}$ Not allocated

$\left.\begin{array}{l}40 \; x' = c\\41 \; x' = x + c\\42 \; x' = -c\\43 \; x' = x - c\\44 \; x' = c - x\\45 \; x' = x \,\&\, c\\46 \; x' = x \not\equiv c\end{array}\right\} c = N2^{-38}$

47 Not allocated

$\left.\begin{array}{ll}50 \; x' = 2^N x & \\51 \; x' = 2^{-N}x \text{ (rounded)} &\end{array}\right\}$ $\begin{array}{l}\text{single-length arith-}\\\text{metical shifts}\end{array}$ $\left.\rule{0pt}{24pt}\right\}$ Note: $x' = x$ if $N = 0$

$\left.\begin{array}{l}52 \; \text{Shift } x \text{ up } N \text{ places}\\53 \; \text{Shift } x \text{ down } N \text{ places}\end{array}\right\}$ $\begin{array}{l}\text{single-length logical}\\\text{shifts}\end{array}$

$\left.\begin{array}{l}54 \; (pq)' = 2^N(pq)\\55 \; (pq)' = 2^{-N}(pq) \text{ (un-}\\\qquad\text{rounded)}\end{array}\right\}$ $\begin{array}{l}\text{double-length arith-}\\\text{metical shifts}\end{array}$ $\left.\rule{0pt}{24pt}\right\}$ $\begin{array}{l}\text{Note: } p' = p\\\text{and } q' = q\\\text{if } N = 0\end{array}$

56 (Normalize) $(pq)' = 2^\mu(pq);$

$$x' = x - 2^{-38}\mu \quad \begin{cases} \text{either (1)} \ \ \tfrac{1}{4} \le (pq)' < \tfrac{1}{2} \ \text{and} \\ \qquad\qquad -1 \le \mu \le N - 1 \\ \text{or (2)} \ -\tfrac{1}{2} \le (pq)' < \tfrac{1}{4} \ \text{and} \\ \qquad\qquad -1 \le \mu \le N - 1 \\ \text{or (3)} \ -\tfrac{1}{4} \le (pq)' < \tfrac{1}{4} \ \text{and} \\ \qquad\qquad \mu = N - 1 \end{cases}$$

57 Not allocated

60 Jump to N if $x = 0$
61 Jump to N if $x \ne 0$
62 Jump to N if $x \ge 0$
63 Jump to N if $x < 0$
64 Jump to N if overflow staticizor clear; clear overflow staticizor.
65 Jump to N if overflow staticizor set; clear overflow staticizor.
66 (Unit-modify) $x'_m = x_m + 1$. Jump to N if $x'_m \not\equiv 0$ (mod. 8)
67 (Unit-count) $x'_c = x_c - 1$. Jump to N if $x'_c \ne 0$

70 Single word read to accumulator 1. $1' = s$
71 Single word write from accumulator 1. $s' = 1$
72 Block read from main store $u' = b$
73 Block write into main store $b' = u$
74 External conditioning
75
76 $\Big\}$ Not allocated
77 Stop

The notation used here is as follows:

N is the first address (the register address) in an order.
X is the accumulator specified in an order.
n is the word in N before obeying the order.
x is the word in X before obeying the order.
p and q are the words in 6 and 7 before obeying the order.
$(pq) = p + 2^{-38}q$, with $q \ge 0$. This is a double-length number.
x', n', p' and q' are the corresponding values after obeying the order.
B is a block in the main store (the drum).
U is a block in the computing store.
P is the position number of a word within a block.
OVR is the overflow indicator.
xm is the modifier in X, i.e. an integer represented by the digits 1 to 13 of x.
xc is the counter in X, i.e. an integer represented by the digits 14 to 38 of x.

Chapter 10

An 8-bit-character computer

Introduction

We present in this chapter the result of an exercise to design an 8-bit computer. Although a rather trivial machine, it is not without interest, either as manipulator of variable-length character strings or as an interpreter of more complex computers in a role similar to a microprogrammed Pc. In the latter role a read-only memory could be used as Mp to speed up the Pc.

This computer is typical of 8-bit character-oriented computers. Among the similar machines are the Interdata Model 3, the RCA 1600, the IBM System/360 Model 25, and the Data Machines Inc. DMI 520/I. A processor of this type rarely stands alone but is used with a fixed program in the following ways: as a control in a larger C, as a control to a laboratory or other complex instrument, and as a microprogrammed processor to interpret an ISP.[1]

The processor must perform fixed-length operations on both 8-bit characters and 16-bit addresses. The address (double length) operations are necessary for performance reasons, because almost all programs operate on address integers. (For example, see the program on page 185.) Thus, extending (generalizing) the operation length to three and four characters is comparatively inexpensive. It should be noted that a processor might allow the operation length to be specified between 1 and perhaps 2^8 (256) characters for a much more general capability. We limit the directly addressable Mp to 2^{16} (or 65,384) characters. An alternative design might allow the maximum addressable Mp to be 2^{24} words, or, alternatively, it could be variable. Although 24-bit operations are defined, their implementation might be expensive. Aligning the 24-bit words on 32-bit-word boundaries would simplify the address calculation hardware.

The ISP

The basic information unit is the 8-bit character. Instructions are, in general, one character in length. However, both instructions and data formats are of variable length, instructions being 1, 2, 3, 4, and 5 characters long, and data being 1, 2, 3, and 4 characters long. The Pc state contains ~35 characters, which are organized to be dealt with as eight 8-, 16-, 24-, or 32-bit registers (shown

in the ISP description in Appendix 1 of this chapter). Of these registers, the first (register 0) is taken to be a special accumulator, A.

The Pc state contains both operands and addresses to operands. The instructions to load or store register A, from or into Mp, with or without incrementing a general register, all use the general registers as a two-character address pointer. Any general register may be loaded or stored direct from or to Mp. The binary arithmetic and logical operations are with a register and the accumulator, and leave the result in the accumulator; i.e., they are of the form

$$A \leftarrow A \ b \ R[r]$$

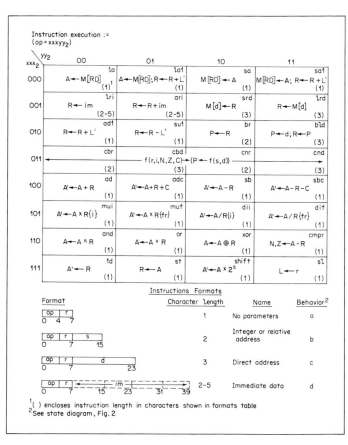

Fig. 1. Instruction coding for an 8-bit-character computer.

[1] The structure should be compared with the elaborate microprogrammed IBM System 360/Model 30 (Chap. 32).

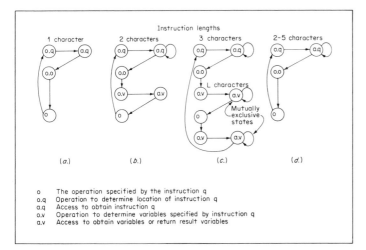

Fig. 2. An 8-bit-character-computer instruction-interpretation state diagram. (*a*) No parameters; (*b*) integer or relative address; (*c*) direct address; (*d*) immediate data.

The general registers discussed above are similar to those of the general register processors. Since it is assumed that this type of processor might be used to interpret another ISP, the +1 and −1 instructions provide for both string and stack memory operations. The instructions for a microprogrammed P and the I/O devices are not defined. For example, a 16-way branch instruction which branched to one of 16 locations based on 4 bits of the accumulator might facilitate writing an interpreter.

The ISP is given in Appendix 1 of this chapter. The Pc state is organized about a small scratch-pad memory, although Mp could be used instead. The instruction formats and the operation code assignments are shown in Fig. 1.

The instructions behave as illustrated in the state diagram (Fig. 2). For example, the instruction "lri 3, A907$_{16}$" is coded

00100,011	1010,1001	0000,0111

and the effect is

$$R[3]\langle 0{:}15\rangle \leftarrow A907_{16}$$

The instruction, xor 3, with L = 2, is coded

11010,011

and the effect is

$$R[0]\langle 0{:}23\rangle \leftarrow R[0]\langle 0{:}23\rangle \oplus R[3]\langle 0{:}23\rangle$$

In these examples, the behavior of lri and xor is specified in the state diagrams of Fig. 1*d* and 1*a*, respectively.

An open subprogram to perform the n-component vector (16-bit) addition[1] $\overleftarrow{A} \leftarrow \overleftarrow{B} + \overleftarrow{C}$ is

start	sl 2 − 1	*set register length = 2*
	lri 4, A	*set up vector pointers to*
	lri 5, B	*locations A, B, C in Mp*
	lri 6, C	
	lri 7, 2 × n	*set up count at 2n*
loop	lal 5	*fetch B*
	st 3	*store B temporarily*
	lal 6	*fetch C*
	ad 3	*add*
	stl 4	*store in A*
	sul 7	*decrement n count*
	cnr 4, loop	*branch if negative n*

The above program loop is nine characters long. A program loop for the IBM System/360 is about 16 characters long. The setup is 13 characters, as opposed to 6 ∼ 16 characters for the 360.

Conclusions

We have violated our principle of showing "real" computers by designing this computer. We think it is typical of a small processor, but slightly more interesting.

---

[1] The length is specified by register L.

APPENDIX 1 AN 8-BIT-CHARACTER COMPUTER ISP DESCRIPTION

Appendix 1

An 8 Bit Character Computer ISP Description

Pc State
The following array of 8 general registers, R, are mapped into the first 8 x (L+1) Mp cells. The register length is
$<0:8 \times (L+1)) - 1>$. The first register of each array, R[0] is an accumulator, and has special properties.

$R[0:7]<0:(8 \times L') -1> := M[0:7][0:L]<0:7>$	*General Registers of length (L+1) x 8 bits*
$A<0:(8 \times L') -1> := R[0]<0:(8 \times L') -1>$	*Accumulator (generally)*
$RQ[0:7]<0:31> := M[0:7][0:3]<0:7>$	*Quadruple Registers*
$AQ<0:31> := RQ[0]<0:31>$	*Quadruple Accumulator*
$RT[0:7]<0:23> := M[0:7][0:2]<0:7>$	*Triple Registers*
$AT<0:23> := RT[0]<0:23>$	*Triple Accumulator*
$RD[0:7]<0:15> := M[0:7][0:1]<0:7>$	*Double Registers*
$AD<0:15> := RD[0]<0:15>$	*Double Accumulator*
$RS[0:7]<0:7> := M[0:7][0:0]<0:7>$	*Single Registers*
$AS<0:7> := RS[0]<0:7>$	*Single Accumulator*

The following flags are set by the result of all arithmetic and logical instructions on the Accumulator, A. These are connected
to A to form A'.

N	*Negative result flag*
Z	*Zero flag, set if the register contains a zero*
C	*Carry flag, set if there is a carry or borrow from bit 0 of the addition*

$A'<N,Z,C,0:(8 \times L') -1> := N\square Z\square C\square A<0:(8 \times L') -1>$

$L<0:1>$ *2 bit register to indicate the character length of operations;*
                          1,2,3,4 for S,D,T,Q

$L'<1>_4 := L+1$

$P<0:15>$ *Program counter*

Mp State
$M[0:177777_8]<0:7>$ *primary memory*

Instruction Format

$i[0:4]<0:7>$	*1 to 5 character instruction*
$op<0:4> := i[0]<0:4>$	*op code*
$r<0:2> := i[0]<5:7>$	*register address*
$s<0:7> := i[1]$	*signed integer for shifts*
$d<0:15> := i[1:2]$	*address integer*
$im<0:(8 \times L') -1> := i[1:L']<0:7>$	*variable length immediate data*

Instruction Interpretation Process

$((\text{instruction}[0:4]<0:7> \leftarrow M[P:P+4]; P \leftarrow P + 1); \text{next}$	*fetch*
$((op = 011\$) \vee (op = 1\$11) \vee (op = 1001)) \rightarrow (P \leftarrow P + 2);$	
$((op = 11\$0) \vee (op = 1010)) \rightarrow (P \leftarrow P + 1);$	
$(op = 010\$) \rightarrow (P \leftarrow P + L+1); \text{next}$	
Instruction_execution)	*execute*

Instruction Set and Instruction Execution Process

```
Instruction_execution := (
la   (:= op = 0) → (A ←M[RD[r]]);                              load A
lal  (:= op = 1) → (A ←M[RD[r]]; next RD[r] ← RD[r] + L');     load A, increment
sa   (:= op = 2) → (M[RD[r]] ←A);                              store A
sal  (:= op = 3) → (M[RD[r]] ←A; next RD[r] ←RD[r] + L');      store A, increment
lri  (:= op = 4) → (R[r] ←im);                                 load register immediate
ari  (:= op = 5) → (R[r] ←im + R[r]);                          add register immediate
srd  (:= op = 6) → (M[d] ←R[r]);                               store register
lrd  (:= op = 7) → (R[r] ←M[d]);                               load register
adl  (:= op = 01000) → (R[r] ←R[r] + L');                      add 1 to register
sul  (:= op = 01001) → (R[r] ←R[r] - L');                      subtract 1 from register
br   (:= op = 01010) → (P ←R[r]);                              branch return
bld  (:= op = 01011) → (P ←d; R[r] ←P);                        branch and link direct
cbr  (:= op = 01100) → ((cond ≠ 0) →P ←P + s);                 conditional branch relative
cbd  (:= op = 01101) → ((cond ≠ 0) →P ←d);                     conditional branch direct
cnr  (:= op = 01110) → ((cond = 0) →P ←P + s);                 conditional not branch relative
cnd  (:= op = 01111) → ((cond = 0) →P ←d);                     conditional not branch direct
     cond := (r ∧ N□Z□C)
ad   (:= op = 10000) → (A' ←A + R[r]);                         add
adc  (:= op = 10001) → (A' ←A + R[r]+ C);                      add with carry
sb   (:= op = 10010) → (A' ←A - R[r]);                         subtract
sbc  (:= op = 10011) → (A' ←A - R[r] - C);                     subtract with carry
mui  (:= op = 10100) → (A' ←A × R[r] {i});                     integer multiply
muf  (:= op = 10101) → (A' ←A × R[r] {fr});                    fraction multiply
dii  (:= op = 10110) → (A' ←A / R[r] {i});                     integer divide
dif  (:= op = 10111) → (A' ←A / R[r] {fr});                    fraction divide
and  (:= op = 11000) → (A ←A ∧ R[r]);                          logical and
or   (:= op = 11001) → (A ←A ∨ R[r]);                          logical or
xor  (:= op = 11010) → (A ←A ⊕ R[r]);                          exclusive or
cmpr (:= op = 11011) → (N□Z ←A - R[r]);                        compare used to N and Z
ld   (:= op = 11100) → (A' ←R[r]);                             load
st   (:= op = 11101) → (R[r] ← A);                             store
shift(:= op = 11110) → (A' ←A × 2^s);                          shift right or left
sl   (:= op = 11111) → (L ← r)                                 set operation length
        )                                                      end Instruction_execution
```

Part 3

The instruction-set processor level: variations in the processor

In this part we discuss computers whose ISP's are variations from the main-line computers in Part 2. These variations represent historical computers that have not remained viable in the judgment of the computer engineering community, responses to particular technology, and explorations that were either too advanced for their time or still exist as open options.

Section 1, Processors with greater than 1 address per instruction, is mostly of historical and comparative interest. The general register organization with large Mp's (hence large addresses) almost surely dominate them.

Section 2, Processors constrained by a cyclic, primary memory, describes a response to a historical feature of Mp technology. The use of a drum, delay line, or disk was a matter of necessity rather than choice. When better random access core memories were available, the drum ceased to be a primary memory component.

Section 3 presents processors for variable string data. These processors are no longer built in their original form. However, they were very successful for a while (IBM 1401). Furthermore, string data-types have been incorporated in later processors.

Section 4 presents two desk calculator computers. Although we too often dismiss these devices as mere desk calculators, they have facilities that qualify them as general purpose stored program computers. Unlike most computers, because of the production cost constraint, these calculator computers are all very cleverly designed.

Section 5, Processors with stack memories, describes an organization that has never reached the main line state. Nevertheless, the idea of a stack memory is gradually being assimilated. For example, the DEC PDP-6 and PDP-10 computers use their general registers for stack pointer control, as suggested in Chap. 3, page 62.

In Sec. 6 the ideas of multiprogramming are presented. These ideas are recent and have not yet been adequately incorporated in main line designs. They undoubtedly will be standard features in the next generation, although the exact form cannot yet be known.

Section 1

Processors with greater than 1 address per instruction

Multiple-address instruction formats exist for several reasons. The addition of an explicit address to determine the next instruction occurs with cyclic Mp's to make them efficient. Section 2 is devoted to this case, and it will not be considered further here. These processors are known as n + 1 address. A second reason is that many operations have more than one operand (as in A + B or A ∨ B), and it seems to be efficient encoding to put them all into an instruction. A third reason is that many operations need to be followed by writing the result in memory, to permit the Pc to be used for operations on other data. Thus, coupling each operation with the address where the result is to be stored seems to be advantageous. However, in evaluating complex arithmetic expressions, more instruction bits and memory references are required than in a single-address computer. Also, for unary operators one address field is unused. It seems fair to say that ISP organizations with two or three addresses have not proved themselves in competition with the main line of 1, (1 + index), or (1 + general register) organizations. However, no definitive demonstration of their inefficiency under all technological conditions exists, and they are worth studying.

For microprogrammed processors, multiple-address instructions allow a high degree of parallelism to be obtained in a single instruction. Multiple-address formats survive in this form.

The Pilot ACE

The National Physics Laboratory's Pilot ACE is the first of several cyclic memory computers which have been designed to provide optimum coding of instructions. Subsequent machines which it influenced include the nearly identical English Electric Deuce, the Bendix G-15, and the Packard Bell PB-250.[1] The PMS structure does not strictly follow our lattice model (page 65). The Deuce PMS structure is given in Fig. 1. A 32-word block in Mp.delay_line can be transferred to Ms.drum in one instruction (transfer time of 1,024 μs). Another capability of

ACE allows it to perform operations on vectors of up to 32 elements in 1 instruction.

The ACE structure (Chap. 11) has a common M which contains much of the processor state and Mp. Many of the locations used for processor state can store programs for direct execution. The diagram on page 198 in Chap. 11 describes the instruction execution process and implementation.

Alan M. Turing is credited with the basic design of ACE (see introduction, page 193, and Turing's biography [Turing, 1959]).

ZEBRA, a simple binary computer

ZEBRA illustrates the organizational details of another serial arithmetic computer with Mp.cyclic. ZEBRA, like ACE, allows the user to construct instructions for the hardware which are almost directly interpreted. In both ACE and ZEBRA very little decoding is built into the machine; a large instruction set is available since the instructions are microcoded. In these computers the programming problem can be as complex as the user wishes, because a large number of different instructions can be micro-

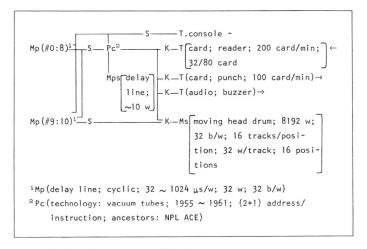

Fig. 1. English Electric Deuce PMS diagram.

---

[1] H. D. Huskey was involved in the design of ACE, G-15, and PB-250; he was undoubtedly the idea carrier.

coded. The LGP-30 (Chap. 16), by contrast, has only a basic instruction set. Hence a problem can be coded only one or two ways. ZEBRA's performance of 60 percent memory-cycle utilization is rather outstanding and raises the possibility that random-access primary memories may not be necessary.

UNIVAC scientific (1103A) instruction logic

The UNIVAC 1103A (Chap. 13) is a two-address computer. The computer was designed initially by Engineering Research Associates (ERA) of St. Paul.[1] UNIVAC acquired ERA in 1952 as a scientific-computer division. The evolution of the 1103A later yielded the 1107 and 1108 general register processors. The reader should compare the 1103A with the IBM 704 series (Chap. 41). At the time both were used, it was not clear which computer was better.

[1] As the third in a series that started with the ERA 1101 and 1102.

The RW-400: a new polymorphic data system

The RW-400 in Chap. 38 is a two-address, binary computer. It is discussed in Part 5, Sec. 4, page 470.

Instruction logic of the MIDAC

The University of Michigan's MIDAC (Michigan Digital Automatic Computer) is based on the National Bureau of Standards' SEAC (Standards' Electronic Automatic Computer). MIDAC, a three-address, binary computer, is presented in Chap. 14.

Instruction logic of the Soviet Strela (Arrow)

The Russian Strela is presented in Chap. 15. Since it is used only to illustrate a three-address organization, the chapter consists of only the instruction set.

Chapter 11

The Pilot ACE[1]

J. H. Wilkinson

Introduction

A machine which was almost identical with the Pilot ACE was first designed by the staff of the Mathematics Division at the suggestion of Dr. H. D. Huskey during his stay at the National Physical Laboratory in 1947. It was based on an earlier design by Dr. A. M. Turing and its principal object was to provide experience in the construction of equipment of this type. It was not intended that it would be used on an extensive programme of computation, but it was hoped that it would give practical experience in the production of subroutines which would serve as a useful guide to the design of a full scale machine. An attempt to build the Pilot Model, during Dr. Huskey's stay, was unsuccessful, but a year later after the formation of an Electronics Section at the NPL a combined team consisting of this section and four members of the Mathematics Division started on the construction of a Pilot Model, the design of which was taken over almost unchanged from the earlier version. The machine first worked, in the sense that it carried out automatically a simple sequence of operations, in May 1950 and by the end of that year it had reached the stage at which a successful Press Demonstration was held. The successful application of the machine to the solution of a number of problems made it apparent that, in spite of its obvious shortcomings, it was capable of being converted into a powerful computer comparable with any then in existence and much faster than most. Accordingly a small programme of modifications was embarked upon early in 1951, but the machine was not functioning satisfactorily again until November of that year. After a month of continuous operation it was transferred from the Electronics Section to Mathematics Division where it has since been in use on a 13-hour day. During its first year of full scale operation it achieved a 65% serviceability figure based on a very strict criterion. Its performance during its second year has so far been considerably better than this.

[1]*Automatic Digital Computation, National Physical Laboratory, Teddington, England*, pp. 5–14, March, 1953.

General description

The Pilot ACE is a serial machine using mercury delay line storage and working at a pulse repetition rate of 1 megacycle/sec. Its high speed store consists of 11 long delay lines each of which stores 32 words of 32 binary digits each, with a corresponding circulation period of 1024 microseconds, 5 short lines storing one word each with a circulation period of 32 microseconds and two delay lines storing two words each. It was inevitable that in the design of a machine originally intended for experimental purposes, overriding consideration should be given to the minimization of equipment rather than to making the machine logically satisfying as a whole. This is reflected to a certain extent in the code adopted for the machine and in its arithmetic facilities, which are in general fairly rudimentary. The design of the machine was also decisively influenced by the attempt to overcome the loss of speed due to the high access time of the long storage units. The machine in fact uses what is usually known as a system of "optimum coding."

Code of Pilot ACE

The Pilot ACE may be said to have a "three-address code" though this form of classification is not particularly appropriate. Each instruction calls for the transfer of information from one of 32 "sources" to one of 32 "destinations" and selects which of eight long delay lines will provide the next instruction. This third address is necessary because consecutive instructions do not occupy consecutive positions but are placed in such relative positions that, in so far as is possible, each instruction emerges during the minor cycle in which the current instruction is completed. An unusual feature of the instructions is that the transfers they describe may last for any number of consecutive minor cycles from one to thirty-two. The instruction word contains three other main elements which are known as the wait number, the timing number and the characteristic which together determine when the transfer starts, when it stops and which instruction in the selected instruction source is the next to be obeyed. The structure of the instruction word is as follows:

Next instruction source	Digits 2–4
Source	Digits 5–9
Destination	Digits 10–14
Characteristic	Digits 15–16
Wait number	Digits 17–21
Timing number	Digits 25–29
Go digit	Digit 32

The remaining digits are spare.

Coding of a problem takes place in two parts, in the first of which only the source, the destination and the period of transfer are specified, the last being a function of the characteristic, wait number and timing number. In the second part, the detailed coding, the other elements are added.

The sources and destinations

Simplest among the sources and destinations are those associated with the short delay lines. The six one-word delay lines are each given numbers and these for reasons associated with the history of the machine are 11, 15, 16, 20, 26 and 27. They are usually referred to as Temporary Stores or TS's because they are used to store temporarily those numbers which are being operated upon most frequently at each stage of a computation. In general TSn has associated with it a source, source n, and a destination, destination n. An instruction of the type

15–16

in the preliminary stage of the coding represents the transfer of a copy of the contents of TS15 via source 15 to TS16 via the destination 16. After it has taken place both stores contain the number originally in TS15. The period of the transfer is not mentioned in the coding because a transfer of more than one minor cycle is irrelevant. Most transfers are for one minor cycle and hence the period of transfer is not specified unless it is greater than one minor cycle. Associated with the TS's are a number of functional sources and destinations. TS16 for instance has two other destinations 17 and 18 associated with it, in addition to destination 16. Any number transferred to destination 17 is added to the contents of TS16 while any number transferred to destination 18 is subtracted from the contents of TS16. TS16 may be said to have some of the functions associated with the accumulator

on an orthodox machine. The period of transfer to destinations 17 and 18 is very important. Thus

15–17   (n minor cycles)

has the effect of adding the contents of TS15, n times to the contents of TS16. This prolonged transfer is used in this way to give small multiples (up to 32) of numbers. Similarly, we may have

15–18   (n mc)

The instruction

16–17   (n mc)

is of special significance because it has the effect of adding the content of TS16 to itself for each minor cycle of the transfer, that is it gives multiplication by 2^n or a left shift of n binary places.

TS26 has associated with it a number of functional sources. Source 17 gives the ones complement of the number in TS26, Source 18, the contents divided by 2, and Source 19, the contents multiplied by 2. The instruction

18–26   (n mc)

thus has the effect of dividing the contents of TS26 by 2^n, that is a right shift of n places. Similarly

19–26   (n mc)

gives a left shift of n places.

There are two functional sources which give composite functions of the numbers in TS26 and TS27. These are Source 21 which gives the number

TS26 & TS27

and Source 22 which gives the number

TS26 $\not\equiv$ TS27

There are a number of sources which give constant numbers which are of frequent use in computation. These are Source 23 which gives the number which has a zero everywhere except in the 17th position, usually known as P17, Source 24 which gives P32, Source 25 which gives P1, Source 28 which gives zero and Source 29 which gives a number consisting of 32 consecutive ones. These sources are valuable because they provide numbers with an access time of one minor cycle and are thus almost as useful as several extra TS's.

The use of a number of TS's with the arithmetic facilities distributed among them makes it possible to take advantage of the placing of instructions in appropriate positions in the long storage units so that they emerge as required. The coding of a trivial example will illustrate the uses of the TS's and their associated sources. It is required to build up the successive natural numbers, their squares and their cubes simultaneously. It is natural to store the values in TS's and we may suppose TS15 contains n, TS20, n^2 and TS26, n^3.

Instruction		Description	
1.	28–15	zero to TS15 *i.e.* 0	These 3 instructions set the
2.	28–20	zero to TS20 *i.e.* 0^2	initial values
3.	28–26	zero to TS26 *i.e.* 0^3	
	—		
4.	26–16	TS16 contains n^3	
5.	20–17 (3mc)	TS16 contains $n^3 + 3n^2$	
6.	15–17 (3mc)	TS16 contains $n^3 + 3n^2 + 3n$	
7.	25–17	TS16 contains $n^3 + 3n^2 + 3n + 1$	
8.	16–26	TS26 contains $(n + 1)^3$	
9.	20–16	TS16 contains n^2	
10.	15–17 (2mc)	TS16 contains $n^2 + 2n$	
11.	25–17	TS16 contains $n^2 + 2n + 1$	
12.	16–20	TS20 contains $(n + 1)^2$	
13.	15–16	TS16 contains n	
14.	25–17	TS16 contains $(n + 1)$	
15.	16–15	TS15 contains $(n + 1)$ Next instruction (4)	

The instructions (1) to (3) set the initial conditions. The instruction (4) − (15) have the effect of changing the contents of 15, 20, 26 from n, n^2, n^3 to $(n + 1)$, $(n + 1)^2$, $(n + 1)^3$. As remarked earlier, each instruction selects the next instruction and here instruction (15) selects instruction (4) as the next instruction. In the preliminary coding this is usually denoted by using an arrow; it must be catered for in the detailed coding by the correct choice of the timing number, as will be shown below.

The branching of a programme is achieved by the use of two destinations, destination 24 and destination 25. If a transfer is made from any source to destination 24 then the next instruction is one or other of two according as the number transferred is positive or negative. Similarly if a transfer is made to destination 25 then the next instruction is one or other of two according as the number transferred is zero or non-zero. In the preliminary coding the bifurcation is denoted by the use of arrows, thus:

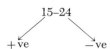

In the detailed coding the effect is that if the number transferred to destination 24 is negative then the timing number is increased

by 1. Similarly for destination 25; the two possible next instructions are consecutive in the store.

The two double word stores are numbered DS12 and DS14. DS12 has only source 12 and destination 12 associated with it, but DS14 has, in addition to source 14 and destination 14, a number of functional sources and destinations. Source 13 gives the contents of DS14 divided by 2, while transfers to destination 13 have the effect of adding the numbers transferred to DS14. In specifying transfers from, and to, the double length stores, the time of the transfer must be specified, *i.e.* whether it takes place in an even or an odd minor cycle or both. Thus the transfer

12–14 (odd minor cycle) usually written

12–14 (o)

represents the transfer of the word in the odd positions of DS12 to the odd position in DS14 while

12–14 (2 minor cycles)

represents the transfer of both words in 12 to the corresponding positions in 14. The operation

13–14 (2n)

gives us a method of shifting the contents of TS14 n places to the right while

14–13 (2n)

produces a shift of n places to the left.

The machine is not equipped with a fully automatic multiplier. To multiply two numbers, a and b, together, a must be sent to TS20, b to DS14 odd, zero to DS14 even and a transfer (source irrelevant) made to destination 19. The product is then produced in DS14 in 2 milliseconds, but a and b are treated as positive numbers. Corrections must be made to the answer if a and b are signed numbers. To make multiplication fast, it has been made possible to perform other operations while multiplication is proceeding. Thus the corrections necessary if a and b are signed numbers may be built up in TS16 during multiplication, and signed multiplication takes only a little over two millisecs. It is, of course, therefore, a subroutine but a very fast one. The amount of equipment associated with the multiplier is very small. The main part of the store consists of the long storage units known as DL1, DL2, . . ., DL11. Each of these has a source and a destination with the same number as the DL number. The words in each DL are numbered 0 to 31 and the nth word in DLM is usually denoted by DLM_n. Transfers to and from long lines in the preliminary coding are denoted thus:

8_n- 16 (transfer nth word of DL8 to TS16)

$8_{m-n}-17$ (add all the words from 8_m to 8_n *i.e.* n − m + 1 consecutive words of DL8^m to TS16)

Detailed coding

In the second stage of the coding the true instruction words are derived from the preliminary coding. This is a fairly automatic process and recent experience has shown that it can be carried out satisfactorily by quite junior staff. The timing of each instruction is given relative to the position of that instruction in the store. This is an incidental feature of the code which arose from the attempts to minimize equipment. It would be dropped in any future machine in favour of an absolute timing system. If an instruction occupies position m in a DL and has a wait number W and timing number T then the transfer always begins in minor cycle (m + W + 2) and the next instruction is always in minor cycle (m + T + 2) of the selected next instruction source. The period of transfer depends on the value of the characteristic. If the characteristic is zero then the transfer lasts for the whole period from (m + W + 2) to (m + T + 2), that is (T − W + 1) minor cycles. If the characteristic is one, then the transfer is for one minor cycle, that is minor cycle (m + W + 2). If the characteristic is three then the transfer is for two minor cycles (m + W + 2) and (m + W + 3). The characteristic value, two, is not used. The characteristic value zero gives a prolonged transfer which is peculiar to the Pilot ACE. The characteristics 1 and 3 are analogous to the facility on EDSAC whereby full length or $\frac{1}{2}$-length words may be transferred. On the Pilot ACE we transfer single or double length words. This facility is invaluable for double length, floating and complex arithmetic. In the above definitions the numbers (m + W + 2) etc. are to be interpreted modulo 32. In general, timing and wait numbers are simpler than they appear from the definitions because they are very frequently both zero, corresponding to a transfer for one minor cycle. The detailed coding of the problem given earlier will illustrate the procedure. All the instructions are in DL1 so that the next instruction source is always one. The key to the headings in the following table is:

m.c.	Minor cycle position of instructions in DL1
N.I.S.	Next instruction source
S	Source
D	Destination
C	Characteristic
W	Wait number
T	Timing number

The last column gives the position of the next instruction in DL1; it is given by (m + T + 2). The first 4 instructions occupy minor cycles, 0, 2 and 4, 6 and each takes two minor cycles, and gives a transfer for one minor cycle only. The next instruction occupies minor cycle number 8 and it requires a transfer lasting 3 minor cycles. The simplest and fastest way of getting this is to have W = 0 and T = 2 giving a transfer of (2 − 0 + 1) minor cycles. The next instruction is in position (8 + 2 + 2), that is minor cycle 12, and so on. When we reach the instruction in minor cycle 31, viz. 25–17, a transfer for one minor cycle is required. The simplest way is to have W = 0 T = 0 and this makes the next instruction occupy position (31 + 0 + 2) i.e. position 33 which is position 1. If position 1 had been already occupied, a value of T could have been chosen in order to land in an unoccupied position. In order to ensure that a transfer of one minor cycle only took place, the characteristic could have been made 1. It should be appreciated that the choice of C, W and T is far from unique. Whenever possible T = 0 and W = 0 are chosen because this gives the highest speed of operation besides being simplest. The instruction occupying position 1 is of special interest because this is the last instruction of the cycle needed to build up a square and cube and it must select as its next instruction the first of the cycle, which is, in position number 6. This is achieved by making T = 3 (giving the next instruction in m.c. 1 + 3 + 2 = 6). This incidentally gives a transfer lasting four minor cycles but since it is a transfer from one TS to another and no functional source or destination is in use, the prolonged transfer produces no harmful effect. If a prolonged transfer had to be avoided then the characteristic could be taken as 1. It is seldom necessary to use any characteristic other than zero for transfers to and from TS's but when transfers are made to and from DL's, characteristic values of 1 or 3 are almost universal. All 12 instructions which comprise the repeated cycle of the computation take a total time of one major cycle exactly (32 minor cycles) the last instruction of the cycle having been specially designed to get back to the beginning of the cycle. This is in contrast to the position in a machine not using optimum coding, where 12 major cycles would be necessary quite apart from the fact that the multiplications by factors of 3 and 2, each of which uses one instruction, would normally need more than one instruction if a prolonged transfer were not available. Figure 1 gives a simplified diagram of the machine. The sequence of events in obeying the instruction

N	S		D	C	W	T
2	16	–	2C	0	8	10

occupying DL1$_2$ for example is as follows. Starting from the time when the last instruction was completed, the instruction from

Minor cycle position of instructions in DL1	Next instruction source	Source	Destination	Charac- teristic	Wait no.	Timing no.	Minor cycle position of next instruction
0	1	28	15	0	0	0	(2)
1	1	16	15	0	0	3	(6)
2	1	28	20	0	0	0	(4)
3							
4	1	28	16	0	0	0	(6)
5							
6	1	26	16	0	0	0	(8)
7							
8	1	20	17	0	0	2	(12)
9							
10							
11							
12	1	15	17	0	0	2	(16)
13							
14							
15							
16	1	25	17	0	0	0	(18)
17							
18	1	16	26	0	0	0	(20)
19							
20	1	20	16	0	0	0	(22)
21							
22	1	15	17	0	0	1	(25)
23							
24							
25	1	25	17	0	0	0	(27)
26							
27	1	16	20	0	0	0	(29)
28							
29	1	15	16	0	0	0	(31)
30							
31	1	25	17	0	0	0	(1)

$DL1_2$ will have passed into the special TS marked TS COUNT during minor cycle number 2. By the end of minor cycle number 3, S switch number 16 will be over and also N switch number 2. The contents of TS16 will be passing into HIGHWAY and those of DL2 into INSTRUCTION HIGHWAY. At the beginning of minor cycle number 12 (*i.e.* $2 + 8 + 2$), D switch number 20 will go over, and TS20 will stop recirculating and the number on the HIGHWAY will pass into TS20. The transfer will continue until minor cycle 14 (*i.e.* $2 + 10 + 2$) when the D switch number 20 will switch back. At the beginning of minor cycle 14, the switch X on COUNT will go over and the number on INSTRUCTION HIGHWAY during this minor cycle, $DL2_{14}$, will pass into COUNT. At the end of minor cycle 14, the X switch will close again and

$DL2_{14}$ will be trapped in COUNT. The cycle of events is now complete. COUNT is associated with a counter and it is this counter which determines from the wait, timing, and characteristic numbers of the trapped instruction, when the D and X switches go over and back.

Input and output

The only part of the instruction word not described is the GO digit. If the GO digit is a one, the instruction is carried out at high speed, but if it is a zero the machine stops and does not proceed until a manual switch is operated. The GO digit is omitted in strategic instructions when a programme is being tested. It also

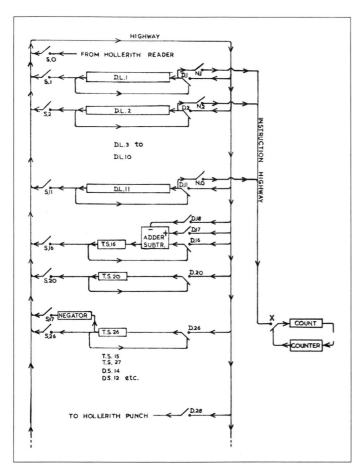

Fig. 1. Simplified diagram showing some sources, destinations, and next-instruction sources.

serves a further purpose in synchronising the input and output facilities with the high speed computer. Input on the machine is by means of Hollerith punched cards. When cards are passed through the reader the numbers on the card may be read row by row as each passes under a set of 32 reading brushes. When a row of a card is under the reading brushes, the number punched on that row, regarded as a number of 32 binary digits, is available on source 0. In order to make certain that reading takes place when a row is in position and not between rows, transfers from source 0, have the GO digit omitted and it is arranged that the Hollerith reader has the same effect as operating the manual switch each time a row comes into position. The passage of a card through the reader is called for by a transfer from any source to destination 31. No transfer of information from the card takes place unless the appropriate instruction using source 0 is obeyed during the passage of the card. Output on the machine is also provided

by a Hollerith punch. The passage of a card through the punch is called for by a transfer from any source to destination 30. While a card is passing through the punch a 32 digit number may be punched on each row by a transfer to destination 28. Again synchronisation is ensured by omitting the GO digit in instructions calling for a transfer to destination 28, and arranging that the Hollerith punch effectively operates the manual switch as each row comes into position. The reader feeds cards at the rate of 200 cards per minute and the punch, at the rate of 100 cards per minute. The speed of input for binary digits is $200 \times 32 \times 12$ per minute or 1280 per second. The output speed is 640 digits per second. Data may be fed in and out in decimal, but it then requires conversion subroutines. The computation involved in the conversion is done between the rows of the card and up to 30 decimal digits per card may be translated. This speed of conversion is only possible because of the use of optimum coding. The facility for carrying out computation between rows of cards is used extensively particularly in linear algebra when matrices exceeding the storage capacity of the machine are involved. The matrices are stored on cards in binary form with one number on each of the 12 rows of each card, all the computation being done either between rows when reading or when punching. Times comparable with those possible with the matrices stored in the memory are often achieved in this way, when the computation uses a high percentage of the available time between rows. Up to 80% of this time may be safely used.

Initial input

The initial input of instructions is achieved by choosing destination 0 in a special manner. When a transfer is made to destination 0, then the instruction transferred becomes the next to be obeyed and the next instruction source is ignored. Source 0 has already been chosen specially since it is provided from a row of a card. The instruction consisting of zeros has the effect of injecting the instruction punched on a row of a card into the machine as the next to be obeyed. The machine is started by clearing the store and starting the Hollerith reader which contains cards punched with appropriate instructions. Destination 0 is also used when an instruction is built up in an arithmetic unit ready to be obeyed.

Miscellaneous sources and destinations

Destination 29 controls a buzzer. If a non-zero number is transferred to destination 29 the buzzer sounds.

Source 30 is used to indicate when the last row of a card is in position in the reader or punch. This source gives a non-zero number only when a last row is in position. The operation of the arithmetic facilities on DS14 may be modified by a transfer to destination 23. If a transfer with an odd characteristic is made from any source to destination 23 then, from then on, DS14 behaves as though it were two single length accumulators in series. This means that carries are suppressed at the end of each of the single words. This condition persists until a transfer is made to destination 23 using an even characteristic, when DS14 behaves as an accumulator for double length numbers with their least significant parts in even minor cycles and more significant parts in odd minor cycles.

The operation TS20 is modified by transfers to destination 21. If a transfer with an odd characteristic is made to destination 21 then TS20 ceases to have an independent existence and from then on is fed continuously from DL10. Source 20 then gives the contents of DL10 one minor cycle later than from source 10. TS20 reverts to its former condition when a transfer with an even characteristic is made to destination 21. The facility is used to move the 32 words in DL10 round one position so that the word in minor cycle n is available in minor cycle (n + 1).

Assessment of optimum coding

A detailed assessment of the value of optimum coding is by no means simple. Roughly speaking, subroutines are on an average about 4 or 5 times as fast as on an orthodox machine using the same pulse repetition rate. In main tables a somewhat lower factor is usually achieved. The factor of 4 or 5 would be exceeded if less of the advantage given by optimum coding were used to overcome disadvantages due to the rudimentary nature of the arithmetic facilities on Pilot ACE. Even so, the bald statement of the average ratio of speeds does not do full justice to the value of optimum coding on the Pilot ACE. Its value springs as much from the fact that it has made possible the programmes in which computing is done between the rows of cards and also the high output speed of decimal numbers. The binary decimal conversion routines for punching out several decimal numbers simultaneously on a card and also decimal-binary conversion routines for reading several numbers, achieve a ratio of something like 14 to 1, and on a machine which is being used extensively for scientific computation on a commercial basis this is of immense importance.

Future programme

Engineered versions of the Pilot Model are now under construction by the English Electric Company. These machines will be similar to the Pilot Model but will have a little more high-speed store, an automatic divider, two quadruple length stores and a subtractive input on the double length accumulator besides several minor modifications including a rationalization of the numbering of the stores! In addition a magnetic drum intermediate store with the equivalent of 32DL's storage capacity will be added. A full scale machine will probably soon be under development employing a 4 address code. Typical instructions will be of the form

$$A \pm B \ C$$

and will select the next source of instruction. This code is more economical in instruction storage space and since all single word stores will then become complete accumulators with all facilities except multiplication on them, it will be possible to take much fuller advantage of optimum coding.

Sources, destination and next instruction sources

Sources		Destinations		Next instr. sources	
0.	Input	0.	INSTRUCTION	0.	DL11
1.	DL1	1.	DL1	1.	DL1
2.	DL2	2.	DL2	2.	DL2
3.	DL3	3.	DL3	3.	DL3
4.	DL4	4.	DL4	4.	DL4
5.	DL5	5.	DL5	5.	DL5
6.	DL6	6.	DL6	6.	DL6
7.	DL7	7.	DL7	7.	DL7
8.	DL8	8.	DL8		
9.	DL9	9.	DL9		
10.	DL10	10.	DL10		
11.	DL11	11.	DL11		
12.	DS12	12.	DS12		
13.	DS14 + 2	13.	DS14 add		
14.	DS14	14.	DS14		
15.	TS15	15.	TS15		
16.	TS16	16.	TS16		
17.	TS26	17.	TS16 add		
18.	TS26 ÷ 2	18.	TS16 subtract		
19.	TS26 × 2	19.†	MULTIPLY		
20.	TS20	20.	TS20		
21.	TS26 & TS27	21.	Modifies Source 20		
22.	TS26 ≢ TS27	22.	—		
23.	P17	23.	Modifies Source 13, Destination 13		
24.	P32	24.	DISCRIMINATE on sign		
25.	P1	25.	DISCRIMINATE on zero		
26.	TS26	26.	TS26		
27.	TS27	27.	TS27		
28.	Zero	28.	Output		
29.	Ones	29.	BUZZER		
30.	Last row of card	30.†	PUNCH		
31.	—	31.†	READ		

† Independent of source used.

References

WilkJ53; TuriS59

Chapter 12

ZEBRA, a simple binary computer[1]

W. L. van der Poel

Summary   The computer ZEBRA is a computer based on the following ideas:

1. The logical structure of the arithmetic and control units of the machine have been simplified as much as possible; there is not even a built-in multiplier nor a divider.

2. The separate bits in an instruction word are used functionally and can be put together in any combination.

3. Conventional two stage operation (set-up, execution) has been abandoned. Each unit time interval can be used for arithmetical operations.

4. A small number of fast access registers is used as temporary storage; at the same time these registers serve as modifier registers (B-lines).

5. Optimum programming is almost automatically done to a very great extent. The percentage of word times effectively used is usually greater than 60%.

6. An instruction can be repeated and modified while repeated by using an accumulator as next instruction source and the address counter as counter. This can be done without any special hardware.

This has resulted in a machine which has a very simple structure and hence contains only a very moderate number of components, giving high reliability and easy maintenance. Because of the functional bit coding, the programming is extremely flexible. In fact the machine code is a sort of micro-programming. Full-length multiplication or half-length multiplication in half the time are just as easy, only require a different micro-programme. The minimum latency programming together with the effective use of word times lost in other systems results in a very high speed of operation compared to the basic clock pulse frequency.

Introduction

In the Dr. Neher Laboratory of the Dutch Postal & Telecommunications Services the logical design of a computer called ZEBRA has been developed, and this computer has been engineered and constructed by Standard Telephones & Cables Ltd, England. The logical system is so different from most computers, that it is worth while to devote a special lecture to it. As time is limited,

[1]*Proc. ICIP, UNESCO*, pp. 361–365, June, 1959.

no technical details nor questions about dimensions or capacity will be discussed. They can all be found in the literature [van der Poel, 1956; van der Poel, 1952].

The main idea of the machine is to economise as far as possible on the number of components by simplifying the logical structure. For example, multiplication and division are not built in but must be programmed. Of course this system can only work with an appropriate internal code which has enough properties to execute basic arithmetic and logical routines effectively. In fact, the internal machine code is more or less a system of microprogramming [Wilkes and Stringer, 1953].

Operation part of the instruction

The most conspicuous, but probably not the most important, characteristic is the functional use of the separate bits in the operation part of an instruction. An instruction word in ZEBRA is composed as follows:

$$
\begin{array}{|c|c|c|c|}
\hline
 & \text{15 bits} & & \text{5 bits} \\
\text{A K Q L R I B C D E} & \text{V } x_4\ x_2\ x_1 & \text{W} & 0\ 0\ 0\ 0\ 0 \\
 & \text{test bits} & & \text{fast store} \\
 & \text{operation part} & & \text{address} \\
\hline
\end{array}
$$

$$
\begin{array}{|c|}
\hline
\text{13 bits} \\
\text{x x x x x x x x x x x x x} \\
\text{drum store address} \\
\hline
\end{array}
$$

It is a binary, two-address machine with one address of 13 bits for the selection of a location in the main store (a drum of 8192 locations divided into 256 tracks of 32 words each), and a second address of 5 bits for the selection of one of 12 fast access store registers and several permanently wired locations (e.g., input, output, accumulators, constants). The operation part has 15 bits, each one having a separate and independent meaning. The most important of these are the A, K, D and E bits.

A- and K-bits

There are four main components in the machine: the drum store, the fast store, the arithmetic unit and the control. The A-bit in the instruction controls the interconnection of the drum and the

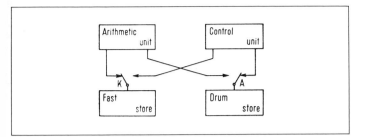

Fig. 1. The main units of the computer.

arithmetic unit or the control. In the same way the K-bit controls the interconnection of the fast store with the arithmetic unit or the control unit. These interconnections can be seen from Fig. 1.

It will be seen that A and K can have 4 possible combinations:

Case 1. A = 0, K = 0. This is called the adding jump (Fig. 2*a*).

While a new instruction is coming into the control from the drum, the arithmetic unit can at the same time do an operation with the operand coming from the fast store. This is the fastest type of operation. When the following instruction is placed in the next location on the drum there is no waiting time, and 32 instructions of this type can be executed per revolution. (One revolution = 10 ms, one word time = 312 μs.)

Case 2. A = 0, K = 1. This is called the double jump (Fig. 2*b*).

Both stores are now used for giving information to the control, i.e., making a jump. Since the fast store is used for the control, the instruction coming in from the drum is modified by the contents of a fast register. In this way the B-line facility, as it is often called, is realised.

Case 3. A = 1, K = 0. This is called the double addition (Fig. 2*c*).

Both stores are now connected to the arithmetic unit. The control must take care of itself using the address counter which is *stepped up by 2 at a time*, thus enabling this type of instruction to reach the number lying between the two successive instructions without any waiting time. Constants in particular will always be taken from optimum places on the drum.

Case 4. A = 1, K = 1. This is called the jumping addition (Fig. 2*d*).

While the drum is used for the arithmetic unit the address counter is modified by a fast register. Control may thus be passed to any instruction, and not only to the next instruction.

D- and E-bits

The functional bits D and E control the direction of flow of information.

D = 0 means: read from the drum.

E = 0 means: read from the fast store.

D = 1 means: write to the drum.

E = 1 means: write to the fast store.

A few possible instructions will be given below. In the written code a drum address will always be written with 3 or more digits and the absence of the A-bit will be indicated by the letter X. (This is necessary for the input programme to recognize the beginning of a new instruction.)

A200.5	Add ⟨200⟩ (the contents of address 200) and ⟨5⟩ to the accumulator. Step the address counter by 2.
X200E5	Take next instruction from 200 (= jump to 200) and store contents of accumulator in 5.
X200KE5	Jump to 200 and store previous contents of address counter in 5. This amounts to placing a link instruction for return from a sub-routine.
X200K5	Take next instruction from 200 but modify it with ⟨5⟩ thus making a variable instruction.

Arithmetic bits

The remainder of the function bits have arithmetic meanings. We shall only briefly indicate their different actions.

B: Do not use the A accumulator (most significant accumulator) but the B accumulator.

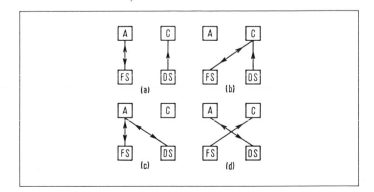

Fig. 2. The possible combinations of the A- and K-bits.

C: Clear the accumulator specified by B after storing, or before addition. (In a serial machine like ZEBRA this is automatically the case, cf. Fig. 3.)

I: Subtract instead of add.

Q: Add one (unit in the least significant place) to the B-accumulator.

L: Shift both accumulators one place to the left.

R: Shift both accumulators one place to the right. The accumulators are always coupled together in shifting except when C is present.

A few more examples will be given.

A200BCE25   Store ⟨B⟩ in 5, clear B and add ⟨200⟩ to B.

X200QLIBCE6   Jump to 200. Store ⟨B⟩ in 6, put −1 in B (because of QIBC) and shift the A accumulator one place to the left. Shifting from B into A is prevented by the presence of C.

X200RBC3   Jump to 200. Shift A to the right. Copy ⟨3⟩ into B. As register 3 is just an address for the B accumulator itself, this means that A is shifted while B is static.

X200K3QIBC   Take the instruction from 200 and modify it with the contents of the B accumulator (= register 3). Put −1 in B afterwards.

As can be seen, many complicated operations can be composed by the elementary possibilities of the separate bits.

The accumulator

A simplified block diagram of one of the accumulators is shown in Fig. 3.

Shifting is effected by looping the accumulator over one place less or one place more. In a double addition the contents of the drum store and the fast store are first added together in the pre-adder (possibly augmented by unity in the B accumulator, if Q is present) and this result is added into the accumulator (or subtracted in case of I). A clearing gate controlled by C interrupts the recirculation of the previous contents.

The control unit

The control unit has two shifting registers, the C-register which receives the next instruction to be executed and the D-register or counter. The block diagram is shown in Fig. 4. After a new instruction has come into C, it is taken over in parallel form into E in the interword time. It remains in E while the next instruction is coming into C. Let us explain the action of this control with a short programme.

Examples of programmes

100 X101E5
101 AC102
102 constant
103 etc.

The actions in the several registers are now:

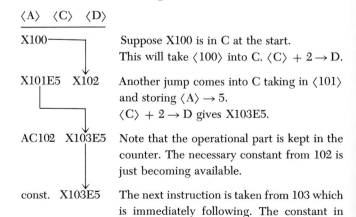

⟨A⟩   ⟨C⟩   ⟨D⟩

X100 —   Suppose X100 is in C at the start. This will take ⟨100⟩ into C. ⟨C⟩ + 2 → D.

X101E5 X102   Another jump comes into C taking in ⟨101⟩ and storing ⟨A⟩ → 5. ⟨C⟩ + 2 → D gives X103E5.

AC102 X103E5   Note that the operational part is kept in the counter. The necessary constant from 102 is just becoming available.

const. X103E5   The next instruction is taken from 103 which is immediately following. The constant in A is stored to 5 by E5, and is still active after coming back from D.

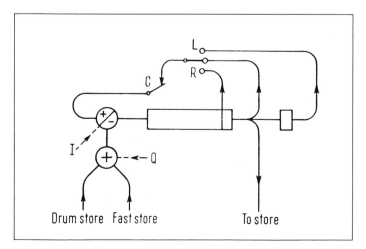

Fig. 3. Accumulator.

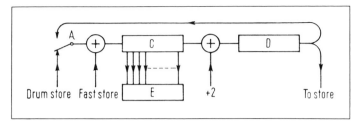

Fig. 4. Control unit.

This is the most important aspect of the machine. An instruction in the address counter comes back after an A-instruction and can do something useful. To our surprise we found that in many more cases than we first suspected, the second action could be used effectively. In most other computers the time of access to the next instruction is lost because nothing can be done concurrently in the arithmetic unit.

Another example of the action of the control is the jump to a sub-routine. Suppose that we have the following piece of pro-gramme:

100 X200KE5	Jump to sub-routine starting in 200. Place return jump in 5.
102 etc.	Sub-routine returns here.

The action is as follows:

$\langle C \rangle$	$\langle D \rangle$	
X100		The instruction is taken from 100.
X200KE5	X102	X200KE5 $\to$ C and X100 + 2 $\to$ D. Now KE5 stores D in 5. Thus $\langle 5 \rangle$ = X102.
(200)		The subroutine at 200 is executed and ends with XK5: jump to 5.
XK5		Take instruction from 5.
X102 (102)		Now the main programme proceeds to 102 etc.

By ending the sub-routine:

220   X221K5
221 − 1

we can return not two but one location further on, i.e., X221K5 takes as next instruction $\langle 5 \rangle - 1 =$ X101. Here 5 contains the instruction and the drum modifier.

The test bits

The digits V x_4 x_2 x_1 will not be dealt with extensively but the different combinations of these 4 digits represent different types of test. When for example V1 is attached to an instruction, this instruction will be executed when $\langle A \rangle$ is negative, but will be skipped altogether when $\langle A \rangle$ is positive or zero. The harmless A-instruction will then be executed instead. The test can be attached to a jump, giving a conditional jump, as well as to an A-instruction, giving a conditional addition.

The W-bit

So far the digit W has not been mentioned. When W is present in an instruction the drum address is not used. The instruction is not kept waiting but is immediately executed and the drum is completely disregarded. With the help of this digit W, jumps can be made to instructions in the fast store, e.g., XK5W takes the instruction from 5 only, and the drum does not deliver any number. The use of this type of instruction has very peculiar consequences. Let us take the following example:

100 X101KE6	$\langle 5 \rangle$ = ARW
101 X8186K5RW	$\langle 6 \rangle$ = filled with return instruction
102 etc.	

The action is as follows:

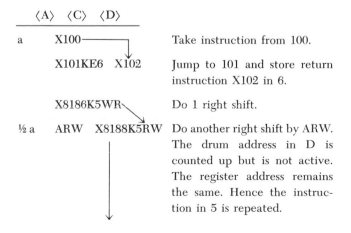

	$\langle A \rangle$	$\langle C \rangle$	$\langle D \rangle$	
a		X100		Take instruction from 100.
		X101KE6	X102	Jump to 101 and store return instruction X102 in 6.
		X8186K5WR		Do 1 right shift.
½ a	ARW	X8188K5RW		Do another right shift by ARW. The drum address in D is counted up but is not active. The register address remains the same. Hence the instruction in 5 is repeated.

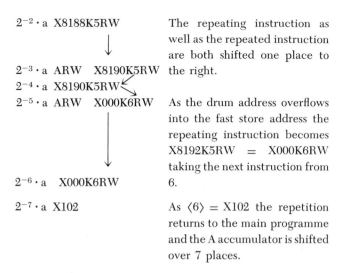

$2^{-2} \cdot$ a X8188K5RW

$2^{-3} \cdot$ a ARW   X8190K5RW

$2^{-4} \cdot$ a X8190K5RW

$2^{-5} \cdot$ a ARW   X000K6RW

$2^{-6} \cdot$ a   X000K6RW

$2^{-7} \cdot$ a X102

The repeating instruction as well as the repeated instruction are both shifted one place to the right.

As the drum address overflows into the fast store address the repeating instruction becomes X8192K5RW = X000K6RW taking the next instruction from 6.

As ⟨6⟩ = X102 the repetition returns to the main programme and the A accumulator is shifted over 7 places.

The instruction ARW has thus been repeated p times when the drum address of the repeating instruction is 8192—2p. This way of repeating an instruction has made it possible to do multiplication, division, block transfers, table look up and many other small basic repetitive processes in a very simple way. There is no special hardware present in the machine to do the counting necessary for the repetition, as this counting is done by the normal address counter.

As a last example we shall give a programme for the summa-

tion of a block of locations from 200 to 300 in the store. This involves 101 locations. The programme reads:

100 A101BC	Put A200Q in B (B has address 3).
101 A200Q	
102 X103KE4C	Put return jump X104 in 4. Clear A in advance.
103 X7990K3W	Repeat A200Q 101 times. Because A200Q is standing in B the Q augments the instruction itself at every repetition. Hence successively ⟨200⟩, ⟨201⟩ etc. are added to A. At the end the sum is left in A and the programme proceeds at 104.
104 etc.	

It is left to the reader to work out the action diagram.

This example is not programmed for minimum waiting, but by supplying the repeating instruction X7990K3W with a Q it will step up the repeated instruction A200Q by 2 every time. Now, once the first instruction has been located, all even locations following are emerging from the drum just at the right time. The odd numbered locations must be summed in a second, similar repetition.

References

VandW59; VandW52, 56; WilkM53a.

Chapter 13

UNIVAC Scientific (1103A) instruction logic[1]

John W. Carr III

The UNIVAC Scientific computer is a $(35, 0, 0)^2$ binary machine, with option of $(27, 8, 0)$. The arithmetic unit contains two 36-bit X (exchange) and Q (quotient) registers and one 72-bit A register (accumulator). Negative numbers are represented in one's complement notation.

Input-output is via high-speed paper tape reader and punch, direct card reader and punch, and Uniservo magnetic tape units, which may be connected to peripheral punched card readers and punches and a high-speed printer. In addition, information may be recorded on magnetic tape directly from keyboards by the use of Unitypers. Communication with external equipment is via an 8-bit (IOA) register and a 36-bit (IOB) register. Information sent to these registers controls magnetic tapes as well as other input-output equipment. The program address counter (PAK) contains the present instruction address. Storage is in up to 12,288 locations of magnetic core storage, along with a directly addressable drum of 16,384 locations. Instructions are of the two-address form, with six bits for the operation code and two fifteen-bit addresses (u and v).

The following information is taken from a Univac Scientific Manual [Univac Scientific Electronic Computing System Model 1103A, Form EL338].

Definitions and conventions

Instruction word

oc 6 bits	u 15 bits	v 15 bits
$1_{35}\cdots$	$1_{29}\cdots$	$1_{14}\cdots\qquad 1_0$

oc   Operation code
u   First execution address
v   Second execution address

For some of the instructions, the form jn or jk replaces the u address; for others the form k replaces the v address.

j   One-digit octal number modifying the instruction
n   Four-digit octal number designating number of times instruction is to be performed
k   Seven-digit binary number designating the number of places the word is to be shifted to the left

Address allocations (octal)

MC	00000–07777	4096
	00000–17777	8192 or
	00000–27777	12,288 36-bit words
Q	31000–31777	1 36-bit word
A	32000–37777	1 72-bit word
MD	40000–77777	16,384 36-bit words

Fixed addresses

F_1	00000 or 40001
F_2	00001
F_3	00002
F_4	00003

Arithmetic section registers

A	72-bit accumulator with shifting properties
A_R	Right-hand 36 bits of A
A_L	Left-hand 36 bits of A
Q	36-bit register with shifting properties
X	36-bit exchange register

Note: Parentheses denote *contents of.* For example, (A) means contents of A (72-bit word in A); (Q) means contents of Q (36-bit word in Q).

---

[1] In E. M. Grabbe, S. Ramo, and D. E. Wooldridge (eds.), "Handbook of Automation, Computation, and Control," vol. 2, chap. 2, pp. 77–83, John Wiley & Sons, Inc., New York, 1959.

[2] Carr's triplet notation for: fractional significant digits, digits in exponent, and digits to left of radix point.

Input-output registers

IOA   8-bit in-out register

IOB   36-bit in-out register

TWR   6-bit typewriter register

HPR   7-bit high-speed punch register

Word extension

D(u)   72-bit word whose right-hand 36 bits are the word at address u, and whose left-hand 36 bits are the same as the leftmost bit of the word at u.

S(u)   72-bit word whose right-hand 36 bits are the word at address u, and whose left-hand 36 bits are zero.

D(Q)   72-bit word—right-hand 36 bits are in register Q, left-hand 36 bits are same as leftmost bit in register Q.

S(Q)   same as D(Q) except left 36 bits are zero.

D(A_R), S(A_R)   are similarly defined.

L(Q)(u)   72-bit word—left-hand 36 bits are zero, right-hand 36 bits are the bit-by-bit product of corresponding bits of (Q) and word at address u.

L(Q′)(v)   72-bit word—left-hand 36 bits are zero, right-hand 36 bits are the bit-by-bit product of corresponding bits of the complement of (Q) and word at address v.

Transmit instructions

11[1]   Transmit Positive TPuv[2]: Replace (v) with (u).

13   Transmit Negative TNuv: Replace (v) with the complement of (u).

12   Transmit Magnitude TMuv: Replace (v) with the absolute magnitude of (u).

15   Transmit U-address TUuv: Replace the 15 bits of (v) designated by v_{15} through v_{29}, with the corresponding bits of (u), leaving the remaining 21 bits of (v) undisturbed.

16   Transmit V-address TVuv: Replace the right-hand 15 bits of (v) designated by v_0 through v_{14}, with the corresponding bits of (u), leaving the remaining 21 bits of (v) undisturbed.

35   Add and Transmit ATuv: Add D(u) to (A). Then replace (v) with (A_R).

36   Subtract and Transmit STuv: Subtract D(u) from (A). Then replace (v) with (A_R).

22   Left Transmit LTjkv: Left circular shift (A) by k places. If j = 0 replace (v) with (A_L); if j = 1 replace (v) with (A_R).

[1] Octal notation.
[2] Mnemonic notation.

Q-controlled instructions

51   Q-controlled Transmit QTuv: Form in A the number L(Q)(u). Then replace (v) by (A_R).

52   Q-controlled Add QAuv: Add to (A) the number L(Q)(u). Then replace (v) by (A_R).

53   Q-controlled Substitute QSuv: Form in A the quantity L(Q)(u) plus L(Q′)(v). Then replace (v) with (A_R). The effect is to replace selected bits of (v) with the corresponding bits of (u) in those places corresponding to 1's in Q. The final (v) is the same as the final (A_R).

Replace instructions

21   Replace Add RAuv: Form in A the sum of D(u) and D(v). Then replace (u) with (A_R).

23   Replace Subtract RSuv: Form in A the difference D(u) minus D(v). Then replace (u) with (A_R).

27   Controlled Complement CCuv: Replace (A_R) with (u) leaving (A_L) undisturbed. Then complement those bits of (A_R) that correspond to ones in (v). Then replace (u) with (A_R).

54   Left Shift in A LAuk: Replace (A) with D(u). Then left circular shift (A) by k places. Then replace (u) with (A_R). If u = A, the first step is omitted, so that the initial content of A is shifted.

55   Left Shift in Q LQuk: Replace (Q) with (u). Then left circular shift (Q) by k places. Then replace (u) with (Q).

Split instructions

31   Split Positive Entry SPuk: Form S(u) in A. Then left circular shift (A) by k places.

33   Split Negative Entry SNuk: Form in A the complement of S(u). Then left circular shift (A) by k places.

32   Split Add SAuk: Add S(u) to (A). Then left circular shift (A) by k places.

34   Split Subtract SSuk: Subtract S(u) from (A). Then left circular shift (A) by k places.

Two-way conditional jump instructions

46   Sign Jump SJuv: If $A_{71} = 1$, take (u) as NI. If $A_{71} = 0$, take (v) as NI. (NI means next instruction.)

47   Zero Jump ZJuv: If (A) is not zero, take (u) as NI. If (A) is zero, take (v) as NI.

44 Q-Jump QJuv: If $Q_{35} = 1$, take (u) as NI. If $Q_{35} = 0$, take (v) as NI. Then, in either case, left circular shift (Q) by one place.

One-way conditional jump instructions

41 Index Jump IJuv: Form in A the difference D(u) minus 1. Then if $A_{71} = 1$, continue the present sequence of instructions; if $A_{71} = 0$, replace (u) with (A_R) and take (v) as NI.

42 Threshold Jump TJuv: If D(u) is greater than (A), take (v) as NI; if not, continue the present sequence. In either case, leave (A) in its initial state.

43 Equality Jump EJuv: If D(u) equals (A), take (v) as NI, if not, continue the present sequence. In either case leave (A) in its initial state.

One-way unconditional jump instructions

45 Manually Selective Jump MJjv: If the number j is zero, take (v) as NI. If j is 1, 2, or 3, and the correspondingly numbered MJ selecting switch is set to "jump," take (v) as NI; if this switch is not set to "jump," continue the present sequence.

37 Return Jump RJuv: Let y represent the address from which CI was obtained. Replace the right-hand 15 bits of (u) with the quantity y plus 1. Then take (v) as NI.

14 Interpret IP: Let y represent the address from which CI was obtained. Replace the right-hand 15 bits of (F_1) with the quantity $y + 1$. Then take (F_2) as NI.

Stop instructions

56 Manually Selective Stop MSjv: If $j = 0$, stop computer operation and provide suitable indication. If j = 1, 2, or 3 and the correspondingly numbered MS selecting switch is set to "stop," stop computer operation and provide suitable indication. Whether or not a stop occurs, (v) is NI.

57 Program Stop PS—Stop computer operations and provide suitable indication.

External equipment instructions

17 External Function EF-v: Select a unit of external equipment and perform the function designated by (v).

76 External Read ERjv: If $j = 0$, replace the right-hand 8 bits of (v) with (IOA); if $j = 1$, replace (v) with (IOB).

77 External Write EWjv: If $j = 0$, replace (IOA) with the right-hand 8 bits of (v); if $j = 1$, replace (IOB) with (v). Cause the previously selected unit to respond to the information in IOA or IOB.

61 PRint PR-v: Replace (TWR) with the right-hand 6 bits of (v). Cause the typewriter to print the character corresponding to the 6-bit code.

63 PUnch PUjv: Replace (HPR) with the right-hand 6 bits of (v). Cause the punch to respond to (HPR). If $j = 0$, omit seventh level hole; if $j = 1$, include seventh level hole.

Arithmetic instructions

71 Multiply MPuv: Form in A the 72-bit product of (u) and (v), leaving in Q the multiplier (u).

72 Multiply Add MAuv: Add to (A) the 72-bit product of (u) and (v), leaving in Q the multiplier (u).

73 Divide DVuv: Divide the 72-bit number (A) by (u), putting the quotient in Q, and leaving in A a non-negative remainder R. Then replace (v) by (Q). The quotient and remainder are defined by: $(A)_i = (u) \cdot (Q) + R$, where $0 \leqq R < |(u)|$. Here $(A)_i$ denotes the initial contents of A.

74 Scale Factor SFuv: Replace (A) with D(u). Then left circular shift (A) by 36 places. Then continue to shift (A) until $A_{34} \neq A_{35}$. Then replace the right-hand 15 bits of (v) with the number of left circular shifts, k, which would be necessary to return (A) to its original position. If (A) is all ones or zeros, $k = 37$. If u is A, (A) is left unchanged in the first step, instead of being replaced by $D(A_R)$.

Sequenced instructions

75 RePeat RPjnw: This instruction calls for the next instruction, which will be called NIuv, to be executed n times, its u and v addresses being modified or not according to the value of j. Afterwards the program is continued by the execution of the instruction stored at a fixed address F_1. The exact steps carried out are:

a Replace the right-hand 15 bits of (F_1) with the address w.

b Execute NIuv, the next instruction in the program, n times.

c　If j = 0, do not change u and v.
　　If j = 1, add one to v after each execution.
　　If j = 2, add one to u after each execution.
　　If j = 3, add one to u and v after each execution.

The modification of the u address and v address is done in program control registers. The original form of the instruction in storage is unaltered.

d　On completing n executions, take (F_1), as the next instruction. F_1 normally contains a manually selective jump whereby the computer is sent to w for the next instruction after the repeat.

e　If the repeated instruction is a jump instruction, the occurrence of a jump terminates the repetition. If the instruction is a Threshold Jump or an Equality Jump, and the jump to address v occurs, (Q) is replaced by the quantity j, (n − r), where r is the number of executions that have taken place.

Floating point instructions

64　Add FAuv: Form in Q the normalized rounded packed floating point sum (u) + (v).

65　Subtract FSuv: Form in Q the normalized rounded packed floating point difference (u) − (v).

66　Multiply FMuv: Form in Q the normalized rounded packed floating point product (u) · (v).

67　Divide FDuv: Form in Q the normalized rounded packed floating point quotient (u) ÷ (v).

01　Polynomial Multiply FPuv: Floating add (v) to the floating product $(Q)_i$ · (u), leaving the packed normalized rounded result in Q.

02　Inner Product FIuv: Floating add to $(Q)_i$ the floating product (u) · (v) and store the rounded normalized packed result in Q. This instruction uses MC location $F_4 = 00003$ for temporary storage, where $(F_4)_f = (Q)_i$. The subscripts i and f represent "initial" and "final."

03　Unpack UPuv: Unpack (u), replacing (u) with $(u)_M$ and replacing $(v)_C$ with $(u)_C$ or its complement if (u) is negative. The characteristic portion of $(u)_f$ contains sign bits. The sign portion and mantissa portion of $(v)_f$ are set to zero. *Note.* The subscripts M and C denote the mantissa and characteristic portions.

04　Normalize Pack NPuv: Replace (u) with the normalized rounded packed floating point number obtained from the possibly unnormalized mantissa in $(u)_i$ and the biased characteristic in $(v)_c$. *Note.* It is assumed that $(u)_i$ has the binary point between u_{27} and u_{26}; that is, that $(u)_i$ is scaled by 2^{-27}.

05　Normalize Exit NEj-: If j = 1 normalize without rounding until a master clear or until the instruction is again executed with j = 0.

References

Univac Scientific Electronic Computing System Model 1103A, Form EL 338

Chapter 14

Instruction logic of the MIDAC[1]

John W. Carr III

The MIDAC, Michigan Digital Automatic Computer [Carr, 1956], was constructed on the basis of the design of the SEAC at the National Bureau of Standards. Its instruction code is particularly of interest because it incorporates the index register concept into a three-address binary instruction. Numbers in this machine are $(44, 0, 0)$[2] fixed points. The word length is 45 binary digits with serial operation.

Word structure

The data or address positions of an instruction are labeled the α, β, and γ positions. Each contains twelve binary digits represented externally as three hexadecimal digits. Four binary digits, or one hexadecimal digit, are used to convey the instruction modification or relative addressing information. The next four binary digits or single hexadecimal digit represents the operation portion of the instruction. The final binary digit is the halt or breakpoint indicator for use with the instruction.

For example, the 45-binary-digit word

000001100100000011001000000100101100000001011

considered as an instruction would be interpreted as

α	β	γ	abcd	Op	halt
000001100100	000011001000	000100101100	0000	0101	1

In external hexadecimal form this would be written

064 0c8 12c 0 5 —

The above binary word is the equivalent machine representation of the following instruction: "Take the contents of hexadecimal address 064, add to it the contents of hexadecimal address 0c8, and store the result in hexadecimal address 12c. There is *no* modification of the 12-binary-digit address locations given by the

instruction. Upon completion of the operation, stop the machine if the proper external switches are energized." The binary combination represented by 5 is the operation code for addition.

Data or addresses

The addresses given by the twelve binary digits in each of the three locations designate in the machine the individual acoustic storage cells and blocks of eight magnetic drum storage cells. The addresses from 0 to 1023 (decimal) or 000 to 3FF (hexadecimal) correspond to acoustic storage cells. The addresses from 1024 to 4095 (decimal) or 400 to FFF (hexadecimal) correspond to magnetic drum storage blocks. In certain operations, however, the addresses 0 to 15 (decimal) or 0 to F (hexadecimal) represent input-output stations rather than storage locations.

These twelve-binary-digit groups will in some cases be modified by the machine in order to yield a final twelve-binary-digit address. The method of processing will depend on the values of the instruction modification digits. After modification, the final result will then be interpreted by the control unit as a machine address.

In some instructions, namely those that perform change of control operations, which involve cycling and counting rather than simple arithmetic operations on numbers, the α and β positions in an instruction are not considered as addresses. In those cases, they are used instead as counters or tallies. In other instructions, which do not require three addresses, but only one or two, the β position is not considered as an address. In these cases, the oddness or evenness of the β address is used to differentiate between two operations having the same operation code digits. That is, the parity of binary digit P22 is used as an extra function designator.

Instruction modification digits

The four binary digits P9-P6 are used as instruction modification or relative addressing digits. Their normal function is relatively simple; nevertheless, the possible exceptions to the general rule can make their behavior complicated. These four digits are labeled

---

[1] In E. M. Grabbe, S. Ramo, and D. E. Wooldridge (eds.), "Handbook of Automation, Computation, and Control," vol. 2, chap. 2, pp. 115–121, John Wiley & Sons, Inc., New York, 1959.

[2] Carr's triplet notation for: fractional significant digits, digits in exponent, and digits to left of radix point.

the a, b, c, and d digits. Ordinarily the a digit is associated with the α position, the b digit with the β position, and the c digit with the γ position in an instruction.

When binary digit P22 (or the β position) is used in an instruction to represent extra operation information, the instruction modification digit b is ignored. In the case of input and output instructions, when the various address positions represent machine address locations on the drum, input-output stations, or block lengths, and modification of these addresses is not desired in any case, the corresponding relative addressing digits are ignored.

The purpose of the instruction modification digits is to tell the machine whether or not to modify the twelve binary digits making up the corresponding address position in an instruction by addition of the contents of one or the other of two counters. In the normal case, if the a, b, or c digit is a zero, the twelve binary digits in the corresponding position are interpreted, unchanged, as the binary representation of the machine address of the number word to be processed by the instruction.

If one or more of the a, b, or c digits is a one, the contents of one of two auxiliary address counters is added to the corresponding twelve binary digits to yield a final address usually different from that given by the original twelve-digit portion of the instruction word. The addresses are then said to be relative to the counter.

The two counters involved in the address modification feature of the MIDAC are known as the instruction counter and the base counter. In the normal case, if the fourth instruction modification or d digit is a zero, the contents of the instruction counter will be added to the contents of the various twelve-digit addresses (dependent on the values of the a, b, and c digits) before further processing of the instruction. If the a digit is one and the d digit zero, the contents of the instruction counter will be added to the α address; similarly for b and d digits and β address, etc.

If the d digit is a one, the contents of the base counter will be normally added to the contents of the twelve digits in the α, β, and γ positions (again dependent on the values of the a, b, and c digits), before further processing of the results. If the a digit is one and the d digit one, the contents of the base counter will be added to the α address, etc.

The effect of the instruction modification digits may be summarized as follows:

The contents of the two counters will be designated by C_d
(d = 0, 1).

C_0 = contents of the instruction counter
C_1 = contents of the base counter

Then the modified addresses α', β', and γ' are related to the α, β, and γ addresses appearing in the instruction by the following:

$$\alpha' = \alpha + aC_d \qquad \beta' = \beta + bC_d \qquad \gamma' = \gamma + cC_d$$
$$(a, b, c, d = 0, 1)$$

In certain instructions addresses relative to one of the two counters may be prohibited. Thus, if in a particular instruction α may be relative only to the instruction counter, then for that instruction

$$\alpha' = \alpha + aC_0$$

no matter whether the d digit is a 0 or a 1.

The notation (α'), (β'), or (γ') is used to indicate the word stored in the location whose address is α', β', or γ'.

Instruction counter

The instruction counter is a twelve-binary digit (modulo 4096) counter which contains the binary representation of the address of the instruction which the control unit is processing or is about to process. In normal operation when no change of control operation is being processed, the contents of the instruction counter is increased by one at the completion of each instruction. Thus, normally the next instruction to be processed is stored in the acoustic storage cell immediately following the cell which contains the present instruction.

A change of control operation is one which selects a next instruction not stored in sequence in the acoustic storage. That is, at the completion of such instructions the contents of the instruction counter is not increased by one, but instead is changed entirely.

Base counter

The base counter is a second twelve-binary-digit counter (modulo 4096), physically identical to the instruction counter, which contains the binary representation of a base number or tally. Unlike the instruction counter, however, the base counter does not sequence automatically, but remains unchanged until a change of base instruction is processed. This counter serves two primary purposes, dependent on the usage to which it is put:

1   It may contain the address of the initial word in a group, thus serving as a base address to which integers representing the relative position of a given word in the group of words may be added by using the address modification digits.

2 It may contain a counter or tally which can be increased by a base instruction. This instruction makes use of the address modification digits to change the counter so as to count the number of traversals of a particular cycle of instructions.

Instruction types

Instructions used in MIDAC can be divided into three categories: change of information, change of control, and transfer of information. The first category can be further subdivided into arithmetic and logical instructions. In the arithmetic instructions are included addition, subtraction, division, various forms of multiplication; power extraction, number shifting; and number conversion instructions. The sole logical instruction is extract, which modifies information in a nonarithmetic fashion.

The transfer of information or data transfer instructions include transfers of individual words or blocks of words into and out of the acoustic storage and drum and magnetic tape control.

The possible change of control instructions includes two comparisons that provide different future sequences dependent on the differences of two numbers. In the compare numbers or algebraic comparison, the difference is an algebraic, signed one. In the compare magnitudes or absolute comparison, the difference is one between absolute values. Two other instructions, file and base, perform other tasks beside transferring control. The file instruction transfers control unconditionally. The file instruction files or stores the contents of the base or instruction counter in a specific address position of a particular word in the storage. The base or tally instruction provides a method for referring addresses automatically relative to the address given by the base counter, irrespective of its contents. The base instruction also gives a conditional transfer of control.

The nineteen MIDAC instructions can be described functionally as follows:

Change of information

1 **Add.** $(\alpha') + (\beta')$ is placed in γ'. Result must be less than 1 in absolute value.

2 **Subtract.** $(\alpha') - (\beta')$ is placed in γ'. Result must be less than 1 in absolute value.

3 **Multiply, Low Order.** The least significant 44 binary digits of $(\alpha') \times (\beta')$ are placed in γ'.

4 **Multiply, High Order.** The most significant 44 binary digits of $(\alpha') \times (\beta')$ are placed in γ'.

5 **Multiply, Rounded.** The most significant 44 binary digits of $(\alpha') \times (\beta') \pm 1 \cdot 2^{-45}$ are placed in γ'. The $1 \cdot 2^{-45}$ is added if $(\alpha') \times (\beta')$ is positive, and subtracted if $(\alpha') \times (\beta')$ is negative.

6 **Divide.** The most significant 44 binary digits of $(\beta')/(\alpha')$ are placed in γ'. (Note the inversion of order of α and β.) Result must be less than 1 in absolute value.

7 **Power Extract.** The number $n \cdot 2^{-44}$ is placed in γ' where n is the number of binary 0's to the left of the most significant binary 1 in (α'). The b digit is ignored; β may be any even number. If (α') is all zeros, zero is placed in γ'.

8 **Shift Number.** The 44 binary digits immediately to the right of the radix point in $(\alpha') \cdot 2^{(\beta') \cdot 2^{44}}$ are placed in γ'. The result, in γ', is the equivalent of shifting (α') n places, where $n \cdot 2^{-44} = (\beta')$ and n positive indicates a shift left, n negative a shift right. If $|n| \geqq 44$, zero is placed in γ'.

9 **Extract or Logical Transfer.** Those binary digits in (γ'), including the sign digit, whose positions correspond to 1's in (β') are replaced by the digits in the corresponding positions of (α').

10 **Decimal to Binary Conversion.** This operation may be interpreted in two ways: (a) (α') is considered as a binary-coded-decimal integer times 2^{-44}. It is converted to the equivalent binary integer times 2^{-37} and the result is placed in γ', or (b) (α') is considered as a binary-coded-decimal fraction, D. It is converted into an intermediate binary fraction, B_i, such that $B_i = D \times 10^{11} \times 2^{-37}$ and the result placed in γ'. To obtain B, the true binary equivalent of D, B_i must be multiplied by $(10^{-11} \times 2^{37})$. However, since this factor is greater than 1 and therefore cannot be represented in the machine, two operations must be performed. For example,

$$B_i \times (10^{-11} \times 2^{37} - 1) = B_j$$
$$B = B_i + B_j$$

Here the b digit is ignored, and β may be any *even* number.

11 **Binary-to-Decimal Conversion.** (α'), considered as a binary fraction, is converted into the equivalent eleven-digit binary-coded-decimal fraction. The result is placed in γ'. The b digit is ignored, and β may be any odd number.

Change of control

12 **Compare Numbers.** γ can be relative only to the instruction counter. If $(\alpha') \geqq (\beta')$, the contents of the instruction counter are increased by one as is normally done at the end of each instruction. If $(\alpha') < (\beta')$, the contents of the instruction counter are set to γ'.

13   **Compare Magnitudes.** γ can be relative only to the instruction counter. If $|(\alpha')| \geqq |(\beta')|$, the contents of the instruction counter are increased by one as is normally done at the end of each instruction. If $|(\alpha')| < |(\beta')|$, the contents of the instruction counter is set to γ'.

14   **Base or Tally.** The d digit is ignored. α and β may be relative only to the base counter, γ only to the instruction counter. If $\alpha' \geqq \beta'$, the contents of the base counter are set to zero and the contents of the instruction counter increased by one as usual. If $\alpha' < \beta'$, the contents of the base counter are set to α' and the contents of the instruction counter to γ'. (*Note.* The comparisons made here are of addresses themselves, not their contents.)

15   **File.** β may be any odd number. α and γ may be relative only to the instruction counter.

If d = 0, the contents of the instruction counter increased by one is placed in the γ position of (α'), and the instruction counter is set to γ'.

If d = 1, the contents of the base counter is placed in the α position of (α'), and the instruction counter is set to γ'. In addition, if b = 1, the contents of the base counter is set to zero; if b = 0, the contents of the base counter is not changed.

Transfer of information

16   **Read In.** The a digit must be 0; the b digit is ignored. If β is in the range 0 to 7 (decimal) or 000 to 007 (hexadecimal) α words are read into the acoustic storage from input-output station β. The first word read in is placed in γ', the second in $\gamma' + 1$, etc. If β is in the range 1024 to 1791 decimal (400 to 6FF hexadecimal), α words are read into the acoustic storage from the drum starting with the first word in the drum block whose address is β. The first word is placed in γ', the second in $\gamma' + 1$, etc.

17   **Read Out.** The a digit must be 0, the c digit is ignored. Starting with (β'), read out α consecutive words from the acoustic storage to input-output station γ, if γ is in the range 0 to 7 decimal (000 to 007 hexadecimal), or to the drum starting at the beginning of the drum block whose address is γ, if γ is in the range 1024 to 1791 decimal (400 to 6FF hexadecimal).

16   **Alphanumeric Read In.** The a digit must be 1; the b digit is ignored. If β is in the range 0 to 7 (decimal) or 000 to 007 (hexadecimal) α characters are read into the acoustic storage from input-output station β. The first character read in is placed in γ', the second in $\gamma' + 1$, etc. Each character occupies the six most significant digit positions of the register into which it is read; the other positions are set to zero. This operation may not be used to read words from the drum into the acoustic storage.

17   **Alphanumeric Read Out.** The a digit must be 1; the c digit is ignored. Starting with (β'), read out α consecutive characters from the acoustic storage to input-output station γ; γ must be in the range 0 to 7 (decimal) or 000 to 007 (hexadecimal). This operation may not be used to read words from the acoustic storage onto the drum.

18   **Move Tape Forward.** (a, b, c and d digits are ignored.) β may be any *even* number; γ must be in the range 0 to 15 decimal (000 to 00F hexadecimal). The magnetic tape at input-output station γ is moved forward n blocks where

$$n = \left[\frac{\alpha - 1}{8} \right] + 1$$

that is, one plus the integral part of $\alpha - \frac{1}{8}$, or the number of blocks that include α words.

19   **Move Tape Backward.** (a, b, c, and d digits are ignored.) β may be any *odd* number; γ must be in the range 0 to 15 decimal (000 to 00F hexadecimal). The magnetic tape at input-output station γ is moved backward n blocks where

$$n = \left[\frac{\alpha - 1}{8} \right] + 1$$

that is, one plus the integral part of $\alpha - \frac{1}{8}$, or the number of blocks that include α words.

References

CarrJ56. SEAC computer references: AinsE52; AlexS51; ElboR53; GreeS52, 53; HaueR52; PikeJ52; SerrR62; ShupP53; SlutR51. DYSEAC computer references: LeinA54.

Chapter 15

Instruction logic of the Soviet Strela (Arrow)[1]

John W. Carr III

A typical general purpose digital computer using three-address instruction logic is the Strela (Arrow) constructed in quantity under the leadership of Iu. Ia. Basilewskii of the Soviet Academy of Sciences, and described in detail by Kitov [1956]. This computer uses a $(35, 6, 0)$[2] binary floating point number system. Its instruction word, of 43 digits, contains a six-digit operation code, and three 12-digit addresses, with one breakpoint bit. In octal notation, two digits represent the operation, four each the addresses, and one bit the breakpoint. This machine operates with up to 2048 words of high-speed cathode ray tube storage.

Input-output is ordinarily via punched cards and punched paper tape. A "standard program library" is attached to the computer as well as magnetic tape units (termed "external accumulators" below). *Note.* This computer is different from both the BESM described by Lebedev [1956] and the Ural reported by Basilewskii [1957]. Apparently, it is somewhat lower in performance than BESM.

Since all arithmetic is ordinarily in floating point, "special instructions" perform fixed point computations for instruction modifications.

Ordinarily instructions are written in an octal notation, but external to the machine operation symbols are written in a mnemonic code. The two-digit numerals are the octal instruction equivalent.

Arithmetic and logical instructions

01.   + α β γ. Algebraic addition of (α) to (β) with result in γ.

02. $+_1$ α β γ. Special addition, used for increasing addresses of instructions. The command (α) or (β) is added to the number (β) or (α) and the result sent to the cell with address γ.

As a rule, the address of the instruction being changed corresponds to the address γ.

03. $-$ α β γ. Subtraction with signed numbers. From the number (α) is subtracted the number (β) and the result sent to γ.

04. $-_1$ α β γ. Difference of the absolute value of two numbers $|(\alpha)| - |(\beta)| = (\gamma)$.

05. $\times$ α β γ. Multiplication of two numbers (α) and (β) with result sent to γ.

06. $\wedge$ α β γ. Logical multiplication of two numbers in cells α and β. This instruction is used for extraction from a given number or instruction a part defined by the special number (β).

07. $\vee$ α β γ. Logical addition of two numbers (α) and (β) and sending the result to cell γ. This instruction is used for forming numbers and commands from parts.

10.   Sh α β γ. Shift of the contents of cell α by the number of steps equal to the exponent of the (β). If the exponent of the (β) is positive then the shift proceeds to the left, in the direction of increasing value; if negative, then the shift is right. In addition, the sign of the number, which is shifted out of the cell, is lost.

11. $-_2$ α β γ. Special subtraction, used for decreasing the addresses of instructions. In the cell α is found the instruction to be transformed, and in cell β the specially selected number. Ordinarily addresses α and γ are identical.

12. $\neq$ α β γ. Comparison of two numbers (α) and (β) by means of digital additions of the numbers being compared *modulo* two. In the cell γ is placed a number possessing ones in those digits in which inequivalence results in the numbers being compared.

Control instructions

13.   C α β 0000. Conditional transfer of control either to instruction (α) or to instruction (β), depending on the results of the preceding operation. With the operations of addition, subtraction, and subtraction of absolute values, it appraises the sign

[1] In E. M. Grabbe, S. Ramo, and D. E. Wooldridge (eds.), "Handbook of Automation, Computation, and Control," vol. 2, chap. 2, pp. 111–115, John Wiley & Sons, Inc., New York, 1959.
[2] Carr's triplet notation for: fractional significant digits, digits in exponent, and digits to left of radix point.

of the result: for a positive or zero result it transfers control to the command (α), for negative results to the command (β).

The result of the operation of multiplication is dependent on the relationship to unity. Transfer is made to the command (α) in the case where the result is greater than or equal to one, and to command (β), if it is smaller than one.

For conditional transfer after the operation of comparison, transfer to the instruction (α) is made in the case of equality of binary digits, and to (β) when there is any inequivalence.

After the operation $\wedge$ (logical sequential multiplication) the conditional transfer command jumps to the instruction (α) when the result is different from zero, and to instruction (β) when it is equal to zero.

A forced comparison is given by

C α α 0000

The third address in this command is not used and in its place is put zero.

14. I-O α 0000 0000. This instruction is executed parallel with the code of the other operations, and guarantees bringing into working position in good time the zone of the external accumulator (magnetic tape unit) with the address α.

15. H 0000 0000 0000. This instruction executes an absolute halt.

Group transfer instructions

Special instructions for group transfer serve for the accomplishment of a transfer of numbers to and from the accumulators. In the second address in these instructions stands an integer, designating the quantity of numbers in the group which must be transferred. Group transfers always are produced in increasing sequence of addresses of cells in the storage.

16. T_1 0000 n γ. The instruction T_1 guarantees transfer from a given input unit (with punched cards, perforated tape, etc.) into the storage. In the third address γ of the instruction is indicated the initial address of the group of cells in the storage where numbers are to be written. With punched paper tape or punched cards the variables are written in sequence, beginning with the first line.

17. T_2 0000 n γ. The instruction T_2 guarantees transfer of a group of n numbers from an input unit into the external accumulator in zone γ.

20. T_3 α n γ. This instruction guarantees a line-by-line sequence of transfers of n numbers from zone α of the external accumulator into the cells of the storage beginning with the cell with address γ.

21. T_4 α n 0000. This instruction guarantees the transfer to the input-output unit (to punched paper tape or punched cards) of a group of n numbers from the storage, beginning with address α. The record on punched paper tape or punched cards as a rule will begin with the first line and therefore a positive indication of the addresses of the record is not required.

22. T_5 α n γ. Instruction T_5 guarantees transfer of a group of n numbers from one place in the storage with initial address α into another place in the storage with initial address γ.

23. T_6 α n γ. Instruction T_6 guarantees transfer of a group of n numbers from the storage with initial address α into the external accumulator with address γ.

24. T_7 α n 0000. Instruction T_7 serves for transfer of n numbers from the zone of the external accumulator with address α into the input-output unit.

Instructions T_2 and T_7 cannot be performed concurrently with other machine operations.

Standard subroutine instructions

Certain instructions in the Strela, although written as ordinary instructions, are actually "synthetic" instructions which call on a subroutine for computation of the function involved. The amount of machine time (number of basic instruction cycles) for an iterative process depends on the required precision of the computed function. The figures given below are based on approximately ten-digit decimal numbers with desired precision one in the tenth place.

25. D α β γ. This standard subroutine serves for execution of the operation of division: The number (α) is divided into the number (β) and the quotient is sent to cell γ.

The actual operation of division is executed in two steps: the initial obtaining of the value of the inverse of the divisor, by which the dividend is then multiplied. The computation of the inverse is given by the usual Newton formula, originally used with the EDSAC [Wilkes et al., 1952].

$$y_{n+1} = y_n(2 - y_n x)$$

For $x = d \cdot 2^p$, where $\frac{1}{2} < d < 1$, the first approximation is taken as 2^{-p}. The standard subroutine takes 8 to 10 instructions and can be executed in 18–20 machine cycles (execution time for one typical command).

26. $\sqrt{}$ α 0000 γ. This instruction guarantees obtaining the value $\sqrt{x}$ from the value $x = (\alpha)$ and sending the result to cell γ. Initially $1/\sqrt{x}$ is computed by the iteration formula

$$y_{n+1} = \frac{1}{2} y_n(3 - x y_n^2)$$

where the first approximation is taken as

$$y_0 = 2^{[p/2]}$$

the bracket indicating "integral part of." After this the result is multiplied by x to obtain $\sqrt{x}$. This standard subroutine contains 14 instructions and is executed in 40 cycles.

27. e^x α 0000 γ. This instruction guarantees formation of e^x for the value $x = (\alpha)$ and sending the result to cell γ. The computation is produced by means of expansion of e^x in a power series. The standard subroutine contains 20 instructions and is executed in 40 cycles.

30. ln x α 0000 γ. This instruction guarantees formation of the function $\ln x$ for the value $x = (\alpha)$ and sending the result to location γ. Computation is produced by expansion of $\ln x$ in series. The subprogram contains 15 instructions and is executed in 60 cycles.

31. sin x α 0000 γ. This instruction guarantees execution of the function $\sin x$ and sending the result to location γ. The computation is produced in two steps: initially the value of the argument is translated into the first quadrant, then the value of the function is obtained by a series expansion. The subroutine contains 18 instructions and is executed in 25 cycles.

32. DB α n γ. This instruction performs conversion of a group of n numbers, stored in locations α, $\alpha + 1, \ldots$ from binary-coded decimal into binary and sending of the result to locations γ, $\gamma + 1, \ldots$. The subroutine contains 14 instructions and is executed in 50 cycles (for each number).

33. BD α n γ. This instruction performs the conversion of a group of n numbers stored in locations α, $\alpha + 1, \ldots$ from the binary system into binary-coded decimal and sends them to locations γ, $\gamma + 1, \ldots$. The subroutine contains only 30 instructions and is executed with 100 cycles (for each number).

34. MS α n γ. This is an instruction for storage summing. This instruction produces the formal addition of numbers, stored in locations beginning with address α, and the result is sent to location γ. Numbers and instructions are added in fixed point. This sum may be compared with a previous sum for control of storage accuracy.

References

Basil57; KitoA56; LebeS56; WilkM52.

Section 2

Processors constrained by a cyclic, primary memory

These processors use one extra (the +1) address to specify the address of the next instruction. Obviously this address is used to allow complete freedom in the location of both operands and next instructions in an optimum manner. The IBM 650, a 1 + 1 address computer, is the most straightforward to understand. ACE and ZEBRA have subtle microcoded instructions to achieve powerful instruction sets. The LGP-30 and LGP-21 have a simple 1 address instruction format; they interlace several logical addresses between the physical addresses to help with the optimum location of operands.

The Olivetti Underwood Programma 101 desk calculator

The Programma 101 is a desk calculator computer implemented with a cyclic Mp. The cyclic memory is not apparent from the user's viewpoint because the response is adequate (less than 0.1 sec for simple arithmetic operations). The Programma 101 is discussed in Part 3, Sec. 4, page 235.

ZEBRA, a simple binary computer

The ZEBRA is presented in Chap. 12 and is discussed in Part 3, Sec. 1, page 190.

The LGP-30 and LGP-21

The LGP-30 (Chap. 16) is a first-generation, 31-bit computer with an Mp.cyclic and a very simple ISP. The computer appears to be characteristic of small-scale drum computers in the first generation. We think of this class of computer as having very little power when compared, for example, with the IBM 701. However, the power is mostly related to the drum-based technology, with 0.26 ~ 16.66 millisecond access times.

The Pilot ACE

The NPL Pilot ACE is presented in Chap. 11. Its relationship in the computer space is discussed in Part 3, Sec. 1, page 190.

The UNIVAC system

The UNIVAC I is described in Chap. 8. A discussion is given in Part 2, Sec. 1, page 91.

The design philosophy of Pegasus, a quantity-production computer

The Pegasus cyclic memory, general register computer (Chap. 9) is discussed in Part 2, Sec. 2, page 170.

IBM 650 instruction logic

The IBM 650 has a 1 + 1 address format and a very complete instruction set. Because of the long word length (10 decimal digits) we would consider it to have general utility. The 650's high performance is achieved by using a fast drum (6 milliseconds/revolution). The characteristics given in Chap. 17 present the machine as it was first introduced in 1954. Later versions provided options for floating point arithmetic and index registers. A 96-word core buffer was also added for disk and magnetic-tape buffering. The machine structure is a simple 1 Pc without concurrent processing and input/output transfer ability. Although the 650 has a large word, it initially processed only fixed point integers.

NOVA: a list-oriented computer

The NOVA (Chap. 26) is a specialized computer for processing array data. It is discussed in Part 4, Sec. 2, page 315.

Chapter 16

The LGP-30 and LGP-21

The LGP-30 is a small computer with an Mp.drum. It is distinct from the first (and succeeding) generation computers using Mp.random_access and can be described by using the PMS diagram in Fig. 1. The LGP-21, a direct descendant of the LGP-30, having the same ISP, is also described by Fig. 1.

Since there is only one address/instruction, a method is needed for the optimal allocation of operands. Otherwise, each instruction might have to wait a complete drum (or disk) revolution each time a data reference is made. The LGP-30 provides for operand-location optimization by interlacing the logical addresses on the drum so that two adjacent addresses (e.g., 00 and 01) are separated by nine physical locations.[1] These spaces allow for operands to be located next to the instructions which use them. There are 64 tracks, each with 64 words (sectors). Each word is accessed by a track address of 6 bits and a word address of 6 bits. The sequence of words (sectors) within a track is 00, 57, 50, 43, 36, 29, 22, 15, 08, 01, 58, 51, 44, 37, . . . , 06, 63, 56, 49, 42, 35, 28, 21, 14, 07, 00. The time between two adjacent physical words is approximately 0.260 millisecond, and the time between two adjacent addresses is 9×0.260 or 2.340 milliseconds. The actual maximum t.access is 16.66 ms.[2]

Half of the instruction (15 bits) is unused. It could be used for extra instructions, indexing, indirect addressing, or a second $(+1)$ address to locate the next instruction, all of which increase the preformance.

[1] The LGP-21 has a space of 18 words.
[2] The later LGP-21 appears to have a lower performance than the LGP-30 by about a factor of 3.

```
C ⎡'LGP-30; technology: (113 vacuum tubes), (1350 diodes);
  ⎢  power: 1500 watts; weight: 800 pounds; number produced:
  ⎢  320 ~ 490; t.delivery: September 1956; descendant: 'LGP-21;
  ⎢  Pc(1 address; 1 instruction/w; data: w,bv,i,fr; Mps(~ 2 w);
  ⎢      operations: (+,-,×,/,∧,× 2))
  ⎢  Mp(drum; t.cycle: 260 µs/w; t.access: (.260 ~ 16.6) ms;
  ⎢      i.rate: 2.34 ms/w contiguous addresses:  4096 w; (31,1
  ⎢      space) b/w)
  ⎣  T(Flexowriter, paper tape)

C ⎡'LGP-21; technology: (460 transistors), (375 diodes); power:
  ⎢  300 watts; weight: 90 pounds; number produced: ~ 150;
  ⎢  t.delivery: December 1962;
  ⎢  Mp(fixed head disk; cyclic; t.cycle: 400 µs/w; t.access:
  ⎢      (0 ~ 52) ms; i.rate: 7.26 ms/w contiguous addresses:
  ⎢      4096 w; (31,1 space) b/w)
  ⎣  T(#1:32; Flexowriter, paper tape, analog, CRT, card)
```

Fig. 1. LGP-30 and LGP-21 PMS diagrams.

The ISP, given in Appendix 1 of this chapter, is about the most straightforward in the book. There are only 16 instructions, and the program state is less than two words. Although the performance is limited because of an Mp.cyclic_access, an Mp.random_access would serve to make the ISP fairly similar to other faster computers, e.g., an IBM 701.

APPENDIX 1 LGP-30 AND LGP-21 ISP DESCRIPTION

Appendix 1

LGP-30 and LGP-21 ISP Description

Pc State

A<0:30>   *Accumulator*

C<18:23,24:29>   *Program Counter register*

Ov   *Overflow, LGP-21 only on LGP-30 machine stops if an overflow*

Run

Pc Console State

BP<4,8,16,32>   *Break Point switches*

TC   *Transfer Control switch*

Mp State

M[0:77$_8$][0:77$_8$]<0:30>   *primary memory; 2^{12} w; track and sector (word)*

K State

The following Input Output devices do not have synchronization description variables. LGP-21 only. LGP-30 has a Flexowriter.

Input device[0:31]<1:6>

stop code   *condition signifying input device has read a special code*

Output device[0:31]<1:6>

Instruction Format

i<0:30>   *instruction*

op<0:3>   := i<12:15>   *operation code*

t<0:5>   := i<18:23>   *track select bit on Mp*

t'<0:4>:= t<1:5>   *input-output select, LGP-21 only*

s<0:5>   := i<24:29>   *sector select bit of Mp*

skip condition := ((t<0:3> ∧ ¬ BP) ≠ 0)

Instruction Interpretation Process

Run → (i ←M[C]; C ←C + 1; next   *fetch*

        Instruction execution)   *execute*

Instruction Set and Instruction Execution Process

Instruction execution := (

Z (:= op = 0) → (

(t = 00000$_8$) → (Run ←0);   *stop*

skip condition → (C ←C + 1);   *sense BP and transfer*

i<0> → (Ov → (Ov ←0; C ←C + 1)));   *sense overflow and transfer*

B (:= op = 1) → (A ←M[t][s]);   *bring from memory*

Y (:= op = 2) → (M[t][s]<18:29>←A<18:29>);   *store address*

R (:= op = 3) → (M[t][s]<18:29> ←C + 1):   *set return address*

I (:= op = 4) → (*shifts, and input*

¬ i<0> ∧ (t=62) → (A ←A x 2^6 {logical});

  i<0> ∧ (t=62) → (A ←A x 2^4 {logical}):

¬ i<0> ∧ (t≠62) → (input 6 bit):

  i<0> ∧ (t≠62) → (input 4 bit)):

APPENDIX 1   LGP-30 AND LGP-21 ISP DESCRIPTION (Continued)

```
    input₆bit := (A ← A x 2⁶ {logical}; next
                   A<25:30> ← Input device[t']; next
                   (¬A<0> ∨ stop code) → input₆bit)
    input₄bit := (A ← A x 2⁴ {logical}; next
                   A<27:30> ← Input device[t']<1:4>; next
                   (¬A<0> ∨ stop code) → input₄bit)
D  (:= op = 5)  → (0v,A ← round(A / M[t][s]));
N  (:= op = 6)  → (A ← A x M[t][s] {s.integer});
M  (:= op = 7)  → (A ← A x M[t][s] {s.fraction});
P  (:= op = 10₈) → (
  ¬ i<0> → (Output device[t']<1:6> ← A<0:5>):
    i<0> → (Output device[t']<1:6> ← A<0:3>□□0));
E  (:= op = 11₈) → (A ← A ∧ M[t][s]);
U  (:= op = 12) → (C ← t□s);
T  (:= op = 13) → (i<0> → ((A<0> ∨ TC) → (C ← t□s));
                  ¬i<0> → (A<0> → (C ← t□s)));
H  (:= op = 14) → (M[t][s] ← A);
C  (:= op = 15) → (M[t][s] ← A; next A ← 0);
A  (:= op = 16) → (0v□A ← A + M[t][s]);
S  (:= op = 17) → (0v□A ← A - M[t][s])
               )
```

```
input processes



wait




divide
multiply, save right
multiply, save left


print 6 bit
print 4 bit
extract
unconditional transfer
transfer control
conditional transfer
hold and store
clear
add
subtract
end Instruction execution
```

Chapter 17

IBM 650 instruction logic[1]

John W. Carr III

The basic IBM 650 is a magnetic drum $(10, 0, 0)$[2] decimal computer with one-plus-one address instruction logic. It has a storage of 1000 or 2000 10-digit words (plus sign) with addresses 0000–0999 or 0000–1999. More extended versions of the equipment have built-in floating point arithmetic and index accumulators, but the basic machine will be described here. There are three arithmetic registers in addition to the standard program register and program counter. All information from the drum to the arithmetic unit passes through a signed 10-digit *distributor*. A twenty-digit *accumulator* is divided into a lower and upper part, each of 10 digits with sign. Each of these is addressable (distributor 8001, lower accumulator 8002, and upper accumulator 8003). Each accumulator may be cleared to zero separately (in IBM 650 terminology, "reset"). The entire 20-digit register can be considered as a unit, or each part separately (but affecting the other in case of carries). The 10-digit instruction is broken down into the following form:

10	9	8	7	6	5	4	2	3	1	0
Op. Code		Data Address			Next Instruction Address					Sign

One particular instruction, Table Look-Up, allows automatic table search for one particular element in a table, which can be stored with a corresponding functional value. Input-output is via 80-digit numerical punched cards. An "alphabetic device" allows limited alphabetical entry on cards. Only certain 10-word groups on the magnetic drum are available for input and output. The following information is taken from an IBM 650 manual [Type 650, Magnetic Drum Data-Processing Machine Manual of Operations]. Much of the input-output is handled via board wiring, which is not described in detail below. The two-digit pair represents the machine code. The BRD (Branch on Digit) operation is used with special board wiring to tell when certain specific card punches exist.

[1] In E. M. Grabbe, S. Ramo, and D. E. Wooldridge (eds.), "Handbook of Automation, Computation, and Control," vol. 2, chap. 2, pp. 93–98, John Wiley & Sons, Inc., New York, 1959.
[2] Carr's triplet notation for: fractional significant digits, digits in exponent, and digits to left of radix point.

Input-output instructions

70 RD (Read). This operation code causes the machine to read cards by a two-step process. First, the contents of the 10 words of read buffer storage are automatically transferred to one of the 20 (or 40) possible 10-word groups of read general storage. The group selected is determined by the D address of the Read instruction. Secondly, a card is moved under the reading brushes, and the information read is entered into buffer storage for the next Read instruction.

71 PCH (Punch). This operation code causes card punching in two steps. First the contents of one of the 20 (or 40) possible 10-word groups of punch storage are transferred to punch buffer storage. The group selected is specified by the D address of the Punch instruction. Secondly, the card is punched with the information from buffer storage.

69 LD (Load Distributor). This operation code causes the contents of the D address location of the instruction to be placed in the distributor.

24 STD (Store Distributor). This operation code causes the contents of the distributor with the distributor sign to be stored in the location specified by the D address of the instruction. The contents of the distributor remain undisturbed.

Addition and subtraction instructions

10 AU (Add to Upper). This operation code causes the contents of the D address location to be added to the contents of the upper half of the accumulator. The lower half of the accumulator will remain unaffected unless the addition causes the sign of the accumulator to change, in which case the contents of the lower half of the accumulator will be complemented. Also, the units position of the upper half of the accumulator will be reduced by one.

15 AL (Add to Lower). This operation code causes the contents of the D address location to be added to the contents of the lower half of the accumulator. The contents of the upper half of the accumulator could be affected by carries.

11 SU (Subtract from Upper). This operation code causes the contents of the D address location to be subtracted from the contents of the upper half of the accumulator. The contents of the lower half of the accumulator will remain unaffected unless the subtraction causes a change of sign in the accumulator, in which case the contents of the lower half of the accumulator will be complemented. Also, the units position of the upper half of the accumulator will be reduced by one.

16 SL (Subtract from Lower). This operation code causes the contents of the D address location to be subtracted from the contents of the lower half of the accumulator. The contents of the upper half of the accumulator could be affected by carries.

60 RAU (Reset and Add into Upper). This operation code resets the entire accumulator to plus zero and adds the contents of the D address location into the upper half of the accumulator.

65 RAL (Reset and Add into Lower). This operation code resets the entire accumulator to plus zero and adds the contents of the D address location into the lower half of the accumulator.

61 RSU (Reset and Subtract into Upper). This operation code resets the entire accumulator to plus zero and subtracts the contents of the D address location into the upper half of the accumulator.

66 RSL (Reset and Subtract into Lower). This operation code resets the entire accumulator to plus zero and subtracts the contents of the D address location into the lower half of the accumulator.

Accumulator store instructions

20 STL (Store Lower in Memory). This operation code causes the contents of the lower half of the accumulator with the accumulator sign to be stored in the location specified by the D address of the instruction. The contents of the lower half of the accumulator remain undisturbed.

It is important to remember that the D address for all store instructions must be 0000–1999. An 8000 series D address will not be accepted as valid by the machine on any of the store instructions.

21 STU (Store Upper in Memory). This operation code causes the contents of the upper half of the accumulator with the accumulator sign to be stored in the location specified by the D address of the instruction. If STU is performed after a division operation, and before another division, multiplication, or reset operation takes place, the contents of the upper accumulator will be stored with the sign of the remainder from the divide operation (Op-Code 14). The contents of the upper half of the accumulator remain undisturbed.

22 STDA (Store Lower Data Address). This operation code

causes positions 8–5 of the distributor to be replaced by the contents of the corresponding positions of the lower half of the accumulator. The modified word in the distributor with the sign of the distributor is then stored in the location specified by the D address of the instruction.

23 STIA (Store Lower Instruction Address). This operation code causes positions 4–1 of the distributor to be replaced by the contents of the corresponding positions of the lower half of the accumulator. The modified word in the distributor with the sign of the distributor is then stored in the location specified by the D address of the instruction. The contents of the lower half of the accumulator remain unchanged, and the sign of the accumulator is not transferred to the distributor. The modified word remains in the distributor upon completion of the operation.

Absolute value instructions

17 AABL (Add Absolute to Lower). This operation code causes the contents of the D address location to be added to the contents of the lower half of the accumulator as a positive factor regardless of the actual sign. When the operation is completed, the distributor will contain the D address factor with its actual sign.

67 RAABL (Reset and Add Absolute into Lower). This operation code resets the entire accumulator to zeros and adds the contents of the D address location into the lower half of the accumulator as a positive factor regardless of its actual sign. When the operation is completed, the distributor will contain the D address factor with its actual sign.

18 SABL (Subtract Absolute from Lower). This operation code causes the contents of the D address location to be subtracted from the contents of the lower half of the accumulator as a positive factor regardless of the actual sign. When the operation is completed, the distributor will contain the D address factor with its actual sign.

68 RSABL (Reset and Subtract Absolute into Lower). This operation code resets the entire accumulator to plus zero and subtracts the contents of the D address location into the lower half of the accumulator as a positive factor, regardless of the actual sign. When the operation is completed, the distributor will contain the D address factor with its actual sign.

Multiplication and division

19 MULT (Multiply). This operation code causes the machine to multiply. A 10-digit multiplicand may be multiplied by a 10-digit multiplier to develop a 20-digit product. The multiplier must be placed in the upper accumulator prior to multiplication. The location of the multiplicand is specified by the D address of the instruction. The product is developed in the accumulator beginning in the low-order position of the lower half of the accumulator and extending to the left into the upper half of the accumulator as required.

14   DIV   (Divide). This operation code causes the machine to divide without resetting the remainder. A 20-digit dividend may be divided by a 10-digit divisor to produce a 10-digit quotient. In order to remain within these limits, the absolute value of the divisor must be *greater than* the absolute value of that portion of the dividend that is in the upper half of the accumulator. The entire dividend is placed in the 20-position accumulator. The location of the divisor is specified by the D address of the divide instruction.

64   DIV   RU   (Divide and Reset Upper). This operation code causes the machine to divide as explained under operation code 14 (DIV). However, the upper half of the accumulator containing the remainder with its sign is reset to zeros.

Branching instructions (decision operations)

44   BRNZU   (Branch on Non-Zero in Upper). This operation code causes the contents of the upper half of the accumulator to be examined for zero. If the contents of the upper half of the accumulator is nonzero, the location of the next instruction to be executed is specified by the D address. If the contents of the upper half of the accumulator is zero, the location of the next instruction to be executed is specified by the I address. The sign of the accumulator is ignored.

45   BRNZ   (Branch on Non-Zero). This operation code causes the contents of the entire accumulator to be examined for zero. If the contents of the accumulator is nonzero, the location of the next instruction to be executed is specified by the D address. If the contents of the accumulator is zero, the location of the next instruction to be executed is specified by the I address. The sign of the accumulator is ignored.

46   BRMIN   (Branch on Minus). This operation code causes the sign of the accumulator to be examined for minus. If the sign of the accumulator is minus, the location of the next instruction to be executed is specified by the D address. If the sign of the accumulator is positive, the location of the next instruction to be executed is specified by the I address. The contents of the accumulator are ignored.

47   BROV   (Branch on Overflow). This operation code causes the overflow circuit to be examined to see whether it has been set. If the overflow circuit is set, the location of the next instruction to be executed is specified by the D address. If the overflow circuit is not set, the location of the next instruction to be executed is specified by the I address.

90-99   BRD   1-10   (Branch on 8 in Distributor Position 1-10). This operation code examines a particular digit position in the distributor for the presence of an 8 or 9. Codes 91–99 test positions 1–9, respectively, of the test word; code 90 tests position 10. If an 8 is present, the location of the next instruction to be executed is specified by the D address. If a 9 is present, the location of the next instruction to be executed is specified by the I address. The presence of other than an 8 or 9 will stop the machine.

Shift instructions

30   SRT   (Shift Right). This operation code causes the contents of the entire accumulator to be shifted right the number of places specified by the units digit of the D address of the shift instruction. A maximum shift of nine positions is possible. A data address with units digit of zero will result in no shift. All numbers shifted off the right end of the accumulator are lost.

31   SRD   (Shift Round). This operation causes the contents of the entire accumulator to be shifted right the number of places specified by the units digit of the D address of the instruction. A 5 is added (-5 if the accumulator is negative) in the twenty-first (blind) position of the amount in the accumulator. A data address units digit of zero will shift 10 places right with rounding.

35   SLT   (Shift Left). This operation code causes the contents of the entire accumulator to be shifted left the number of places specified by the units digit of the D address of the instruction. A maximum shift of nine positions is possible. A data address with a units digit of zero will result in no shift. All numbers shifted off the left end of the accumulator are lost. However, the overflow circuit will not be turned on.

36   SCT   (Shift Left and Count). This operation code causes (1) the contents of the entire accumulator to be shifted to the left until a nonzero digit is in the most significant place, (2) a count of the number of places shifted to be inserted in the two low-order positions of the accumulator. This instruction is to aid fixed-point scaling.

Table look-up instructions

84   TLU   (Table Look-up). This operation code performs an automatic table look-up using the D address as the location of the first table argument and the I address as the address of the next instruction to be executed. The argument for which a search is to be made must be in the distributor. The address of the table argument equal to, or higher than (if no equal exists) the argument given is placed in positions 8–5 of the lower accumulator. The search argument remains, unaltered, in the distributor.

Miscellaneous instructions

00 No-Op (**No Operation**).   This code performs no operation. The data address is bypassed, and the machine automatically refers to the location specified by the instruction address of the **No-Op** instruction.

01 Stop.   This operation code causes the program to stop provided the *programmed* switch on the control console is in the stop position. When the *programmed* switch is in the run position the 01 code will be ignored and treated in the same manner as **00** (**No-Op**).

References

Type 650 Magnetic Drum Data-Processing Machine Manual of Operations; HughE54; SerrR62.

Section 3

Processors for variable-length-string data

Although only two computers are described in this section, the reader might refer to other computers in the book which handle variable-length strings. The IBM System/360 processes a string whose length is specified in the instruction. The Burroughs B 5000 has a very nice string data ISP (both simple and powerful).

Variable-length strings imply some method to specify at instruction execution time the actual length of the character strings being processed. Which method is used has a substantial effect on the ISP of the resulting machine, and it is noteworthy that a wide variety of devices has been tried without any apparent consensus yet on the appropriate mechanism:

1 An extra bit in each character to mark the string boundary (IBM 1401)

2 A special terminal character to mark the string boundary (IBM 702)

3 A field variable in the instruction to specify the string length (IBM System/360)

4 A register variable in the processor to specify the string length (an 8-bit-character computer—Chap. 10)

5 A fixed number of characters at the head of the string

to specify the length (and data type) of the string (used extensively for variable-length records on tape and disk, though we know of no ISP that uses it)

The IBM 1401

The 1401 was IBM's most popular computer, measured by quantity produced, prior to the 1130/1800 and System/360. However, the authors of this book were unable to find any technical papers on its design or design philosophy. The 1401 is based on earlier business-oriented computers (Fig. 1, page 225). It evolved a great deal, as can be seen from the number of "features" which can be appended to improve it. Successors, the 1440 and 1460, are also improvements. It is assumed that early computers mainly influence successor computers within the same organization.

An 8-bit-character computer

An 8-bit-character computer (Chap. 10) has been suggested by the authors. It is a very restricted computer for processing string data and illustrates another approach to string definitions; the string length is specified by a variable in the processor.

Chapter 18

The IBM 1401

The second-generation transistor-technology IBM 1401 has been included both because a large number[1] have been produced and because it differs from common fixed word length binary and decimal computers. IBM 1401s are used in business data-processing applications requiring variable-length character strings or fields and rather limited calculating ability. Two specific applications are as a card processor in making a transition from plugboard programmed calculators to full-scale automatic computations and for converting data from one medium to another, for example, from card to tape. The 1401 was little used by the scientific, engineering, and scientific business data-processing communities, probably because of the limited Mp size, the low overall processing speed, and the lack of concurrent I/O operation in the smaller configurations. However, it did achieve considerable use as a stand-alone Cio in C('7090) installations, perhaps because of the speed and quality of the T('1403; line; printer).

Although undoubtedly influenced by machines outside the IBM organization, the IBM 1401 is derived primarily from the IBM 702 and 705, which are variable word length decimal machines. The relationship of the various IBM decimal computers to one another is shown in Fig. 1. (RCA's early computers[2] also use a combination of fixed-length and variable-length 7-bit character strings and may have influenced the 1401.)

The IBM 1401's ISP was the first to be adopted by another company. Honeywell defined its H-200 ISP to be a superset of the IBM 1401 ISP. The ISP of the H-200 is more complex and increases performance by organizing Mp by both characters and words.

The IBM 1401, 1440, and 1460 are the only IBM computers to be completely character-string oriented. That is, both instructions and data are stored in variable-length character strings; these strings are addressed by a pointer register to the string. The address integer is fixed at three characters. The encoding process for addresses is given in Appendix 1 of this chapter. The 3-character address (3 × 6 bits) is assigned as 3 × 4 bcd characters for encoding addresses 0:999; 2 × 2 bits for selecting 16 × 1,000 addresses; and 2 bits for selecting one of the three index registers.

The IBM 1620 processes variable-length data strings, although the instruction length is a fixed 12-digit string corresponding to a word in Mp. The 1620, though not identical to the 1401, is almost a member of the same family.

The 1401 evolved. Figure 1 shows the evolution of "features" which have created new computers. The 1401's optional features are mainly design afterthoughts; they sometimes increase performance, sometimes make certain operations possible, and sometimes provide substantive change. There are approximately 19 features in the 1401: memory expansion beyond the anticipated 4,000 characters and index registers required encoding the field bits of the A and B addresses; store A-Address and store B-Address register

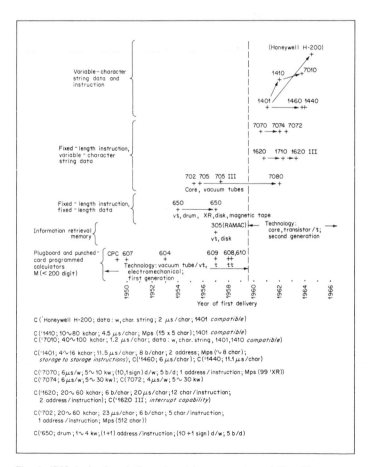

Fig. 1. IBM decimal and character-string computer relationships.

C ('Honeywell H-200; data: w, char. string; 2 μs/char; 1401 *compatible*)

C('1410; 10∿80 kchar; 4.5 μs/char; Mps (15 x 5 char); 1401 *compatible*)
C('7010; 40∿100 kchar; 1.2 μs/char; data: w, char. string, 1401,1410 *compatible*)

C('1401; 4∿16 kchar; 11.5 μs/char; 8 b/char; 2 address; Mps (∿8 char); *storage to storage instructions*); C('1460; 6 μs/char; C('1440; 11.1 μs/char)

C('7070; 6μs/w; 5∿10 kw; (10,1sign) d/w; 5 b/d; 1 address /instruction; Mps (99 'XR))
C('7074; 6μs/w; 5∿30 kw); C('7072; 4μs/w; 5∿30 kw)

C('1620; 20∿60 kchar; 6 b/char; 20μs/char; 12 char/instruction; 2 address/instruction); C('1620 III; *interrupt capability*)

C('702; 20∿60 kchar; 23μs/char; 6 b/char; 5 char/instruction; 1 address/instruction; Mps (512 char))

C('650; drum; 1∿4 kw; (1+1) address/instruction; (10 +1 sign) d/w; 5 b/d)

---

[1] Up to 1966, more 1401s were produced than any other model. An estimated 7,500 1401s, 1,500 1401 G's (card-only system), 3,600 1440s, and 1,500 1460s were produced. About 1,800 1620s were produced.

[2] RCA 301, 501, and 601.

instructions are necessary for subroutines—the Store Address Register Feature; Indexing Feature; Multiply-Divide Feature; High-Low-Equal Compare Feature; Read Release and Punch Release Feature; the Column Binary Feature; Early-Card-Read Feature; Processing Overlap Feature, etc.

PMS structure

The 1401 PMS structure (Fig. 2) is an early 1 Pc structure. The diagram does not show the S(fixed) Pc interconnection structure with the Ms and T. The Pc-(Ms|T) interconnection restricts the concurrency of T and Ms. The optional processing overlap feature provides a link to Mp to allow the T(card; read, punch) to be run concurrently with Pc processing. When any of the peripheral devices are operating without the processing overlap feature, the Pc is dedicated to be a data transmission link or K (as in earlier computers). The device K is connected directly to Pc. For example, Ms(disk, magnetic tape) data transfers use the main registers of the Pc and can tie it up full time during data transmission. By careful programming, several devices can be synchronized and thus run concurrently for communicating with Pc from a K. The Pc does not have an interrupt system. Thus the peripherals have no way of communicating with Pc. Subsequent models, the 1440 and 1460, added interrupt capability and made it easier to control multiple simultaneous data transfers among the peripheral K's and Pc.

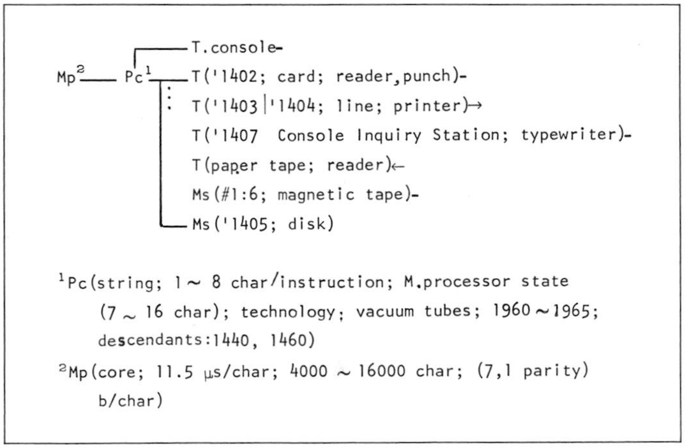

$$Mp^2 \longrightarrow Pc^1 \left\{ \begin{array}{l} \text{T.console}\leftarrow \\ \text{T('1402; card; reader,punch)-} \\ \text{T('1403}|\text{'1404; line; printer)}\rightarrow \\ \text{T('1407 Console Inquiry Station; typewriter)-} \\ \text{T(paper tape; reader)}\leftarrow \\ \text{Ms(\#1:6; magnetic tape)-} \\ \text{Ms('1405; disk)} \end{array} \right.$$

[1]Pc(string; 1 ~ 8 char/instruction; M.processor state
   (7 ~ 16 char); technology; vacuum tubes; 1960 ~ 1965;
   descendants:1440, 1460)

[2]Mp(core; 11.5 µs/char; 4000 ~ 16000 char; (7,1 parity)
   b/char)

Fig. 2. IBM 1401 PMS diagram.

ISP structure

The IBM 1401 ISP is given in Appendix 1 of this chapter. Instruction strings and data strings are delimited by the special F bit in a character. A character in Mp is of the form[1]

C⟨check,F,B′,A′, 8, 4, 2, 1⟩
An n-character string is C[0], C[1], . . . C[n − 1]
and would be stored in Mp[j:j + n − 1]

The first character (or head) of an instruction must contain the word-mark flag or F bit. The head of the instruction, which is to be interpreted next, is held at Mp[I], and succeeding characters of the instruction are at Mp[I + 1], Mp[I + 2], etc. Correctly defined instructions are 1, 2, 4, 5, 7, and 8 characters long. Undefined instruction lengths of up to 8 characters are also interpreted without an error condition. The interpretation algorithm presented in the ISP description does not explain the action of instructions which have an incorrect length. Actually, the 1401 Reference Manual does not go into details of general instruction interpretation but dwells on "correct" operation. Table 1 presents the correct instruction lengths and formats. If we take the instructions in the table, the set is not variable in length but is fixed at these six sizes. The instruction set (not including the input/output instructions) is presented in Table 2. This table also provides a hint of the implementation, since the execution times are given in terms of memory cycles.

   The ISP state, unlike that of more conventional processors, has no temporary operand storage (e.g., accumulators). The ISP state has registers which point to operands. The state of the machine (see Appendix 1) is basically: Mp, the Instruction Location Counter, Indicators or miscellaneous bits, three 3-character blocks of Mp reserved for Index registers, and the two registers A_address and B_address which point to data operands.

Instruction interpretation

There are three principal state types in processing an instruction: o.q., when the instruction is being formed; o.v., when the operands are being accessed or the results are being stored in Mp; and o, when the operation specified by the instruction is being carried out. Each state transition corresponds essentially to a memory access. The three instruction types of Fig. 3 each have their own particular states. Only types 1 and 2 process the variable-length

[1]See Appendix 1 of this chapter for the meaning of the bits in a character. We have renamed the A and B bits A′ and B′ to avoid confusion with the registers.

Table 1 IBM 1401 instruction formats

Length (char)	Location: M[I]	M[(I + 1):(I + 3)]	M[(I + 4):(I + 6)]	M[I + 7]	Types
1	C[0]				no-op, halt, or single character to specify a chained instruction
2	C[0]	C[1]			the d_character is used to specify additional instruction information (e.g., select, card stacker)
4	C[0]	C[1, 2, 3]			unconditional branch instruction or single address arithmetic; M[A] ← f(M[A])
5	C[0]	C[1, 2, 3]	C[4]		conditional branch instruction; C[4] selects a specific test
7	C[0]	C[1, 2, 3]	C[4, 5, 6]		two address instruction; M[B] ← M[B] b M[A]; (e.g., add, subtract)
8	C[0]	C[1, 2, 3]	C[4, 5, 6]	C[7]	conditional branch based on Mp[B] character; d_character is test character; (e.g., branch if character equal)

Function of instruction characters:

C[0] op code; always contains a word-mark flag or F bit.

C[1, 2, 3] = branch address for I_Address register or first operand address for the A_Address register.

C[1] or C[4] or C[7] d_character; used as a single character for additional operation code information or a character for comparison, or to select a test.

C[4, 5, 6] primary operand (B_Address register specification).

character strings, {char.string}, and the state diagram accounts for strings on a character-at-a-time basis. For an add instruction Fig. 3 oversimplifies the execution because it implies that each character of the A and B operand is accessed, the addition is performed, and the result is restored according to the B_address register. A more complex description must account for A and B strings of unequal length, and the case of getting a number which must be recomplemented because it is the wrong sign. The recomplementation process requires a reverse scan to find the end of the B string and then a forward scan to recomplement each character of B. Figure 4 is a detailed state diagram of the add execution process.

The states in the ISP description (Appendix 1) within the instruction-interpretation process correspond to the three state types just described: the single-instruction character-fetch operation, the fetch-operand-addresses for the remainder of the instruction, and Instruction_execution. Instruction_execution is not given in any detail. For example, the execution of add is defined as "A"(:= op = 110001) → Ov□M[B] ← M[B] + M[A] {char.string};. The state diagram (Fig. 4) presents this execution in detail. Note that in the ISP description we omit telling the reader that the A and B

address registers point to the next lowest variable-length string in M after an operation is performed. We allow the definition of a variable-string operation, for example, + {char.string}, to imply the action on the processor state.

Some instructions can be defined with a single character, and these are called chained instructions. Chained instructions take the previous values of the pointer registers, the A and B address registers, as the operand addresses. The add instruction, for example, can be either 1 (chained), 4, or 7 characters; the forms of all instructions appear in Table 1. The 4-character add instruction places the A address field in both the A and B address registers; thus the effect is an instruction to double a string (add it to itself).

Data

An n-decimal-digit numeric data string is represented as

$$\underline{C[n - 1]}, C[n - 2], \ldots, C[1], C[0], \underline{C[M]}$$

The underlined characters, $C[n - 1]$ and $C[M]$, have the flag bit present, that is, $(C[n - 1]\langle F \rangle = 1)$ and $(C[M]\langle F \rangle = 1)$. The n characters are stored in locations Mp[j], Mp[j + 1], ..., Mp[j +

Table 2   IBM 1401 instruction set (excluding input, output)

Instruction	Op Code†	Execution time in memory cycles‡	Length (char.)	Data type
Add (no recomplement)	A	$L_I + 3 + L_A + L_B$	1, 4, 7	char. string
Add (recomplement)	A	$L_I + 3 + L_A + 4L_B$	1, 4, 7	char. string
Branch	B	$L_I + 1$	4	3 char
Branch if Bit Equal§	W	$L_I + 2$	8	1, 3 char
Branch if Character Equal	B	$L_I + 2$	8	1, 3 char
Branch if Indicator On	B	$L_I + 1$	5	1, 3 char
Branch if Word Mark and/or Zone	V	$L_I + 2$	8	1, 3 char
Clear Storage	/	$L_I + 1 + L_x$	1, 4, 7	char. string
Clear Word Mark	⌑	$L_I + 3$	1, 4, 7	1 char
Compare	C	$L_I + 1 + L_A + L_B$	1, 7	char. string
Divide (aver.)§	%	$L_I + 2 + 7L_R L_Q + 8L_Q$	7	char. string
Halt	.	$L_I + 1$	1	
Load Characters to A Word Mark	L	$L_I + 1 + 2L_A$	4, 7	char. string
Modify Address§	#	$L_I + 9$	4, 7	3 char
Move Characters to A or B Word Mark	M	$L_I + 1 + 2L_w$	4, 7	char. string
Move Characters and Edit	E	$L_I + 1 + L_A + L_B + L_y$	7	char. string
Move Characters to Record or Word Mark§	P	$L_I + 1 + 2L_A$	7	char. string
Move Characters and Suppress Zeros	Z	$L_I + 1 + 3L_A$	7	char. string
Move and Insert Zeros§	X	$L_I + 1 + 2\Sigma L_A + \Sigma L_z$	7	char. string
Move Numeric	D	$L_I + 3$	1, 7	1 char
Move Zone	Y	$L_I + 3$	1, 7	1 char
Multiply (aver.)§	@	$L_I + 3 + 2L_C + 5L_C L_M + 7L_M$	7	char. string
No operation	N	$L_I + 1$	1	
Set Word Mark	,	$L_I + 3$	4, 7	1 char
Store A-Address Register§	Q	$L_I + 5$	4	3 char
Store B-Address Register§	H	$L_I + 4$	4	3 char
Subtract (no recomplement)	S	$L_I + 3 + L_A + L_B$	1, 4, 7	char. string
Subtract (recomplement)	S	$L_I + 3 + L_A + 4L_B$	1, 4, 7	char. string
Zero and Add	?	$L_I + 1 + L_A + L_B$	1, 4, 7	char. string
Zero and Subtract	!	$L_I + 1 + L_A + L_B$	1, 4, 7	char. string

†Alphanumeric code used to specify instruction.
‡M(t.cycle: 11.5 μs/char)
§Optional-feature instructions.

Abbreviations for symbols used in timing:
L_A = length of the A-field (in characters)
L_B = length of the B-field
L_C = length of multiplicand field
L_I = length of instruction
L_M = length of multiplier field
L_Q = length of quotient field
L_R = length of divisor field
L_S = number of significant digits in divisor (excludes highorder Os and blanks)
L_W = length of A- or B-field, whichever is shorter
L_X = number of characters to be cleaned
L_Y = number of characters back to rightmost 0 in control field
L_Z = number of Os inserted in a field
Σ = number of fields included in an operation

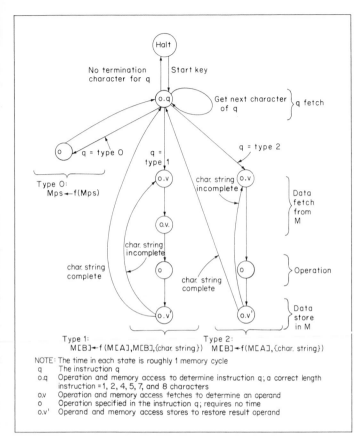

Fig. 3. IBM 1401 instruction-interpretation state diagram.

character-by-character scan with string A and B being added together; the result string is placed in B. States 4 and 5 define the string addition, when string A is terminated; i.e., it is considered to be zero. States 7, 8, 9, and 10 define the recomplementation process in which the B string has to be recomplemented. This condition occurs when the operand signs differ, and the A-field result is greater than the B field; the results are in ten's complement form. States 7 and 8 define the B-field scan (to return to find the least digit of B), and states 9 and 10 define the recomplementation of each character. Thus an add operation may require up to three scans of the B string.

The 1401 ISP (Appendix 1 of this chapter) has four parts: State Declaration, Instruction-interpretation process, Instruction-execution process, and Operand address-register calculation process. The Operand address-register calculation process is analogous to the Effective-address calculation in more conventional Pc's and is the most elaborate part of the instruction interpretation. The operand address registers A_address and B_address are part of the Pc state and must be retained between instructions. At the end of an instruction, these registers point to the character of the next lowest data string in Mp, that is, the character at C[n].

Implementation

The 1401 has a small Pc state, and there are only a few registers in the implementations. Figure 5 shows the registers, interregister transfer paths, and data operations that make up the register-

n − 1]. The values of the string are based on the bcd value of the 8, 4, 2, 1 bits of each digit. The magnitude of the integer is

$$C[n-1] \times 10^{n-1} + C[n-2] \times 10^{n-2} + \cdots + C[0] \times 10^0$$

and the sign is

$$\text{Sign} := ((\neg C[0]\langle A'\rangle \wedge C[0]\langle B'\rangle) \to -;$$
$$\neg(\neg C[0]\langle A'\rangle \wedge C[0]\langle B'\rangle) \to +)$$

A string is addressed (or accessed) via the A_address or B_address pointer registers. These point to the tail (or least significant digit), that is, C[0], of the string. The instruction-execution state diagram of a variable-string add is shown in Fig. 4. The state diagram assumes that A and B address registers are set up according to Fig. 3. Thus Fig. 4 is a more detailed description of states o.v, o.v, o, and o.v'. Each horizontal pair of states (Fig. 4) corresponds to a single scan of the states of type 1 instruction o.v, o.v, o, o.v' in Fig. 3. Transitions among states 2 and 3 correspond to the

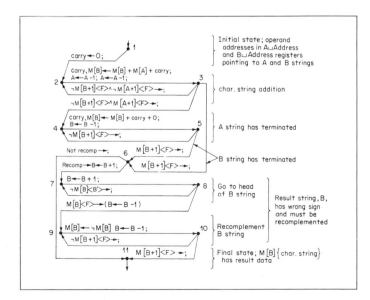

Fig. 4. IBM 1401 add-instruction-execution state diagram.

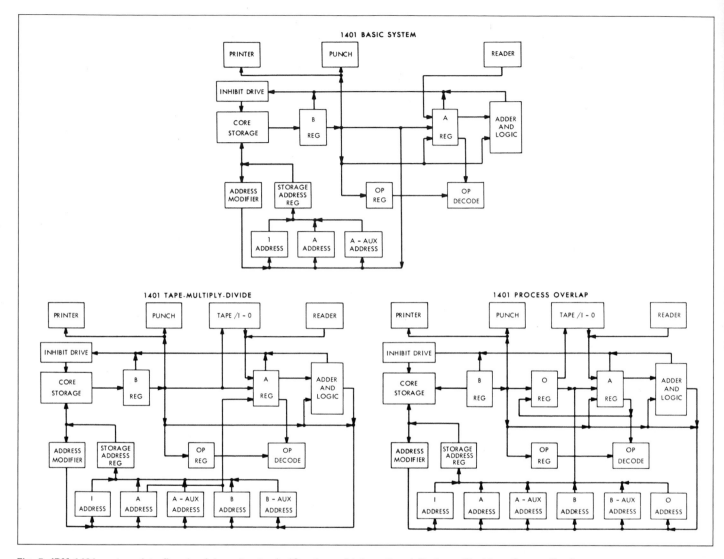

Fig. 5. IBM 1401 system data flow (registers structure). (*Courtesy of International Business Machines Corporation.*)

transfer level primitives of the complete computer together with several options. The options, of course, increase the complexity (and concurrency). Without the overlap feature, for example, all data are accessed in Mp via Pc's address registers.

There are register pairs consisting of a 3-character memory address (access) register, and a 1-character data register. The memory-address, memory-data register pairs are A_address, A_data; B_address, B_data; I_address, Operation/Op; Overlap-_address, Overlap_data/O.

The implementation is straightforward, and the instruction times (Table 2) show the implementation at the register-transfer level. For example, as an instruction is being read by Pc, prior to instruction execution, each new character is taken in and examined for the instruction-terminating flag bit. When the flag bit is present, the instruction is complete and ready to be executed. The character of the next instruction is not saved but is picked up again after the previous instruction has been executed.

APPENDIX 1   IBM 1401 ISP DESCRIPTION

Appendix 1

IBM 1401 ISP Description

Pc, Pc Console, and IO Device Control States

The following description is a highly simplified description of the IBM 1401. For example, the edit instruction given below in one line corresponds to a three page description in the Reference Manual for the 1401. It does not include the input-output instructions which transfer character strings to fixed blocks of primary memory. The character strings are denoted as character.string/ ch.string/ch.s. For the character.string operations the A_address/A and B_address/B registers contain a pointer to the next A and B strings at the end of the operations; this aspect of the operation is not described--but implied in the string operations.

I $[1:3]$<B',A',8,4,2,1>   *I_address register, the instruction location pointer*

A $[1:3]$<B',A',8,4,2,1>   *A_address register*

B $[1:3]$<B',A',8,4,2,1>   *B_address register*

String Data pointer registers A and B point to the least significant digit end of a variable length string in memory (see Mp State definition below). Normally A and B are decreased by one and move to the more significant end for variable length string {ch.s} operations. B is normally the result string, and the length is defined by a word mark, F, the last character of the B string. If A string has a word mark, and is shorter than the B string, then the remaining A string is taken to be a zero. I is a pointer to the most significant digit of the instruction. Although Pc register characters have the B',A',8,4,2,1 bits, the M has two additional bits check, and field. The bits of Mp are:

Check/Parity_bit. The sum (modulo 2) + 1, of the F,B',A',8,4,2,1, bits.

WM/Word_Mark/F/Field_bit. This bit defines the beginning of each instruction. The F bit also defines the most significant digit (the last digit) of a variable length numeric integer string.

B',A',8,4,2,1 bits. A 6 bit character is encoded in these bits. If numeric data is represented, the 8,4,2,1 bits are used as a bcd digit. The sign is encoded with the least significant digit. For numeric data, a minus sign, -, is encoded by (A' = 0) ∧ (B' = 1). All other combinations of A',B' represent a plus sign, +.

XR $[1:3][1:3]$<B',A',8,4,2,1>₁= M$[87:89,92:94,97:99]$<B',A',8,4,2,1>   *3 three character optional index registers stored in Mp*

Indicators $[0:63]$ *logical bit array encoding Pc State (not including I,A, and B)*

There are a set of 31 status bits of the possible 64. They can be cleared or set under instruction control. Some Indicators are used by external Pc status or I/O status. The indicators can be selected for testing by the d character of an instruction. The Pc indicators assignment to Pc State is:

Unconditional := 1   *always a 1*

Sense_switch<A,B,C,D,E,F,G>   *a set of 7 console keys*

Unequal_compare $B \neq A$

Equal_compare $B = A$

Low_compare $B < A$

High_compare $B > A$

Overflow   *set by arithmetic overflow, cleared by a branch instruction if it is set*

The indicator array is partially encoded below:

Indicator $[000000]$:= Unconditional

Indicator $[110001]$:= Sense_switch<A>
⋮

Indicator $[010001]$:= Unequal_compare
⋮

Indicator $[011001]$:= Overflow

Mp State

M$[0:15999]$<Check,F,B',A',8,4,2,1>   *primary memory*

   address[X$[1:3]$<B',A',8,4,2,1>]<1:5>₁₀ := (*Address encoding for 1 of 16000 from a 3 char value of register X. Indexing described below.*

APPENDIX 1   IBM 1401 ISP DESCRIPTION (Continued)

$$X[3]<B',A'> \times 4000_{10} \quad +$$
$$X[1]<B',A'> \times 1000_{10} \quad +$$
$$X[1:3]<8,4,2,1>\{bcd.string\})$$

Instruction Format

op<F,B',A',8,4,2,1>	*instruction register specifying the operation*
d_char<F, B',A',8,4,2,1>	*additional character used in some instructions*
d_char_present	*indicates a d_char is used in the current instruction*
active	*indicates an instruction string is still being fetched*
A_address_present	*indicates there is an A address part of an instruction*
B_address_present	*indicates there is a B address part of an instruction*

Move, load, and store instruction types control the initialization of A and B.

move or load or store A or B/mls := ((move characters and edit = op) ∨ (load characters to A word mark = op) ∨ (move characters to A or B words mark = op) ∨ (move characters and suppress zeros = op) ∨ (move numerical = op) ∨ (move zone = op) ∨ (store A address register = op) ∨ (store B address register = op))

Instruction Interpretation Process

Run → (op ←M[I]; I ← I + 1; next	*fetch operation*
Fetch_operand_addresses; next	*fetch addresses for A and B*
Instruction_execution)	*execute*

Address Calculation Process

The 1401 calculates explicit effective addresses by first setting up the A, and B address registers. Operands are not fetched in Instruction_execution. There are 1,2,4,5,7 and 8 character instructions which have the op and the following operands (respectively): no char, d char, the I or A address, the I or A address and d char, the A and B address, and the I or A address and B address and d char. The following process defines the operation for correct length instructions.

Fetch_operand_addresses := (
d_char_present ← 0;

M[I]<F> → (active ← 0);	*1 char instruction*
¬M[I]<F> → (active ← 1; ¬ mls → B ← 0); next	*proceed to get an I or A address*
active → (d_char ← get_char; next A[1] ← d_char;	*I or A address set up or d_char*
d_char_present ← 1; next	
¬mls → (B[1] ← A[1])); next	
active → (A[2] ← get_char; next ¬ mls → B[2] ← A[2]); next	
active → (A[3] ← get_char; next ¬ mls → B[3] ← A[3]): next	
active → (A_address_present ←1);	*record whether I or A address is present*
¬ active → (A_address_present ← 0); next	
A_address_present → (d_char_present ← 0;	*add index register to I or A*
(A[2]<B',A'> ≠ 0) → (A ← A + XR[A[2]<B',A'>] {3.ch}));	
¬ M[I]<F> → (B ← 0); next	*B address set up or d_char*
active → (d_char ← get_char; next B[1] ← d_char;	
d_char_present ← 1);	
active → (B[2] ← get_char); next	
active → (B[3] ← get_char); next	
active → (B_address_present ← 1);	*record whether B address is present*
¬ active → (B_address_present ← 0); next	
B_address_present → (	*add index register to B*
d_char_present ← 0;	
(B[2]<B',A'> ≠ 0) → (B ← B + XR[B[2]<B',A'>]{ 3.ch}));	

APPENDIX 1   IBM 1401 ISP DESCRIPTION (Continued)

```
      (¬ M[I]<F> ∧ active) → (d_char ← get_char;               final d_char
                          d_char_present ← 1);  next
      (¬ M[I]<F> ∧ active) → Run ← 0;                          halt if more than 8 char instruction
          )                                                    end Fetch_operand_addresses
get_character:
   A sub-process used to fetch each new character in the instruction.  If F is found in a character, the process terminates.
      get_char<B',A',8,4,2,1> := (
          ¬ M[I]<F> ∧ active → (M[I];  I ← I + 1);             value is present character
          M[I]<F> → active ← 0);                               no value, terminate
Instruction Set and Instruction Execution Process
   Instruction_execution := (
      character string/ch.s movement and clear memory:
          "M" (:= op = 100100) → (M[B] ←M[A] {ch.s});          move characters to A or B word mark - character string (ch.s)
          "Z" (:= op = 011001) → (M[B] ←M[A] {ch.s}; next      move characters and suppress zeros
                               M[B] ← f(M[B]) {ch.s});
          "L" (:= op = 100011) → (M[B] ←M[A] {ch.s});          load characters to A word mark
          "E" (:= op = 110101) → (M[B] ← f(M[A],M[B], {ch.s}));  move characters and edit
      This instruction moves the A field string to the B field string under control of an edit character string in the original B field.
          "/" (:= op = 010001) → (M[B] ←0 {ch.s.mod.100};      clear storage, ignores the F mark and moves to next modulo
                                                                   100 address
                      ¬ B_address_present → ;                  clear storage
                        B_address_present → I ←A);             clear storage and branch
      character string, {ch.s}, arithmetic:
          "A" (:= op = 110001) → (Ov,M[B] ←M[B] + M[A] {ch.s}):  add
          "S" (:= op = 010010) → (Ov,M[B] ←M[B] - M[A] {ch.s});  subtract
          "!" (:= op = 101010) → (M[B] ←0       - M[A] {ch.s});  zero and subtract
          "?" (:= op = 111010) → (M[B] ←0       + M[A] {ch.s});  zero and add
          "@" (:= op = 001100) → (Ov,M[B] ←M[B] × M[A] {ch.s});  multiply; full length product in M[B], special hardware option
          "%" (:= op = 011100) → (Ov,M[B] ←M[B] / M[A] {ch.s});  divide; quotient and remainder both end up in M[B].
          "#" (:= op = 001011) → (M[B] ←M[B] + M[A] {3.ch};    modify address
                               B ←B - 3; A ←A - 3);
      branches, halt, no-operation:
          "N" (:= op = 100101)→  ;                             no operation
          "." (:= op = 111011)→ (Run ←0;
                      ¬ A_address_present → ;                  halt
                        A_address_present → I ← A);            halt and branch
          "B" (:= op = 110010) → (
             (¬ B_address_present ∧ ¬ d_char_present) → I ← A;     branch
             (¬ B_address_present ∧   d_char_present) → (         branch if indicator on
                Indicator [f(d_char)] → (I ←A);
                Indicator [f(d_char)] ←0);
             (B_address_present ∧ d_char_present) → (            branch if char equal
                B ←B - 1;
                (M[B] = d_char) → I ← A)):
```

APPENDIX 1 IBM 1401 ISP DESCRIPTION (Continued)

```
        "V" (:= op = 010101) → (B ← B - 1;                          branch if word mark and/or zone
                            M[B]<f(d⌣char)> → (I ←A));
        "C" (:= op = 110011) → (                                    compare
            Indicators ←M[A] = M[B] {ch.s});
    subroutine calling:
        "Q" (:= op = 101000) → (                                    store A address register
          M[A - 2:A] ←A[1:3]; A ← A - 3);
        "H" (:= op = 111000) → (                                    store B address register
          M[A - 2:A] ← B[1:3]; A ← A - 3);
    single character operations
        "," (:= op = 011011) → (M[A]<F> ← 1; M[B]<F> ← 1;           set word mark
                            A ← A - 1; B ← B - 1);
        "⬚" (:= op = 111100) → (M[A]<F> ← 0: M[B]<F> ← 0;           clear word mark
                            A ← A - 1; B ← B - 1);
        "D" (:= op = 110100) → (M[B]<8,4,2,1> ← M[A]<8,4,2,1>;      move numerical
                            A ← A - 1; B ← B - 1);
        "Y" (:= op = 011000) → (M[B]<B',A'> ← M[A]<B',A'>;          move zone
                            A ← A - 1; B ← B - 1);
            )                                                       end Instruction⌣execution
```

Section 4

Desk calculator computers: keyboard programmable processors with small memories

These stored program computers have interesting features. For example, the keyboard is utilized several ways:

1 T.console mode; a conventional console for entering data in response to a stored program

2 Program entry mode; a device for creating stored programs

3 Desk calculator mode; a part of the arithmetic (data) element by issuing direct instructions and thus obtaining results directly independent of a program

Uses 2 and 3 are both internally and externally programmed. The data types are decimal (both fixed and floating) because of the intimate interface they require to the user. Some calculators interpret nested (parenthesized) algebraic expressions.

These calculators easily meet the definition for a stored-program computer. It is apparent their designers know a great deal about general purpose stored-program computers. The machines are cleverly designed and make efficient use of the hardware they possess. Eventually there may be more of these computers than conventional stored program computers. The reader should note that not all "electronic desk calculators" are computers; most are electronic versions of their mechanical or electromechanical ancestors.

The OLIVETTI UNDERWOOD PROGRAMMA 101 desk calculator

The Programma 101 (Chap. 19) is at the limit of what we call a stored program computer. It has a sufficient instruction set to be classified as a computer, but the storage for temporary data, constants, and programs is limited. The machine's instruction set is interesting because memory is not addressed explicitly. A jump, for example, is executed by scanning the program for a particular marker which was named in the jump instruction. The Programma 101 uses an Mp.cyclic.

The program library for the Programma 101 is extensive and provides an indication of its capability.

The Hewlett-Packard Model 9100A computing calculator

The HP 9100A (Chap. 20), like the Programma 101 (Chap. 19), is a desk calculator. They are both stored program computers. Programma is designed for simpler accounting and statistical-tabulation tasks and has fixed-point decimal data. (Programma 101 costs somewhat less.) The HP 9100A operates on both fixed- and floating-point decimal data with scalar, rectangular, and polar coordinate vectors and is designed for engineering and scientific calculations. Thus, according to a measure based on data types and operators, the HP 9100A is about the most complete computer in the book. Its operations are given in the PMS diagram of Fig. 1.

```
Mp(read,write; core; 368 w; 6 b/w)
 ├ T.console(keyboard)←
 ├ T.console(CRT; display; numeric; decimal; mixed, floating)→
Pc[data:(scalar, rectangular co-ordinate vector, polar co-      ]¹
   ordinate vector); fixed, floating; decimal; operations:(+,
   -, X, /, cos, sin, tan, sin⁻¹, cos⁻¹, tan⁻¹, sinh, cosh,
   tanh, sinh⁻¹, cosh⁻¹, tanh⁻¹, ln, log₁₀, abs, e, sqrt,
   integer part,{rectangular co-ordinate vector} ← {polar co-
   ordinate vector}, {polar co-ordinate vector} ← {rectangular
   co-ordinate vector})
 ├ T.numeric printer→
 ├ T.plotter→
 └ L.external device -
 └ T─┌M[magnetic card; 2 programs; 196 program steps/program;]
     └ 6 b/program step

¹Pc := [Mp(read only; 512 w, 64 b/w)                    ]
       [P.microprogrammed(M.processor state(40 b))²     ]

²P.microprogrammed := P.microprogrammed
                      |
                      Mp(control; read only; 800 ns/w;
                         64 w; 29 b/w)
```

Fig. 1. Hewlett-Packard Model 9100A Computing Calculator PMS diagram.

The implementation has approximately 36.2 kb of memory, including the read-only and read-write parts. The design is physically outstanding, and its use of microprogramming is superb. The reader should note there are two levels of M(read only). We could draw the PMS structure of Pc as a P.microprogrammed within a P.microprogrammed. HP rightfully regards the two ISP's (29-bit and 64-bit word) as proprietary and carefully avoids discussing these points in the article (Chap. 20). It might be noted that an IBM System/360 Model 30 requires about 2.9 milliseconds for a floating-point square root, whereas the HP 9100A requires 19 milliseconds. By way of evidence of its outstanding packaging, its cost is about five-eighths that of a PDP-8/I for about the same amount of physical hardware. The cost difference, though truly difficult to compare, is partially the result of a design from an instrument maker (Hewlett-Packard) versus a design from a computer manufacturer (DEC). The TV-like construction of the HP 9100A is an important lesson that computer manufacturers have not learned. In other words, a Henry Ford has yet to emerge from the computer field. (Our guess is that he may come from Japan.)

Whereas many computers in this book are included because they are typical of points in the computer space, the HP 9100A is included because it is innovative. It is worthy of note that only one of the engineers had some computer design experience; Cochran, who did the programming, had prior experience with circuitry and instrumentation. Had he been a programmer by training, a larger Mp might have been required. By way of comparative evidence, the IBM 1800 floating-point arithmetic functions $+$, $-$, $\times$, $/$, sin, cos, $\tan^{-1}$, $\sqrt{\ }$, log, exponential, tanh, binary to decimal, and decimal to binary require approximately 1,425 16-bit words, or 23 kb. On the other hand, the FOCAL[1] interactive calculator program for a 4,096-word PDP-8 (49 kb) provides the user with all but polar-rectangular coordinates and hyperbolic functions, but it does have a complete program editing capability, text handling, control structure, and 1,600-character Mp.

[1] Similar in scope to Dartmouth's BASIC.

Chapter 19

The OLIVETTI Programma 101 desk calculator[1]

The Programma 101 is manufactured by the Olivetti Underwood Corporation. The cost of Programma 101 is about $3,500 (in 1968). Several thousand are currently in use. Unlike conventional stored program computers it has instructions which can be executed directly as commands from a keyboard or instructions which can be stored in a program and interpreted by the processor. The processor uses the decimal representation for mixed numbers. The decimal point location is controlled manually. Although information is stored in character strings, the maximum length is 22 digits or 24 instructions for a register. A program can be up to 120 characters long and is stored as a continuous string. The internal encoding of a character is 8 bits. There are no absolute addresses for instructions, and jump instructions are programmed by placing labels or references in the string to transfer to. The Programma 101 is composed of the following elements.

Memory. The memory stores numeric data and program instructions.

Keyboard. The keyboard has four functions: It is used for operator control of the calculator (power on, off, etc); in manual mode the instructions are executed immediately as in a conventional desk calculator (e.g., add); the keys write a program's instructions in the memory, and the instructions are executed when the program is run; and numeric data may be entered to a running program.

Printing unit. Serial printing is from right to left, at 30 characters per second; this unit prints all keyboard entries, programmed output, and instructions.

Magnetic-card reader/recorder. This device permits instructions and constants for a program to be stored and retrieved from magnetic cards.

Control and arithmetic units. The *control unit* is the administrative section of the computer. It receives the incoming information, determines the computation to be performed, and directs the

[1] The description is partially taken from the Programma 101 Programming Manual.

arithmetic unit where to find the information and what operation to perform.

The PMS diagram shown below is, of course, very simple. It conforms closely to the classic diagram of what a digital computer looks like:

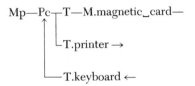

Primary memory and processor memory

The memory has 10 registers; eight are for general storage and two are used exclusively for instructions. A character can have several meanings, depending on the register and its use.

The two instruction registers, 1 and 2, each store 24 instructions. An instruction is one character long.

The eight storage registers, M, A, R, B, C, D, E, and F, have a capacity of 22 decimal digits, plus decimal point and sign. The sign and decimal point do not require character space. Alternatively, D, E, and F hold 24 instructions. M, A, and R are operating registers and take part in all arithmetic operations. They are considered to be the arithmetic unit.

The M register is the Median (or distributive) register. All keyboard figure entries are held in the M register and distributed to the other registers as instructed.

The A register functions with the arithmetic unit to form the Accumulator. Arithmetic results are developed and retained in the A register. A result of up to 23 digits can be produced in the A register.

The R register retains the complete results in addition and subtraction, the complete product in multiplication, the remainder in division, and a remainder in square root. B, C, D, E, and F are storage registers. Each can be split into two registers, each with a capacity of 11 digits, plus decimal point and sign. When storage registers are split, the right portion of the split register retains its original designation, and the left side is identified with the corresponding lowercase letter. Thus these registers become b, B, c, C, d, D, e, F, f and F. The lowercase designation is obtained by first entering the corresponding uppercase letter and then depressing the "/" key, for example, c ≡ C/.

The registers D, E, and F or their splits have the additional capability of storing either instructions or constants to be used within programs. Thus they can store 1 signed 22-digit number, 2 signed 11-digit numbers, 1 signed 11-digit number, and 11 instructions, or 24 instructions. Programs of up to 120 instructions can be stored internally (Fig. 1). When registers D, E, and F and their splits are not used for instructions, they are free to store constants or intermediate results.

The relationship of memory, keyboard, printer, and magnetic card is shown in Fig. 1. Registers are referenced explicitly. Programs do not use explicit addresses in instruction. Thus, special marker characters are placed in the instructions to serve as jump reference addresses (program labels).

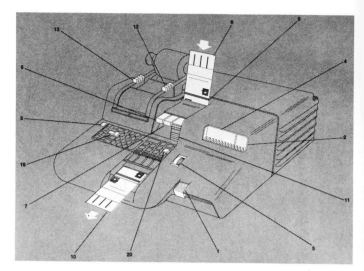

Fig. 2. Programma 101. (*Courtesy of Olivetti Underwood Corporation.*)

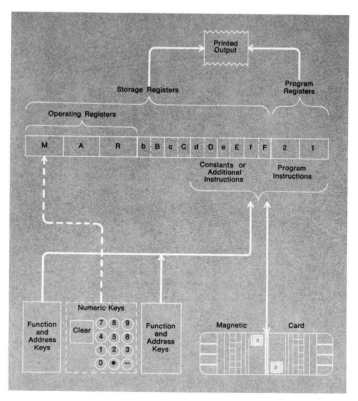

Fig. 1. Programma 101 functional block diagram. (*Courtesy of Olivetti Underwood Corporation.*)

Structure

The calculator parts are described briefly below. The parts correspond to both the numbers (Fig. 2) and the lettered keyboard (Fig. 3). The following parts are, in effect, the console. Some of the keys are used for control of the calculator, and some can be used either as programmed instructions or as commands which are executed directly. The following section discusses their instruction function.

The on-off key (1). This is a dual-purpose switch for both the on and off positions. (Note: The OFF position automatically clears all stored data and instructions.)

The error (red) light (2). This lights when the computer is turned on and whenever the computer detects an operational error, e.g., exceeding capacity, division by zero.

The general reset key (3). This key erases all data and instructions from the computer and turns off the error light.

The correct-performance (green) light (4). This light indicates the computer is functioning properly. A steady light indicates that the computer is ready for an operator decision; a flickering light indicates that the computer is executing programmed instructions and that the keyboard is locked.

The decimal wheel (5). This determines the number of decimal places (0, 1, . . . , 15) to which computations will be carried out in the A register and the decimal places in the printed output, except for results from the R register. Up to 22 decimal digits may be developed in, and printed from, the R register.

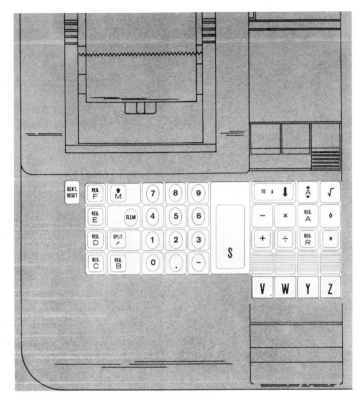

Fig. 3. Programma 101 keyboard. (*Courtesy of Olivetti Underwood Corporation.*)

The record program switch (6). When this switch is off, the commands pressed on the keyboard are executed directly. When this switch is on, it directs the computer to store instructions either in the memory from the keyboard or onto a magnetic program card from the memory.

The record program switch must be off to load instructions from a magnetic program card into the memory.

The print program switch (7). When this switch is on (in), it directs the computer to print out the instructions stored in memory from its present location in the program to the next Stop instruction (S), whenever the print key (20) is depressed.

The magnetic program card (8). This is a plastic card with a ferrous oxide backing, used to record programs for external storage. The card is inserted into a magnetic reader/writer (9) to record instructions and/or constants into or from the computer memory. Once inserted, the card may be removed from the computer (10) without disturbing the stored instructions.

(*Note:* The magnetic-card reader/writer uses only half the magnetic card at a time; consequently, two sets of 120 instructions and/or constants may be stored on a single card.)

The keyboard release key (11). This key reactivates a locked keyboard. If two or more keys are depressed simultaneously, the keyboard will lock to indicate a misoperation. Because the operator does not know what entry was accepted by the computer, after touching the keyboard release key, the clear entry key (16) must be depressed and the complete figure reentered.

Tape advance (12). This advances the printing paper tape.

Tape release lever (13). This enables adjustment when changing tape rolls.

The routine selection (keys V, W, Y, and Z). These keys direct the computer to the proper program or subroutine.

The numeric keyboard (keys 0, 1, . . . , 9, . , −). This keyboard allows entry of a signed, mixed decimal number. Keyboard entries are automatically stored in the M register.

The clear entry key. This key clears the entire keyboard entry. When keying in the program, a depression of the clear key will erase the last instruction that has been entered into the memory. The printing tape will be spaced.

The start key (S). This key restarts the computer in programmed operation; it is used to code a stop instruction when keying in programs.

The register address (keys A, B, C, D, E, F, and R). These keys identify the corresponding registers. The operating register M has no keyboard identification since the computer automatically relates all instructions to the M register unless otherwise instructed.

The split key (/). This key combined with a register (for example, C/) divides that register into two equal parts. When storage registers are split, the right portion of the split register retains the original designation, and the left side is identified on the tape with the corresponding lowercase letter (for example, C/ ≡ c).

The print key (◊). This key prints the contents of an addressed register.

The clear key (°). This key clears the contents of an addressed register. When the computer is operated manually, a depression of this key will print the number in the register and clear it.

The transfer keys (↓, ↑, ↕). These keys perform transfer operations between the storage registers and the operating registers.

The arithmetic keys (−, +, ×, ÷, √). These keys perform their indicated arithmetic function.

Keyboard and stored-program operations

All the following keys can be used as direct instructions (i.e., manually) if the record program switch is off. Alternatively, if the record program switch is on, the keys specify the instruction to be recorded in the program memory. Finally, the descriptions specify the instruction's behavior as it is executed within a program.

Start S. The instruction S (used in creating a program) directs the computer to stop and release the keyboard for the entry of figures or the selection of a subroutine. After figure entry, the program is restarted by touching the start key (S).

The program can also be restarted by touching a routine selection key. When the S instruction stops the program, the computer may also be operated in the manual mode without disturbing the program instructions in the memory. Any figures entered on the keyboard before depression of start or an operation key will be printed automatically.

Clear °. The clear operation ° directs the computer to clear the selected register. The M and R registers cannot be cleared with this instruction.

When the computer is operated manually this key will cause it to print the contents of the selected register, r. $(r \leftarrow 0)$

Data-transfer operations

To A ↓. An instruction containing the operation ↓ directs the computer to transfer contents of the addressed register, r, to A while retaining them in the original register. The contents of M and R are not affected. The previous contents of A are destroyed. $(A \leftarrow r)$

From M ↑. An instruction containing the operation ↑ directs the computer to transfer the contents of M to the addressed register while retaining them in M. The contents of registers A and R are unaffected by this instruction. The original contents of the addressed register are destroyed. $(r \leftarrow M)$

Exchange ↕. An instruction containing the operation ↕ directs the computer to exchange the contents of the A register with the contents of the addressed register. The contents of M are not affected except by the exchange between A and M. The contents of the R register are not affected. $(A \leftarrow r; r \leftarrow A)$

D-R exchange RS. The instruction RS directs the computer to exchange the contents of D (both D and d registers) with the contents of the R register. $(D \leftarrow R; R \leftarrow D)$

This instruction has a special use in multicard programs to store temporarily the contents of the D (d,D) register in R, when a new card has to be read to continue the program. During this temporary storage no instruction affecting the R register should be executed.

Decimal part to M /↕. The instruction /↑ directs the computer to transfer the decimal portion of the contents of A to the M

register while retaining the entire contents in A. The original contents of the M register are destroyed. The R register is not affected by this instruction. $(M \leftarrow \text{fraction\_part}(A))$

Arithmetic operations

All arithmetic operations are performed in the operating registers M, A, and R. An arithmetic operation is performed in two phases:

1 The contents of the selected register are automatically transferred to the M register. The M register is selected automatically if no other register is indicated.

2 The operation is carried out in the M, A, and R registers.

Programma 101 can perform these arithmetic operations: $+$, $-$, $\times$, $\div$, $\sqrt{\ }$, and absolute value. Figures are accepted and computed algebraically. A negative value is entered by depressing the negative key at any time during the entry of a figure. If there is no negative indication, the computer will accept the figure as positive.

The subtract operation key is separate from the numeric keyboard and is used exclusively for subtraction (not negation).

Addition $+$. An instruction containing the operation $+$ directs the computer to add the contents of the selected register (addend) to the contents of the A register (augend). Addition is executed in two phases:

1 Transfer the contents of the selected register (addend) to M.

2 Add the contents of M to the contents of A (augend) obtaining in A the sum truncated according to the setting of the decimal wheel. The complete sum is in R. M contains the addend. $(M \leftarrow r; \text{next } R \leftarrow A + M; \text{next } A \leftarrow f(R,\text{decimal\_wheel}))$

Multiplication $\times$. An instruction containing the operation $\times$ directs the computer to multiply the contents of the selected register (multiplicand) by the contents of the A register (multiplier).

1 Transfer the contents of the addressed register to M.

2 Multiply the contents of M by the contents of A, obtaining in A the product truncated according to the setting of the decimal wheel. The complete product is in R. M contains the multiplicand. $(M \leftarrow r; \text{next } R \leftarrow A \times M; \text{next } A \leftarrow f(R, \text{decimal\_wheel}))$

Subtraction −. An instruction containing the operation − directs the computer to subtract the contents of the selected register (subtrahend) from the contents of the A register (minuend).

1 Transfer the contents of the selected register (subtrahend) to M.

2 Subtract the contents of M from the contents of A (minuend), obtaining in A the difference truncated according to the setting of the decimal wheel. The complete difference is in R. M contains the subtrahend. (M ← r; next R ← A − M; next A ← f(R,decimal_wheel))

Division ÷. An instruction containing the operation ÷ directs the computer to divide the contents of the selected register (divisor) into the contents of the A register (dividend).

1 Transfer the contents of the addressed register to M.

2 Divide the contents of M into the contents of A, obtaining in A the quotient truncated according to the setting of the decimal wheel. The decimally correct fractional remainder is in R. M contains the divisor. (M ← r; next A ← A ÷ M; R ← A mod M)

Square Root √. An instruction containing the operation √ directs the computer to:

1 Transfer the contents of the selected register to M.

2 Extract the square root of the contents of M, as an absolute value, obtaining in A the result truncated according to the setting of the decimal wheel. The R register contains a nonfunctional remainder. At the end of the operation, M contains double the square root. (M ← r; next M,R ← sqrt(abs(M)) × 2; next A ← f(M/2, decimal_wheel))

Absolute Value A↕. The absolute-value instruction A↕ changes the contents of the A register, if negative, to positive. (A ← abs(A))

Jump operations

The jump operation directs the computer to depart from the normal sequence of step-by-step instructions and jump to a preselected point in the program.

These instructions provide both internal and external (manual) decision capability and are useful to create "loops" that allow repetitive sequences in a program to be executed; routines or subroutines to be performed at the discretion of the operator; and automatically to "branch" to alternate routines or subroutines according to the value in the A register.

The jump process consists of two related instructions or characters:

1 The reference point or label, l, is where the program begins or where the jump is to start. The sequence is restarted at this point. This label has no effect when interpreted.

2 The jump instruction specifies the label for the instruction sequence.

There are two types of jump instructions: unconditional jumps and conditional jumps.

Unconditional jumps. These jumps are executed whenever the instruction is read. The labels or reference points for unconditional jumps, L, and the corresponding jump instructions, j, are given as (L,j). The permissible jump labels and jump constructions are:

(AV,V), (AW,W), (AY,Y), (AZ,Z), (BV,CV), . . . ,
(BZ,CZ), (EV,DV), . . . , (EZ,DZ), (FV,RV), . . . , (FZ,RZ)

All programs must begin with reference parts of an unconditional jump instruction. Reference points AV, AW, AY, AZ are used so that these program sequences can be started by touching the routine selection keys V, W, Y, or Z.

Conditional Jumps. If the contents of the A register are:

Greater than zero: the program jumps to the corresponding reference point (label).

Zero or less: the program continues with the next instruction in sequence.

The labels or reference points for conditional jumps, L, and the corresponding conditional jump instruction, cj, are given as (L,cj). The permissible jump labels and jump instructions are

(aV,/V), . . . , (aZ,/Z), (bV,cV), . . . ,
(bZ,cZ), (eV,dV), . . . , (eZ,dZ), (fV,rV),
. . . , (fZ,rZ)

Constants as instructions A/↑. A one-digit constant can be generated by a special instruction. The results of the instruction place the digit in M. The digit value of the constant must follow A/↑.

Instructions and data in the same register. An instruction can be considered to be data and, therefore, used as both a constant and an instruction. Another technique allows the computer to interpret data as null instructions so that both data (for reading and writing) and instructions can be stored in the same register.

Examples. A program to take values for the numbers A, B, C, and D from the keyboard and then print the value of the expression $[(A + B) \times C]/D$ would be written as follows:

instruction	comments
AV	label to allow the program to be started by key, V
S	wait; enter A from keyboard into M
↓ or ↓M[1]	A value goes to A register
S	wait, enter B from keyboard
+M	a register contains A + B
S	wait, enter C from keyboard
×M	a register ×C or (A + B) × C
S	wait, enter D from keyboard
÷M	a register has expression
A◇	print A register
V	jump back to beginning label to recalculate expression for new variables

[1] M is implied if left blank.

The following program computes and prints n!. n is entered from the keyboard, where $n \geq 1$, and an integer. The program is started by pressing key Z.

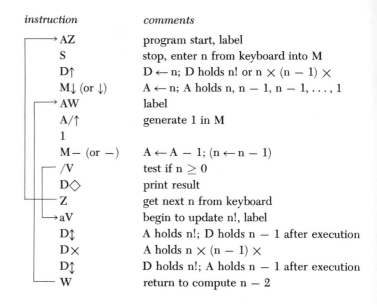

instruction	comments
AZ	program start, label
S	stop, enter n from keyboard into M
D↑	D ← n; D holds n! or n × (n − 1) ×
M↓ (or ↓)	A ← n; A holds n, n − 1, n − 1, . . . , 1
AW	label
A/↑	generate 1 in M
1	
M− (or −)	A ← A − 1; (n ← n − 1)
/V	test if n ≥ 0
D◇	print result
Z	get next n from keyboard
aV	begin to update n!, label
D↕	A holds n!; D holds n − 1 after execution
D×	A holds n × (n − 1) ×
D↕	D holds n!; A holds n − 1 after execution
W	return to compute n − 2

Conclusion

Many algorithms have been written for Programma 101, being coded in impressively small space. The techniques have sometimes been borrowed from conventional computer programming. For example, multiple card programs operate by using chains in the same way as large FORTRAN programs. The significant fact to the reader is that the Programma 101 calculator is a nicely designed stored program computer.

Chapter 20

The HP Model 9100A computing calculator[1]

Richard E. Monnier / *Thomas E. Osborne* /
David S. Cochran

A new electronic calculator with computerlike capabilities

Many of the day-to-day computing problems faced by scientists and engineers require complex calculations but involve only a moderate amount of data. Therefore, a machine that is more than a calculator in capability but less than a computer in cost has a great deal to offer. At the same time it must be easy to operate and program so that a minimum amount of effort is required in the solution of typical problems. Reasonable speed is necessary so that the response to individual operations seems nearly instantaneous.

The HP Model 9100A Calculator, Fig. 1, was developed to fill this gap between desk calculators and computers. Easy interaction between the machine and user was one of the most important design considerations during its development and was the prime guide in making many design decisions.

CRT display

One of the first and most basic problems to be resolved concerned the type of output to be used. Most people want a printed record, but printers are generally slow and noisy. Whatever method is used, if only one register is displayed, it is difficult to follow what is happening during a sequence of calculations where numbers are moved from one register to another. It was therefore decided that a cathode-ray tube displaying the contents of three registers would provide the greatest flexibility and would allow the user to follow problem solutions easily. The ideal situation is to have both a CRT showing more than one register, and a printer which can be attached as an accessory.

Figure 2 is a typical display showing three numbers. The X register displays numbers as they are entered from the keyboard one digit at a time and is called the keyboard register. The Y register is called the accumulator since the results of arithmetic operations on two numbers, one in X and one in Y, appear in the Y register. The Z register is a particularly convenient register to use for temporary storage.

Numbers

One of the most important features of the Model 9100A is the tremendous range of numbers it can handle without special attention by the operator. It is not necessary to worry about where to place the decimal point to obtain the desired accuracy or to avoid register overflow. This flexibility is obtained because all numbers are stored in 'floating point' and all operations performed using 'floating point arithmetic.' A floating point number is expressed with the decimal point following the first digit and an exponent representing the number of places the decimal point should be moved—to the right if the exponent is positive, or to the left if the exponent is negative.

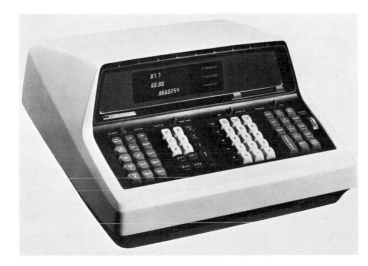

Fig. 1. This new HP Model 9100A calculator is self-contained and is capable of performing functions previously possible only with larger computers.

[1] This chapter is a compilation of three articles [Monnier, 1968; Osborne, 1968; Cochran, 1968], reprinted from *Hewlett-Packard Journal*, vol. 20, no. 1, pp. 3–9, 10–13, 14–16, September, 1968.

Fig. 2. Display in fixed point with the decimal wheel set at 5. The Y register has reverted to floating point because the number is too large to be properly displayed unless the digits called for by the DECIMAL-DIGITS setting are reduced.

$$4.398\,364\,291 \times 10^{-3} = .004\,398\,364\,291$$

The operator may choose to display numbers in FLOATING POINT or in FIXED POINT. The FLOATING POINT mode allows numbers, either positive or negative, from 1×10^{-99} to $9.999\,999\,999 \times 10^{99}$ to be displayed just as they are stored in the machine.

The FIXED POINT mode displays numbers in the way they are most commonly written. The DECIMAL DIGITS wheel allows setting the number of digits displayed to the right of the decimal point anywhere from 0 to 9. Figure 2 shows a display of three numbers with the DECIMAL DIGITS wheel set at 5. The number in the Y register, $5.336\,845\,815 \times 10^5 = 533\,684.5815$, is too big to be displayed in FIXED POINT without reducing the DECIMAL DIGITS setting to 4 or less. If the number is too big for the DECIMAL DIGITS setting, the register involved reverts automatically to floating point to avoid an apparent overflow. In FIXED POINT display, the number displayed is rounded, but full significance is retained in storage for calculations.

To improve readability, 0's before the displayed number and un-entered 0's following the number are blanked. In FLOATING POINT, digits to the right of the decimal are grouped in threes.

Pull-out instruction card

A pull-out instruction card, Fig. 3, is located at the front of the calculator under the keyboard. The operation of each key is briefly

explained and key codes are listed. Some simple examples are provided to assist those using the machine for the first time or to refresh the memory of an infrequent user. Most questions regarding the operation of the Model 9100A are answered on the card.

Data entry

The calculator keyboard is shown in Fig. 4. Numbers can be entered into the X register using the digit keys, the π key or the ENTER EXP key. The ENTER EXP key allows powers of 10 to be entered directly which is useful for very large or very small numbers. 6.02×10^{23} is entered ⬚6 ⬚0 ⬚ENTER EXP ⬚2 ⬚3 . If the ENTER EXP key is the first key of a number entry, a 1 is auto-

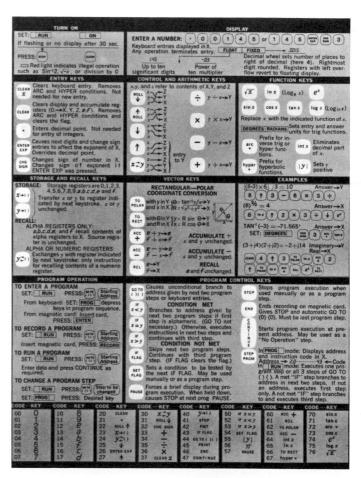

Fig. 3. Pull-out instruction card is permanently attached to the calculator and contains key codes and operating instructions.

Fig. 4. Keys are in four groups on the keyboard, according to their function.

matically entered into the mantissa. Thus only two keystrokes [ENTER EXP] [6] suffice to enter 1,000,000. The CHG SIGN key changes the sign of either the mantissa or the exponent depending upon which one is presently being addressed. Numbers are entered in the same way, regardless of whether the machine is in FIXED POINT or FLOATING POINT. Any key, other than a digit key, decimal point, CHG SIGN or ENTER EXP, terminates an entry; it is not necessary to clear before entering a new number. CLEAR X sets the X register to 0 and can be used when a mistake has been made in a number entry.

Control and arithmetic keys

ADD, SUBTRACT, MULTIPLY, DIVIDE involve two numbers, so the first number must be moved from X to Y before the second is entered into X. After the two numbers have been entered, the appropriate operation can be performed. In the case of a DIVIDE, the dividend is entered into Y and the divisor into X. Then the [÷] key is pressed causing the quotient to appear in Y, leaving the divisor in X.

One way to transfer a number from the X register to the Y register is to use the double sized key, [↑], at the left of the digit keys. This repeats the number in X into Y, leaving X unchanged; the number in Y goes to Z, and the number in Z is lost. Thus, when squaring or cubing a number, it is only necessary to follow [↑] with [×] or [×] [×]. The [↓] key repreats a number in Z to Y leaving Z unchanged, the number in Y goes to X, and the number in X is lost. The [↑ ROLL] key rotates the number in the X and Y registers up and the number in Z down into X. [ROLL ↓] rotates the numbers in Z and Y down and the number in X up into Z. [x⇌y] interchanges the numbers in X and Y. Using the two ROLL keys and [x⇌y], numbers can be placed in any order in the three registers.

Functions available from the keyboard

The group of keys at the far left of the keyboard, Fig. 4, gives a good indication of the power of the Model 9100A. Most of the common mathematical functions are available directly from the keyboard. Except for [|y|] the function keys operate on the number in X replacing it with the function of that argument. The numbers in Y and Z are left unchanged. [√x] is located with another group of keys for convenience but operates the same way.

The circular functions operate with angles expressed in RADIANS or DEGREES as set by the switch above the keyboard. The sine, cosine, or tangent of an angle is taken with a single keystroke. There are no restrictions on direction, quadrant or number of revolutions of the angle. The inverse functions are obtained by using the [arc ↓] key as a prefix. For instance, two key depressions are necessary to obtain the arc sin x: [arc ↓] [TAN x]. The angle obtained will be the standard principal value. In radians:

$$-\frac{\pi}{2} \le \text{Sin}^{-1} x \le \frac{\pi}{2}$$

$$0 \le \text{Cos}^{-1} x \le \pi$$

$$-\frac{\pi}{2} < \text{Tan}^{-1} x < \frac{\pi}{2}$$

The hyperbolic sine, cosine, or tangent is obtained using the [hyper ↓] key as a prefix. The inverse hyberbolic functions are obtained with three key depressions. $\text{Tanh}^{-1} x$ is obtained by [arc ↓] [hyper ↓] [TAN x]. The arc and hyper keys prefix keys below them in their column.

Log x and ln x obtain the log to the base 10 and the log to the base e respectively. The inverse of the natural log is obtained with the e^x key. These keys are useful when raising numbers to odd powers as shown in one of the examples on the pull-out card, Fig. 3.

Two keys in this group are very useful in programs. [int x] takes the integer part of the number in the X register which deletes the part of the number to the right of the decimal point. For example int$(-3.1416) = -3$. [|y|] forces the number in the Y register positive.

Storage registers

Sixteen registers, in addition to X, Y, and Z, are available for storage. Fourteen of them, 0, 1, 2, 3, 4, 5, 6, 7, 8, 9, a, b, c, d, can be used to store either one constant or 14 program steps per register. The last registers, e and f, are normally used only for constant storage since the program counter will not cycle into them. Special keys located in a block to the left of the digit keys are used to identify the lettered registers.

To store a number from the X register the key ⌈x→()⌉ is used. The parenthesis indicates that another key depression, representing the storage register, is necessary to complete the transfer. For example, storing a number from the X register into register 8 requires two key depressions: ⌈x→()⌉ ⌈8⌉ . The X register remains unchanged. To store a number from Y register the key ⌈y→()⌉ is used.

The contents of the alpha registers are recalled to X simply by pressing the keys a, b, c, d, e, and f. Recalling a number from a numbered register requires the use of the ⌈x⇄y⌉ key to distinguish the recall procedure from digit entry. This key interchanges the number in the Y register with the number in the register indicated by the following keystroke, alpha or numeric, and is also useful in programs since neither number involved in the transfer is lost.

The CLEAR key sets the X, Y, and Z display registers and the f and e registers to zero. The remaining registers are not affected. The f and e registers are set to zero to initialize them for use with the ⌈ACC +⌉ and ⌈ACC −⌉ keys as will be explained. In addition the CLEAR key clears the FLAG and the ARC and HYPER conditions, which often makes it a very useful first step in a program.

Coordinate transformation and complex numbers

Vectors and complex numbers are easily handled using the keys in the column on the far left of the keyboard. Figure 5 defines the variables involved. Angles can be either in degrees or radians. To convert from rectangular to polar coordinates, with y in Y and x in X, press ⌈TO POLAR⌉. Then the display shows θ in Y and R in X. In

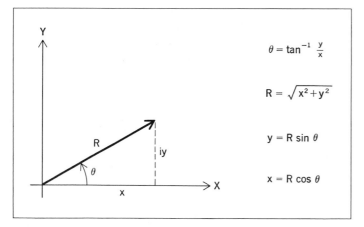

$$\theta = \tan^{-1} \frac{y}{x}$$

$$R = \sqrt{x^2 + y^2}$$

$$y = R \sin \theta$$

$$x = R \cos \theta$$

Fig. 5. Variables involved in conversions between rectangular and polar coordinates.

converting from polar to rectangular coordinates, θ is placed in Y, and R in X, ⌈TO RECT⌉ is pressed and the display shows y in Y and x in X.

ACC+ and ACC− allow addition or subtraction of vector components in the f and e storage registers. ACC+ adds the contents of the X and Y register to the numbers already stored in f and e respectively; ACC− subtracts them. The RCL key recalls the numbers in the f and e registers to X and Y.

Illegal operations

A light to the left of the CRT indicates that an illegal operation has been performed. This can happen either from the keyboard or when running a program. Pressing any key on the keyboard will reset the light. When running a program, execution will continue but the light will remain on as the program is completed. The illegal operations are:

Division by zero
$\sqrt{x}$ where $x < 0$
ln x where $x \le 0$; log n where $x \le 0$
$\sin^{-1} x$ where $|x| > 1$; $\cos^{-1} x$ where $|x| > 1$
$\cosh^{-1} x$ where $x < 1$; $\tanh^{-1} x$ where $|x| > 1$

Accuracy

The Model 9100A does all calculations using floating point arithmetic with a twelve digit mantissa and a two digit exponent. The two least significant digits are not displayed and are called guard digits.

The algorithms used to perform the operations and generate the functions were chosen to minimize error and to provide an extended range of the argument. Usually any inaccuracy will be contained within the two guard digits. In certain cases some inaccuracy will appear in the displayed number. One example is where the functions change rapidly for small changes in the argument, as in tan x where x is near 90°. A glaring but insignificant inaccuracy occurs when an answer is known to be a whole number, but the least significant guard digit is one count low: 2.000 000 000 ≃ 1.999 999 999.

Accuracy is discussed further in the 'Internal Programming' section in this chapter. But a simple summary is: the answer resulting from any operation or function will lie within the range of true values produced by a variation of ±1 count in the tenth digit of the argument.

Programming

Problems that require many keyboard operations are more easily solved with a program. This is particularly true when the same operations must be performed repeatedly or an iterative technique must be used. A program library supplied with the Model 9100A provides a set of representative programs from many different fields. If a program cannot be found in the library to solve a particular problem, a new program can easily be written since no special experience or prior knowledge of a programming language is necessary.

Any key on the keyboard can be remembered by the calculator as a program step except STEP PRGM. This key is used to 'debug' a program rather than as an operation in a program. Many individual program steps, such as 'sin x' or 'to polar' are comparatively powerful, and avoid the need of sub-routines for these functions and the programming space such sub-routines require. Registers 0, 1, 2, 3, 4, 5, 6, 7, 8, 9, a, b, c, d can store 14 program steps each. Steps within the registers are numbered 0 through d just as the registers themselves are numbered. Programs can start at any of the 196 possible addresses. However 0-0 is usually used for the first step. Address d-d is then the last available, after which the program counter cycles back to 0-0.

Registers f and e are normally used for storage of constants only, one constant in each register. As more constant storage is required, it is recommended that registers d, then c, then b, etc., are used starting from the bottom of the list. Lettered registers are used first, for the frequently recalled constants, because constants stored in them are more easily recalled. A register can be used to store one constant or 14 program steps, but not both.

Branching

The bank on the far right of the keyboard, Fig. 4, contains program oriented keys. $\binom{GO\ TO}{(\ )\ (\ )}$ is used to set the program counter. The two sets of parentheses indicate that this key should be followed by two more key depressions indicating the address of the program step desired. As a program step, 'GO TO' is an unconditional branch instruction, which causes the program to branch to the address given by the next two program steps. The 'IF' keys in this group are conditional branch instructions. With $\binom{IF}{x<y}$ $\binom{IF}{x=y}$, and $\binom{IF}{x>y}$ the numbers contained in the X and Y registers are compared. The indicated condition is tested and, if met, the next two program steps are executed. If the first is alphameric, the second must be also, and the two steps are interpreted as a branching address. When the condition is not met, the next two steps are skipped and the program continues. $\binom{IF}{FLAG}$ is also a very useful conditional branching instruction which tests a 'yes' or 'no' condition internally stored in the calculator. This condition is set to 'yes' with the SET FLAG from the keyboard when the calculator is in the display mode or from a program as a program step. The flag is set to a 'no' condition by either asking IF FLAG in a program or by a CLEAR instruction from the keyboard or from a program.

Data input and output

Data can be entered for use in a program when the machine is in the display mode. (The screen is blank while a program is running.) A program can be stopped in several ways. The $\boxed{\text{STOP}}$ key will halt the machine at any time. The operation being performed will be completed before returning to the display mode. As a program step, STOP stops the program so that answers can be displayed or new data entered. END must be the last step in a program listing to signal the magnetic card reader; when encountered as a program step it stops the machine and also sets the program counter to 0-0.

As a program step, PAUSE causes a brief display during program execution. Nine cycles of the power line frequency are counted—the duration of the pause will be about 150 ms for a 60 Hz power line or 180 ms for a 50 Hz power line. More pauses can be used in sequence if a longer display is desired. While a program is running the PAUSE key can be held down to stop the machine when it comes to the next PAUSE in the program. PAUSE provides a particularly useful way for the user and the machine to interact. It might, for instance, be used in a program so that the convergence to a desired result can be observed.

Other means of input and output involve peripheral devices such as an X-Y Plotter or a Printer. The PRINT key activates the printer, causing it to print information from the display register. As a program step, PRINT will interrupt the program long enough for the data to be accepted by the printer and then the program will continue. If no printer is attached, PRINT as a program step will act as a STOP. The FMT key, followed by any other keystroke, provides up to 62 unique commands to peripheral equipment. This flexibility allows the Model 9100A to be used as a controller in small systems.

Sample program—N!

A simple program to calculate N! demonstrates how the Model 9100A is programmed. Figure 6 (top) shows a flow chart to compute N! and Fig. 6 (bottom) shows the program steps. With this program, 60! takes less than $\frac{1}{2}$ second to compute.

Program entry and execution

After a program is written it can be entered into the Model 9100A from the keyboard. The program counter is set to the address of

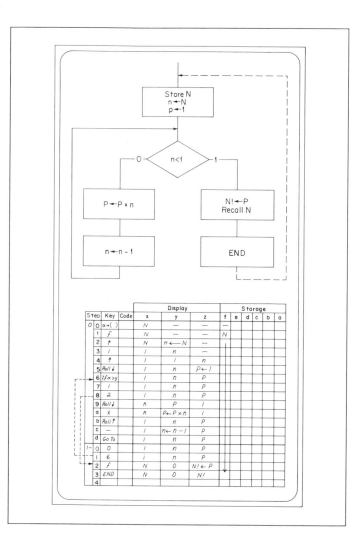

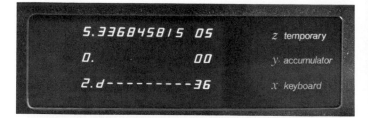

Fig. 7. Program step address and code are displayed in the X register as steps are entered. After a program has been entered, each step can be checked using the STEP PRGM key. In this display, step 2-d is 36, the code for multiply.

is made in a step, it can be corrected by using the (GO TO 0 0) key without having to re-enter the rest of the program.

To run a program, the program counter must be set to the address of the first step. If the program starts at 0-0 the keys (GO TO 0 0) (0) (0) are depressed, or simply just (END) since this key automatically sets the program counter to 0-0. CONTINUE will start program execution.

Magnetic card reader-recorder

One of the most convenient features of the Model 9100A is the magnetic card reader-recorder, Fig. 8. A program stored in the Model 9100A can be recorded on a magnetic card, Fig. 9, about

Fig. 6. Flow chart of a program to compute N! (top). Each step is shown (bottom) and the display for each register. A new value for N can be entered at the end of the program, since END automatically sets the program counter back to 0-0.

The flow chart and table (Fig. 6):

Step		Key	Code	Display			Storage					
				x	y	z	f	e	d	c	b	a
0	0	α→( )		N	—	—	—					
	1	f		N	—	—	N					
	2	↑		N	n←N	—						
	3	/		l	l	—						
	4	↑		l	l	n						
	5	Roll↓		l	n	P←l						
	6	If x≠>y		l	n	P						
	7	l		l	n	P						
	8	2		l	n	P						
	9	Roll↓		n	P	l						
	a	X		n	P←Pxn	l						
	b	Roll↑		l	n	P						
	c	—		l	n←n-l	P						
	d	Go To		l	n	P						
1-	0	0		l	n	P						
	1	6		i	n	P						
	2	f		N	0	N!←P						
	3	END		N	0	N!						
	4											

Flow chart boxes:
Store N / n←N / p←1 → n<1 → (0) P←Pxn → n←n-1 ; (1) N!←P Recall N → END

the first program step by using the GO TO ( ) ( ) key. The RUN-PROGRAM switch is then switched from RUN to PROGRAM and the program steps entered in sequence by pushing the proper keys. As each step is entered the X register displays the address and key code, as shown in Fig. 7. The keys and their codes are listed at the bottom of the pull-out card, Fig. 3. Once a program has been entered, the steps can be checked using the STEP PRGM key in the PROGRAM mode as explained in Fig. 7. If an error

Fig. 8. Programs can be entered into the calculator by means of the magnetic program card. The card is inserted into the slot and the ENTER button pressed.

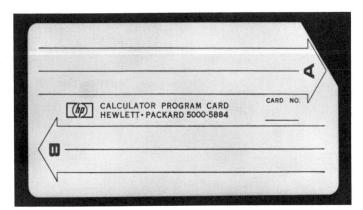

Fig. 9. Magnetic programming card can record two 196-step programs. To prevent accidental recording of a new program over one to be saved, the corner of the card is cut as shown.

the size of a credit card. Later when the program is needed again, it can be quickly re-entered using the previously recorded card. Cards are easily duplicated so that programs of common interest can be distributed.

As mentioned earlier, the END statement is a signal to the reader to stop reading recorded information from the card into the calculator. For this reason END should not be used in the middle of a program. Since most programs start at location 0-0 the reader automatically initializes the program counter to 0-0 after a card is read.

The magnetic card reader makes it possible to handle most programs too long to be held in memory at one time. The first entry of steps can calculate intermediate results which are stored in preparation for the next part of the program. Since the reader stops reading at the END statement these stored intermediate results are not disturbed when the next set of program steps is entered. The stored results are then retrieved and the program continued. Linking of programs is made more convenient if each part can execute an END when it finishes to set the program counter to 0–0. It is then only necessary to press CONTINUE after each entry of program steps.

Hardware design of the Model 9100A calculator

All keyboard functions in the Model 9100A are implemented by the arithmetic processing unit, Figs. 10 and 11. The arithmetic unit operates in discrete time periods called clock cycles. All

---

Specifications of HP Model 9100A*

The HP Model 9100A is a programmable, electronic calculator which performs operations commonly encountered in scientific and engineering problems. Its log, trig and mathematical functions are each performed with a single key stroke, providing fast, convenient solutions to intricate equations. Computer-like memory enables the calculator to store instructions and constants for repetitive or iterative solutions. The easily-readable cathode ray tube instantly displays entries, answers and intermediate results.

Operations
Direct keyboard operations include:
Arithmetic: addition, subtraction, multiplication, division and square-root.
Logarithmic: log x, ln x and e^x.
Trigonometric: sin x, cos x, tan x, $\sin^{-1}x$, $\cos^{-1}x$ and $\tan^{-1}x$ (x in degrees or radians).
Hyperbolic : sinh x, cosh x, tanh x, $\sinh^{-1}x$, $\cosh^{-1}x$, and $\tanh^{-1}x$.

Coordinate transformation: polar-to-rectangular, rectangular-to-polar, cumulative addition and subtraction of vectors.
Miscellaneous: other single-key operations include—taking the absolute value of a number, extracting the integer part of a number, and entering the value of π. Keys are also available for positioning and storage operations.

Programming
The program mode allows entry of program instructions, via the keyboard, into program memory. Programming consists of pressing keys in the proper sequence, and any key on the keyboard is available as a program step. Program capacity is 196 steps. No language or code-conversions are required. A self-contained magnetic card reader/recorder records programs from program memory onto wallet-size magnetic cards for storage. It also reads programs from cards into program memory for

repetitive use. Two programs of 196 steps each may be recorded on each reusable card. Cards may be cascaded for longer programs.

Speed
Average times for total performance of typical operations, including decimal-point placement:
add, subtract: 2 milliseconds
multiply: 12 milliseconds
divide: 18 milliseconds
square-root: 19 milliseconds
sin, cos, tan: 280 milliseconds
ln x: 50 milliseconds
e^x: 110 milliseconds
These times include core access of 1.6 microseconds.

General
Weight: Net 40 lbs, (18,1 kg.); shipping 65 lbs. (29,5 kg.).
Power: 115 or 230 V $\pm$ 10%, 50 to 60 Hz, 400 Hz, 70 watts.
Dimensions: 8¼'' high, 16'' wide, 19'' deep.

*Courtesy of Loveland Division.

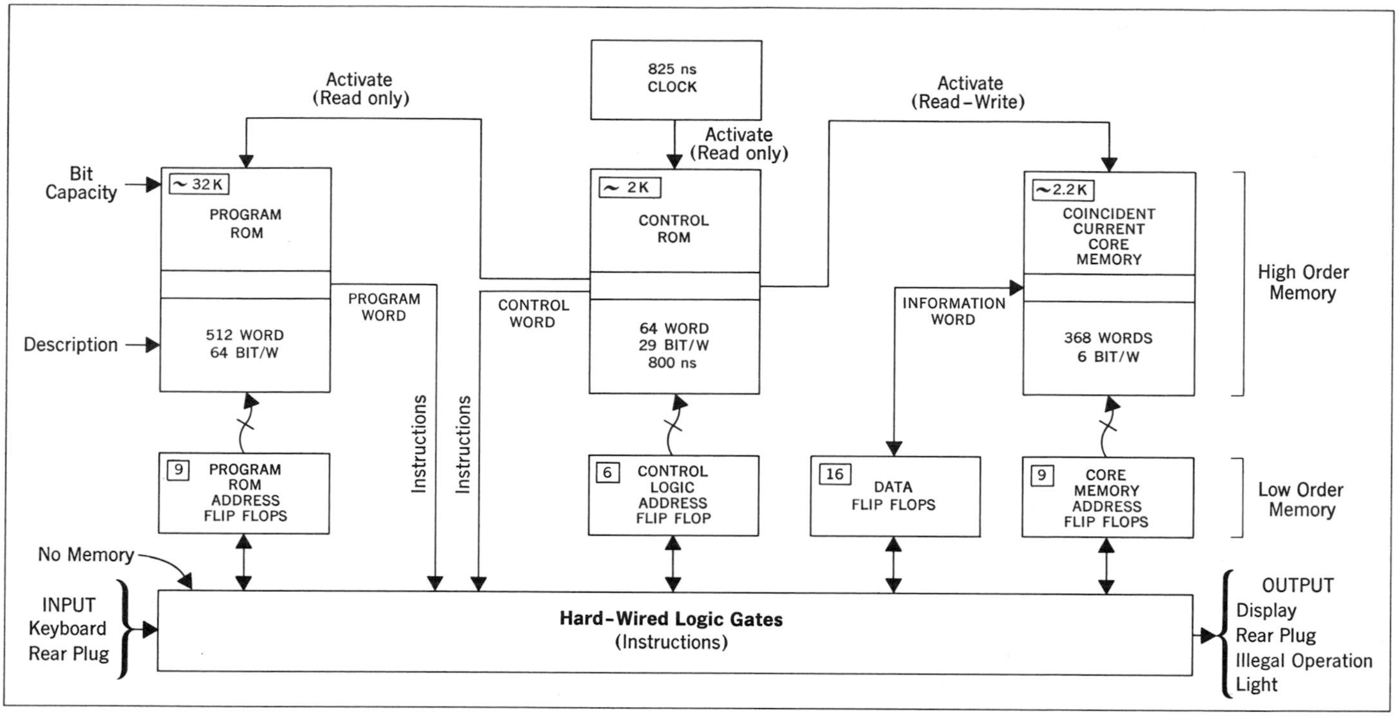

Fig. 10. Arithmetic processing unit block diagram. This system is a marriage of conventional, reliable diode-resistor logic to a 32,000-bit read-only memory and a coincident current core memory.

operations are synchronized by the clock shown at the top center of Fig. 10.

The clock is connected to the control read only memory (ROM) which coordinates the operation of the program read only memory and the coincident current core read/write memory. The former

Fig. 11. Arithmetic unit assembly removed from the calculator.

contains information for implementing all of the keyboard operations while the latter stores user data and user programs.

All internal operations are performed in a digit by digit serial basis using binary coded decimal digits. An addition, for example, requires that the least significant digits of the addend and augend be extracted from core, then added and their sum replaced in core. This process is repeated one BCD digit at a time until the most significant digits have been processed. There is also a substantial amount of 'housekeeping' to be performed such as aligning decimal points, assigning the proper algebraic sign, and floating point normalization. Although the implementation of a keyboard function may involve thousands of clock cycles, the total elapsed time is in the millisecond region because each clock cycle is only 825 ns long.

The program ROM contains 512 64-bit words. When the program ROM is activated, signals (micro-instructions) corresponding to the bit pattern in the word are sent to the hard wired logic gates shown at the bottom of Fig. 10. The logic gates define the changes to occur in the flip flops at the end of a clock cycle. Some of the micro-instructions act upon the data flip flops while others change the address registers associated with the program ROM, control ROM and coincident current core memory. During the next clock cycle the control ROM may ask for a new set of micro-instructions from the program ROM or ask to be read from or written into the coincident current core memory. The control ROM also has the ability to modify its own address register and to issue micro-instructions to the hard wired logic gates. This flexibility allows the control logic ROM to execute special programs such as the subroutine for unpacking the stored constants required by the keyboard transcendental functions.

Control logic

The control logic uses a wire braid toroidal core read only memory containing 64 29-bit words. Magnetic logic of this type is extremely reliable and pleasingly compact.

The crystal controlled clock source initiates a current pulse having a trapezoidal waveform which is directed through one of 64 word lines. Bit patterns are generated by passing or threading selected toroids with the word lines. Each toroid that is threaded acts as a transformer to turn on a transistor connected to the output winding of the toroid. The signals from these transistors operate the program ROM, coincident current core, and selected micro-instructions.

Coincident current core read/write memory

The 2208 (6 $\times$ 16 $\times$ 23) bit coincident current memory uses wide temperature range lithium cores. In addition, the X, Y, and inhibit drivers have temperature compensated current drive sources to make the core memory insensitive to temperature and power supply variations.

The arithmetic processing unit includes special circuitry to guarantee that information is not lost from the core memory when power is turned off and on.

Power supplies

The arithmetic processing unit operates from a single -15 volt supply. Even though the power supply is highly regulated, all circuits are designed to operate over a voltage range of -13.5 to -16.5.

Display

The display is generated on an HP electrostatic cathode ray tube only 11 inches long. The flat rectangular face plate measures $3\frac{1}{4} \times 4\frac{13}{16}$ inches. The tube was specifically designed to generate a bright image. High contrast is obtained by using a low transmissivity filter in front of the CRT. Ambient light that usually tends to 'wash out' an image is attenuated twice by the filter, while the screen image is only attenuated once.

All the displayed characters are 'pieces of eight.' Sixteen different symbols are obtained by intensity modulating a figure 8 pattern as shown in Fig. 12. Floating point numbers are partitioned into groups of three digits and the numeral 1 is shifted to improve readability. Zeros to the left of the most significant digit and insignificant zeros to the right of the decimal point are blanked to avoid a confusing display. Fixed point numbers are automatically rounded up according to the decimal wheel setting. A fixed point display will automatically revert to floating point notation if the number is too large to be displayed on the CRT in fixed point.

Multilayer instruction logic board

All of the hard wired logic gates are synthesized on the instruction logic board using time-proven diode-resistor logic. The diodes and resistors are located in separate rows, Fig. 13. All diodes are oriented in the same direction and all resistors are the same value. The maze of interconnections normally associated with the back plane wiring of a computer are located on the six internal layers of the multilayer instruction logic board. Solder bridges and accidental shorts caused by test probes shorting to leads beneath components are all but eliminated by not having interconnections on the two outside surfaces of this multilayer board. The instruction logic board also serves as a motherboard for the control logic board, the two coincident core boards and the two flip flop boards, the magnetic card reader, and the keyboard. It also contains a connector, available at the rear of the calculator, for connecting peripherals.

Flip flops

The Model 9100A contains 40 identical J-K flip flops, each having a threshold noise immunity of 2.5 volts. Worst case design techniques guarantee that the flip flops will operate at 3 MHz even though 1.2 MHz is the maximum operating rate.

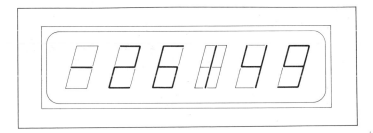

Fig. 12. Displayed characters are generated by modulating these figures. The digit 1 is shifted to the center of the pattern.

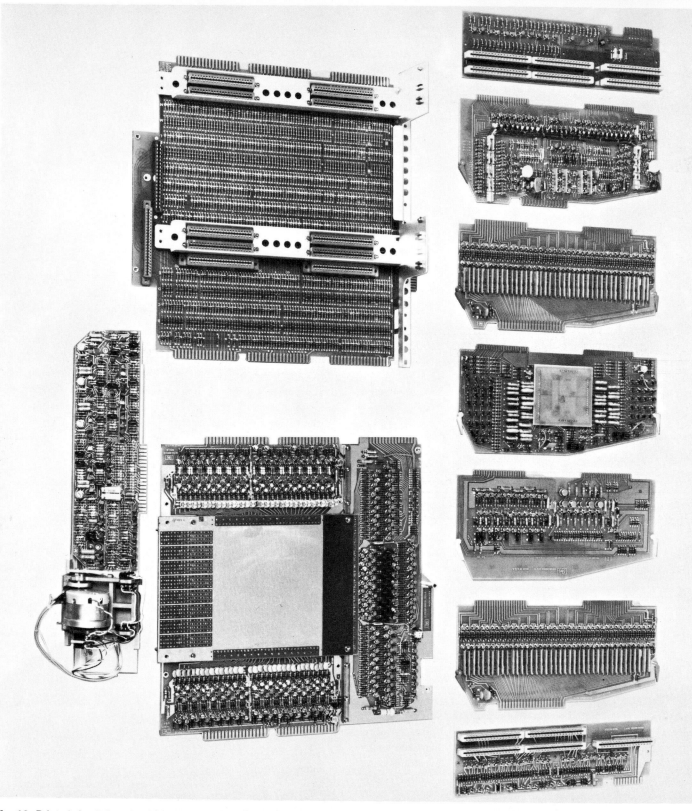

Fig. 13. Printed-circuit boards which make up the arithmetic unit are, left to right at top, side board, control logic, flip flop, core and drivers, core sense amplifiers and inhibit, flip flop, and side board. Large board at the lower left is the multilayer instruction board, and the program ROM is at the right. The magnetic card reader and its associated circuitry are at the bottom.

Program read only memory

The 32,768 bit read only program memory consists of 512 64-bit words. These words contain all of the operating subroutines, stored constants, character encoders, and CRT modulating patterns. The 512 words are contained in a 16 layer printer-circuit board having drive and sense lines orthogonally located. A drive line consists of a reference line and a data line. Drive pulses are inductively coupled from both the reference line and data line into the sense lines. Signals from the data line either aid or cancel signals from the reference line producing either a 1 or 0 on the output sense lines. The drive and sense lines are arranged to achieve a bit density in the ROM data board of 1000 bits per square inch.

The program ROM decoder/driver circuits are located directly above the ROM data board. Thirty-two combination sense amplifier, gated-latch circuits are located on each side of the ROM data board. The outputs of these circuits control the hard wired logic gates on the instruction logic board.

Side boards

The program ROM printed circuit board and the instruction logic board are interconnected by the side boards, where preliminary signal processing occurs.

The keyboard

The keyboard contains 63 molded plastic keys. Their markings will not wear off because the lettering is imbedded into the key body using a double shot injection molding process. The key and switch assembly was specifically designed to obtain a pleasing feel and the proper amount of tactile and aural feedback. Each key operates a single switch having gold alloy contacts. A contact closure activates a matrix which encodes signals on six data lines and generates an initiating signal. This signal is delayed to avoid the effects of contact bounce. An electrical interlock prevents errors caused by pressing more than one key at a time.

Magnetic card reader

Two complete 196 step programs can be recorded on the credit card size magnetic program card. The recording process erases any previous information so that a card may be used over and over again. A program may be protected against accidental erasure by clipping off the corner of the card, Fig. 9, page 249. The missing corner deactivates the recording circuitry in the magnetic card reader. Program cards are compatible among machines.

Information is recorded in four tracks with a bit density of 200 bits per inch. Each six-bit program step is split into two time-multiplexed, three-bit codes and recorded on three of the four tracks. The fourth track provides the timing strobe.

Information is read from the card and recombined into six bit codes for entry into the core memory. The magnetic card reading circuitry recognizes the 'END' program code as a signal to end the reading process. This feature makes it possible to enter subroutines within the body of a main program or to enter numeric constants via the program card. The END code also sets the program counter to location 0-0, the most probable starting location. The latter feature makes the Model 9100A ideally suited to 'linking' programs that require more than 196 steps.

Packaging and servicing

The packaging of the Model 9100A began by giving the HP industrial design group a volume estimate of the electronics package, the CRT display size and the number of keys on the keyboard. Several sketches were drawn and the best one was selected. The electronics sections were then specifically designed to fit in this case. Much time and effort were spent on the packaging of the arithmetic processing unit. The photographs, Figs. 11 and 14, attest to the fact that it was time well spent.

The case covers are die cast aluminum which offers durability, effective RFI shielding, excellent heat transfer characteristics, and convenient mechanical mounts. Removing four screws allows the case to be opened and locked into position, Fig. 14. This procedure exposes all important diagnostic test points and adjustments. The keyboard and arithmetic processing unit may be freed by removing four and seven screws respectively.

Any component failures can be isolated by using a diagnostic routine or a special tester. The faulty assembly is then replaced and is sent to a service center for computer assisted diagnosis and repair.

Reliability

Extensive precautions have been taken to insure maximum reliability. Initially, wide electrical operating margins were obtained by using 'worst case' design techniques. In production all transistors are aged at 80% of rated power for 96 hours and tested before being used in the Model 9100A. Subassemblies are computer tested and actual operating margins are monitored to detect trends that could lead to failures. These data are analyzed and corrective action is initiated to reverse the trend. In addition, each calculator is operated in an environmental chamber at 55°C for 5 days prior to shipment to the customer. Precautions such as these allow Hewlett-Packard to offer a one year warranty in a field where 90 days is an accepted standard.

Fig. 14. Internal adjustments of the calculator are easily accessible by removing a few screws and lifting the top.

Internal programming of the 9100A calculator

Extensive internal programming has been designed into the HP Model 9100A Calculator to enable the operator to enter data and to perform most arithmetic operations necessary for engineering and scientific calculation with a single key stroke or single program step. Each of the following operations is a hardware subroutine called by a key press or program step:

Basic arithmetic operations
    Addition
    Subtraction
    Multiplication
    Division

Extended arithmetic operations
    Square root
    Exponential—e^x
    Logarithmic—$\ln x$, $\log x$
    Vector addition and subtraction

Trigonometric operations
    Sin x, cos x, tan x
    Arcsin x, arccos x, arctan x
    Sinh x, cosh x, tanh x
    Arcsinh x, arccosh x, arctanh x
    Polar to rectangular and rectangular to
        polar coordinate transformation

Miscellaneous
    Enter π
    Absolute value of y
    Integer value of x

In the evolution of internal programming of the Model 9100A Calculator, the first step was the development of flow charts of each function. Digit entry, Fig. 15, seemingly a trivial function, is as complex as most of the mathematical functions. From this functional description, a detailed program can be written which uses the microprograms and incremental instructions of the calculator. Also, each program must be married to all of the other programs which make up the hard-wired software of the Model 9100A. Mathematical functions are similarly programmed defining a step-by-step procedure or algorithm for solving the desired mathematical problem.

The calculator is designed so that lower-order subroutines may be nested to a level of five in higher-order functions. For instance, the 'Polar to Rectangular' function uses the sin routine which uses multiply which uses add, etc.

Addition and subtraction

The most elementary mathematical operation is algebraic addition. But even this is relatively complex—it requires comparing signs and complementing if signs are unlike. Because all numbers in the Model 9100A are processed as true floating point numbers, exponents must be subtracted to determine proper decimal alignment. If one of the numbers is zero, it is represented in the calculator by an all-zero mantissa with zero exponent. The difference between the two exponents determines the offset, and rather than shifting the smaller number to the right, a displaced digit-by-digit addition is performed. It must also be determined if the offset is greater than 12, which is the resolution limit.

Although the display shows 10 significant digits, all calculations are performed to 12 significant digits with the two last significant digits (guard digits) absorbing truncation and round-off errors. All registers are in core memory, eliminating the need for a large number of flip-flop registers. Even with the display in 'Fixed Point' mode, every computed result is in storage in 12 digits.

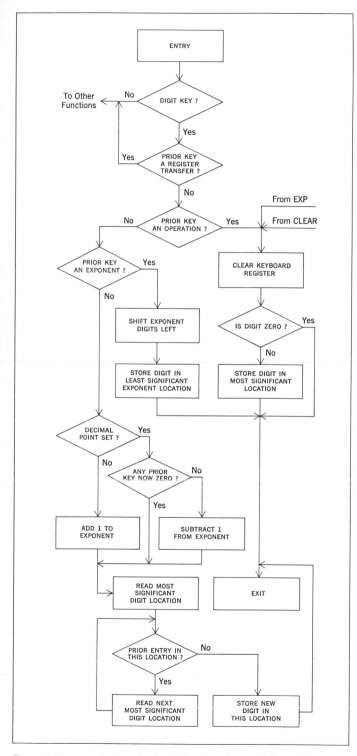

Fig. 15. Flow chart of a simple digit entry. Some of these flow paths are used by other calculator operations for greater hardware efficiency.

Multiplication

Multiplication is successive addition of the multiplicand as determined by each multiplier digit. Offset in the digit position flip-flops is increased by one after completion of the additions by each multiplier digit. Exponents are added after completion of the product. Then the product is normalized to justify a carry digit which might have occurred.

Division

Division involves repeated subtraction of the divisor from the dividend until an overdraft occurs. At each subtraction without overdraft, the quotient digit is incremented by one at the digit position of iteration. When an overdraft occurs, the dividend is restored by adding the divisor. The division digit position is then incremented and the process continued. Exponents are subtracted after the quotient is formed, and the quotient normalized.

Square root

Square root, in the Model 9100A, is considered a basic operation and is done by pseudo division. The method used is an extension of the integer relationship.

$$\sum_{i=1}^{n} 2i - 1 = n^2$$

In square root, the divisor digit is incremented at each iteration, and shifted when an overdraft and restore occurs. This is a very fast algorithm for square root and is equal in speed to division.

Circular routines

The circular routines (sin, cos, tan), the inverse circular routines (arcsin, arccos, arctan) and the polar to rectangular and rectangular to polar conversions are all accomplished by iterating through a transformation which rotates the axes. Any angle may be represented as an angle between 0 and 1 radian plus additional information such as the number of times $\pi/2$ has been added or subtracted, and its sign. The basic algorithm for the forward circular function operates on an angle whose absolute value is less than 1 radian, but prescaling is necessary to indicate quadrant.

To obtain the scaling constants, the argument is divided by 2π, the integer part discarded and the remaining fraction of the circle multiplied by 2π. Then $\pi/2$ is subtracted from the absolute value until the angle is less than 1 radian. The number of times $\pi/2$ is subtracted, the original sign of the argument, and the sign upon completion of the last subtraction make up the scaling constants. To preserve the quadrant information the scaling constants are stored in the core memory.

The algorithm produces $\tan \theta$. Therefore, in the Model 9100A, $\cos \theta$ is generated as

$$\frac{1}{\sqrt{1 + \tan^2\theta}}$$

and $\sin \theta$ as

$$\frac{\tan \theta}{\sqrt{1 + \tan^2\theta}}$$

Sin θ could be obtained from the relationship $\sin \theta = \sqrt{1 - \cos^2\theta}$, for example, but the use of the tangent relationship preserves the 12 digit accuracy for very small angles, even in the range of $\theta < 10^{-12}$. The proper signs of the functions are assigned from the scaling constants.

For the polar to rectangular functions, $\cos \theta$ and $\sin \theta$ are computed and multiplied by the radius vector to obtain the X and Y coordinates. In performing the rectangular to polar function, the signs of both the X and Y vectors are retained to place the resulting angle in the right quadrant.

Prescaling must also precede the inverse circular functions, since this routine operates on arguments less than or equal to 1. The inverse circular algorithm yields arctangent functions, making it necessary to use the trigonometric identity.

$$\sin^{-1}(x) = \tan^{-1}\frac{x}{\sqrt{1 - x^2}}$$

If $\cos^{-1}(x)$ is desired, the arcsin relationship is used and a scaling constant adds $\pi/2$ after completion of the function. For arguments greater than 1, the arccotangent of the negative reciprocal is found which yields the arctangent when $\pi/2$ is added.

Exponential and logarithms

The exponential routine uses a compound iteration algorithm which has an argument range of 0 to the natural log of 10 (ln 10). Therefore, to be able to handle any argument within the dynamic range of the calculator, it is necessary to prescale the absolute value of the argument by dividing it by ln 10 and saving the integer part to be used as the exponent of the final answer. The fractional part is multiplied by ln 10 and the exponential found. This number is the mantissa, and with the previously saved integer part as a power of 10 exponent, becomes the final answer.

The exponential answer is reciprocated in case the original argument was negative, and for use in the hyperbolic functions. For these hyperbolic functions, the following identities are used:

$$\sinh x = \frac{e^x - e^{-x}}{2}$$

$$\cosh x = \frac{e^x + e^{-x}}{2}$$

$$\tanh x = \frac{e^x - e^{-x}}{e^x + e^{-x}}$$

Natural logarithms

The exponential routine in reverse is used as the routine for natural logs, with only the mantissa operated upon. Then the exponent is multiplied by ln 10 and added to the answer. This routine also yields these $\log_{10}$ and are hyperbolic functions:

$$\mathrm{Log}_{10}x = \frac{\ln x}{\ln 10}$$

$$\sinh^{-1}(x) = \ln(x + \sqrt{x^2 + 1})$$
$$\cosh^{-1}(x) = \ln(x + \sqrt{x^2 - 1})$$
$$\tanh^{-1}(x) = \ln\sqrt{\frac{1 + x}{1 - x}}$$

The $\sinh^{-1}(x)$ relationship above yields reduced accuracy for negative values of x. Therefore, in the Model 9100A, the absolute value of the argument is operated upon and the correct sign affixed after completion.

Accuracy

It can be seen from the discussion of the algorithms that extreme care has been taken to use routines that have accuracy commensurate with the dynamic range of the calculator. For example; the square root has a maximum possible relative error of 1 part in 10^{10} over the full range of the machine.

There are many algorithms for determining the sine of an angle; most of these have points of high error. The sine routine in the Model 9100A has consistent low error regardless of quadrant. Marrying a full floating decimal calculator with unique mathematical algorithms results in accuracy of better than 10 displayed digits.

Section 5

Processors with stack memories
(zero addresses per instruction)

This section contains only computers which use a stack memory in their Pc and hence are denoted Pc.stack. Although the implementation details differ, they are based on the common idea of a stack as described in Chap. 3, page 62. Several theory or language-based processors—IPL-VI and EULER—use a stack in Mp. However, for these language-based machines the stack is not the main design theme as it is with the other computers in Table 1. In fact, data in IPL-VI are organized (Chap. 30) about lists, which are a more general data structure than stacks. A stack permits push and pop operations to be performed on the top of the stack; a list permits push and pop operations to be performed on each cell of the list (they are then called insert

Table 1   Pc.stack computers

Company or basis computer name	Disclosure date[a]	Delivery date	Ancestry	Relative power	References
English Electric KDF 9	/60	4/63	George[c]	. . .	AllmR62, DaviG60, HambC62
Burroughs (Paoli, Pa.)					
D825[d]	/61				AndeJ62
D830[d]			extended performance D825		
B 8500[e]	4/66[b]	1/67[f]	developed at laboratory producing D825, D830	20–30	
Burroughs (Pasadena, Calif.)					
B 5000	/62	2/63		1/2	AllmR62, BartR61, BockR63, CarlC63,
B 5500		11/64	successor to B 5000	1–1.7[g]–1.9[g]	LoneW61, HaucE68
B 6500		1/68[f]	B 5500 based with improved multi- and shared-programmed mapping	5–6	
B 7500			extended performance B 6500	10	
Theory or language-based:					
IPL-VI	/58		language: IPL-IV, V		ShawJ58
EULER	/67	/67	language: EULER(ALGOL+)		WebeH67, WirtN66a,b
ALGOL			language: ALGOL		AndeJ61
Argonne Laboratory					
IPL-VC			language: IPL-V		HodgD64

[a] First edition of manual, or a paper, or the appearance in *Adams Computing Characteristics Quarterly*.
[b] Still evolving. B 8501 was discontinued in 1968.
[c] George, University of New South Wales, interpreter using Polish notation and a stack. Circa 1957 [Hamblin, 1962].
[d] Produced for command and control (military) applications.
[e] B 8500 is a system name; the Pc is a B 8501.
[f] Reported. Actual delivery unknown.
[g] Dual processor.

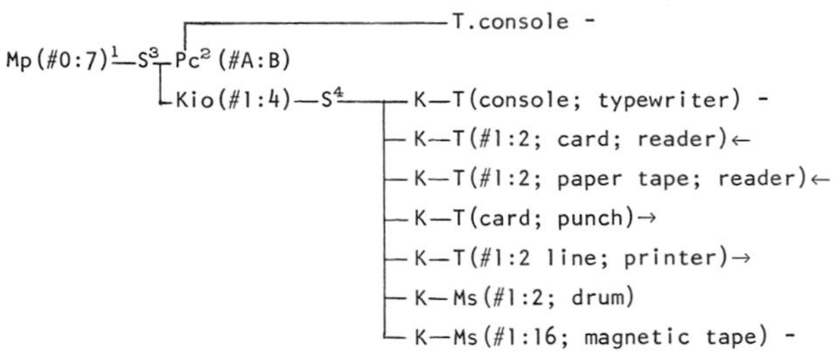

```
                                ┌──────────────── T.console -
Mp(#0:7)¹─S³─Pc²(#A:B)
              └Kio(#1:4)─S⁴──┬─K─T(console; typewriter) -
                             ├─K─T(#1:2; card; reader)←
                             ├─K─T(#1:2; paper tape; reader)←
                             ├─K─T(card; punch)→
                             ├─K─T(#1:2 line; printer)→
                             ├─K─Ms(#1:2; drum)
                             └─K─Ms(#1:16; magnetic tape) -

¹Mp(core; 4 μs/w; 4096 w; (48,3) b/w)
²Pc(stack; 12 b/syllable; 6 b/char; data: si,sf,bv,w,char.
    string; (1 ~ 2) syllable/instruction; Mps(~ 4 w) ante-
    cedents: 'ALGOL language; descendants; 'B 5000, B 6500,
    B 7500; technology: transistor; ~(1961 ~ 1963)))
³S(from: 2 Pc,4 K; to: 8 Mp; concurrency: 4)
⁴S(from: 4 Kio; to: KT,KMs; concurrency: 4)
```

Fig. 1. Burroughs B 5000 PMS diagram.

and delete, respectively). Thus a list is like a nested set of overlapping stacks. EULER (Chap. 32) uses a stack to store temporary data and subroutine calls both when compiling and when interpreting the compiled program. However, the language-based machines can still be studied profitably with the stack in mind.

The following comments will be directed to the P.stack computers manufactured by both English Electric and Burroughs. There are three basic P.stack computer families: B 5000 → B 5500 → B 6500/B 7500; D825 → D830 → B 8500; and KDF9. Each root member was made available at about the same time by Burroughs (Pasadena, Calif.), Burroughs (Paoli, Pa.), and English Electric. The IBM Corporation later responded with a proposed Pc.stack, but the machine never entered the production phase.

The Pc.stack is a major alternative to the main line organization of 1 address per instruction (augmented with index registers or general registers). It tries to capitalize on the hierarchical character of computation to avoid having to give memory shuffling instructions explicitly. In Chap. 3, page 64, we gave a comparison of a trivial computation using a stack and a general-register organization, in order to make clear the case

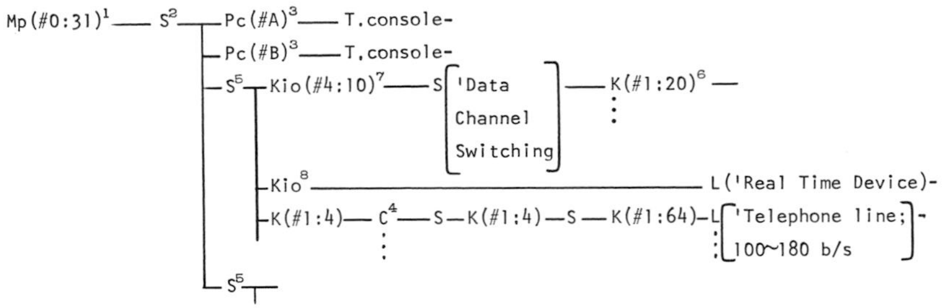

```
Mp(#0:31)¹──── S²────┬──Pc(#A)³──── T.console-
                     ├──Pc(#B)³──── T.console-
                     ├─S⁵─┬Kio(#4:10)⁷────S┌'Data       ┐──K(#1:20)⁶──
                     │    │                │ Channel     │    ⋮
                     │    │                │ Switching   │
                     │    │                └─────────────┘
                     │    ├─Kio⁸ ─────────────────────────── L('Real Time Device)-
                     │    └─K(#1:4)── C⁴──S─K(#1:4)─S ─ K(#1:64)─L┌'Telephone line;┐-
                     │       ⋮                                    └:  100~180 b/s   ┘
                     └─S⁵─┬

¹Mp((core; 1.2 μs/w)│(thin film; .6 μs/w); 16 kw; 51 b/w)
²S(32 Mp; 4(Pc,K,S); concurrency: 4)
³Pc(stack; technology: integrated circuits; ~ 1969; data: sf,df,i,char.string,
    boolean vector, address integer; 4,6,8 b/char)
⁴C('Data Communications Processor)
⁵Identical peripheral structures possible with two switches
⁶See Figures 3, 4, and 5.
⁷Kio('Input/Output Multiplexor)
⁸Kio('Real Time Adapter)
```

Fig. 2. B 6500, B 7500 PMS diagram.

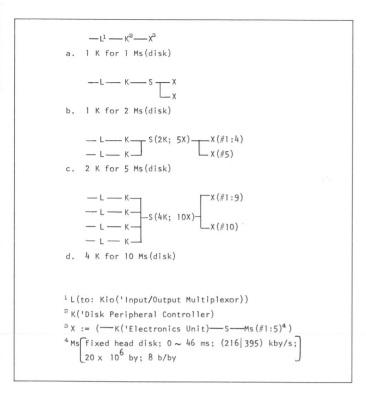

```
        —L¹—K²—X³
  a.  1 K for 1 Ms(disk)

        —L—K—S ⊤X
                  └X
  b.  1 K for 2 Ms(disk)

        —L—K ⊤S(2K; 5X) ⊤X(#1:4)
        —L—K ┘          └X(#5)
  c.  2 K for 5 Ms(disk)

        —L—K ⊤            ⊤X(#1:9)
        —L—K ┤            │
             ├S(4K; 10X) ─┤
        —L—K ┤            │
        —L—K ┘            └X(#10)
  d.  4 K for 10 Ms(disk)
```

¹L(to: Kio('Input/Output Multiplexor))
²K('Disk Peripheral Controller)
³X := (—K('Electronics Unit)—S—Ms(#1:5)⁴)
⁴Ms[fixed head disk; 0 ~ 46 ms; (216|395) kby/s; 20 x 10⁶ by; 8 b/by]

Fig. 3. Burroughs B 6500, B 7500 Ms (disk) PMS diagrams.

for stacks. However, we did not there attempt any analysis. It has been asserted [Amdahl et al., 1964a] that the Pc.stack derives its power only from its having some fast-working memory in the Pc, thus that it is dominated by the general-register organization. Our own feeling is that the compile and compiled program execution times for the Pc.stack are indeed impressive. However, no definitive analysis has been published, as far as we know. Pc.stack is certainly an organization that rates serious study by any computer designer.

The PMS structure of the examples

The PMS structure diagram of the B 5000 and B 6500/B 7500 (Figs 1 to 5) should be compared with Burroughs own structure representation (Chap. 22, page 268). The D825 structure is similar; it is given in Chap. 36, page 447. All the Burroughs computers in Table 1 have the multiprocessor structure.

Burroughs was probably the first computer company to take matters of the structure and organization seriously. The D825 hardware and software were designed for military command

and control applications which demand very high uptime and availability. As various computer components in the structures fail, continuous operation is possible at a reduced level through the fail-soft design. However, to our knowledge, no published account exists on how well this design works in practice from a performance and reliability viewpoint. The philosophy and details of the D825 software and hardware are discussed in Chap. 36.

The structures in the B 6500, especially, allow Kio's to be freely assigned to any T or Ms, thereby achieving better equipment utilization. The S(16 Mp; 16 P) is probably overdesigned in the Burroughs B 6500 computers. These structures generally have a maximum 4(P + Kio), although the design is based on 16(P + Kio). The Kio's (Chap. 22) may be overdesigned, too, since a K capable of controlling a simple T.card_reader can also control a complex Ms.disk or Ms.magnetic_tape.

The PMS structure of the English Electric KDF9 (Fig. 6) is fairly simple. The 16 K's for direct memory access appear

```
  —L¹—K²—S³—Ms[#0:7; magnetic tape;
               9 ~ 144 kchar/s; 6|8
               b/char; 200|556|800|
               1600 char/in; forward
               and reverse motion]
  a.  1 K for 8 Ms(magnetic tape)

  —L—K ⊤S(2 K; 10 Ms)——Ms(#0:9; magnetic tape)
  —L—K ┘
  b.  2 K for 10 Ms(magnetic tape)

  —L—K ⊤S(4 K; 16 Ms)——Ms(#0:15; magnetic tape)
  —L—K ┤
  —L—K ┤
  —L—K ┘
  c.  4 K for 16 Ms(magnetic tape)
```

¹L(to: Kio('Input/Output Multiplexor))
²K('Peripheral Controller)
³S(1K; 8 Ms; bus)

Fig. 4. Burroughs B 6500, B 7500 Ms (magnetic tape) PMS diagrams.

```
      —L¹ ——— K ——— T(console; keyboard, printer)—
      — L ——— K ——— T(card; reader) ←
      — L ——— K ——— T(card; punch) →
      — L ——— K ——— T(paper tape; reader) ←
      — L ——— K ——— T(paper tape; punch) →
      — L ——— K ——— T(CRT; display) →
      — L ——— K ——— T(line; printer) →


      ¹L(to: Kio('Small Peripheral Control))
```

Fig. 5. Burroughs B 6500, B 7500 peripheral K——T PMS diagrams.

to be both overdesigned (or overly general) and there are too few of them. The limit of only 16(T + Ms) components is small, especially considering that the KDF9 is to be time-shared from several consoles.

The ISP of the examples

The comparison of Pc.stack, Pc.1address, and Pc.general‿registers (page 64) makes the assumption that an unlimited

```
                              ┌————————————— T.console—
Mp(#0:15)¹ — S² — Pc³ —┐      │
                        └——K(#1) — S —— Ms(magnetic tape)—
                        │ ├—K——————————— T(typewriter)—
                        │ ├—K——————————— T(paper tape)—
                        │ ⋮
                        └—K⌐#16; data vector
                           └transmission to Mp┘
```

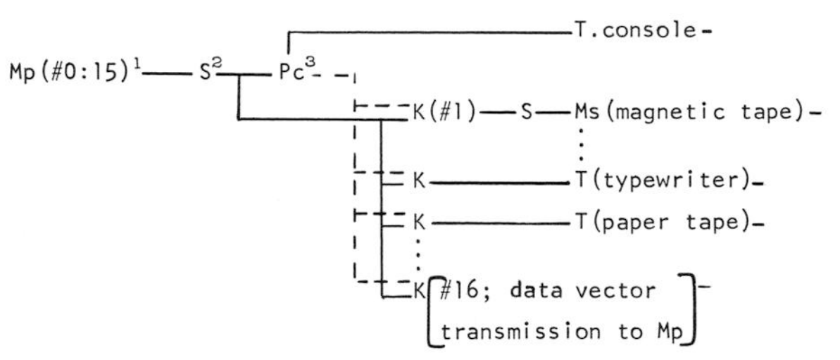

¹Mp(core; 6 μs/w; 4 ~ 32 kw; 48 b/w)
²S(16 Mp; 16(P,K); concurrency: 1)
³Pc(stack; 8 b/syllable; 0 ~ 1 address/instruction; 6 b/char;
   technology: transistor; data: syllable, char, w, bv, si,
   di, sf, df, hw; 1-3 syllables/instruction; operators: +,
   -, ×, /, ∧, ∨, ⊕,←{char.string}, Mp ← stack, stack ← Mp;
Mps('Subroutine Jump Nesting Store[0:7]<0:17> *stack*;
   'Nesting Store[0:15]<0:47> *arithmetic stack*;
   'Q-store[0:15]<0:17,18:31,32:48> *Q-store is used for*
   indexing, and contains a counter, an increment, and a
   modifier))

Fig. 6. English Electric KDF9 PMS diagram.

hardware stack resides in Pc. The B 5500 has a local M.stack in Pc of 4 words. The size and number of stacks, and their use by software, are most important. The IPL-VI machine has any number of stacks since the front of each list is a stack. The KDF9 (Fig. 6) has two independent stacks: one for arithmetic expression evaluation and one for holding subroutine return addresses. The DEC 338 P.display (Chap. 25) uses a stack for storing subroutine return addresses.

Unfortunately, we have not been able to include a discussion of the "cactus stack" of the B 6500, which is a data structure more like a list [Hauck and Dent, 1968]. The Hauck and Dent paper describes both the relationship to a Pc.stack and its relevance to program mapping and memory management for multiprogramming.

The C('D825) parameters are given in Fig. 7. The D825 ISP differs from other Pc.stack computers in that the data, d, for operations can be in either of two places, the stack or Mp. Consider the unary or binary operations:

```
C('Burroughs D825; multiprocessor structure;
   S(cross-point; 16 M; 16(Pc,Kio))
   Mp(4.33 μs/w; 65 kw; (48,1 parity) b/w);
   S(cross-point; 4 Kio; 64 (T,Ms));
   T(console, paper tape, printer, card, time, communication
     link);
   Ms(drum, disk, magnetic tape);
   Kio(#1:4);
   Pc(#1:2; 12 b/syllable; stack; 0 ~ 3 addresses/instruction;
      multiprogrammed; data: (integer, floating, single char-
      acter, fractional precision word, boolean vector); opera-
      tions: (+, -, ×, /, ∧, ∨, ⊕, ¬, round, {si} ← {sf}, abs,
      negate,-abs);
      instruction-size: (1 ~ 7) syllable;
      operation-code-size: 5/12 syllable;
      address-size: (7/12 + 0 ~ 6) syllable;
      operation forms: (d3 ← d1 b d2, d2 ← u d1);
      variable addresses: (stack, Mp[syllable + BAR],Mp[syllable
                           + BAR + X[A] + X [B] + X [C]]);
   Mps('Stack/S, Index Registers[1:15]/X[1:15],
      'Index Comparison Limit Registers[1:15],
      'Base Address Registers/BAR,
      'Program Address Register/PAR,
      'Program Counter/PC)))
```

Fig. 7. Burroughs D825 PMS diagram.

$$d_2 \leftarrow u \; d_1$$
$$d_3 \leftarrow d_1 \; b \; d_2$$

In either of these cases d_1, d_2, or d_3 can be the top of Stack/S; or Mp[Address + Base Address + [Σindex registers [A,B,C]]]. This flexibility allows the Pc to behave as a 0, 1, 2, or 3 address per instruction processor.

The B 5000 is more conventional than the D825 in its use of stacks (see references, Table 1). There are only load and store (that is, push and pop instructions) to transfer data between Mp and one stack. Actually, the B 5000 has several important features that make it worthy of study:

1   The stacks.

2   Data-type specification. A data type is declared by placing a type identifier with the data. Thus, for example, there is one add operation for both fixed and floating point, the data telling which addition is to take place.

3   Multiprogram mapping. Descriptors are used to access variables (scalars, vectors, and arrays). This indirect addressing technique allows multiprogramming; however, the reader should note that the data are not protected against other accesses (corrected in the B 6500).

4   Failure of the Pc.stack for character processing. The B 5000 has a character mode to allow processing of string data, and the stack is not used in this mode. In effect, a separate string processing ISP is incorporated in the Pc.

5   Multiprocessing. A B 5000 can have two Pc's.

A command structure for complex information processing

The IPL-VI (Chap. 30) is discussed in Part 4, Sec. 4 page 348 as a language-based processor.

Microprogrammed implementation of EULER on IBM System/360

EULER (Chap. 32) is discussed in Part 4, Sec. 4 page 348 as a microprogrammed, language-based processor.

Chapter 21

Design of an arithmetic unit incorporating a nesting store[1]

R. H. Allmark / J. R. Lucking

Summary This paper describes the arithmetic unit of a computer whose order code is based on the *Reverse Polish* algebraic notation. The order code has been realised by causing the arithmetic unit to operate on data stored in the most accessible registers of a nesting store; these registers are of the transistor flip-flop type but are backed up by sixteen fast magnetic core registers. The functions are performed as micro-programmes of transfers between the registers in the arithmetic unit, and the necessary arrangement of transfer paths, logical gates and arithmetic circuits is described. The number system is binary, using the two's-complement representation of negative numbers. Automatic floating-point operations are included which use an autonomous unit to perform the shifts required.

Introduction

The arithmetic unit of a general purpose digital computer contains circuits to perform at least the basic operations of addition, subtraction, multiplication and division. In many machines it is possible to use some of the registers in the arithmetic unit as temporary storage for the partial results arising during a calculation; thus the accumulator of a one-address machine is used to store the result of the last arithmetic operation. The arithmetic unit described in this paper uses a nesting store, operating on the last-in-first-out principle, for the storage of its data and partial results. The nesting store consists of a stack of cells, of which only the most accessible supply data to the arithmetic unit, the results are automatically returned to the most accessible cells and the original operands erased, less accessible information being moved into the cells made vacant by the operation.

The computer and its order code

The arithmetic unit is part of a general purpose synchronous system, working in the parallel mode, with main core storage of (up to) 32, 768 48-bit words, and provision for the time sharing of up to 4 programmes. The order code of the computer is based

[1]*Proc. IFIP Congr. 62*, pp. 694–698, 1962.

on the *Reverse Polish* algebraic notation, and contains four groups of operations:

a Transfers between the arithmetic unit and the main store.

b Arithmetic, logical and manipulative functions on data in the nesting store.

c Conditional and unconditional jump instructions used to interrupt the normal sequencing of instructions.

d Instructions for controlling the operation of the various peripheral devices which may be attached to the machine.

Main store transfers include instructions for transferring half and full-length words to the most accessible cell of the nesting store, information already in the stack being retained by transfer to the less accessible cells. The contents of the most accessible cell of the stack may be stored in the main store; they are then automatically erased from the stack while information is moved from the less accessible cells to a more accessible position.

Arithmetic operations also feature the transfer of data in the nesting store so that the operands are destroyed, the results are left in the most accessible cell (or cells), and data not involved in the operation are moved to fill any vacated cells.

Thus the programme for evaluating

$$f = (a - b)/(c + de)$$

may be written:

fetch a,
fetch b,
subtract (forming $a - b$ in the most accessible cell
    and erasing both a and b from the stack),
fetch d,
fetch e,
multiply (forming de in the most accessible cell,
    erasing d and e, and thus leaving $a - b$ in the
    second most accessible cell),

fetch c,
add (forming $c + de$)
divide (forming f),
store as f(leaving the nesting store in the same state
    as before the *fetch a* instruction).

For instructions, the 48-bit word has been divided into 6 *syllables* of eight bits each, and these are then treated as a continuous sequence of variable length instructions. Arithmetic operations are specified by single syllable instructions, but main store transfers require three syllables to accommodate both the address and the address modifying information of the word to which they refer; jump instructions also have three syllables. Two-syllable instructions include the peripheral transfers, and instructions for processing address modifiers and performing shifts. The first syllable of every instruction contains two bits whose values specify the length of the instruction; the redundant case being used to differentiate between main store transfers and jump instructions. The first syllable of an instruction contains enough information to specify any arithmetic unit operation required; thus in the machine, each instruction is treated by two controls; the first or *Store Control* organising the fetching and storing of information in advance of the second or *Arithmetic Unit Control* which completes the instruction on the information in the first syllable.

Range of functions

The allocation of bits to the instructions described above allows 64 possible functions, of which 59 are used to specify the wide range of operations needed in a general purpose computer.

As well as the normal single-length fixed-point arithmetic operations, functions have been provided for the addition and subtraction of double-length numbers. These simplify the programming of multi-length operations as well as giving increased accuracy. For normal scientific and engineering calculations automatic floating-point facilities are available. A single length word may represent a floating-point number with a 40-bit fractional part f, and an 8-bit characteristic c; the value of the number is then $f2^{c-128}$. The fractional part is limited to the range $-1 \leq f < -\frac{1}{2}$, or $1 > f \geq \frac{1}{2}$, or $f = 0$ when c is also zero. All floating-point operations assume that operands are in this standard form and give correctly rounded results in standard form. Functions for the addition and subtraction of double-length floating-point numbers have been provided, as these give increased accuracy and stability in many matrix operations.

An increase in operating speed and a saving of instructions are effected by the use of instructions which re-order the position of information in the most accessible cells of the nesting store, including reversing and cycling operations. The normal logical operations are provided.

All arithmetic operations in the arithmetic unit are carried out on binary numbers using the two's-complement notation for negative numbers; instructions being provided for the conversion to and from binary of information stored as 6-bit characters in other radix systems. For the convenience of the programmer, double-length numbers are stored in the arithmetic unit with their more significant half in a more accessible cell; the sign of the less significant half is ignored and is set positive after all double-length operations.

The nesting store

Although the concept of a nesting store is similar to that of a rifle magazine where the addition of a cartridge displaces those already there, movement of information only occurs in the three most accessible cells of the nesting store, which are transistor flip-flop registers forming part of the arithmetic unit. The less accessible cells are core registers which are addressed in a sequential manner by a reversible counter. Reading from these cores reduces the count by one, thus selecting the next word; the read-out is destructive so that the cores are in the correct state for a subsequent writing operation, which is the reverse of a read. The access time of the cores is reduced by providing separate counters and reading and writing mechanisms for the odd and even numbered rows of cores; thus when reading or writing from odd rows the addressing mechanism for the next even row is set, so that it is available for immediate use. Thus with a simple one core per bit system successive *reads* can be made at 1 μsec intervals and *writes* at 2 μsec intervals; as these operations are performed in parallel with the functioning of the arithmetic unit, their times do not increase the time required to complete the functions.

The arithmetic unit

As shown in Fig. 1, there are six full length transistor flip-flop registers in the arithmetic unit; there are also two 8-bit registers used when performing floating-point operations. The main facilities associated with these registers are as follows.

$W1$, $W2$ and $W3$ are the three most accessible cells of the nesting store; transfers to the core part of the nesting store, being

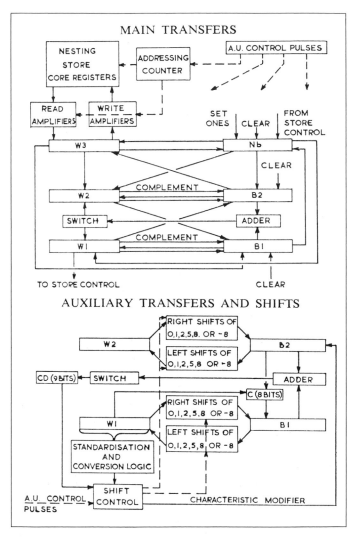

Fig. 1. Block diagram of the arithmetic unit. Full lines represent information transfers; dotted lines represent control pulses. All registers are 48-bits long unless otherwise stated.

made via $W3$. $W1$ and $W2$, together with $B1$ and $B2$, form a double-length shifting register which may be used as two independent single-length shifting registers.

$B1$ and $B2$ are the inputs to the 48-bit adder whose output may be routed to $W1$, $W2$, or to the characteristic difference register CD.

The adder contains 13 carry-skip stages which reduce the carry propagation time to a maximum of 150 nsec. Subtraction is performed by adding the minuend's complement to the subtrahend with a carry inserted into the right-most adder stage.

Nb acts as a buffer between store control and the arithmetic unit, and together with $B1$ and $B2$, is used in nearly every function.

Arithmetic unit control interprets each instruction as a sequence of timed pulses along lines which activate the various transfers etc., between the registers. The sequences have been constructed so that many operations are performed simultaneously, reducing the overall time to a minimum; thus the function *single-length fixed-point add* is performed by:

   i   Transferring $W1$, $W2$, $W3$ to $B2$, $B1$ and Nb respectively, simultaneously commencing a read from the nesting store, clearing the carry inserted into the right-most adder stage and switching the adder's output to $W1$.

   ii   Adding and simultaneously transferring Nb to $W2$.

Each step takes 0.5 μsec and by the end of the last step, $W3$ has been refilled from the core nesting store.

To speed up multiplication and division, these functions are carried out in a separate unit employing the stored carry principle, but the results are finally assimilated within the arithmetic unit.

A similar arithmetic unit operating only on single-length numbers could be designed using only four full-length registers. At least five registers are required to perform the function which interchanges the contents of the two most accessible cells in the nesting store with those of the next most accessible pair. The sixth register enables all double-length arithmetic operations to be performed without writing information back into the nesting during the function; this would have complicated the sequences and increased the time for the functions.

When determining the arrangement of transfer paths between the various registers, it was found sufficient to consider only the double-length functions which required complicated or lengthy sequences; in particular the function for adding two double-length floating numbers had great influence.

An overflow indication is set on fixed-point addition and subtraction if the sign of the result differs from that expected, and on floating-point operations if the characteristic exceeds the maximum allowable; shifting may also cause overflow.

Shift control

Shifting operations are effected by transfers between $W1$ (and/or $W2$) and $B1$ (and/or $B2$), and back again. The shift transfer paths from the W to the B registers provide right shifts of 0, 1, 2, 5 or 8 places, and a left shift of 8 places; the paths from the B to the W registers provide the same shifts in the reverse direction. The two sets of shift paths are used alternately, those from the W registers being used first; all shifts are terminated using a path into the W registers. Shifts of a large number of places are accomplished by a series of shifts of eight places in the appropriate direction until the number of places remaining is less than eight; if necessary the number is then transferred back into the W registers: the remaining shifts, or the whole shift if the number of places is less than eight, is then completed by a transfer to the B registers and back again using two appropriate paths. With the shifts available, extension of the B registers by two bits at the right-most end enables any shift to be performed without loss of accuracy. In double-length arithmetic shifts, the sign digit of the less significant word is by-passed. When a shift is to be performed, the number of places and the type of shift are transferred into a semi-autonomous unit, called the shift control, which is then supplied with a string of command pulses by the arithmetic unit control; shift control then re-routes these pulses to perform the transfers necessary to obtain the shift.

When performing floating-point addition and subtraction, shifts are required to equalize the characteristics of the two numbers; the amount of shift is calculated by a modified subtraction, operating on the characteristic positions of the two numbers. After the addition, the shift required to restore the result to standard form is determined by logical circuits which interpret the pattern of bits in $W1$ into shift information. The number of shifts performed during this standardising operation is made available to the arithmetic unit control for use in forming the correct characteristic of the result.

The character conversion operations to, and from, binary are accomplished by shift control, using a method involving successive shifting of the character word, and adding or subtracting portions of the radix word.

Examples of sequences

To illustrate the working of the arithmetic unit, two sequences are described.

a $-\mathbf{D}$, (i.e. subtract the double-length fixed-point number in $W1$ and $W2$ from the number in $W3$ and the most accessible core register of the nesting store).

   i Transfer $W1$, $W2$, $W3$ to $B2$, $B1$ and Nb respectively, simultaneously reading from the core nesting store.

  ii A dummy pulse.

  iii Transfer the complement of $W2$ to $B2$ (but setting the sign of $B2$ positive), transfer $W3$ directly to $B1$ ($W3$ has by now been filled with fresh data), switch the adder's output to $W2$, inserting a carry into the right-most adder stage, and read from the nesting store.

  iv Add.

  v Transfer the complement of $W1$ to $B1$ and Nb to $B2$, switch the adder's output to $W1$ and insert a carry into the right-most adder stage if $W2$ is negative.

  vi Add, simultaneously clearing the sign of $W2$.

b $+\mathbf{F}$ (i.e. add the two single-length floating numbers in $W1$ and $W2$).

   i Transfer the complement of $W1$ to $B1$, transfer $W2$ to $B2$ and switch the adder's output to register CD.

  ii Store the characteristic of $W1$ in the eight-bit register C and add.

  iii Clear the characteristic positions of $W1$, simultaneously transferring CD into the shift number register in shift control. This latter operation is such that the shift register contains minus the difference in characteristics.

  iv Clear the characteristic of $W2$, and if $W1$ is about to be shifted, determined by the sign digit of CD, replace the contents of C by the characteristic of $B2$; thus C contains the larger characteristic.

  v Supply control pulses to shift control and thus perform the required right-shift of eight $W1$ or $W2$.

  vi Having completed the shift, transfer $W1$, $W2$ and $W3$ to $B2$, $B1$ and Nb respectively, simultaneously switching the adder's output to $W1$, clearing the carry into the right-most adder stage and reading from the core-nesting store.

  vii Add the fractional parts, simultaneously transferring Nb to $W2$.

  viii Supply control pulses to shift control so as to cause it to enter the standardization procedure and perform the shifts required.

  ix Store the complement of the number of left-shifts performed in (viii) in the characteristic position of $B2$, transfer C to the characteristic position of $B1$, switch the adder to $W1$.

  x Perform a special add operation which only affects the characteristic positions of $W1$.

The sum is thus formed in $W1$. Rounding the answer is carried out using two special control pulses which complete all floating-point operations, these call up logic to deal with the cases when the rounding operation necessitates re-standardization of the result.

Conclusions

The advantages of a machine incorporating a nesting store in the arithmetic unit are:—

i   The machine is simple to programme using the *machine language*.

ii Programmes are faster, since many main store transfers are eliminated, and the access time of the nesting store is virtually zero. They are more compact because less information is required to specify many instructions.

iii As the operation of the arithmetic unit is largely independent of the main store, their controls may readily be separated. This allows store control to process instructions whilst the arithmetic unit control processes a prior instruction, thereby leading to faster execution of the programme.

The main disadvantage is an increase in the order of complexity involved.

References

AllmR62; DaviG60; HaleA62

Chapter 22

Design of the B 5000 system[1]

William Lonergan / Paul King

Computing systems have conventionally been designed via the 'hardware' route. Subsequent to design, these systems have been handed over to programming systems people for the development of a programming package to facilitate the use of the hardware. In contrast to this, the B 5000 system was designed from the start as a total hardware-software system. The assumption was made that higher level programming languages, such as ALGOL, should be used to the virtual exclusion of machine language programming, and that the system should largely be used to control its own operation. A hardware-free notation was utilized to design a processor with the desired word and symbol manipulative capabilities. Subsequently this model was translated into hardware specifications at which time cost constraints were considered.

Design objectives

The fundamental design objective of the B 5000 system was the reduction of total problem through-put time. A second major objective was facilitation of changes both in programs and system configurations. Toward these objectives the following aspects of the total computer utilization problem were considered:

Statement of problems in higher-level machine-independent languages; efficiency of compilation of machine language; speed of compilation of machine language; program debugging in higher-level languages; problem set-up and load time; efficiency of system operation; ease of maintaining and making changes in existing programs, and ease of reprogramming when changes are made in a system configuration.

Design criteria

Early in the design phase of the B 5000 system the following principles were established and adopted:

Program should be independent of its location and unmodified as stored at object time; data should be independent of its location; addressing of memory within a program should take advantage of contextual addressing schemes to reduce redundancy; provisions

[1]*Datamation*, vol. 7, no. 5, pp. 28–32, May, 1961.

should be made for the generalized handling of indexing and subroutines; a full complement of logical, relational and control operators should be provided to enable efficient translation of higher-level source languages such as ALGOL and COBOL; program syntax should permit an almost mechanical translation from source languages into efficient machine code; facilities should be provided to permit the system to largely control its own operation; input-output operations should be divorced from processing and should be handled by an operating system; multi-programming and true parallel processing (requires multiple processors) should be facilitated, and changes in system configuration (within certain broad limitations) should not require reprogramming.

System organization

The B 5000 system achieves its unique physical and operational modularity through the use of electronic switches which function logically like telephone crossbar switches. Figure 1 depicts the basic organization of the system as well as showing a maximum system.

Master control program

A master control program will be provided with the B 5000 system. It will be stored on a portion of the magnetic drum. During normal operations, a small portion of the MCP will be contained in core memory. This portion will handle a large percentage of recurrent system operations. Other segments of the MCP will be called in from the magnetic drum, from time to time, as they are required to handle less frequently-occurring events, or system situations. Whenever the system is executing the master control program, it is said to be in the Control State. All entries to the Control State are made via 'interrupts.' A special operation is provided, which can only be executed when the system is in the Control State, to permit control to return to the object program it was executing at the time the 'interrupt' occurred.

The following are a few typical occurrences which cause an automatic 'interrupt' in the system: An input-output channel is

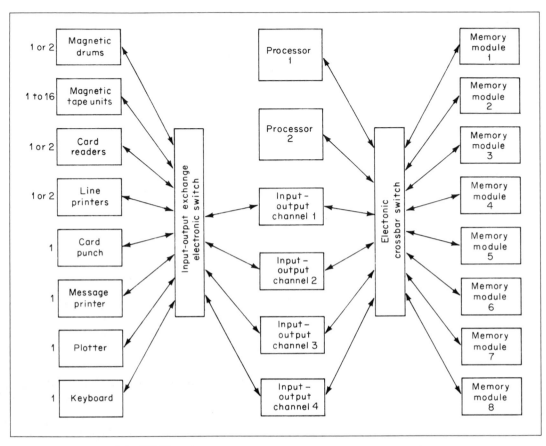

Fig. 1. Organization of the B5000 system.

available, an input-output operation has been completed or an indexing operation was attempted which violated the storage protection features built into the system.

In addition to processing interrupt conditions, the master control program handles fundamental parts of the total system operation such as the initiation of all input-output operations, tanking of input-output areas when required, file control, allocation of memory, scheduling of jobs (priority ratings, system requirements of each object program, and the present system configuration are considered), maintenance of an operations log and maintenance of a system description.

Operating modes

The B 5000 can either operate with fixed-length words or with variable-length fields. These two modes of operation are called the word mode and the character mode. For certain operations, a processor operating on words is most desirable and for other operations, a variable field length mode of operation is most desirable. By combining both abilities in one processor, a processor can operate in the mode most desirable for the operation at hand. In a B 5000 system, it is even possible for one processor to be operating in the word mode and the other in the character mode.

When operating in the word mode, a standard format for the data word is used as illustrated in Fig. 2.

Note that the standard word is an octal floating point word. However, the mantissa is treated as an integer rather than as a fraction (heretofore the reverse has been common practice). This provides two benefits: first, an integer has the same internal representation as its unnormalized floating point correspondent; and, second, the range of numbers that can be expressed, rather than being from 8^{+64} to 8^{-63}, is 8^{+76} to 8^{-51}. The first feature eliminates

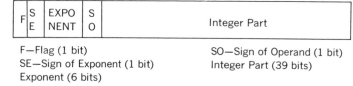

F—Flag (1 bit)
SE—Sign of Exponent (1 bit)
Exponent (6 bits)

SO—Sign of Operand (1 bit)
Integer Part (39 bits)

Fig. 2. Data word — word mode.

the need for fixed-to-floating point conversion; integers and floating point numbers can be mixed in arithmetic calculations. The second expands the range where trouble with range is most often encountered, namely, in numbers with extremely large magnitude.

The flag serves a dual purpose. The function of the flag depends on how the program references the data word. If the data word is a single variable and not an element of an array, the flag identifies the word as being operand, that is, a data word. If the word is an element of an array, the flag may be used to identify this particular element as an element of data which is not to be processed by the normal program (for example, a boundary point in mesh calculations).

When operating in the character mode, each data word consists of eight alphanumeric characters as illustrated in Fig. 3. Programs in the character mode can address any character in a word. Fields can start at any position in a word. A processor in a single operation can operate on fields of any length up to 63 characters long; operations on fields of greater length can easily be programmed. For example, two 57 character fields could be compared in a single operation.

There are two instances when the character mode operates with words of the type used in the word mode. Operations are provided in the character mode for converting numeric information in the alphanumeric representation to the standard word type of the word mode and vice versa. In both of these instances, the length of the alphanumeric fields being converted to or from the word mode type of word can be no greater than eight characters long. Again, conversion of fields of greater length can easily be programmed.

The purpose of the word mode is to provide the advantages of high-speed parallel operations, floating-point abilities and the inherent information density possible in a binary machine. In the first case, it is economically feasible to provide parallel operations in a word machine; the cost of parallel operations on variable length fields would be prohibitive. In the last case, a given size memory can contain over twenty percent more numeric information if that information is expressed in binary rather than binary-

coded decimal, and over eighty percent more information than can be expressed in six-bit alphanumeric representation.

The purpose of the character mode is to provide editing, scanning, comparison and data manipulative abilities (although addition and subtraction are also provided). The type of editing facilities provided obviate the need for the artificial "add-shift-extract-store" type of editing. For example, operations are provided for generalized insertion of editing symbols (such as blanks, decimal points, floating dollar signs, etc.) and for the substitution or suppression of any unwanted characters. For those interested in the new area of Information Processing Languages, the character mode is particularly well suited to list structures.

Program organization

Programs in the B 5000 are composed of strings of syllables. A syllable is the basic unit of the program and is twelve bits in length. The term "syllable" is used rather than instruction to distinguish it from conventional single-address or multi-address instructions. Each program word contains four syllables and they are executed sequentially in a left-to-right order within the program word, and sequentially by word. Branching is allowed to any syllable within a word. Before delving into some of the details of the internal operation of the B 5000 processor, it is necessary to discuss stacks, Polish notation, and the Program Reference Table.

The stack

The internal organization of single-address computers forces the wasting of both programming and running time for the storage and recall of the intermediate results in the sequence of computation. The data must be placed into the proper registers and memory cells before the operation can be executed, and their contents must often be completely rearranged before the next operation can be performed. Multi-address computers are constructed to make the execution of a few selected operations more efficient, but at the expense of building inefficiencies into all the rest. Automatic programming aids attack this problem indirectly: they relieve the programmer of the need to laboriously code his

First Character	Second Character	Third Character	Fourth Character	Fifth Character	Sixth Character	Seventh Character	Eighth Character

Fig. 3. Data word — character mode.

way around machine design, but they still must provide object coding to accomplish the storage and recall functions. In brief, conventionally designed computers, with or without automatic programming aids, require the wasteful expenditure of programming effort, memory capacity, and running time to overcome the limitations of their internal organization.

The problem is attacked directly in the B 5000 by incorporation of a "pushdown" stack, which completely eliminates the need for instructions (coded or compiled) to store or recall intermediate results.

In a B 5000 processor, the stack is composed of a pair of registers, the A and B registers, and a memory area. As operands are picked up by the programs, they are placed in the A register. If the A register already contains a word of information, that word is transferred to the B register prior to loading the operand into the A register. If the B register is also occupied by information, then the word in B is stored in a memory area defined by an address register S. Then the word in A can be transferred to B and the operand brought into the A register. The new word coming into the stack has pushed down the information previously held in the registers. As each pushdown occurs, the address in the S register is automatically increased by one. The information contained in the registers is the last information entered into the stack; the stack operates on a "last in–first out" principle. As information is operated on in the stack, operands are eliminated from the stack and results of operations are returned to the stack. As information in the stack is used up by operations being performed, it is possible to cause "pushups," i.e., a word is brought from the memory area addressed by the S register, and the address in the S register is decreased by one.

To eliminate unnecessary pushdowns and pushups, the A and B registers both have indicators used for remembering whether the registers contain information or are empty. When an operand is to be placed in the stack and either of the registers is empty, no pushdown into memory occurs. Also, when an operation leaves one or both of the registers empty, no automatic pushup occurs.

Polish notation

The Polish logician, J. Lukasiewicz, developed a notation which allows the writing of algebraic or logical expressions which do not require grouping symbols and operator precedence conventions. For example, parentheses are necessary as grouping symbols in the expression A(B+C) to convey the desired interpretation of the expression. In the expression A+B/C, the normal interpretation is A+(B/C), rather than (A+B)/C, because of the convention that

the / operator is of higher precedence than the + operator. The right-hand Polish notation used in the B 5000 is based on placing the operators to the right of their operands: A + B becomes AB+ in Polish notation. A+B+C can be written either as AB+C+, or as ABC+ +. In the expression ABC+ +, the first + operator says to add the operands B and C. The second + operator says to add A to the sum of B and C. Returning to the first examples above, A(B+C) can be written as BC+A× or ABC+× in Polish. The second example is written as BC/A+ or ABC/+. The extension of Polish notation to handle equations is shown in the following example:

Conventional notation $Z = A(B-C)/(D+E)$
Polish notation $ABC-\times DE+/Z=$

The stack in use

To illustrate the functioning of the stack, two simple examples are shown in Figs. 4 and 5. In the examples, the letters P, Q and R represent syllables in the program that cause the operands P, Q, and R to be picked up and placed in the stack. The symbols + and × represent syllables that cause the add and multiply operations to occur. The two examples represent different ways of writing P(Q+R) in Polish notation. The first example in Fig. 4 does not require pushdowns or pushups. The second example, shown in Fig. 5, requires a pushdown in the execution of the syllable R, and a pushup in the execution of the syllable ×. The columns in the table represent the contents of the various registers after execution of the syllable listed in the first column.

Independence of addressing

One of the goals set in the design of the B 5000 was to make the programs independent of the actual memory locations of both the program itself and the data, in order to provide really automatic

Polish Notation QR+P×

Syllable Executed	Contents of	
	Register A	Register B
Q	Q	Empty
R	R	Q
+	Empty	R+Q
P	P	R+Q
×	Empty	P(R+Q)

Fig. 4

Polish Notation PQR$+\times$

Syllable Executed		Contents of			
		Register A	Register B	Register S	Cell 101
P		P	Empty	100	—
Q		Q	P	100	—
R	Pushdown	Empty	Q	101	P
	Execute	R	Q	101	P
+		Empty	Q−R	101	P
$\times$	Pushup	Q−R	P	100	—
	Execute	Empty	P(Q−R)	100	—

Fig. 5

program segmentation. Through automatic program segmentation, it is possible to have program size practically independent of the size of core memory. The systems analyst or programmer intending to do multi-processing is then no longer faced with the difficult task of planning what jobs are to be run together in order that system storage capacities are not exceeded.

In achieving independence of addressing, a solution requiring large contiguous areas of memory was not deemed satisfactory. Each segment of the program and each data area should be completely relocatable without modification to the program. It is then possible to load all the segments of a program or programs onto the drum at load time and call in the segments to any available space in core memory as needed during run time. If some segment of a program is overlaid by a subsequent segment of a program, the segment of the program destroyed in core memory is still available on the drum to be called in again if needed.

Due to the very high program densities in the B 5000, the availability of high capacity drum storage on every system and automatic segmentation, a minimum B 5000 system has the capacity for a program or programs equivalent to approximately 40,000 to 60,000 single address instructions. Of course, if an installation normally ran such large programs, the system would very likely not be a minimum system. However, the installation having an occasional need to run very large programs is not prevented from doing so by storage capacity.

Processing speed now becomes a function of the size of core memory. If large programs are run in a system with small core memory, time will be consumed in recalling program segments from drum to core. If the core memory is expanded, less time will be spent in such activity and the program or programs will be speeded up, and no reprogramming is required.

Program reference table

The means of achieving independence of addressing in the B 5000 is called a Program Reference Table (PRT). The PRT is a 1,025 word relocatable area in memory used primarily for storing control words that locate data areas or program segments. There are also control words for describing input-output operations. These control words, called descriptors, contain the base address and size of data areas, program segments and input-output areas. A descriptor specifying an input-output operation also contains the designation of the unit to be used and the type of operation to be performed. Operands may also be stored in the PRT, providing direct access to single values such as indices, counts, control totals, etc.

In the word mode of the B 5000, every item of data is considered to be either a single value or an element of an array of data. If it is a single value, it will be obtained directly by indexing a descriptor contained in the PRT.

Program segments are described by program descriptors. In addition to core base address, the program descriptor contains the location in drum storage of the program segment and an indication if the program segment is currently in core memory starting at the address specified in the descriptor. Entry to a program segment is made via its program descriptor contained in the PRT. If the program segment is in core memory, entry will be made to the program segment. However, when entry is attempted to a program segment whose descriptor indicates that the segment is not in core memory, automatic entry to the Master Control Program will occur and the desired segment will then be brought in from the drum. Notice that in moving from one segment to another, it is not necessary to know whether the segment to be entered is currently in core memory. Branching within a program segment is self-relative, i.e., the distance to jump either forward or backward is specified, not the address to be jumped to.

As a result of keeping all actual addresses of data and program in the PRT, the program itself does not contain any addresses, but only references to the PRT. To specify one of the 1,024 positions in the PRT requires only 10 bits which contributes greatly to the high program density achieved in the B 5000. Since the PRT is relocatable, references to the PRT contained in the program are to relative locations, thus completely freeing the program from any dependence whatsoever on actual memory locations.

Word mode program

The word mode of the B 5000 processor has four types of syllables. The syllable type is distinguished by the two high-order bits of each 12-bit syllable. The types of syllable and the identification bits are:

00—Operator Syllable
01—Literal Syllable
10—Operand Call Syllable
11—Descriptor Call Syllable

The first of these, the operator syllable, causes operations to be performed. The remaining ten bits of the operator syllable are the operation codes. There are approximately sixty different operations in the word mode. For those operations requiring an operand or operands, the processor checks for sufficient operands in the registers; if they are not there, pushups from the stack in memory occur automatically.

The literal syllable is used for placing constants in the stack to be used as operands. The ten bits of the literal syllable are transferred to the stack. This allows the program to contain integers less than 1,024 as constants.

The operand call syllable, and the descriptor call syllable address locations in the program reference table. The purpose of the operand call syllable is to place an operand in the stack; the purpose of the descriptor call syllable is to place the address of an operand, a descriptor, in the stack. There are four situations that arise, depending on the word read from the program reference table.

1 The word is an operand.

2 The word is a descriptor containing the address of the operand.

3 The word is a descriptor containing the base address of the data area in which the operand resides.

4 The word is a program descriptor containing the base address of a subroutine.

For (1), the operand call syllable has completed its action by placing an operand in the stack. The descriptor call syllable will cause the construction of a descriptor of the operand, replacing the operand by the constructed descriptor.

For (2), the operand call syllable then reads the operand from the cell addressed. The descriptor call syllable has completed its action.

For (3), indexing of the descriptor by the item that is now the second item in the stack occurs. For an operand call syllable, the operand is obtained from the indexed address; for the descriptor call syllable, action is complete after the indexing.

In the case of (4), subroutine entry occurs to the subroutine addressed. A word of the three previous types may be left in the registers upon return from the subroutine, in which instance the actions described above will take place, depending upon the type of syllable which initiated the subroutine.

Essentially, the four types of action that occur for an operand call syllable are obtaining an operand directly, indirectly, from an array, or by computation. Sometimes in the use of the call syllables, it is not known which type of action will occur for a particular syllable when the program is created. This is particularly true for call syllables in subroutines.

Programs in the word mode consist of strings of syllables which follow the rules of Polish notation. Variable length strings of call syllables and literal syllables, which place items of information in the stack, are followed by operator syllables which perform their operations on information in the stack.

The indexing features of the B 5000 allow generalized indexing and at the same time provide complete storage protection. Data areas and program segments of different programs may be intermingled, but a program is prevented from storing outside of its data areas. The method of indexing allows any of the 1,024 words of the program reference table to be considered index registers. Multilevel indexing is provided, i.e., indices of arrays can themselves be elements of arrays.

The subroutine control provided in the B 5000 allows nesting of subroutines—even recursive nesting (a subroutine is a subroutine of itself)—arbitrarily deep. Dynamic allocation of storage for parameter lists and temporary working storage simplify the use of subroutines. Storage is automatically allocated and deallocated as required.

Character mode program

In the character mode of the B 5000 Processor, there is only one type of syllable, called the operator syllable. Program segments in the character mode are constructed of strings of these syllables. The character mode is designed to provide editing, formatting, comparison, and other forms of data manipulation. In doing so, the processor uses two areas of memory—the source and destination areas. When a program switches from word mode to character mode, two descriptors containing the base addresses of these areas are supplied. The source area or destination area may be changed at any time during character mode so that the program may act on several areas.

The character mode operator syllable is split into two 6-bit parts; the last part specifies the operation to be performed and the first part specifies the number of times the operation is to be performed. Operations are provided for the transferring, deletion, comparison, and insertion of characters or bits. Also, there are operations which allow the repetition of syllable strings. This is quite useful for complex table look-up operations and for editing information which contains repeated patterns.

Conclusion

The Burroughs B 5000 system has been designed as an integrated hardware-software package which offers such benefits as savings in the memory space required to store equivalent object programs; multi-processing and parallel processing; and running identical programs on systems with different size memories and different system configurations with no loss in individual system efficiency.

References

LoneW61; BartR61; BockR63; CarlC63; MaheR61

Section 6

Processors with multiprogramming ability

The processors in this section have features which allow multiple programs to exist in the primary memory at the same time. The programs can be executed alternately by a single processor without having to wait for new programs to be input. The cost is only that of changing the processor state, which involves only a few instructions at most (and only one instruction on some systems, such as the CDC 6600). Since programs are subject to numerous unpredictable delays within a single run for interchange with the external environment (either via Ms or T), substantial increases in Pc utilization can be achieved by multiprogramming. If more than a single processor has access to Mp, the system is called a multiprocessor system.

Time-shared computers are generally multiprogrammed. Alternatively, time-shared systems can be implemented by swapping programs, one at a time, into primary memory for interpretation. The Berkeley Time-Sharing System (Chap. 24) uses both multiprogramming and program swapping. The Burroughs B 5000 (Chap. 22) is an early computer to have multiprogram capability. The idea of multiprogramming is so fundamental that it should be among the first concepts to be understood by the student of computing systems. A very nice review of memory mapping and storage allocation is presented in the paper Dynamic Storage Allocation Systems [Randell and Kuehner, 1968].

Atlas

The Atlas is one of the most important machines described in this book. The prototype was originally designed and constructed at Manchester University. The Atlas 1 and Atlas 2 were produced by Ferranti Corp. (prior to becoming part of I.C.T.[1]). Atlas 1 is the most interesting; it incorporates most of the features of the Atlas prototype. The Lincoln Laboratory TX-2 [Clark, 1957] influenced some Atlas features: multiple index registers and interrupt processing of input/output devices. Atlas' detailed internal structure is described in a paper [Sumner et al., 1962].

[1] International Computers and Tabulators, U. K.

Two original features, one-level storage and extracodes, have been copied in many other machines. A one-level store is common to most new computers which are time-shared or multiprogrammed; the scheme for memory paging in the SDS 940 is essentially that of Atlas.

The extracodes feature allows ordinary machine operation codes to be used to call subroutines. Commonly used complex instructions (such as sin, cos, and monitor calls) can be written in a common operating system accessible to all users. Initially these subroutines were stored in a read-only memory.

The ISP is straightforward and extremely nice. The extracode idea appears in the SDS 900 series and was used in the SDS 940 system for defining common-user instructions. The IBM System/360 SVC (supervisor call) instruction is an adaptation of the extracode.

Atlas was about the earliest computer to be designed with a software operating system and the idea of user machine in mind. The operating system has been nicely described [Kilburn et al., 1961] and evaluated [Morris et al., 1967].

In a letter to the authors of this book, F. H. Sumner makes the following comments on Atlas.

The initial ideas and the preliminary research on the Atlas computer system started in the Department of Computer Science of the University of Manchester in 1956. The team, under the direction of Professor T. Kilburn, was later supplemented by several members of the I.C.T. Computer Research Department, and the prototype machine was working in the department by the Autumn of 1961. The first production model became operational in January 1963.

The significant features of the system can be summarised as:

1 The provision of a virtual address field greater than the real address space.

2 The implementation of a "one-level" store using a mixture of core store and drum store.

3 The interrupt system and the method of peripheral control.

4 The realisation at the design stage that there would be a complex operating system and the provision in the hardware of specific features to assist such an operating system.

The method of peripheral control permitted the attachment of a large number of on-line peripherals with rapid response and entry into the operating system for a peripheral requiring attention. This, together with the multiprogramming features, makes the design ideal for the attachment of keyboards for the provision of multi-access operation. In the original design, provision for several such on-line typewriters was made, but at the production stage it was decided to remove these as an economy measure. In view of the subsequent development of on-line operation, this was rather an unfortunate decision.

The Atlas computer at the University has now been in continuous operation for four years and it is expected to provide for the major part of the University's computing needs until 1971.

During the period of its operation the provision of extensive monitoring and logging information has permitted the behaviour of the system to be studied in detail. The results of these studies have been extremely valuable in the design of a successor to the Atlas.

Design of the B 5000 System

The Burroughs B 5000 computer is described in Part 3, Sec. 5, page 257, Chap. 22.

A user machine in a time-sharing system

The Berkeley Time-Sharing Computer (Fig. 1) is based on the SDS 930 (Chap. 24). The hardware modifications to the SDS 930, together with the operating system software, were sold by Scientific Data Systems as the SDS 940. The operating system and hardware modifications for multiprogramming make the 940 one of the first commercially available combined hardware-software time-sharing computers.[1]

The description in Chap. 24 is concerned with the machine as it appears to the user. That is, the hardware and the operating system software are both presented in the context in which they contribute to form a user machine.

The 940 uses a memory map which is almost a subset of that of Atlas but is more modest than that of the IBM 360/67 [Arden et al., 1966] and GE 645 [Dennis, 1965; Daley and Dennis, 1968]. A number of instructions are apparently built in via the programmed operator calling mechanism, based on Atlas extracodes (Chap. 23). The software-defined instructions emphasize the need for hardware features. For example, floating-point arithmetic is needed when several computer-bound programs are run. The SDS 945 is a successor to the 940, with slightly increased capability but at a lower cost.

[1] Time-shared computers consist of both hardware and a complex software operating system. *Adams Computer Characteristics Quarterly* lists the deliveries of general-purpose time-shared computers as DEC PDP-6 hardware, October, 1964 (software in early 1965); SDS 940 hardware (and Berkeley software) April, 1966; GE 635, 645 hardware, May, 1965 (M.I.T.'s project MULTICS software, around 1969); IBM System/360 Model 67 hardware, March, 1966 (software, around 1968).

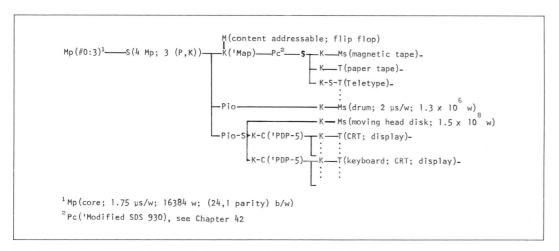

Fig. 1. University of California (Berkeley) time-shared-computer PMS diagram.

Chapter 23

One-level storage system[1]

T. Kilburn / D. B. G. Edwards / M. J. Lanigan
F. H. Sumner

Summary After a brief survey of the basic Atlas machine, the paper describes an automatic system which in principle can be applied to any combination of two storage systems so that the combination can be regarded by the machine user as a single level. The actual system described relates to a fast core store-drum combination. The effect of the system on instruction times is illustrated, and the tape transfer system is also introduced since it fits basically in through the same hardware. The scheme incorporates a "learning" program, a technique which can be of greater importance in future computers.

1. Introduction

In a universal high-speed digital computer it is necessary to have a large-capacity fast-access main store. While more efficient operation of the computer can be achieved by making this store all of one type, this step is scarcely practical for the storage capacities now being considered. For example, on Atlas it is possible to address 10^6 words in the main store. In practice on the first installation at Manchester University a total of 10^5 words are provided, but though it is just technically feasible to make this in one level it is much more economical to provide a core store (16,000 words) and drum (96,000 words) combination.

Atlas is a machine which operates its peripheral equipment on a time division basis, the equipment "interrupting" the normal main program when it requires attention. Organization of the peripheral equipment is also done by program so that many programs can be contained in the store of the machine at the same time. This technique can also be extended to include several main programs as well as the smaller subroutines used for controlling peripherals. For these reasons as well as the fact that some orders take a variable time depending on the exact numbers involved, it is not really feasible to "optimum" program transfers of information between the two levels of store, *i.e.*, core store and drum, in order to eliminate the long drum access time of 6 msec. Hence a system has been devised to make the core drum store combination appear to the programmer as a single level of storage, the

[1] *IRE Trans., EC-11*, vol. 2, pp. 223–235, April, 1962.

requisite transfers of information taking place automatically. There are a number of additional benefits derived from the scheme adopted, which include relative addressing so that routines can operate anywhere in the store, and a "lock out" facility to prevent interference between different programs simultaneously held in the store.

2. The basic machine

The arrangement of the basic machine is shown in Fig. 1. The available storage space is split into three sections; the private store which is used solely for internal machine organization, the central store which includes both core and drum store, in which all words are addressed and is the store available to the normal user, and finally the tape store, which is the conventional backing-up large capacity store of the machine. Both the private store and the main core store are linked with the main accumulator, the *B*-store, and the *B*-arithmetic unit. However the drum and tape stores only have access to these latter sections of the machine via the main core store.

The machine order code is of the single address type, and a comprehensive range of basic functions are provided by normal engineering methods. Also available to the programmer are a number of extra functions termed "extracodes" which give automatic access to and subsequent return from a large number of built-in subroutines. These routines provide

1   A number of orders which would be expensive to provide in the machine both in terms of equipment and also time because of the extra loading on certain circuits. An example of this is the order:
Shift accumulator contents $\pm n$ places where n is an integer.

2   The more complex mathematical operations, *e.g.*, $\sin x$, $\log x$, etc.,

3   Control orders for peripheral equipments, card readers, parallel printers, etc.,

4   Input-output conversion routines,

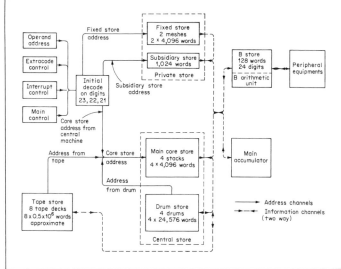

Fig. 1. Layout of basic machine.

5 Special programs concerned with storage allocation to different programs being run simultaneously, monitoring routines for fault finding and costing purposes, and the detailed organization of drum and tape transfers.

All this information is permanently required and hence is kept in part of the private store termed the "fixed store" [Kilburn and Grimsdale, 1960a] which operates on a "read only" basis. This store consists of a woven wire mesh into which a pattern of small "linear" ferrite slugs are inserted to represent digital information. The information content can only be changed manually and will tend to differ only in detail between the different versions of the Atlas computer. In Muse this store is arranged in two units each of 4096 words, a unit consisting of 16 columns of 256 words, each word being 50 bits. The access time to a word in any one column is about 0.4 μsec. If a change of column address is required, this figure increases by about 1 μsec due to switching transients in the read amplifiers. Subsequent accesses in the new column revert to 0.4 μsec. The store operates in conjunction with a subsidiary core store of 1024 words which provides working space for the fixed store programs, and has a cycle time of about 1.8 μsec. There are certain safeguards against a normal machine user gaining access to addresses in either part of the private store, though in effect he makes use of this store through the extracode facility.

The central store of the machine consists of a drum and core store combination, which has a maximum addressable capacity of about 10^6 words. In Muse the central store capacity is about 96,000 words contained on 4 drums. Any part of this store can be transferred in blocks of 512 words to/from the main core store, which consists of four separate stacks, each stack having a capacity of 4096 words.

The tape system provides a very large capacity backing store for the machine. The user can effect transfers of variable amounts of information between this store and the central store. In actual fact such transfers are organized by a fixed store program which initiates automatic transfers of blocks of 512 words between the tape store and the main core store. The system can handle eight tape decks running simultaneously, each producing or demanding a word on average every 88 μsec.

The main core store address can thus be provided from either the central machine, the drum, or the tape system. Since there is no synchronization between these addresses, there has to be a priority system to allocate addresses to the core store. The drum has top priority since it delivers a word every 4 μsec, the tape next priority since words can arise every 11 μsec from 8 decks and the machine uses the core store for the rest of the available time. A priority system necessarily takes time to establish its priority, and so it has been arranged that it comes into effect only at each drum or tape request. Thus the machine is not slowed down in any way when no drum or tape transfers take place. The effect of drum and tape transfers on machine speed is given in Appendix 1.

To simplify the control commands given to the drum, tape, and peripheral equipment in the machine, the orders all take the form $b \rightarrow S$ or $s \rightarrow B$ and the identification of the required command register is provided by the address S. This type of storage is clearly widely scattered in the machine but is termed collectively the V-store.

In the central machine the main accumulator contains a fast adder [Kilburn et al., 1960b] and has built-in multiplication and division facilities. It can deal with fixed or floating point numbers and its operation is completely independent of the B-store and B-arithmetic unit. The B-store is a fast core store (cycle time 0.7 μsec) of 120 twenty-four bit words operating in a word selected partial flux switching mode [Edwards et al., 1960]. Eight "fast" B lines are also provided in the form of flip-flop registers. Of these, three are used as control lines, termed main, extracode, and interrupt controls respectively. The arrangement has the advantage that the control numbers can be manipulated by the normal B-type orders, and the existence of three controls permits the machine to switch rapidly from one to another without having to transfer control numbers to the core store. Main control is used when the central machine is obeying the current program, while the extra-code control is concerned with the fixed store subroutines. The interrupt control provides the means for handling numerous peripheral equipments which "interrupt" the machine when they either require or are providing information. The remaining "fast" B lines are mainly used for organizational procedures, though $B124$ is the floating point accumulator exponent.

The operating speed of the machine is of the order of 0.5×10^6 instructions per second. This is achieved by the use of fast transistor logic circuitry, rapid access to storage locations, and an extensive overlapping technique. The latter procedure is made possible by the provision of a number of intermediate buffer storage registers, separate access mechanisms to the individual units of core store and parallel operation of the main accumulator and B-arithmetic units. The word length throughout the machine is 48 bits which may be considered as two half-words of 24 bits each. All store transfers between the central machine, the drum and tape stores are parity checked, there being a parity digit associated with each half-word. In the case of transfers within the central store (*i.e.*, between main core store and drum) the parity digits associated with a given word are retained throughout the system. Tape transfers are parity checked when information is transferred to and from the main core store, and on the tape itself a check sum technique involving the use of two closely spaced heads is used.

The form of the instruction, which allows for two B-modifications, and the allocation of the address digits is shown in Fig. 2a. Half of the addressable store locations are allocated to the central store which is identified by a zero in the most significant digit of the address. (See Fig. 2b.) This address can be further subdivided into block address, and line address in a block of 512 words. The least significant digits, 0 and 1, make it possible to address 6 bit characters in a half word and digit 2 specifies the half word.

The function number is split into several sections, each section relating to a particular set of operations, and these are listed in Fig. 2c. The machine orders fall into two broad classes, and these are

1 *B codes:* These involve operations between a B line specified by the B_A digits in the instruction and a core store line whose address can be modified by the contents of a B line determined by the B_m digits. There are a total of 128 B lines, one of which, B_0, always contains zero. Of the other lines 90 are available to the machine user, 7 are special registers previously mentioned, and a further 30 are used by extracode orders.

2 *A codes:* These involve operations between the Accumulator and a core store line whose address can now be doubly

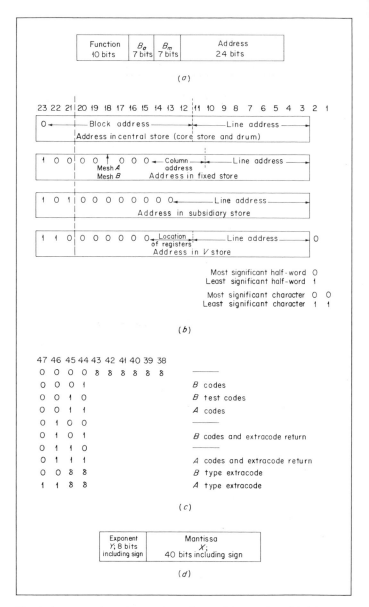

Fig. 2. Interpretation of a word. (a) Form of instruction. (b) Allocation of address digits. (c) Function of decoding. (d) Floating-point number $X8^Y$.

modified first by contents of B_m and then by the contents of B_A. Both fixed and floating point orders are provided, and in the latter case numbers take the form of $X8^Y$, the digit allocation of X and Y being shown in Fig. 2d. When fixed point working occurs, use is made only of the X digits.

3.   One-level store concept

The choice of system for the fast access store in a large scale computer is governed by a number of conflicting factors which include speed and size requirements, economic and technical difficulties. Previously the problem has been resolved in two extreme cases either by the provision of a very large core store, *e.g.*, the 2.5 megabit [Papian, 1957] store at M.I.T., or by the use of a small core store (40,000 bits) expanded to 640,000 bits by a drum store as in the Ferranti Mercury [Lonsdale and Warburton, 1956; Kilburn et al., 1956] computer. Each of these methods has its disadvantages, in the first case, that of expense, and in the second case, that of inconvenience to the user, who is obliged to program transfers of information between the two types of store and this can be time consuming. In some instances it is possible for an expert machine user to arrange his program so that the amount of time lost by the transfers in the two-level storage arrangement is not significant, but this sort of "optimum" programming is not very desirable. Suitable interpretative coding [Brooker, 1960] can permit the two-level system to appear as one level. The effect is, however, accompanied by an effective loss of machine speed which, in some programs and depending on details of machine design, can be quite severe, varying typically, for example, between one and three.

The two-level storage scheme has obvious economic advantages, and inconvenience to the machine user can be eliminated by making the transfer arrangements completely automatic. In Atlas a completely automatic system has been provided with techniques for minimizing the transfer times. In this way the core and drum are merged into an apparent single level of storage with good performance and at moderate cost. Some details of this arrangement on the Muse are now provided.

The central store is subdivided into blocks of 512 words as shown by the address arrangements in Fig. 2*b*. The main core store is also partitioned into blocks of this size which for identification purposes are called pages. Associated with each of these core store page positions is a "page address register" (P.A.R.) which contains the address of the block of information at present occupying that page position. When access to any word in the central store is required the digits of the demanded block address are compared with the contents of all the page address registers. If an "equivalence" indication is obtained then access to that particular page position is permitted. Since a block can occupy any one of the 32 page positions in the core store it is necessary to modify some digits of the demanded block address to conform with the page positions in which an equivalence was obtained.

These processes are necessarily time consuming but by providing a by-pass of this procedure for instruction accesses (since, in general, instruction loops are all contained in the same block) then most of this time can be overlapped with a useful portion of the machine or core store rhythm. In this way information in the core store is available to the machine at the full speed of the core store and only rarely is the over-all machine speed affected by delays in the equivalence circuitry.

If a "not equivalence" indication is obtained when the demanded block address is compared with the contents of the P.A.R.'s then that address, which may have been *B*-modified, is first stored in a register which can be accessed as a line of the *V*-store. This permits the central machine easy access to this address. An "interrupt" also occurs which switches operation of the machine over to the interrupt control, which first determines the cause of the interrupt and then, in this instance, enters a fixed store routine to organize the necessary transfers of information between drum and core store.

A.   Drum transfers

On each drum, one track is used to identify absolute block positions around the drum periphery. The records on these tracks are read into the θ registers which can be accessed as lines of the *V*-store and this permits the present angular drum position to be determined, though only in units of one block. In this way the time needed to transfer any block while reading from the drums can be assessed. This time varies between 2 and 14 msec since the drum revolution time is 12 msec and the actual transfer time 2 msec.

The time of a writing transfer to the drums has been reduced by writing the block of information to the first available empty block position on any drum. Thus the access time of the drum can be eliminated provided there are a reasonable number of empty blocks on the drum. This means, however, that transfers to/from the drum have to be carried out by reference to a directory and this is stored in the subsidiary store and up-dated whenever a transfer occurs.

When the drum transfer routine is entered the first action is to determine the absolute position on a drum of the required block. The order is then given to carry out the transfer to an empty page position in the core store. The transfer occurs automatically as soon as the drum reaches the correct angular position. The page address register in the vacant position in the core store is set to a specific block number for drum transfers. This technique simplifies the engineering with regard to the provision of this number from the drum and also provides a safeguard against transferring to the wrong block.

As soon as the order asking for a read transfer from the drum has been given the machine continues with the drum transfer program. It is now concerned with determining a block to be transferred back from the core store to the drum. This is necessary to ensure an empty core store page position when the next read transfer is required. The block in the core store to be transferred has to be carefully chosen to minimize the number of transfers in the program and this optimization process is carried out by a learning program, details of which are given in Sec. 5. The operation of this program is assisted by the provision of the "use" digits which are associated with each page position of the core store.

To interchange information between the core store and drums, two transfers, a read from and a write to the drum are necessary. These have to be done sequentially but could occur in either order. The technique of having a vacant page position in the core store permits a read transfer to occur first and thus allows the time for the learning program to be overlapped either into the waiting period for the read transfer or into the transfer time itself. In the time remaining after completion of the learning program an entry is made into the over-all supervisor program for the machine, and a decision is taken concerning what the machine is to do until the drum transfer is completed. This might involve a change to a different main program.

A program could ask for access to information in a page position while a drum or tape transfer is taking place to that page. This is prevented in Atlas by the use of a "lock out" (L.O.) digit which is provided with each Page Address Register. When a lock out digit is set at 1, access to that page is only permitted when the address has been provided either by the drum system, the tape system, or the interrupt control. The latter case permits all transfers from paper tape, punched card, and other peripheral equipments, to be handled without interference from the main program. When the transfer of a block has been completed the organizing program resets the L.O. digit to zero and access to that page

position can then be made from the central machine. It is clear that the L.O. digit can also be used to prevent interference between programs when several different ones are being held in the machine at the same time.

In Sec. 3 it was stated that addresses demanding access to the core store could arise from three distinct sources, the central machine, the drum, and the tape. These accesses are complicated because of (1) the equivalence technique, and (2) the lock out digit. The various cases and the action that takes place are summarized in Table 1.

The provision of the Page Address Registers, the equivalence circuitry, and the learning program have permitted the core store and drum to be regarded by the ordinary machine user as a one-level store, and the system has the additional feature of "floating address" operation, i.e., any block of information can be stored in any absolute position in either core or drum store. The minimum access time to information in this store is obviously limited by the core store and its arrangement and this is now discussed.

B. Core store arrangement

The core store is split into four stacks, each with individual address decoding and read and write mechanisms. The stacks are then combined in such a way that common channels into the machine for the address, read and write digits are time shared between the various stacks. Sequential address positions occur in two stacks alternately and a page position which contains a block of 512 sequential addresses is thus arranged across two stacks. In this way it is possible to read a pair of instructions from consecutive addresses in parallel by increasing the size of the read channel. This permits two instructions to be completely obeyed in three store "accesses." The choice of this particular storage arrangement is discussed in Appendix 2.

The coordination of these four stacks is done by the "core stack coordinator" and some features of this are now discussed, starting with the operation of a single stack.

Table 1 Comparison of demanded block address with contents of the P.A.R.'s resultant state of equivalence and lock out circuits

Source of address	$\begin{cases}\textit{Equivalence} \\ \textit{Lock out} = 0\end{cases}$ [E.Q.]	Not equivalence [N.E.Q.]	$\begin{cases}\textit{Equivalence} \\ \textit{Lock out} = 1\end{cases}$ [E.Q. & L.O.]
1. Central Machine	Access to required page position	Enter drum transfer routine	Not available to this program
2. Drum System	Access to required page position	Fault condition indicated	Fault condition indicated
3. Tape System	Access to required page position	Fault condition indicated	Fault condition indicated

C. Operation of a single stack of core store

The storage system employed is a coincident current M.I.T. system arranged to give parallel read out of 50 digits. The reading operation is destructive and each read phase of the stack cycle is followed by a write phase during which the information read out may be rewritten. This is achieved by a set of digit staticizors which are loaded during the read phase and are used to control the inhibit current drivers during the write phase. When new information is to be written into the store a similar sequence is followed, except that the digit staticizors are loaded with the new information during the read phase. A diagram indicating the different types of stack cycle is shown in Fig. 3.

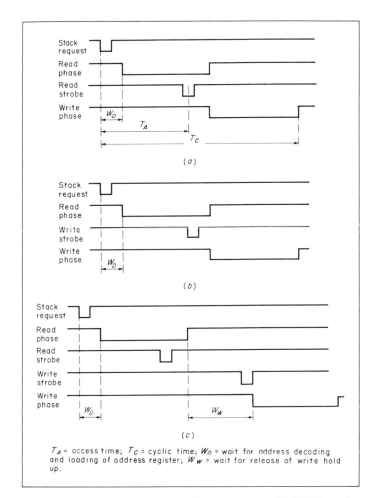

T_A = access time; T_C = cyclic time; W_D = wait for address decoding and loading of address register; W_W = wait for release of write hold up.

Fig. 3. Basic types of stack cycle. (a) Read order (s → A). (b) Write order (a → s). (c) Read-write order (b + s → S).

There is a small delay W_D ($\simeq 100$ mμsec) between the "stack request" signal, SR, and the start of the read phase to allow for setting of the address state and the address decoding. The output information from the store appears in the read strobe period, which is towards the end of the read phase. In general, the write phase starts as soon as the read phase ends. However, the start of the write phase may be held up until the new information is available from the central machine. This delay is shown as W_w in Fig. 3c. The interval T_A between the stack request and the read strobe is termed the stack access time, and in practice this is approximately one third of the cycle time T_C. Both T_A and T_C are functions of the storage system and assuming that W_w is zero have typical values of 0.7 μsec and 1.9 μsec respectively. A holdup gate in the request channel prevents the next stack request occurring before the end of the preceding write phase.

D. Operation of the main core store with the central machine

A schematic diagram of the essentials of the main core store control system is shown in Fig. 4. The control signals SA_1 and SA_2 indicate whether the address presented is that of a single word or a pair of sequentially addressed instructions. Assuming that the flip-flop F is in the reset condition, either of these signals results in the loading of the buffer address register (B.A.R.). This loading is done by the signal B.A.B.A. which also indicates that the buffer register in the central machine has become free.

In dealing with the first request the block address digits in the B.A.R. are compared with the contents of all the page address registers. Then one of the indications summarized in Table 1 and indicated in Fig. 4 is obtained. Assuming access to the required store stack is permitted then a set C.S.F. signal is given which resets the flip-flop F. If this occurs before the next access request arises, then the speed of the system is not store-limited. In most cases SET CSF is generated when the equivalence operation on the demanded block address is complete, and the read phase of the appropriate stack (or stacks) has started. Until this time the information held in the B.A.R. must not be allowed to change. In Fig. 5 a flow diagram is shown for the various cases which can arise in practice.

When a single address request is accepted it is necessary to obtain an "equivalence" indication and form the page location digits before the stack request can be generated. The SET CSF signal then occurs as soon as the read phase starts. If a "not equivalent" or "equivalent and locked out" indication is obtained a stack request is not generated, and the contents of the B.A.R. are copied in to a line of the V-store before SET CSF is generated.

When access to a pair of addresses is requested (*i.e.*, an instruc-

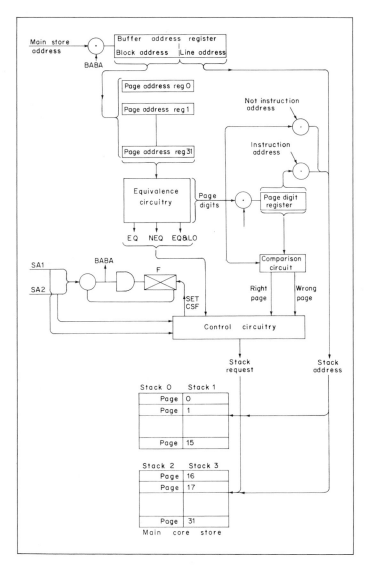

Fig. 4. Main core store control.

tion pair) the stack requests are generated on the assumption that these instructions are located in the same page position as the last pair requested, *i.e.*, the page position digits are taken from the page digit register. (See Fig. 4.) In this way the time required to obtain the equivalent indication and form the page location digits is not included in the over-all access time of the system. The assumption will normally be true, except when crossing block boundaries. The latter cases are detected and corrected by comparing the true position page digits obtained as a result of the

equivalence operation with the contents of the page digit register and a "right page" or "wrong page" indication is obtained. (See Fig. 4.) If a wrong page is accessed this is indicated to the central machine and the read out is inhibited. The true page location digits are copied into the page digit register, so that the required instruction pair will be obtained when next requested. The read out to the central machine is also inhibited for "not equivalent" or "equivalent and locked out" indications.

In Fig. 5 the waiting time indicated immediately before the stack request is generated can arise for a number of reasons.

1   The preceding write phase of that stack has not yet finished.

2   The central machine is not yet ready either to accept information from the store, or to supply information to it.

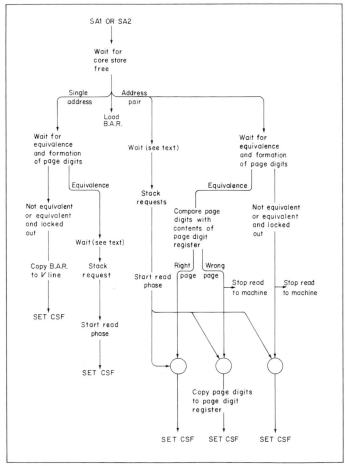

Fig. 5. Flow diagram of main core store control.

3 It is necessary to ensure a certain minimum time between successive read strobes from the core store stacks to allow satisfactory operation of the parity circuits, which take about 0.4 μsec to check the information. This time could be reduced, but as it is only possible to get such a condition for a small part of the normal instruction timing cycle it was not thought to be an economical proposition.

The basic machine timing is now discussed.

4. Instruction times

In high-speed computers, one of the main factors limiting speed of operation is the store cycle time. Here a number of techniques, *e.g.*, splitting the core store into four separate stacks and extracting two instructions in a single cycle, have been adopted despite a fast basic cycle time of 2 μsec in order to alleviate this situation. The time taken to complete an instruction is dependent upon

1 The type of instruction (which is defined by the function digits)

2 The exact location of the instruction and operand in the core or fixed store since this can affect the access time

3 Whether or not the operand address is to be modified

4 In the case of floating point accumulator orders, the actual numbers themselves

5 Whether drum and/or tape transfers are taking place

The approximate times for various instructions are given in Table 2. These figures relate to the times between completing instructions when a long sequence of the same type of instruction is obeyed. While this method is not ideal, it is necessary because in practice obeying one instruction is overlapped in time with some part of three other instructions. This makes the detailed timing complicated, and so the timing sequence is developed slowly by first considering instructions obeyed one after another. It is convenient to make these instructions a sequence of floating point additions with both instruction and operand in the core store and with the operand address single *B*-modified.

To obey this instruction the central machine makes two requests to the core store, one for the instruction and the second for the operand. After the instruction is received in the machine the function part has to be decoded and the operand address modified by the contents of one of the *B* registers before the operand request can be made. Finally, after the operand has been obtained the actual accumulator addition takes place to complete the instruction. The time from beginning to end of one instruction is 6.05 μsec and an approximate timing schedule is as follows in Table 3.

If no other action is permitted in the time required to complete the instruction (steps 1 to 8 in Table 3), then the different sections of the machine are being used very inefficiently, *e.g.*, the accumulator adder is only used for less than 1.1 μsec. However, the organization of the computer is such that the different sections such as store stacks, accumulator and *B*-arithmetic unit, can operate

Table 2 Approximate instruction times

Type of instruction	Number of modifications of address	Instruction in core store. Operands in core store. Time in μsec	Instructions in fixed store. Operands in core store. Time in μsec	Instructions in fixed store. Operands in fixed store. Time in μsec
Floating Point Addition	0	1.4	1.65	1.2
	1	1.6	1.65	1.2
	2	2.03	1.9	1.9
Floating Point Multiplication	0, 1 or 2	4.7	4.7	4.7
Floating Point Division	0, 1 or 2	13.6	13.6	13.6
Add Store Line to an Index Register	0	1.53	1.65	1.15
	1	1.85	1.85	1.85
Add Index Register to Store Line and Rewrite to Store Line	0	1.63	1.65	—
	1	1.8	1.7	—

Table 3† Timing sequence for floating point addition (instructions and operands in the core store)

Sequence	Time interval between steps μsec	Total time μsec
1. Add 1 to Main Control		0
(Addition time)	0.3	
2. Make Instruction Request		0.3
(Transfer times, equivalence time and stack access time)	1.75	
3. Receive Instruction in Central Machine		2.05
(Load register and decode)	0.2	
4. Function decoding complete		2.25
(Single address modification)	0.85	
5. Request Operand		3.10
(Transfer times, equivalence time and stack access time)	1.75	
6. Receive Operand in Central Machine		4.85
(Load register)	0.1	
7. Start Addition in Accumulator		4.95
(Average floating point addition, including shift round and stand-ardise)	1.1	
8. Instruction complete		6.05

† In step 4, time is for single address modification. Times for no modification and two modifications are 0.25 μsec and 1.55 μsec respectively.

at the same time. In this way several instructions can be started before the first has finished, and then the effective instruction time is considerably reduced. There have, of course, to be certain safeguards when for example an instruction is dependent in any way on the completion of a preceding instruction.

In the time sequence previously tabulated, by far the longest time was that between a request in the central machine for the core store and the receipt in the central machine of the information from that store. This effective access time of 1.75 μsec is made up as shown in Table 4. It has been reduced in practice by the provision of two buffer registers, one in the central machine and the other in the core stack coordinator. These allow the equivalence and transfer times to be overlapped with the organization of requests in the central machine.

In this way, provided the machine can arrange to make requests fast enough, then the effective access time is reduced to 0.8 μsec. Further, since three accesses are needed to complete two instructions (one for an instruction pair and one for each of the two operands) the theoretical minimum time of an instruction is 1.2 μsec $3 \times 0.8/2$ and it then becomes store limited. Reference to

Table 3 shows that the arithmetic operation takes 1.2 μsec to complete so that, on the average, the capabilities of the store and the accumulator are well matched.

Another technique for reducing store access time for instructions has also been adopted. This permits the read cycles of the two stacks to start assuming that the same page will be referred to as in the previous instruction pair. This, of course, will normally be true and there is sufficient time to take corrective procedures should the page have been changed. The limit of 1.2 μsec per instruction is not reduced by this technique, but the possibility of reaching this limit under other conditions is enhanced.

A schematic diagram of the practical timing of a sequence of floating point addition orders is shown in Fig. 6. The overlapping is not perfect and in the time between successive instruction pairs the computer is obeying four instructions for 25 per cent of the time, three for 56 per cent and two for 19 per cent. It is therefore to be expected that the practical time for the complete order is greater than the theoretical minimum time; it is in fact approximately 1.6 μsec.

For certain types of functions the reading of the next pair of instructions before completing both instructions of the first pair would be incorrect, e.g., functions causing transfer of control. Such situations are recognized during the function decoding, and the request for the next instruction pair is held up until a suitable time.

In a sequence of floating point addition orders with the operand addresses unmodified the limit is again 1.2 μsec while the time obtained is 1.4 μsec. For accumulator orders in which the actual accumulator operation imposes a limit in excess of 2 μsec then the actual time is equal to this limit.

Perhaps a more realistic way of defining the speed of the computer is to give the time for a typical inner loop of instructions. A frequently occurring operation in matrix work in the formation of the scalar product of two vectors, this requires a loop of five instructions:

Table 4 Effective store access time

Sequence	Total time μsec
1. Request in Central Machine	0
2. Request in Core Stack Coordinator	0.25
3. Equivalence complete and request made to selected stack	0.95
4. Information in Core Stack Coordinator	1.65
5. Information in Central Machine	1.75

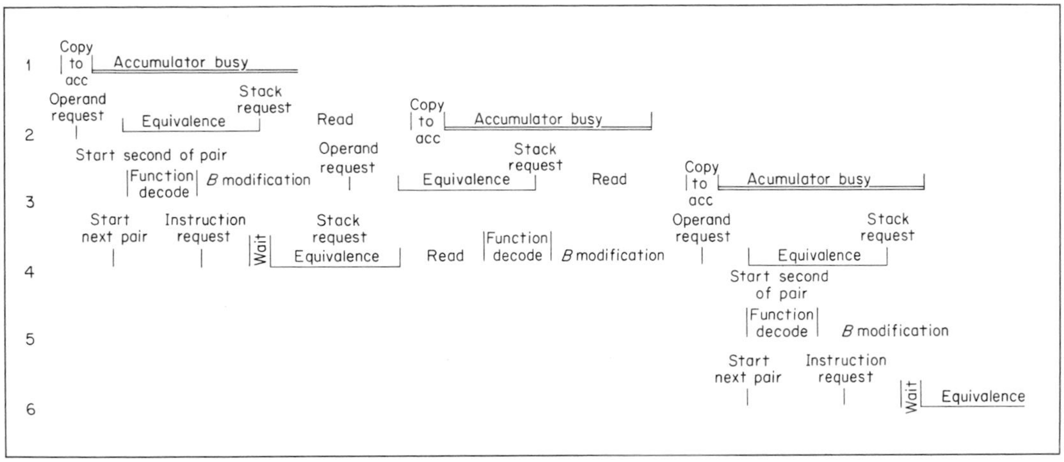

Fig. 6. Timing diagram for a sequence of floating point addition orders. (Single-address modification.)

1 Element of first vector into accumulator. (Operand *B*-modified.)

2 Multiply accumulator by element of second vector. (Operand *B*-modified.)

3 Add partial product to accumulator.

4 Copy accumulator to store line containing partial product.

5 Alter count to select next elements and repeat.

The time for this loop with instructions and operands on the core store is 12.2 μsec. The value of the overlapping technique is shown by the fact that the time from starting the first instruction to finishing the second is approximately 10 μsec.

When the drum or tape systems are transferring information to or from the core store then the rate of obeying instructions which also use the core store will be affected. The affect is discussed in more detail in Appendix 1. The degree of slowing down is dependent upon the time at which a drum or tape request occurs relative to machine requests. It also depends on the stacks used by the drum or tape and those being used by the central machine. The approximate slowing down is by a factor of 25 per cent during a drum transfer and by 2 per cent for each active tape channel. (See Appendix 1.)

5.   The drum transfer learning program

The organization of drum transfers has been described in Sec. 2A. After the transfer of the required block from the drum to the core

store has been initiated, the organizing program examines the state of the core store, and if empty pages still exist, no further action is taken. However, if the core store is full it is necessary to arrange for an empty page to be made available for use at the next nonequivalence. The selection of the page to be transferred could be made at random; this could easily result in many additional transfers occurring, as the page selected could be one of those in current use or one required in the near future. The ideal selection, which would minimize the total number of transfers, could only be made by the programmer. To make this ideal selection the programmer would have to know (1) precisely how his program operated, which is not always the case, and (2) the precise amount of core store available to his program at any instant. This latter information is not generally available as the core store could be shared by other central machine programs, and almost certainly by some fixed store program organizing the input and output of information from slow peripheral equipments. The amount of core store required by this fixed store program is continuously varying [Kilburn et al., 1961]. The only way the ideal pattern of transfers can be approached is for the transfer program to monitor the behavior of the main program and in so doing attempt to select the correct pages to be transferred to the drum. The techniques used for monitoring are subject to the condition that they must not slow down the operation of the program to such an extent that they offset any reduction in the number of transfers required. The method described occupies less than 1 per cent of the operating time, and the reduction in the number of transfers is more than sufficient to cover this.

That part of the transfer program which organizes the selection of the page to be transferred has been called the "learning" program. In order for this program to have some data on which to operate, the machine has been designed to supply information about the use made of the different pages of the core store by the program being monitored.

With each page of the core store there is associated a "use" digit which is set to "1" whenever any line in that page is accessed. The 32 "use" digits exist in two lines of the V-store and can be read by the learning program, the reading automatically resetting them to zero. The frequency with which these digits are read is governed by a clock which measures not real time but the number of instructions obeyed in the operation of the main program. This clock causes the learning program to copy the "use" digits to a list in the subsidiary store every 1024 instructions. The use of an instruction counter rather than a normal clock to measure "time" for the learning program is due to the fact that the operations of the main program may be interrupted at random for random lengths of time by the operation of peripheral equipments. With an instruction counter the temporal pattern of the blocks used will be the same on successive runs through the same part of the program. This is essential if the learning program is to make use of this pattern to minimize the number of transfers.

When a nonequivalence occurs and after the transfer of the required block has been arranged, the learning program again adds the current values of the "use" digits to the list and then uses this list to bring up to date two sets of times also kept in the subsidiary store. These sets consist of 32 values of t and T, one of each for each page of the core store. The value of t is the length of time since the block in that page has been used. The value of T is the length of the last period of inactivity of this block. The accuracy of the values of t and T is governed by the frequency with which the "use" digits are inspected.

The page to be written to the drum is selected by the application in turn of three simple tests to the values of t and T.

1 Any page for which $t > T + 1$, or

2 That page with $t \neq 0$ and $(T - t)$ max, or

3 That page with T_{max} (all $t = 0$).

The first rule selects any page which has been currently out of use for longer than its last period of inactivity. Such a page has probably ceased to be used by the program and is therefore an ideal one to be transferred to the drum. The second rule ignores all pages with $t = 0$ as they are in current use, and then selects the one which, if the pattern of use is maintained, will not be required by the program for the longest time. If the first two rules fail to select a page the third ensures that if the page finally selected is wrong, in that it is immediately required again, then, as in this case, T will become zero and the same mistake will not be repeated.

For all the blocks on the drum a list of values of τ is kept. The values of τ are set when the block is transferred to the drum:

$$\tau = \text{time of transfer—value of } t \text{ for transferred page}$$

When a block is transferred to the core store the value of τ is used to set the value of T.

$$T = \text{time of transfer—value of } \tau \text{ for this block}$$
$$= \text{length of last period of inactivity}$$

For the block transferred from the drum t is set to 0.

In order to make its decision the learning program has only to update two short lists and apply at the most three simple rules; this can easily be done during the 2 msec transfer time of the block required as a result of the nonequivalence. As the learning program uses only fixed and subsidiary store addresses it is not slowed down during the period of the drum transfer.

The over-all efficiency of the learning program cannot be known until the complete Atlas system is working. However, the value of the method used has been investigated by simulating the behavior of the one-level store and learning program on the Mercury computer at Manchester University. This has been done for several problems using varying amounts of store in excess of the core store available. One of these was the problem of forming the product A of two 80th order matrices B and C. The three matrices were stored row by row each one extending over 14 blocks, only 14 pages of core store were assumed to be available. The method of multiplication was

$$b_{11} \times \text{1st row of } C = \text{partial answer to 1st row of } A$$
$$b_{12} \times \text{2nd row of } C + \text{partial answer} = \text{second partial answer,}$$
$$\text{etc.}$$

Thus matrix B was scanned once, matrix C 80 times and each row of matrix A 80 times.

Several machine users were asked to spend a short time writing a program to organize the transfers for a general matrix multiplication problem. In no case when the method was applied to the above problem were fewer than 357 transfers required. A program written specifically for this problem which paid great attention to the distribution of the rows of the matrices relative to block divisions required 234 transfers. The learning program required 274 transfers; the gain over the human programmer was chiefly due to the fact that the learning program could take full advantage of the occasions when the rows of A existed entirely within one block.

Many other problems involving cyclic running of single or multiple sets of data were simulated, and in no case did the learning program require more transfers than an experienced human programmer.

A. *Prediction of drum transfers*

Although the learning program tends to reduce the number of transfers required to a minimum, the transfers which do occur still interrupt the operation of the program for from 2 to 14 msec as they are initiated by nonequivalence interrupts. Some or all of this time loss could be avoided by organizing the transfers in advance. A very experienced programmer having sole use of the core store could arrange his own transfers in such a way that no unnecessary ones ever occurred and no time was ever wasted waiting for transfers to be completed. This would require a great deal of effort and would only be worthwhile for a program that was going to occupy the machine for a long time. By using the data accumulated by the learning program it is possible to recognize simple patterns in the use made by a program of the various blocks of the one-level store. In this way a prediction program could forecast the blocks required in the near future and organize the transfers. By recording the success or failure of these forecasts the program could be made self-improving. For the matrix multiplication problem discussed above the pattern of use of the blocks containing matrix C is repeated 80 times, and a considerable degree of success could be obtained with a simple prediction program.

6. Conclusions

A specific system for making a core-drum store combination appear as a single level store has been described. While this is the actual system being built for the Atlas machine the principles involved are applicable to combinations of other types of store. For example, a tunnel diode–fast core store combination for an even faster machine. An alternative which was considered for Atlas, but which was not as attractive economically, was a fast core–slow core store combination. The system too can be extended to three levels of storage, and indeed if 10^6 words of total storage had to be provided then it would be most economical to provide it on a third level of store such as a file drum.

The automatic system does require additional equipment and introduces some complexity, since it is necessary to overlap the time taken for address comparison into the store and machine operating time if it is not to introduce any extra time delays. Simulated tests have shown that the organization of drum transfers are reasonably efficient and other advantages which accrue, such as efficient allocation of core storage between different programs and store lock out facilities are also invaluable. No matter how intelligent a programmer may be he can never know how many programs or peripheral equipments are in operation when his program is running. The advantage of the automatic system is that it takes into account the state of the machine as it exists at any particular time. Furthermore if as in normal use there is some sort of regular machine rhythm even through several programs, there is the possibility of making some sort of prediction with regard to the transfers necessary. This involves no more hardware and will be done by program. However, this stage will probably be left until results on the actual system are obtained.

It can be seen that the system is both useful and flexible in that it can be modified or extended in the manner previously indicated. Thus despite the increase in equipment, the advantages which are derived completely justify the building of this automatic system.

APPENDIX 1   ORGANIZATION OF THE ACCESS REQUESTS TO THE CORE STORE

There are three sources of access requests to the core store, namely the central machine, the drum, and the tape systems. In deciding how the sequence of requests from all three sources are to be serialized and placed in some sort of order, a number of facts have to be considered. These are

1   All three sources are asynchronous in nature.

2   The drum and tape systems can make requests at a fairly high rate compared with the store cycle time of approximately 2 μsec. For example, the drum provides a request every 4 μsec and the tape system every 11 μsec when all 8 channels are operative.

3   The drum and tape systems can only be stopped in multiples of a block length, *i.e.*, 512 words. This means that any system devised for accessing the core store must deal with both the average rates of drum and tape requests specified in 2. Only the central machine can tolerate requests being stopped at any time and for any length of time. From these facts a request priority can be stated which is
  a   Drum request.
  b   Tape request.
  c   Central machine request.

4   A machine request can be accepted by the core store, but because there is no place available to accept the core store information, its cycle is inhibited and further requests held up. In the case of successive division orders this time can be as long as 20 μsec, in which case 5 drum requests could be made. To avoid having an excessive amount of buffer storage for the drum two techniques are possible:

a When drums or tapes are operative do not permit machine requests to be accepted until there is a place available to put the information.

b Store the machine request and then permit a drum or tape request.

The latter scheme has been adopted because it can be accommodated more conveniently and it saves a small amount of time.

5   If the central machine is using the private store then it is desirable for drum and tape transfers to the core store not to interfere with or slow down the central machine in any way.

6   When the central machine, drum and tape are sharing the core store then the loss of central machine speed should be roughly proportional to the activity of the drum or tape systems. This means that drum or tape requests must "break" into the normal machine request channel as and when required.

The system which accommodates all these points is now discussed. Whenever a drum or tape request occurs inhibit signals are applied to request channel into the core stack coordinator and also to the stack request channels from this coordinator. This results in a "freezing" of the state of flip-flop F (Fig. 5) and this state is then inspected (Fig. 7, point X). If the state is "busy" this means that a machine order has been stopped somewhere between the loading of the buffer address register (B.A.R.) and the stack request. Normally this time interval can vary from about 0.5 μsec if there are no stack request holdups, to 20 μsec in the case of certain accumulator holdups. In either case sufficient time is allowed after the inspection to ensure that the equivalence operation has been completed. If an equivalence indication is obtained all the information relevant to this machine order (*i.e.*, the line address, page digits, stack(s) required and type of stack order) are stored for future reference. Use is made here of the page digit register provided to allow the by-pass on the equivalence circuitry for instruction accesses. The core store is then made free for access by the drum or the tape. If the core store had been found to be free on inspection, the above procedure is omitted.

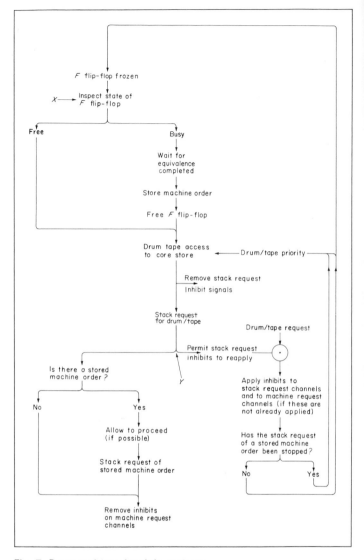

Fig. 7. Drum and tape break in systems.

A drum or tape access (as decided by the priority circuit) to the core store then occurs, which removes the inhibits on the stack request channels. When the stack request for the drum or tape cycle is initiated these inhibits are allowed to reapply. At this stage (Fig. 7, point Y), if there is a stored machine order it is allowed to proceed if possible. The inhibits on the machine request channels are removed when the stack request for the stored machine order occurs. If there is no stored machine order this is done immediately, and the central machine is again allowed access to the core store. However, another drum or tape request can arise before the stack request of the stored machine order occurs, in particular because this latter order may still be held up by the central machine. If this is the case the drum or tape is allowed immediate access and a further attempt is made to complete the stored machine order when this drum or tape stack request occurs.

If the stored machine order was for an operand, the content of the page digit register will correspond to the location of this operand. The next machine request for an instruction pair will then almost certainly result in a "wrong page" indication. This is prevented by arranging that the next instruction pair access does not by-pass the equivalence circuitry.

The effect on the machine speed when the drum or tapes are transferring information to or from the core store is dependent upon two factors. First, upon the proportion of time during which the buffer register in the core coordinator is busy dealing with machine requests, and secondly, upon the particular stacks being used by the central machine and the drum or tape. If the computer is obeying a program with instructions and operands on the fixed or subsidiary store then the rate of obeying instructions is unaffected by drum or tape transfers. A drum or tape interrupt occurring when the B.A.R. is free prevents any machine address being accepted onto this buffer for 1.0 μsec. However, if the B.A.R. is busy then the next machine request to the core store is delayed until 1.8 μsec after the interrupt if different stacks are being used, or until 3.4 μsec after the interrupt if the stacks are the same.

When the machine is obeying a program with instructions and operands on the core store the slowing down during drum transfers can be by a factor of two if instructions, operands, and drum requests use the same stacks. It is also possible for the machine to be unaffected. The effect on a particular sequence of orders can be seen by considering the one discussed in Sec. 4 and illustrated in Fig. 6. In this sequence the instructions are on stacks 0 and 1 while the operands are on stacks 2 and 3. If the drum or tape is transferring alternately to stacks 0 and 1 then the effect of any interrupt within the 3.2 μsec of an instruction pair is to increase this time by between 0.5 and 3.4 μsec depending upon where the interrupt occurred. The average increase is 1.8 μsec and for a tape transfer with interrupts every 88 μsec the computer can obey instructions at 98 per cent of the normal rate. During drum transfers the interrupts occur every 4 μsec which would suggest a slowing down to 60 per cent of normal. However, for any regular sequence of orders the requests to the core store by the machine and by the drum rapidly become synchronized with

the result in this particular case that the machine can still operate at 80 per cent of its normal speed.

APPENDIX 2   METHODS OF DIVISION OF THE MAIN CORE STORE

The maximum frequency with which requests can be dealt with by a single stack core store is governed by the cycle time of the store. If the store is divided into several stacks which can be cycled independently then the limit imposed on the speed of the machine by the core store is reduced. The degree of division which is chosen is dependent upon the ratio of core store cycle time to other machine operations and also upon the cost of the multiple selection mechanisms required.

Considering a sequence of orders in which both the instruction and operand are in the core store, then for a single stack store the limit imposed on the operating speed by the store is two cycle times per order, i.e., 4 μsec in Atlas. This is significantly larger than the limits imposed by other sections of the computer (Sec. 4). If the store is divided into two stacks and instructions and operands are separated, then the limit is reduced to 2 μsec which is still rather high. The provision of two stacks permits the addressing of the store to be arranged so that successive addresses are in alternate stacks. It is therefore possible by making requests to both stacks at the same time to read two instructions together, so reducing the number of access times to three per instruction pair. Unfortunately such an arrangement of the store means that operands are always on the same stacks as instruction pairs, and the limit imposed by the cycle time is still 2 μsec per order even if the two operand requests in the instruction pair are to different stacks and occur at the same time.

Division into any number of stacks with the addressing system working through each stack in turn cannot reduce the limit below 2 μsec since successive instructions normally occur in successive addresses and are therefore in the same stack. However, four stacks arranged in two pairs reduces the limit to 1 μsec as the operands can always be arranged to be on different stacks from the instruction pairs. In order to reduce the limit to 0.5 μsec it is necessary to have eight stacks arranged in two sets of four and to read four instructions at once, which would increase the complexity of the central machine.

The limit of 1 μsec is quite sufficient and further division with the stacks arranged in pairs only enables the limit to be more easily obtained by suitable location of the instructions and operands.

The location of instructions and operands within the core store is under the control of the drum transfer program; thus when there

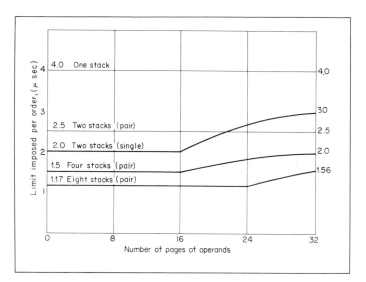

Fig. 8. Limit imposed by cycle time on operating speed for different divisions of the core store.

are several stacks instructions and operands are separated wherever possible. Under these conditions it is possible to calculate the limit imposed on the operating speed by the cycle time for different divisions of the core store. The results are shown in Fig. 8, for stacks arranged in pairs instructions are read in pairs and in all cases both instructions and operands are assumed to be on the core store. Operands are assumed to be selected at random from the operand space, for instance in the case of two stacks arranged as a pair, successive operand requests have equal probability of being to the same stack or to alternate stacks.

The limit imposed by a four stack store is never severe compared with other limitations, for example the sequence of floating point addition orders discussed in Sec. 4 required 1.6 μsec per order with ideal distribution of instructions and operands. Division into eight stacks, although it reduces the limit, will not have an equivalent effect on the over-all operating speed, and such a division was not considered to be justified.

References

KilbT62; BrooR60; EdwaD60; KilbT56; 60a, 60b, 61; LonsK56; PapiW57; FothJ61; HartD68; HowaD61; 62, 63; MorrD67; SumnF62

Chapter 24

A user machine in a time-sharing system[1]

B. W. Lampson / *W. W. Lichtenberger* / *M. W. Pirtle*

Summary This paper describes the design of the computer seen by a machine-language programmer in a time-sharing system developed at the University of California at Berkeley. Some of the instructions in this machine are executed by the hardware, and some are implemented by software. The user, however, thinks of them all as part of his machine, a machine having extensive and unusual capabilities, many of which might be part of the hardware of a (considerably more expensive) computer.

Among the important features of the machine are the arithmetic and string manipulation instructions, the very general memory allocation and configuration mechanism, and the multiple processes which can be created by the program. Facilities are provided for communication among these processes and for the control of exceptional conditions.

The input-output system is capable of handling all of the peripheral equipment in a uniform and convenient manner through files having symbolic names. Programs can access files belonging to a number of people, but each person can protect his own files from unauthorized access by others.

Some mention is made at various points of the techniques of implementation, but the main emphasis is on the appearance of the user's machine.

Introduction

A characteristic of a time-sharing system is that the computer seen by the user programming in machine language differs from that on which the system is implemented [Bright, 1964; Comfort, 1965; Forgie, 1965; McCullogh et al., 1965; Schwartz, 1964]. In fact, the *user machine* is defined by the combination of the time-sharing hardware running in user mode and the software which controls input-output, deals with illegal actions which may be taken by a user's program, and provides various other services. If the hardware is arranged in such a way that calls on the system have the same form as the hardware instructions of the machine [Lichtenberger and Pirtle, 1965], then the distinction becomes irrelevant to the user; he simply programs a machine with an unusual and powerful instruction set which relieves him of many of the problems of conventional machine-language programming [Lampson, 1965; McCarthy et al., 1963].

[1]*Proc. IEEE, 54*, vol. 12, pp. 1766–1774, December, 1966.

In a time-sharing system which has been developed by and for the use of members of Project Genie at the University of California at Berkeley [Lichtenberger and Pirtle, 1965], the user machine has a number of interesting characteristics. The computer in this system is an SDS 930, a 24 bit, fixed-point machine with one index register, multi-level indirect addressing, a 14 bit address field, and 32 thousand words of 1.75 μs memory in two independent modules. Figure 1 shows the basic configuration of equipment. The memory is interleaved between the two modules so that processing and drum transfers may occur simultaneously. A detailed description of the various hardware modifications of the computer and their implications for the performance of the overall system has been given in a previous paper [Lichtenberger and Pirtle, 1965].

Briefly, these modifications include the addition of monitor and user modes in which, for user mode, the execution of a class of instructions is prevented and replaced by a trap to a system routine. The protection from unauthorized access to memory has been subsumed in an address mapping scheme: both the 16 384 words addressable by a user program (logical addresses) and the 32 768 words of actual core memory (physical addresses) have been divided into 2048-word *pages*. A set of eight six-bit hardware registers defines a *map* from the logical address space to the real memory by specifying the real page which is to correspond to each of the user's logical pages. Implicit in this scheme is the capability of marking each of the user's pages as unassigned or read-only, so that any attempt to access such a page improperly will result in a trap.

All memory references in user mode are mapped. In monitor mode, all memory references are normally absolute. It is possible, however, with any instruction in monitor mode, or even within a chain of indirect addressing, to specify use of the user map. Furthermore, in monitor mode the top 4096 words are mapped through two additional registers called the monitor map. The mapping process is illustrated in Fig. 2.

Another significant hardware modification is the mechanism for going between modes. Once the machine is in user mode, it can get to monitor mode under three circumstances:

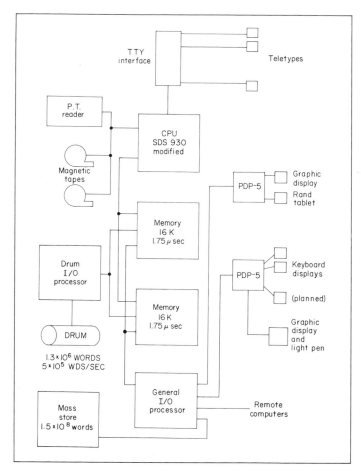

Fig. 1. Configuration of equipment.

1 If a hardware interrupt occurs

2 If a trap is generated by the user program as outlined.

3 If an instruction with a particular configuration of two bits is executed. Such an instruction is called a system programmed operator (SYSPOP).

In case 3, the six-bit operation field is used to select one of 64 locations in absolute core. The current address of the instruction is put into absolute location zero as a subroutine link, the indirect address bit of this link word is set, and another bit is set, marking the memory location in the link word as having come from user-mapped memory. The system routine thus invoked may take a parameter from the word addressed by the SYSPOP, since its address field is not interpreted by the hardware. The routine will

address the parameter indirectly through location zero and, because of the bit marking the contents of location zero as having come from user mode, the user map will be applied to the remainder of the address indirection. All calls on the system which are not inadvertent are made in this way.

A monitor mode program gets into user mode by transferring to an address with mapping specified. This means, among other things, that a SYSPOP can return to the user program simply by branching indirect through location zero.

As the above discussion has perhaps indicated, the mode-changing arrangements are very clean and permit rapid and natural transfers of control between user and system programs. Advantage has been taken of this fact to create a rather grandiose machine for the user. Its features are the subject of this paper.

Basic features of the machine

A user in the Berkeley time-sharing system, working at what he thinks of as the hardware language level, has at his disposal a machine with a configuration and capability which can be conveniently controlled by the execution of machine instruction sequences. Its simplest configuration is very similar to that of a

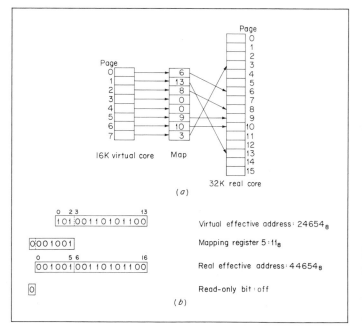

Fig. 2. The hardware memory map. (a) Relation between virtual and real memory for a typical map. (b) Construction of a real memory address.

standard medium-sized computer. In this configuration, the machine possesses the standard 930 complement of arithmetic and logic instructions and, in addition, a set of software interpreted monitor and executive instructions. The latter instructions, which will be discussed more fully in the following, do rather complex input-output of many different kinds, perform many frequently used table lookup and string processing functions, implement floating point operations, and provide for the creation of more complex machine configurations. Some examples of the instructions available are:

1   Load A, B, or X (index) registers from memory or store any of the registers. Indexing and indirect addressing are available on these and almost all other instructions. Double word load and store are also available.

2   The normal complement of fixed-point arithmetic and logic operations.

3   Skips on various arithmetic and logic conditions.

4   Floating point arithmetic and input-output. The latter is in free format or in the equivalent of Fortran E or F format.

5   Input a character from a teletype or write a block of arbitrary length on a drum file.

6   Look up a string in a hash-coded table and obtain its position in the table.

7   Create a new process and start it running concurrently with the present one at a specified point.

8   Redefine the memory of the machine to include a portion of that which is also being used by another program.

It should be emphasized that, although many of these instructions are software interpreted, their format is identical to the standard machine instruction format, with the exception of the one bit which specifies a system interpreted instruction. Since the system interpretation of these instructions is completely invisible to the machine user, and since these instructions do have the standard machine instruction format, the user and his program make no distinction between hardware and software interpreted instructions.

Some of the possible 192 operation codes are not legal in the user machine. Included in this category are those hardware instructions which would halt the machine or interfere with the input-output if allowed to execute, and those software interpreted instructions which attempt to do things which are forbidden to the program. Attempted execution of one of these instructions will result in an *illegal instruction* violation. The effect of an illegal instruction violation is described later.

Memory configuration

The memory size and organization of the machine is specified by an appropriate sequence of instructions. For example, the user may specify a machine which has 6K of memory with addresses from 0 to 13777_8; alternatively, he may specify that the 6K should include addresses 0 to 3777_8, 14000_8 to 17777_8, and 34000_8 to 37777_8. The user may also specify the size and configuration of the machine's secondary storage and, to a considerable extent, the structure of its input-output system. A full discussion of this capability will be deferred to a later section.

The next few paragraphs discuss the mechanism by which the user's program may specify its memory size and organization. This mechanism, known as the *process map* to distinguish it from the hardware memory address mapping, uses a (software) mapping register consisting of eight 6-bit bytes, one byte for each of the eight 2K blocks addressable by the 14 bit address field of an instruction. Each of these bytes either is 0 or addresses one of the 63 words in a table called the private memory table (*PMT*). Each user has his own private memory table. An entry in this table provides information about a particular 2K block of memory. The block may be either *local* to the user or it may be *shared*. If the block is local, the entry gives information about whether it is currently in core or on the drum. This information is important to the system but need not concern the user. If the block is shared, its *PMT* entry points to an entry in another table called the shared memory table (*SMT*). Entries in this table describe blocks of memory which are shared by several users. Such blocks may contain invariant programs and constants, in which case they will be marked as *read-only*, or they may contain arbitrary data which is being processed by programs belonging to two different users.

A possible arrangement of logical or virtual memory for a process is shown in Fig. 3. The nature of each page has been noted in the picture of the virtual memory; this information can also be obtained by taking the corresponding byte of the map and looking at the *PMT* entry specified by that byte. The figure shows a large amount of shared memory, which suggests that the process might be a compilation, sharing the code for the compiler with other processes translating programs written in the same source language. Virtual pages one and two might hold tables and temporary storage which are unique to each separate compilation. Note that, although the flexibility of the map allows any block of code or data to appear anywhere in the virtual memory, it is certainly not true that a program can run regardless of which pages

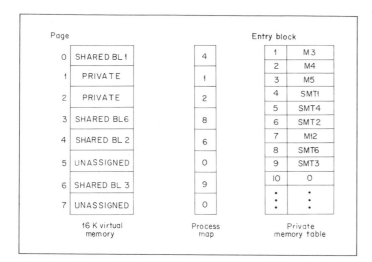

Fig. 3. Layout of virtual memory for a typical process.

it is in. In particular, if it contains references to itself, such as branch instructions, then it must run in the same virtual pages into which it was loaded.

Two instructions are provided which permit the user to read and modify his process map. The ability to read the process mapping registers permits the user to obtain the current memory assignment, and the ability to write the registers permits him to reassign memory in any way which suits his fancy. The system naturally checks each new map as it is established to ensure that the process is not attempting to obtain unauthorized access to memory which does not belong to it.

When the user's process is initiated, it is assigned only enough memory to contain the program data as initially loaded. For instance, if the program and constants occupy 3000_8 words, two blocks, say blocks 0 and 1, will be assigned. At this point, the first two bytes of the process mapping register will be nonzero; the others will be zero. When the program runs, it may address memory outside of the first 4K. If it does, and if the user has specified a machine size larger than 4K, a new block of memory will be assigned to him which makes the formerly illegal reference legal. In this way, the user's process may obtain more memory. In fact, it may easily obtain more than 16K of memory simply by addressing 16K, reading and preserving the process mapping register, setting it with some of the bytes cleared to zero, and grabbing some more memory. Of course, only 16K can be addressed at one time; this is a limitation imposed by the address field of the machine.

There is an instruction which allows a process to specify the maximum amount of memory which it is allowed to have. If it attempts to obtain more than this amount, a *memory violation* will occur. A memory violation can also be caused by attempts to transfer into or indirect through unassigned memory, or to store into read-only memory. The effect of this violation is similar to the effect of an illegal instruction violation and will be discussed.

The facilities just described are entirely sufficient for programs which need to reorganize the machine's memory solely for internal purposes. In many cases, however, the program wishes to obtain access to memory blocks which have been created by the system or by other programs. For example, there may be a package of mathematical and utility routines in the system which the program would like to use. To accommodate this requirement, there is an instruction which establishes a relationship between a name and a certain process mapping function. This instruction moves the PMT entries for the blocks addressed by the specified process mapping function into the shared memory table so that they are generally accessible to all users. Once this correspondence has been established, there is another instruction which allows a different user to deliver the name and obtain in return the associated process map. This instruction will, if necessary, make new entries in the second user's PMT. Various subsystems and programs of general interest have names permanently assigned to them by the system.

The user machine thus makes it possible for a number of processes belonging to independent users to run with memory which is an arbitrary combination of blocks local to each individual process, blocks shared between several processes, and blocks permanently available in the system. A complex configuration is sketched in Fig. 4. Process 1.1 was shown in more detail in Fig. 3. Each box represents a process, and the numbers within represent the eight map bytes. The arrows between processes show the process hierarchy, which is discussed in the next section. Note that the PMT's belong to the users, not to the processes.

From the above discussion, it is apparent that the user can manipulate the machine memory configuration to perform simple memory overlays, to change data bases, or to perform other more complex tasks requiring memory reconfiguration. For example, the use of common routines is greatly facilitated, since it is necessary only to adjust the process map so that (1) memory references internal and external to the common routine are correct, and (2) the memory area in which the routine resides is read-only. In the simplest case, in which the common routine and the data base fit into 16K of memory, the map is initially established and remains static throughout the execution of the routine. In other cases where the routine and data base do not fit into 16K, or where several common routines are concurrently employed, it may be necessary to make frequent adjustment to the map during execution.

Multiple processes

An important feature of the user machine allows the user program, which in the current context will be referred to as the controlling process, to establish one or more subsidiary processes. With a few minor exceptions, to be discussed, each subsidiary process has the same status as the controlling process. Thus, it may in turn establish a subsidiary process. It is therefore apparent that the user machine is in fact a multi-processing machine. The original suggestion which gave rise to this capability was made by Conway [Conway, 1963], more recently the Multics system has included a multi-process capability [Corbato and Vyssotsky, 1965; Dennis and Van Horn, 1966; Saltzer, 1966].

A *process* is the logical environment for the execution of a program, as contrasted to the physical environment, which is a hardware *processor*. It is defined by the information which is required for the program to run; this information is called the *state vector*. To create a new process, a given process executes an instruction which has arguments specifying the state vector of the new process. This state vector includes the program counter, the central registers, and the process map. The new process may have a memory configuration which is the same as, or completely different from, that of the originating process. The only constraint placed on this memory specification is that the total memory available to the multi-process system is limited to 128K by the process mapping mechanism, which is common to all processes. Each user, of course, has his own 128K.

This facility was put into the system so that the system could control the user processes. It is also of direct value, however, for many user processes. The most obvious examples are input-output buffering routines, which can operate independently of the user's main program, communicating with it through memory and with interrupts (see the following). Whether the operation being buffered is large volume output to a disc or teletype requests for information about the progress of a running program, the degree of flexibility afforded by multiple processes far exceeds anything which could have been built into the input-output system. Furthermore, the overhead is very low: an additional process requires about 15 words of core, and process switching takes about 1 ms under favorable conditions. There are numerous other examples of the value of multiple processes; most, unfortunately, are too complex to be briefly explained.

A process may create a number of subsidiary processes, each

of which is independent of the others and equivalent to them from the point of view of the originating process. Figure 4 shows two simple multi-process structures, one for each of two users. Note that each process has associated with it pointers to its controlling process and to one of its subsidiary processes. When a process has two immediate descendants, as in the case of processes 1.2 and 1.3, they are chained together on a ring. Thus, three pointers, up, down, and ring, suffice to define the process structure completely. The up pointers are, of course, redundant, but are convenient for the implementation. The process is identified by a *process number* which is returned by the system when it is created.

A complex structure such as that in Fig. 5 may result from the creation of a number of subsidiary processes. The processes in Fig. 5 have been numbered arbitrarily to allow a clear description of the way in which the pointers are arranged. Note that the user need not be aware of these pointers; they are shown here to clarify the manner in which the multiple process mechanism is implemented.

A process may destroy one of its subsidiary processes by executing the appropriate instruction. For obvious reasons this operation is not legal if the process being destroyed itself has subsidiary

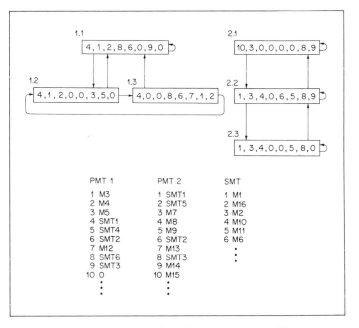

Fig. 4. Process and memory configuration for two users. (The processes are numbered for each user and are represented by their process mapping registers. Memory blocks are identified by drum addresses, which are written M1, M2, ...)

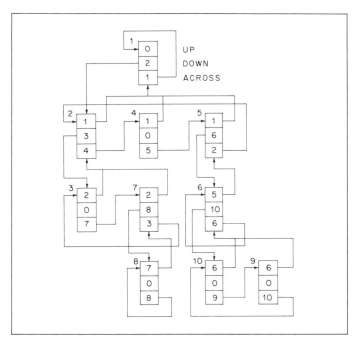

Fig. 5. Hierarchy of processes.

processes. It is possible to find out what processes are subsidiary to any given one; this permits a process to destroy an entire tree of sub-processes by reading the tree from the top down and destroying it from the bottom up.

The operations of creating and destroying processes are entirely separate from those of starting and stopping their execution, for which two more operations are provided. A process whose execution has been stopped is said to be *suspended.*

To assure that these various processes can effectively work together on a common task, several means of interprocess communication exist. The first allows the controlling process to obtain the current status of each of its subsidiary processes. This status information, which is read into a table by the execution of the appropriate system instruction, includes the current state vector and operating status. The operating status of any process may be

1   Running

2   Dismissed for input-output

3   Terminated for memory violation

4   Terminated for illegal violation, or

5   Terminated by the process itself

A second instruction allows the controlling process to become dormant until one of its subsidiary processes terminates. Termination can occur in the following four ways:

1   Because of a memory violation

2   Because of an illegal instruction violation

3   Because of self-termination

Interactions described previously provide no method by which a process can attract the attention of another process which is pursuing an independent course. This can be done with a program interrupt. Associated with each process is a 20-bit interrupt mask. If a mask bit is set, the process may, under certain conditions (to be described in the following), be interrupted; i.e., a transfer to a fixed address will be simulated. The program will presumably have at this fixed address the location of a subroutine capable of dealing with the interrupt and returning to the interrupted computation afterwards. The mechanism is functionally almost identical to many hardware interrupt systems.

A process may cause an interrupt by delivering the number of the interrupt to the appropriate instruction. The process causing the interrupt continues undisturbed, but the nearest process which is either on the same level as the one causing the interrupt or above it in the hierarchy of processes, and which has the appropriate interrupt armed, will be interrupted. This mechanism provides a very flexible way for processes to interact with each other without wasting any time in the testing of flags or similar frivolous activities.

Interrupts may be caused not only by the explicit action of processes, but also by the occurrence of several special conditions. The occurrence of a memory violation, attempted execution of an illegal instruction, an unusual input-output condition, the termination of a subsidiary process, or the intervention of a user at a console (by pushing a reserved button) all may cause unique interrupts (if they have been previously armed). In this way, a process may be notified conveniently of any unusual conditions associated with other processes, the process itself, or a console user.

The memory assignment algorithm discussed previously is slightly modified in the presence of multiple processes. When a process is activated, one of three options may be specified:

1   Assign new memory to the process entirely independently of the controlling process.

2   Assign no new memory to the process. Any attempt to obtain new memory will cause a memory violation.

3   If the process attempts to obtain new memory, scan upward through the process hierarchy until the topmost process is reached. If at any time during this scan a process is found for which the address causing the trap is legal, propagate the memory assigned to it down through the hierarchy to the process causing the trap.

Option 3 permits a process to be started with a subset of memory and later to reacquire some of the memory which was not given to it initially. This feature is important because the amount of memory assigned to a process influences the operating efficiency of the system and thus the speed with which it will be able to respond to teletypes or other real-time devices.

The input-output system

The user machine has a straightforward but unconventional set of input-output instructions. The primary emphasis in the design of these instructions has been to make all input-output devices interface identically with a program and to provide as much flexibility in this common interface as possible. Two advantages result from this uniformity: it becomes natural to write programs which are essentially independent of the environment in which they operate, and the implementation of the system is greatly simplified. To the user the former point is, of course, the important one.

It has been common, for example, for programs written to be controlled from a teletype to be driven instead from a file on, let us say, the drum. A command exists which permits the recognizer for the system command language and all of the subsystems to be driven in this way. This device is particularly useful for repetitive sequences of program assemblies and for background jobs which are run in the absence of the user. Output which normally goes to the teletype is similarly diverted to user files. Another application of the uniformity of the file system is demonstrated in some of the subsystems, notably the assembler and the various compilers. The subsystem may request the user to specify where he wishes the program listing to be placed. The user may choose anything from paper tape to drum to his own teletype. In the absence of file uniformity each subsystem would require a separate block of code for each possibility. In fact, however, the same input-output instructions are used for all cases.

The input-output instructions communicate with *files*. The system in turn associates files with the various physical devices. Programs, for the most part, do not have to account for the peculiarities of the various actual devices. Since devices differ widely in characteristics and behavior, the flexibility of the operations available on files is clearly critical. They must range from single-character input to the output of thousands of words.

A file is *opened* by giving its name as an argument to the appropriate instruction. Programs thus refer to all files symbolically, leaving the details of physical location and organization to the system. If authorized, a program may refer to files belonging to other users by supplying the name of the other user as well as the file name. The owner of a file determines who is authorized to access it. The reader may compare this file naming mechanism with a more sophisticated one [Daley and Neumann, 1965], bearing in mind the fact the file names can be of any length and can be manipulated (as strings of characters) by the program.

Access to files is, in general, either sequential or random in nature. Some devices (like a keyboard-display or a card reader) are purely sequential, while others (like a disk) may be either sequentially or randomly accessed. There are accordingly two major I/O interfaces to deal with these different qualities. The interface used in conjunction with a given file depends on whether the file was declared to be a *random* or a *sequential* file. The two major interfaces are each broken down into other interfaces, primarily for reasons of implementation. Although the distinction between sequential and random files is great, the subinterfaces are not especially visible to the user.

Sequential files

The three instructions CIO (character input-output), WIO (word input-output), and BIO (block input-output) are used to communicate with a sequential file. Each instruction takes as an operand a *file number*. This number is given to the program when it opens a file. At the time of opening a file it must be specified whether the file is to be read from or written onto. Whether any given device associated with the file is character-oriented or word-oriented is unimportant; the system takes care of all necessary character-to-word assembly or word-to-character disassembly.

There are actually three separate, full-duplex physical interfaces to devices in the sequential file mechanism. Generally, these interfaces are invisible to programs. They exist, of course, for reasons of system efficiency and also, because of the way in which some devices are used. The interfaces are:

1   Character-by-character (basically for low-speed, character-oriented devices used for man-machine interaction)

2   Buffered block I/O (for medium-speed I/O applications)

3   Block I/O directly from user core (for high-speed situations)

It should be pointed out that there is no particular relation between these interfaces and the three instructions CIO, WIO, and BIO. The interface used in a given situation is a function of the device involved and, sometimes, of the volume of data to be transmitted, not of the instruction.

Any interface may be driven by any instruction.

Of the three subinterfaces under discussion, the last two are straightforward. The character-by-character interface is, however, somewhat different and deserves some elaboration. Devices associated with this interface are generally (but not necessarily) used for man-machine interaction. Consider the case of a person communicating with a program by means of a keyboard-display (or a teletype). He types on the keyboard and the information is transmitted to the computer. The program may wish to make an immediate response on the display screen. In many cases this response will consist of an echo of the same character, so that the user has the feeling of typing directly onto the screen (or onto the teleprinter).

So that input-output can be carried out when the program is not actually in main memory, the character-by-character input interface permits programs a choice of a number of *echo tables;* it further permits programs a choice of grade of service by permitting them to specify whether a given character is an attention (or *break*) character. Thus, for example, the program may specify that each character typed is to be echoed immediately and that all control characters are to result in activation of the program regardless of the number of characters in the input buffer. Alternatively, the program may specify that no characters are echoed and every character is a break character. By changing the specification the program can obtain an appropriate (and varying) grade of service without putting undue load on the system. Figure 6 shows the components of the character-by-character interface; responsibility for its operation is split between the interrupt called when the device signals for attention and the routine which processes the user's I/O request.

The advantage of the full-duplex, character-by-character mode of operation is considerable. The character-by-character capability means that the user can interact with his program in the smallest possible unit—the character. Furthermore, the full-duplex capability permits, among other things (1) the program to substitute characters on strings of characters as echoes for those received, (2) the keyboard and display to be used simultaneously (as, for example, permitting a character typed on a keyboard to pre-empt the operation of a process. In the case of typing information in during the output of information, a simple algorithm prevents the random admixture of characters which might otherwise result), and (3) the ready detection of transmission errors.

Instructions are included to enable the state of both input and output buffers to be sensed and perhaps cleared (discarding unwanted output or input). Of course, it is possible for a program to use any number of authorized physical devices; in particular, this includes those devices used as remote consoles. A mechanism is provided to permit output which is directed to a given device to be copied on all other devices which are *output linked* to it (and similarly for input). This is useful when communication among users is desired and in numerous other situations.

The sequential file has a structure somewhat similar to that of an ordinary magtape file. It consists of a sequence of *logical records* of arbitrary length and number. On some devices, such as a card reader or the teletype, a file may have only one logical record. The full generality is available for drum files, which are the ones most commonly used. The logical record is to be contrasted with the variable length physical record of magtape or the fixed length record of a card. Instructions are provided to insert or delete logical records and increase or decrease them in length. Other instructions permit the file to be "positioned" almost instantaneously to a specified logical record. This gives the sequential file greater flexibility than one which is completely unaddressable. This flexibility is only possible, of course, because the file is on a random-access device and the sequential structure is maintained by pointers. The implementation is discussed in the following.

When reading a sequential file, CIO and WIO return certain unusual data configurations when they encounter an end of record or end of file, and BIO terminates transmission on either of the conditions and returns the address of the last word transmitted. In addition, certain flag bits are set by the unusual conditions, and an interrupt may be caused if it has been armed.

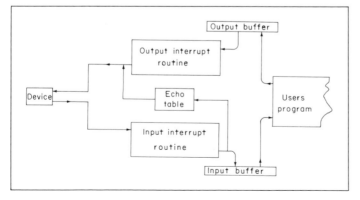

Fig. 6. The character-oriented interface.

The implementation of the sequential file scheme for auxiliary storage is illustrated in Fig. 7. Information is written on the drum in 256-word physical records. The locations of these records are kept track of in 64-word index blocks containing pointers to the data blocks. For the file shown, the first logical record is more than 256 words long but ends in the second 256-word block. The second logical record fits in the third 256-word block and the third logical record—in the 4th data block—is followed by an end of file. If a file requires more than 64 index words, additional index blocks are chained together, both forward and backward. Thus, in order to access information in the file it is necessary only to know the location of the first index block. It may be worthwhile to point out that all users share the same drum. Since the system has complete control over the allocation of space on the drum, there is no possibility of undesired interaction among users.

Available space for new data blocks or index blocks is kept track of by a bit table, illustrated in Fig. 8. In the figure, each column represents one of the 72 physical bands on the drum allocated for the storage of file information. Each row represents one of the 64 256-word sectors around a band. Each bit in the table thus represents one of the 4608 data blocks available. The bits are set when a block is in use and cleared when the block becomes available. Thus, if a new data block is required, the system has only to read the physical position of the drum, use this position to index in the table, and search a row for the appearance of a 0. The column in which a 0 is found indicates the physical track on which a block is available. Because of the way the row was chosen, this block is immediately accessible. This scheme has two advantages over its alternative, which is to chain unused blocks together:

1 It is easy to find a block in an optimum position, using the algorithm just described.

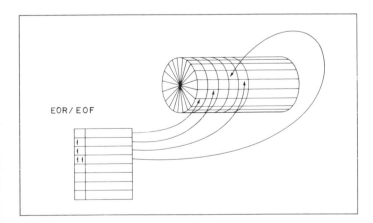

Fig. 7. Index blocks and pointers to data blocks.

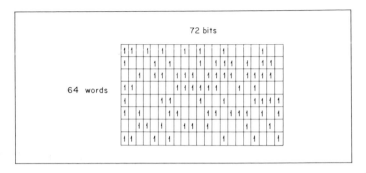

Fig. 8. Bit table for allocation of space on the drum.

2 No drum operations are required when a new block is needed or an old one is to be released.

It may be preferable to assign the new block so that it becomes accessible immediately after the block last assigned for the file. This scheme will speed up subsequent reading of the file.

Random files

Auxiliary storage files can also be treated as extensions of core memory rather than as sequential devices. Such files are called *random files*. A random file differs from a sequential file in that there is no logical record structure to the file and that information is extracted from or written into the random file by addressing a specific word or block of words. It may be opened like a sequential file; the only difference is that it need not be specified as an output or an input file.

Four instructions are used to input and output words and blocks of words on a random file. To permit the random file to look even more like core memory, an instruction enables one of the currently open random files to be specified as the *secondary memory* file. Two instructions, LAS (load A from secondary memory) and SAS (store A in secondary memory), act like ordinary load and store instructions with one level of indirect addressing (see Fig. 9) except, of course, that the data are in a random file instead of in core memory.

Random files are implemented like sequential files except that end of record indicators are not meaningful. Although as many index blocks are used up as required by the size of a random file, only those data blocks which actually contain information will be attached to a random file. As new locations are accessed, new data blocks are attached.

Subroutine files

Whereas it makes little sense to associate, say, a card reader with a random file, a sequential file can be associated with any physi-

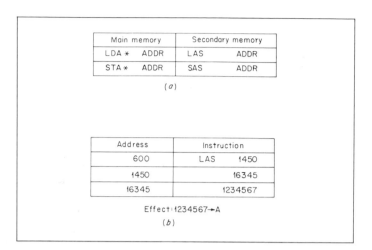

Fig. 9. Load and store form main and secondary memory. (a) Instructions. (b) Addressing.

cal device in the system. In addition, a sequential file may be associated with a subroutine. Such a file is called a *subroutine file*, and the subroutine may thus be thought of as a "nonphysical" device. The subroutine file is defined by the address of a subroutine together with information indicating whether it is an input or an output file and whether it is word or character oriented. An input operation from a subroutine file causes the subroutine to be called. When it returns, the contents of the A register is taken to be the input requested. Correspondingly, an output operation causes the subroutine to be called with the word or character being output in A. The subroutine is completely unrestricted in the kinds of processing it can do. It may do further input or output and any amount of computation. It may even call itself if it preserves the old return address.

Recall that for sequential files the system transforms all information supplied by the user to the format required by the particular file; hence, the requirement that the user, in opening a subroutine file, must specify whether the file is to be character or word oriented. The system will thereafter do all the necessary packing and unpacking.

Subroutine files are the logical end-product of a desire to decouple a program from its environment. Since they can do arbitrary computations, they can provide buffers of any desired complexity between the assumptions a program has made about its environment and the true state of things. In fact, they make it logically unnecessary to provide an identical interface for all the input-output devices attached to the system; if uniformity did not exist, it could be simulated with the appropriate subroutine files. Considerations of convenience and efficiency, of course, militate against such an arrangement, but it suggests the power inherent in the subroutine file machinery.

Summary

The user machine described was designed to be a flexible foundation for development and experimentation in man-machine systems. The user has been given the capability to establish configurations of multiple processes, and the processes have the ability to communicate conveniently with each other, with central files, and with peripheral devices. A given user may, of course, wish only to use a subsystem of the general system (e.g., a compiler or a debugging routine) for his particular job. In the course of using the subsystem, however, he may become dissatisfied with it and wish to revise or even rewrite the subsystem. The features of the user machine not only permit this activity but make it easier.

References

BrigH64; ComfW65; ConwM63; CorbF65; DaleR65; DennJ66; ForgJ65; LampB65; LichW65; McCaJ63; McCuJ65; SaltJ66; SchwJ64

Part 4

The instruction-set processor level: special-function processors

This part contains descriptions of processors that do not interpret general programming languages; that is, they are not Pc's. They are all P's, however, since they have an interpreter that determines not only the operations to be taken, given the current instruction, but the next instruction to be obtained.

A Pio (Sec. 1) is a processor that controls T and Ms components. It manages block or vector transmission between Ms or T and Mp.

A P.array (Sec. 2) processes both vectors and two-dimensional matrices. By recognizing these data as fundamental units, programs (or algorithms) can be expressed efficiently in terms of primitive operators. The chief advantage of these P's is their ability to take advantage of the data structure for parallel interpretation, thereby increasing processing speed.

A microprogram processor (Sec. 3) is designed to interpret and process a datatype which is a program. In effect, this processor is a computer within another computer, programmed to act as an interpreter.

A language processor (Sec. 4) interprets a data-type derived from the primitives of a programming language. In contrast, a conventional processor interprets a language based on fundamental hardware implementation primitives. The difference is clearly apparent as increased complexity of the language processors.

Section 1

Processors to control terminals and secondary memories (input-output processors)

The first three chapters of this section show the evolution of the IBM Data Channels (io processors) from 1958 (the 7094 II) to the present (the 1800, which came after the 360). The processor approach for controlling T and Ms components, while more general, should be contrasted with the specialized one-instruction controls in the B 5000 (Chap. 22) and Burroughs D825 (Chap. 36).

The fourth chapter, on the DEC 338, shows a processor that controls cathode-ray-tube display consoles. The graphic terminals are the first T's of sufficient complexity to utilize a processor of their own. The first CRT displays used the Pc (e.g., on Whirlwind); then small Pc's were adapted to the task; the DEC 338 is one of the earliest special P.display's that appeared.

There is no example in this section of a specialized P for message concentration and switching. For computer systems multiple remote inputs are still recent enough so that either the main Pc handles the task, via specialized K, or small Pc's are committed to it. However, in the telephone industry there has been a very substantial development by the Bell System of the Electronic Switching System (ESS), which uses specialized C's to control switching (routing). In computer systems, we can expect the use of such specialized processors to increase in the near future.

The IBM 7094 II

The IBM 709, a member of the IBM 701-7094 II family, is one of the first computers to have an io processor (IBM name: Data Channel) in its structure. Chapter 41 discusses the two Data Channel types: the early 7607 and the later 7909. The 7909 Data Channel ISP, and a K which it controls, are given in Appendix 2 and 3 of Chap. 41. The principal difference is that Pc controls the Pio ('7909) which in turn controls the K, which in turn controls a T or Ms; the Pc controls the Pio ('7607) and the K; the K controls the T or Ms. The series is discussed in Part 6, Sec. 1, page 515.

The structure of System/360
Part I—outline of the logical structure

The io processors (Selector and Multiplexor Channels) in the System/360 have evolved from the IBM 701-7094 II Series. Part 6, Sec. 3 presents the ISP and PMS structures for these processors. Depending on the computer model, the implementations are realized by a microprogrammed processor interpreting a shared control program for both Pio's and Pc, or by a hardwired Pio. The multiple Pio's in a 360 Multiplexor Channel, though logically independent, are implemented as a single, shared physical processor.

The IBM 1800

The Pio's in this structure are presented in Chap. 33, and the structure is discussed in Part 5, Sec. 2, page 396.

The Digital Equipment Corporation DEC 338 display processor

The DEC 338 is an early P.display. It directly interprets a stored program to control a T.display. Earlier T.displays were controlled by Pc (Whirlwind, Chap. 6), or by a special K.display without stored-program capability, or by a general-purpose Pio. The last method outputs fixed length blocks containing data to be interpreted by T.display as points, vectors, characters, curved line segments, etc. The control of T.display first by Pc, then by a K, then by a Pio, and finally by a P.display has been observed as an evolution [Myer and Sutherland, 1968]. Myer and Sutherland also observe that the evolution is about to become a closed cycle because the generality of a Pc is needed to control a T.display.

Note that the 338 has a very extensive ISP. In fact, the P.display's ISP is more extensive than the companion Pc of the PDP-8 (Chap. 5). There are some display tasks which require Pc, for example, compiling programs (pictures), calculating elaborate light-pen tracking figures, making coordinate and curved lines to straight-line vector approximation transformations, and communicating with other system components.

Another approach to the design of a P.display is based on a P.microprogram which is shared among many T.displays [Rose, 1967]. Yet another alternative, which has not yet been tried, is to incorporate a Pio (P.display) as a special mode in a conventional Pc. Thus the P would interpret either conventional Pc instructions or P.display instructions.

P.display is the interpreter for the output of pictures or graphics. The 338 utilizes data space efficiently simply because the data are long variable-length strings (word vectors). The instruction requires almost no space to specify the data operations and addresses; data are interpreted directly or immediately in the instruction rather than via instruction addresses.

Another feature which allows a program to be efficiently encoded is the stack mechanism for storing subroutine linkages. Subroutines in P.displays are actually programs which form part of a more complete picture. Subroutines are actually subpictures. Although the stack mechanism allows for recursive picture calls, the stack is used principally to save space and to allow multiple T.displays to use common picture programs.

A problem in the 338 which is common to all multi-P structures is intercommunication among the P's. Pc is the controlling P, as is the case with most Pc-Pio structures. The P('338) has no trap to itself but relies on an interrupt signal to Pc. The Pc processes both tasks which P.display might process, given an interrupt system, and other tasks beyond P.display's capability.

A clock should be built into the 338. The brightness or intensity of a picture is determined both electronically (see the mode instructions for controlling intensity) and by the rate at which the pictures are repeated. A clock would allow the time when pictures are started or drawn to be specified; thus the intensity would be independent of picture length.

The 338 requires more hardware than a simpler Pc. However, a large amount of this hardware is used to control the generation of characters and lines. The lines (vectors) are drawn using a DDS (Digital Differential Analyzer) technique. Perhaps one-half of the registers could be eliminated if the 338 were not a P. A simpler alternative was constructed about a similar computer, the PDP-9, by Bell Telephone Laboratories and DEC, using the approach of making the display only a K.

A more elaborate Pc interrupt system with reduced overhead time would enable Pc to take on the specialized program control functions in the 338. Such a scheme might pass the program or instruction counter parameter directly from P.display to Pc. In this way, Pc or P.display would alternatively process part of a single instruction stream, depending on the task.

Despite the problems of this early P.display, it has a sophistication which successors appear to be following.

Chapter 25

The DEC 338 display computer

Introduction

The C(display; 'DEC 338) is a C('DEC PDP-8) with a P.display which can connect to T(#1:8; CRT; display; area: 9.375 × 9.375 in.²). The PMS structure is shown in Fig. 1, Chap. 5, describing the PDP-8. The Pc ISP is given in Appendix 1 of Chap. 5.

The C('338), although designed to stand alone, is generally used as a satellite to a larger C, via an L(Dataphone). The rationale for using a C as a T is based on the bandwidth and storage requirements needed to maintain graphical picture displays. A human being manipulating pictures (rotation, scale change, and conversion of internal linked data structure to a picture structure) requires short response time; this requirement places high processing demands on larger C's. Thus this C(display) is a preprocessor for larger, more general C's.

The actual T(CRT) is a 16-inch CRT with a $9\frac{3}{8}$-inch square viewing area covered by 1,024 × 1,024 (XY) points. The diameter of the points is ∼0.015 inch. The spot is magnetically deflected and focused. All eight T(CRT)'s can be driven together or used independently. A photomultiplier connected through a fiber-optic bundle link is used as a light pen (a photosensitive sensor) to detect spots on the T. The light pen allows the P.display to detect whether a user has "pointed to" a displayed spot.

Pc and P.display access the same Mp; the total data rate available from Mp is one 12-bit word/1.5 microseconds. The instruction times of P.display are a function of the point plotting times of the T(CRT):0.3 microsecond to the next incremental unintensified point (approximately 0.010 inch away); 1.2 microseconds to an incremental intensified point; and 35 microseconds to a point plotted at a random position.

The state (registers) of C.display is given in the ISP description of Appendix 1 of this chapter. There are four parts of the state: the control registers for Program Flow State, the Picture State (or position of beam), Console and Light-pen State, and Mp State. The instruction interpreter is fairly simple and is best described by the state diagram (Fig. 1). The instructions are given in Tables 1 and 2. The remainder of the chapter discusses the P.display instructions and the Pc instructions for communicating with P.display.

Principle of operation

The actual picture is held stationary by repeatedly displaying (intensifying) a particular point, line, etc. The number of times a figure has to be displayed so that it appears stationary and does not flicker depends on the CRT phosphor, the figure, and environmental parameters. The generally accepted range is a plotting rate of 20 ∼ 50 plots/second; thus a complete picture has to be drawn in 50 ∼ 20 milliseconds. If we assume a 30-Hz plot rate, about 28,000 points can be plotted in vector mode (or 280 ∼ 1120 inches, depending on the spacing). About 1,000 characters can be displayed in 30 milliseconds using character mode.

When the light pen is used, a display program is required to "track" the pen. The pen's position is determined by displaying known points. The pen, of course, detects the points when it is present at the displayed points position; therefore the program knows the location of the pen.

The parameters of interest for a display vary, depending on the application. However, the general parameters are:

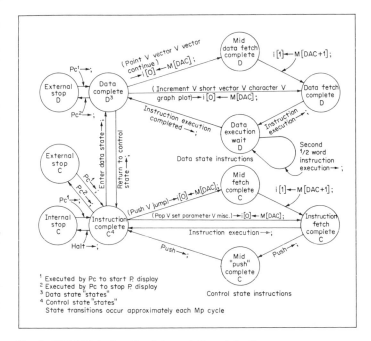

Fig. 1. DEC 338 instruction-interpretation state diagram.

Table 1   DEC 338 control-mode instruction set

Instruction Op Code

	Bits 0:2	3	4	5	6	7	8	9	10	11
Parameters	0	set† Scale	Scale ⟨0:1⟩		set lt pen	lt pen	set Intensity	Intensity ⟨0:2⟩		
Mode	1	stop	clear flags	set mode	Data_Mode ⟨0:2⟩			clear sector	clear X, Y	enter Data_State
Jump‡	2	set Scale	Scale ⟨0:1⟩		set lt pen	lt pen	push	Memory field ⟨0:2⟩		
Pop	3	set Scale	Scale ⟨0:1⟩		set lt pen	lt pen	inh§ Data_Mode	inh Scale, lt pen	inh intensity	enter Data_State
Conditional skip	4	reverse¶ test	clear bits after test	complement after test	Push_Buttons ⟨0:5⟩/PB ⟨0:5⟩					
Conditional skip	5	reverse¶ test	clear bits after test	complement after test	Push_Buttons ⟨6:11⟩/PB ⟨6:11⟩					
Arithmetic compare PB	6	0	0	0	Push_Buttons ⟨0:5⟩					
Arithmetic compare PB	6	0	0	1	Push_Buttons ⟨6:11⟩					
Skip on flags	6	0	1	0	skip	skip if not in sector	skip if PB ⟨0:5⟩ = 0	skip if PB ⟨5:11⟩ = 0		
Count	6	0	1	1	count scale	0 → +1 1 → −1	count Intensity	0 → +1 0 → −1		
Set slaves	6	1	Group number ⟨0:1⟩		set unit 0	lt pen	Intensity	set unit 1	lt pen	Intensity
Spare	7									

† Set; allow instruction bits to specify new value.

‡ A two-word instruction, second word contains low-order 12 bits for DAC (jump address).

¶ Skip can be for true or false.

§ Inhibit restoration of bits.

1   Picture
   a   Display area
   b   Phosphor type (intensity and color as function of time)
   c   Spot size
   d   Resolution
   e   Linearity
   f   Short-term and long-term stability

2   Figure plotting (generation) characteristics
   a   Data types: points, lines (vectors), graphs, characters (from a fixed set), characters (from a defined set), curved-line segments, etc.
   b   Plotting time

3   Transformation and internal representations
   a   Space to encode (specify) a figure
   b   Scale change, rotation, coordinate-system transformation abilities
   c   Ability to communicate between a displayed data structure and an internal representation of a picture

4   Light-pen or graphic input capability

Instructions and their interpretation in P(display)

Two instruction-set types are interpreted in the P.display: Data State, in which instructions specify display information; and Control State, in which instructions specify program control information (e.g., jumps, modes, etc.). A state diagram for the interpretation process is given in Fig. 1.

Data-state instructions

There are seven instructions (which DEC calls modes) that can be executed while P.display is in data state. The instructions (modes) are really substates of data state. The instructions (actually more like data) are interpreted for the mode. When all the data-mode instructions have been interpreted, an escape instruction returns the P.display to control state. A control instruction is issued to select a mode and simultaneously place the display in data state.

Increment mode. This mode is used to draw curves and alphanumeric characters and other small symbols. Two instructions are stored per word. An instruction will cause the beam position to be moved one, two, or three times, in 0.010-inch increments, in one of eight directions. Direction 0 is to the right, direction 1 is up and to the right, etc.

Table 2   DEC 338 data-mode instruction set

				Instruction bits:											
Mode	Function	Time (μs)	Word	0	1	2	3	4	5	6	7	8	9	10	11
0	point	6 ~ 35	1 of 2	int[a]	inh[b]	Y	coordinate								
			2 of 2	esc[c]	inh	X	coordinate								
1	increment	1.5 + 2 × (.9 ~ 3.6)	1	int	move count[d]		move direction[e]			same as bits 0 ~ 5					
2	vector	1 ~ 150	1 of 2	int	±	Delta Y									
			2 of 2	esc	±	Delta X									
3	vector continue	1 ~ 1,200	1 of 2	int	±	Delta Y									
			2 of 2	esc	±	Delta X									
4	short vector	1.8 ~ 24	1	int	±	Delta Y				±	esc	Delta X			
5	6-bit character	3.75+	1	character 1						character 2					
5	7-bit character	4.5+	1	blank					character						
6	graph plot	6 ~ 35	1	esc	X/Y[f]	Y or X coordinate									
7	spare														

[a] Intensify; turn on beam.

[b] Inhibit; do not set value into Y or X coordinate.

[c] Escape; enter control state.

[d] 0 → move 1 and escape; 1, 2, 3, → move 1, 2, 3.

[e] 8 directions.

[f] 0 → set Y and increment X; 1 → set X and increment Y.

Vector mode. The vector mode is used to draw straight-line segments. This two-word instruction causes the beam position to be moved along a line represented by an 11-bit delta y and an 11-bit delta x.

Vector continue mode. This mode is used to draw a straight line to the edge of the screen. It is similar to vector mode but causes the line to be extended until an "edge" is encountered.

Short vector mode. The short vector mode is used to draw figures composed of short line segments. A one-word instruction specifies a 5-bit delta y and a 5-bit delta x quantity. It is transformed within the display to the same format as vector mode and operates in the same manner.

The preceding modes move the beam by counting the X and Y position registers. The counting is done at 1.2 microseconds per step on an intensified move and at 0.30 microsecond per step on a nonintensified move.

Point mode. Point mode is used for random point plotting. A two-word instruction specifies new Y and/or X coordinates to be placed into the Y and X position registers.

Graph-plot mode. This is used to draw curves of mathematical functions. A one-word instruction has data for the Y or X position register; at the same time, X or Y, respectively, is incremented by a count of one, two, four, or eight, depending on the scale factor.

Point and graph-plot modes operate at a rate depending upon the position of the new point with respect to the previous point. If a point is only one-eighth of the screen away, the delay for beam-settling time is 6 microseconds; otherwise the settling time is 35 microseconds.

Character generation option instructions. The alphanumeric characters or special symbols which make up a character set are stored in Mp in increment mode or short vector mode. These characters can be arbitrarily defined. A 6-bit (or 7-bit) character code in the instruction is used to locate a word in a table in Mp called the dispatch table. The base address of the table is specified by the Starting Address Register/SAR⟨0:5⟩.

SAR may be loaded by instructions from the Pc. The SAR represents the most significant 6 bits of a 15-bit memory address. The character code represents the least significant 6 (or 7) bits. A seventh SAR bit, corresponding to the octal position 100, is used with 6-bit characters as a case bit (i.e., uppercase or lowercase characters) and may be set or cleared with a control character.

A word in the dispatch table has the following format:

Bit 0:   If bit 0 is a 1, bits 1 to 11 are used to perform a control function as specified by particular control instructions. If bit 0 is a 0, bits 2 to 11 are combined with SAR to specify the address at which the character definition program starts. (The address bit 2 is common to both the SAR and bit 2 of the dispatch word and so may be specified in either place or in both places.)

Bit 1:   Determines the mode in which the character is to be displayed. If bit 1 is a 0, the increment mode is used to plot the character used; if bit 1 is a 1, the short vector mode is used to plot the character.

Control-state instructions

There are six control-state instructions.

Parameter. Parameter is used to set values in scale, light-pen, and intensity registers.

Mode. Mode is used to set up the data-state mode (or data-mode instruction). Mode also is used to stop the display.

Conditional skip. The skip instruction tests the state of the P.display and the pushbuttons.

Miscellaneous. These instructions include both tests and additional parameter control.

Display jump and push-jump subroutine instructions. The display jump instruction has 15 address bits, so that a jump may be executed to any location in the display file within the 32-kw memory.

The display subroutine instructions are push-jump (an extension of the jump instruction) and pop, the return from subroutine. The push-jump works as follows: The current state of the display (Light Pen Enable, Data Mode, Scale, and Intensity) is stored, along with the return address, in two successive locations in the first 4,096 words of memory. The locations are determined by the pushdown pointer, PDP. This pointer is initially set by a Pc instruction. The normal jump is then executed.

To return from a subroutine, the pop instruction is executed. It has no address bits. Its function is to return the display to a previous state by sending the last words on the push-down stack back to the display.

The stack approach to subroutining as implemented on the 338 has certain advantages over the jump to subroutine instruction normally used in Pc's:

1 Memory space is conserved since return address locations are not required in each subroutine in memory.

2 A subroutine can be called any number of times before return to the main routine.

3 Since the state of the display is saved on the stack and subsequently restored, subroutines are truly transparent; that is, after the return they leave the state of the display program the same as before the subroutine call.

4 The subroutines can either retain the same state or change the state of the display by using one or more of the "inhibit restore" bits available in the pop instruction. The programmer can elect independently to inhibit restoration of mode, light pen, and scale, or intensity information.

Instructions in Pc for communicating with P(display)

Instructions in Pc communicate with P.display. The physical connection is by the S('I/O Bus). The in-out transfer instructions in Pc are used to initialize and read the state of P.display.

P.display state initialization from Pc instructions
  Set Push Down Pointer from AC
  Set Display Address Counter from AC
  Set Push Button contents from AC
  Set miscellaneous flag and status bits from AC
  Set character generator SAR address

P.display status to Pc instructions
  Read Push Down Pointer into AC
  Read X register into AC
  Read Y register into AC
  Read Display Address Counter into AC
  Read Status words 1, 2, 3, 4, 5 into AC (60 miscellaneous bits of flags, modes, etc.)

Picture debugging modes. These modes aid programmed and picture debugging. A bit can be set to override the nonintensify bit in data-mode instructions. When this bit is a 1, all points and vectors are plotted, whether they are to be intensified or not. The search enable instruction forces the display to run until a particular instruction type is found. The instruction type is specified by the search enable instruction.

APPENDIX 1   DEC 338 DISPLAY PROCESSOR ISP DESCRIPTION

Appendix 1

DEC 338 Display Processor ISP Description (partially complete)

P.display State

  Program Flow State

    DAC<0:14>   *Display Address Counter; holds memory address of display instruction*

    PDP<0:11>   *Push Down Pointer to stack holding subroutine return addresses*

    Internal_Stop   *denotes halt by a P.display instruction*

    External_Stop   *denotes a request by Pc for P.display to halt*

Data_State and Control_State are two mutually exclusive states. Data_State instructions are interpreted by P.display as points, lines, and characters to be displayed on T. There are 7 modes for specifying the data types. The Data_Mode register holds the data type being interpreted. Control_State instructions include jump to subroutines using the stack, controlling P.display state registers and switching to a specific data mode.

    Data_State

    Control_State := ¬ Data_State

    Data_Mode/DM<0:2>   *specifies interpretation of Data_State instructions*

    SAR<0:5>   *Starting Address Register; base register of a dispatch table for calling character display subroutines*

  Picture State

    X<0:12>   *beam position; only integers in range $0 \le X|Y \le 2^{10+dimension-1}$*
    Y<0:12>   *are plotted*

    Vertical_edge_flag/Vef   *denotes if beam is within a displayable area*

    Horizontal_edge_flag/Hef   *set when beam moves outside the display area*

    Edge_Interrupt/EI

    CHSZ   *Character Size, 0 indicates 6 bit character set 1 indicates 7 bit character set*

    Scale<0:1>   *used to set increment size for Data_Mode instructions, incre-*
      Scale'<0:2> := (Scale<0:1> + 1)   *ments are X 2^{scale}*

    Intensity<0:2>   *brightness of displayed points*

    X_dimension<0:1>   *maximum dimension of plotting area, 9.375, 18.75, 37.5, 75.0 in*
    Y_dimension<0:1>   *on, to display a point or line; automatically turned off at*
    Beam   *instruction completion*

  Console and Light Pen State

    Push_Buttons/PB<0:11>   *register with lights: can be complemented manually or by processor*

    Push_Button_Hit/PBH   *flag is set by manually striking any push button*

    Manual_Interrupt/MI   *key which is used to interrupt Pc and becomes one when struck*

    Light_Pen_Find/LPF   *stops the display and interrupts Pc whenever the Light Pen has seen a displayed spot and the Light_Pen_Enable is a one*

    Light_Pen_Enable/LPE   *a bit to enable the Light_Pen_Find flag to cause an interrupt*

  Mp State

    M[0:7] [0:4095]<0:11>   *primary memory for P.display and Pc*

  Instruction Format

    instruction/i<0:11>   *The individual instructions fields are defined below. Each instruction type has its own bit field assignments.*

      enter_data_state   := i<11>   *common bits for several instructions*

      pb_sense   := i<3>   *push button control bits*

APPENDIX 1   DEC 338 DISPLAY PROCESSOR ISP DESCRIPTION (Continued)

```
        pb⌣clear                    := i<4>
        pb⌣complement               := i<5>
        pb⌣select<0:5>              := i<6:11>
        scale⌣change/sc             := i<3>                    scale (size) control bits
        scale⌣value/sv<0:1>         := i<4:5>
        light⌣pen⌣change/lpc        := i<6>                    light pen test control bits
        light⌣pen⌣bit/lpb           := i<7>
```

Instruction Interpretation Process

```
    (¬ Internal⌣Stop V ¬ External⌣Stop) →                      fetch
        (instruction[0:1] ← M[DAC:DAC+1] ; DAC ← DAC + 1; next
        (Control⌣State ∧ (instruction<0:1> = 2)) → (DAC ← DAC + 1);    2 w instruction
        (Data⌣State ∧ ((Data Mode = 0) V (Data⌣Mode = 2) V          2 w data
            (Data Mode = 3))) → (DAC ← DAC + 1);
        next Instruction⌣execution)                            execute
```

Instruction Set and Instruction Execution Process
The following instruction set definition is not complete. It does not include the complete character instruction definition or the miscellaneous and conditional skip instructions. Most of the instructions are microcoded.

```
    Instruction⌣execution := (

    Control Instructions
    parameter<0:11> := i[0]<0:11>                              set parameter instruction format
        parameter⌣opcode           := (i<0:2> = 000)
        parameter⌣intensity⌣change := parameter<8>            set parameter execution
        parameter⌣intensity<0:2>   := parameter<9:11>
    parameter⌣opcode ∧ Control⌣State → (
        scale⌣change → (Scale ← scale⌣value);
        light⌣pen⌣change → (Light⌣Pen⌣Find ← ¬ light⌣pen⌣bit);
        intensity⌣change → (Intensity ← parameter⌣intensity));

    mode<0:11> := i<0:11>                                      set mode instruction format
        mode⌣opcode                := (i<0:2> = 001)
        mode⌣stop⌣code             := mode<3>
        mode⌣clear⌣push⌣button⌣flag:= mode<4>
        mode⌣data⌣mode⌣change      := mode<5>
        mode⌣set<0:2>              := mode<6:8>
        mode⌣clear⌣sector          := mode<9>
        mode⌣clear⌣coordinate      := mode<10>
    mode⌣opcode ∧ Control⌣State → (                            set mode execution
        mode⌣stop⌣code → (Internal⌣Stop ← 1);
        mode⌣clear⌣push⌣button⌣flag → (Push⌣Button⌣Hit ← 0);
        mode⌣data⌣mode⌣change → (Data⌣Mode ← mode⌣set);
        mode⌣clear⌣sector → (X<0:2> ← 0; Y<0:2> ← 0);
        mode⌣clear⌣coordinate → (X<3:12> ← 0; Y<3:12> ← 0);
        enter⌣data⌣state → (Data⌣State ← 1)));
```

APPENDIX 1   DEC 338 DISPLAY PROCESSOR ISP DESCRIPTION (Continued)

```
    PB_1<0;11> := i<0;11>                                    group 1 push button test and set instruction format for
       PB_1_opcode := (PB_1<0:2> = 100)                        Push Buttons 0 to 5
                                                             group 2 (not defined) is for Push Buttons 6 to 11
       PB_1_opcode ∧ Control_State → (                       PB_1 instruction execution
          pb_sense ⊕ (pb_select<0:5> ≡ (PB<0:5> ∧ pb_select<0:5>)) → ( skip test
             DAC ← DAC + 2);


          pb_clear → (PB<0:5> ← PB<0:5> ∧ pb select<0:5>); next
          pb_complement → (PB<0:5> ← PB<0:5> + pb_select<0:5>));
    jump[0:1]<0:11> := i[0:1]<0:11>                          jump and stack push down (subroutine calling) instruction
       jump_op          := (i[0]<0:2> = 010)                   format
       jump_push        := i[0]<8>
       jump_field<0:2>  := i[0]<9:11>
    jump_op ∧ Control_State → (
       scale_change → (Scale ← scale_value);                jump and push down execution
       light_pen_change → (Light_Pen_Find ← light_pen_bit);
       DAC ← jump_field□i[1];
       jump_push → (
          M[PDP + 1] ← DAC<0:2>□LPF□Scale□Data_Mode□Intensity:
          M[PDP + 2] ← DAC<3:14>;
          PDP ← PDP + 2);
    pop<0:11> := i[0]<0:11>                                  stack pop instruction format; subroutine return
       pop_op_code           := (i<0:2> = 011)
       pop_inhibit_mode      := pop<8>
       pop_inhibit_scale_pen := pop<9>
       pop_inhibit_intensity := pop<10>
    pop_op_code ∧ Control_State → (                          pop execution
       DAC<3:14> ← M[PDP];
       DAC<0:2> ← M[PDP-1];
       ¬ pop_inhibit_intensity → (Intensity ← M[PDP-1]<9:11>);
       ¬ pop_inhibit_mode → (Data_Mode ← M[PDP-1]<6:8>);
       ¬ pop_inhibit_scale_change → (
          Scale ← M[PDP-1]<4:5>
          LPF ← M[PDP-1]<3>);
       PDP ← PDP - 2; next
       scale_change → (Scale ← scale_value);
       light_pen_change → (LPF ← light_pen_bit);
       enter_data_mode → (Data_Mode ← 1));
Data Mode Instructions                                       point data instruction format
point[0:1]<0:11> := i[0:1]<0:11>
   point_intensity     := point[0]<0>
   point_inhibit_y     := point[0]<1>
   point_y<0:9>        := point[0]<2:11>
   point_x<0:9>        := point[1]<2:11>
   point_escape        := point[1]<0>
   point_inhibit_x     := point[1]<1>
```

APPENDIX 1   DEC 338 DISPLAY PROCESSOR ISP DESCRIPTION (Continued)

```
(Data_Mode = 000) ∧ Data_State → (                          point data execution
  ¬ point_inhibit_x → (X ← point_X);
  ¬ point_inhibit_y → (Y ← point_Y);
    point_intensify → (Beam ← 1);
    point_escape → (Data_State ← 0));
vector[0]<0:11>:= i[0:1]<0:11>                              vector data instruction format
  vector_intensify      := vector[0]<0>
  vector_escape         := vector[1]<0>
  vector_dy<0:10>       := vector[0]<1:11>
  vector_dx<0:10>       := vector[1]<1:11>
(Data_Mode = 010) ∧ Data_State → (                          vector data execution
  Y ← Y + vector_dy;                                        not correct, since the vector from point Y,X to Y+ vector_dy,
  X ← X + vector_dx;                                         X+ vector_dx is plotted
  vector_intensify → (Beam ← 1);
  vector_escape → Data_State ← 0);

vector continue[0:1]<0:11> := i[0:1]<0:11>                  vector continue instruction format same as vector
(Data_Mode = 011) ∧ Data_State → (                          vector continue execution
  Y ← Y + sign_extend(vector_dy);
  X ← X + sign_extend(vector_dx);                           not correct, as vector continues plotting until edge is found
  vector_intensify → (Beam ← 1);
  vector_escape → (Data_State ← 0));
short_vector<0:11> := i[0]<0:11>                            short vector instruction format
  short_vector_intensify   := short_vector<0>
  short_vector_escape      := short_vector<6>
  short_vector_dx          := short_vector<8:11>
  short_vector_dy          := short_vector<1:5>
(Data_Mode = 100) ∧ Data_State → (                          short vector execution
  X ← X + sign_extend(short_vector_dx);
  Y ← Y + sign_extend(short_vector_dy);
  short_vector_intensify → (Beam ← 1);
  short_vector_escape → (Data_State ← 0));

increment<0:5>                                              increment instruction format; 2 increment/instruction
  increment_intensify         := increment<0>
  increment_direction/id<0:2> := increment<3:5>            1 of 8 directions
  increment_count/ic<0:1>     := increment<1:2>
      ic1e   := (ic = 0)                                    count 1 and escape to Control_State
      ic1    := (ic = 1)                                    count 1
      ic2    := (ic = 2)                                    count 2
      ic3    := (ic = 3)                                    count 3
(Data_Mode = 001) ∧ Data_State → (                          increment instruction execution
  increment ← i<0:5>; next plot_increment_vector; next
  increment ← i<6:11>; next plot_increment_vector)
```

APPENDIX 1 DEC 338 DISPLAY PROCESSOR ISP DESCRIPTION (Continued)

```
plot increment vector := (
   icle → (move 1 position; Control State ← 1);          move 1 and escape
   icl → (move 1 position);                              move 1
   ic2 → (move 1 position; next move 1 position)         move 2
   ic3 → (move 1 position; next move 1 position; next    move 3
       move 1 position)
Move 1 position := (                                     sub process for moving beam
   (id = 0) → (X ← X + Scale);                           1 of 8 positions
   (id = 1) → (X ← X + Scale; Y ← Y + Scale);
   (id = 2) → (Y ← Y + Scale);
   (id = 3) → (Y ← Y + Scale; X ← X - Scale);
   (id = 4) → (X ← X - Scale);
   (id = 5) → (Y ← Y - Scale; X ← X - Scale);
   (id = 6) → (Y ← Y - Scale);
   (id = 7) → (Y ← Y - Scale; X ← X + Scale);
   increment intensify → Beam ← 1)
character<0:11> := i<0:11>                               character instruction format
   6 bit [0:1]<0:5> := character<0:11>
   7 bit<5:11>      := character<5:11>
(Data Mode = 101) ∧ Data State → (                       character instruction execution;
   (CHSZ = 0) → (
       X,Y ← f(M[SAR☐6 bit[0]],M);                       plot function;
       X,Y ← f(M[SAR☐6 bit[1]],M));
   (CHSZ = 1) → (X,Y ← f(M[SAR☐7 bit],M)));              see text
graph plot<0:11> := i [0]<0:11>                          graph data instruction format
   graph plot escape<0>  := graph plot<0>
   graph plot x y<0>     := graph plot<1>
   graph plot data<0:9>  := graph plot<2:11>
(Data mode = 110) ∧ Data State → (                       graph data execution
   ¬ graph plot x y → (X ← X + Scale'; Y ← graph plot data; Beam ← 1);
     graph plot x y → (Y ← Y + Scale'; X ← graph plot data; Beam ← 1);
     graph plot escape → (Data State ← 0))
                                     )                   end Instruction execution
```

Section 2

Processors for array data

Two array processors are discussed in this section. Conceptually, they are an outgrowth of both the parallel, distributed computer [Holland, 1959], and the matrix-interpreter-based programs for general-purpose computers. NOVA is a very low cost special processor. ILLIAC IV is a very general array processor. Another approach, the ILLIAC III [McCormick, 1963] stores information on photographic media, so that optical processing (inherently parallel) can be used.

NOVA

NOVA is a proposed, non-general-purpose machine based on the belief that efficient, special-function processors can be built to solve particular problems.

It is reasonable to assume that there are problems for which NOVA, with its cyclic memory, would perform no worse than a processor with a random-access memory. Unless the operations performed on the arrays were extremely simple or restricted, a single system might not always work very efficiently. By using a variable-speed cyclic memory to match the operation time in the form of an address transformation or renaming mechanism, the access problems might be avoided.

NOVA represents a particular idea for effective utilization of hardware and is presented to remind us that a memory now considered obsolete may perform nicely for a restricted application.

The ILLIAC IV computer

D. L. Slotnick is responsible for the ILLIAC IV computer. The idea for a computer with a number of parallel data operators or processing elements appeared some time ago in the SOLOMON computer [Gregory and McReynolds, 1963]. The technology of the first and second generation made SOLOMON impractical to build. ILLIAC IV was designed at the University of Illinois under a contract to the Department of Defense's Advanced Research Projects Agency.[1] The processing elements are constructed from third-generation technology although some medium- and large-scale integrated circuits are used in the design.

The design is about the most ambitious ever undertaken. The direct and indirect effects should be numerous.

[1] The University of Illinois monitored the contract to the Burroughs Corporation, Paoli, Pa.

Chapter 26

NOVA: a list-oriented computer[1]

Joseph E. Wirsching

Since the advent of the internally-stored program computer, those of us concerned with problems involving massive amounts of computation have taken a one-operation, one-operand approach. But there is a very large class of problems involving massive amounts of computation that may be thought of as one-operation, many-operand in nature. Some familiar examples are numerical integration, matrix operations, and payroll computation.

This article proposes a computer, called NOVA, designed to take advantage of the one-operation, many-operand concept. NOVA would use rotating memory instead of high-cost random access memory, reduce the number of program steps, and reduce the number of memory accesses to program steps. In addition it is shown that NOVA could execute typical problems of the one-operation, many-operand type in times comparable to that of modern high-speed random access computers.

Rotating memories were used in early computers because of low cost, reliability, and ease of fabrication. These machines have been replaced by machines with more costly random access memories primarily to increase computing speed as the result of a decrease in access time to both operands and instructions.

The NOVA approach

Let us take two simple examples and use them to compare conventional computing techniques with those proposed for NOVA.

Example 1. Consider two lists (a's and b's) of which the corresponding pairs are to be added. With a conventional computer this is done with a program that adds the first a to the first b, the second a to the second b, etc., and counts the operations. The working part of such a program might consist of the following instructions:

Fetch a
Add b
Store $(a + b)$
Count, Branch, and Index

[1] *Datamation*, vol. 12, no. 12, pp. 41–43, December, 1966.

In general, the four or more instructions must be brought from the memory to the instruction register once for each pair in the lists. This seems to be a great waste when only one arithmetic operation is involved. Indeed it is, when one considers that the majority of computing work consists of the performance of highly repetitive operations that are merely combinations of the simple example given. Attempts have been made to alleviate this waste by incorporating "instruction stacks" and "repeat" commands into the instruction execution units of more recent computers.

Example 2. Consider three lists (a's, b's and c's), where we wish to compute $(a + b) \times c$ for each trio. There are two distinct methods by which this can be accomplished: first, by forming $(a + b) \times c$ for each trio of numbers in the list, or second, by forming a new list consisting of $(a + b)$ for each a and b, and then multiplying each c by the corresponding member of the new list. Clearly the second method is wasteful of memory space and wasteful of programming steps.

Next, let us take a look at the memory requirements for these two examples. First, the instructions are kept in a high-speed random access memory, and while the bulk of the variables need not be kept in a random access memory, they must be brought to one before the algorithm can be performed. This extra transfer may entail more instructions to perform the logistics. Thus the simplicity of the overall program is directly related to the size of the memory. The variables (a's, b's, etc.) are usually stored in consecutive memory locations. Except for indexing this ordering of the data is not exploited.

In NOVA, lists of variables are kept on tracks of a rotating bulk memory. When called for, the lists of variables are streamed through an arithmetic unit and the results immediately replaced on another track for future use. This process takes maximum advantage of the sequential ordering of the variables. Instructions need only be brought to the instruction execution unit *once* for each pair of lists rather than once for each operand; thus the instructions need not be stored in a random access memory but may also be stored on the rotating bulk memory. This departure from the requirement for random access memory significantly reduces the cost of the computer, without sacrificing speed of problem solution.

Solution of a network problem

Before going further into the structure of NOVA, let us consider a significant example, which shows that NOVA is well suited to the solution of differential equations using difference methods over a rectangular network.

Let Fig. 1 represent an artificial network used as a model for some physical process. Generally speaking, the method of advancing the variables at a mesh point (j, k) from one time step to the next involves only information from the neighboring mesh points. A typical hydrodynamics problem will require a list of 10 to 20 variables (physical quantities) at each mesh point. The traditional computer solution involves listing these variables to each point in a contiguous fashion and in a regular sequence with respect to the rows and columns of the array. If the total array does not fit into the fast memory, three adjacent columns (or rows) are brought to the fast memory; as a new column is calculated, the next column in sequence is brought in from bulk memory and the oldest of the three is written to bulk memory. In this fashion one proceeds across the array. This process is then repeated until some significant physical occurrence happens and the problem is ended.

In NOVA, the variables are organized into separate lists rather than by mesh point. From a computational standpoint this is possible since the main memory of NOVA may be essentially unlimited in size, at least exceeding the size of the largest present network problems. One then proceeds to execute operations on

Original Lists	V Shifted Down by 1	V Shifted Down By 2	V Shifted Down By K
$U_{0,0}$ $V_{0,0}$	—	—	—
$U_{0,1}$ $V_{0,1}$	$V_{0,0}$	—	—
$U_{0,2}$ $V_{0,2}$	$V_{0,1}$	$V_{0,0}$	—
. .	$V_{0,2}$	$V_{0,1}$	—
. .	.	.	—
$U_{1,0}$ $V_{1,0}$	$V_{0,K}$	$V_{0,K-1}$	$V_{0,0}$
$U_{1,1}$ $V_{1,1}$	$V_{1,0}$	$V_{0,K}$	$V_{0,1}$
$U_{1,2}$ $V_{k,2}$	$V_{1,1}$	$V_{1,0}$	$V_{0,2}$
. .	.	.	.
. .	.	.	.
$U_{j,k}$	$V_{j,k}$	$V_{j,k-1}$	$V_{j-1,k}$
. .	.	.	.
. .	.	.	.
$U_{J,K}$ $V_{J,K}$	$V_{J,K-1}$	$V_{J,K-2}$	$V_{J-1,K}$
	$V_{J,K}$	$V_{J,K-1}$	.
		$V_{J,K}$	.

Fig. 2. Lists of variables.

lists of variables rather than single variables, performing a single operation for all mesh points in the array in sequence.

Let us look more closely at the variables and their possible combinations. Let $U_{j,k}$ and $V_{j,k}$ be variables associated with the array of Fig. 1. These variables are listed sequentially by column in Fig. 2, along with further lists of the V column shifted by various increments.

With some concentration, one discovers in Fig. 2 that an arithmetic operation between $U_{j,k}$ and $V_{j,k}$ is simply a matter of taking the two columns as they exist and operating on them in pairs. To combine $U_{j,k}$ with a nearby neighbor, $V_{j,k-1}$, the V column is shifted down one place, at which time the proper neighboring variables are found opposite one another for the entire network. At certain boundaries of the array some elements have no proper neighbors. In NOVA these boundary elements must be handled separately in the same way as they must be handled separately in a conventional machine. In NOVA, calculations at boundaries may be temporarily inhibited by having a third input to the arithmetic unit which allows the calculation of a result for a pair of operands to proceed or not, as appropriate. This third input is defined as "conditions," and is brought as a bit string to the arithmetic unit concurrently with the operands. This bit string may contain any number from one to several bits for each pair of operands.

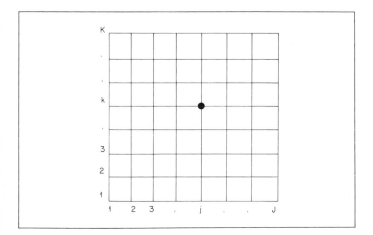

Fig. 1. Two-dimensional array.

Further observation shows not only that it is possible to obtain the nearest neighbors easily by shifting the columns of variables with respect to one another, but that any neighbor relationship can be obtained. In general, for an operation with a neighbor $\pm n$ rows away and $\pm m$ columns away, the lists are offset by $\pm n \pm m \cdot K$, where K is the number of rows in the array.

Many problems (for example, payroll and inventory records) are essentially list-structured but do not require offsetting of variables. Clearly the NOVA structure is well suited for the solutions of these problems also.

Structure

The most difficult problem to be solved in the proposed computer is to synchronize movement of the columns of data that require offset. Buffers of various types could be used to solve this problem; they could range all the way from rotating memory devices or delay lines to core memories. The former are simple, direct, and low in cost but are limited in their general capabilities. On the other hand, a number of small random access buffer memories could be used for offsetting lists of variables and for facilitating special functions such as boundary calculations but at a higher equipment cost.

Figure 3 shows a block diagram of the organization of NOVA. The rotating memory, which might be a disc or drum, would be

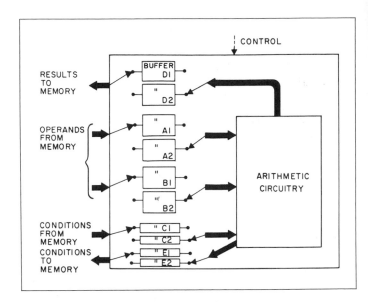

Fig. 4. Buffering in arithmetic unit.

composed of several hundred tracks, each storing several thousand words, with a total capacity between one and two million words. Each track would have an individual read-write head. The heads would be organized in such a way as to attain a high word-transfer rate, perhaps as high as one million words per second. With this in mind an ideal execution time for one addition would be the time required to move two operands from the disc to the arithmetic unit; i.e., 1-2 microseconds. The disc synchronizer would be capable of simultaneously reading two lists of operands, writing one list of results, and reading one list and writing one list of conditional control information. In addition, instructions would be read from another channel in small blocks.

The bit string of conditions coming from the memory is used to control individual operations on pairs of operands in the lists, and in essence each bit (or bits) is a subordinate part of the individual operations. Conditions going to the memory are the subsidiary result of the operation of one list upon another. These bit strings may be used later as control during another list operation. They want also to contain information on the occurrence of an overflow or underflow, or on the presence of an illegal operand, etc.

Figure 4 shows a suggested organization for the arithmetic unit that incorporates five sets of alternating buffers. Two sets are for lists of operands coming from the memory, one set for lists of results going to the memory, and two sets for "conditions" (conditional control information) coming from and going to the memory.

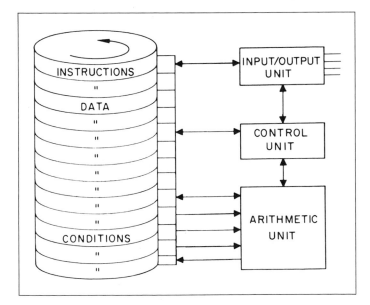

Fig. 3. Block diagram of NOVA computer.

These buffers should be equivalent in length to the number of words on a track of the rotating memory.

The loading and unloading of the buffers to and from the rotating memory is dependent on the timing of the rotating memory, whereas the loading and unloading of the buffers to and from the arithmetic unit is guided solely by the rate at which the arithmetic can be performed. Here again it may also be possible to take advantage of the streaming nature of the operands by designing an "assembly-line" arithmetic unit in which more than one pair of operands could be in process at the same time. With this kind of unit it may be possible to execute additions at a rate equal to the word-transfer rate from the rotating memory; however, a multiplication or division of two lists may require several revolutions of the memory. The timing diagram of Fig. 5 shows several typical instructions being carried out. A certain amount of look-ahead is required, but there is ample time for this, since instructions are prepared for execution at an average rate of less than one per revolution of the rotating memory.

While a detailed cost estimate has not been made for a simple prototype NOVA, a quick estimate would be $50,000 for a head-per-track disc and $50,000 for the arithmetic and control section, making a total of $100,000. For a buffering scheme such as the one shown in Fig. 4 the cost would be considerably higher but would be offset by increased versatility.

Conclusions

In the previous paragraphs we have demonstrated that NOVA is capable of handling network problems at a significantly lower cost than contemporary computers, and at a comparable speed. The availability of such a machine as NOVA would stimulate further

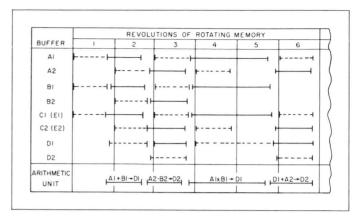

Fig. 5. Timing diagram of buffers, rotating memory, and arithmetic unit. Dotted line shows movement of data into a device; solid line shows movement out.

interest in the one-operation, many-operand approach to computation and no doubt would uncover many other problems to which it could be applied.

Because NOVA makes it possible to easily establish neighbor-relationships between mesh points that are further away than nearest neighbors, it may be possible to develop new differencing techniques for the solution of coupled sets of differential equations. This may increase the accuracy or shorten the time required for their solution.

The memory, arithmetic, and other units needed for NOVA are commercially available now. No new technology would be required to fabricate a prototype model. In view of the potential advantages of such a machine, it seems clear that construction of a model would justify the minimal development costs.

Chapter 27

The ILLIAC IV computer[1]

George H. Barnes / Richard M. Brown / Maso Kato
David J. Kuck / Daniel L. Slotnick / Richard A. Stokes

Summary The structure of ILLIAC IV, a parallel-array computer containing 256 processing elements, is described. Special features include multiarray processing, multiprecision arithmetic, and fast data-routing interconnections. Individual processing elements execute 4×10^6 instructions per second to yield an effective rate of 10^9 operations per second.

Index terms Array, computer structure, look-ahead, machine language, parallel processing, speed, thin-film memory.

Introduction

The study of a number of well-formulated but computationally massive problems is limited by the computing power of currently available or proposed computers. Some involve manipulations of very large matrices (e.g., linear programming); others, the solution of sets of partial differential equations over sizable grids (e.g., weather models); and others require extremely fast data correlation techniques (phased array signal processing). Substantive progress in these areas requires computing speeds several orders of magnitude greater than conventional computers.

At the same time, signal propagation speeds represent a serious barrier to increasing the speed of strictly sequential computers. Thus, in recent years a variety of techniques have been introduced to overlap the functions required in sequential processing, e.g., multiphased memories, program look-ahead, and pipeline arithmetic units. Incremental speed gains have been achieved but at considerable cost in hardware and complexity with accompanying problems in machine checkout and reliability.

The use of explicit parallelism of operation rather than overlapping of subfunctions offers the possibility of speeds which increase linearly with the number of gates, and consequently has been explored in several designs [Slotnick et al., 1962; Unger, 1958; Holland, 1959; Murtha, 1966]. The SOLOMON computer [Slotnick et al., 1962], which introduced a large degree of overt parallelism into its structure, had four principal features.

1 A large array of arithmetic units was controlled by a single control unit so that a single instruction stream sequenced the processing of many data streams.

2 Memory addresses and data common to all of the data processing were broadcast from the central control.

3 Some amount of local control at the individual processing element level was obtained by permitting each element to enable or disable the execution of the common instructions according to local tests.

4 Processing elements in the array had nearest-neighbor connections to provide moderate coupling for data exchange.

Studies with the original SOLOMON computer indicated that such a parallel approach was both feasible and applicable to a variety of important computational areas. The advent of LSI circuitry, or at least medium-scale versions, with gate times of the order of 2 to 5 ns, suggested that a SOLOMON-type array of potentially 10^9 word operations per second could be realized. In addition, memory technology had advanced sufficiently to indicate that 10^6 words of memory with 200 to 500-ns cycle times could be produced at acceptable cost. The ILLIAC IV Phase I design study during the latter part of 1966 resulted in the design discussed in this paper. The machine, to be fabricated by the Defense Space and Special Systems Division of Burroughs Corporation, Paoli, Pa., is scheduled for installation in early 1970.

Summary of the ILLIAC IV

The ILLIAC IV main structure consists of 256 processing elements arranged in four reconfigurable SOLOMON-type arrays of 64 processors each. The individual processors have a 240-ns ADD time and a 400-ns MULTIPLY time for 64-bit operands. Each processor requires approximately 10^4 ECL gates and is provided with 2048 words of 240-ns cycle time thin-film memory.

Instruction and addressing control

The ILLIAC IV array possesses a common control unit which decodes the instructions and generates control signals for all

[1] *IEEE Trans., C-17*, vol. 8, pp. 746–757, August, 1968.

processing elements in the array. This eliminates the cost and complexity for decoding and timing circuits in each element.

In addition, an index register and address adder are provided with each processing element, so that the final operand address a_i for element i is determined as follows:

$$a_i = a + (b) + (c_i)$$

where a is the base address specified in the instruction, (b) is the contents of a central index register in the control unit, and (c_i) is the contents of the local index register of the processing element i. This independence in operand addressing is very effective for handling rows and columns of matrices and other multidimensional data structures [Kuck, 1968].

Mode control and data conditional operations

Although the goal of the ILLIAC IV structure is to be able to control the processing of a number of data streams with a single instruction stream, it is sometimes necessary to exclude some data streams or to process them differently. This is accomplished by providing each processor with an ENABLE flip-flop whose value controls the instruction execution at the processor level.

The ENABLE bit is part of a test result register in each processor which holds the results of tests conditional on local data. Thus in ILLIAC IV the data conditional jumps of conventional computers are accomplished by processor tests which enable or disable local execution of subsequent commands in the instruction stream.

Routing

Each processing element i in the ILLIAC IV has data routing connections to 4 of its neighbors, processors $i + 1$, $i - 1$, $i + 8$, and $i - 8$. End connection is end around so that, for a single array, processor 63 connects to processors 0, 62, 7, and 55.

Interprocessor data transmissions of arbitrary distance are accomplished by a sequence of routings within a single instruction. For a 64-processor array the maximum number of routing steps required is 7; the average overall possible distances is 4. In actual programs, routing by distance 1 is most common and distances greater than 2 are rare.

Common operand broadcasting

Constants or other operands used in common by all the processors are fetched and stored locally by the central control and broadcast to the processors in conjunction with the instruction using them. This has several advantages: (1) it reduces the memory used for storage of program constants, and (2) it permits overlap of common operand fetches with other operations.

Processor partitioning

Many computations do not require the full 64-bit precision of the processors. To make more efficient use of the hardware and speed up computations, each processor may be partitioned into either two 32-bit or eight 8-bit subprocessors, to yield 512 32-bit or 2048 8-bit subprocessors for the entire ILLIAC IV set.

The subprocessors are not completely independent in that they share a common index register and the 64-bit data routing paths. The 32-bit subprocessors have separate enabled/disabled modes for indexing and data routing; the 8-bit subprocessors do not.

Array partitioning

The 256 elements of ILLIAC IV are grouped into four separate subarrays of 64 processors, each subarray having its own control unit and capable of independent processing. The subarrays may be dynamically united to form two arrays of 128 processors or one array of 256 processors. The following advantages are obtained.

1　Programs with moderately dimensioned vector or matrix variables can be more efficiently matched to the array size.

2　Failure of any subarray does not preclude continued processing by the others.

This paper summarizes the structure of the entire ILLIAC IV system. Programming techniques and data structures for ILLIAC IV are covered in a paper by Kuck [1968].

ILLIAC IV structure

The organization of the ILLIAC IV system is indicated in Fig. 1. The individual processing elements (PEs) are grouped in four arrays, each containing 64 elements and a control unit (CU). The four arrays may be connected together under program control to permit multiprocessing or single-processing operation. The system program resides in a general-purpose computer, a Burroughs B 6500, which supervises program loading, array configuration changes, and I/O operations internal to the ILLIAC IV system and to the external world. To provide backup memory for the ILLIAC IV arrays, a large parallel-access disk system (10 bits, 10^9 bit per second access rate, 40-ms maximum latency) is directly coupled to the arrays. There is also provision for real-time data connections directly to the ILLIAC IV arrays.

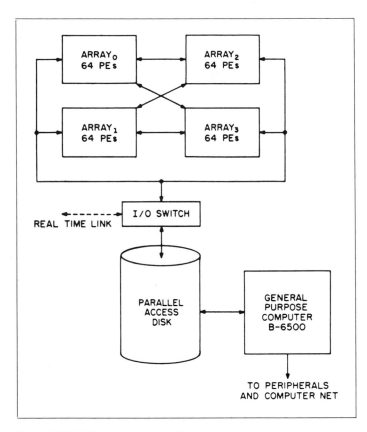

Fig. 1. ILLIAC IV system organization.

Array organization

The internal structure of an array is indicated in Fig. 2. The 64 processing elements in each array are arranged in a string and are controlled by the control unit (CU) which receives the instruction string, generates the appropriate control signals and address parameters of the instructions, and transmits them to the individual processing elements for execution. In addition, each CU can broadcast via the common data bus operands for common use (e.g., constant).

Full word length (64 bits) communication exists between the processing elements for exchange of information by organized routing of words along the string array. Direct routing connections exist for nearest neighbors and also for processing elements 8 units away. Routing for intermediate distances are generated via sequences of routes of $+1$, -1, $+8$, or -8. The end connections of the string are circular, but can be broken and connected to the ends of other arrays when the system is organized in one of the multiarray configurations.

All processing elements of an array execute, of course, the same instruction in unison under the control of the CU; local control is provided by the mode bit in each processing element which enables or disables the execution of the current instruction. The control unit is able to sense the mode bits of all processing elements under its control and thereby monitor the state of operation.

Multiarray configurations

To permit more optimal matching of array size to problem structure, the four arrays may be united in three different configurations, as shown in Fig. 3. To enlarge the arrays, the end connections of the PE strings are decoupled and attached to the ends of the other arrays to form strings of 128 or 256 processors. For multiarray configurations all CUs receive the same instruction string and any data centrally accessed. The control units execute the instructions independently, however, with inter-CU synchronization occurring only on those instructions in which data or control information must cross array boundaries. This simplifies and speeds up the instruction execution in multiarray configurations. The multiplicity of array configurations introduces complexities in memory addressing which will be discussed in a later section.

Control unit

The array control unit (CU) has the following five functions.

1   To control and decode the instruction streams

2   To generate the control pulses transmitted to the processing elements for instruction execution

3   To generate and broadcast those components of memory addresses which are common to all processors

4   To manipulate and broadcast data words common to the calculations of all the processors

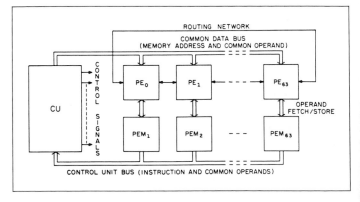

Fig. 2. Array structure.

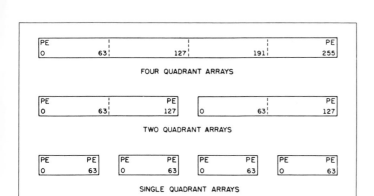

Fig. 3. **Multiarray configurations.**

5   To receive and process trap signals arising from arithmetic faults in the processors, from internal I/O operations, and from the B 6500.

The structure of the control unit is shown in Fig. 4. Principal components of the CU are two fast-access buffers of 64 words each, one associatively addressed, which holds current and pending instructions (PLA), and the other a local data buffer (LDB). The four 64-bit accumulator registers (CAR) are central to communication within the CU and hold address indexing information and active data for logical manipulation or broadcasting. The CU arithmetic unit (CULOG) performs addition, subtraction, and Boolean operations; more complex data manipulations are relegated to the PE's. To specify and control array configurations, there are three 4-bit configuration control registers whose use will be described in another section.

Instruction processing

All instructions are 32 bits in length and belong to one of two classes: CU instructions, which generate operations local to the CU (e.g., indexing, jumps, etc.), and PE instructions, which are decoded in the CU and then transmitted via control pulses to all the processing elements. Instructions flow from the array memory upon demand in blocks of 8 words (16 instructions) into the instruction buffer. As the control advances, individual instructions are extracted from the instruction buffer and sent to the advanced instruction station (ADVAST) which decodes them and executes those instructions local to the CU. In the case of PE instructions, ADVAST constructs the necessary address or data operands and stacks the result in a queue (FINQ) to await transmission to the PEs. PE instructions are taken from the bottom of the stack to

the final instruction station (FINST) which controls the broadcast of address or data and holds the PE instruction during the execution period.

The use of the PE instruction queue permits overlap between the CU and PE instruction executions; the amount of overlap depends, of course, on the distribution of CU and PE instructions. As in all overlap strategies, careful attention to the instruction sequence by the programmer or compiler can result in considerable speedup of program execution.

The instruction buffer holds a maximum of 128 instructions, sufficient to hold the inner loop of many programs. For such loops, after initial loading, instructions are fetched from the buffer with minimal delay.

A variety of strategies for instruction buffer loading were examined, and the following straightforward approach was taken. When the instruction counter is halfway through a block of 8

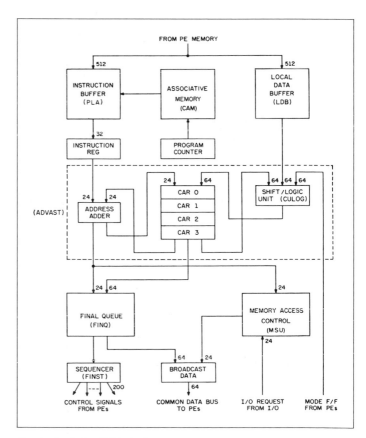

Fig. 4. **Control-unit block diagram.**

words (16 instructions), fetch of the next block is initiated; the possibility of pending jumps to different blocks is ignored. If the next block is found to be already resident in the buffer, no further action is taken; else fetch of the next block from the array memory is initiated. On arrival of the requested block, the instruction buffer is cyclically filled; the oldest block is assumed to be the least required block in the buffer and is overwritten. Jump instructions initiate the same procedures.

Fetch of a new instruction block from memory requires a delay of approximately three memory cycles to cover the signal transmission times between the array memory and the control unit. On execution of a straight line program, this delay is overlapped with the execution of the 8 instructions remaining in the current block.

In a multiple-array configuration, instructions are fetched from the array memory specified by the program counter, and broadcast simultaneously to all the participating control units. Instruction processing thereafter is identical to that for single-array operation, except that synchronization of the control units is necessary whenever information, in the form of either data or control signals, must cross array boundaries. CU synchronization must be forced at all fetches of new instruction blocks, upon all data routing operations, all conditional program transfers, and all configuration-changing instructions. With these exceptions, the CUs of the several arrays run independently of one another. This simplifies the control in the multiple-array operation; furthermore, it permits I/O transactions with the separate array memories without stealing memory cycles from the nonparticipating memories.

Memory addressing

Both data and instructions are stored in the combined memories of the array. However, the CU has access to the entire memory, while each PE can only directly reference its own 2,048-word PEM. The memory appears as a two-dimensional array with CU access sequential along rows and with PE access down its own column. In multiarray configurations the width of the rows is increased by multiples of 64.

The resulting variable-structure addressing problem is solved by generating a fixed-form 20-bit address in the CU as shown in Fig. 5. The lower 6 bits identify the PE column within a given array. The next 2 bits indicate the array number, and the remaining higher-order bits give the row value. The row address bits actually transmitted to the PE memories are configuration-dependent and are gated out as shown.

Addresses used by the PE's for local operands contain three components: a fixed address contained in the instruction, a CU

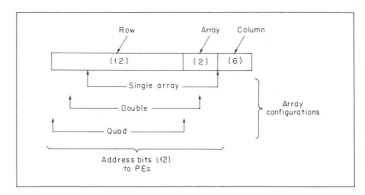

Fig. 5. Memory address structure.

index value added from one of the CU accumulators, and a local PE index value added at the PE prior to transmission to its own memory.

CU data operations

The control unit can fetch either individual words or blocks of 8 words from the array memory to the local data buffer. In addition, it can fetch 1 bit selected from the 8-bit mode register of each processing element to form a 64-bit word read into the CU accumulator. The CU program counter (PCR) and the configuration registers are also directly addressable by the CU. Data manipulations $(+, -,$ Boolean$)$ are performed on a selected CAR and the result returned to the CAR. Data to be broadcast to the processing elements is inserted into the FINQ along with the accompanying instruction and transmitted to the PEs at the appropriate time.

Configuration control

With the variety of array configurations for ILLIAC IV, it is necessary to specify and control the subarrays which are conjoined and to designate the instruction and data addressing. For this purpose each CU has three configuration control registers (CFC), each of 4-bit length, where each bit corresponds to one of the four subarrays. The CFC registers may be set by the B 6500 or a CU instruction.

CFC0 of each CU specifies the array configuration in which it is participating by means of a 1 in the appropriate bits of CFC0. CFC1 specifies the instruction addressing to be used within the array. In a united configuration it is thus possible for the instruction stream to be derived from any subset of the united arrays. CFC2 specifies the CU data addressing form in a manner similar to the CFC1 control of instruction addressing.

The addressing indicated by both CFC1 and CFC2 must be consistent with the actual configuration designated by CFC0, else a configuration interrupt is triggered.

Trap processing

Because external demands on the arrays will be preprocessed through the B 6500 system computer, the interrupt system for the control units is relatively straightforward. Interrupts are provided to handle B 6500 control signals and a variety of CU or array faults (undefined instructions, instruction parity error, improper configuration control instruction, etc.). Arithmetic overflow and underflow in any of the processing elements is detected and produces a trap.

The strategy of response to an interrupt is an effective FORK to a single-array configuration. Each CU saves its own status word automatically and independently of other CU's with which it may previously have been configured.

Hardware implementation consists of a base interrupt address register (BIAR) which is dedicated as a pointer to array storage into which status information will be transferred. Upon receipt of an interrupt, the contents of the program counter and other status information and the contents of CAR 0 are stored in the block pointed to by the BIAR. In addition, CAR 0 is set to contain the block address used by BIAR so that subsequent register saving may be programmed. Interrupt returns are accomplished through a special instruction which reloads the previous status word and CAR 0 and clears the interrupt.

Interrupts are enabled through a mask word in a special register. The interrupt state is general and not unique to a specific trigger or trap. During the interrupt processing, no subsequent interrupts are responded to, although their presence is flagged in the interrupt state word.

The high degree of overlap in the control unit precludes an immediate response to an interrupt during the instruction which generates an arithmetic fault in some processing element. To alleviate this it is possible under program control to force non-overlapped instruction execution permitting access to definite fault information.

Processing element (PE)

The processing element, shown in Fig. 6, executes the data computations and local indexing for operand fetches. It contains the following elements.

1 Four 64-bit registers (A, B, R, S) to hold operands and results. A serves as the accumulator, B as the operand register, R as the multiplicand and data routing register, and S as a general storage register.

2 An adder/multiplier (MSG, PAT, CPA), a logic unit (LOG), and a barrel switch (BSW) for arithmetic, Boolean, and shifting functions, respectively.

3 A 16-bit index register (RGX) and adder (ADA) for memory address modification and control.

4 An 8-bit mode register (RGM) to hold the results of tests and the PE ENABLE/DISABLE state information.

As described earlier, the PEs may be partitioned into subprocessors of word lengths of 64, 2×32, or 8×8 bits. Figure 7 shows the data representations available. Exponents are biased and relative to base 2. Table 1 indicates the arithmetic and logical operations available for the three operand precisions.

PE mode control

Two bits of the mode register (RGM) control the enabling or disabling of all instructions; one of these is active only in the 32-bit precision mode and controls instruction execution on the second operand. Two other bits of RGM are set whenever an arithmetic fault (overflow, underflow) occurs in the PE. The fault bits of all PEs are continuously monitored by the CU to detect a fault condition and initiate a CU trap.

Data paths

Each PE has a 64-bit wide routing path to 4 of its neighbors (± 1, ± 8). To minimize the physical distances involved in such routing, the PEs are grouped 8 to a cabinet (PUC) in the pattern shown in Fig. 8. Routing by distance ± 8 occurs interior to a PUC; routing by distance ± 1 requires no more than 2 intercabinet distances.

CU data and instruction fetches require blocks of 8 words, which are accessed in parallel, 1 word per PUC, into a CU buffer (CUB) 512-bit wide, distributed among the PUCs, 1 word per

Table 1 PE data operations

| Operation | 64 bit | Operation time per element | |
		2×32 bit	8×8 bit
+, −	200 ns	240 ns	80 ns
×	400 ns	400 ns	
÷	2200 ns	3040 ns	
Boolean	80 ns		
Shift	80/240 ns†	160 ns	

† (Single length)/(double length)

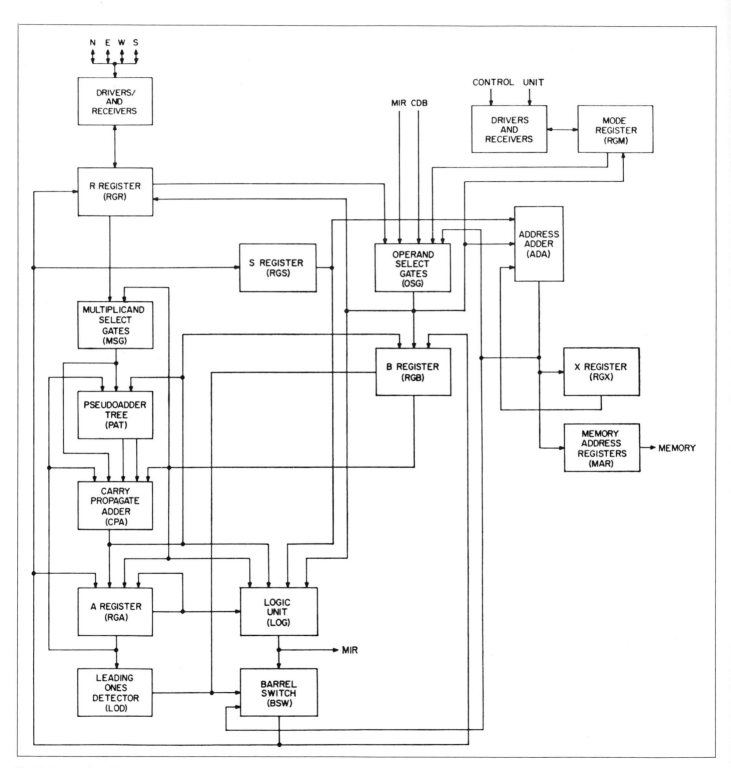

Fig. 6. Processing-element block diagram.

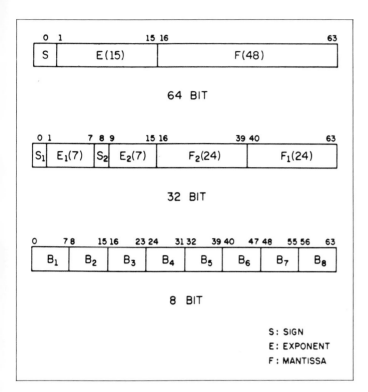

Fig. 7. ILLIAC IV data representation.

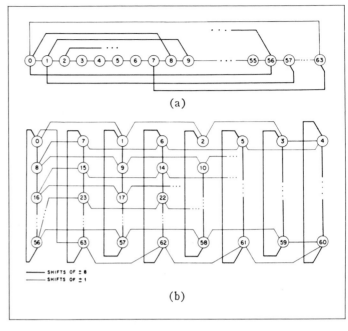

Fig. 8. (a) Electrical connectivity for routing. (b) Physical layout.

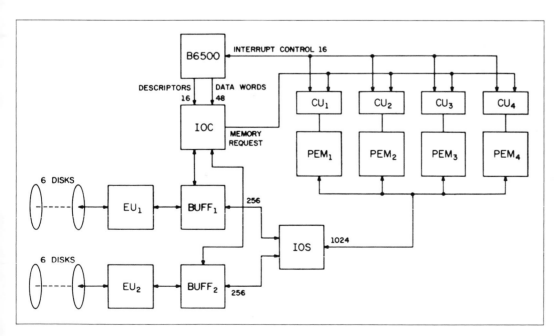

Fig. 9. I/O data path.

cabinet. Data is transmitted to the CU from the CUB on a 512-line bus.

Disk and on-line I/O data are transmitted on a 1024-line bus which can be switched among the arrays. Within each array, parallel connection is made to a selected 16 of 64 PEs, 2 per PUC. Maximum data rate is one I/O transaction per microsecond or 10^9 bits per second. The I/O path of 1024 lines is expandable to 4096 lines if required.

Processing element memory (PEM)

The individual memory attached to each processing element is a thin-film DRO linear select memory with a cycle time of 240 ns and access time of 120 ns. Each has a capacity of 2048 64-bit words. The memory is independently accessible by its attached PE, the CU, or I/O connections.

Disk-file subsystem

The computing speed and memory of the ILLIAC IV arrays require a substantial secondary storage for program and data files as well as backup memory for programs whose data sets exceed fast memory capacity. The disk-file subsystem consists of six Burroughs model IIA storage units, each with a capacity of 1.61×10^8 bits and a maximum latency of 40 ms. The system is dual; each half has a capacity of 5×10^8 bits and independent electronics capable of supporting a transfer rate of 500 megabits per second. The data path from each of the disk subsystems becomes 1024 bits wide at its interface with the array. Figure 9 shows the organization of the disk-file system.

B 6500 control computer

The B 6500 computer is assigned the following functions.

1 Executive control of the execution of array programs

2 Control of the multiple-array configuration operations

3 Supervision of the internal I/O processes (disk to arrays, etc.)

4 External I/O processing and supervision

5 Processing and supervision of the files on the disk file subsystem

6 Independent data processing, including compilation of ILLIAC IV programs

To control the array operations, there is a single interrupt line and a 16-bit data path both ways between the B 6500 and each of the control units. In addition, the B 6500 has a control and data

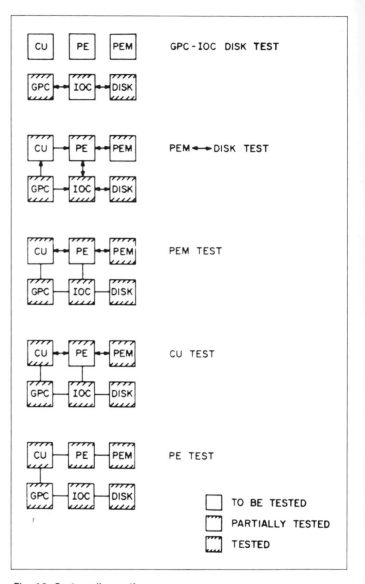

Fig. 10. System diagnostic sequence.

path to the I/O controller (IOC) which supervises the disk, and also direct connections to the array memories.

Reliability and maintenance of the ILLIAC IV

The progress in computer components from vacuum tubes to semiconductors over several generations has improved the mean-time-between-failures for computers from tens of hours to several thousand hours. By using larger scale integration, a tenfold increase in number of gates per system should be possible with comparable reliability.

It is only by virtue of high-density integration (50- to 100-gate package) that the design of a three-million-gate system can be contemplated. Reliability of the major part of the system, 256 processing elements and 256 memory units, is expected to be in the range of 10^5 hours per element and 2×10^3 hours per memory unit.

The organization of the ILLIAC IV as a collection of identical units simplifies its maintenance problems. The processing elements, the memories, and some part of power supplies are designed to be pluggable and replaceable to reduce system down time and improve system availability.

The remaining problems are (1) location of the faulty subsystem, and (2) location of the faulty package in the subsystem.

Location of the faulty subsystem assumes the B 6500 to be fault-free, since this can be determined by using the standard B 6500 maintenance routines. The steps to follow are shown in Fig. 10.

The B 6500 tests the control units (CU) which in turn test all PEs. PEMs are tested through the disk channel. This capability for functional partitioning of the subsystems simplifies the diagnostic procedure considerably.

References

HollJ59; KuckD68; MurtJ66; SlotD62; UngeS58

APPENDIX 1
A1. CLASSIFIED LIST OF CU INSTRUCTIONS

A1.1 *Data transmission*

ALIT	Add literal (24 bit) to CAR.
BIN	Block fetch to CU memory.
BINX	Indexed (by PE index) block fetch.
BOUT	Block store from CU memory.
BOUTX	Indexed block store.
CLC	Clear CAR.
COPY	Copy CAR into CAR of other quadrant.
DUPI	Duplicate inner half of CU memory address contents into both halves of CAR.
DUPO	Duplicate outer half of CU memory address contents into both halves of CAR.
EXCHL	Exchange contents of CAR with CU memory address contents.
LDL	Load CAR from CU memory address contents.
LIT	Load CAR with 64-bit literal following the instruction.
LOAD	Load CU memory from contents of PE memory address found in CAR.
LOADX	Load CU memory from contents of PE memory address found in CAR, indexed by PE index.
ORAC	OR all CARS in array and place in CAR.
SLIT	Load CAR with 24-bit literal.
STL	Store CAR into CU memory.
STORE	Store CAR into PE memory.
STOREX	Store CAR into PE memory, indexed by PE index.
TCCW	Transmit CAR counterclockwise between CUs in array.
TCW	Transmit CAR clockwise between CUs in array.

A1.2 *Skip and test*

$\text{CTSB}\begin{Bmatrix} T, A \\ F \end{Bmatrix}$ — Skip on nth bit of CAR. If T is present, skip if 1; if F is present, skip if 0. If A is present, AND together bits from all CUs in array before testing; if absent, OR together bits from all CUs in array before testing.

4 Instructions: CTSBT, CTSBTA, CTSBF, CTSBFA.

$\text{EQL}\begin{Bmatrix} T, A \\ F \end{Bmatrix}$ — Skip on CAR equal to CU memory address contents. The letters T, F, and A have the same meaning as in CTSB above.

4 Instructions: EQLT, EQLTA, EQLF, EQLFA.

$\text{EQLX}\begin{Bmatrix} T, A \\ F \end{Bmatrix}$ — Skip on index portion of CAR (bits 40 through 63) equal to bits 40 through 63 of CU memory address contents. The letters T, F, and A have the same meaning as in CTSB above.

4 Instructions: EQLXT, EQLXTA, EQLXF, EQLXFA.

$\text{GRTR}\begin{Bmatrix} T, A \\ F \end{Bmatrix}$ — Skip on index part of CAR (bits 40 through 63) greater than bits 40 through 63 of CU memory address contents. The letters T, F, and A have the same meaning as in CTSB above.

4 Instructions: GRTRT, GRTRTA, GRTRF, GRTRFA.

$\text{LESS}\begin{Bmatrix} T, A \\ F \end{Bmatrix}$ — Skip on index part of CAR (bits 40 through 63) less than bits 40 through 63 of CU memory address contents. The letters T, F, and A have the same meaning as in CTSB above.

4 Instructions: LESST, LESSTA, LESSF, LESSFA.

$\text{ONES}\begin{Bmatrix} T, A \\ F \end{Bmatrix}$ — Skip on CAR equal to all 1's. The letters T, F, and A have the same meaning as in CTSB above.

4 Instructions: ONEST, ONESTA, ONESF, ONESFA.

$\text{ONEX}\begin{Bmatrix} T, A \\ F \end{Bmatrix}$ — Skip on bits 40 through 63 of CAR equal to all 1's. The letters T, F, and A have the same meaning as in CTSB above.

4 Instructions: ONEXT, ONEXTA, ONEXF, ONEXFA.

$\text{SKIP}\begin{Bmatrix} T, A \\ F \end{Bmatrix}$ — Skip on T–F flip-flop previously set. The letters T, F, and A have the same meaning as in CTSB above.

4 Instructions: SKIPT, SKIPTA, SKIPF, SKIPFA.

SKIP — Skip unconditionally.

$\text{TXL}\begin{Bmatrix} T, A, I \\ F \end{Bmatrix}$ — Skip on index portion of CAR (bits 40 through 63) less than limit portion (bits 1 through 15). The letters T, F, and A have the same meaning as in CTSB above. If I is present, the index portion of CAR is incremented by the increment portion of CAR (bits 16 through 39) while the test is in progress; if I is not present, no incrementing takes place.

8 Instructions: TXLT, TXLTI, TXLTA, TXLTAI, TXLF, TXLFI, TXLFA, TXLFAI.

$\text{TXE}\begin{Bmatrix} T, A, I \\ F \end{Bmatrix}$ — Skip on index portion of CAR (bits 40 through 63) equal to limit portion of CAR (bits 1 through 15). See CTSB for the meaning of T, F, and A; see TXL above for the meaning of I.

8 Instructions: TXET, TXETI, TXETA, TXETIA, TXEF, TXEFI, TXEFA, TXEFIA.

$\text{TXG} \begin{Bmatrix} T, A, I \\ F \end{Bmatrix}$ Skip on index portion of CAR (bits 40 through 63) greater than limit portion of CAR (bits 1 through 15). See CTSB for the meaning of T, F, and A; see TXL above for the meaning of I.

8 Instructions: TXGT, TXGTI, TXGTA, TXGTAI, TXGF, TXGFI, TXGFA, TXGFAI.

$\text{ZER} \begin{Bmatrix} T, A \\ F \end{Bmatrix}$ Skip on CAR all 0's. See CTSB for the meaning of T, F, and A.

4 Instructions: ZERT, ZERTA, ZERF, ZERFA.

$\text{ZERX} \begin{Bmatrix} T, A \\ F \end{Bmatrix}$ Skip on index portion of CAR (bits 40 through 63) all 0's. See CTSB for the meaning of T, F, and A.

4 Instructions: ZERXT, ZERXTA, ZERXF, ZERXFA.

A1.3 Transfer of control

EXEC	Execute instruction found in bits 32 through 63 of CAR.
EXCHL	Exchange contents of CAR with contents of CU memory address.
HALT	Halt ILLIAC IV.
JUMP	Jump to address found in instruction.
LOAD	Load CU memory address contents from contents of PE memory address found in CAR.
LOADX	Load CU memory address contents from contents of PE memory address found in CAR, indexed by PE index.
STL	Store CAR into CU memory.

A1.4 Route

RTE	Route. Routing distance is found in address field (CAR indexable), and register connectivity is found in the skip field.

A1.5 Arithmetic

ALIT	Add 24-bit literal to CAR.
CADD	Add contents of CU memory address to CAR.
CSUB	Subtract contents of CU memory address from CAR.
INCRXC	Increment index word in CAR.

A1.6 Logical

CAND	AND CU memory to CAR.
CCB	Complement bit of CAR.
CEXOR	Exclusive OR CU memory to CAR.

CLC	Clear CAR.
COR	OR CU memory to CAR.
CRB	Reset bit of CAR.
CROTL	Rotate CAR left.
CROTR	Rotate CAR right.
CSB	Set bit of CAR.
CSHL	Shift CAR left.
CSHR	Shift CAR right.
LEADO	Detect leading ONE in CAR of all quadrants in array.
LEADZ	Detect leading ZERO in CAR of all quadrants in array.
ORAC	OR all CARS in array and place in CAR.

A2. CLASSIFIED LIST OF PE INSTRUCTIONS

A2.1 Data transmission

LDA	Load A register.
LDB	Load B register
LDR	Load R register.
LDS	Load S register.
LDX	Load X register.
LDC0	Load CAR 0 from PE register.
LDC1	Load CAR 1 from PE register.
LDC2	Load CAR 2 from PE register.
LDC3	Load CAR 3 from PE register.
LEX	Load exponent of A register.
ONES	Load all ONES into A register.
STA	Store A register.
STB	Store B register.
STC	Store C register.
STR	Store R register.
STS	Store S register.
STX	Store X register.
SWAPA	Interchange inner and outer contents of A register.
SWAP	Interchange the contents of A register and B register.
SWAPX	Interchange outer operand of A register and inner operand of B.

A2.2 Index operations

$\text{IX} \begin{Bmatrix} L \\ E, I \\ G \end{Bmatrix}$ Set I on comparison of X register and operand. The presence of L means set I if X is less than operand; the presence of E means set I if X is equal to operand; the presence of G means set I if X is greater than operand. If I is present, increment X while performing test; if I is absent, do not increment X.

6 Instructions:	IXL, IXLI, IXE, IXEI, IXG, IXGI.
JX $\begin{Bmatrix} L \\ E, I \\ G \end{Bmatrix}$	Set J on comparison of X register and operand. See above for meaning of L, E, G, and I.
6 Instructions:	JXL, JXLI, JXE, JXEI, JXG, JXGI.
XI	Increment PE index (X register) by bits 48 through 63 of operand.
XIO	Increment PE index of bits 48 through 63 of operand plus one.

A2.3 Mode setting/comparisons

EQB	Test A and B for equality bytewise.
GRB	Test B register greater than A register bytewise.
LSB	Test B register less than A register bytewise.
CHWS	Change word size.
I $\begin{Bmatrix} L \\ A \\ M \end{Bmatrix}$ L	Set I if A register is less than operand. L means test logical; A means test arithmetic; M means test mantissa.
3 Instructions:	ILL, IAL, IML.
I $\begin{Bmatrix} L \\ A \\ M \end{Bmatrix}$ E	Set I if A register is equal to operand. See above for meaning of L, A, and M.
3 Instructions:	ILE, IAE, IME.
I $\begin{Bmatrix} L \\ A \\ M \end{Bmatrix}$ G	Set I if A register is greater than operand. See above for meaning of L, A, and M.
3 Instructions:	ILG, IAG, IMG.
I $\begin{Bmatrix} L \\ A \\ M \end{Bmatrix}$ Z	Set I if A register is equal to all zeros.
3 Instructions:	ILZ, IAZ, IMZ.
I $\begin{Bmatrix} L \\ A \\ M \end{Bmatrix}$ O	Set I if A register is equal to all ONES.
3 Instructions:	ILO, IAO, IMO.
J $\begin{Bmatrix} L & L \\ A, & E \\ M & G \end{Bmatrix}$ Z O	Set J under conditions specified in set of instructions immediately above.
15 Instructions:	JLL, JAL, JML, JLE, JAE, JME, JLG, JAG, JMG, JLZ, JAZ, JMZ, JLO, JAO, JMO.
IX $\begin{Bmatrix} L \\ E, I \\ G \end{Bmatrix}$	Set I on comparison of X register and operand. See Section A2.2 for meaning of L, E, G, and I.

6 Instructions:	IXL, IXLI, IXE, IXEI, IXG, IXGI.
JX $\begin{Bmatrix} L \\ E, I \\ G \end{Bmatrix}$	Set J on comparison of X register and operand. See Section A2.2 for meaning of L, E, G, and I.
6 Instructions:	JXL, JXLI, JXE, JXEI, JXG, JXGI.
IS $\begin{Bmatrix} L \\ E \\ G \end{Bmatrix}$	Set I on comparison of S register and operand. See Section A2.2 for meaning of L, E, and G.
3 Instructions:	ISL, ISE, ISG.
JS $\begin{Bmatrix} L \\ E \\ G \end{Bmatrix}$	Set J on comparison of S register and operand. See Section A2.2 for meaning of L, E, and G.
3 Instructions:	JSL, JSE, JSG.
ISN	Set I from the sign bit of A register.
JSN	Set J from the sign bit of A register.
SETE	Set E bit as a logical function of other bits.
SETEO	Set $E1$ bit similarly.
SETF	Set F bit similarly.
SETFO	Set $F1$ bit similarly.
SETG	Set G bit similarly.
SETH	Set H bit similarly.
SETI	Set I bit similarly.
SETJ	Set J bit similarly.
SETC0	Set Pth bit of CAR 0 similarly.
SETC1	Set Pth bit of CAR 1 similarly.
SETC2	Set Pth bit of CAR 2 similarly.
SETC3	Set Pth bit of CAR 3 similarly.
IBA	Set I from Nth bit of A register; bit number is found in address field.
JBA	Set J from Nth bit of A register; bit number is found in address field.

A2.4 Arithmetic

ADB	Add bytewise.
SBB	Subtract operand from A register bytewise.
ADD	Add A register and operand as 64-bit operands.
SUB	Subtract operand from A register as 64-bit quantities.
AD{R, N, M, S}	Add operand to A register. The R, N, M, S specify all possible variants of the arithmetic instruction. The meaning of each letter, if present in the mnemonic, is

R	round result
N	normalize result
M	mantissa only
S	special treatment of signs.

16 Instructions:	ADM, ADMS, ADNM, ADNMS, ADN, ADNS, ADRM, ADRMS, ADRM, ADRNMS, ADRN, ADRNS, ADR, ADRS, AD, ADS.
ADEX	Add to exponent.
DV{R, N, M, S}	Divide by operand. See AD instruction for meaning of R, N, M, and S.
16 Instructions:	DVM, DVMS, DVNM, DVNMS, DVN, DVNS, DVRM, DVRMS, DVRNM, DVRNS, DVRN, DVRNS, DVR, DVRS, DV, DVS.
EAD	Extend precision after floating point ADD.
ESB	Extend precision after floating point SUBTRACT.
LEX	Load exponent of A register.
ML{R, N, M, S}	Multiply by operand. See AD instruction for meaning of R, N, M, and S.
16 Instructions:	MLM, MLMS, MLNM, MLNMS, MLN, MLNS, MLRM, MLRMS, MLRNM, MLRNMS, MLRN, MLRNS, MLR, MLRS, ML, MLS.
SAN	Set A register negative.
SAP	Set A register positive.
SBEX	Subtract exponent of operand from exponent of A register.
SB{R, N, M, S}	Subtract operand from A register. See AD instruction for meaning of R, N, M, and S.
16 Instructions:	SBM, SBMS, SBNM, SBNMS, SBN, SBNS, SBRM, SBRMS, SBRNM, SBRNMS, SBRN, SBRNS, SBR, SB, SBS.
NORM	Normalize A register.
MULT	In 32-bit mode, perform MULTIPLY and leave outer result in A register and inner result in B register, with both results extended to 64-bit format.

A2.5 Logical

$$\begin{Bmatrix} N \\ Z \\ O \end{Bmatrix} \text{AND} \begin{Bmatrix} N \\ Z \\ O \end{Bmatrix}$$ AND A register with operand. The left-hand set of letters specifies a variant on the A register, the right-hand set, on the operand. The meaning of these variants is

not present	use true
N	use complement
Z	use all ZEROS
O	use all ONES.

16 Instructions:	AND, ANDN, ANDZ, ANDO, NAND, NANDN, NANDZ, NANDO, ZAND, ZANDN, ZANDZ, ZANDO, OAND, OANDN, OANDZ, OANDO.
CBA	Complement bit of A register.
CHSA	Change sign of A register.
$\begin{Bmatrix} N \\ Z \\ O \end{Bmatrix}$ EOR $\begin{Bmatrix} N \\ Z \\ O \end{Bmatrix}$	Exclusive OR A register with operand.
16 Instructions:	EOR, EORN, EORZ, EORO, NEOR, NEORN, NEORZ, NEORO, ZEOR, ZEORN, ZEORZ, ZEORO, OEOR, OEORN, OEORZ, OEORO.
LEX	Load exponent of A register.
$\begin{Bmatrix} N \\ Z \\ O \end{Bmatrix}$ OR $\begin{Bmatrix} N \\ Z \\ O \end{Bmatrix}$	OR A register with operand.
16 Instructions:	OR, ORN, ORZ, ORO, NOR, NORN, NORZ, NORO, ZOR, ZORN, ZORZ, ZORO, OOR, OORN, OORZ, OORO.
RBA	Reset bit A register to ZERO.
RTAL	Rotate A register left.
RTAML	Rotate mantissa of A register left.
RTAMR	Rotate mantissa of A register right.
RTAR	Rotate A register right.
SAN	Set A register negative.
SAP	Set A register positive.
SBA	Set bit of A register to ONE.
SHABL	Shift A and B registers double-length left.
SHABR	Shift A and B registers double-length right.
SHAL	Shift A register left.
SHAML	Shift A register mantissa left.
SHAR	Shift A register right.
SHAMR	Shift A register mantissa right.

Section 3

Processors defined by a microprogram

Processors defined by a microprogram have only recently come into existence, although Wilkes suggested the idea in 1951. The discussion in Chap. 3 (page 71) suggests reasons why this controversial idea has taken so long to be adopted.

Microprogramming and the design of the control circuits in an electronic computer

Chapter 28 is an extension of an earlier paper by Wilkes. It includes an example of a microprogrammed processor (page 337). In the earlier paper, The Best Way to Design an Automatic Computing Machine [Wilkes, 1951a], the essential ideas of microprogramming were first outlined.

The observation that an instruction set, or ISP, should be looked at as a program to be interpreted is the basis of microprogramming. The idea of an ISP is our acknowledgment that we, too, view a processor as a program.

There is little to say about this chapter; it is historical, yet timely and well written. Microprogramming, like other of Wilkes' ideas, is present in many of our computers.

The design of a general-purpose microprogram-controlled computer with elementary structure

The SD-2 computer (Chap. 29) is described by Kampe in a casual but highly communicative fashion. Most engineers tend to be somewhat formal and stuffy when describing the machines they have designed. This formal ruse can be used to make the design seem difficult but well founded—certainly not arbitrary. Kampe truthfully admits to making decisions in a somewhat arbitrary fashion.

The SD-2 microprogram structure, unlike that of the IBM System 360 models, has a P.microprogram which is similar to the external Pc which it defines. As such, the main question about this design is whether it is cheaper to have a single, hard-wired Pc rather than a computer within a computer. The Packard Bell 440 [Boutwell and Hoskinson, 1963] is an example of a better-known Pc whose internal P resembles the SD-2.

The authors of this book feel that, when the internal and external P's are so similar, it may be better to have a single P which suits both needs. To gain speed and still define powerful functions, Mp could be made up of both the conventional Mp and a small, fast Mp.

The Hewlett-Packard HP 9100A computing calculator

The HP 9100A (Chap. 20) is discussed in Part 3, Sec. 4, page 235.

Microprogrammed implementation of EULER on the IBM System 360/Model 30

This microprogrammed processor in Chap. 32 is also discussed as a language processor in Part 4, Sec. 4, page 348.

Chapter 28

Microprogramming and the design of the control circuits in an electronic digital computer[1]

M. V. Wilkes / *J. B. Stringer*

1. Introduction

Experience has shown that the sections of an electronic digital computer which are easiest to maintain are those which have a simple logical structure. Not only can this structure be readily borne in mind by a maintenance engineer when looking for a fault, but it makes it possible to use fault-locating programmes and to test the equipment without the use of elaborate test gear. It is in the control section of electronic computers that the greatest degree of complexity generally arises. This is particularly so if the machine has a comprehensive order code designed to make it simple and fast in operation. In general, for each different order in the code some special equipment must be provided, and the more complicated the function of the order the more complex this equipment. In the past, fear of complicating unduly the control circuits of the machines has prevented the designers of electronic machines from providing such facilities as orders for floating-point operations, although experience with relay machines and with interpretive subroutines has shown how valuable such orders are. This paper describes a method of designing the control circuits of a machine which is wholly logical and which enables alterations or additions to the order code to be made without *ad hoc* alterations to the circuits. An outline of this method was given by one of us [Wilkes, 1951a] at the Conference on Automatic Calculating Machines at the University of Manchester in July 1951.

The operation called for by a single machine order can be broken down into a sequence of more elementary operations; for example, shifting a number in the accumulator one place to the right may involve, first, a transfer of the number to an auxiliary shifting register, and secondly, the transfer of the number back to the accumulator along an oblique path. These elementary operations will be referred to as *micro-operations*. Basic machine operations, such as addition, subtraction, multiplication, etc., are thought of as being made up of a *micro-programme* of micro-operations, each micro-operation being called for by a *micro-order*. The process of writing a micro-programme for a machine order is very similar to that of writing a programme for the whole calculation in terms of machine orders.

For the method to be applicable it is necessary that the machine should contain a suitable permanent rapid-access storage device in which the micro-programme can be held—a diode matrix is proposed in the case of the machine discussed as an example below—and that means should be provided for executing the micro-orders one after the other. It is also necessary that provision should be made for conditional micro-orders which play a role in micro-programming similar to that played by conditional orders in ordinary programming.

Since the only feature of the machine which has to be designed specially for any particular set of machine orders is the configuration of diodes in the matrix, or the corresponding configuration in whatever equivalent device is used, there is no difficulty in making changes to the order code of the machine if experience shows them to be desirable; in fact, the design of the machine in the first place can be carried out completely without a firm decision on the details of the order code being taken, as long as care is taken to provide accommodation for the greatest number of micro-orders that are likely to be required. It would even be possible to have a number of interchangeable matrices providing for different order codes, so that the user could choose the one most suited to his particular requirements.

2. Description of the proposed system

The system will be described in relation to a parallel machine having an arithmetical unit designed along conventional lines. This will contain a set of registers and an adder together with a switching system which enables the micro-operations in the various machine orders to be performed. Some of the micro-operations will be simple transfers of a number from one register to another with or without shifting of the number one place to the left or

[1]*Proc. Cambridge Phil. Soc.*, pt. 2, vol. 49, pp. 230–238, April, 1953.

the right, while others will also involve the use of the adder. Any particular micro-operation can be performed by applying pulses simultaneously to the appropriate gates of the switching system. In certain cases it may be possible for two or more micro-operations to take place at the same time.

It will be convenient to regard the control system as consisting of two parts. A register is needed to hold the address of the next order due to be executed, and another to hold the current order while it is being executed, or at any rate during part of that time. Some means of counting the number of steps in a shifting operation or a multiplication must also be provided. One method of meeting these requirements is to provide a group of registers and an adder together with a switching system which enables transfers of numbers, with or without addition, to be made. This part of the control system will be called the *control register unit*. In any case the operations which need to be performed on the numbers standing in the control register unit during the execution of an order are, like the operations performed in the arithmetical unit, regarded as being made up of a sequence of micro-operations, each of which is performed by the application of pulses to appropriate gates.

The other part of the control system is concerned with control of the sequence of micro-orders required to carry out each machine order, and with the operation of the gates required for the execution of each micro-order. This will be called the *micro-control unit;* it consists of a decoding tree, two rectifier matrices and two registers (additional to those of the control register unit) connected as indicated in Fig. 1, which shows how the pulses used to operate the gates in the arithmetical unit and control register unit are generated. A series of control pulses from a pulse generator are applied to the input of the decoding tree. Each pulse is routed to one of the output lines of the tree, according to the number standing in register I. The output lines all pass into a rectifier matrix A and the outputs of this matrix are the pulses which operate the various gates associated with micro-operations. Thus one input line of the matrix corresponds to one micro-order. The *address* of the micro-order is the number which must be placed in register I to cause the control pulse to be routed to the corresponding line. The output lines from the tree also pass into a second matrix B, which has its outputs connected to register II. This matrix has wired on it the address of the micro-order to be performed next in time so that the address of this micro-order is placed in register II. Just before the next control pulse is applied to the input of the tree a connexion is established between register II and register I, and the address of the micro-order due to be executed next is transferred into register I. In this way the decoding tree is prepared to route the next incoming control pulse

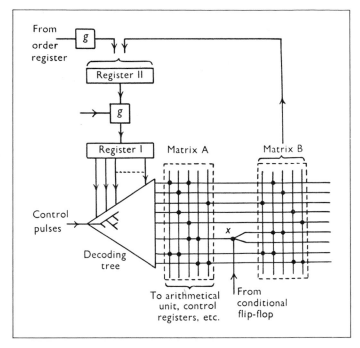

Fig. 1. Micro-control unit.

to the correct output line. Thus application of pulses alternately to the input of the tree and to the gate connecting registers I and II causes a predetermined sequence of micro-orders to be executed.

It is necessary to have means whereby the course of the micro-programme can be made conditional on whether a given digit in one of the registers of the arithmetical unit or control register unit is a 1 or a 0. The means of doing this is shown at X in Fig. 1. A two-way switch, controlled by a special flip-flop called a *conditional flip-flop*, is inserted between matrix A and matrix B. The conditional flip-flop can be set by an earlier micro-order with any digit from any one of the registers. Two separate addresses are wired into matrix B, and the one which passes into register I, and thus becomes the address of the next micro-order, is determined by the setting of the conditional flip-flop.

Conditional micro-orders play the same part in the construction of micro-programmes as conditional orders play in the construction of ordinary programmes; apart from their obvious uses in micro-programmes for such operations as multiplication and division, they enable repetitive loops of micro-orders to be used.

If desired, two branchings may be inserted in the connexions between matrix A and matrix B, so that any one of four alternative addresses for the next micro-order may be selected according to the settings of two conditional flip-flops. Another possibility is to make the output from the decoding tree branch before it enters matrix A so that the nature of the micro-operation that is performed depends on the setting of the conditional flip-flop.

The micro-programme wired on to the matrices contains sections for performing the operations required by each order in the basic order code of the machine. To initiate the operation it is only necessary that control in the micro-programme should be sent to the correct entry point. This is done by placing the function digits of the order in the least significant part of register II, the other digits in this register being made zero. The micro-programme is constructed so that when this number passes into register I, control in the micro-programme is sent to the correct entry point.

The switching system in the arithmetical unit may either be designed to permit a large variety of micro-operations to be performed, or it may be restricted so as to allow only a small number of such operations. In a machine with a comprehensive order code there is much to be said for having the more flexible switching system since this will enable an economy to be made in the number of micro-orders needed in the micro-programme.

A similar remark applies in connexion with the degree of flexibility to be provided when designing the switching system for the control register unit. If the specification of the machine allows the same number of registers to be used in the arithmetical and control sections, the construction of these two sections may be identical except as far as the number of digits is concerned. In a new machine under construction in the Mathematical Laboratory, Cambridge, the registers are being constructed in basic units each containing five registers and an adder-subtractor together with the associated switching system. It is hoped that it will be possible to use identical units in the arithmetical unit and in the control register unit.

3.   Example

An example will now be given to show the way in which a micro-programme can be drawn up for a machine with a single-address order code covering the usual operations. It is supposed that the arithmetical unit contains the following registers:

A   multiplicand register

B   accumulator (least significant half)

C   accumulator (most significant half)

D   shift register

The registers in the control register unit are as follows:

E   register connected to the access circuits of the store; the address of a storage location to which access is required is placed here

F   sequence control register; contains address of next order due to be executed

G   register used for counting

It was assumed when drawing up the micro-programme that there was an adder-subtractor in the arithmetical unit with one input permanently connected to register D, and a similar adder-subtractor in the control register unit with one input permanently connected to register G. For convenience it was assumed that the switching systems in each case were comprehensive enough to provide any micro-operation required. It was further supposed that the arithmetical unit provided for 20 digits and that the numbers 0, 1 and 18 could be introduced at will into one of the registers or the adder of the control register unit. Two conditional flip-flops are used. All micro-operations including those involving access to the store are supposed to take the same amount of time. Reference will be made to this point in §4.

Table 1 gives the order code of the machine, and Table 2 the micro-programme. Each line of Table 2 refers to one micro-order; the first column gives the address of the micro-order, the second column specifies the micro-operations called for in the arithmetical unit of the machine, and the third column specifies the micro-

Table 1

Notation: Acc = accumulator
Acc_1 = most significant half of accumulator
Acc_2 = least significant half of accumulator
n = storage location n
$C(X)$ = contents of X (X = register or storage location)

Order	Effect of order
$A\ n$	$C(Acc) + C(n)$ to Acc
$S\ n$	$C(Acc) - C(n)$ to Acc
$H\ n$	$C(n)$ to Acc_2
$V\ n$	$C(Acc_2) \cdot C(n)$ to Acc, where $C(n) \geq 0$
$T\ n$	$C(Acc_1)$ to n, 0 to Acc
$U\ n$	$C(Acc_1)$ to n
$R\ n$	$C(Acc) \cdot 2^{-(n+1)}$ to Acc
$L\ n$	$C(Acc) \cdot 2^{n+1}$ to Acc
$G\ n$	If $C(Acc) < 0$, transfer control to n; if $C(Acc) \geq 0$, ignore (i.e., proceed serially)
$I\ n$	Read next character on input mechanism into n
$O\ n$	Send $C(n)$ to output mechanism

Table 2

Notation: $A, B, C, \ldots$ stand for the various registers in the arithmetical and control register units (see §3 of the text). 'C to D' indicates that the switching circuits connect the output of register C to the input of register D; '$(D+A)$ to C' indicates that the output of register A is connected to the one input of the adding unit (the output of D is permanently connected to the other input), and the output of the adder to register C. A numerical symbol n in quotes (e.g., 'n') stands for the source whose output is the number n in units of the least significant digit.

		Arithmetical unit	Control register unit	Conditional flip-flop Set	Use	Next micro-order 0	1
	0		F to G and E			1	
	1		$(G+\text{'}1\text{'})$ to F			2	
	2		Store to G			3	
	3		G to E			4	
	4		E to decoder			—	
A	5	C to D				16	
S	6	C to D				17	
H	7	Store to B				0	
V	8	Store to A				27	
T	9	C to Store				25	
U	10	C to Store				0	
R	11	B to D	E to G			19	
L	12	C to D	E to G			22	
G	13		E to G	$(1)C_s$		18	
I	14	Input to Store				0	
O	15	Store to Output				0	
	16	$(D+\text{Store})$ to C				0	
	17	$(D-\text{Store})$ to C				0	
	18				1	0	1
	19	D to B (R)†	$(G-\text{'}1\text{'})$ to E			20	
	20	C to D		$(1)E_s$		21	
	21	D to C (R)			1	11	0
	22	D to C (L)‡	$(G-\text{'}1\text{'})$ to E			23	
	23	B to D		$(1)E_s$		24	
	24	D to B (L)			1	12	0
	25	'0' to B				26	
	26	B to C				0	
	27	'0' to C	'18' to E			28	
	28	B to D	E to G	$(1)B_l$		29	
	29	D to B (R)	$(G-\text{'}1\text{'})$ to E			30	
	30	C to D (R)		$(2)E_s$	1	31	32
	31	D to C			2	28	33
	32	$(D+A)$ to C			2	28	33
	33	B to D		$(1)B_l$		34	
	34	D to B (R)				35	
	35	C to D (R)			1	36	37
	36	D to C				0	
	37	$(D-A)$ to C				0	

† Right shift. The switching circuits in the arithmetic unit are arranged so that the least significant digit of register C is placed in the most significant place of register B during right shift micro-operations, and the most significant digit of register C (sign digit) is repeated (thus making the correction for negative numbers).

‡ Left shift. The switching circuits are similarly arranged to pass the most significant digit of register B to the least significant place of register C during left shift micro-operations.

operations called for in the control register unit. The fourth column shows which conditional flip-flop, if any, is to be set and the digit which is to be used to set it; for example, $(1)C_s$ means that flip-flop number 1 is set by the sign digit of the number in register C, while $(2)G_l$ means that flip-flop number 2 is set by the least significant digit of the number in register G. In the case of unconditional micro-orders columns 5 and 7 are blank and column 6 contains the address of the next micro-order to be executed. In the case of conditional micro-orders column 5 shows which flip-flop is used to operate the conditional switch and columns 6 and 7 give the alternative addresses to which control is to be sent when the conditional flip-flop contains a 0 or a 1 respectively.

Micro-orders 0 to 4 are concerned with the extraction of orders from the store. They serve to bring about the transfer of the order from the store to register E and then cause the five most significant digits of the order to be placed in register II with the result that control is transferred to one of the micro-orders 5 to 15, each of which corresponds to a distinct order in the machine order code. In this way the sequence of micro-orders needed to perform the particular operation called for is begun.

The way in which the various operations are performed can be followed from Table 2. In the section dealing with multiplication, it is assumed that numbers lie in the range $-1 \leq x < 1$ and that negative numbers are represented in the machine by their complements with respect to 2. It will be noted that the process of drawing up a micro-programme is very similar to that of drawing up an ordinary programme for an automatic computing machine and the problems involved are very much alike.

4. The timing of micro-operations

The assumption that all micro-operations take the same length of time to perform is not likely to be borne out in practice. In particular in a parallel machine it may not be possible to design an adder in which the carry propagation time is sufficiently short to enable an addition to be performed in substantially the same length of time as that taken for a simple transfer. It will be necessary, therefore, to arrange that the wave-form generator feeding the decoding tree should, when suitably stimulated by a pulse from one of the outputs from matrix A, supply a somewhat longer pulse than that normally required. Other operations may take many times as long to perform as an ordinary micro-order; for example, access to and from the store (particularly if a delay store is used) and operation of the input and output devices of the machine. The sequence of operations in the micro-programme must therefore be interrupted. One way of doing this is to prevent pulses from

the wave-form generator reaching the decoding tree during the waiting period. This method, although quite feasible, appears to involve just the kind of complication which the present system is designed to avoid. A more attractive system is to make the machine wait on a conditional micro-order which transfers control back to itself unless the associated conditional flip-flop is set. Setting of this flip-flop takes place when the operation is completed, and control then goes to the next micro-order in the sequence. The machine is thus in a condition of 'dynamic stop' while waiting for the operation to be completed. This system has the advantage that no complication is introduced into the units supplying the wave-forms to the decoding tree and that the control equipment required is similar to that already provided for other purposes.

5. Discussion

It will be seen that the equipment needed to execute a complicated order in the machine order code is of the same form as that required for a simple one, namely outlets from the decoding tree and diodes in the matrices. Quite complicated orders can, therefore, be built into the machine without difficulty. In particular, arithmetical operations on numbers expressed in floating binary form and other similar operations can be micro-programmed and it is found that they do not involve very large numbers of micro-orders. For example, a micro-programme providing for the floating-point operations of addition, subtraction, and multiplication needs about 70 micro-orders. The switching system in the arithmetical unit must, of course, be designed with these operations in view. The decoding tree and matrices of a parallel machine with 40 digits in the arithmetical unit and provision for 256 micro-orders would only amount to about 15% of the total equipment in the machine, so that it appears that such a machine can well be provided with built-in facilities of considerable complexity.

The number of micro-orders needed in a complicated micro-programme can sometimes be reduced by making use of what might be called *micro-subroutines*. For example, when two numbers have to be added together in a floating binary machine, some shifting of one of them is usually necessary before the addition can take place. By making the micro-orders for this shifting operation serve also when a multiplication is called for, considerable saving is effected.

Four registers is the bare minimum needed in the arithmetical unit in order to enable the basic arithmetical operations to be performed. If any extension or refinement of the facilities provided is required, it may be necessary to increase the number of registers.

For example, four registers are not sufficient to enable a succession of products to be accumulated without the transfer of intermediate results to the store, since the accumulator must be clear at the beginning of a multiplication. The addition of one register enables the accumulation of products to be provided for in the micro-programme. If this register is associated with the outlet from the store, it also enables some of the waiting time for storage access to be eliminated. To do this the micro-programme is arranged to call for a number from the store as soon as it is known that the number will be required and to continue with other necessary micro-operations before finally proceeding to use the number. The 'dynamic stop' would occur just before the number is required for use. Another way of saving time is to arrange, in the case of those orders which permit it, for the next order to be extracted from the store before the operation currently being performed has been completed.

The minimum number of registers required in the control register unit of the machine for the simplest mode of operation is three. If extra registers are provided facilities similar to those provided by the B-lines in the machine at Manchester University could be included in the micro-programme.

6. Microprogramming applied to serial machines

All the discussion so far has been with reference to parallel machines because the technique described in this paper is most adapted to that type of machine. It is, however, possible to design a serial machine along the same lines. In a parallel computer with an asynchronous arithmetical unit every gate requires only one kind of wave-form to operate it and the timing of that wave-form is not critical. In a serial machine, on the other hand, different gates require different wave-forms and the same gate may require different wave-forms at different times; further, all these wave-forms must be critically timed. These complications may be handled by including in the micro-control unit a third matrix, C, for selecting the appropriate wave-form for each micro-order. The main wave-form, routed by the decoding tree and matrix A, opens a gate which is fed by a wave-form selected by matrix C. This enables a wave-form of correct duration to be applied to any selected gate in the arithmetical or control sections of the machine.

References

WilkM51a; BoutE63; FlynM67; GreeJ64, 66; MercR57; Patz67; RosiR69; TuckS67; WilkM58b, 69; WebeH67

Chapter 29

The design of a general-purpose microprogram-controlled computer with elementary structure[1]

Thomas W. Kampe

Summary   This paper presents the design of a parallel digital computer utilizing a 20-μsec core memory and a diode storage microprogram unit. The machine is intended as an on-line controller and is organized for ease of maintenance.

A word length of 19 bits provides 31 orders referring to memory locations. Fourteen bits are used for addressing, 12 for base address, one for index control, and one for indirect addressing. A 32nd order permits the address bits to be decoded to generate special functions which require no address.

The logic of the machine is resistor-transistor; the arithmetic unit is a bus structure which permits many variants of order structure.

In order to make logical decisions, a "general-purpose" logic unit has been incorporated so that the microcoder has as much freedom in this area as in the arithmetic unit.

Introduction

This paper discusses the logical design of a binary, parallel, real-time computer. Only those aspects of packaging and circuitry which bear directly on this topic will be considered.

Since the specifications for the job a computer is to perform are not enough to fix the design, the logical designer is faced with an undetermined system. One of his main functions is to analyze the system in its natural environment, *i.e.*, with malfunctions, operator errors, etc., and to supply the remainder of the side conditions which do fix the design.

In this discussion, the exposition will be directed toward the design philosophy which led to a machine now being built. In order to accomplish this, we shall consider the functional requirements, their analysis in terms of the state of the art, the basic design decisions, and, finally, a description of the computer as it stands.

[1] *IRE Trans.*, EC-9, vol. 2, pp. 208–213, June, 1960.

Functional requirements

The design of the computer (known, for a variety of reasons, as the SD-2) was undertaken to supply a computer capable of moderately fast arithmetic with perhaps five decimal places of accuracy and 3000 or more words of storage. Furthermore, the computer must reside in a hostile environment (a small house, 0° to 85°C temperature), withstand severe shocks, and be maintained by men with only two weeks training on the system. The volume limitation is 40 cubic feet. Within this space must reside the control computer, memory, power supplies, complete maintenance facilities, and sufficient input/output equipment to handle 20 shaft position outputs, 30 such inputs, numerous switch settings, and 20 or more display or relay signals.

The final specification (or blow) was that 15 months were available from the start of preliminary design to the delivery of an operating instrument with debugged program.

Design analysis

The maintenance requirement was evidently the major problem. In order to achieve the simplicity required, two design criteria were necessary.

First, the computer had to be readily understood. This implied that the usual clever logical tricks such as intensive time sharing of control and arithmetic were undesirable.

Second, if built-in maintenance facilities were to be kept simple, the machine must be designed with this in mind.

Since temperature and reliability were important, an extremely conservative approach had to be taken with respect to component performance.

With the schedule requirements, a machine which could be designed and released in pieces was needed. Since the control system is usually the most troublesome part of a computer to design, a simple control was needed.

The volume available, together with the schedule, required a logical design with natural packaging properties in the sense that it should break, in a natural way, into logical packages of a reasonable size having a minimum of interpackage communication.

Design decisions

The need for 2000 operations per second poses a serious access problem with a serial memory, unless one resorts to several simultaneously operating control units which are neither small nor simple. Hence, a random access memory seemed advisable. Magnetic core memories at 85°C are a problem, but they can be built, provided memory cycle time is not too short. The memory was chosen as 4096 words of core storage, with a 20-μsec cycle time.

The requirement for training a man in two weeks to maintain the machine argues for a simple-structured parallel machine. Providing that much use is made of asynchronous transfer, there are a variety of simple maintenance methods, particularly if a bus structure is adopted. Also, asynchronous, or semi-asynchronous, parallel machines require only average performance of a set of components, not of any particular component; the central limit theorem of statistics can come to the aid of reliability. This approach was finally adopted.

The simplicity of both design and understanding is aided by the use of a microprogram control system. Further, maintenance is made rather simple by two provisions on the maintenance console.

The first of these is a manner of going through the microprogram on a step-by-step basis. While this tests little of the dynamics, it can often locate totally defective parts, and it helps factory checkout immeasurably.

The second is a means of taking out the microprogram unit and substituting a set of switches. This permits a maintenance man to exercise specific registers, or the memory, at will.

This is a powerful tool, and is almost free with a microprogram control. Finally, and rather pragmatically, microprogramming permits "last minute" changes in machine operation without serious hardware modifications. This approach was chosen.

Regardless of the control used, at various times in the process of executing orders, decisions must be made. Occasionally these are on a single bit, more often on two, and occasionally on more than two. If one excludes order decoding, only such functions as zero detection require the use of more than two bits. At this point, the logical designer is faced with a rather sticky decision: whether to design a specific set of decision logic, which is cheap to build

but sometimes messy, or to use some microcontrolled logic-generating scheme.

In this case, the latter alternative was taken. A unit, called (for several obscure reasons) the alteration unit, was designed which amounted to a three-address, one-bit unit. It can generate any Boolean function of two binary variables and transmit this value to another variable. A special set of logic was needed for detecting zeros.

Because of the rather wild nature of the inputs, it seemed desirable to include a trapping mode. The logic for this was made an adjunct to the alteration unit.

The circuitry chosen was resistor-transistor logic, which yields either Sheffer stroke or NOR logic, as one prefers, high or low true logic, and p-n-p or n-p-n transistors. In this case, the combination was high true logic and p-n-p transistors, so that the logical operation is Sheffer stroke. Because of temperature and reliability requirements, the maximum frequency available was a 250-kc square wave. This gave a cycle time of 4 μsec available for asynchronous transfer in any sequence of logic.

An index register seemed advisable because of the amount of data processing. Thus, additions were needed for indexing, arithmetic, and counter advance. It seemed undesirable to have more than one parallel adder, so that an adder accessible to all registers was chosen. This was another argument for a bus structure.

Because of the multiplicity of problems being handled simultaneously, one index register was not really enough. Rather than add another register, indirect addressing was chosen.

At this point, one needs 12 bits for address, one for index tagging, and one to specify whether the address is direct or indirect, or 14 bits for operand selection. Thirty-two orders was a tight minimum, so the minimum word length was 19 bits. Since this was consistent with five decimal place accuracy, it was tentatively chosen. It was decided, however, to design a structure basically suited to any length word.

Shifting is necessary to multiply and divide and is required on two registers, yet shift registers for asynchronous operation are complex. Hence, it was decided to put the shift facility on the data transfer bus. By providing complementing here, subtraction could be generated.

It was decided to use two-complement arithmetic, first because of the simplicity of the multiply-divide logic, and second because it avoids the whole negative zero question.

The precise number of microsteps needed was determined by a trial microprogram. The machine was designed for up to 512 microsteps although only 384 are now used. Eight bits were in a register, called *J*, and one was a flip-flop, *TO*, in the alteration unit, thus allowing fixed sequence with a one-bit microprogrammed choice. This, incidentally, is the genesis of the name "alteration unit."

The SD-2 computer

Figure 1 is a block diagram of the computer. There will be, presently, a block-by-block description of the computer.

The two boxes on the left were added to facilitate input and output. The output buffer holds 20 words, and outputs all values in a 4.8-msec cycle, thus providing for nearly continuous outputs. The output distributor is a selection system which allows the programmer to transmit the contents of the accumulator onto one of eight channels to control external devices. The "inputs" line represents up to 32 channels which can be read into the accumulator. The numbers 8 and 32 are purely arbitrary; the upper limit of 32 is a microcode convenience only.

The alteration unit, in addition to its decision making duties, has several other functions. It has a five bit counter, used for microsubroutines, which can be set to any value chosen or to any number on the arithmetic unit. The alteration unit can sense when it goes from all zeros to all ones. In addition, the flip-flops con-

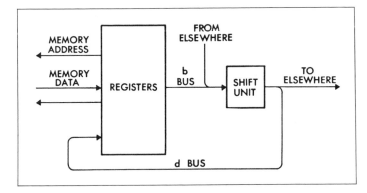

Fig. 2. Arithmetic flow.

trolling initial carry in the adder, end carry in shifting, and memory read or write control are in this unit.

Figure 2 is a block diagram of the arithmetic unit. Information may be put onto the *b* bus from any register, or from outside sources, such as inputs, or constants from the microprogram unit; thence to the shift unit, and finally to the *d* bus. From the *d* bus, it may be sent to other places, such as the output distributor, microprogram register, etc., or to an arithmetic register.

Data and addressing between memory and the arithmetic unit have their own private channels, leaving the bus free during memory operation. The memory buffer and address register are a part of the arithmetic unit.

Figure 3 is an expanded view of this unit. Capital letters stand for registers, small letters for logical entities. Registers *A*, *B*, *C* and *E* are simply storage registers, and are used as the Accumulator, *B*-line, Counter and Extension (least significant arithmetic) register. The Distributor, *D*, is the memory buffer, and is often used as working storage. Registers *F* and *G* are the inputs to the adder logic. The *a* logic is the algebraic sum of $(F) + (G)$; *e* is a rather weird logic, ($e = \bar{F} + G$, which is used in generating the extract order); *f*, which yields $\bar{F}G + F\bar{G}$, is used for the "exclusive" or generation; *c* is the carry logic; *g* is a constant emitter, under microprogram control; and *h* is a set of gates used for input.

As a number moves from *b* to *d*, one of five operations may be performed; *viz.*, normal, shift left one bit, shift right one bit, complement or shift left 5 bits. The last is used for automatic fill and in connection with the microprogram unit control.

As an example, to add the number in the *A* and *D* registers, three microprogram steps would be needed. First, transfer *A* to *G*, *D* to *F*, and finally *a* to *A*; 12 μsec would be required.

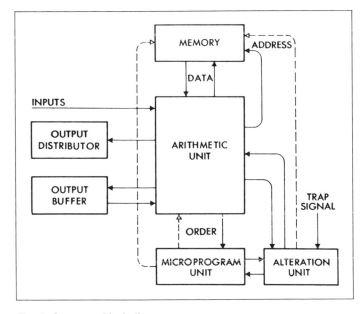

Fig. 1. Computer block diagram.

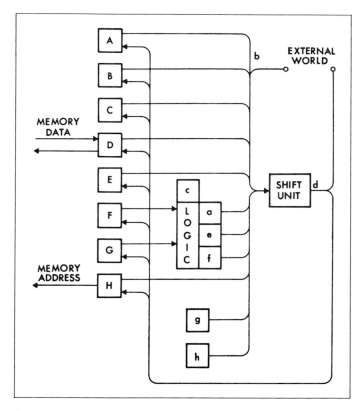

Fig. 3. Arithmetic unit detail.

Figure 4 is a diagram of the microprogram unit. The eight-bit *J* register, augmented by the *TO* flip-flop of the alteration unit, is decoded for up to 512 steps. Students of microprogramming will recognize the Wilkes model in its pure form [Wilkes and Stringer, 1953]. The "next" value of the microprogram register may be chosen in one of three ways.

First, the value may be controlled by the microprogram itself.

Second, five bits of the bus, corresponding to the order portion of the word, may be entered; the other three bits are set to zero. In this manner, the order decoding is accomplished.

Third, all eight bits of the *J* register may be filled from the *d* bus. In practice, the order is shifted five bits to the left, presenting eight bits of the address to get the *J* register. In this manner, one may generate "no address" commands.

In principle, the programmer may start on any microstep which amuses him; in practice, only a limited number of these will yield no-address orders, the other steps being used for parts of add, subtract, order procure, etc. The author has no doubt, however,

that someone will find a useful reason for popping into the middle of divide or some other command. There is no feature of a machine, however pathological, which cannot be exploited by a programmer.

The actual decoding of these nine bits is accomplished partly by logic, and partly by current switching of the clock pulse. A diode matrix is used to convert the microsteps into control signals.

No more than 15 micro operations may be called out on a single step, including selection of the next microorder.

When stepping the microregister, a ploy is used to reduce the number of diodes. Instead of specifying the next step, the microcoder specifies the bits of *J* which he wishes to reverse. Instead of the minimum latency coding of earlier days, the microcoder of the SD-2 must do minimum diode coding. This is roughly analogous to asking for a fast, efficient computer program containing a minimum of 1's. The author, as well as others, has spent endless hours trying to devise a computer program to do such microcoding, with no results.

One may note in passing that the man who wrote the microcode, Tomo Hayata, has for several years specialized in advanced programming problems. Wilkes' views,[1] that logical design will in the future be done by programmers, seem to be verified here. Because of the limited microarithmetic available here, microcoding of the highest order is a must, since each microstep is 4 μsec of time.

For simple orders (*e.g.*, extract), the processes of order procure, indexing (but not indirect addressing), operand procure and execution can be compressed into the time for two memory cycles, *i.e.*, 40 μsec. Each indirect reference adds another memory cycle

[1] Private communication; Aug. 17, 1959.

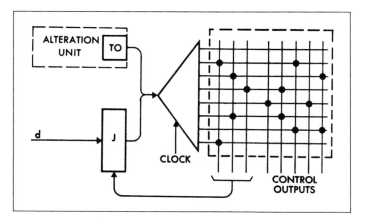

Fig. 4. Microprogram unit.

to this time. Only on multiply, divide, and shift does the ultra-simple structure begin to be expensive in time.

If the temperature requirement were not imposed, the clock frequency could be doubled, materially improving the performance of the machine on multicycle orders.

Figure 5 is a block diagram of the alteration unit. It consists of gates which permit entry of conditions within the computer or the outside world, flip-flops used as working storage, flip-flops, including *TO*, to make its conclusions known to all and sundry, a five-bit tally register (*I*), a circuit to detect a zero on the *d* bus, and the trap logic. There are as many as 20 input gates, 9 storage flip-flops and 10 output flip-flops, exclusive of *TO*.

The *I* register can change its contents in one of two ways, *viz.*, counting down by one, or by accepting an entry from the *d* bus. It may transmit intelligence in two ways, *viz.*, to the *b* bus, or by notifying the input gate system that, should anyone care, it has just counted past zero.

The zero detector signals the truth of the statement that *d* is identically zero. In practice, it checks only the lower digits, not the sign. This is related to the existence of the number -1 in a two-complement system, which is the system's answer to the negative zero of a one's complement logic.

The trap logic is as follows: one of the output signals of the alteration unit signals whether or not the system is receiving trap signals; if it is not, the trap logic makes a note of callers. When the system is again accepting those signals, it transmits whether or not signals have been received, and resets its memory to zero. The timing is such that no trap signal will ever be lost.

The lines going into the logic unit are actually two busses. Any logic source may read to either bus. The logic unit has four control wires from the microprogram unit, specifying which of the 16 Boolean functions of the two busses is to be put on the output bus. This value is then routed to the appropriate logic destination.

The output flip-flops have inputs from the logic unit, and their outputs go to various control points in the machine. Three major points are: (1) establishing whether a memory cycle is read/restore or erase/write; (2) setting the initial carry in the adder; and (3) determining what value shall shift into the vacant spot on a left or right shift.

The initial carry is used for more than simply adding one to a value; since the logic is two complement, but the one complement one is transmitted on the bus, the initial carry is, in general, one during subtraction and zero during addition.

Microprogram details

Figure 6 gives circuit details of the microprogram decode system. The nine flip-flops used are broken into two groups, one of four, the other of five flip-flops. These are decoded into, respectively, 16 and 32 wires. In each group, one and only one wire goes negative. When the clock signal, of 2 μsec width, is applied to the emitters of the first set of 16 gates, it is passed by the selected gating transistor. From the collector of this transistor, it is routed to the emitter of a set of 32 transistors; again, only one can pass current. Thus, the clock signal is routed to one of $16 \times 32 \times 512$ lines. Diodes on the selected line then cause this signal to be routed to appropriate gates in the arithmetic or alteration unit.

By appropriate placement of diodes, a microstep can operate a variety of gates, the number of which is limited by the current available.

Some of the microcontrol wires return to the *J* register so that the microcoder may control the selection of the next microstep. This register is so designed that the actual change of state is inhibited until the clock goes negative.

While each output of the decoding trees may go to 16 bases, only one transistor of the 16 will have a signal on the emitter; thus only one must be driven.

From an engineering point of view, the control of a computer is an elaborate timing system. A microprogram unit is thus a programmable timing generator. The gating transistor/diode decoding system is but one of many ways to achieve this.

Wilkes has observed[1] that, with the diode system, one has an

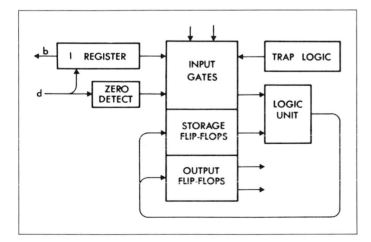

Fig. 5. Alteration unit.

[1] M. V. Wilkes, private communication; Aug. 17, 1959.

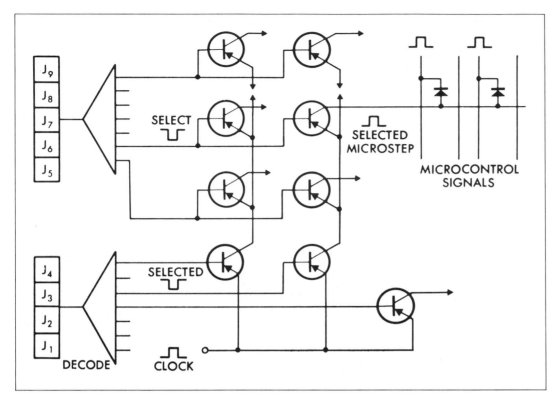

Fig. 6. Details of the microdecode system.

acute packaging problem. He and his co-workers have been led to consider the use of switch-core decoding [Wilkes et al., 1958a].

Eachus[1] and his co-workers have evolved yet another switch-core system which does not depend on coincident current switching.

Order code

Since the order code is only a small problem in the design of a microprogrammed machine (GOTT SEI DANKE), there is little need to dwell on it. There are several comments of design interest, however.

We were unable, with this structure, to get the multiplication below five microsteps per iteration, nor the divide below six, thus costing respectively 20 and 24 μsec per bit dealt with. Moreover, division required some precalculations (overflow detect) and some

[1] Dr. Joseph Eachus of Minneapolis-Honeywell, private conversation; September, 1959.

postcalculation (obtaining a rounded quotient with a correct remainder) which further boosted its time.

Because of the asynchronous nature of transfer, it is not possible to read into and out of a register simultaneously. Hence, shifting one register requires two steps, or 8 μsec per bit, and double-length shifting requires 16 μsec. This is painful.

Because of the short words, four double-length orders were microprogrammed: add, subtract, clear and add, and store. These take a total of 60 μsec to execute.

A rich collection of branch orders was included. BRanch Unconditionally, BRanch Negative, and BRanch Zero are self-explanatory. BRanch on B is the tally loop order which decreases (B) by one, and branches if it does not go negative. BR1, BR2, BR3, and BR4 are sense toggle branch; if the toggle is set, it is turned off and the program branches. These sense toggles are actually storage flip-flops T1, T2, T3, and T4 of the alteration unit. These may be set by other orders. T1 is also used as an overflow mark.

The machine has a "dynamic" idle. When it is halted, either externally or by order, this fact is observed by the microprogram, through the alteration unit, whereupon the microprogram goes into a tight loop, continuously asking, "Can I go? Can I go? Can I go? . . . ." Two forms of halting are provided. In "Halt and Display," registers are presented; in the other halt, the console lights are left unaltered. A manual halt is equivalent to halt and display.

For an addressed order, bit positions one through five are sent into the microprogram unit. During order procure, the microprogram examines bits zero and six for indirect addressing and index modification.

A nonaddress order is recognized by the binary equivalent of 31 in the order bits; the microprogram unit causes the order word to shift left 5 bits, and the 8 high bits of the "address" field enter the *J* register.

Conclusion

This paper is not intended to be an argument in favor of the general acceptance of the SD-2 structure as an ideal. Like all computers, the SD-2 is a state-of-the-art device, intended not only to meet the needs of the problems at hand, but also, more importantly, to meet the side conditions of its use. In a vague analogy, the computer specification is like a partial differential equation. The logical designer must choose the boundary conditions and solve the problem, or at least approximate the solution.

With today's emphasis on system speed performance, some serious mental gear-shifting on the designer's part is required in order to design a simple machine. It goes against the grain of instinct and experience. *A posteriori*, the SD-2 could have been made even simpler, particularly with respect to several peripheral areas not discussed in the paper.

Several conclusions can be drawn here, however. The bus structure is easy to fabricate and maintain; this has been proven on the MILSMAC, a breadboard for the SD-2. It is a highly flexible structure, permitting wide variation in order code with no change in arithmetic unit. At the same time, the components are cascaded to a point where one has the absurd situation of fast-switching in a relatively slow computer. A designer of a bus-structured machine would do well to consider alternatives, such as multiple busses, accumulators, etc., to permit more parallelism when speed is important.

The use of a special-purpose logic unit, such as the alteration unit of the SD-2, gives a freedom of design not possible with a special-purpose logic. At the same time, it uses more parts, is slow in handling multiple variable problems, and requires a great deal of control input. It appears to be a weapon of opportunity.

The use of microprogramming is much the same as the general logic unit. Its flexibility and speed of design are unquestionable. Also, it uses more parts than a special-purpose control.

There is no real substitute for a special-purpose design. The use of generalized elements in computer design can be justified only by the side conditions, never by the basic specification. Where simplicity and speed of design are major items, their use seems indicated.

Wilkes once presented a paper on the best way to design a computer and launched the microprogramming notions. The author would like to comment that if ease and reliability of design are criteria, he was absolutely correct.

References

KampT60; WilkM53a; WilkM58a

Section 4

Processors based on a programming language

Programming-language-based processors are described in Chap. 3 (page 73). Three examples are presented in this section. Two of the languages, FORTRAN and EULER, are algebraic languages operating on conventional data types, whereas IPL-VI is more like a conventional machine language operating on unconventional data types (i.e., list structures). A peculiar feature of IPL-VI is its conception of data as program (as well as of program as data) and the multiprogramming organization to which this led.

A command structure for complex information processing

The IPL-VI processor (Chap. 30) discussed in Part 3, Sec. 5, is an outgrowth of the IPL series of programming languages by Newell, Shaw, and Simon. The paper seriously treats both the language and the merits of casting a language in a hardware processor. IPL-VI was never implemented in hardware. (A partial IPL-V processor for the CDC 3600 was built at the Argonne National Laboratory.) A hardware processor for IPL-VI in the third generation would undoubtedly exist as an interpreter in a microprogrammed processor.

System design of a FORTRAN machine

This paper (Chap. 31) presents a way to map a software program into hardware. The machine's passes (or modes) corre-spond to activities one would see when compiling, loading, and executing a FORTRAN program.

BCD format is used for the arithmetic. The symbol table is simply organized and, therefore, has to be searched. A more serious approach for the actual implementation of such a machine might follow the lines of EULER (Chap. 32).

A microprogrammed implementation of EULER on IBM System 360/Model 30

This very clearly written paper describes a processor to implement an ALGOL-like language [Wirth and Weber, 1966]. An earlier processor was proposed to directly execute ALGOL [Anderson, 1961]. It is implemented using the Model 30 IBM System/360 P.microprogrammed. We include the paper both because it describes the Model 30 and because of EULER.

The P.language operates like a conventional compiler and operating system. The description presents clearly the process of compiling before execution.

The microprogramming aspects of the Model 30 are typical of other IBM System/360 models. The IBM approach to a P.microprogrammed is significantly different from that in Kampe's SD-2 (Chap. 29). In the 360 a microprogram instruction is encoded in a long word (60 to 100 bits, depending on the model) with a number of microcoded operations which can be selected in parallel. The SD-2 uses a short word, and only one operation is encoded in a single instruction.

Chapter 30

A command structure for complex information processing[1]

J. C. Shaw / A. Newell / H. A. Simon / T. O. Ellis

The general-purpose digital computer, by virtue of its large capacity and general-purpose nature, has opened the possibility of research into the nature of complex mechanisms per se. The challenge is obvious: humans carry out information processing of a complexity that is truly baffling. Given the urge to understand either how humans do it, or alternatively, what kinds of mechanisms might accomplish the same tasks, the computer is turned to as a basic research tool. The varieties of complex information processing will be understood when they can be synthesized: when mechanisms can be created that perform the same processes.

The last few years have seen a number of attempts at synthesis of complex processes. These have included programs to discover proofs for theorems [Newell et al., 1956, 1957b], programs to synthesize music [Brooks et al., 1957b], programs to play chess [Bernstein et al., 1958; Kister et al., 1957], and programs to simulate the reasoning of particular humans [Newell et al., 1958]. The feasibility of synthesizing complex processes hinges on the feasibility of writing programs of the complexity needed to specify these processes for a computer. Hence, a limit is imposed by the limit of complexity that the human programmer can handle. The measure of this complexity is not absolute, for it depends on the programming language he uses. The more powerful the language, the greater will be the complexity of the programs he can write. The authors' work has sought to increase the upper limit of complexity of the processes specified by developing a series of languages, called information processing languages (IPL's), that reduce significantly the demands made upon the programmer in his communication with the computer. Thus, the IPL's represent a series of attempts to construct sufficiently powerful languages to permit the programming of the kinds of complex processes previously mentioned.

The IPL's designed so far have been realized interpretively on current computers [Newell and Shaw, 1957a]. Alternatively, of course, any such language can be viewed as a set of specifications for a general-purpose computer. An IPL can be implemented far

more expeditiously in a computer designed to handle it than by interpretation in a computer designed with a quite different command structure. The mismatch between the IPL's designed and current computers is appreciable: 150-machine cycles are needed to do what one feels should take only 2 or 3 machine cycles. (It will become apparent that the difficulty would not be removed by "compiling" instead of "interpreting," to resurrect a set of well-worn distinctions. The operations that are mismatched to current computers must go on during execution of the program, and hence cannot be compiled out.)

The purpose of this paper is to consider an IPL computer, that is, a computer constructed so that its machine language is an information processing language. This will be called language *IPL-VI*, for it is the sixth in the series of IPL's that have been designed. This version has not been realized interpretively, but has resulted from considering hardware requirements in the light of programming experience with the previous languages.

Some limitations must be placed on the investigation. This paper will be concerned only with the central computer, the command structure, the form of the machine operations, and the general arrangements of the central hardware. It will neglect completely input-output and secondary storage systems. This does not mean these are unimportant or that they present only simple problems. The problem of secondary storage is difficult enough for current computing systems; it is exceedingly difficult for IPL systems, since in such systems initial memory is not organized in neat block-like packages for ease of shipment to the secondary store.

Nor is it the case that one would place an order for the IPL computer about to be described without further experience with it. Results are not entirely predictable. IPL's are sufficiently different from current computer languages that their utility can be evaluated only after much programming. Moreover, since IPL's are designed to specify large complicated programs, the utility of the linguistic devices incorporated in them cannot be ascertained from simple examples.

One more caution is needed to provide a proper setting for

[1] *Proc. WJCC*, pp. 119–128, 1958.

this paper. Most of the computing world is still concerned with essentially numerical processes, either because the problems themselves are numerical or because nonnumerical problems have been appropriately arithmetized. The kinds of problems that the authors have been concerned with are essentially nonnumerical, and they have tried to cope with them without resort to arithmetic models. Hence the IPL's have not been designed with a view to carrying out arithmetic with great efficiency.

Fundamental goals and devices

The basic aim, then, is to construct a powerful programming language for the class of problems concerned. Given the amount and kind of output desired from the computer, a reduction in the size and complexity of the specification (the program) that has to be written in order to secure this output is desired.

The goal is to reduce programming effort. This is not the same as reducing the computing effort required to produce the desired output from the specification. Programming feasibility must take precedence over computing economics; since it is not yet known how to write a program that will enable a computer to teach itself to play chess, it is premature to ask whether it would take such a computer one hour or one hundred hours to make a move. This is not meant as an apology, but as support for the contention that, in seeking to write programs for very large and complicated tasks, the overriding initial concerns must be to attain enough flexibility, abbreviation, and automation of the underlying computing processes to make programming feasible. And these concerns have to do with the power of the programming language rather than the efficiency of the system that executes the program.

In the next section a straightforward description of an IPL computer is begun. To put the details in a proper setting, the remainder of this section will be devoted to the basic devices that *IPL-VI* uses to achieve a measure of power and flexibility. These devices include: organization of memory into list structure, provision for breakouts, identity of data with program, two-stage interpretation, invariance of program during execution, provision for responsibility assignments, and centralized signalling of test results.

List structure

The most fundamental and characteristic feature of the IPL's is that they organize memory into list structures whose arrangement is independent of the actual physical geometry of the memory cells and which undergo continual change as computation proceeds. In all computing systems the topology of memory, the character-

istics of hardware and program that determine what memory cells can be regarded as "next to" a given cell, plays a fundamental role in the organization of the information processing. This is obviously true for serial memories like tape; it is equally true from random access memories. In random access memories the topological structure is derived from the possibility of performing arithmetic operations on the memory addresses that make use of the numerical relations among these addresses. Thus, the cell with address 1435 is next to cell 1436 in the specific sense that the second can be reached from the first by adding one to the number in a counter.

In standard computers use is made of the static topology based on memory addresses to facilitate programming and computation. Index registers and relative addressing schemes, for example, make use of program arithmetic and depend for their efficacy upon an orderly matching of the arrangement of information in memory with the topology of the addressing system.

When memory is organized in a list structure, the relation between information storage and topology is reversed. The topology of memory is continually modified to adapt to the changing needs of organization of memory content. No arithmetic operations on memory addresses are permitted; the topology is built on a single, asymmetric, modifiable, ordinal relation between pairs of memory cells which is called adjacency. The system contains processes that make use of the adjacency relations in searching memory, and processes that change these relations at will inexpensively in the course of processing.

A list structure can be established in computer memory by associating with each word in memory an address that determines what word is adjacent to it, as far as all the operations of the computer are concerned. Memory space of an additional address associated with each word is given up, so that the adjacency relation can be changed as quickly as a word in memory can be changed. Having paid this price, however, many of the other basic features of IPL's are obtained almost without cost: unlimited hierarchies of subroutines; recursive definition of processes; variable numbers of operands for processes; and unlimited complexity of data structure, capable of being created and modified to any extent at execution time.

Breakouts

Languages require grammar-fixed structural features so that they can be interpreted. Grammar imposes constraints on what can be said, or said simply, in a language. However, the constraints created by fixed grammatical format can be alleviated at the cost of introducing an additional stage of processing by devices that allow one to "break out" of the format and to use more general modes of specification than the format permits. Devices for breakouts exchange processing time for flexibility. Several devices achieve this in *IPL-VI*. Each is associated with some part of the format.

As an illustrative example, *IPL-VI* has a single-address format. Without breakout devices, this format would permit an information process to operate on only a single operand as input, and would permit the operand of a process to be specified only by giving its address. Both of these limitations are removed: the first by using a special communication list to store operands, the second by allowing the address for an operand to refer either to the operand itself or to any process that will determine the operand.

The latter device, which allows broad freedom in the method of specifying an operand, illustrates another important facet of the flexibility problem. Breakouts are of great importance in reducing the burden of planning that is imposed on the programmer. It is certainly possible, in principle, to anticipate the need for particular operands at particular stages of processing, and to provide the operands in such a way that their addresses are known to the programmer at the appropriate times. This is the usual way in which machine coding is done. However, such plans are not obtained without cost; they must be created by the programmer. Indeed, in writing complex programs, the creation of the plan of computation is the most difficult part of the job; it constitutes the task of "programming" that is sometimes distinguished from the more routine "coding." Thus, devices that exchange computing time for a reduction in the amount of planning required of the programmer provide significant increases in the flexibility and power of the language.

Identity of data with programs

In current computers, the data are considered "inert." They are symbols to be operated upon by the program. All "structure" of the data is initially developed in the programmer's head and encoded implicitly into the programs that work with the data. The structure is embodied in the conventions that determine what bits the processes will decode, etc.

An alternative approach is to make the data "active." All words in the computer will have the instruction format: there will be "data" programs, and the data will be obtained by executing these programs. Some of the advantages of this alternative are obvious: the full range of methods of specification available for programs is also available for data; a list of data, for example, may be specified by a list of processes that determine the data. Since data are only desired "on command" by the processing programs, this approach leads to a computer that, although still serial in its

control, contains at any given moment a large number of parallel active programs, frozen in the midst of operation and waiting until called upon to produce the next operation or piece of data. This identity of data with program can be attained only if the processing programs require for their operation no information about the structure of the data programs, only information about how to receive the data from them.

Two-stage interpretation

To identify the operand of an *IPL-VI* instruction, a designating operation operates on the address part of the instruction to produce the actual operand. Thus, depending on what designating operation is specified, the address part may itself be the operand, may provide the address of the operand, or may stand in a less direct relation to the operand. The designating operation may even delegate the actual specification of the operand to another designating operation.

Invariance of program during execution

In order to carry out generalized recursions, it is necessary to provide for the storage of indefinite amounts of variable information necessary for the operation of such routines. In *IPL-VI* all the variable information is stored externally to the associated routine, so that the routine remains unmodified during execution. The name of a routine can appear in the definition of the routine itself without causing difficulty at execution time.

Responsibility assignments

The automatic handling of such processes as erasing a list, or searching through a list requires some scheme for keeping track of what part of the list has been processed, and what part has not. For example, in erasing a program containing a local subroutine that appears more than once within the program, care must be taken to erase the subroutine once and only once. This is accomplished by a system for assigning responsibility for the parts of the list. In general, the responsibility code in *IPL-VI* handles these matters without any explicit attention from the programmer, except in those few situations where the issue of responsibility is the central problem.

Centralized signalling of test results

The structure of the language is simplified by having all conditional processes set a switch to symbolize their output instead of producing an immediate conditional transfer of control. Then, a few specialized processes are defined that transfer control on the basis of the switch setting. By symbolizing and retaining the conditional information, the actual transfer can be postponed to the most convenient point in the processing. The flexibility obtained by this device proves especially useful in dealing with the transmission of conditional information from subroutines to the routines that call upon them.

General organization of the machine

The machine that is described can profitably be viewed as a "control computer." It consists of a single control unit with access to a large random-access memory. This memory should contain 10^5 words or more. If less than 10^4 words are available in the primary memory, there will probably be too frequent occasions for transfer of information between primary and secondary storage to make the system profitable.

The operation of the computer is entirely nonarithmetic, there being no arithmetic unit. Since arithmetic processes are not used as the basis of control, as they are in standard computers, such a unit is inessential, although it would be highly desirable for the computer to have access to one if it is to be given arithmetic tasks. The computer is perfectly capable of proving theorems in logic or playing chess without an arithmetic adjunct.

Memory

The memory consists of cells containing words of fixed length. Each word is divided into two parts, a symbol and a link. The entire memory is organized into a list structure in the following way. The link is an address; if the link of a word a is the address of word b, then b is adjacent to a. That is, the link of a word in a simple list is the address of the next word in the list.

The symbol part of a word may also contain an address, and this may be the address of the first word of another list. As indicated earlier, the entire topology of the memory is determined by the links and by addresses located in the symbol parts of words. The links permit the creation of simple lists of symbols; the links and symbol parts together, the creation of branching list structures.

The topology of memory is modified by changing addresses in links and symbol parts, thereby changing adjacency relations among words. The modification of link addresses is handled directly by various list processes without the attention of the programmer. Hence, the memory can be viewed as consisting of symbol occurrences connected together by mechanisms or structure whose character need not be specified.

The basic unit of organization is the list, a set of words linked together in a particular order by means of their link parts, in the way previously explained. The address of the first word in the sequence is the name of the list. A special terminating symbol T, whose link is irrelevant, is in the last word on every list. A simple list is illustrated in Fig. 1; its name is L_{100}, and it contains two symbols, S_1 and S_2.

The symbols in a list may themselves designate the names of other lists. (The symbols themselves have a special format, so that they are not names of lists but designate the names in a manner that will be described.) Thus, a list may be a list of lists, and each of its sublists may be a list of lists.

An example of a list structure is shown in Fig. 2. The name of the list structure is the name of the main list, L_{200}. L_{200} contains two sublists, L_{300} and L_{500}, plus an item of information, I_4, that is not a name of a list. L_{300} in its turn consists of item I_1 plus another sublist, L_{400}, while L_{500} contains just information, and is not broken out further into sublists. Each of these lists terminates in a word that holds the symbol T.

Available space list

A list uses a certain number of cells from memory. Which cells it uses is unimportant as long as the right linkages are set up. In executing programs that continually create new lists and destroy old ones, two requirements arise. When creating a list, cells in memory must be found that are not otherwise occupied and so are available for the new list. Conversely, when a list is destroyed (when it is no longer needed in the system) its cells become available for other uses, but something must be done to gain access to these available cells when they are needed.

The device used to accomplish these two logistic functions is the available space list. All cells that are available are linked together into the single long list. Whenever cells are needed, they are taken from the front of this available space list: whenever cells are made available, they are inserted on the front of the available space list just behind the fixed register that holds the link to the first available space. The operations of taking cells from the available space list and returning cells to the available space list involve, in each case, only changes of addresses in a pair of links.

Fig. 1. A simple list.

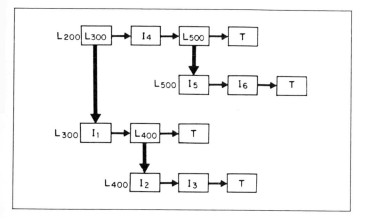

Fig. 2. A list structure.

Organization of central unit

Figure 3 shows the special registers of the machine and the main information transfer paths. Four addressable registers accomplish fixed functions. These are shown as part of the main memory, but would be fast access registers.

Communication list, L_0. The system allows the introduction of unlimited numbers of processes with variable numbers of inputs and outputs. The communication of inputs and outputs among processes is centralized in a communication list with known name, L_0. All subroutines find their inputs on this list, and all subroutines put their outputs on the same list.

Available space list, L_1. All cells not currently being used are on the available space list: cells can be obtained from it when needed and are returned to it when they are no longer being used.

List of current instruction addresses (CIA), L_2. At any given moment in working sequentially through a program, there will be a whole hierarchy of instructions that are in process or interpretation, but whose interpretation has not been completed. These will include the instruction currently being interpreted, the routine to which this instruction belongs, the superroutine to which this routine belongs, and so on. The CIA list is the list of addresses of this hierarchy of routines. The first symbol on the list gives the address of the instruction currently being interpreted; the second symbol gives the address of the current instruction in the next higher routine, etc. In this system it proves to be preferable to keep track of the current instruction being interpreted, rather than the next one.

List of current CIA lists, L_3. The control sequence is complicated in this computer by the existence of numerous programs which become active when called upon, and whose processing may be interspersed among other processes. Hence, a single CIA list does not suffice; there must be such a list for each program that has not been completely executed. Therefore, it is necessary also to have a list that gives the names of the CIA lists that are active. This list is L_3.

Besides these special addressable registers, three nonaddressable registers are needed to handle the transfers of information. Two of these, R_1 and R_2, are each a full word in length, and transfer information to and from memory. Register R_1 receives input from memory; R_2 transmits output to memory. The comparator that provides the information for all tests takes as its input for comparison the symbols in R_1 and R_2. This pair of registers also performs a secondary function in regenerating words in memory: the basic "read" operation from memory is assumed to be destructive; a nondestructive "read" merely shunts the word received from memory in R_1 to R_2 and back, by means of a "write" operation, to the same memory cell.

A register, A, which holds a single address, controls references to the memory, that is, specifies the memory address at which a "read" or "write" operation is to be performed. References to the four addressable registers, L_0 to L_3, can be made either by A or directly by the control unit itself; other memory cells can be referred to only by A. Finally, the computer has a single bit register which is used to encode and retain test results.

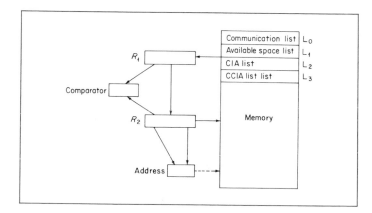

Fig. 3. Machine information transfer paths.

The environment

How input-output, secondary storage, and high-speed arithmetic could be handled with such a machine will be indicated. The machine manipulates symbols: it can construct complex structures, search them, and tell when two symbol occurrences are identical. These processes are sufficient to play chess, prove theorems, or do most other tasks. The symbols it manipulates are not "coded"; they simply form a set of arbitrary distinguishable entities, like a large alphabet.

This computer can manipulate things outside itself if hardware is provided to make some of its symbols refer to outside objects, and other symbols refer to operations on these objects. It could do high-speed arithmetic, for example, if some of its symbols were names of words in memory encoded as numbers as in the usual computer fashion, and others were names of the arithmetic operations. In such a scheme these words would not be in the IPL language; they would have some format of their own, either fixed or floating-point, binary or decimal. They might occupy the same physical memory as that used by the control computer. Thus the IPL language would deal with numbers at one remove, by their names, in much the same manner as the programmer deals with numbers in a current computer. A similar approach can be used for manipulating printers, input devices, etc.

The word and its interpretation

All words in IPL have the same format, shown in Fig. 4. The word a is divided into two major parts: the symbol part, $bcde$, and the link, f. It has been observed that the programmer never deals explicitly with the link, although it will be frequently represented explicitly to show how manipulations are being accomplished. Since the same symbol can appear in many words, the symbol occurrence of the symbol in the word a will be discussed.

A symbol occurrence consists of an operation, b, a designation

operation, c, an address, d, and a responsibility code, e. The operation, b, takes as operand a single symbol occurrence, which is called s. The operand, s, is determined by applying the designation operation, c, to the address, d. Thus, the process determined by a word is carried out in two stages: the first-stage operation (the designation operation) determines an operand that becomes the input to the second-stage operation.

The responsibility bit

The single bit, e, is an essential piece of auxiliary information. The address, d, in a symbol may be the address of another list structure. The responsibility code in a symbol occurrence indicates whether this occurrence is "responsible" for the structure designated by d. If the same address, d, occurs in more than one word, only one of these will indicate responsibility for d.

The main function of the responsibility code is to provide a way of searching a branching list structure so that every part of the structure will, sooner or later, be reached, and so that no part will be reached twice. The need for a definite assignment of responsibility for the various parts of the structure can be seen by considering the process of erasing a list. Suppose that a list has a sublist that appears twice on it, but that does not appear anywhere else in memory. When the list is erased, the sublist must be erased if it is not to be lost forever, and the space it occupies with it. However, after the sublist has been erased when an occurrence of its name is encountered on the other list, it is imperative that it not be erased again on the second encounter. Since the words used by the sublist would have been returned to the available space list prior to the second encounter, only chaos could result from erasing it again. The responsibility code would indicate responsibility, in erasing, for one and only one of the two occurrences of the name of the sublist.

Detailed consideration of systems of responsibility is inappropriate in this paper. It is believed that an adequate system can be constructed with a single bit, although a system that will handle merging lists also requires a responsibility bit on the link f. The responsibility code is essentially automatic. The programmer does not need to worry about it except in those cases where he is explicitly seeking to modify structure.

Interpretation cycle

A routine is a list of words, that is, a list of instructions. Its name is the address of the first word used in the list. The interpretation of a program proceeds according to a very simple cycle. An instruction is fetched to the control unit. The designation operation is decoded and executed, placing the location of s in the address

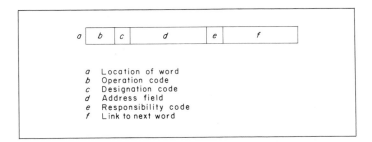

| a | b | c | d | e | f |

a Location of word
b Operation code
c Designation code
d Address field
e Responsibility code
f Link to next word

Fig. 4. IPL word format.

register, A, of Fig. 3. Then operation b is decoded and performed on s. The cycle is then repeated using f to fetch the next instruction.

The operation codes

The simple interpretation cycle previously described provides none of the powerful linguistic features that were outlined at the beginning of the paper: hierarchies of subroutines, data programs, breakouts, etc. These features are obtained through particular b and c operations that modify the sequence of control. The operation codes will be explained under the following headings: the designation code, sequence-controlling operations, save and delete operations, communication list operations, signal operations, list operations, and other operations.

The designation code

The designation operation, c, operates on the address, d, to designate a symbol occurrence, s, that will serve as input, or operand, for the operation b. The designation operation places the address of the designated symbol, s, in the address register.

The designation codes proposed, based on their usefulness in coding with the IPL's, are shown in Appendix 1. The first four, $c = 0$, 1, 2, or 3, allow four degrees of directness of reference. They are usable when the programmer knows in advance where the symbol, s, is located. To illustrate their definition, consider an instruction a_1, with parts b_1, c_1, d_1, and e_1, which can collectively be called s_1. The address part, d_1, of this instruction may be the address of another instruction $d_1 = a_2$; the address part, d_2, of a_2 may be the address of a_3, etc.

The code $c_1 = 1$ means that s is the symbol whose address is d_1, that is, the symbol s_2. In this case the designating operation puts d_1, the address of s_2, in the address register. The code $c_1 = 2$ means that s is s_3; hence, the operation puts d_2, the address of s_3, in the address register. The code $c_1 = 3$ puts d_3, the address of s_4, in the address register. Finally, $c_1 = 0$ designates as s the actual symbol in a_1 itself; hence, this means that b is to operate on s_1. Therefore, this operation places a_1 in the address register.

The remaining two designation operations, $c = 4$ and 5, introduce another kind of flexibility, for they allow the programmer to delegate the designation of s to other parts of the program. When $c_1 = 4$, the task of designating s is delegated to the symbol of the word $d_1 = a_2$. In this case, s is found by applying the designation operation, c_2 of word a_2, to the address, d_2, of word a_2. An operation of this kind permits the programmer to be unaware of the way in which the data are arranged structurally in memory. Notice that the operation permits an indefinite number of stages of delegation, since if $c_2 = 4$, there will be a further delegation of the designation operation to c_3 and d_3 in word a_3.

The last designation operation, $c = 5$, provides both for delegation and a breakout. With $c_1 = 5$, d_1 is interpreted as a process that determines s. Any program whatsoever, having its initial instruction at d_1, can then be written to specify s. When this program has been executed, an s will have been designated, and the interpretation will continue by reverting to the original cycle, that is, by applying b_1 to the s that was just designated. It is necessary to provide a convention for communicating the result of process d_1 to the interpreter. The convention used is that d_1 will leave the location of s in L_0, the standard communication cell.

Sequence-controlling operations

Appendix 2 lists the 35 b operations. The first 12 of these are the ones that affect the sequence of control. They accomplish 5 quite different functions: executing a process ($b = 1$, 10), executing variable instructions ($b = 2$), transferring control within a routine ($b = 3$, 4, 5), transferring control among parallel program structures ($b = 0$, 6, 7, 8, 9,), and, finally, stopping the computer ($b = 11$).

A routine is a list of instructions; its name is the address of the first word in the list. To execute a routine, its name (i.e., its name becomes the s of the previous section) is designated and to it is applied the operation $b = 1$, "execute s." The interpreter must keep track of the location of the instruction that is being executed in the current routine and return to that location after completing the execution of the instruction (which, in general, is a subroutine). All lists end in a word containing $b = 10$, which terminates the list and returns control to the higher routine in which the subroutine just completed occurred. (The symbol T is really any symbol with $b = 10$.)

Figure 5 provides a simple illustration of the relations between routines and their subroutines. In the course of executing the routine L_{10} (i.e., the instructions that constitute list L_{10}), an instruction, $(1, 0, L_{20})$, is encountered that is interpreted as "execute L_{20}." In the course of executing L_{20}, an instruction is encountered that is interpreted as "execute L_{30}." Assuming that L_{30} contains no subroutines, its instructions will be executed in order until the terminate instruction is reached. Because of the 10 in its b part, this instruction returns control to the instruction that follows L_{30} in L_{20}. When the final word in L_{20} is reached, the operation code 10 in its b part returns control to L_{10}, which then continues with the instruction following L_{20}. (Only the b part, $b = 10$, of the terminal word in a routine is used in the interpretation; the c and

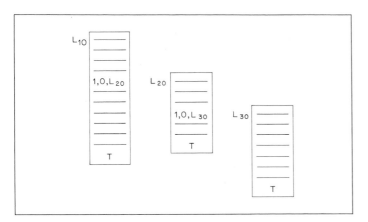

Fig. 5. A simple subroutine hierarchy.

d parts are irrelevant.) This is a standard subroutine linkage, but with all the sequence control centralized.

The operation code $b = 2$, "interpret s," delegates the interpretation to the word s. The effect of an instruction containing $b = 2$ is exactly the same as if the instruction contained, instead, the symbol, s, that is designated by its c and d parts. One can think of the instruction with $b = 2$ as a variable whose value is s. Thus, a routine can be altered by modifying the symbol occurrence s, without any modification whatsoever in the words belonging to the routine itself.

The three operations, $b = 3$, 4, and 5, are standard transfer operations. The first is an unconditional transfer; the two others transfer conditionally on the signal bit. As mentioned earlier, all binary conditional processes set the signal either "on" or "off." In order to describe operations $b = 0$, 6, 7, 8, 9 the concept of program structure must be defined. A program structure is a routine together with all its subroutines and designation processes. Such a structure corresponds to a single, although perhaps complex, process. The computer is capable of holding, at a given time, any number of independent program structures, and can interrupt any one of these processes, from time to time, in order to execute one of the others. All of these structures are coordinate, or parallel, and the operations $b = 0$, 6, 7, 8, 9, are used to transfer control, perhaps conditionally, from the one that is currently active to a new one or to the previously active one. In this sense, the computer being described may be viewed as a serial control, parallel program machine.

The execution of a particular routine in program structure A will be used as an example. Operation $b = 6$ will transfer control to an independent program structure determined by s; call it B.

The machine will then begin to execute B. When it encounters a "stop interpretation" operation ($b = 0$) in B, control will be returned to the program structure, A, that was previously active. But the "stop interpretation" operation, unlike the ordinary termination, $b = 10$, does not mark the end of program structure B. At any later point in the execution of A, control may again be transferred to B, in which case execution of the latter program will be resumed from the point where it was interrupted by the earlier "stop interpretation" command. The operation that accomplishes the second transfer of control from A to B is $b = 7$, "continue parallel program s." Thus, $b = 0$ is really an "interrupt" operation, which returns control to the previous structure, but leaves the structure it interrupts in condition to continue at a later point. There can be large numbers of independent program structures all "open for business" at once, with a single control passing from one to the other, determining which has access to the processing facilities, and gradually executing all of them. Operations $b = 8$ and 9 simply allow the interruption to be conditional on the test switch.

Notice that the passage of control from one structure to another is entirely decentralized; it depends upon the occurrence of the appropriate b operations in the program structure that has control.

When control is transferred to a parallel program structure, either of two outcomes is possible. Either a "stop interpretation" instruction is reached in the structure to which control has been transferred, or execution of that structure is completed and a termination reached. In either case, control is returned to the program structure that had it previously, together with information as to whether it was returned by interruption or by termination. Thus, $b = 0$ turns the signal bit on when it returns control; $b = 10$ in the topmost routine of a structure turns the signal off.

The operation, $b = 11$, simply halts. Processing continues from the location where it halted upon receipt of an external signal, "go."

Save and delete operations

The two operations, $b = 12$ and 13, are sufficiently fundamental to warrant extended treatment. For example, consider a word, L_{100}, that contains the symbol I_1:

Location	Symbol	Link
L_{100}.....................	I_1.....................	t

The link of L_{100}, t, indicates that the next word holds the termination operation, $b = 10$. The "save" operation ($b = 12$)

provides a copy of I_1 in such a way that I_1 can later be recalled, even if in the meantime the symbol in L_{100} has been changed. After the "save" operation has been performed on $s = L_{100}$, the result is:

Location	Symbol	Link
L_{100}	I_1	L_{200}
L_{200}	I_1	t

A new cell, which happened to be L_{200}, was obtained during the "save" operation from the available space list, L_1, and a copy of I_1 was put in it. The symbol in L_{100} can now be changed without losing I_1 irretrievably. Suppose a different symbol is copied, for example, I_2, into L_{100}. Then:

Location	Symbol	Link
L_{100}	I_2	L_{200}
L_{200}	I_1	t

Although I_1 has been replaced in L_{100}, I_1 can be recovered by performing the "delete" operation, $b = 13$. Before the "delete" operation is explained, it will be instructive to show what happens when the "save" operation on L_{100} is interated. If it is executed again, it will make a copy of I_2. Therefore:

Location	Symbol	Link
L_{100}	I_2	L_{300}
L_{300}	I_2	L_{200}
L_{200}	I_1	t

Notice that the cell L_{200}, in which the copy of symbol I_1 is retained, was not affected at all by this second "save" operation. Only the top cell in the list and the new cell from the available space list are involved in the transaction of saving. The same process is performed no matter how long the list that trails out below L_{100}; thus, the save operation can be applied as many times as desired with constant processing time.

The "delete" operation, $b = 13$, applied to the symbol I_2 in L_{100}, will now be illustrated. This operation puts the symbol and link of the second word in the list, L_{300}, into the first cell, L_{100}, and puts L_{300} back on the available space list, with the following result:

Location	Symbol	Link
L_{100}	I_2	L_{200}
L_{200}	I_1	t

The result is the exact situation obtained before the last "save" was performed.

In the description of the "delete" operation up to this point, only the changes it makes in the "push-down" list, in this case L_{100}, have been considered. The operation does more than this, however; "delete s" also erases all structures for which the symbol s (I_1 and I_2 in the examples) is responsible. When a copy of a symbol is made, e.g., the operation that initially replaced I_1 by I_2 in L_{100}, the copy is not assigned responsibility for the symbol ($e = 0$ was set in the copy). Thus, no additional erasing would be required in the particular "delete" operation illustrated. If, on the other hand, the I_2 that was moved into L_{100} had been responsible for the structure that could be reached through it (if it were the name of a list, for example), then a second "delete" operation, putting I_1 back into L_{100}, would also erase that list and put all its cells back on the available space list. Thus "delete" is also equivalent to "erase" a list structure.

Communication list operations

In describing a process as a list of subprocesses, the question of inputs and outputs from the processes has been entirely by-passed. Since each subroutine has an arbitrary and variable number of operands as input, and provides to the routine that uses it an arbitrary number of outputs, some scheme of communication is required among routines. The communication list, L_0, accomplishes this function in IPL.

That the inputs and outputs to a routine be symbols is required. This is no real restriction since a symbol can be the name of any list structure whatever. Each routine will take as its inputs the first symbols in the L_0 list. That is, if a routine has three inputs, then the first three symbols in L_0 are its inputs. Each routine must remove its inputs from L_0 before terminating with $b = 10$, so as to permit the use of the communication list by subsequent routines. Finally, each routine leaves its outputs at the head of list L_0.

The b operations 14 through 19 are used for communication in and out of L_0. Their one common feature is that, whenever they put a symbol in L_0, they save the symbol already there, that is, they push down the symbols already "stacked" in L_0. Likewise, whenever a symbol is moved from L_0 to memory, the symbol below it in L_0 "pops up" to become the top one. (To be precise, the responsibility bit travels with a symbol when it is moved. Hence for example, $b = 16$ and 17, do not, unlike the "delete" operation, erase the structure for which $1L_0$ is responsible.)

The four operations, $b = 14$, 15, 16, and 17, are the main in-out operations for L_0. Two options are provided, depending on whether the programmer wishes to retain the s in memory ($b = 14$ and 16) or destroy it ($b = 15$ and 17). (The move in operation 15 has the same significance as in 16 and 17; the responsibility bit moves with the symbol, and the symbol previously in the location of s, is recalled.)

Operation $b = 18$ is a special input to aid in the breakout designation operation, $c = 5$. Recall that the latter operation requires d to place the location of s, the symbol it determines, in L_0. Operation 18 allows the process d to accomplish this.

Operation $b = 19$ provides the means for creating structures. It takes a cell, for example, L_{200}, from available space, and puts its name, as the symbol $(0, 0, L_{200})$, in the location of the designated symbol, s. The symbol s, previously in this location is pushed down and saved.

Signal operations

Ten b operations are primarily involved in setting and manipulating the signal bit. Observe that the test of equality ($b = 20$ and 21) is identity of symbols. Since there is nothing in the system that provides a natural ordering of symbols, inequality tests like $s > 1L_0$, are impossible. ($1L_0$ means the symbol in L_0.) It is necessary to be able to detect the responsibility bit ($b = 22$), since there are occasions when the explicit structure of lists is important, and not just the information they designate. Finally, although the signal bit is just a single switch, it is necessary to have two symbols, one corresponding to "signal on" and the other to "signal off" ($b = 26$ and 27), so that the information in the signal can be retained for later use ($b = 28$ and 29).

The sense of the signal is not arbitrary. In general "off" is used to mean that a process "failed," "did not find," or the like. Thus, in operations $b = 6$ and 7, the failure to find a "stop interpretation" operation sets the signal to "off." Likewise, the end of a list will by symbolized by setting the signal to "off."

List operations

Both the "save" and "delete" operations are used to manipulate lists, but besides these, several others are needed. The three operations, $b = 30$, 31, 32, allow for search over list structures. They can be paraphrased as: "get the referent," "turn down the sublist," and "get the next word of the list." They all have in common that they replace a known symbol with an unknown symbol. This unknown symbol need not exist; that is, the symbol referred to may contain a $b = 10$ operation, which means that the end of the list has been reached. Consequently, the signal is always set "on" if the symbol is found, and "off" if the symbol is not found. One of the virtues of the common signal is apparent at this point, since, if the programmer knows that the symbol exists, he will simply ignore the signal. Instruction formats that provide for additional addresses for conditional transfers would force the programmer to attend to the condition even if it only meant leaving a blank space in the program.

To illustrate how these search operations work, Fig. 6 shows a list of lists, L_{300}, and a known cell, L_{100}. Cell L_{100} contains the reference to the list structure. The programmer does not know how the list, L_{300}, is referenced. He wants to find the last symbol on the last list of the structure. His first step is $(30, 1, L_{100})$ which replaces the reference by the name of the list, L_{300}. He then searches down to the end of list L_{300} by doing a series of operations: $(32, 1, L_{100})$. Each of these replaces one location on the list by the next one. In fact, a loop is required, since the length of the list is unknown. Hence, after each "find the next word" operation, he must transfer, on the basis of the signal, back to the same operation if the end of the list hasn't been reached. The net result, when the end of the list is reached, is that the location of the last word on list L_{300} rests in L_{100}. Since in this example he wants to go down to the end of the sublist of the last word on the main list, he next performs $(31, 1, L_{100})$. This operation replaces the location of the last word with the name of the last list, L_{700}. Now the search down the sublist is repeated until the end is again reached, at this point the location of the last symbol on the last list is in L_{100}, as desired. The sequence of code follows:

Location	Symbol			Link
	b	c	d	
	30	1	L_{100}	
L_{888}	32	1	L_{100}	
	4	0	L_{888}	
	31	1	L_{100}	
L_{999}	32	1	L_{100}	
	4	0	L_{999}	

The operations, $b = 33$ and 34, allow for inserting symbols in a list either before or after the symbol designated. The lists in this system are one-way: although there is always a way of finding the symbol that follows a designated symbol, there is no way of finding the symbol that precedes a designated symbol. The "insert before" operation does not violate this rule. In both operations,

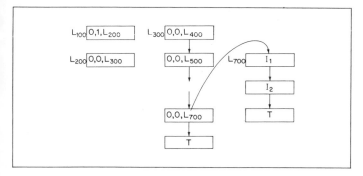

Fig. 6. Example of finding last item of last sublist.

33 and 34, a cell is obtained from the available space list and inserted after the word holding the designated symbol. (This is identical with the first step of the "save" operation.) In the "insert before" operation ($b = 33$) the designated symbol, s, is copied into the new cell, and $1L_0$ is moved into the previous location of s. In "insert after" ($b = 34$), the designated symbol is left unchanged, and $1L_0$ is moved into the new cell. In both cases $1L_0$ is moved, that is, it no longer remains at the head of the communication list.

Other operations

This completes the account of the basic complement of operations for the IPL computer. These form a sufficient set of operations to handle a wide range of nonnumerical problems. To do arithmetic efficiently, one would either add another set of b's covering the standard arithmetic operations or deal with these operations externally via a breakout operation on b (not formally defined here) that would move a full symbol into a special register for hardware interpretation relative to external machines: adders, printers, tapes, etc.

The set of operations has not been described for reading and writing the various parts of the word: b, c, d, e, and f (although it may be possible to automatize this last completely). These operations rarely occur, and it seemed best to ignore them as well as the input-output operations in the interest of simple presentation.

Interpretation

This section will describe in general terms the machine interpretation required to carry out the operation codes prescribed. There is not enough space to be exhaustive, therefore selected examples will be discussed.

Direct designation operations

Figure 7 shows the information flows for $c = 2$, an operation that is typical of the first four designation operations. These flows follow a simple, fixed interpretation sequence. Assume that instruction $(-, 2, L_{100})$ is inside the control unit. The contents of L_{100} are brought into R_1, the input register, then transferred to R_2, the output register, and back to L_{100} again. The d part of R_2 now contains the location of s, and this location is transferred from R_2 to the address register.

Execute subroutine ($b = 1$)

When "execute s" is to be interpreted, the address register already contains the location of s, which was brought in during the first stage of the interpretation cycle. L_2, the current instruction address list (CIA), holds the address of the instruction containing the "execute" order. A "save" operation is performed on L_2, and s is transferred into L_2, which ends the operation. The result is to have the interpreter interpret the first instruction on the next sublist, and to proceed down it in the usual fashion. Upon reaching the terminate operation, $b = 10$, the delete operation is performed on $1L_2$, thus bringing back the original instruction address from which the subroutine was executed. Now, when the interpretation cycle is resumed, it will proceed down the original list. Thus, the two operations, save and delete, perform the basic work in keeping track of subroutine linkage.

Parallel programs

A single program structure, that is, a routine with all its subroutines, and their subroutines etc., requires a CIA list in order to keep track of the sequence of control. In order to have a number of independent program structures, a CIA list is required for each. L_3 is the fixed register which holds the name of the current CIA

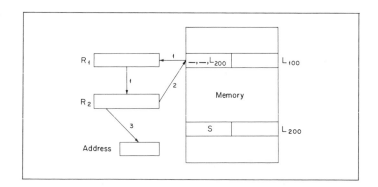

Fig. 7. Information transfers in c = 2 operation.

list. The name of the CIA list for the program structure which is to be reactivated on completion or interruption of the current program structure is the second item on the L_3 list, etc. Therefore, the L_3 list is appropriately called the current CIA list. The "save" and "delete" operations are used to manipulate L_3 analogously to their use with L_2 previously described.

Appendix 3 gives a more complete schematic representation of the interpretation cycle. It has still been necessary to represent only selected b operations.

Data programs

In the section on list operations a search of a list was described. There the data were passive; the processing program dictated just what steps were taken in covering the list. Consider a similar situation, shown in Fig. 8, where there is a working cell, L_{100}, which contains the name of a list, L_{300}. L_{300} is a data program. There is a program that wants to process the data of L_{300}, which is a sequence of symbols. This program knows L_{100}. To obtain the first symbol of data, it does $(6, 1, L_{100})$, that is, "execute the parallel program whose name is in L_{100}." The result is to create a CIA list, L_{500}, put its name in L_{100}, and fire the program. Some sort of processing will occur, as indicated by the blank words of L_{300}. Presumably this has something to do with determining what the data are, although it might be some bookkeeping on L_{300}'s experience as a data file. Eventually L_{700} is reached, which contains $(0, 1, L_{800})$. This operation stops the interpretation, and returns control to the original processing program. The first symbol of data is defined to be $1L_{800}$. The processing program can designate this by $4L_{100}$, since the sequence of $c = 4$ prefixes in L_{100} and L_{500} pass along the interpretation until it ultimately becomes $1L_{800}$. Now the processing program can proceed with the data. It remains

completely oblivious to the processing and structure that were involved in determining what was the first symbol of data. Similarly, although it is not shown, the processing program is able to get the second symbol of data at any time simply by doing a "continue parallel program $1L_{100}$" ($b = 7$).

One virtue of the use of data programs is the solution it offers for "interpolated" lists. In working on a chess program, for example, one has various lists of men: pawns, pieces, pieces that can move more than one square, such as rooks, queens, etc. One would like a list of all men. There already exists a list of all pieces and a list of all pawns. It would be desirable to compose these lists into a single long list without losing the identity of either of the short lists, since they are still used separately. In other words form a list whose elements are the two lists, but such that, when this list of lists is searched it looks like a single long list. Further, and this is the necessary condition for doing this successfully, one cannot afford to make the program that uses this list of lists know the structure. The operation "execute s" ($b = 1$) is precisely the operation needed to accomplish this task in a data program. It says "turn aside and go down the sublist s." Since it does not have the operation $b = 0$, it is not "data." It is simply "punctuation" that describes the structure of the data list, and allows the appropriate symbols to be designated. Figure 9 shows a data list of the kind just described. The authors have taken the liberty of writing in the names of the chessmen.

The stretch of code that follows shows the use of a data program for a "table look up" operation. The table has arbitrary arguments, each of which has a symbol for its value. A_1, A_2, etc. have been used to represent the arguments. To find the value corresponding to argument A_5, for example, A_5 is put in the communication cell with $(14, 0, A_5)$. Then the data program is executed with $(6, 0, L_{100})$. Control now lies with the table, which tests each argument against the symbol in the communication lists: i.e., A_5, and sets the signal accordingly. The program stops interpreting ($b = 8$) at the word holding the value only if the arguments are the same. In this case it would stop, designating L_{350}. If no entry was found, of course, control would return to the inquiring program with the signal off.

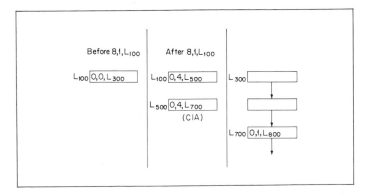

Fig. 8. Example of a data program.

Location	Symbol	Link
L_{100}	20,0,A_1	
	8,0,L_{300}	
	20,0,A_2	
	8,0,L_{320}	
	20,0,A_5	
	8,0,L_{350}	t

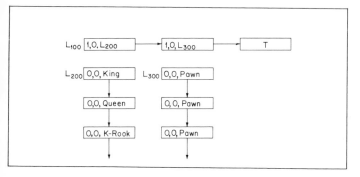

Fig. 9. Application of a data program to chess.

Conclusions

The purpose of this paper has been to outline a command structure for complex information processing, following some of the concepts used in a series of interpretive languages, called IPL's. The ultimate test of a command structure is the complex problems it allows one to solve that would not have been solved if the coding language were not available.

At least two different factors operate to keep problems from being solved on computers: the difficulty of specification, and the effort required to do the processing. The primary features of this command structure have been aimed at the specification problem. The authors have tried to specify the language requirements for complex coding, and then see what hardware organization allowed their mechanization. All the features of delegation, indirect referencing, and breakout imply a good deal of interpretation for each machine instruction. Similarly, the parallel program structure requires additional processing to set up CIA lists, and when a data symbol is designated, there is delegated interpreting through several words, each of which exacts its toll of machine time. If one were solely concerned with machine efficiency, one would require the programmer to so plan and arrange his program that direct and uniform processes would suffice. Considering the size of current computers and their continued rate of growth toward megaword memories and microsecond operations, it is believed that the limitation already lies with the programmer with his limited capacity to conceive and plan complicated programs. The authors certainly know this to be true of their own efforts to program theorem proving programs and chess playing programs, where the IPL languages or their equivalent in flexibility and also in power have been a necessary tool.

Considering the amount of interpretation, and the fact that interpretation uses the same operations as are available to the programmer; e.g., the save and delete operations, one can think of alternative ways to realize an IPL computer. At one extreme are interpretive routines on current computers, the method that the authors have been using. This is costless in hardware, but expensive in computing time. One could also add special operations to a standard repertoire to facilitate an interpretive version of the language. Probably much more fruitful is the addition of a small amount of very fast storage to speed up the interpreter. Finally, one could wire in the programs for the operations to get even more speed. It is not clear that there is any arrangement more direct than the wired in program because of the need of the interpreter to use the whole capability of its own operation code.

References

ShawJ58; BernA58; BrooF57b; KistJ57; NeweA56, 57a, 57b, 58

APPENDIX 1 *c* OPERATIONS (DESIGNATING OPERATIONS)

c Nature of operation for (a) = b c d e.

0 (*a*) is the symbol *s*.
1 *d* is the address of the symbol *s*.
2 *d* is the address of the address of the symbol *s*.
3 *d* is the address of the address of the address of the symbol *s*.
4 *d* is the address of the designating instruction that determines *s*.
5 *d* is the address (name) of a process that determines *s*.

APPENDIX 2 *b* OPERATIONS

b Nature of operation

SEQUENCE-CONTROL OPERATIONS

0 Stop interpreting; return to previous program structure.
1 Execute process named *s*.
2 Interpret instruction *s*.
3 Transfer control to location *s*.
4 Transfer control to location *s*, if signal is on.
5 Transfer control to location *s*, if signal is off.
6 Execute parallel program *s*; turn signal on if stops; off if not.
7 Continue parallel program *s*; turn signal on if stops; off if not.
8 Stop interpreting, if signal is on.
9 Stop interpreting, if signal is off.
10 Terminate.
11 Halt; proceed on go.

SAVE AND DELETE OPERATIONS

12 Save *s*.
13 Delete *s* (and everything for which *s* is responsible).

COMMUNICATION LIST OPERATIONS

14   Copy s into communication list, saving $1L_0$.
15   Move s into communication list, saving $1L_0$.
16   Move $1L_0$ into location of s, saving s.
17   Move $1L_0$ into location of s, destroying s.
18   Copy location of s into communication list, saving $1L_0$.
19   Create a new symbol in location of s, saving s.

SIGNALLING OPERATIONS

20   Turn signal on if $s = 1L_0$, off if not.
21   Turn signal on if $s = 1L_0$, off if not; delete $1L_0$.
22   Turn signal on if s is responsible, off if not.
23   Turn signal on.
24   Turn signal off.
25   Invert signal.
26   Copy signal into location of s.
27   Copy signal into location of s, saving s.
28   Set signal according to s.
29   Set signal according to s; delete s.

LIST OPERATIONS

30   Replace s by the symbol designated by s, and turn signal on; if symbol doesn't exist ($b = 10$), leave s and turn signal off.
31   Replace s by the symbol in d of s and turn signal on; if symbol doesn't exist, leave s and turn signal off.
32   Replace s by the location of the next symbol after d of s and turn signal on (s replaced by "0, 4, (f, part of d of s)"); if next symbol does not exist, leave s and turn signal off.
33   Insert $1L_0$ before s (move symbol from communication list).
34   Insert $1L_0$ after s (move symbol from communication list).

APPENDIX 3 · THE INTERPRETATION CYCLE

1. Fetch the current instruction according to the current instruction address (CIA) of the current CIA list.

2. Decode and execute the c operation:
    If $c = 3$ replace d by d part of the word at address d, reduce c to $c = 2$ and continue. If $c = 2$ replace d by d part of the word at address d, reduce c to $c = 1$ and continue. If $c = 1$ put d in the address register and go to step 3.
    If $c = 0$ put CIA in the address register and go to step 3.
    If $c = 4$ replace c, d by the c, d parts of the word at address d and go to step 2.
    If $c = 5$ mark CIA "incomplete," save it, set a new CIA $= d$, and go to step 1.

3. Decode and execute the b operation: (Some of the b operations which affect the interpretation cycle follow.)
    If $b = 0$ turn the signal on, delete CIA and go to step 4.
    If $b = 1$ save CIA, set a new CIA $= d$ part of s and go to step 1.
    If $b = 2$ replace b, c, d by s and go to step 2.
    If $b = 3$ replace CIA by the d part of s and go to step 1.
    If $b = 10$ delete CIA.
        If no CIA "pops up" turn signal off, delete CIA and go to step 4.
        If "popped up" CIA is marked "incomplete" fetch the current instruction again, move $1L_0$ into address register and go to step 3.
        Otherwise go to step 4.

4. Replace CIA by the f part of the current instruction and go to step 1.

Chapter 31

System design of a FORTRAN machine[1]

Theodore R. Bashkow / Azra Sasson / Arnold Kronfeld

Summary   A system design is given for a computer capable of direct execution of FORTRAN language source statements. The allowed types of statements are the FORTRAN DO, GO TO, computed GO TO, Arithmetic, READ, PRINT, arithmetic IF, CONTINUE, PAUSE, DIMENSION and END statements. Up to two subscripts are allowed for variables and no FORMAT statement is needed. The programmer's source program is converted to a slightly modified form while being loaded and placed in a Program Area in lower memory. His original variable names and statement numbers are retained in a Symbol Table in upper memory, which also serves as the data storage area. During execution of the program each FORTRAN statement is read and interpreted at basic circuit speeds since the machine is a hardware interpreter for these statements. The machine corresponds therefore to a "one-pass, load-and-go" compiler except, of course, that there is no translation to a different machine language. It is estimated that the control circuitry for this machine will require on the order of 10,000 diodes and 100 flip-flops. This does not include arithmetic circuitry.

Index Terms   Digital computer system, digital machine design, direct execution of FORTRAN, FORTRAN computer system, FORTRAN language machine, hardware interpreter.

Introduction

The algebraic languages, in particular FORTRAN in this country, have had enormous impact on the utilization of computers for scientific and engineering computation. They were designed in large part to overcome the annoyance of lengthy learning time and the laborious attention to detail needed to use a basic machine language.

These annoyances are overcome by providing a language which is closer to English in form, and freer of "bookkeeping" details, than the usual machine languages, and by providing a machine language program, called a compiler or translator, to convert from the source program written by a user to an object program executable by a computer. Thus the original drawbacks are overcome but the discrepancy between the external language of the user and the internal language of the machine leads to at least two others. The compilation run of the machine, during which the language translation is accomplished, is a waste of time and money to the user since he must pay for this time though he gets no problem answers from it. Secondly, the user has specified the logical flow and arithmetic details of his solution in the source language. However, when the machine "hangs up" or when he attempts to debug his program, all he finds displayed on the machine console is the machine language. (On large machines he gets equivalently an esoteric print-out in a symbolic form of machine language.) To overcome these difficulties one could use an interpretive translator of the source language instead, but the historical deficiencies of interpreters, loss of memory space and loss of speed of execution have caused this solution to be shunned.

Another solution is also possible—design a machine which executes an algebraic language directly as its "machine language." This approach is based on a recognition that once the allowable syntax and associated semantics of language statements have been firmly specified it is a matter of choice whether to write a compiler, to write an interpreter or to build an interpreter out of hardware. The software choice has been almost overwhelmingly to write a compiler. Since the choice of hardware interpreter, or machine, has not been made, and in fact has hardly been explored to any great extent, a study has been made in order to see if this choice leads to a system which is competitive with the usual software system. It should be understood that such a machine has *not* been constructed. However, the design[2] is sufficiently complete that construction seems feasible.

Language—design philosophy

Since the machine language is to be an algebraic one it seemed reasonable to choose a simple subset of the most commonly used one, FORTRAN. This eliminates the necessity for inventing still another such language and allows attention to be focused on machine design. In fact, the subset chosen is quite close to that known as "Preliminary FORTRAN for the IBM 1620," which is complete enough to be quite useful, but which does not include

[1] *IEEE Trans., EC-16*, vol. 4, pp. 485–499, August, 1967.

[2] See final technical report for Contract AF 19(628)-2798.

such innovations as subroutines, etc. In addition, the usual "built in" subroutines SIN (x), COS (x), etc., are not included. Their inclusion would require additional effort for their hardware implementation which did not appear to be worth expending at this time.

The FORTRAN statement types which are accepted by the machine as machine language are in the table that follows.[1]

Statement	Comment
$a = b$	The value of the arithmetic expression b is stored in the memory location referenced by the variable name a, which may have up to two subscripts.
GO TO n	Program control is transferred to the statement numbered n.
GO TO $(n_1, n_2, \ldots, n_m)$, i	Program control is transferred to *one* of the statements numbered $n_1, n_2, \ldots, n_m$ depending on the value of i at the time this statement is executed.
IF(e) n_1, n_2, n_3	Program control is transferred to the statement numbered n_1 if the algebraic expression e is negative, to that numbered n_2 if e is zero, and to that numbered n_3 if e is positive.
PAUSE	Program execution is halted until restarted by console switch.
DO n $i = m_1, m_2, m_3$	All statements following this one in the program, including the statement numbered n, are executed repeatedly. The first execution is with i equal m_1, i is incremented by the value of m_3 before each succeeding execution. This continues until i is greater than m_2 at which time program control is transferred either to the statement following n or to that statement required by the DO sequencing rules for DO nests. If m_3 is not given it is understood to be 1.
CONTINUE	This statement has the effect of the "no operation" instruction in conventional machines. Program control goes to the next statement in the program unless the CONTINUE is the last statement in the range of a DO. In this case normal DO sequencing takes place.
END	This statement generates a control signal to start execution of the program.

[1] Some familiarity with the FORTRAN language is assumed.

READ, List PRINT, List	These statements cause data to be read or printed, respectively, in accordance with the specified list of variables which may be subscripted; however, the "implied DO" feature has not been implemented. No FORMAT control is available with this machine, therefore no statement number need be given.
DIMENSION $v, v, \ldots$	This statement has the effect of reserving memory space for the subscripted variables v. Each v stands for a variable name followed by parentheses enclosing one or two constants.

No distinction is made in this machine between fixed (integer) and floating point (real) variables. These may have names of any length, starting with any alphabetic character.

Fixed point constants may be specified, in a program or as data, as any combination of one to four numeric characters preceded by a + or − sign. however, these are converted to an internal decimal floating point number and so there are no restrictions on "mixed mode" expressions. Statement numbers must be unsigned fixed point constants, which are not so converted since they only affect program control and not arithmetic processing.

Floating point constants are specified in the form of a mantissa of one to four numeric symbols *preceded* by a decimal point (and a + or − sign). These are followed by the character E and a single (positive or negative) digit representing the power of ten in the usual scientific notation.

These constraints on number size and format are made to simplify certain circuits and could easily be relaxed if desired. The restriction to a two-subscript maximum for subscripted variables is similarly motivated.

Internally, all numerical data require three 8-bit words (Fig. 1). The first two words contain the four-digit mantissa, packed two per word in a 4-bit code for each digit. A decimal point is assumed to exist to the left of the most significant digit. The most significant two bits of the third word are zero. The third bit is 0 if the mantissa is positive, or 1 if it is negative, and similarly the fourth bit is 0 or 1 if the exponent is, respectively, positive or negative. The single exponent digit occupies the least significant four bits of this word. All other characters occupy a full 8-bit word of which the two most significant are 1's. Any numeric characters which are symbols of a variable, e.g., the "2" in $AB2X$, also occupy a full word of this type. Statement numbers are simply packed 2 digits per word and always occupy 2 full words.

Before proceeding with the description of the overall charac-

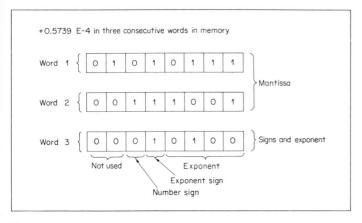

Fig. 1. Data format in memory.

teristics of a machine that loads and executes the language specified above, it may be well to indicate two basic design goals.

1   The card deck or tape containing the Hollerith or BCD version of the English language form of a source program should be the only deck or tape required at any time to execute the program.

2   Once this program is loaded into memory and execution started, any look "into the machine" should reveal information in the same form in which it was entered. Thus if the program is executing $X = A + B$, then one should find "X", "$=$", "A", "$+$", "B", at least in their BCD form.

The second goal has been compromised somewhat as far as the internal representation of the program is concerned in the interest of execution speed. However, all such compromises have been kept to a minimum. In addition, the mechanisms by which one can take such looks "into the machine" are such as to conceal these compromises.

Memory organization

The machine is, in effect, a hardware version of on "one-pass-load-and-go" compiler and it operates in two modes. In the load mode FORTRAN statements are read. They are analyzed as required and stored in memory. When the last statement has been stored, the execution mode is entered and program execution begins at the first executable statement that was read. The input/output device for the machine design is a Flexowriter Model SPD. Programs are assumed to be punched onto a paper tape, one

statement per line, followed by a "carriage return" which generates a paper tape symbol to separate statements. When this tape is read into memory, blanks are automatically "squeezed out."

The memory around which the machine is designed is a 4096-word, 8-bit-per-word, random-access core memory.[1] It is treated by the control circuits as though it consisted of three distinct regions.

1   *Input/output (I/O) buffer:* One statement at a time is loaded sequentially into memory locations 0–99. The six-bit paper tape codes are first converted to internal (often different) six-bit memory codes and stored in the six least significant positions of the 8-bit words. The carriage return symbol is encoded into a special "end-of-statement" symbol represented in the paper as "‡." When this symbol is read the tape is also automatically stopped.

2   *Symbol table area:* Memory locations 4095 and sequentially downward in memory hold the programmer's names for variables, statement numbers, etc., as well as "pointers" to machine addresses, plus empty (before execution) locations for data.

3   *Program area:* Memory locations 100 and sequentially upward hold the FORTRAN program, in a slightly modified form.

Operating modes

The load mode circuits control the input of FORTRAN statements. They place certain information in the Symbol Table Area and the modified form of the FORTRAN statements in the Program Area. It is while in this mode that the necessary searches for variable names take place and machine addresses are assigned. These addresses replace portions of the variable names in the statement as it appears in the Program Area. Similar processing replaces programmer-assigned statement number references in the Program Area with various internal "pointers" for control of GO TO, DO, and IF statements. This modification is done so that statement execution in the execute mode can proceed at high speed. In short, the FORTRAN statement in the Program Area is modified to the extent that variable names are replaced by actual data addresses and statement number references are replaced by actual addresses of statement locations in the Program Area. This translation is done once only, when the statement is analyzed in the load mode. It might be noted here that because of the "one-pass" nature of the translation (a given statement is analyzed only once), certain

[1] 5-μs cycle time, EE Co Model 781.

of the pointers correspond to indirect addresses. Figure 2 shows a sketch of the overall system control and Tables 2 to 7 show to what extent the original statements have been altered.

Loading a program

A program, which is punched in a paper tape, is loaded into memory by energizing the tape read circuit which reads a statement on the tape, including the end-of-statement symbol ±, into the I/O buffer. The read circuit is then de-energized. The least significant 6 bits of each word of the buffer hold the internal BCD representation of each symbol.

A scan circuit (Fig. 3) now picks up each symbol in the statement from left to right and as each symbol is decoded it reacts as follows.

1 If the first symbol is a digit, control is turned over to a Statement Number Load circuit. This circuit shifts the statement number digit by digit into a register (SHR). The maximum allowable length of a statement number is 4 digits and all statement numbers are carried internally in this form, i.e., a programmer's statement number 13 is carried in 2 words as 0013. A search is now made of the Symbol Table area. One of three possibilities exists:

   a   The statement number is not found in the Symbol Table.

It is put into the Symbol Table followed by the value of the current Program location. The statement number is also put into the Program Area starting at this location and the Program Counter incremented appropriately, i.e., by 2 since two 8-bit words are used.

   b   The statement number is found in the Symbol Table because it has been previously referred to by an IF or GO TO. The current value of the Program Counter is placed into the two memory locations following the statement number. (These were left blank when the statement number was previously processed.) The statement number is put into the Program Area and the Program Counter is incremented.

   c   The statement number is found in the Symbol Table because it has been previously referred to by a DO statement. A description will be deferred until the DO statement loading is described since the circuit's behavior is more meaningful in that context.

2 After a statement number has been processed in this fashion or if the first symbol in the statement was not a digit (no statement number was assigned) then the scan circuit continues to pick up each symbol from left to right until it is able to classify the statement as to type. It then turns over control to the appropriate loading circuit as indicated in Fig. 3.

All of these loading circuits put the statements into the Program Area after replacing variable names and statement number references in the program with addresses or pointers. They also replace reserved names such as GO TO or CONTINUE with a single 8-bit code (token). Each unique variable name in the program, however, is also stored in the Symbol Table *once* using an 8-bit code for each symbol. For nonsubscripted variables the three words following the name are reserved for the data that will be associated with this name when the program is executed. Subscripted variable names are found in DIMENSION statements which must precede the use of these variables in the program. In this case as many locations following the name are reserved as have been computed from the DIMENSION statement. The name in the Symbol Table is preceded by a special symbol α, to indicate that it is a subscripted variable. In addition, the first of the two subscript values in the DIMENSION statement is also stored immediately following the name. This number is needed during program execution for constructing the proper element of the array specified by a subscripted variable.[1] The address of

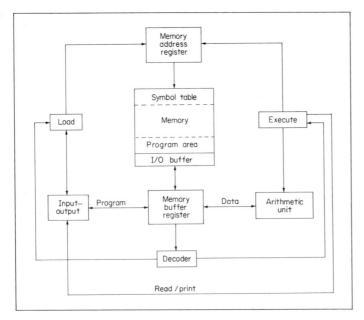

Fig. 2. FORTRAN computer system.

[1] A pointer to the next available location in the Symbol Table is also stored for speed in Symbol Table searching.

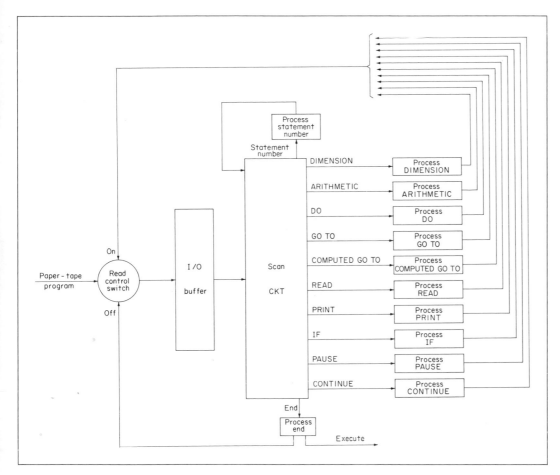

Fig. 3. Load processing sequence and control.

the data location replaces all symbols of the variable name in the Program Area except for the first. This symbol, which must be alphabetic, is retained in the Program Area as an indicator that this is indeed a variable. All special symbols such as (,), +, −, etc. are simply stored sequentially in the Program Area in the 8-bit BCD form as they appear in the original statement.

Statement numbers in IF and GO TO statements are similarly replaced by the address in the Symbol Table which holds the address in the Program Area of the statement having that number. Note that this is an indirect address to the statement. Statement numbers in DO statements are dealt with somewhat differently as will be explained later. Because variable names and statement number references can appear many times in a program, these searches of the Symbol Table are controlled by two special circuits, the Variable Match Unit (VMU) and the Statement Match Unit

(SMU). These circuits indicate either that the name or statement number is already in the Symbol Table or it is not. Thus the first appearance of a variable name, statement number, or reference to a statement number causes it to be put into the Symbol Table. Subsequent references merely utilize these previously assigned data or Program addresses. Therefore each name or statement number is stored in the Symbol Table only once with an exception noted below. In general, the programmer's statement is altered only in the above described fashion. However, for ease of execution the computed GO TO has its index parameter name, i.e., the "i" in GO TO $(n_1, n_2. \cdots, n_m), i$, changed from the position following the parenthesis to a position preceding the parenthesis.

The DO statement requires the most complex loading algorithm. Basically, the idea is to place the DO statement itself, essentially unchanged, into the Program Area but to extract the range statement number (which specifies the last statement in the range of the DO) and put it into the Symbol Table. It is there preceded by a special symbol Δ, designating it as being referenced by a DO, and followed by the Program Area address of the corresponding DO statement. The DO statement in the Program Area has its original statement number replaced by a special symbol, λ, and an internal address which is determined as follows (see Table 6).

a   If this DO is one of a nest of DO's, the internal address is the Program Area address of the λ token of the next preceding DO statement. This is easily found by a Symbol Table search for the range statement number since there is an entry in the Symbol Table corresponding to every DO statement. Thus for a DO nest three deep all ending in statement number 100, for example, there will be three entries in "DO nest order" of the number 0100 each followed by the corresponding DO statement Program Area address.

b   If this DO is the *first* of a nest of DO's, or if it is the only DO specifying a particular range statement number, then this internal address is the program address of the next statement *outside* the DO range, i.e., the address to which control should go if this DO or DO nest is satisfied.

This *outside* address is found by the Statement Number Load circuit at the time the last statement *in* the range appears in the I/O buffer for loading. The circuit first detects that a matching statement number in the Symbol Table is preceded by a Δ. It then extracts and saves the Program Area address of the first DO and the last DO, if there is a nest, or simply the only address if there is just one. The statement number is put in the Program Area as always. In addition, the Program Area address of the λ token of the last DO in the nest is also put in the Program Area immediately following it. In addition, a special flip-flop, the LSFF, is set. The loading circuit for each statement type allowed to be the last statement in a DO range, tests this LSFF *after* it has loaded the statement into the Program Area. If it is on, the current contents of the Program Counter, the address of the next statement *outside* the DO range are used as the internal address in the first (or only) DO of the nest.

It should be noted that this DO range statement number together with its own Program Area location will also appear in the Symbol Table without a preceding Δ. This is necessary because it is possible (and even legal in some cases!) to have an IF or GO TO refer to it also.

The method used to design the circuits which implement these

functions is the same in each case. From the English language description of the function a sequential circuit state diagram is constructed. The circuit is then synthesized from the state diagram using established methods. The state diagrams of the Arithmetic Statement Loading circuits and the Variable Match Unit, which are used during Loading, are shown in the Appendix.

The hardware implementation of the state diagram of the Variable Match Unit is also described there.

Executing a program

When the END statement signaling the end of a source program is encountered by the scan unit, the machine leaves its load mode, executes an automatic RESET, and enters the execution mode. (Reset forces the address 100 into the Program Counter.) Pressing the console *start* button causes statement execution to begin at the first executable statement which is always found at memory address 100. There is a separate statement execution circuit for each statement type. In addition, the Statement Number processing circuit reacts to a digit as the first symbol in a statement. Each of these circuits is in an initial state when execution begins. *One and only one* can leave its initial state when the first symbol of a statement is read from memory. The responding circuit then retains control as it executes the statement until the ‡ (end of statement symbol) is read from memory. It then returns to its initial state. The first symbol of the next statement, as indicated by the Program Counter, is read and causes some circuit to leave its initial state, etc. Thus the first symbol of a statement acts like the "operation code" portion of a conventional computer instruction word. The first symbol must be (since the load circuitry causes this) one of the 8-bit tokens for the various statement types, or a digit of a statement number, or the alphabetic character of the variable on the left of the "=" symbol of an arithmetic statement. The tokens are represented in this paper shown in Table 1.

Table 1

Statement type	Token
GO TO n	*GO TO*
GO TO $(n_1, n_2, \ldots, n_m), i$	*COMGOTO*
IF $(e)\ n_1, n_2, n_3$	*IF*
PAUSE	*PAUSE*
DO $n\ i = m_1, m_2, m_3$	*DO*
CONTINUE	*CONTINUE*
READ	*READ*
PRINT	*PRINT*

It is possible, however, for the DO execution circuitry to leave its initial state either by reading of the *DO* or by reading of the λ token immediately following it. The former causes DO initialization, the latter causes DO indexing and testing as will be described later.

The action of the execution circuits is briefly given below.

Statement number processing

When the first symbol of a statement is a digit this circuit is energized. If there are only four digits (packed into two memory words) the circuit returns to its initial state and the remainder of the statement is executed. If there are eight digits (packed into four memory words), the last four digits (the address of the λ of the last, or only, DO in a nest) are saved in a register, SSAR. The LSFF is turned on, the circuit returns to its initial state and the remainder of the statement is executed. If the remainder of the statement is *not* an IF, GO TO, or DO statement, the execution circuitry in control executes the statement and then tests for the LSFF being on. If it is on, the Program Counter contents are replaced with the SSAR contents, the LSFF is reset, and the circuit returns to its initial state. In this case the SSAR holds the program address of the λ token of the innermost DO. When this λ is read, DO indexing and testing take place. If the LSFF is off, the circuit returns to its initial state.

GO TO n

The *GOTO* token energizes this circuit. The four-digit address (packed into two memory words) immediately following the token is extracted. The contents of this address are put into the Program Counter and the circuit returns to its initial state.

Example.[1] GO TO 15‡ (Table 2).

GO TO $(n_1, n_2, \cdots, n_m)$, i

The *COMGOTO* token energizes this circuit. The initial alphabetic symbol of i, now immediately following the token, is read and discarded and the four-digit address immediately following is extracted. The contents of this address (the current value of i) are put into a register and decremented by one.

1   If the result is zero, the four-digit address following the left parenthesis is extracted. The contents of this address are put into the Program Counter and the circuit returns to its initial state.

[1] All examples are written as though this statement or statements were the first in the program.

Table 2

Symbol table			Program area		
Address	contents		Address	contents	
4095	00	⎫Machine form for	0100	*GOTO*	
4094	15	⎭Statement 15	0101	40	⎫Address of the address
4093	02	⎫Address of	0102	93	⎭of Statement 15
4092	50	⎭Statement 15	0103	‡	
·			·		
·			·		
·			·		
·			0250	00	⎫Statement 15
·			0251	15	⎭in the program

2   If the result is nonzero, the four-digit address following the left parenthesis is read and discarded. The register is decremented by one again.

3   If the result is zero, the four-digit address following the next comma is treated as in 1 above.

4   If the result is nonzero, the four-digit address following the next comma is read and discarded. The register is decremented by one again.

Steps 3 and 4 above are repeated until the register is zero. If the right parenthesis is read while the register is nonzero an error condition has been found and will be indicated.

Example. GO TO (5, 10, 150), *ITALY*± (Table 3).

IF(e)n_1, n_2, n_3

The *IF* token energizes this circuit. The left parenthesis immediately following the token is read. Control is then given temporarily to the Arithmetic Statement execution circuit. The latter circuit is forced to the state in which it would be if it were ready to evaluate an expression to the right of the equal sign in an Arithmetic Statement. A special F/F, the IFFF, is also set to 1. The expression e of the IF statement is read and evaluated until the final right parenthesis of the IF statement is read. Since the Arithmetic Statement circuit was not allowed to read the initial left parenthesis, it would normally go to an error condition under these circumstances of "unbalanced" parentheses. However, sensing that the IFFF is set to 1, it resets the IFFF, places the value of the expression e just evaluated into the accumulator, returns to its own initial state, and re-energizes the IF statement circuit. The accumulator is equipped to sense its own contents and energizes one

Table 3

Symbol table		Program area	
Address	contents	Address	contents
4095	I	0100	COMGOTO
4094	T	0101	I
4093	A	0102	40 ⎫ Address of the
4092	L	0103	90 ⎭ data for ITALY
4091	Y	0102	(
4090		0103	40 ⎫ Address of the address
4089		0104	85 ⎭ of Statement 5
4088		0105	,
4087	00 ⎫ Representation of	0106	40 ⎫ Address of the address
4086	05 ⎭ Statement 5	0107	81 ⎭ of Statement 10
4085	02	0108	,
4084	50	0109	40 ⎫ Address of the address
4083	00	0110	77 ⎭ of Statement 150
4082	10	0111	)
4081	03 ⎫ Address of	0112	‡
4080	50 ⎭ Statement 10	.	
4079	01	.	
4078	50	.	
4077	05 ⎫ Address of	0250	00
4076	53 ⎭ Statement 150	0251	05
.		.	
.		.	
.		.	
		0350	00
		0351	10
		.	
		.	
		.	
		0553	01
		0554	50

of three signal lines depending on whether the number is zero, positive, or negative. The IF circuit senses these lines and reacts as follows.

1 If the accumulator signal is negative, the next four-digit address (n_1) is extracted. The contents of this address are put into the Program Counter and the circuit returns to its initial state.

2 If the accumulator signal is zero, the next four-digit address is skipped over. The four-digit address following the next commas (n_2) is treated as in 1 above.

3 If the accumulator signal is positive, the next 2 four-digit addresses and the intervening comma are skipped over. The four-digit address following the next comma (n_3) is treated as in 1 above.

Example. IF(A − B) 10, 20, 20± (Table 4).

PAUSE

The *PAUSE* token energizes this circuit. The end of statement symbol, ‡, is read and discarded. All execution circuits are forced to a state O' and automatic reading of the memory ceases. A START signal, initiated by a console switch, is required to return these circuits to state O and to initiate memory reading at the location specified by the current contents of the Program Counter.

Example. PAUSE ‡ (Table 5).

DO n i = m_1, m_2, m_3 (or DO n i = m_1, m_2)

This circuit is energized (i.e., caused to leave its initial state) either by a *DO* token or by the λ token. Its action is different in these two cases and will be described separately.

Table 4

Symbol table		Program area	
Address	contents	Address	contents
4095	A	0100	IF
4094		0101	(
4093		0102	A
4092		0103	40
4091	B	0104	94
4090		0105	—
4089		0106	B
4088		0107	40
4087	00	0108	90
4086	10	0109	)
4085	03 ⎫ Address of	0110	40 ⎫ Address of the address
4084	50 ⎭ Statement 10	0111	85 ⎭ of Statement 10
4083	00	0112	,
4082	20	0113	40 ⎫ Address of the address
4081	04 ⎫ Address of	0114	81 ⎭ of Statement 20
4080	41 ⎭ Statement 20	0115	,
.		0116	40 ⎫ Address of the address
.		0117	81 ⎭ of Statement 20
.		0118	‡
		.	
		.	
		.	
		0350	00
		0351	10
		.	
		.	
		.	
		0441	00
		0442	20

Table 5

Symbol table	Program area	
	Address contents	
(not applicable)	0100	*PAUSE*
	0101	‡

1 *The circuit is energized by the DO token:* The λ token and the four-digit address immediately following are read and discarded. The initial alphabetic symbol of *i* is read and discarded and the four-digit address immediately following is extracted and saved in a register called SAR. The = symbol is read and discarded. The initial value m_1 of this statement can be either purely numeric or it may be the name of a variable.

    a If it is purely numeric the load circuitry will have replaced it with the internal machine representation of the number. Therefore this number is simply read and stored in the Symbol Table starting at the address given in the SAR register.

    b If it is the name of a variable, the initial alphabetic symbol is read and discarded. The four-digit address following is extracted. The contents of this address are treated as in *a* above.

    In either event then, *i*, is given the value m_1 as required. The remainder of the DO statement including the ‡ symbol is read and discarded and the circuit returns to its initial state.

2 *The circuit is energized by the λ token:* The four-digit address immediately following is extracted and saved in the SSAR. The initial alphabetic symbol of *i* is read and discarded, the four-digit address immediately following is put into the SAR and the contents of this address are placed in the accumulator. (This is the current value of *i*.) The = symbol and all symbols up to and including the next comma are read and discarded. The final value, m_2, may be numeric or the name of a variable.

    a If it is numeric, this value is placed in a numeric register, SHR.

    b If it is the name of a variable the initial symbol is read and discarded. The contents of the four-digit address following is extracted and placed in the numeric register SHR. The next symbol is read. This will be a comma if m_3 has been specified or ‡ if m_3 has not been specified.

    c If it is a comma either the following purely numeric value is added to the contents of the accumulator or the contents of the following four-digit address is added.

    d If it is the ‡ symbol then the contents of the accumulator are incremented by one. In either event, after the current value of *i* has been incremented by either m_3 or one, the contents of the accumulator are put in the Symbol Table starting at the address given in the SAR.

Now the final value, saved in the SHR, is subtracted from the accumulator. If the accumulator signal is positive then the value of *i* must be greater than the final value of m_2. Therefore the address in the SSAR is placed in the Program Counter and the circuit returns to its initial state. The address in the SSAR will either be the address of the λ token of a preceding DO in the nest or it will be the address of the next statement outside the DO nest depending on which DO statement is being executed. If the accumulator signal is not positive then the value of *i* is less than or equal to m_2 and the circuit just returns to its initial state. Thus the next statement after the DO statement will be executed.

Example. (See Table 6.)

DIMENSION *B*(20, 10)‡
DO 5   *IT* = 1, 100, *L*‡
DO 5   *J* = *N, M*‡
5   *A* = *B*(*IT, J*)‡

CONTINUE

The *CONTINUE* token energizes this circuit. The ‡ symbol is read. If the LSFF is not on, the circuit returns to its initial state. If the LSFF is on, it is turned off. The contents of the SSAR replace the contents of the Program Counter and the circuit returns to its initial state. Thus if this statement is either not labeled or is not the last statement in a DO range, its execution has no effect on the program. The example assumes the usual case where it is the last statement in a DO range.

Example. (Table 7)

DO 5 *I* = 1, 150‡
5   CONTINUE‡

READ, list. (PRINT, list.)

The *READ* token energizes this circuit which then energizes the Flexowriter read circuits. Data from paper tape is read into the I/O buffer until the end-of statement symbol, ‡, is stored. The data must be punched as one to four decimal digits for fixed point numbers or one to four decimal digits preceded by a decimal point for floating point numbers. The latter may also be followed by

Table 6

Symbol table		Program area	
Address contents		Address contents	
4095	α	0100	DO
4094	B	0101	λ
4093	34 } Next free symbol	0102	01 } Address of Statement
4092	88 } Table Address	0103	57 } following the DO nest
4091	00 } Machine form of	0104	I
4090	20 } the constant 20	0105	34 } Address of data
4089	04 }	0106	81 } for IT
·		0107	=
·		0108	00
·		0109	01
3488	Δ	0110	04
3487	00 } Machine form	0111	,
3486	05 } Statement 5	0112	01
3485	01 } Address of 1st	0113	00
3484	01 } DO in nest	0114	04
3483	I	0115	,
3482	T	0116	L
3481		0117	34
3480		0118	77
3479		0119	‡
3478	L	0120	DO
3477		0121	λ
3476		0122	01 } Address of preceding
3475		0123	01 } DO in the nest
3474	Δ	0124	J
3473	00	0125	34
3472	05	0126	68
3471	01 } Address of 2nd	0127	=
3470	21 } DO in nest	0128	N
3469	J	0129	34
3468		0130	64

Symbol table		Program area	
Address contents		Address contents	
3467		0131	,
3466		0132	M
3465	N	0133	34
3464		0134	60
3463		0135	‡
3462		0136	00
3461	M	0137	05
3460		0138	01 } Address of last
3459		0139	21 } DO in the nest
3458		0140	A
3457	00	0141	34
3456	05	0142	52
3455	01 } Address of	0143	=
3454	36 } Statement 5	0144	B
3453	A	0145	40
3452		0146	91
3451		0147	(
3450		0148	I
·		0149	34
·		0150	81
·		0151	,
		0152	J
		0153	34
		0154	68
		0155	)
		0156	‡
		0157	
		·	
		·	
		·	

the letter E and a single positive or negative digit indicating a power of ten. Numbers must be separated by a comma to distinguish them, since no FORMAT information is available and the read circuits "squeeze out" blanks.

The first set of digits starting at the beginning of I/O buffer, memory address 0, is read into a 24-bit register (which is the size of the three 8-bit memory words required for data). Numerical information in the I/O buffer is in a 6-bit code. The two most significant bits are 0 if the code is for a numeric character. The placing of information into the 24-bit register is easier to understand if we consider it as a 16-bit mantissa register M, which can hold four decimal digits, and an 8-bit sign and exponent register X, which can hold 2 bits of sign information and an exponent digit.

Both registers are set to zero initially. If the first character is a minus sign, the bit in the mantissa sign position of X is set to one. (The internal form of data representation was described earlier in the section on Language-Design Philosophy.) If it is a plus sign no action is required since a zero in the mantissa sign position indicates a positive mantissa. Further action depends on the next character.

1   If the next character is numeric (or if there was no sign given and the first character is numeric) this must be a fixed point constant. The four bits of numeric information are gated to the *least* significant four positions of register M. If the next character is numeric, M is shifted left four positions and this character is also gated to the least significant

Table 7

Symbol table		Program area	
Address	contents	Address	contents
4095	Δ	0100	DO
4094	00	0101	λ
4093	05	0102	01
4092	01	0103	22
4091	01	0104	I
4090	I	0105	40
4089		0106	89
4088		0107	=
4087		0108	00
4086	00	0109	01
4085	05	0110	04
4084	01	0111	,
4083	16	0112	01
		0113	50
		0114	04
		0115	±
		0116	00
		0117	05
		0118	01 ⎫ Address of the
		0119	01 ⎭ DO statement
		0120	CONTINUE
		0121	±
		0122	

position. This continues until the comma is read. The numeric code for four is now gated to the least significant four positions of X. Since the arithmetic unit assumes a decimal point at the left of all data, this action insures that a fixed point number is properly interpreted.

2 If the next character after the sign (if there is one) is a decimal point this must be a floating point number. In this case the following digits are stored into M as indicated above, but three shifts of M are always taken, whether or not four digits are stored in M. This is required to insure proper interpretation of the number. If a comma follows the series of digits no further action is taken. If an E follows then the digit following it is placed in the least significant 4 positions of X. If a minus sign is found following the E a setting of the exponent sign position of X precedes this action. The comma is then read.

After this first piece of data has been placed in M and X, the alphabetic character following the READ token is read and discarded. The next 4 digits are used as the address in which the

most significant two digits in M are stored and it is then decremented appropriately to store the remainder of the data.

The remaining data in the I/O buffer are then stored one by one in sequence at the addresses given by the remainder of the READ list. A subscripted variable on this list requires additional arithmetic operations to compute the correct address from the current index values and the original DIMENSION information stored in the Symbol Table. These operations will be given later in the Arithmetic Statement description.

When the ‡ token in the I/O buffer is reached, the next character in the READ list is read. If this character is also the ‡ token then the circuit returns to its initial state. If, however it is not, then the Flexowriter is again energized such as to read data into the I/O buffer, and processing proceeds as before until reading of the ‡ of the READ statement returns the circuit to its initial state.

The PRINT statement circuit operates in almost exactly inverse fashion and will not be described in detail. The list variables are used in sequence to extract data from the proper memory locations and place it in the M and X registers. The contents of these registers are then put sequentially into the I/O buffer, together with 6-bit codes for the decimal point, plus and minus signs, commas, and the E symbol at appropriate places. All data are thus output in floating point form. When the ‡ token is read, the Flexowriter print circuits are energized and the circuit returns to its initial state.

Example.

READ, A, B, C(I, J)‡
PRINT, B, C(I, J)±‡

The appearance of the Symbol Table and Program Area should be apparent from previous examples. Since this would add little to the description of circuit action they will be omitted.

a = b

The Arithmetic Statement execution unit is energized by any 8-bit alphabetic character code. This first character of the variable name represented above as "a" is discarded. Then either the following four-digit data address is saved or the data address of a subscripted variable is computed and saved in a register. After reading and discarding the = symbol, the circuit executes the expression b in accordance with the given sequence of arithmetic operator symbols, +, −, *, /, which are used to control the arithmetic unit. The partial results at any time during the execution are stored in the I/O buffer area which is, of course, otherwise unused during

Arithmetic Statement execution. These storage areas for partial results are called d_{i0}, d_{i1}, where i specifies the "level" at which computation is taking place,[1] i is equal to zero until a left parenthesis is encountered which increases the current value of i by 1. An exception occurs if the left parenthesis immediately follows the = symbol. In this case the level remains at zero. It is also necessary to store control information which relates to these partial results.

Two control values are required at every level. The count of left parentheses at any i level is stored as a number, l_i. Before i is incremented, the incompleted arithmetic operations still required at the current level are indicated by giving an indicator t_i the value 1, 2, or 3. Also needed are indicators $t +_i$ and $t *_i$ to distinguish + from − and * from /. To clarify the significance of these control values an analysis will be made of the following expression, which contains some unneeded but legitimate sets of parentheses:

$$A = ((B + (C/((D + E^*(F)))) + G))\ddagger$$

1 The circuit reads and saves the address of A, then reads and discards the = which puts the circuit at the level $i = 0$. The first two left parentheses cause l_0 to be set to 2. The value of B is stored in d_{00}. The plus sign followed by a left parenthesis cause the indicator t_0 to be set to 1 to indicate the condition "$B + ($". Since we might in other cases find "$B - ($", t_0 is set to zero to indicate the plus sign.

2 The left parenthesis also causes i to be incremented to one and since it is the only one at this level, l_1 is also set to 1. The value of C is stored in d_{10}. The division symbol followed by a left parenthesis causes t_1 to be set to 2 to indicate the condition "$C/($". Since we might find "$C*($" in other cases, $t *_1$ is set to 1 to indicate the division.

3 The left parenthesis also causes i to be incremented to 2 and the next left parenthesis increments l_2 to 2. The value of D is stored in d_{20} and the value of E put into d_{21}, respectively. The multiplication symbol followed by a left parenthesis causes t_2 to be set to 3 to indicate the condition "$D + E * ($". $t +_2$ and $t *_2$ are each set to zero to indicate the plus and multiplication symbols, respectively.

4 The left parenthesis before the F causes i to be incremented to 3 and l_3 to be set to 1. The value of F is placed in d_{30}.
   The Arithmetic Statement circuit always puts the final value computed at any level into the arithmetic unit register, SR. It does this whenever $l_i = 0$ for any i. Clearly l_i must be decremented by one for each right parenthesis.

[1] Basic circuit operation at any level is described in the earlier report. See page 363, footnote 2.

Therefore the first right parenthesis after the F causes l_3 to equal zero. This condition causes the value stored in d_{30} to be placed in the SR. The value of i is decremented to 2.

5 t_2 being 3 (and $t +_2 = t *_2 = 0$) causes the computation, $d_{20} + d_{21} *$ SR to be stored in d_{20}. The next two parentheses after F cause l_2 to equal zero. Therefore, this result is placed in the SR. The value of i is decremented to 1.

6 Since t_1 is equal to 2 and $t *_1$ is equal to 1 the computation $d_{10}/$SR is made and stored in d_{10}. The final parenthesis after the F causes l_1 to equal zero. Therefore this result goes to SR. i is decremented to zero.

7 Since t_0 is one and $t +_0$ is zero the computation, $d_{00} +$ SR, is made and the result is stored in d_{00}.

8 The $+G$ causes the computation $d_{00} + G$ to be made and stored in d_{00}. The final two parentheses cause l_0 to be zero; therefore the value in d_{00} is placed in SR. (If *another* right parenthesis were found, this would cause an error condition to be indicated.) The $\ddagger$ symbol causes the contents of SR to be stored at the previously saved memory address for A.

Any subscripted variable addresses are computed easily from the initial DIMENSION statement information, saved in the Symbol Table, and the current value of the subscripts. Assume the first data location for an array $A(I, J)$ is stored at a location $A_{\text{base}} + 1$. If the DIMENSION statement read DIMENSION A(5, 10) then the computation, $A_{\text{base}} + 5 * (J - 1) + I$, gives the correct data address for any nonzero value of I and J. (This is true only if a complete data word is stored per memory word; in this machine the expression is slightly more complicated.)

In this machine the partial result locations d_{i0} and d_{i1} are actually 3 words long, of course, to accommodate the data. An additional word is used to store control information where 4 bits are used for t_i, $t +_i$, and $t *_i$ and the remaining 4 bits for the l_i count. The i counter therefore is actually incremented or decremented by 7 instead of one. Thus at any level, of which there can be 14 since the I/O buffer is 100 words long, the l_i count can be as great as 15. This is more than adequate since it allows for 210 left parentheses, which is much longer than the I/O buffer length.

Since the appearance of the Symbol Table and Program Area would add little to this discussion, an example will be omitted.

Conclusion

We have illustrated in some detail that a machine for direct translation of a simple algebraic language is possible. It would therefore seem that further investigation be made of the economic position of this solution vis-à-vis the software compiler solution. Unfortunately, the present authors are not sufficiently versed in compiler construction to make such a comparison.

The actual construction of such a machine as an independent unit is probably not reasonable except under particular circumstances in which only small one-shot scientific problems form the bulk of the computing. However, as an adjunct to a larger general purpose machine, it may well serve a need as a hardware interpreter for widely used higher level languages.

As a result of a fairly complete design of the control circuits of this machine, it is estimated that 10,000 diodes and 100 flip-flops would be needed for these alone (not including arithmetic circuits). The design techniques used are simple and straightforward but rather expensive. These designs should probably only be considered for use with integrated circuitry.

References

AndeJ61; BashT64; International Business Machines Corporation, General Information Manual; FORTRAN, Form F28-807401, December, 1961; IBM 1620 FORTRAN: Preliminary Specifications, Form J29-4200-2, April, 1960

APPENDIX[1]

The variable match unit (*VMU*) (*Fig. 4*)

The Symbol Table at the end of the load mode should contain all variable names used by the program, together with empty locations reserved for data associated with these names. The Program Area at the end of the load mode should have a program in which all variable names have been modified in that only the first letter is retained, followed by the *Symbol Table address* of the data associated with this name. Since any variable name may appear many times in a program, a search is required, during the loading, to see if the name already exists in the Symbol Table. The search of the Symbol Table (ST) consists of comparing each name there with the variable name in the statement being loaded. All statements are loaded by an appropriate circuit of Fig. 3 from the I/O buffer and into the Program Area of the memory. Therefore the variable name in the Statement exists physically in the I/O buffer.

It is the function of the VMU to make this search when energized or "called" by the loading circuits for DIMENSION, DO, computed GO TO, READ, PRINT, IF and Arithmetic statements in which variable names appear. The output action of the VMU

[1]Symbols used in this Appendix are described in Table 8.

is to set either the OK, AOK or EOL flip-flops. These flip-flops respectively indicate that the ST either:

1   holds the variable in question as a result of previous loading, or

2   that the variable is subscripted and has been previously loaded by the DIMENSION statement loading circuit, or

3   that the End-of-List (EOL) token was found, indicating the absence of the variable in the ST.

The state diagram for this circuit is shown in Fig. 4. When triggered by the START VMU signal in state 0, the circuit goes to state 1, the next clock pulse sends it to state 2 from which it starts its search of the ST. In going from 1 to 2, the I/O Counter (CIO) contents are saved in register SCIO since the name may have to be scanned again. The Symbol Table Counter (STC) is initialized to 4095 since the ST is scanned sequentially downward.

If a character of a variable name in the I/O buffer is found in the corresponding position of a name in the ST, the character is said to be matched. The VMU proceeds from state 2 to state 3 if the first character of the name under scan matches. Otherwise the state changes from 2 to 8, if the NO MATCH signal is given. The MATCH or NO MATCH signals are generated as a result of comparing the contents of the ST location undergoing the scan (the contents reside in the Memory Buffer Register, MBR), with the contents of the register COMP which has the character from the I/O buffer. The first character is put into COMP by the calling circuit, thereafter the VMU picks them up in the 3–4 transition. The CIO and STC counters are incremented and decremented, respectively, and the VMU oscillates between states 3 and 4 as long as matching continues. This comparison process will terminate when, either an arithmetic operator S_0, is read from the I/O buffer sending the circuit to state 6 from state 3, or the ST contents cause a NO MATCH signal with respect to the contents of the COMP unit causing the transition from state 4 to 5.

In state 6, if a digit is next read from the ST, corresponding in position to the appearance of the operator from the I/O buffer clearly the names are the same and the OKFF is set to 1, and the transition from 6 to 0 is made. On the other hand, if another alphameric character in the ST corresponds to an operator, S_0, in the I/O buffer, the names are not the same and the transition from 6 to 5 is made. In state 5 the circuit just reads to the end of the nonmatching name in the ST. A digit at the end of this name causes the transition 5–7 during which the STC is stepped over the 3 data locations to the next ST entry and the CIO reini-

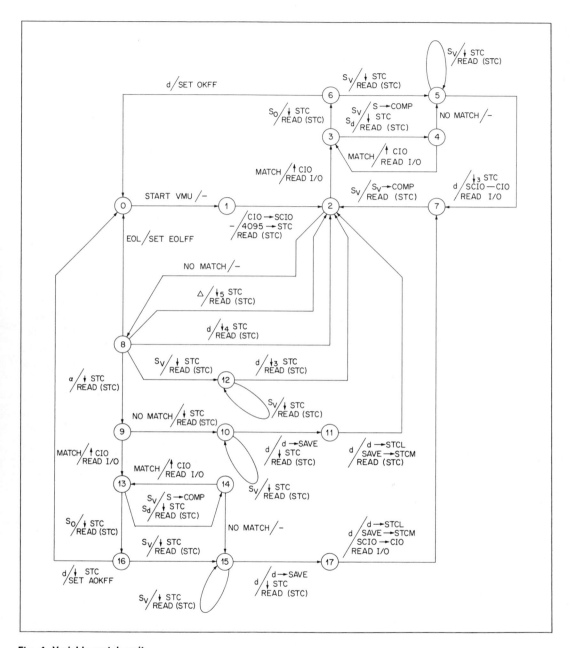

Fig. 4. Variable match unit.

tialized to the start of the name being sought. The first character in this name is read and placed in COMP as circuit goes to 2.

As stated earlier, when the *first* character from the I/O buffer does not match the contents of ST, the state becomes 8. If the mismatch was caused by the EOL token in the ST the EOLFF

is set to 1 and state 0 is reached. If the mismatch was due to a Δ at the present ST location, the STC is decremented by 5 which steps over the 2 four-digit numbers stored after a Δ and the circuit returns to 2 to try a match on the next ST entry. If the mismatch is caused by a digit then this is statement number information

Table 8

CIO	Counter for the input output buffer, 4 BCD numeric character (4 bits each), counts up. Can be set to any given number.
CP	Program Counter. (During execution it points to the statement to be executed, during loading it points to the location where the program is to be loaded.) 4 BCD numeric characters, counts up. Can be set to any given value.
COMP	Comparator register, 8 bits. During loading holds a character to be matched with some other character in the memory, during execution saves the input symbol that drives the execution circuits. (Acts as second rank of Memory Buffer Register.)
SAR	Save Address register, 4 BCD numerics. Counts down. During loading holds the address of the last DO in a nest. During execution it is an auxiliary counter.
SAVE	2 BCD (8 bits total) auxiliary register, each bit can be set independently of the others.
SCIO	4 BCD numeric register, holds temporarily the value of CIO.
SHR	Special Shift register, 4 BCD character, can be shifted to the left 1 BCD character (4 bits) at a time.
SR	24-bit register, used with the accumulator in the arithmetic unit. Bits 1–8, 9–16, 17–24 can be gated independently.
SSAR	Special Save register, 4 BCD numeric (used as auxiliary register in loading and execution).
STC	Symbol Table counter, 4 BCD character, counts down.
S_v	The 8 bits in the MBR are decoded as a single alphabetic character (A–Z).
S_d	The 8 bits in the MBR are decoded as a digit (0–9) and bits 1–4 represent in BCD the value of the digit.
S_o	The bits in the MBR are decoded as one of the following operators. $+ - * / ( )$,
α	8-bit character that precedes a subscripted variable name in the Symbol Table.
Δ	8-bit character that precedes the statement number of a last statement of a DO nest in the symbol table.
λ	An 8-bit character that follows the *DO* token in the program area.
d	The 8 bits in the MBR are decoded as 2 BCD digits of 4 bits each.
EOL	An 8-bit character that is placed at the current end of the Symbol Table.
MATCH	Signal that is generated when the content of the MBR is identical to the content of the COMP.

which requires a decrement of STC by 4 to get to the next entry. If an unmatched alphabetic character in the ST was the reason for the mismatch, this variable is read to its end in state 12 as was done in state 5.

The only other ST symbol which could have caused a mismatch is an α, the array symbol. This symbol sends the VMU to state 9. If a match is now to occur, it will be with a subscripted variable name. Thus a match causes a transition from 9 to 13 and states 13 and 14 correspond to state 3 and 4 for a simple variable as matching proceeds.

Reading an arithmetic operator in the I/O buffer causes transition to 16 where a corresponding digit in the ST causes the AOKFF to be set and the circuit returns to 0, during which time it decrements the STC. This is necessary in order for the STC to hold the address of the first constant given in the DIMENSION statement which caused this ST entry. The transition 16 to 15 corresponds to the 6 to 5 transition, the ST name is longer than the I/O buffer name, and in state 15 the rest of the name is stepped over. Now, however, the next two words in the ST hold the address of the next ST entry. Therefore, these are saved and put into the STC during transition 15-17-7, which otherwise corresponds to the transition 5-7 for a single variable.

If, however, there was no match in state 9, the circuit steps over the rest of the name in the ST in state 10 and initializes the STC to the next ST entry in the transition 10-11-12.

Note that when the VMU returns to its 0 state after setting either EOL or OK or AOK flip-flops, the STC holds precisely the address needed for further action. An EOL needs to be replaced, starting at this STC address, with the new variable name. In the case of OK or AOK this STC address is the one to be placed in the program since it holds the data address for simple variables or the address of the required indexing constant for subscripted variables.

After the calling circuit has used the VMU it has received one of the 3 signals from the VMU. For certain statements these signals can be used to detect syntax errors. If there are none then the calling circuit takes whatever further action is necessary on the variable name being scanned.

The arithmetic statement loading circuit (*Fig. 5*)

An arithmetic statement consists of a string of alphameric symbols, $S_v S_d$, grouped to form variable names, of numeric symbols, S_d, grouped to form constants, and of arithmetic or other operator symbols, S_o, which separate them. The Arithmetic Statement loading circuit calls on the VMU circuit to find the variable names as has been described. It then puts a new name into the ST (if required)

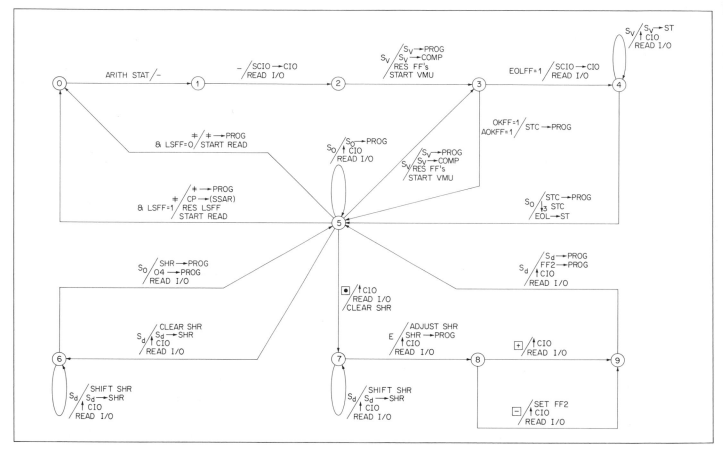

Fig. 5. Arithmetic statement loading.

or it puts the data address into the program. The 8-bit BCD forms of the operator symbols are simply put into the program. The constants are put into the program after conversion to machine form. The state diagram of this circuit is shown in Fig. 5. The scan circuit signal ARITH STAT sends the circuit from 0 through 1 to 2. The scan circuit has saved the address of the beginning of this statement in a register SCIO. This is used to initialize the CIO so that this statement can be read from the beginning.

The first symbol of an arithmetic statement, which must be a variable and not a digit, takes the circuit to state 3 after this symbol has been put into the program ($S_v \rightarrow$ PROG) and the VMU initialized and started. Any one of the VMU signals is possible and valid and simply forces the circuit to state 5. During the 3-5 transition the circuit loads the appropriate address into the program when the name has matched. If it has not matched any existing name the circuit first goes to state 4 and puts the name

into the Symbol Table before going to state 5. State 5 is that from which all further loading is accomplished. Variable names are separated by operators, which are loaded into the program by the cycle in state 5 ($S_0 \rightarrow$ PROG). Note the convention that S_0 represents any operator symbol not explicitly specified on another exit from 5. Any variable names cause a transition to state 3 with the same output action as from state 2. Floating point constants are loaded via states 5-9-5. A decimal point indicates a floating point constant and takes the circuit to state 7. (Note that a minus sign preceding a constant is simply an operator and is processed in state 5.) The SHR is cleared in preparation for the storing of the following digits in state 7. When E is received the digits of the fraction in the SHR are left adjusted (ADJUST SHR), if there are less than four of them, and placed in the program area. The exponent sign is found in the transition 8 to 9. The exponent digit together with the exponent sign bit is stored in the program area during the

9 to 5 transition. Fixed point constants are handled in state 6. The important difference is that the digits are *not* left adjusted in the SHR and a 04 is put into the program as the exponent since a decimal point is assumed to precede the first data word. See Fig. 1.

The ‡ takes the circuit to its initial state. If this statement happens to be the last in a DO nest, the Statement Number Load circuit has set the LSFF to 1. It has also put the ST address of the word following the λ symbol of the first DO of the nest into the SSAR register. Since the program counter (CP) now holds the correct exit address for this DO statement it is placed at the address given by the SSAR during the transition to state 0. During the transition the signal START READ is also sent to the paper tape reader in order for it to put the next Statement into the I/O buffer.

Hardware implementation of the VMU state diagram

Each function mentioned in the paper plus some other auxiliary ones are initially represented in a state diagram form, such as the state diagram for the loading of the Arithmetic Statement (Fig. 5) and the Variable Match Unit (VMU) (Fig. 4).

We will describe the method used to realize a circuit which will perform the function defined by a given state diagram (SD). As an example we will use the VMU. All the information needed is present on the SD. The operations on the right-hand side of the "/,, in the SD are the output operations required to be performed. In order to implement these operations we must specify the actual register gating signals, memory read and write signals, arithmetic unit signals, etc., required by them. We will call these various signals the microsteps of an output operation. Therefore to realize the SD of a given function we must implement the microsteps corresponding to the output operations.

We begin by listing from the state diagram some output operations and their corresponding microsteps. For example, in state 2 of Fig. 4, if a MATCH signal is present we are supposed to increment the CIO counter and then read the I/O buffer.

Consequently the microsteps required are:

↑CIO This signal causes the CIO to be incremented by one.
CIO → MAR This signal causes the CIO to be gated to the memory address register.
READ This signal initiates a memory read cycle.
CHANGE STATE This signal causes the VMU to go from state 2 to state 3.

Therefore the execution of the above microsteps, in that order, would implement the 2–3 transition of Fig. 4. Some microsteps

for the VMU are listed at the end of this Appendix. The largest number of microsteps for a transition from one state to another is 8, which occurs in the transition from state 8 to state 2. Once this maximum number of microsteps is determined, a control cycle counter is constructed, which can count as high as this maximum. Since in this case the number is 8 we need 3 flip-flops to realize it. In addition, a "one hot line" decoder is needed such that at each count one and only one line of the decoder has a "one" at its output. Also needed is a state diagram counter which realizes the "skeleton" of the state diagram. This skeletal counter tells us which state we are in and which to change to, given the present input signal or symbol. Thus the skeletal counter "knows" that if the circuit is in state 2 and a MATCH signal is present, it should change to state 3 upon receipt of a change state signal. The realization of such a skeletal counter has been described [Bashkow, 1964]. Now we use the outputs of the skeletal counter which will indicate to us the state we are in, the outputs of the decoder of the control cycle counter, and the input lines (S_v, S_o, MATCH, NO MATCH) and connect them as shown in Fig. 6. Each AND gate in this figure has 3 inputs except those not requiring input line information. One input comes from the input set (S_o, S_v, MATCH, etc.). The second input comes from the state diagram skeletal counter which indicates a unique state of the state diagram, and finally the third comes from the control cycle counter. The output of each AND gate is a line indicating a unique microstep. The AND's feed OR gates, which actually energize the given microstep. For example the output lead of the "READ" Or gate is connected to the "READ" terminal of the memory.

If we assume that the control cycle counts in sequence 1, 2, etc., then the lead numbered 1 will go to the first microstep of each sequence. The one numbered 2 will go the second, etc. Therefore we see that the following microsteps should be executed in the order listed below for states 0, 1, 2, 5 of Fig. 4. The circuit which causes the execution is shown in Fig. 6.

State 0 and START VMU
        CHANGE STATE
State 1 CIO → SCIO
        0100 0000 1001 0101 → STC
        STC → MAR
        READ
        CHANGE STATE
State 2 and MATCH
        INCREASE CIO
        CIO → MAR
        READ

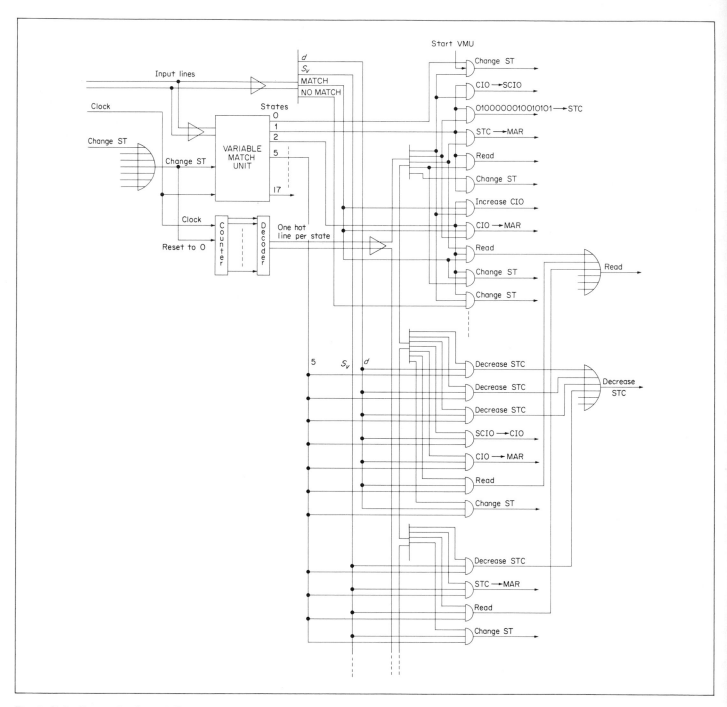

Fig. 6. State diagram implementation.

	CHANGE STATE
State 2	and NO MATCH
	CHANGE STATE
State 5	and S_v
	DECREASE STC
	STC → MAR
	READ
	CHANGE STATE
State 5	and d
	DECREASE STC
	DECREASE STC
	DECREASE STC
	SCIO → CIO
	CIO → MAR
	READ
	CHANGE STATE

In state 0 of Fig. 4 a START VMU signal takes it to state 1. This is accomplished by the top AND of Fig. 6. The only microstep needed is CHANGE STATE. In state 1 of Fig. 4, the next clock pulse (after reaching state 1) causes a transition to state 2. In this case we need to save CIO contents in register SCIO, (CIO → SCIO) set the STC to 4095 (4095 → STC shown above in BCD form) and get the contents of the address now in the Symbol Table Counter (READ(STC)). This latter is implemented by the two microsteps STC → MAR followed by a READ command to the core memory. This transition from 1 to 2 of Fig. 4 is accomplished by the next 5 AND gates shown in Fig. 6. The next AND gates shown accomplish the transition from state 2 to 3 if there is a MATCH. The next AND accomplishes the transition from 2 to 8 if there is NO MATCH (in this case nothing need be done). Finally the lowest two groups of AND gates implement the required microsteps as the circuit changes from state 5 to 7 if a 4-bit digit code is sensed or causes the circuit to remain in state 5 after decrementing the STC if an 8-bit variable code is read.

Chapter 32

A microprogrammed implementation of EULER on IBM System/360 Model 30[1]

Helmut Weber

Summary An experimental processing system for the algorithmic language EULER has been implemented in microprogramming on an IBM System/360 Model 30 using a second Read-Only Storage unit. The system consists of a microprogrammed compiler and a microprogrammed String Language Interpreter, and of an I/O control program written in 360 machine language.

The system is described and results are given in terms of microprogram and main storage space required and compiler and interpreter performance obtained. The role of microprogramming is stressed, which opens a new dimension in the processing of interpretive code. The structure and content of a higher level language can be matched by an appropriate interpretive language which can be executed efficiently by microprograms on existing computer hardware.

Introduction

Programs written in a procedure-oriented language are usually processed in two steps. They are first translated into an equivalent form which is more efficiently interpretable; then the translated text is interpreted ("executed") by an interpretation mechanism. The translation process is a data-invariant and flow-invariant operation. It consists of two parts—an analytical part, which analyzes the higher level language text, and a generative part, which builds up a string of instructions that can be directly interpreted by a machine. The analytical part of the translator depends on the higher level language; the generative part depends on a set of instructions interpretable by a machine. Historically there was only one set of instructions which could be interpreted efficiently by a machine, its "machine language." Figure 1 outlines this scheme.

Some of the processors of the IBM System/360 family are microprogrammed machines. On them the "360 machine language" is interpreted not by wired-in logic but by an interpretive microprogram, stored in control storage, which in turn is interpreted by wired-in logic. Therefore, in a certain sense the 360 language is not the "machine language" of these processors but the (efficiently interpretable) language in which the processors of

the System/360 family are compatible. The true "machine language" of these processors is their microprogram language. This language is on a lower level than the "360 language"; it contains the elementary operations of the machine as operators and the elements of the data flow and storage as operands.

Now it is conceivable to compile a program written in a higher level language into a microprogram language string. This string would undoubtedly contain substrings which occur over and over in the same sequence. We could call these substrings procedures and move them out of the main string, replacing their occurrence by a procedure call symbol, followed by a parameter designator pointing to the particular procedure. Our object program then takes on the appearance of a sequence of call statements. From here it is only a final step to eliminate the call symbols and furnish an interpreting mechanism which interprets the remaining sequence of "procedure designators."

The process just described will result in the definition of a string language and the development of a microprogrammed interpretation system to interpret texts in this string language. The situation is similar to the System/360 case: the string language corresponds to the 360 language. Programs written in a higher level language are compiled into string language text to be stored in main storage. The string language interpreter corresponds to the microprogram

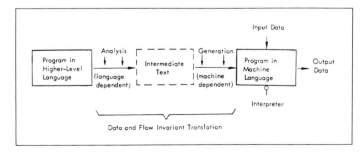

Fig. 1. Processing programs written in higher level languages via translation to machine language.

[1]*Comm. ACM*, vol. 10, no. 9, pp. 549–558, September, 1967.

which interprets 360 language texts. It consists of a recognizing part to read the next consecutive string element and to branch to an appropriate action routine and of action routines to execute the particular procedure called for by the string element.

The essential difference between our situation and the 360 case is that the string language reflects the features of the particular higher level language as well as the features of the particular hardware better than the general purpose 360 language.

What is gained by defining this string language and by providing a microprogrammed interpreter for it? From the method of definition described, it can be seen that the elements of the string language correspond directly to the elements of the higher level language after all simplifying data-invariant and flow-invariant transformations have been performed. But the elements of the string language are also well-adapted to the microprogram structure of the machine. Therefore, during the compiling process (see Fig. 2) only a minimum of generation is necessary to produce the string language text. The compiler is shorter and runs faster.

But the more important aspect is that object code execution is also faster. The string language interpreter in case 2 will be coded to take care of all necessary operations in a concise form, whereas in case 1 it will be necessary to compile a whole sequence of machine language instructions for an elementary operation in the higher level language. Examples of this are the compilation of 360 code for an add operation in COBOL of two numbers with different scaling factors or the compilation of machine instructions for table lookup or search operations, etc. In these cases the string language interpreter of Fig. 2 will execute a function much faster than the machine language interpreter of Fig. 1 will execute the equivalent sequence of machine language instructions. Therefore, object code execution will be faster in scheme 2.

If object code performance is not as much in demand as object storage space economy, the string language interpreter can also be written such that the string language is as tightly packed as possible so that the translated program is as compact as possible and will take up less storage space than the equivalent machine language program under the scheme of Fig. 1.

These ideas are applied in an experimental microprogram system for the higher level language EULER [Wirth and Weber, 1966a and 1966b] described below. Problem areas in this approach are indicated and some ideas for future development are offered.

Special considerations for EULER

The higher level language EULER [Wirth and Weber, 1966a and 1966b] is a dynamic language. This means that for programs written in it many things have to be done at object code execution time which can be done at compile time for other languages. EULER also contains basic functions which do not have comparable basic counterparts in the machine languages of most machines. To compile machine code for these dynamic properties and for those special functions would require rather lengthy sequences of machine language instructions, which would consume considerable object code space and require high object code execution time. Therefore, for a language like EULER, interpretation at the string language level by an interpreter into which the dynamic features and special functions are included by microcode will yield much higher object code economy and object code performance than compilation to machine language and interpretation of this machine language.

Three examples from EULER are given here.

1. Dynamic type handling. To a variable in EULER, constants of varying type can be assigned dynamically. For example in

$$A \leftarrow 3; \cdots; A \leftarrow 4.5_{10}{-}5; \cdots; A \leftarrow \textbf{true}; \cdots; A \leftarrow \text{‘}\cdots\text{’};$$

the quantities assigned to the variable A have the types: integer, real, logical, procedure. Therefore, in EULER each quantity has to carry its type indicator along and each operator operating on a variable has to perform a dynamic type test. The adding operator $+$ for instance in $A + B$ has to test dynamically whether both operands are of type number (integer or real). This type testing is done by the String Language Interpreter in minimum time, whereas it would require extra instructions if the program were to be compiled to 360 machine language.

2. Recursive procedures and dynamic storage allocation. In EULER, procedures can be called recursively, e.g.,

$$F \leftarrow \text{‘formal } N; \text{ if } N = 0 \text{ then } 1 \text{ else } N * F(N - 1)\text{’};$$

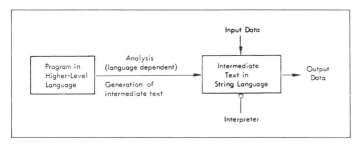

Fig. 2. Processing programs written in higher level languages via translation to interpretive language.

and storage is allocated dynamically, e.g.,

new N; $\cdots$; $N \leftarrow 4$; $\cdots$; **begin new** A; $A \leftarrow$ **list** N;

In order to cope with these problems the EULER execution system uses a run time stack. Each operation is accompanied by stack pointer manipulations which by the microprogram can be accomplished in minimum time (in general, even without extra time because they are overlapped with the operation proper), whereas extra instructions would be required, if the program were compiled.

3. List processing. EULER includes a list processing system, and lists are of a general tree structure, e.g.,

$A \leftarrow (3, 4, (5, 6, 7),$ **true,** '$\cdots$');

List operators are provided like **tail** and **cat** and subscripting:

$B \leftarrow A[3]$; $C \leftarrow B$ **cat** A; $C \leftarrow$ **tail** C;

The string language interpreter handles list operations directly and efficiently by special microprograms. If the program would be compiled to 360 machine language, a sequence of instructions would be required for each list operation.

EULER system on IBM System/360 Model 30

An experimental processing system for the EULER language has been written to demonstrate the validity of these ideas. It is a system running under the IBM Basic Operating System and consists of three parts:

1 A translator, written in Model 30 microcode.[1] This translator is a one-pass syntax-driven compiler which translates EULER source language programs into a reverse polish string form.

2 An interpreter, written in Model 30 microcode,[1] which interprets string language programs.

3 An I/O Control Program written in 360 machine language.[2] This IOCP links the translator and interpreter to the operating system and handles all I/O requests of the translator and interpreter.

---

[1]Stored in the second Read-Only Storage (Compatibility ROS) of Model 30.

[2]The 360 microprograms are stored in the first Read-Only Storage (360 ROS) of the Model 30.

The system is an experimental system. Not all the features of EULER are included,—only the general principles that are to be demonstrated. The restrictions are:

1 Real numbers are not included; only integers are recognized.

2 The interpreter microprograms for the operators Divide, Integer Divide, Remainder, and Exponentiation have not been coded.

3 The type 'symbol' is not included.

4 No garbage collector is provided. Therefore, the system comes to an error stop if a list processing program has used up all available storage space (32K bytes).

Also for reasons of simplicity, the system is written only for a 64K System/360 Model 30 and the storage areas for tables, compiled programs, stacks and free space are assigned fixed addresses.

The string language into which source programs are translated is defined as closely as possible to the interpretive language used in the definition of EULER [Wirth and Weber, 1966a and 1966b]. The question whether this is the ideal directly interpretable language corresponding to the EULER source language given the Model 30 hardware is left open. Also no attempt is made to define the string language so that it becomes relocatable for use in time sharing or conversational processing mode.

The three storage areas used by the execution system are:

1 Program area
2 Stack
3 Variable area

Program area. A translated program in string language consists of a sequence of one-byte symbols for the operators ($+$, $-$, **begin, end,** $\leftarrow$, **go to,** etc.). Some of the symbols have trailer bytes associated with them; for instance, the symbol $+number$ has three trailer bytes for a 24-bit absolute value of the integer constant.

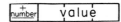

The symbol *reference* (@) has two trailer bytes, one containing the block number (*bn*), the second one the ordinal number (*on*).

The operators **then, else, and, or** and ' have two trailer bytes containing a 16-bit absolute program address, e.g.,

Other operators with trailer bytes are **label** and the list-building operator.

Stack. The execution time stack consists of a sequence of 32-bit words. It contains block and procedure marks to control the processing of blocks and procedures and temporary values of the various types. The first 4-bit digit of a word in stack always is a type indicator. The format of these words is given in Fig. 3.

Variable area. The variable area is an area (32K bytes long) of 32-bit words used for the storage of values assigned to variables and lists (and also for auxiliary words in procedure descriptors; see type procedure in Fig. 3). The format of the entries is exactly the same as the format of the stack entries (see Fig. 3), the only exception being that a mark can never occur in the variable area.

Microprogramming the IBM System/360 Model 30 [Fagg et al., 1964]

Microprograms are sequences of microprogram words. A microprogram word is composed of 60 bits and contains various fields which control the basic functions in the IBM System/360 Model 30 CPU. These basic functions are storage control, control of the

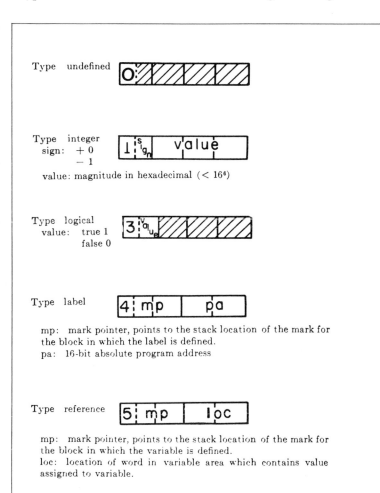

Fig. 3. Format of words in stack and variable area.

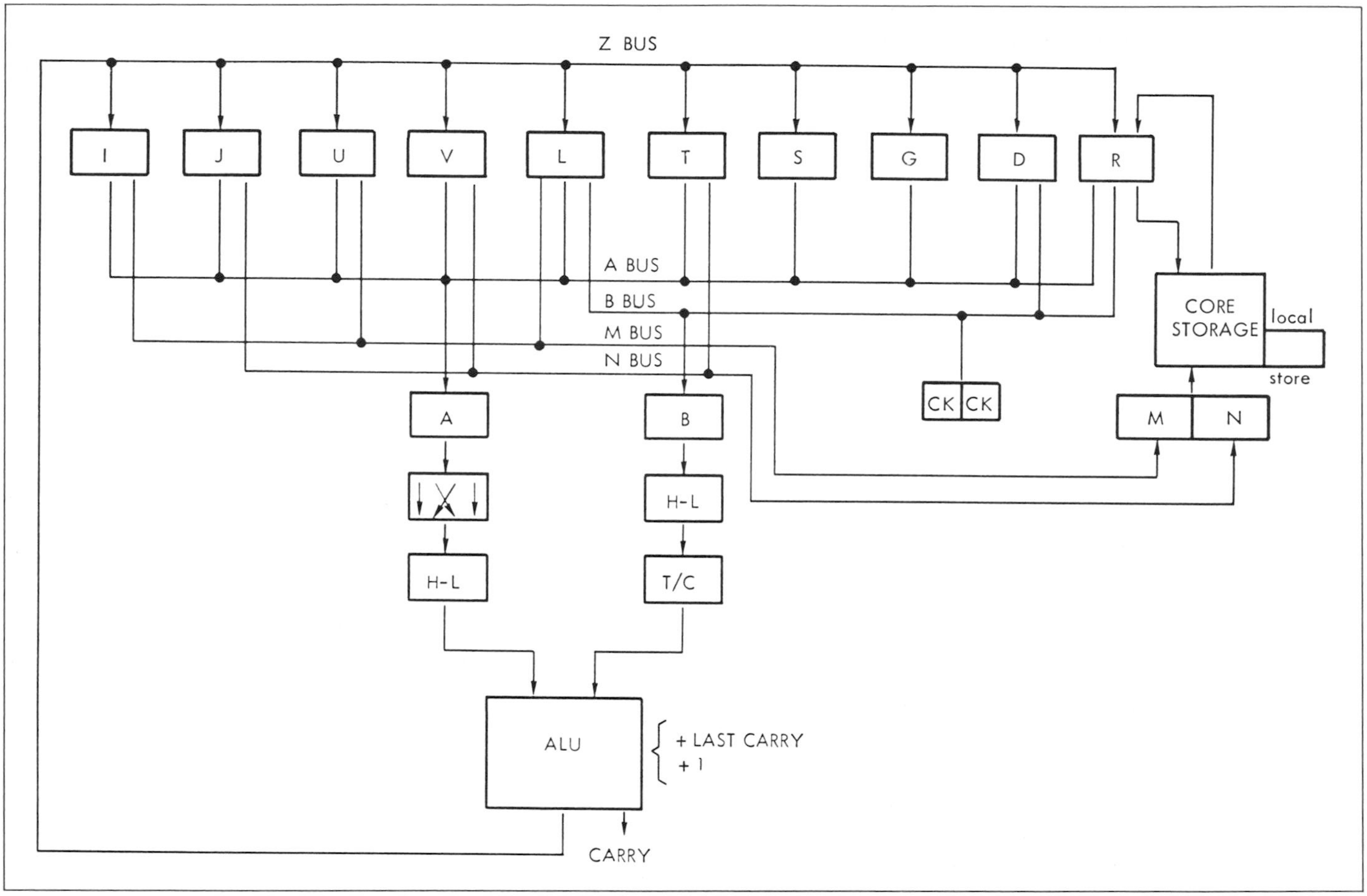

Fig. 4. Simplified data flow of the IBM System/360 Model 30.

data flow registers and the Arithmetic-Logic-Unit (ALU), microprogram sequencing and branching control, and status bit-setting control. Microprogram words are stored in a Card Capacitor Read-Only Storage (CCROS). Fetching one microprogram word and executing it takes 750 nsec, the basic machine cycle.

Figure 4 shows in simplified form the data flow of the IBM System/360 (IBM 2030 CPU). It consists of a core storage with up to 65,536 8-bit bytes and a local storage (accessible by the microprogrammer but not explicitly by the 360 language programmer), a 16-bit storage address register (M, N), a set of 10 8-bit data registers (I, J, $\cdots$, R), an arithmetic-logic-unit (ALU), connecting 8-bit wide buses (Z, A, B, M, N-bus), temporary registers (A, B), switches and gates.

Figure 5 shows the more important fields of a microprogram word. Only 47 bits are shown. Other fields contain various parity bits and special control bits. The field interpretation given in Fig.

5 is as for microprogram words in the second Read-Only Storage unit (Compatibility ROS) if the machine is equipped with the 1620 Compatibility Feature. The meaning of the microprogram word fields is explained in connection with Fig. 6 which shows the symbolic representation of a microprogram word together with an example as it appears on a microprogram documentation sheet.

The fields of the microprogram word can be grouped in five categories:

1 ALU control fields: CA, CF, CB, CG, CV, CD, CC

2 Storage control fields: CM, CU

3 Microprogram sequencing and branching fields: CN, CH, CL

4 Status bit setting field: CS

5 Constant field: CK

CN	CH	CL	CM	CU	CA	CB	CK	CD	CF	CG	CV	CC	CS
0000	O	O	Write	MS	*	R	O	Z	O	O	+	+O	No status setting
0001	⊥	⊥	No access	LS	*	L	⊥	*	L	L	−	+⊥	LZ→S5
0010	RO	*	Store	*	*	D	2	*	H	H	*	And	HZ→S4
0011	S⊥	*	IJ→MN	*	*	K	3	*	Through	Thr.	@	Or	HZ→S4,LZ→S5
0100	*	G⊥	UV→MN		S		4	*	*			+O,save C	O→S4,O→S5
0101	*	R=Valid dec	LT→MN		*		5	*	XL			+⊥,save C	⊥→S⊥
0110	ALU carry	R⊥	*		*		6	S	XH			+C,save C	O→SO
0111	SO	Z=O	*		R		7	R	X			XOR	⊥→SO
1000	R2	G7			D		8	D					O→S2
1001	S2	S3			L		9	L					ANSNZ→S2
1010	S4	S5			G		X'A'†	G					O→S6
1011	S6	S7			T		X'B'	T					⊥→S6
1100	GO	R3			V		X'C'	V					O→S7
1101	G2	G3			U		X'D'	U					⊥→S7
1110	G4	G5			J		X'E'	J					*
1111	G6	Interrupt			I		X'F'	I					O→S⊥

†X'A' means hexadecimal digit A=1010

Fig. 5. IBM System/360 Model 30 microprogram word. (Detailed explanation is provided in text.) The field interpretation is given for microprogram words in compatibility ROS if the machine is equipped with the 1620 compatibility feature. Fields marked "*" contain designators not explained here in order not to confuse the basic principles.

ALU control fields. On the line designated "ALU" in Fig. 6, an ALU statement can appear. It will specify an A-source and a B-source, possibly an A-source modifier and a B-source modifier, an operator, a destination, and possibly a carry-in control and a carry-out control.

CA is the A-source field. It controls which one of the 10 8-bit data registers is connected to the transient A-register and therefore to the A-input of the ALU.

CB is the B-source field. It controls whether the R, L, or D-register or the CK-field is connected to the transient B-register and therefore to the B-input of the ALU. If "K" (CB = 3) is specified in this field, the 4-bit constant field CK is doubled up; i.e., the same four bits are used as the high digit and the low digit.

Between the A-register and the ALU input is a straight/cross switch and a high/low gate. Its function is controlled by the CF-field. Depending on the value of this field, no input is gated into the ALU (0) or only the low (L) or high digit (H) is admitted. CF = 3 gates all eight bits straight through, whereas the codes CF = 5, 6, and 7 cross over the two digits of the byte before admitting the low (XL) or high digit (XH) or both digits (X).

Between the B-register and the ALU input is a high/low gate and a true/complement control. The high/low gate is controlled by the CG-field in the same manner as the high/low gate in the A-input. The true/complement control is operated by the CV-field. It admits the true byte to the ALU (+) or the inverted byte (−) or controls a six-correct mechanism for decimal addition (@).

The operator and carry controls are given by the CC-field. This field specifies binary addition without carry handling (+0), addi-

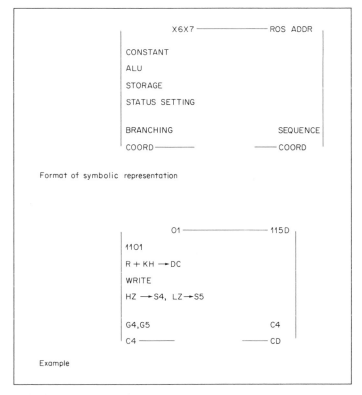

Fig. 6. Symbolic representation of a System/360 Model 30 microprogram word.

tion with injection of a 1 (+1) (for instance, to simulate subtraction in connection with the B-input inverter), addition with saving the carry in bit 3 of register S (+0,Save C, and +1,Save C), and addition using an old carry stored in bit 3 of register S and saving the new carry in this same bit (+C,Save C). Other codes specify logical operations (AND, OR, XOR).

The CD-field specifies into which register the result of the ALU operation is gated. Any one of the 10 data registers can be specified. Z means that the ALU output is gated nowhere and will be lost.

Storage control fields. On the line designated "storage" in Figure 6, a storage statement can appear. It will specify whether this microcycle is a ready cycle, a write cycle, a store cycle or a no-storage access cycle, and from where the storage address is supplied (CM-field) and whether storage access is to main storage or local storage (CU-field). Note that a full storage cycle (1.5 μsec) corresponds to two read-only storage cycles (750 nsec).

The codes CM = 3, 4, or 5 specify read cycles. The addresses are supplied from the register pairs IJ, UV, and LT, respectively. A read cycle reads one byte of data from core storage into the storage data register R.

A write cycle regenerates the data from the storage data register R at the address supplied in the last read cycle.

A store cycle acts exactly as a write cycle except that it inhibits in the read cycle immediately preceding it the insertion of the data byte from storage into the R-register.

The CU-field specifies whether storage access should be to main storage (MS) or to a local storage of 256 bytes not explicitly addressable by the 360 language programmer.

Microprogram sequencing and branching. Each microprogram word is stored at a unique address in ROS. A 13-bit ROS address register (W3···W7, X0···X7) holds the address of the word being executed. For the symbolic representation of a microprogram (Fig. 6) the ROS address is given in hexadecimal in the upper right corner, and the last two bits of this address are repeated in binary on the upper margin.

After execution of a microprogram step, the next sequential word will not be executed. Instead the address of the next word to be executed is derived as follows. The high five bits (W) remain the same, unless they are changed by a special command in the microword, not explained here (so-called module switching). The next six bits (X0···X5) are supplied from the CN-field (written in hexadecimal in the symbolic representation of Fig. 6). The low two bits are set according to conditions specified in the CH and CL fields. X6 is set according to the condition specified by CH.

For instance, if CH = 8, then the bit R2 is transferred to X6; if CH = 6, then X6 is set to one if in the last ALU operation a carry had occurred. It is set to zero if no carry had occurred. X7 is controlled by CL. If, for instance, CL = 0, then X7 is set to zero; if X7 = 5, then X7 is set to one if both digits in R are valid decimal digits (i.e., R0···R3 $\leq$ 9 and R4···R7 $\leq$ 9), X7 is set to zero if either digit in R is not a valid decimal digit (i.e., R0···R3 > 9 or R4···R7 > 9). This microprogram sequencing scheme allows a four-way branch after the execution of each microprogram word.

Status bit setting. The CS-field allows the unconditional or conditional setting of certain status bits to be specified, combined in Register S. If, for instance, CS = 3, then S4 is set to one if the result of the ALU operation performed in this microprogram cycle shows a zero in the high digit (i.e., Z0 = Z1 = Z2 = Z3 = 0); S4 is set to zero otherwise. At the same time, S5 is set to one if the result of the ALU operation shows a zero in the low digit (i.e., Z4 = Z5 = Z6 = Z7 = 0); S5 is set to zero otherwise. If CS = 9, then S2 is set to one if the result of the ALU operation is not zero (i.e., at least one of the bits Z0···Z7 is equal to 1). If the result of the ALU operation is zero, then S2 is not changed.

Constant field. The 4-bit CK-field is used for various purposes. One instance explained in the ALU statement is to supply a constant B-source for an ALU operation. Other examples not explained here any further are the addressing of a few specific scratchpad local storage locations, module switching (replacement of the high part W of the ROS address), and the control of certain special functions.

Symbolic representation of microprograms. Microprograms are symbolically represented as a network of boxes (Fig. 6) each representing a microword, connected by nets indicating the possible branching ways. Figure 7 gives an example of a microprogram (to be explained in the next section). There exist programming systems to aid in the development of microprograms. They contain symbolic translators to translate the contents of a box according to Fig. 6 into the contents of the actual fields of the microprogram word according to Fig. 5. A drawing program generates documentation (Fig. 7 is drawn with such a program). These systems usually also contain programs for simulation and generation of the actual ROS cards.

String language interpreter for EULER

The string language interpreter for EULER is entirely written in Model 30 microcode. It consists of a few microprogram steps to read the next sequential symbol from the program string and to

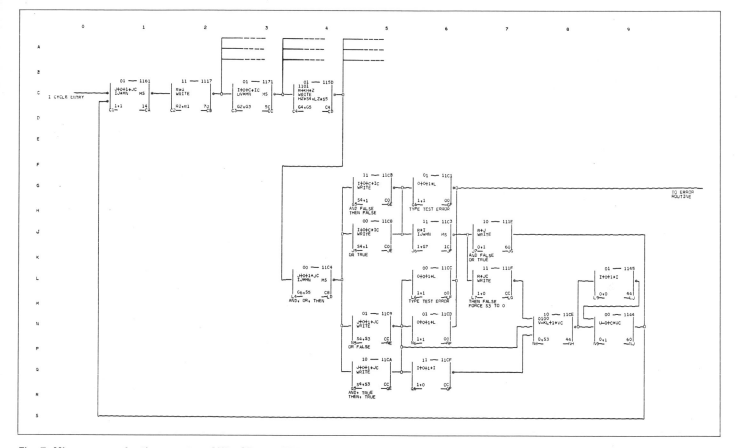

Fig. 7. Microprogram for the operators AND, OR, and THEN.

do a function branch on the symbol and of a group of microprogram routines which perform the necessary operations for the
program byte read. These routines also take care of dynamic type
testing and stack pointer manipulations. The routines are equivalent to the routines described in the definition of the string language for EULER [Wirth and Weber, 1966a and 1966b].

Figure 7 shows, as an example, the microprogram to interpret
the program string symbols **and** (internal representation X'52'[1]),
or X'50' and **then** X'53'. These operators test if the highest entry
in the stack is a value of type logical. The logical operators in
EULER work in the FORTRAN sense, not in the ALGOL sense:
if after the evaluation of the first operand the result is determined
(**false** for **and, true** for **or**), then the second operand is not evaluated but skipped over. If an **and** operator finds the value **false,**
then a branch occurs to the program address given in the two

trailer bytes. If an **and** finds the value **true,** then it deletes this
value from the stack and proceeds to the next symbol in the program string (to evaluate the second operand of **and**). Similarly if
an **or** operator finds the value **true,** then a branch occurs to the
program address given in the two trailer bytes. If an **or** finds the
value **false,** then it deletes this value from the stack and proceeds
to the next symbol in the program string. The **then** operator is a
conditional branch code: it deletes the logical value from the
stack. If this value was **false,** then a branch is taken to the program
address given in the two trailer bytes. If this value was **true,** then
the next symbol in the program string is executed.

The pointer to the symbol in the program string (the instruction
counter) is located in the functionally associated pair of registers
I and J in the Model 30. The pointer to the left-most byte of the
highest entry in the stack (the stack pointer) is located in the two
registers U and V in the Model 30.

In the following the individual steps in this microprogram are
explained in more detail.

[1]X '*nn*' represents the hexadecimal number composed of the digits *n*
($n = 0, \ldots, 9, A, \ldots, F$).

Address	Location in Figure	Description
1161:	C1:	The instruction counter IJ addresses main storage. The addressed byte in main storage is read out into the storage data register R. The instruction counter is updated by adding 1 to register J. A possible carry is saved to be added to 1.
1117:	C2:	The operator has been read out from main storage into R. It is also transferred (through the ALU) to register G. A four-way branch occurs on the two highest bits R0 and R1 of the operator. For the operators 52, 53, and 50 this branch goes to ROS word 1171, whereas other operators cause a branch to 1170, 1172, or 1173, indicated by the three lines not continued.
1171:	C3:	To complete the updating of the instruction counter, the carry from 1161 is added into I. The first byte of the highest entry of the stack is addressed by UV and read out into R. A further four-way branch on the operator is made (G2, G3). For our operators the branch goes to 115D.
115D:	C4:	The high order byte of the highest stack entry has been read out of storage into R. It contains the type of entry in the high digit and if this type was logical then it contains the value **true** (1) or **false** (0) in the second digit. This byte is tested by adding X'D0' to it and observing the result, ignoring the carry. S4 is set to 1 when the type was 3 (logical) otherwise to 0. S5 is set to 1 when the low digit of this byte was 0 (value **false**), S5 is set to 0 when the low digit of this byte was 1 (value **true**). Another four-way branch occurs on the bits G4 and G5 of the operator. If the operator is 50(**or**), 51 (cannot occur), 52 (**and**), or 53(**then**), then a branch to 11C4 occurs.
11C4:	L4:	The next byte is read from the program string, it is the high byte of the two-byte program address trailing the operator. The instruction counter is updated again by adding a 1 to J, saving a possible carry. Another four-way branch occurs on the bit G6 of the operator and the value of the stack entry. If the operator was **and** or **then** (G6 = 1) and the value was **false** (S5 = 1), then branching to 11CB occurs; if the operator was **or** (G6 = 0) and the value was **true** (S5 = 0), then branching to 11C8 occurs. If the operator was **or** (G6 = 0) and the value was **false** (S5 = 1), then branching
		to 11C9 occurs. If the operator was **and** or **then** (G6 = 1) and the value was **true** (S5 = 0), then branching to 11CA occurs.
11CB:	G5:	This word is executed for the operators **and** and **then** when the value was **false**. Here the type test is made. If the type was not logical (S4 = 0), then a branch to 11C1 occurs. If the type was correct, then the microprogram proceeds to fetching the trailing program address (two bytes) to store it as the new instruction counter in IJ. This is done for the **and** operator (G7 = 0) in this word and the following two words 11C3 and 111E; for the **then** operator (G7 = 1) it is done in this word and the words 11C3 and 111F.
11C3, 111E:	J6, J7:	The two bytes trailing of the operators **and** or **or** are stored as the new instruction counter IJ. The operation is completed. The microprogram branches back to 1161 to read out the next operator.
11C3, 111F:	J6, L7:	The two bytes trailing of the operator **then** are stored as the new instruction counter IJ. The carry-saving bit S3 is forced to zero.
11CE, 1144:	N8, N9:	The stackpointer is decremented by four (the operator '−' means complement add) which in effect deletes the highest entry from the stack. Observe that when these two words are entered from 111F (**then** operator with value **false**) the microprogram will not go through 1145 because we have forced S3 to zero in 111F. The operation is completed, and the microprogram branches back to 1161 to read out the next operator.
11C8:	J5:	This word is executed for the operator **or** when the value was **true**. Similarly as in 11CB, the typetest is taken. For types not logical a branch to 11C1 occurs. If the type was correct, then the microprogram proceeds to fetching the trailing program address (two bytes) to store it as the new instruction counter in IJ (words 11C3, 111E).
11C9:	N5:	This word is executed for the operator **or** when the value was **false**. A typetest is made. If the type was correct, then the trailing program address is skipped and IJ is updated by 1 twice in 11C4, 11C9 (possible carries out of J handled in 11CF or 1145). The stackpointer is decremented by four in 11CE, 1144.
11CA:	Q5:	This word is executed for the operators **and** and **then** when the value was **true**. A typetest is made. If the type was correct then the trailing

Address	Location in Figure	Description
		address is skipped, IJ is updated by 1 twice in 11C4, 11CA (possible carries out of J handled in 11CF or 1145). The stackpointer is decremented by four in 11CE, 1144.
11C1, 11CC, 11CD:	G6, L6, N6	These words are executed when a typetest occurs. An error code 01 is set up in L and a branch occurs to the error routine not drawn here.

It can be seen from Fig. 7 that the execution times of the microprograms including the readout of the operator (I-Cycle) are the following:

and   6 μsec[1] (8 microprogram steps)
or   6 μsec (8 microprogram steps)
then 6 μsec for value **true** (8 microprogram steps)
          7.5 μsec for value **false** (10 microprogram steps)

In order to compare this with a hypothetical EULER system for System/360 language, let us assume that the compiler produces in-line code (which probably will give the highest performance although it will be very wasteful with respect to storage space). Then a reasonable sequence for **and** might be:

```
CLI   0 (STACK), LOGFALSE
BE    ANDFALSE
CLI   0 (STACK), LOGTRUE
BNE   TYPEERR
SH    STACK, = '4'
```

Timing: **true**: 90 μsec; **false**: 32 μsec.

This comparison seems to indicate that the microprogram interpreter is about an order of magnitude faster than the equivalent program in 360 language. However, this comparison will only yield such a high factor for functions of EULER which do not have simple System/360 language counterparts (as for instance the list-operators, begin-, end-, and procedure-call-operator) or where the overhead for dynamic testing and stackpointer manipulation is heavy as in the above example of the logical operations. For functions which do have System/360 language counterparts and which are slower so that the overhead is relatively lighter as, for instance, arithmetic operations (especially for real numbers), the microprogrammed interpreter will still be faster than the System/360 language program, but not by a factor of 10.

[1] The cases where carries occur in the IJ and UV updating are disregarded for timing purposes.

The total ROS space requirement for the String Language Interpreter is:

Coded routines	1000 microwords	
Routines for real number handling	500 microwords	(estimated)
Divide, Exponentiation, etc.	400 microwords	(estimated)
Garbage collector	600 microwords	(estimated)
	2500 microwords	

EULER compiler

The translator to translate EULER source language into the Reverse Polish String Language is a one-pass, syntax-driven compiler. The syntax of the language and the precedence functions F and G over the terminal and nonterminal symbols are stored in table form in Model 30 main storage. There is also main storage space reserved for translation tables for character delimiters and word delimiters and for a compile time stack, a name table, and, of course, for the compiled code. All these areas are at fixed storage locations because of the experimental nature of the system.

The microprogram consists of the following parts:

1   A routine reads the next input character from the input buffer to translate it to a 1-byte internal format, if it is a delimiter, or to collect it into a name buffer if it is part of an identifier, or to convert it to hexadecimal if it is part of a numeric constant and to collect the number into a buffer. This "prescan" requires 100+ microwords.

2   As soon as an input unit is collected (delimiter, identifier, number) the main parsing loop is entered which makes use of the precedence tables and the syntax table in main storage. This syntactic analyzer loop requires 100− microwords.

3   When the parsing loop identifies a syntactic unit to be reduced, it calls the appropriate generation routine which performs essentially the functions described as the semantic interpretation rules in the EULER definition. The microprogram space required for these programs amounts to approximately 250 ROS words.

4   If a syntactic error is detected, the system signals an error and does not try to continue with the compilation process. Though this procedure is totally inadequate for a practically useful system, it was deemed sufficient to prove the essential point. For this minimum error analysis and for linkage to the 360 microprograms (IOCP), approximately 60 microwords are required.

The total compiler microprogram space is therefore approximately 500 ROS words. The total main storage space required is approximately 1200 bytes.

The speed of this compiler is limited by the speed of the card-reader of the system (1000 cards/minute). This excellent performance has three main reasons: (1) EULER as a simple precedence language is a language extremely easy to compile. (2) The functions of a compiler are mainly of a table lookup and bit and byte-testing type. Microprogramming is extremely well-suited for these kinds of operations. (3) Since the target language is String Code and not, for example, 360 Machine Language, the generative part of the compiler is relatively short.

It is very difficult to assess the individual contributions of these three main reasons to the high compiler performance. Therefore, it is not possible at this stage to make a statement as to whether the nature of the language EULER or the fact that the compiler is microprogrammed is the dominant factor.

Development of the microprogram

Since there is no higher level language to express microprogram procedures and no compiler to compile microcode, the microprograms were written in the symbolic language explained in Fig. 6. Actually the process was a hand translation of the algorithms in the EULER definition to the symbolic microprogram language. The microprograms were translated into actual microcode and simulated before they were put on the System/360 Model 30 by means of a general microprogram development system.

Outlook and general discussion

It is hoped that the development of this experimental system for EULER shows that with the help of microprogramming we can create systems for higher level languages or special applications, which utilize existing computer hardware to a much higher degree than conventional programming systems.

Among the thoughts which are raised by this scheme are the following:

1   There should be an investigation to determine the ideal directly interpretable languages which correspond to higher level languages. Although several attempts have been made to define string languages for interpretive systems (for instance in Wirth and Weber [1966a and 1966b] and Melbourne and Pugmire [1965]), to the author's knowledge no work has been published which attacks this question in a general and theoretically founded manner.

2   A proliferation of interpretive languages and the development of microprogrammed interpreters can be justified when better tools are developed to reduce the cost of microprogramming. It is necessary that we be able to express microprogramming concepts (and also machine design concepts) in a higher level language form and that we develop compilers which translate the microprograms from higher level language form to actual microcode. Also, good microprogram simulation and debugging tools are called for.

3   The whole relationship between programming, microprogramming, and machine design should be viewed with a common denominator: how should the tradeoffs be made such that the ultimate goal can be reached more effectively, ... how to solve a user's problem? Green [1966] offers some thinking in this direction but the state of the art has to progress further before we will have a complete understanding of what these relationships and tradeoffs are.

References

WebeH67; FaggP64; GreeJ66; HainL65; MelbA65; WirtN66a, 66b; FORTRAN Specifications and Operating Procedures, IBM1401, IBM Systems Ref. Lib. C24-1455-2.

Part 5

The PMS level

This part presents the PMS structure dimension of the computer space. The sections are arranged in order of increasing organizational structure complexity. The sections are as follows; 1 Pc; 1 Pc with multiple Pio; multiprocessing with n Pc; parallel processing with n Pc; computers which are networks; and networks of computers.

In Chap. 37 Lehman defines the terms multiprogramming, multiprocessing, and parallel processing.

Section 1

Computers with one central processor

The computers with one Pc and no Pio's control T and Ms in either of two ways. First, the Pc contains the K for T and Ms; second, a separate K controls a data transmission while Pc initializes the K. In the latter case, a K is like a P where each instruction is received from Pc instead of being fetched automatically by K itself.

The Whirlwind I computer

Whirlwind (Chap. 6) controls data transmissions between Ms or T and Mp by using Pc. Thus, arithmetic and input/output processing concurrency is difficult to achieve. The structure is first discussed in Part 2, Sec. 1, page 90.

The SDS 910-9300 series

The SDS 910-9300 series is presented in Chap. 42 and is discussed in Part 6, Sec. 2, page 542. The input/output and the interrupt system are especially interesting.

Section 2

Computers with one central processor and multiple input/output processors

The computer structures discussed in this section are manufactured mainly by IBM. The reason for this bias toward IBM is that only fairly elaborate or very specialized structures have Pio's; computers of other manufacturers which have Pio's tend to have also the more general multiprocessing capability[1] that would place them in Sec. 3.

The DEC PDP-8

The PDP-8 is presented in Chap. 5, and its 338 P.display appears in Chap. 25. Discussions are given in Part 2, Sec. 1 and Part 4, Sec. 1, respectively. For this section, the reader should look at the methods for transmitting data between Ms or T and Mp. Three methods are used: Pio or P.display is used to control T.displays (Chap. 25); Pc directly transmits a word to the buffer of a K for low-data-rate devices, here a K may request data, using the program interrupt; and a K transmits data directly to Mp.

The IBM 1800

Chapter 33 describes the 1Pc-9Pio IBM 1800 computer. There are five Pio types, depending on the components they control. Although we classify them as Pio's, they are barely processors since the instruction counter has a very restricted behavior. Unless the data channel has "data chaining" capability (in effect a jump instruction), it is not a processor.

The IBM 7094 II

The IBM 7094 II computer is discussed in Part 6, Sec. 1, page 515; its description appears in Chap. 41. The earlier 709 was about the first computer to use independent Pio's. UNIVAC (Chap. 8) has a very extensive K for data transmission concurrent with processing, whereas the 701 and 704 both required Pc to control each data word transmitted. The Pio's of the 7094 II might be looked at as an overreaction or overdesign inspired by the 701-704.

[1] For example, the CDC-3600 [Casale, 1962], and the SDS Sigma 7 [Mendelson and England, 1966].

The structure of System/360, Part I—outline of the logical structure

The structure of the 360 is presented in Part 6, Sec. 3. A discussion of an alternative implementation of the 360 by the authors of this book, using multiprocessors, is given (page 585). Chapter 43 gives an overview of the ISP, and Chap. 44 presents the implementations of various 360 models. The implementations of physical processors to give multiple logical processors using microprogramming are interesting. IBM is rather conservative in regard to providing structures convenient for multiprogramming; and a multiprocessing design appears too complex for them to attempt outside a research environment.

The engineering design of the Stretch computer

Stretch (also known as Model 7030) and the UNIVAC LARC [Eckert, et al., 1959] are perhaps the first computers with the principal design goal of maximizing numerical computing power. Stretch, aptly named because of its influence on the technology (and on the IBM organization), was initiated by the Atomic Energy Commission at Los Alamos. It was designed to interpret large-scale scientific programs for nuclear engineering. Like a number of other high-risk major developmental efforts in the computer field, Stretch was not outstandingly successful as a computer system. Only a few ($5 \sim 10$) were built at a cost substantially exceeding their contract price and with performance only modestly better than the art at the time of their production. However, again in common with other similar efforts, they had a substantial positive effect on the state of the art. In the Stretch case, in particular, the 2.18-microsecond Mp core technology developed for Stretch was transferred to the 7090. In fact, this was a major contribution to why Stretch was only modestly better than 7090. The design goal was performance 100 times an IBM 704. The computer is described at a high level in Chap. 34. Buchholz's book on Project Stretch [Buchholz, 1962] is outstanding as a text on computer structures and as a description of Stretch. It should be read by all computer designers.

Computers built to maximize numerical computing power also include, besides the UNIVAC LARC for the Lawrence Radiation Laboratory at Livermore, the Control Data 6600 (Chap. 39), and the IBM System/360, Models 91 and 85.

Stretch derives its power through:

1 Compound and complex ISP instructions

2 A PMS structure with Mp(2.18 μs/w),Pc(0.25 $\sim$ 1 μs/w), Pio's, and a satisfactory switch between P's and Mp

3 Many data-types

4 Parallelism within the Pc, involving concurrent interpretation of the instruction stream using the "Instruction look-ahead" mechanism

The last of these, internal Pc parallelism, is the most novel. Stretch was possibly the earliest computer to make use of it; each of the other "maximum" power C's listed above also uses some version of instruction look-ahead, for each of these "maximum" systems is faced with how to obtain computing power that goes beyond the basic logic and memory technology available at the time the system is designed. The conclusion, reached in all these cases, is to move toward internal parallelism.

In Stretch the instruction look-ahead mechanism fetches the next several instructions and partially interprets each future instruction. The mechanism is elaborate compared with the straightforward instruction stack in the CDC 6600 (Chap. 39, page 489). The Stretch look-ahead complexity stems from partially interpreting instructions which may later have to be undone.

Stretch uses a basic Mp(core; 16384 w; (64 + 8 parity) b/w; tc:2.18 μs). Sixteen Mp's can be connected to the P's via the S('Memory Bus; time multiplexed). The 8 parity bits are used to give single-error correction and double-error detection, which is a very substantial amount of error protection compared with standard design practice. This is the memory that was incorporated in the IBM 7090 and became operational even before Stretch was delivered. Thus, as is often the case with large development efforts, the by-products are as important as the main product.

There is a single well-designed physical Pio, called the Exchange, consisting of several logical Pio's. Its ability to have the state of all the logical Pio's accessible in Mp is useful and important. This design seems better than the data channels in the IBM 709-7094 series. It is almost a prototype for the IBM System/360 Pio's.

The Stretch word length is 64 bits. It has operations on the following data-types: binary integers, decimal integers, address integers, variable-length integers, boolean vectors, single and double floating point. The length of the variable integer is specified by parameters in the instruction. Noisy-mode floating-point data provide a method of introducing a roundoff error in the least significant bit under program control. Thus a problem can be run in conventional and noisy modes and the results compared. An instruction is either 32 or 64 bits.

The ISP processor state has an instruction counter, a double-length accumulator, 15 index registers, about 6 registers, and about 100 miscellaneous bits. Computing power is obtained by having an instruction set with complex instructions. Hence, there is an instruction for almost every possible operation, though inverse subtract and inverse divide instructions are lacking. However, there is a "multiply and add" instruction. Stretch has the complete set of 16 operators for boolean vectors. Compound instructions, formed from a sequence of simpler instructions, also increase power. These instructions specify the array element to be accessed, an operation on the element, and a calculation to get the next element, in a single instruction. Notice that several of these instructions are oriented toward operations on arrays (i.e., matrices), which are the type of numerical-analysis tasks for which the system was built.

Multiprogramming was done with Stretch [Codd et al., 1959] and undoubtedly had some influence within IBM. Stretch has a pair of bounds registers to relocate and protect a single program. The interrupt scheme for Stretch [Brooks, 1957a] was better than that of existing IBM computers, though it is not described in Chap. 34.

The importance of Stretch lies in the by-products it inspired and its influence on IBM, encouraging a concern with hardware project management. The elaborate ISP and the complex implementation of Stretch may not have been worth the effort, especially when one compares this computer with the later, larger but elegant CDC 6600. It is, however, interesting to note that Stretch was used as a central component in an early specialized multiprocessor system called the IBM Harvest [Herwitz and Pomerene, 1960], which provides extremely powerful data-processing capabilities.

PILOT, the NBS multicomputer system

The National Bureau of Standards' PILOT computer (Chap. 35) was first described in 1959. At that time it was a multiple computer; by our criteria, we classify it as a multiple-processor computer, as shown by its PMS structure (Fig. 1). However,

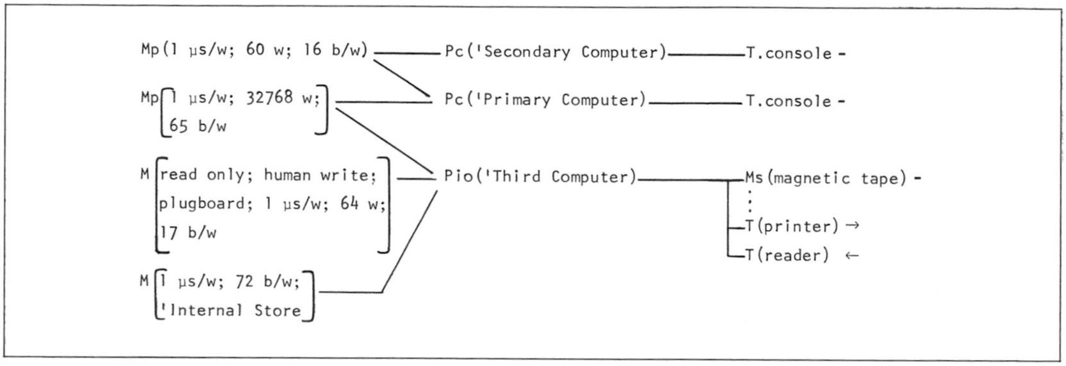

```
    Mp(1 μs/w; 60 w; 16 b/w) ————————— Pc('Secondary Computer)————————T.console -

    Mp⎡1 μs/w; 32768 w; ⎤————————→ Pc('Primary Computer)————————T.console -
      ⎣65 b/w          ⎦

    M⎡read only; human write; ⎤——→ Pio('Third Computer)————————Ms(magnetic tape) -
     ⎢plugboard; 1 μs/w; 64 w;⎥                                :
     ⎣17 b/w                  ⎦                              ┌─T(printer) →
                                                             └─T(reader) ←
    M⎡1 μs/w; 72 b/w;  ⎤
     ⎣'Internal Store  ⎦
```

Fig. 1. National Bureau of Standards' PILOT computer PMS diagram.

unlike present multiprocessors with several identical processors, each PILOT processor is different.

PILOT is a good example of an early attempt to use multiprocessors; successors look little like it. It has one of the best analytical discussions of any computer [Leiner et al., 1957]. With this machine there was an attempt to resolve the controversy between the short-word EDSAC (17 bits) and the long-word Institute for Advanced Studies computers (40 bits) by providing a processor and memory (i.e., computers) for each problem. Only the first computer had substantial Mp, and the other computers, or processors, could be concerned only with the first computer. The third computer was introduced to process devices such as Ms(magnetic tape) and used a plugboard program memory. The idea of an independent processor (IBM 7094) or computer (CDC 6600) for input/output processing is used now, though it is doubtful that PILOT inspired these designs.

The capacitor-diode store is novel and daring for the technology. Two- and three-address computers are used in the primary and secondary computers. The secondary computer, with 16-bit words, is not very useful; its memory is very limited, and it is essentially used only for address calculations. The bookkeeping operation for a three-address computer could easily keep a small processor busy.

Chapter 33

The IBM 1800

Introduction

This third-generation computer is constructed with hybrid-circuit technology (semiconductors bonded to ceramic substrates) known as SLT (Solid Logic Technology). It has a core primary memory.

The 1800 is designed for process control and real-time applications. It is nearly identical to the IBM 1130, which is designed for small-scale, general-purpose, and scientific calculation applications. The two C's perform about the same for computation bound problems. The 1130 and 1800 are not program compatible with the "universal" IBM System/360 series, though introduced at about the same time. However, the 1800 uses terminals and secondary memories similar or identical to the System/360. These are organized about the standard IBM System/360 8-bit byte. Thus their common information media provide a link between the two. Hence an 1800 is sometimes connected to the System/360 as a preprocessor. The relative performance of the IBM 1130, 1800, and the IBM System/360 can be seen on page 586. The 1800 has a better cost/performance ratio than a System/360, Model 40 and has the performance of a Model 30. From now on we will refer only to the IBM 1800, although much applies to the IBM 1130.

The 1800's interface facilities include a large number of T's which can connect to different physical processes; a multiple priority interrupt facility with fast response; multiple Pio's which can transfer information at high data rates;[1] and a complete instruction set for real-time, nonarithmetic processing.

We include the 1800 because it is a typical, 16-bit, real-time, process control computer. The ISP is the most straightforward of the IBM computers in the book (and perhaps the nicest). The several different Pio's and their implementations are unusual and should be carefully studied. Important aspects of the 1800 include the PMS structure as it links to real-time processes, e.g., analog processes; the straightforward Pc ISP (Appendix 1 of this chapter); the specialized Pio's for real-time T's; the Pc implementation; and the Pio implementation. The chapter is written to expose and explain these aspects.[2]

By comparing the 1800 with Whirlwind, an evolutionary progression can be seen. Their ISP's are similar but, because of better technology, the 1800 shows an increase in capability. The 1800 Pc has a medium-sized state (ISP has six registers) including three index registers. The implementation is not elegant; a single register array and adder would provide the basis for a straightforward Pc implementation. The 1800 has features which facilitate higher information processing rates compared with Whirlwind. The major change between Whirlwind and the 1800 machines was brought about by the decreasing cost of registers and primary memory. In the 1800, all K's have independent memory (usually $1 \sim 2$ words or characters) so that concurrent operation of almost all the T and Ms via their K's is possible. In contrast, Whirlwind has only a single, shared register in Pc, and only one device can operate at a time.

Lower hardware costs allow multiple Pio's in the 1800. The Pio's represent an unusual approach to information processing in this period. The Pio's which process standard disk, magnetic tape, and card reader are conventional, but the Pio's for analog and process signals are novel and interesting. The latter Pio's are the most unusual part of the 1800, and they allow independent programs in each Pio to do some very trivial processing tasks such as alarm-condition monitoring independent of Pc. However, the Pio's are limited; for example, it is difficult to transmit or receive a data block between Ms and Mp (using a Pio) without surrounding the data block with Pio control words (thereby transmitting the control words).

The interrupt system is typical of second- and third-generation computers and is comparable to the SDS 900 series (Chap. 42). In later computers interrupt conditions are used to determine a fixed address to which the processor interrupts. There are generally many conditions (100 to 1,000), but only a few discrete levels (8 to 20). The 1800 depends on program polling within a discrete interrupt level; each level has a unique, fixed address.

A principal ISP design problem is the addressing of the 65,536-word Mp. Thus, a 16-bit number has to be generated within Pc for an address. In this regard the 1800 behaves like the 12-bit machines which have to address a 2^{12} (4,096) word memory, and the modes or methods the 1800 uses for addressing are reasonable. It should be noted that it is relatively difficult to write programs which do not modify themselves. For example, the instruction, Store Status, is changed by its execution.

---

[1] Although we refer to the data channels as Pio's, they have a very limited ISP for a Pio; in fact, they might better be called K's.

[2] Some of the material in the chapter has been abstracted from the IBM 1800 Functional Characteristics Manual.

A peculiar feature of the 1800 is its storage protection (see page 408). This feature should provide program relocation capability in addition to protection, but it does not.

PMS structure

A simplified picture of the IBM 1800 structure is given in Fig. 1, without Pio('Data Channel')'s and K('Device Adapter')'s. Each T and Ms have a K which connects Pc's In and Out Bus, the S('Pc to K). Some K's attach to Pio's and some directly to Pc. Information can be transferred between Mp and K via Pio at rates up to 0.5 megaword/s or 8 megabits/s. The IBM Configurator (Fig. 2) gives the restrictions on the possible structures, together with minute L details. It is presented as an alternative to the PMS structure (Fig. 1). The Configurator is intended to show the "permissible structures" but does not show the logical or physical structure. The PMS diagram (Fig. 3) alternatively shows the physical-logical hardware structure and performance parameters. It should be noted that a PMS diagram with the information of the computer component Configurator (Fig. 2) would require slightly more details (and space).

The central processor[1]—primary memory

The IBM 1800 is a fixed-word-length, binary computer with 4, 8, 16, or 32-kword memories of $16 + 1 + 1$ bits, and a memory cycle time of 2 or 4 microseconds. Of the 18 bits 1 bit is used as a parity check (P bit) and 1 bit is used for storage protection (S bit). The Pc instruction set operates on 16-bit and 32-bit words. Indirect addressing and three index registers are used in address modification. The Pc has a 24-level interrupt system, three interval timers, and a console.

The Pc interrupt is a forced branch (jump) in the normal program sequence based upon external or internal Pc conditions. The devices and conditions that cause interrupts are hardwired in fixed priority levels. An interrupt request is not honored while the level of the request itself or any higher level is being serviced, or if the level requested is masked. Examples of interrupt conditions are:

1   An external process condition that requires attention is detected.

[1] IBM name: the Processor-Controller or PC.

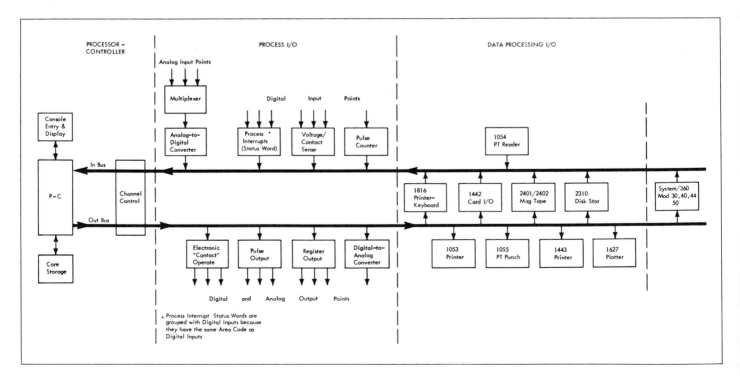

Fig. 1. IBM 1800 data acquisition and control system. (*Courtesy of International Business Machines Corporation.*)

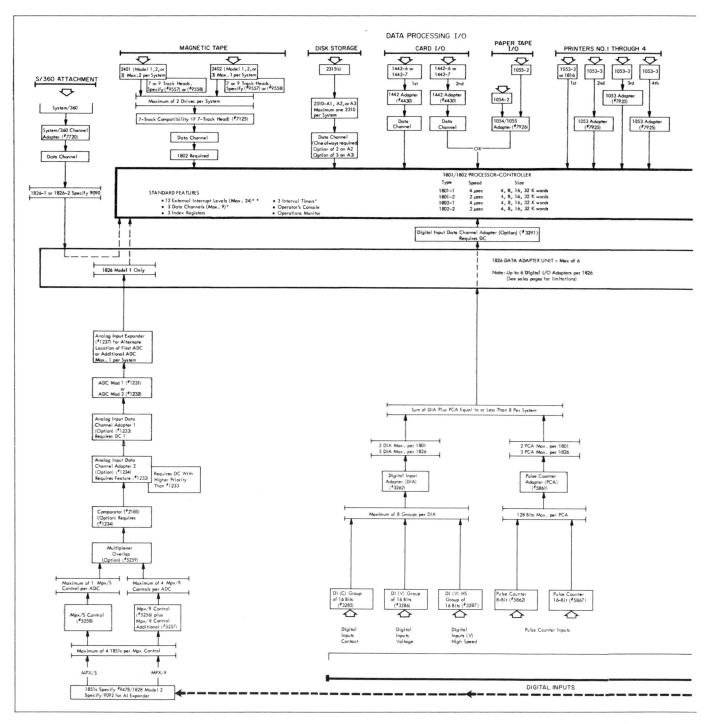

Fig. 2. IBM 1800 data-acquisition and control-system configurator.
(Courtesy of International Business Machines Corporation.)

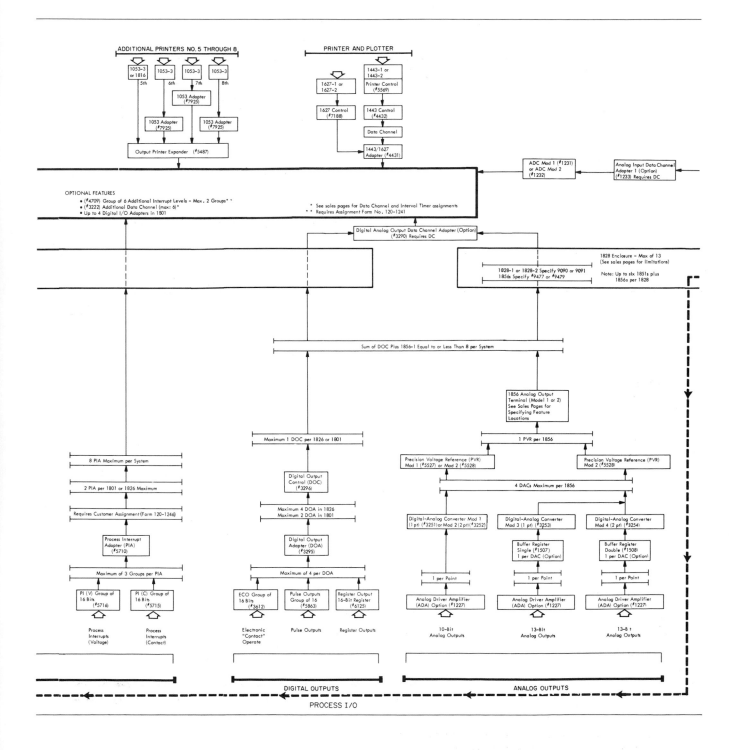

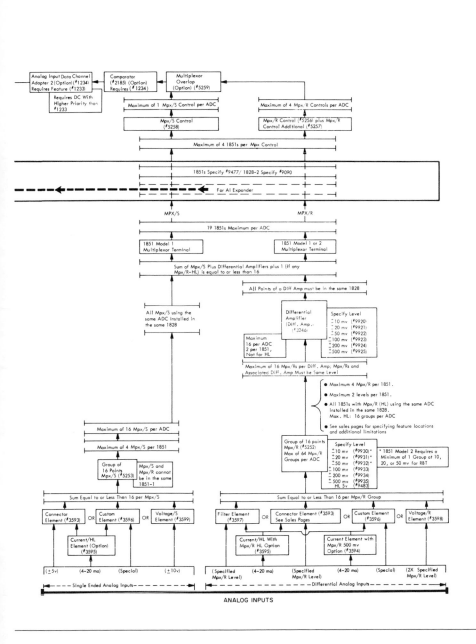

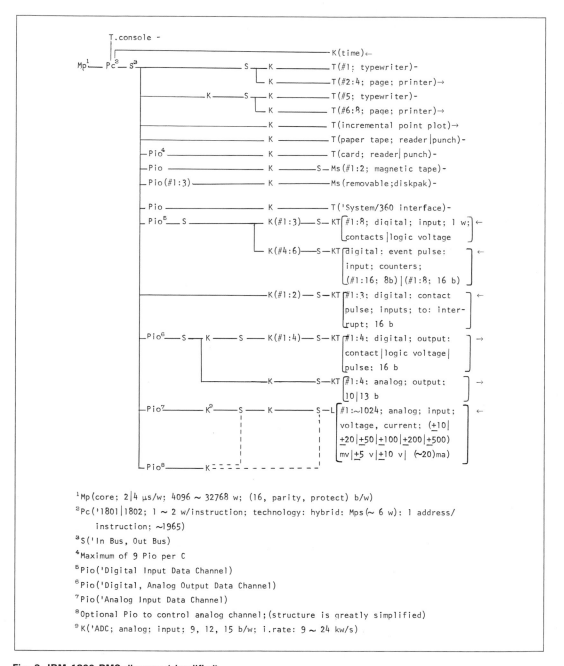

^{1}Mp(core; 2|4 μs/w; 4096 ~ 32768 w; (16, parity, protect) b/w)

^{2}Pc('1801|1802; 1 ~ 2 w/instruction; technology: hybrid: Mps(~ 6 w): 1 address/
        instruction; ~1965)

^{3}S('In Bus, Out Bus)

^{4}Maximum of 9 Pio per C

^{5}Pio('Digital Input Data Channel)

^{6}Pio('Digital, Analog Output Data Channel)

^{7}Pio('Analog Input Data Channel)

^{8}Optional Pio to control analog channel;(structure is greatly simplified)

^{9}K('ADC; analog; input; 9, 12, 15 b/w; i.rate: 9 ~ 24 kw/s)

Fig. 3. IBM 1800 PMS diagram (simplified).

2 An interval timer has counted a previously set time interval.

3 A magnetic-tape drive has completed a data transfer previously requested and is ready for another request.

4 An operator has initiated an interrupt from the Pc console.

5 A device such as a typewriter has just printed a character and is ready to receive the next one.

Primary-memory communication and data transmission with terminals and secondary memory

Two methods are used to transmit data between Mp and Ms, or Mp and T. First, low-speed devices are controlled directly by the program. Each character or word of data is transmitted to or from the Pc and onto T by means of an Execute I/O(XIO) instruction. The Pc program and device synchronization are accomplished by using the interrupt mechanism. Devices operating under direct program control include typewriter, printer, plotter, paper tape reader and punch, analog-to-digital converters, contact sense, voltage-level sense, pulse counters, etc.

The second method of transferring data is via the Pio('Data Channel)'s. The Pio program is started by the XIO instruction of the Pc. The transfer of data words then proceeds under control of the specified Pio, completely asynchronous to and in parallel with Pc program operation. The Pio gains Mp access independent of Pc (Pc operation is suspended for one Mp cycle). During the Mp cycle, the data are taken from or placed into core storage by Pio (via internal Pc control and registers). As soon as the Pio has been satisfied, which normally takes one cycle, the Pc proceeds. The logical state of the Pc, or the Instruction-set Processor, is not changed by Pio's access to Mp. This method of access is referred to as "cycle stealing." Devices (Ms and T) operating under Pio control include magnetic tapes, disks, line printer, card reader-punch, and the link to the IBM System/360.

Some devices can operate under both Pc and Pio control, depending on their characteristics and the configuration, e.g., analog input, analog output, digital input, and digital output.

Process I/O, controls and transducers

Analog inputs. Analog-input equipment includes analog-to-digital converters, multiplexors, amplifiers, and signal conditioning equipment to handle various analog-input signals. The data input rates are up to 20,000 16-bit samples per second, with program selectable resolution and external synchronization. There can be 1,024 (via relay) and 256 (via high-speed solid state) multiplexed analog-input channels connected to a single K (analog-to-digital converter). The Configurator (Fig. 2) shows the allowable inputs.

Digital inputs. The Digital Input provides up to 384 process interrupts; up to 1,024 bits of contact sense, digital input, or parallel register input; and 128 bits of event input counters as 1-, 8-, and 16-bit counting registers.

Analog outputs. Up to 128 analog outputs can be provided.

Digital outputs. Digital Outputs provide up to 2,048 bits of pulse output, contacts, and registers.

IO processors (data channels)

Pio('Data Channels) give a T or Ms the ability to communicate directly with Mp. For example, if an input unit requires a primary memory cycle to store data that it has collected, the Pio communicates directly with Mp and stores the data.

The Pio's run even if Pc is waiting. The Pio's have two registers: a Word Count which is used to count the number of words being transferred in a block between a device and Mp memory; and a Channel Address which points to the next word transferred in a block. The Channel Address is also used to select the next instruction in the program for the next block transfer task.

Two basic types of Pio's are used, nonchaining and chaining.[1] The Pio's provide the ability to transfer either a single block (nonchaining) or multiple blocks (chaining) directly to Mp independent of Pc.

The central processor

Registers in the physical processor

Figure 4 shows the relationship of the registers in Pc, together with those in the Instruction-set Processor. Those registers accessible by the program are shown with an °. All the registers are accessible from the console. A description of the functions of each register is given below.

Storage address register (SAR). All Pc references to Mp are selected or accessed by this 16-bit register. Pio references to Mp use the Channel Address Register (CAR) of the active Pio.

Instruction register (I)°. This 16-bit counter register holds the address of the next instruction.

Storage buffer register (B). This 16-bit register is used for buffering all word transfers with Mp.

[1] A descriptive name undoubtedly concocted by one of IBM's marketing departments.

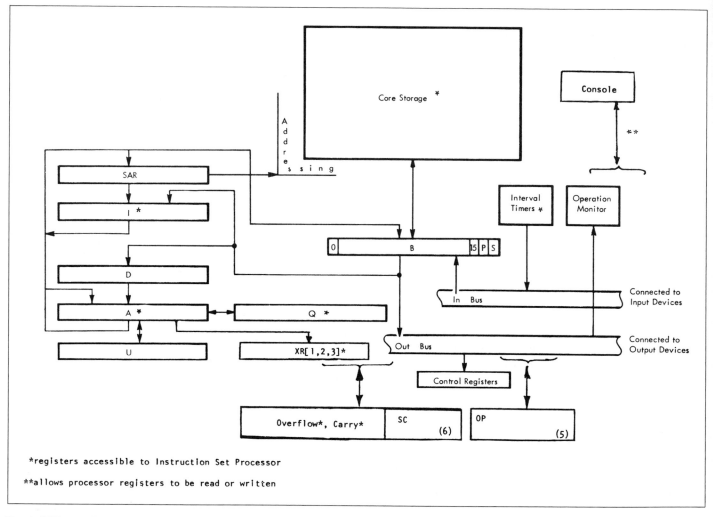

Fig. 4. IBM 1800 Pc data flow. (*Courtesy of International Business Machines Corporation.*)

Arithmetic factor register (D). This 16-bit register is used to hold one operand for arithmetic and logical operations. The Accumulator provides the other factor.

Accumulator (A)°. This 16-bit register contains the results of any arithmetic operation. It can be loaded from or stored into core storage, shifted right or left, and otherwise manipulated by specific arithmetic and logical instructions.

Accumulator extension (Q)°. This register is a 16-bit low-order extension of the Accumulator. It is used during multiply, divide, shifting, and double-precision arithmetic.

Shift control counter (SC). This 6-bit counter is used primarily to control shift operations.

Accumulator temporary (U). The U register is used to store A temporarily during an instruction or an operation which requires the A's facilities.

OP register (OP). This 5-bit register is used to hold the operation code portion of an instruction.

Index registers°. The three 16-bit registers are used in effective-address calculations.

Overflow and carry indicators°. The two indicator bits associated with the Accumulator are Overflow and Carry. The Overflow indicator can be turned on by Add, Subtract, or Divide instruction and indicates a result larger than can be represented in the Accumulator. The Overflow indicator can also be turned on by a Load-status instruction. Once Overflow is on, it will not be changed except by testing the indicator, or by a Load-status or Store-status instruction. The Carry indicator provides the information that a carry (or borrow) from the high-order position of the Accumulator has occurred.

The Carry indicator is used with the Add, Subtract, Shift-left, Load-status, Store-status, and Compare instructions.

In-bus. This 18-bit bus is a link(L) used to carry information from a K to Pc. Generally only 16 of the 18 bits are used, although transfers to magnetic tape can be made three 6-bit characters.

Out-bus. This 18-bit bus is used to carry information from Pc to a K.

Instruction-set processor

The operation of the Pc from a program viewpoint follows. The ISP registers were declared (°) in the previous section and in Fig. 4. The ISP registers are the 16-bit I, A, Q, XR [1, 2, 3], and the 1-bit Overflow and Carry.

An ISP description of the 1800 appears in Appendix 1 of this chapter. It is incomplete in the following respects: The memory protect bit checking is not described; the illegal (undefined) instruction action is not described; double word data must be aligned on even and odd address word boundaries or else a fault occurs; and the IO instruction and interrupt operation are not given.

Instruction formats. Two basic instruction-word formats are used, one word (Fig. 5) and two word (Fig. 6). The bits within the instruction words are used in the following manner:

OP   Operation Code. These 5 bits define the instruction.

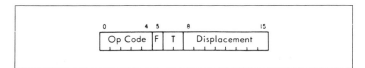

Fig. 5. IBM 1800 one-word-instruction format. (*Courtesy of International Business Machines Corporation.*)

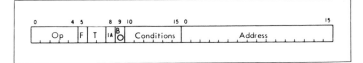

Fig. 6. IBM 1800 two-word-instruction format. (*Courtesy of International Business Machines Corporation.*)

F   Format bit. A 0 indicates a single-word instruction, and 1 a two-word instruction.

T   Tag. These 2 bits specify which of the three index registers is used in address modification or the shift count.

DISP   Displacement. These 8 bits are usually added to the instruction register or the index register specified by T for one-word instructions. The modified address is defined as the Effective Address (EA). If T is 00, the displacement is added to the instruction register (then EA = I + DISP). The displacement is in two's complement form if negative, with the sign in bit 8. The bit in position 8 is automatically extended to the higher-ordered bits (0 to 7) when the displacement is used in EA generation.

IA   Indirect addressing. This bit is used only in the two-word-instruction format. If 0, addressing will be direct. If a 1, addressing will be indirect. Only one level of indirect addressing is permitted. (The Load Index and Modify Index and Skip instructions have exceptions, as shown in the ISP description.)

BO   Branch Out. This bit is used to specify that the Branch or Skip on Condition (BSC) instruction is to be interpreted as a Branch Out (BOSC) when used in an interrupt routine.

COND   Conditions. These 6 bits select the indicators that are to be interrogated on a BSC or BSI instruction. The bit assignments for conditions are:

Cond⟨10⟩   A = 0
Cond⟨11⟩   A < 0
Cond⟨12⟩   A > 0
Cond⟨13⟩   (A⟨15⟩ = 0) *that is, A is even*
Cond⟨14⟩   (Carry = 0)
Cond⟨15⟩   (Overflow = 0)

ADDRESS   These 16 bits usually specify a core storage address

Table 1 Determining effective addresses

	$F = 0$ (direct addressing)†	$(F = 1) \wedge (IA = 0)$ (direct addressing)	$(F = 1) \wedge (IA = 1)$ (indirect addressing)
$T = 00$	$EA \leftarrow I + Disp$‡	$EA \leftarrow Address$	$EA \leftarrow C(Address)$§
$T = 01$	$EA \leftarrow XR[1] + Disp$	$EA \leftarrow Address + XR[1]$	$EA \leftarrow C(Address + XR[1])$
$T = 10$	$EA \leftarrow XR[2] + Disp$	$EA \leftarrow Address + XR[2]$	$EA \leftarrow C(Address + XR[2])$
$T = 11$	$EA \leftarrow XR[3] + Disp$	$EA \leftarrow Address + XR[3]$	$EA \leftarrow C(Address + XR[3])$

† Contents of instruction register (I) or index register (XR[1], XR[2], XR[3]).

‡ May be true positive quantity or negative two's complement quantity.

§ C specifies "contents" at location specified by Address or Address + XR[1], XR[2], or XR[3].

in a two-word instruction. The address can be modified by the contents of an index register or used as an indirect address if the IA bit is on.

Effective-address generation. The Effective Address (EA) is developed as shown in Table 1. The instruction set is divided into five classes as shown in Table 2.

Storage protection. The storage-protection facility protects the contents of specified individual locations of Mp from change due to the erroneous storing of information during the execution of a program. The status of each location is identified as "read only" or "read/write" by the condition of the Storage Protect Bit, S.

The Store-status instruction is used to write and clear Storage Protect Bits. The execution of this instruction is under control of the Write Storage Protect Bits switch on the console. Any attempt by the program to write into a read-only protected location results in a storage-protect violation which causes the Internal Interrupt (the highest priority interrupt).

Instruction interpretation process

The simplified Pc data-flow block diagram (Fig. 4) shows instructions and data entering and leaving memory via the B register. Additional bits in Pc hold the P and S bits for Mp. Input devices send data and instructions to the B register via the 18-bit In-bus. Output devices receive data from the B register via the 18-bit Out-bus. Eighteen bits can be transferred between Pc and K(magnetic tape). As each stored-program instruction is selected, its various parts (op code, format bit, etc.) are directed to the control registers via the B register and the Out-bus. The control registers decode and interpret each instruction before the instruction is executed.

Except for Pio operations, all instructions and data in memory are addressed by the Storage Address Register (SAR). SAR obtains the memory address from the I register or the A register. The

Table 2 Instruction set

Class	Instruction	Indirect addressing	Mnemonic
Load and store	Load accumulator	Yes	LD
	Double load	Yes	LDD
	Store accumulator	Yes	STO
	Double store	Yes	STD
	Load index	‡	LDX
	Store index	Yes	STX
	Load status	No	LDS
	Store status	Yes	STS
Arithmetic	Add	Yes	A
	Double add	Yes	AD
	Subtract	Yes	S
	Double subtract	Yes	SD
	Multiply	Yes	M
	Divide	Yes	D
	And	Yes	AND
	Or	Yes	OR
	Exclusive Or	Yes	EOR
Shift	Shift Left instructions:		
	Shift left logical (A)†	No	SLA
	Shift left logical (AQ)†	No	SLT
	Shift left and count (AQ)†	No	SLC
	Shift left and count (A)†	No	SLCA
	Shift Right instructions:		
	Shift right logical (A)†	No	SRA
	Shift right arithmetically (AQ)†	No	SRT
	Rotate right (AQ)*	No	RTE
Branch	Branch and store I	Yes	BSI
	Branch or skip on condition	Yes	BSC (BOSC)
	Modify index and skip	‡	MDX
	Wait	No	WAIT
	Compare	Yes	CMP
	Double compare	Yes	DCM
I/O	Execute I/O	Yes	XIO

† Letters in parentheses indicate registers involved in shift operations.

‡ See the section for the individual instruction (MDX and LDX).

contents of the I register are developed by one of the following means, depending on the Pc operation:

1 The I register is incremented for each instruction.

2 The effective address of each instruction is developed in the accumulator (A register) and then transferred to SAR. The contents of the accumulator are saved in an auxiliary (U) register during effective-address computation. If the instruction was a branch, the contents of SAR is transferred to the I register.

The following examples illustrate the data flow or instruction interpretation process for the Load Accumulator (LD) instruction.

One-word load instruction
Instruction Cycle

1 A register transfers to U register.

2 I register transfers to SAR (I register is then incremented).

3 SAR addresses the memory location containing the instruction.

4 Memory location transfers to the B register and Out-bus.

5 Control registers store various parts of the instruction (op code, format, and tag).

6 Displacement is stored in the D register.

7 *a* If tag = 00, I register transfers to A register.
   b If tag ≠ 00, the specified XR transfers to A register.

8 Displacement (D register) is added to A register.

Execute Cycle

9 A register transfers to SAR (effective address).

10 U register transfers to A register.

11 SAR addresses data word.

12 Data word transfers to B register.

13 B register loads into A register (via D register).

Two-word load instruction, direct addressing
Instruction Cycle 1

1 A register transfers to U register.

2 I register transfers to SAR (I register is then incremented).

3 SAR addresses the memory location containing the instruction (first word).

4 Memory location transfers to B register and Out-bus.

5 Control registers store various parts of the instruction (op code, format, and tag).

6 If tag ≠ 00, the specified XR transfers to A register.

Instruction Cycle 2

7 I register transfers to SAR (I register is then incremented).

8 SAR addresses second word of instruction.

9 Second word of instruction (address) is read into B register.

10 Address (from B register) is stored in D register.

11 *a* If tag = 00, D register transfers to A register.
   b If tag ≠ 00, D register is added to A register (A register contains contents of XR).

Execute Cycle

12 A register transfers to SAR (effective address).

13 U register transfers to A register.

14 SAR addresses memory at effective address (data word).

15 Data word transfers to B register.

16 B register loads into A register (through D register).

Central-processor communication with the controls[1]

Direct program control of the controls

Pc direct programmed control of I/O devices is on the basis of single-word or character-at-a-time transfers for each XIO instruction executed. One data word or character is transferred to or from Mp to K. The XIO instruction specifies an I/O Control Command (IOCC) with a function of Control, Sense, Read, or Write to a controlled device. This command is either directly to a device or to a Pio.

It is possible for the program sequence to execute an XIO instruction to a device that is busy responding to a previous XIO instruction. Each device has a Busy indicator, which signals whether or not the device can accept data or control information. (Incorrect program sequence timing may cause undetected errors.)

[1]IBM name: Adapter or Device Adapter.

It is possible for a device operating synchronously with the program to request a data word transfer before the program sequence is ready to service the request. Devices with this potential have a "program check" indicator to signal when data have been lost (that is, Pc has not kept up with the device).

Execute I/O instruction (XIO)

This instruction is used for programmed I/O operations and to initialize Pio; it may be either one or two words in length, as specified by the F bit. In the two-word instruction the address is either a direct or indirect address, as specified by the IA bit. For proper operation the effective address must be an even address. The effective address is used to select a two-word I/O Control Command (IOCC) from storage.

The IOCC specifies the I/O operation, I/O device, and core storage address. The format of the two-word IOCC follows, with an explanation of the assigned fields:

Area := $IOCC[1]\langle 0{:}4\rangle$. The area field specifies a unique segment of I/O which may be a single device (1442 Card Read-Punch, 1443 Printer, etc.) or a group of several units (magnetic-tape drives, serial I/O units, contact sense units, etc.). (Area 00000 is used to address system devices such as the console and the Interrupt Mask Register.)

Function := $IOCC[1]\langle 5{:}7\rangle$. The primary I/O functions are specified by the 3-bit function code of the IOCC:

000   Removes an I/O device from on-line status and places it in a "free" mode.

001   Write
Transfers a single word from storage to an I/O unit. The address of the storage location is provided by the Address field of the I/O Control Command.

010   Read
Transfers a single word from an I/O unit to storage. The address of the storage location is provided by the Address field of the I/O Control Command.

011   Sense Interrupt Level
Directs the selected I/O device to make its status available in the Accumulator as the Interrupt Level Status Word (ILSW).

100   Control
Causes the selected device to interpret the address and/or Modifier of the IOCC as a specific control action. Examples are feed card and load interrupt mask register.

101   Initialize Write
Initiates a Write operation on a device or unit which will subsequently make data transfers from storage via a Pc.

110   Initialize Read
Initiates a Read operation from a device or unit which will subsequently make data transfers to storage via a Data Channel.

111   Sense Device
Reads the selected device status word into the Accumulator. A Device Status Word (DSW) and the Process Interrupt Status Word (PISW) are sensed with this instruction.

If Area 00000 is specified, the Console status and Interval Timer status may be brought into the Accumulator as specified by a unit address code in the Modifier field.

The current contents of the Accumulator are destroyed by the execution of Sense Interrupt Level, Sense Device, Initialize Read, Initialize Write, Read, or Write.

Modifier := $IOCC[1]\langle 8{:}15\rangle$. This 8-bit field provides additional detail for either Function or Area. For example, if the Area specifies a disk and if the Function specifies Control (100) then a particular modifier code specifies the direction of the Seek operation. In this case, the Modifier serves to extend the function.

If, however, the Area specifies a group of I/O devices, and if the Function specifies Write (001), then the particular unit address is specified by the modifier.

Address := $IOCC[0]\langle 0{:}15\rangle$. The meaning prescribed for this 16-bit field is dependent upon the Function specified by this I/O Control Command:

1   If Function is Initialize Write (101) or Initialize Read (110), then Address specifies the starting address of a table in storage (an I/O block). The contents of this table are data words and control information.

2   If Function is Control (100) and if, for example, Area specifies the 1443 Printer, the Address may specify a specific control action.

3   If Function is Sense (011 or 111), the Address field is ignored. Instead, an increment of time equivalent to a memory cycle is taken, during which the selected I/O device or Interrupt Level places its status word in the accumulator.

4 If Function is Write (001) or Read (010), the Address specifies the storage location of the data word.

XIO execution interpretation process

1 The EA of the XIO is developed in the accumulator (A) and routed to the Storage Address Register (SAR) to locate the IOCC (as for any EA).

2 Bit position 15 of SAR is forced on to select the EA + 1 where the IOCC Area, Function, and Modifier are found.

3 The Area, Function, and Modifier are routed through the B register to the Out-bus to the control of the device specified by the Area.

4 Bit position 15 of SAR is turned off to allow the address portion of the IOCC word to be transferred from the Mp location specified by the Effective Address (EA) to the B register.

5 If the Function is an Initialize Read, Initialize Write, or Control, the address part of the IOCC is routed through the B register to the Out-bus. The address part of the Initialize Read/Write IOCC goes to the Channel Address Register (CAR) of Pio. If the Function is Read or Write, the address is routed from the B register through the A register to the SAR. SAR addresses the memory location to or from which the data are transmitted.

Interval timers

Three timers are provided to supply real-time information to the program. They are in core-storage locations 0004 (Timer A), 0005 (Timer B), and 0006 (Timer C). Each timer is incremented according to its associated or permanent time base and can be hardwired to be 0.125, 0.250, 0.5, 1, 2, 4, 8, 16, 32, 64, or 128 milliseconds.

The timers can be started or stopped under program control. When the count reaches zero, an interrupt is requested on the level assigned to the timers.

Interrupt

The interrupt feature provides an automatic branch from the normal program sequence, based upon an external condition. A maximum of 24 external interrupt levels (groups) are available, arranged in order of priority. Twelve external interrupt levels are standard. Each interrupt level has a unique core-storage address assigned to it. Several devices may be connected to a single interrupt level, and program polling can be used to differentiate the possible signals causing the interrupt. The Interrupt Level Status Word, ILSW, is used to identify the specific condition causing its interrupt level to request service.

Internal interrupt. When any one of the following error conditions occur, there is an internal interrupt in Pc: an invalid op code; a Mp parity error (an even number of bits); a storage-protect violation; and Channel Address Register check error. The internal interrupt takes priority over all external interrupts and cannot be masked.

A mask register exists for the masking and unmasking of interrupt levels. An interrupt level that is masked cannot initiate a request for service until it has been unmasked.

Device status word (DSW). DSW indicators usually fall into three general categories:

1 Error or exception interrupt conditions

2 Normal data or service-required interrupts

3 Routine status conditions

Process interrupt status word indicators (PISW). The PISW indicators are physically located in Pc and are turned on by events external to the computer, e.g., contact closures or voltage shifts.

IO processors[1]

The Pc initializes each Pio with an XIO instruction. The Pio has priority to the extent that, when the I/O device is ready to send or receive a data word, the Pc is stopped while the word transfers to or from core storage. Pc data and conditions are undisturbed except for the memory locations that receive data from an input device.

I/O devices that are to be operated concurrently must be on separate Pio's.

The XIO instruction for a Pio specifies an I/O Control Command (IOCC) with a function of Initialize Read or Initialize Write. However, even though a device operates with a Pio, the XIO instructions in Pc are used to sense device status and for control.

Registers

Channel address register. The Channel Address Register (CAR) is a 16-bit register used to store the Mp address of the next word that will be addressed by the Pio. Each Pio has a CAR. Pio and its associated CAR are selected when their assigned I/O device is selected by the Area Code and Modifier of an IOCC word. CAR is incremented by 1 after each transfer of its contents to CAB.

[1]IBM name: Data Channel (DC).

Channel address buffer. A common Channel Address Buffer (CAB) is used by all Channel Address Registers to address Mp. When a cycle steal request occurs, the CAR for the requesting Pio is transferred into the Channel Address Buffer.

Channel-address-register check bit. Channel Address Register (CAR) checking is provided to ensure that the first word addressed by a selected CAR is the first word of the correct data table. Thus the check determines if a Pc program has set up the Pio program correctly.[1] A CAR check is made for all devices after the address from the IOCC word is transferred to the selected CAR. A bit-by-bit comparison is made between the contents of the selected CAR and the contents of the B register. If any of the corresponding bits are not equal, a CAR check error has occurred. This CAR check error terminates the Pio task and initiates an internal interrupt.

Word count register. A Word Count Register is provided in each Pio. The Word Count Register is loaded with the contents of the word-count portion of the data table, $\langle 2:15 \rangle$. This register is decremented each time a data word is transferred from (to) the data table.

Scan control register. A Scan Control Register is provided in each Pio that has chaining ability. Scan Control register bits are stored in the first word of the first data table (bit positions 0 and 1) and in the second word (bit positions 0 and 1) of the second data table and all subsequent data tables in a chain.

The Scan Control Register controls the I/O device and the Pio operation at the end of the data table as follows: single scan of data table and stop with an interrupt; single scan of data table and stop (no interrupt); continuous scan of this data table or a different data table with an interrupt at the end of this table; and continuous scan of this data table or a different data table with no interrupt.

The IO processor program operation

The sequence of steps for a Pio program is given below. The memory map or format of the program is shown in Fig. 7.

1   Pc issues an XIO instruction which references the IOCC word and initializes Pio.

2   The Area Code and Modifier of the IOCC select the I/O device. Function specifies the type of operation (Initialize Read or Initialize Write, etc.).

[1] Not a completely arbitrary program fault to check, since processors are involved.

3   *a*   The address portion of the IOCC word is stored in CAR for the selected Data Channel and I/O device.

    b   A CAR check is made between the selected CAR and the B register.

4   A cycle steal is requested by Pio; CAR transfers to CAB.

5   CAB addresses core storage for the first word of the data table while CAR is being incremented by 1.

6   The first word of the data table contains
    a   Scan Control bits (bit positions 0 and 1)
    b   Word Count (bit position 2 to 15)
These are transferred to their respective registers in the I/O device. This is the end of the first cycle steal.

7   When another cycle-steal request from Pio occurs, CAR, which was incremented in step 5, now transfers the next higher address to CAB. CAB then addresses core storage while CAR is being incremented.

8   The first data word is transferred to or from the I/O device via the B register and Data Channel. The Word Count Register in the I/O device is decremented by 1. This is the end of the second cycle-steal cycle.

Steps 7 and 8 now continue on a cycle-steal basis; that is, they occur as the I/O device requests data transfers. The CAR is incremented with each data transfer and the WCR is decremented. This sequence continues until the last data word of the data table is transferred. The last word transfer is sensed by the WCR reaching zero or through some indicator in the device. If the device does not have chaining ability, no more demands for data transfer are made until the device is reinitialized with another XIO instruction.

Chaining. These steps are for the second and all subsequent data tables. See above for steps 1 through 8.

9   The contents of the word following the last data word in the first data table are transferred to CAR. This word must contain the address of the next data table.

10   *a*   When the next cycle is requested, CAR is transferred to CAB to address core storage. The contents of the first word of the next data table is transferred to the B register. This word must contain the address of itself.

10   *b*   CAR check is performed and CAR is incremented by 1.

11   When the next cycle steal is requested, CAR is transferred to CAB and CAB addresses Mp. The Scan-control bits and Word-count bits are transferred from the second word of

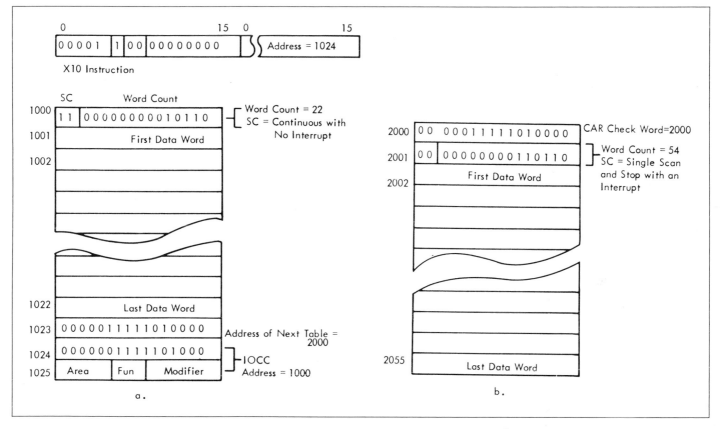

Fig. 7. IBM 1800 data-channel tables for chaining memory maps. (a) First data table; (b) second data table. (Courtesy of International Business Machines Corporation.)

the data table to their respective registers. CAR is incremented by 1.

12 Data are transferred to (from) the I/O device on a cycle-steal basis via the B register and the Data Channel. CAB addresses core storage to transfer a data word to the B register. Each time CAB addresses core storage, CAR is incremented by 1. When the next cycle-steal request occurs, CAR is transferred to CAB. The Word-count Register is decremented for each word transferred.

13 When the last data character is transferred (word count is decremented to zero), operation will continue as specified by the Scan Control Register. (See above section for *Scan-Control Register.*)

Special data channels

The four Pio types for special functions are:

1 Analog input (block data transfers, and comparisons of analog inputs for limits)

2 Digital input/output

3 Analog output

4 Digital output

Analog-input data channels. Memory maps (Fig. 8*a* and *b*) illustrate the command formats interpreted in the Analog Data Channel programs. A list of limit values is placed in a table (Fig. 8*a*), and each analog input is compared with the limits. The operation sequence is: Read a specific addressed analog voltage, called the multiplex[1] point (mpx); compare the input voltage with the limits stored in the table following the analog address (the limit word contains a high and low value in bits $\langle 0:7 \rangle$ and $\langle 8:15 \rangle$, respec-

---

[1] The IBM multiplexor is an S which allows multiple inputs to be read into the T(Analog to Digital Converter) sequentially.

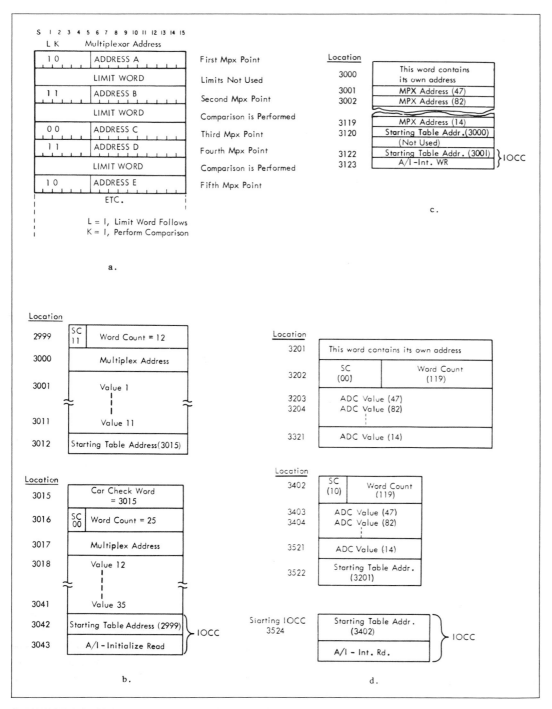

Fig. 8. IBM 1800 data-channel analog-input instruction format and memory maps. (*a*) Multiplexor address table with limit words for comparisons. (*b*) Data table, chained sequential control. (*c*) Multiplexor address table, random addressing. (*d*) Analog-to-digital converter storage tables, random addressing (used with a second data channel).

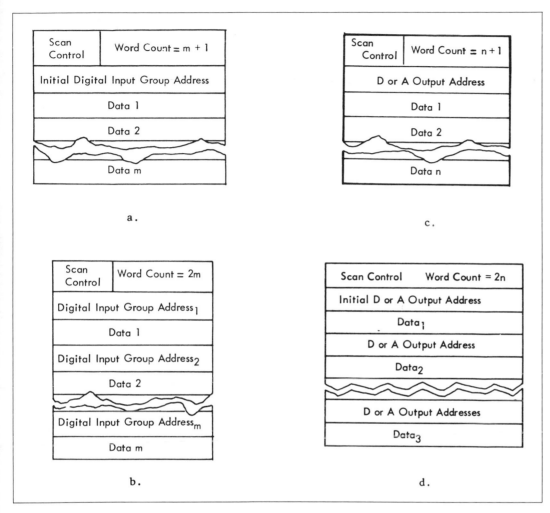

Fig. 9. IBM 1800 data-channel digital or analog-output instruction formats and memory maps. (*a*) Digital input, sequential; (*b*) digital input, random addressing; (*c*) digital or analog output, sequential; (*d*) digital or analog output, random addressing. (*Courtesy of International Business Machines Corporation.*)

tively); and if the analog-input value lies outside the limit range, initiate an interrupt.

Figure 8*b* describes a second use of this data channel. Pio accepts a sequence of analog inputs and packs them into a table following the address initiation instruction. The analog inputs from the T's are either fixed or selected in a cyclic fashion from a Multiplexor.

Two Pio's can be used concurrently: One Pio controls the input from a series of analog-input addresses (Fig. 8*c*); the second Pio packs the corresponding analog values in a second table (Fig. 8*d*).

Digital-input data channels. Digital parameters or events can be read into Mp under the control of a Digital-input Data Channel. The memory map (Fig. 9*a*) shows the control format for selecting and inputting a block or sequence of external data. The memory map (Fig. 9*b*) illustrates a more general ability to address inputs at random and read them into succeeding Mp locations.

Digital- and analog-output data channels. Memory maps (Fig. 9*c* and *d*) show the program format used by the Digital- or Analog-output Data Channels. These channels output selected data points to external analog or digital K's. This Pio is similar to the Digital-input Data Channel.

Conclusions

We have tried to show a typical, third-generation computer used for process control. Many of the facilities the 1800 possesses are general. The Pio's are rather special, designed to monitor and control a process, independent of Pc. Although the Pio's are powerful (by providing parallel data transmission), their use, like other multiprocessing systems, is nontrivial. The Pc ISP is fairly straightforward, and one should write a program using it to appreciate its simplicity.

APPENDIX 1   THE IBM 1800 ISP DESCRIPTION

Appendix 1

IBM 1800 ISP Description

Pc State

A<0:15>	*Accumulator*
Q<0:15>	*Accumulator Extension for multiplier, quotient and double length*
I<0:15>	*Instruction Location Counter*
XR[1:3]<0:15>	*Index Registers*
Ov	*Overflow Indicator*
C	*Carry Indicator*
Run	*denotes running computer*

Mp State

M[0:FFFF$_{16}$]<P,S,0:15>	*Mp with Parity and Protect bits*

Pc Console State

Check Stop Switch	*Pc stops if storage protect violation occurs*
WSPB Switch	*Write Storage Protect Bits; enables the writing of bits in a word*
SPV Indicator	*Storage Protect Violation indicator; set to 1 if a memory reference is made to a protected word*

Instruction Format

instruction/i[0:1]<0;15>

op<0:4>	:= i[0]<0:4>	*operation code*
shop<0:7>	:= op□i[0]<5,8,9>	*shift operation code count*
f	:= i[0]<5>	*format; specifies a 1 or 2 word instruction*
t<0:1>	:= i[0]<6:7>	*tag; index register specification*
d<8:15>	:= i[0]<8:15>	*displacement or short address*
dsgn<0:15>:= sign_extend(d<8>□d<9:15>)		
a<0:15>	:= i[1]<0:15>	*address*
ia	:= i[0]<8>	*indirect address bit*
bo	:= i[0]<8>	*branch out bit*
cond<0:5> := i[0]<10:15>		*conditions for test*

Effective Address Calculation Process

z<0:15>:= (	*effective address*
(t = 0)∧ ¬ f → (dsgn + I);	*1 word, relative*
(t ≠ 0)∧ ¬ f → (dsgn + XR[t]);	*1 word, relative, indexed*
(t = 0) ∧ f ∧ ¬ ia→ a;	*2 word, direct*
(t ≠ 0) ∧ f ∧ ¬ ia→(a + XR[t]);	*2 word, direct, indexed*
(t = 0) ∧ f ∧ ia → M[a];	*2 word, indirect*
(t ≠ 0) ∧ f ∧ ia → (M[a + XR[t]]))	*2 word, indirect, indexed*
z'<0:15> := (¬ f → (dsgn + I);	*effective address for index register instructions*
f ∧ ¬ ia →a;	
f ∧ ia →M[a])	

APPENDIX 1   THE IBM 1800 ISP DESCRIPTION (Continued)

$zd<0:15> := (\neg z<15> \to z + 1;$ *process for locating second operand for double length*
                    $z<15> \to z)$

$xi<0:15> := (\neg f \to dsgn;$ *index increment*
                $f \wedge \neg ia \to a;$
                $f \wedge ia \to M[a])$

$s<0:5> := ($ *shift count calculation*
      $(t = 0) \to d<10:15>$
      $(t \neq 0) \to XR[t]<10:15>)$

Instruction Interpretation Process

  $Run \to (instruction[0:1] \leftarrow M[I:I + 1]; next$ *fetch*
        $\neg f \to (I \leftarrow I + 1); f \to (I \leftarrow I + 2); next$ *1 or 2 word instruction*
        $Instruction\_execution)$ *execute*

Instruction Set and Instruction Execution Process

  $Instruction\_execution := ($

  Load and Arithmetic

LD $(:= op = 11000) \to (A \leftarrow M[z]);$		*load accumulator*
LDD $(:= op = 11001) \to (A\square Q \leftarrow M[z]\square M[zd]);$		*double load*
STO $(:= op = 11010) \to (M[z] \leftarrow A);$		*store accumulator*
STD $(:= op = 11011) \to (M[z]\square M[zd] \leftarrow A\square Q);$		*double store*
A $(:= op = 10000) \to (Ov, C \square A \leftarrow A + M[z]);$		*add*
AD $(:= op = 10001) \to (Ov, C\square A\square Q \leftarrow A\square Q + M[z]\square M[zd]);$		*double add*
S $(:= op = 10010) \to (Ov, C\square A \leftarrow A - M[z]);$		*subtract*
SD $(:= op = 10011) \to (Ov, C\square A\square Q \leftarrow A\square Q - M[z]\square M[zd]);$		*double subtract*
M $(:= op = 10100) \to (A\square Q \leftarrow A \times M[z]);$		*multiply*
D $(:= op = 10101) \to (Ov, Q \leftarrow A\square Q / M[z];$		*divide*
$A \leftarrow A\square Q \bmod M[z]);$		

  Logical instructions

AND $(:= op = 11100) \to (A \leftarrow A \wedge M[z]);$	*logical and*
OR $(:= op = 11101) \to (A \leftarrow A \vee M[z]);$	*logical or*
EOR $(:= op = 11110) \to (A \leftarrow A \oplus M[z]);$	*logical exclusive or*

  Compare

CMP $(:= op = 10110) \to ((A < M[z]) \to (I \leftarrow I + 1);$	*compare*
$(A = M[z]) \to (I \leftarrow I + 2));$	
DCM $(:= op = 10111) \to ((A\square Q < M[z]\square M[zd]) \to (I \leftarrow I + 1);$	*double compare*
$(A\square Q = M[z]\square M[zd]) \to (I \leftarrow I + 2));$	

  Shifts

SLA $(:= shop = 00010\square0\square00) \to ($	*shift left logical*
$A \leftarrow A \times 2^S \{logical\}; C \leftarrow A<s-1>);$	
SLT $(:= shop = 00010\square0\square10) \to ($	*shift double left logical*
$A\square Q \leftarrow A\square Q \times 2^S \{logical\}; C \leftarrow A<s-1>);$	
SRA $(:= shop = 00011\square0\square00) \to (A \leftarrow A / 2^S \{logical\});$	*shift right logical*
SRT $(:= shop = 00011\square0\square10) \to (A\square Q \leftarrow A\square Q / 2^S);$	*shift right A and Q*
RTE $(:= shop = 00011\square0\square11) \to (A\square Q \leftarrow A\square Q / 2^S \{rotate\});$	*rotate right A and Q*
SLCA $(:= shop = 00010\square0\square01) \to ($	*shift left and count A*

APPENDIX 1   THE IBM 1800 ISP DESCRIPTION (Continued)

```
         (t = 0) → (A ← A × 2ˢ; C ← A<s-1>);
         (t ≠ 0) → (A ← normalize(A);
                   C☐XR[t]<10:15> ← normalize_exponent(A);
                   XR[t]<8,9> ← 0));
SLC (:= shop = 00010☐0☐11) → (¬ ((s = 0) ∨ A<0>) → (        shift left and count
   (t = 0) → (A☐Q ← A☐Q × 2ˢ; C ← A<s-1>);
   (t ≠ 0) → (A☐Q ← normalize(A☐Q);
             C☐XR[t] ← normalize_exponent(A☐Q))));
LDX (:= op = 01100) → ((t = 0) → (I ← z');                   load index or instruction counter
                      (t ≠ 0) → (XR[t] ← z'));
STX (:= op = 01101) → ((t = 0) → (M[z'] ← I);                store index or instruction counter
                      (t ≠ 0) → (M[z'] ← XR[t]));
STS (:= op = 00101) → (                                      store status
   (f ∧ bo) → M[z]<P> ← cond<15>;
   ¬bo → (M[z]<8:15> ← 00000☐C☐0v; C☐0v ← 00));
LDS (:= i[0] = 00100☐0☐00☐000000ΦΦ) → (C ← i[0]<14>;         load status
                                      0v ← i[0]<15>);
BSC (:= (op = 01001) ∧ ¬ i<9>) → (                           branch or skip on condition
   ( skip_condition ∧ ¬ f) → (I ← I + 1);
   (¬skip_condition ∧ f) → (I ← z);
   d<15> → 0v ← 0);
   skip_condition := (
      (¬0v ∧ d<15>) ∨                                        overflow off
      (¬ C ∧ d<14>) ∨                                        carry off
      (A<15> ∧ d<13>) ∨                                      Accumulator even
      ((A > 0) ∧ d<12>) ∨                                    Accumulator greater than zero
      (A<0> ∧ d<11>) ∨                                       Accumulator negative
      ((A=0) ∧ d<10>))                                       Accumulator zero
BOSC (:=(op = 01001) ∧ i<9>) → (                             branch out of interrupts
   (skip_condition ∧ ¬ f) → (I ← I + 1; Interrupt ←1);
   (¬ skip_condition ∧ f) → (I ← z; Interrupt ←1);
   d<15> → (0v ← 0));
BSI (:= op = 01000) → (                                      branch and store instruction register
   ¬f → (I ← z + 1; M[z] ←I);
    f → (d<15> → 0v ← 0);
   ¬skip_condition → (I ← z + 1; M[z] ←I));
MDX (:= op = 01110) → (                                      modify index and skip
   (t = 0) ∧ ¬ f → (I ← I + dsgn);                           local branch
   (t = 0) ∧ f → (M[a] ←M[a] + dsgn;
      (Msum=0) ∨ (M[0]<0> ⊕ Msum<0>) → (I ← I + 1));         result zero or sign change
         Msum<0:15> := (M[a] + dsgn)
   (t ≠ 0) → (XR[t] ←XR[t] + xi;
      (Xsum=0) ∨ (XR[t]<0> ⊕ Xsum<0>) → (I ← I + 1)));       result zero or sign change
         Xsum<0:15> := (XR[t] + dsgn)
Wait (:= i = 3000₁₆) → (I ← I - 1);
```

APPENDIX 1   THE IBM 1800 ISP DESCRIPTION (Continued)

```
IO Control Instruction:
    XIO (:= op = 00001) → (                              Execute I/O, not defined
        IOCC[0:1] ← M[z]□M[zd]; next
        Execute⌄IO⌄instruction)
               )                                         end Instruction⌄execution
IO Instruction Format:
    IO Address<0:15>        := IOCC[0]                    address if IO data
    IO Device or Area<0:4>  := IOCC[1]<0:4>              io device name
    IO Function<5:7>        := IOCC[1]<5:7>
    IO Modifier<8:25>       := IOCC[1]<8:15>             device function details
        Device mode off line := (IO Function = 0)
        Device mode write    := (IO Function = 1)
        Device mode read     := (IO Function = 2)
        Device mode sense interrupt level := (IO Function = 3)
        Device mode control := (IO Function = 4)
        Device mode initialize write := (IO Function = 5)
        Device mode initialize read  := (IO Function = 6)
        Device mode sense    := (IO Function = 7)
```

Chapter 34

The engineering design of the Stretch computer[1]

Erich Bloch

Summary The Stretch computer is an advanced scientific computer with variable facilities for floating-point, fixed-point, and variable-field-length arithmetic and data-handling facilities.

The performance goal of 100×704 speed is achieved by high-speed circuits, multiplexing, and simultaneous-operation technique of instruction and data-fetching, as well as overlap within the execution units. This massive overlap and multiplexing results in complicated recovery routines between the look-ahead and instruction units. These units are described in detail, as are the arithmetic units and significant algorithms used in the floating-point arithmetic.

A flexible set of circuits using a current-switching technique with overriding-level facility is described, as well as the packaging of circuits on printed cards. The frame and gate concept is also shown. Performance figures and hardware count illustrate the size, complexity, and performance of the system.

Introduction

The Stretch computer [Dunwell, 1956] project was started in order to achieve two orders of magnitude of improvement in performance over the then existing 704. Although this computer, like the 704, is aimed at scientific problems such as reactor design, hydro-dynamics problems, partial differential equation etc., its instruction set and organization are such that it can handle with ease data-processing problems normally associated with commercial applications, such as processing of alphanumeric fields, sorting, and decimal arithmetic.

In order to achieve the stated goal of performance, all factors that go into the computer design must contribute towards the performance goal; this includes the instruction set [Buchholz, 1958], the internal system organization, the data and instruction word length, and auxiliary features such as status-monitoring devices, the circuits, packaging, and component technology. No one of them by itself can give this hundred-fold increase in speed; only by the combining and interacting of these contributing factors can this performance be obtained.

[1] *Proc. EJCC*, pp. 48–59, 1959.

This paper reviews the engineering design of the Stretch System with primary concentration on the central computer as the main contributor to performance. In it, these new techniques, devices, and instructions have been pushed to the limit set by the present technology and, therefore, its analysis will convey best the problems encountered and the solutions employed.

The Stretch system

Early in the system design, it appeared evident that a six-fold improvement in memory performance and a ten-fold improvement in basic circuit speed over the 704 was the best one could achieve. To meet the proposed performance criteria, the system had to be organized in such a way that it took advantage of every possible overlap of systems function, multiplexing of the major portion of the system, processing of operations simultaneously, and anticipation of occurrences, wherever possible. The system had to be capable of making assumptions based on the probability that certain events might occur, and means had to be provided to retrace the steps when the assumption proved to be wrong.

This simultaneity and multiplexing of operations reflects itself in the Stretch System at all levels, from overall systems organization to the cycle of specific instructions. In the following description, this will be discussed in more detail.

If one considers the Stretch System (Fig. 1) from an overall point of view it becomes apparent that the major parts of the system can operate simultaneously:

a The 2-μsec, 16,384-word core memories are self-contained, with their own clocks, addressing circuits, data registers and checking circuits. The memories themselves are interleaved so that the first two memories have their addresses distributed *modulo* 2 and the other four are interleaved *modulo* 4. The *modulo*-2-interleaved memories are used primarily for instruction storage; since, for high-performance instructions, halfword formats are used, the average rate of obtaining instructions is one per $\frac{1}{2}$ μsec. Similarly, a 0.5-μsec

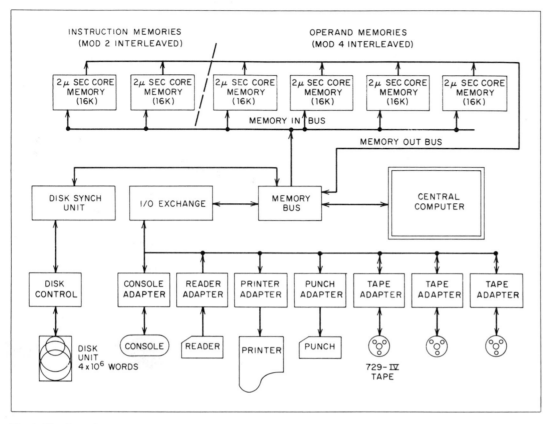

Fig. 1. The Stretch system.

data-word rate is achieved by the use of four *modulo*-4 organized memories. The addressing of the memories and the transfer of information from and to the memories by a memory bus permits new addresses, information, or both to pass through the bus every 200 mμsec.

b   The simultaneously-operating Input/Output units are linked with the memories and the computer through the Exchange, which, after initial instruction by the computer, coordinates the starting of the I/O equipment, the checking and error-correction of the information, the arrangement of the information into memory words, and the fetching and storing of the information from and to memory. All these functions are executed without the use of the computer, so it can in the meantime continue its data processing and computation.

c   The central computer processes and executes the stored program. Here, now, the simultaneity and multiplexing of functions has reached its ultimate.

Before discussing the computer organization, a few general features must be mentioned for completeness:

a   Word length: 64 bits plus eight bits for parity checks and error-correction codes.

b   Memory capacity and addressing: A possible 256,000 words can be randomly addressed. These storage positions are all in external memory, except for the 32 first addresses. These positions consist of the internal registers (accumulators, time clocks, index registers).

c   The instructions are single-address instructions with the exception of a number of special codes that imply the second address explicitly.

The instruction set (Fig. 2) is generalized and contains a full set for single- and double-precision floating-point arithmetic, and a full set for variable-field-length integer arithmetic (binary and decimal). It also has a generalized set for index modification and a branching set, as well as a set of

I/O instructions. All told, 765 different types of instructions are used in the system.

d   The instruction format (Fig. 3) makes use of both half and full words; half words accommodate indexing and floating-point instructions (for optimum performance these two sets of instructions use a rigid format), and full-word formats are used by the variable-field-length instructions. Notice that the latter specifies the operand field by the address of its left-most bit, the length of the field, and the byte[1] size, as well as the starting point (offset) of the implied operand

[1]*Byte:* a generic term to denote the number of bits to be operated on as a unit by a variable-field-length instruction.

(accumulator). Both halves of the word are independently indexable.

e   A general monitoring device used for important status triggers is called the Interrupt [Brooks, 1957] System. This system monitors the flip-flops which reflect internal mal-functions, result significance (exponent range, mantissa zero, overflow, underflow), program errors (illegal instruction, protected memory area), and input/output conditions (unit not ready, etc.). The status of these flip-flops can cause a break in the normal progression of the stored program for fix-up purposes. Their status is automatically interrogated at all times.

COMPUTER VOCABULARY

INSTRUCTION CATEGORY	CLASS	MODIFIER	EXAMPLES	NUMBER OF INSTR
VARIABLE FIELD LENGTH ARITHMETIC	BINARY DECIMAL	SIGNED UNSIGNED SAME SIGN NEGATIVE SIGN	ADD (TO MEMORY) LOAD/STORE MPY DIVIDE CUMULATIVE MPY	280
RADIX CONVERSION	BIN/DEC			32
LOGIC CONNECTS			16 LOGIC STATEMENT	48
FLOATING POINT ARITHMETIC	NORMALIZED UNNORMALIZED	SAME SIGN OPPOSITE SIGN NEGATIVE SIGN NOISY MODE	ADD (SINGLE & DOUBLE) LOAD/STORE MPY/(SINGLE & DOUBLE) DIV (WITH REMAINDER) INTERCHANGE DIVIDE CUMULATIVE MPY SQUARE ROOT	240
INDEXING ARITHMETIC	DIRECT IMMEDIATE PROGRESSIVE			43
BRANCHES	UNCONDITIONAL INDEXING INDICATOR BIT	IF {1 {0 SET 0 LEAVE BIT INVERT BIT		
STORE INST CTR				68
TRANSMIT/SWAP I/O INSTRUCTION				24
			TOTAL	735

Fig. 2. The instruction set.

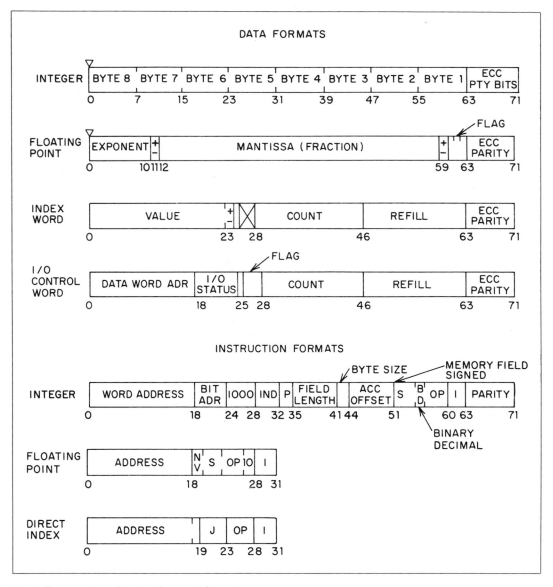

Fig. 3. Data word—and instruction word formats.

The Stretch computer

If one considers the internal organization of the majority of computers that have been produced during the last eight years (and the 704 is a case in point), the organization looks as shown in Fig. 4a. There is a sequential flow of instructions into the computer, and after due processing and execution, the next instruction is called from memory. Compare this with Fig. 4b, showing the organization of Stretch, where two instruction words and four operands can be fetched simultaneously. In addition, the execution of the instruction is done in parallel and simultaneously with the described fetching functions.

All the units of the computer are loosely coupled together, each one controlled by its own clock system, which in turn is synchronized by a master oscillator. This multiplexing of the units of the computer results in a large number of registers and adders, since time-sharing of the major computer organs is no longer possible. All in all, the computer has 3,000 register positions and about 450 adder positions.

Despite the multiplexing and simultaneous operation of successive instructions, the result appears as if sequential step-by-step internal operation were utilized. This has made the design of the interlocks quite complex.

Data flow

The data flow through the computer is shown in Fig. 5 and is comparable to a pipeline which in a steady state (namely, once filled) has a large output rate no matter what its length. The same is true here; after start-up the execution of the instructions is fast and bears no relation at all to the stages it must progress through.

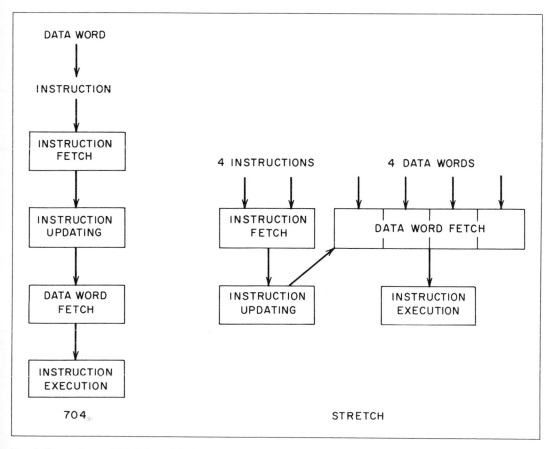

Fig. 4. Comparison of Stretch and 704 organization.

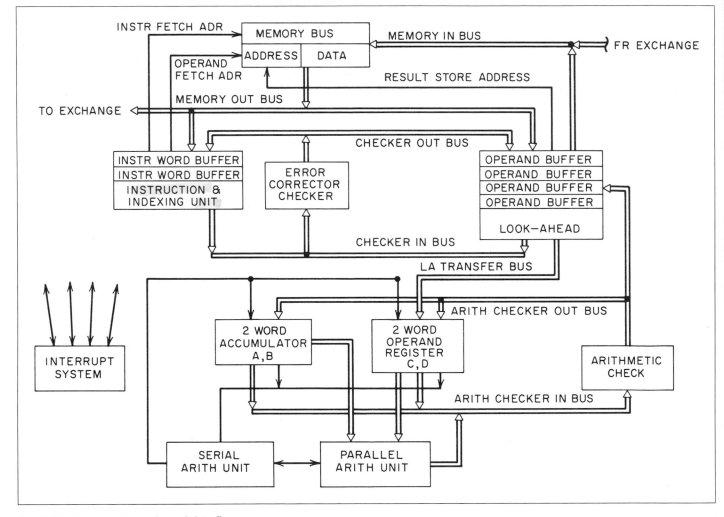

Fig. 5. Stretch computer—units and data flow.

The *Memory Bus* is the communication link between the memories on one side and the exchanges and the computer on the other. It monitors the requests for storage to, or fetches from, memory, and sets up a priority scheme. Since I/O units cannot hold up their requests, the exchange will get highest priority, followed by the computer. In the computer the instruction-fetch mechanism has priority over the operand-fetch mechanism. All told, the memory bus gets requests from and assigns priority to eight different channels.

Since memory can be accessed from multiple sources, and once accessed it is on its own to complete its cycle, a busy condition can exist. Here again, the memory bus tests for busy conditions and delays the requesting unit until memory is ready to be inter-

rogated on data fetches. The return address is remembered and the requesting unit receives the information when it becomes available. To accomplish this, from the time information is requested the receiving data register is in a reserved status.

Requests for stores and fetches can be processed at a 200 mμsec rate and the time, if no busy or priority conditions exist, to return the word to the requesting unit is 1.6 μsec, a direct function of the memory read-out time.

The *Instruction Unit* [Blaauw, 1959] is a computer of its own. It has its own instruction set, its own small memory for index word storage, and its own arithmetic unit. During its operation as many as six instructions can be at various stages of execution.

The Instruction Unit fetches the instruction words from memory, it steps the instruction counter, and performs the indexing of instructions and the initiation of data fetches. After a preliminary decoding of the class of instruction, it recognizes its own instructions and executes indexing instructions. On branches, conditional or unconditional, the instruction unit executes these. In the case of conditional branches, it makes the assumption that the branch will not be successful.

This assumption and the availability of two full-word buffer registers keep the flow of instruction to the computer continuous. Therefore, the rate of instructions entering the instruction unit is for all practical purposes independent of the memory cycle.

Since, for high speed instructions, half-word formats are used, four of these at any one time can be in buffer storage. As soon

as the instruction unit starts processing an instruction, it is removed from the buffer, thus making room for the next memory-word access (Fig. 6). Incidentally, half-word instructions and full-word instructions can be intermixed within the same word, and therefore the latter can cross a word boundary. This permits maximum packing of instructions in memory and also serves as a facility for automatic program assemblers and compilers.

The adder path, index registers, and transfer bus to look-ahead complete the instruction unit system (Fig. 6). It should be noted that the index registers are part of the instruction-unit data path, therefore permitting fast access (no long transmission lines) to an index word. There are 16 index words available to the programmer. The index registers, consisting of multi-aperture cores, are oper-

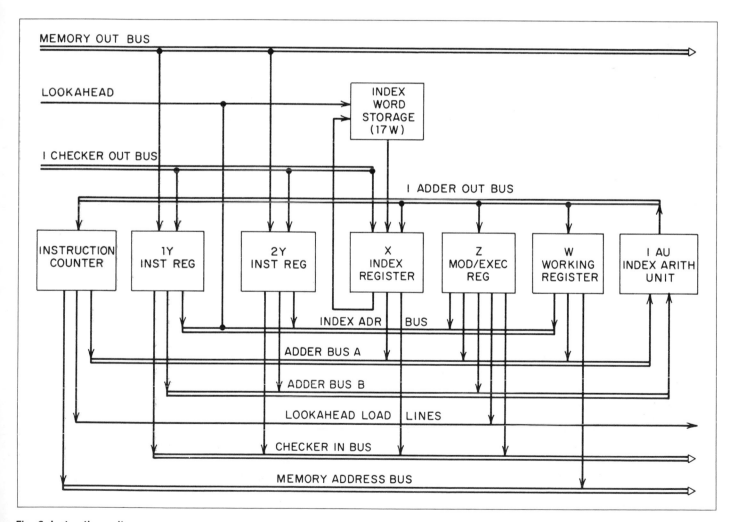

Fig. 6. Instruction unit.

ated in a non-destructive fashion, since in a representative program, the index word is used nine out of ten times without modifying it. This permits fast operation under these conditions, and additional time is only applied where modification is involved.

After processing through the instruction unit, the updated (indexed) instruction enters a level of the *Look-ahead* (Fig. 5). Besides the instruction, all necessary information, its associated instruction counter value, and certain tag information are also stored in the same level. The operand, already requested by the instruction unit, will enter this level directly and will be checked and error-corrected while awaiting transfer to the arithmetic units for execution.

An interlocked counter mechanism in the look-ahead keeps its four levels in step, preventing out-of-sequence execution of instructions, even if all information for a succeeding one is available, before the previous instruction has been started.

The pre-accessing of operands by the look-ahead and of instructions by the instruction unit leads sometimes to embarrassing positions, for which a fix-up routine must be provided. Consider the program

(n)	STORE Accumulator m
(n + 1)	LOAD R
(n + 2)	ADD m

and assume instruction (n) is in look-ahead, waiting for execution. If (n + 2) now enters the look-ahead, a reference to m cannot be made, since the data stored in that position is subject to change by the STORE instruction. The look-ahead must recognize this and "forward" the result of instruction (n), when received, to the level where (n + 2) is stored.

Another example is the case where the instruction unit assumed that a conditional branch would not be executed. This instruction is stored in look-ahead and, when it is recognized that the branch was successful, all modifications of addressable registers made by the instruction unit in the meantime must be restored. Look-ahead in this case acts as a recovery memory for this information. A similar condition exists when interrupts occur due to arithmetic results. The look-ahead here again has the data stored pertaining to registers which were modified erroneously in the meantime. The restoring and recovery routines described break into the instruction unit processing, interrupting temporarily the flow of instruction and their indexing.

The arithmetic units described later are slaves to the look-ahead, receiving not only operands and instruction codes but also the start-execution signal. Conversely, the arithmetic units signal to the look-ahead the termination of an operation and, in the case of "To Memory" operations, place into the look-ahead the result word for transfer to the proper memory position.

Arithmetic units

The design of the arithmetic units was established along lines similar to the design of look-ahead and the instruction unit. Every attempt was made to speed up the execution of arithmetic operations by multiplexing techniques and overlapping of the algorithm, where mathematically permissible.

The arithmetic units, consisting of the Serial Unit and the Parallel Unit, use the same arithmetic registers, namely a double-length accumulator (A,B) consisting of 128 bits and a double-length operand register (C,D) consisting of 128 bits. The reason for the use of the same arithmetic registers is the fact that at any time, a shift from floating-point to variable-field-length operation (or *vice versa*) can be made by the program. Therefore, the result obtained by a floating-point operation can serve as the starting operand for a variable-field-length operation. The chief reason for the double-length registers is the definition of maximum field length to be 64 bits. The field can start with any bit position, and therefore can cross the word boundary.

The executions of floating-point mantissa operations and variable-field-length binary multiply and divide operations are performed by the parallel unit, whereas the floating-point exponent operation and the variable-field-length binary and decimal add-type operations are executed by the serial unit. The square-root operation and the binary-to-decimal conversion algorithm are executed in unison by both units. Salient features of the two units will now be described.

The serial arithmetic unit [Brooks et al., 1959] (Fig. 7). The serial arithmetic consists of a switch matrix which can extract 16 consecutive bits from A,B and C,D. These 16 bits then can be aligned in such a way that the low-order bit of a field as specified by the instruction is at the right end of the field. This wrap-around circuit then feeds into a carry-propagate adder or, in case of logical-connect instructions, into the logic unit. At the adder output, a true complement unit and a binary-to-decimal correction unit are used for subtract and decimal operations. The inverse process of extracting is used to insert the processed byte back into the register without disturbing any neighboring positions. Notice that in one clock cycle, the information is extracted, the arithmetic is performed and the result inserted back into the registers. In addition, the arithmetic information is checked by parity checks on the switch matrices and by duplication and comparison of the arithmetic procedure in a duplicate unit.

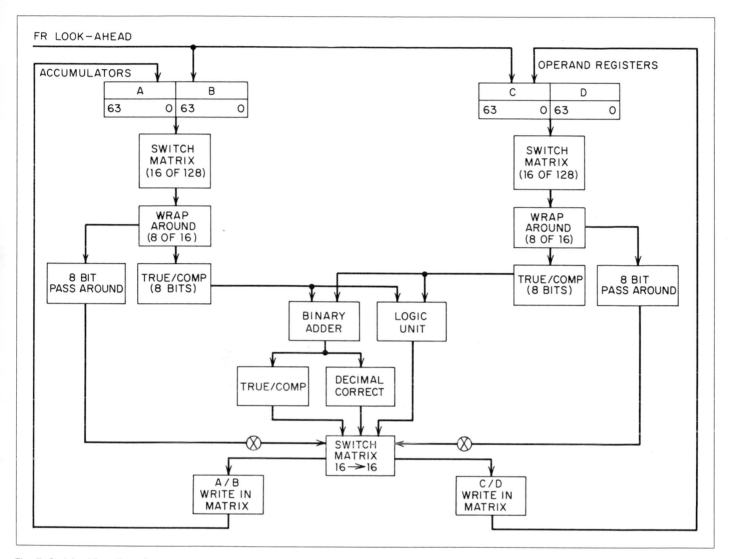

Fig. 7. Serial arithmetic unit.

Parallel arithmetic unit. The parallel arithmetic unit (Fig. 8) is designed to execute floating-point operations with a maximum of efficiency. Since both single- and double-precision arithmetic is performed, the shifter and adder exist in a double-length format of 96 bits. This insures almost the same performance for single- and double-precision arithmetic. The adder is of a carry-propagation type with look-ahead over 4 bits at a time to reduce the delay that normally results in a ripple-carry adder. This carry look-ahead results in a delay time of 150 mμsec for 96-bit binary-number additions. All additions and subtractions are made in one's complement form with automatic end-around carry.

The shifter is capable of shifting up to 4 positions to the right and up to 6 positions to the left. This shifter arrangement takes care of the majority of shifting operations encountered under normal operation. Where higher-order shifts are required, a successive operation is set up between the parallel unit register and the shifter.

To expedite the execution of the multiply instruction, 12 bits

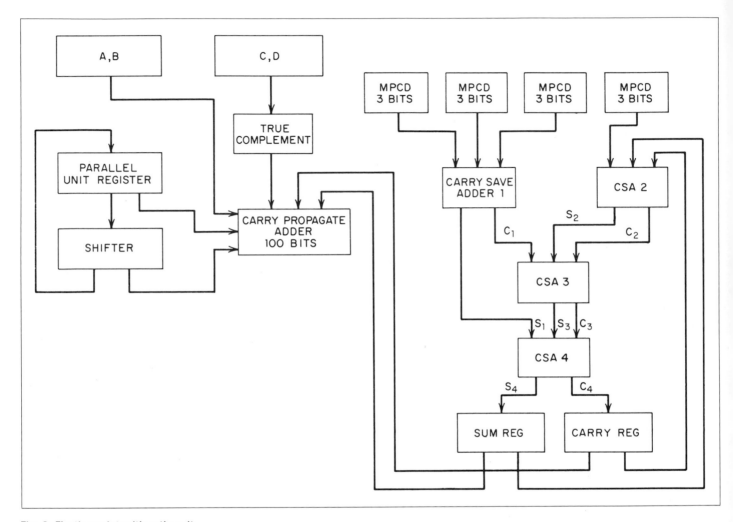

Fig. 8. Floating-point arithmetic unit.

of the multiplier are handled within one cycle. This is accomplished by breaking the 12 bits into groups of three bits each. The action is from right to left and consists of decoding each group of three bits. By observing the lowest-order bit of the next higher group, a decision is made as to what multiple of the multiplicand one must add to the partial product. Since only even multiples of the multiplicand are available, subtraction and addition of the multiples can result. The following example will elaborate this point: (MCD means multiplicand)

		Groups		
$n + 4$	$n + 3$	$n + 2$	$n + 1$	n

Multiplier, 12 bit group

xx0	011	110	101	010

Octal value

3	6	5	2

If two additions of multiples were permitted

$4 \times MCD$	$6 \times MCD$	$6 \times MCD$	$2 \times MCD$
$-1 \times MCD$		$-1 \times MCD$	

Instead of subtracting $1 \times MCD$ in $n + 1$, subtract $8 \times MCD$ in n.

$4 \times MCD$	$6 \times MCD$	$6 \times MCD$	$2 \times MCD$
	$-8 \times MCD$		$-8 \times MCD$

Resulting decoding

$4 \times MCD$	$-2 \times MCD$	$6 \times MCD$	$-6 \times MCD$

The four multiple multiplicand groups and the partial product of the previous cycle are now fed into carry-save adders of the form,

Sum $\quad S = A \mathbin{\not\vee} B \mathbin{\not\vee} C$

Carry $C' = AB + AC + BC$

There are four of these adders, two in parallel followed by two more in series (Fig. 8). The output of Carry-Save Adder 4 then results in a double-rank partial product, the product sum and the product carry. For each cycle this is fed into Carry-Save Adder 2, and, during the last cycle, into the carry-propagate adder, for accumulation of the carries. Since no propagation of carries is required in the four cycles, where multiple multiplicands are added, this operation is fast and is the main contributor to the fast multiply-time of Stretch.

The divide scheme [Robertson, 1958] has a similarity to the multiply scheme. Multiples of the divisor are used, namely, $3/2 \times$ divisor, $3/4 \times$ divisor and $1 \times$ divisor. This, plus shifting over strings of ones and zeros, results in the generation of the required 48 quotient bits within thirteen machine cycles. Most machines using a nonrestoring divide method require 48 cycles for 48 quotient bits. The following example explains this technique. This scheme depends on the use of normalized divisors:

DIVIDEND $\quad$ (DD) $= 101000000000000$

DIVISOR $\quad$ (DR) $= 1100011$

2's COMP DR $(\overline{\text{DR}}) = 0011101$

3/4 DR $\quad\quad\quad = 100101001$

(a) *Using skip over 1/0 only:*

Step 1:
$$\begin{array}{ll} 101000000000000 & \text{DIVIDEND} \\ \underline{0011101} & \text{ADD } \overline{\text{DR}} \\ 1101101 \end{array}$$

Remainder negative, 1st quotient bit $= 0$; shift one position. Leading 1 indicates that next quotient bit must be 1; $Q_1Q_2 = 01$

Step 2:
$$\begin{array}{ll} 011010000 & \text{REMAINDER} \\ \underline{1100011} & \text{ADD DR} \\ 10010111 \end{array}$$

Overflow: Remainder positive and $Q_3 = 1$, leading zero indicates $Q_4 = 0$

Step 3:
$$\begin{array}{ll} 1011100 & \text{REMAINDER} \\ \underline{0011101} & \text{ADD DR} \\ 1111001 \end{array}$$

Negative remainder; $Q_5 = 0$; leading 1's indicate $Q_6Q_7Q_8 = 111$

Number of quotient bits per cycle:

Cycle 1: $\quad 01 = 2$

Cycle 2: $\quad 10 = 2$

Cycle 3: $\quad 0111 = 4$

(b) *The same problem with both skip over 1/0 and 3/4 - 3/2 complement:*

Step 1:
$$\begin{array}{l} 101000000000000 \\ \underline{0011101} \\ 11011010000 \end{array}$$

Same as before, $Q_1Q_2 = 01$

Step 2:
$$\begin{array}{ll} \underline{100101001} & \text{Add 3/4 DR} \\ 111111001 \end{array}$$

This (by table look-up) indicates $Q_3Q_4Q_5Q_6Q_7Q_8 = 100111$

Quotient bits generated per cycle:

Cycle 1: $\quad 01 = 2$

Cycle 2: $\quad 100111 = 6$

In general, this method results in the generation of 3.7 quotient bits per subtraction. While the mantissa operations of multiply and divide are performed by the parallel unit, the serial arithmetic unit executes the exponent arithmetic. Here again is a case where overlap and simultaneity of operation is used to special advantage.

Checking. The operation of the computer is checked in its entirety and correction codes are employed where data transfers from memory and input-output units are involved. In particular, all information sent to memory has a correction code associated with it, which is checked for accuracy on its way from memory. If a single error is indicated, then correction is made and the error is recorded via a maintenance output device. Within the machine, all arithmetic operations are checked, either by parity, duplication, or a "casting out three" process. These checks are overlapped with the execution of the next instruction.

Hardware count. Figure 9 shows the percentage of transistors used in the various sections of the machine. It becomes obvious that the parallel unit and the instruction unit use the highest percentage of transistors. In case of the parallel unit this is due to the extensive circuits for multiply and to the additional hardware to achieve speed up of the divide scheme. In the instruction unit, the controls consume the majority of the transistors, because of the high multiplexed operation encountered.

Performance. The performance comparisons in Fig. 10 show the increase in speed achieved, especially in floating-point operations,

UNIT	NO. OF TRANSISTORS	% OF TOTAL	NO. OF FRAMES
MEMORY CONTROLS	10,500	6.0	2
INSTRUCTION UNIT			
DATA PATH	17,700	22.0	2
CONTROLS	19,500		3–1/2
LOOK–AHEAD			
DATA PATH	17,900	15.6	1
CONTROLS	8,600		1–1/2
ARITH REGISTERS	10,000	5.9	1
SERIAL ARITH UNIT			
DATA PATH	10,000	10.5	1–1/2
CONTROLS	8,700		1
FLOATING PT UNIT			
DATA PATH	32,700	21.0	2–1/2
CONTROLS	3,000		1/2
CHECKING	24,500	14.5	1
INTERRUPT SYSTEM	6,000	3.5	1/2
TOTAL	169,100	100.0	18

DOUBLE CARDS   4,025
SINGLE CARDS   18,747
POWER   21 KW

Fig. 9. Component count.

over the 704. It should be noted that for a large number of problems this particular increase in all arithmetic speeds is almost proportional to the performance increase of the problem as a whole, since the instruction execution-times are overlapped to a great extent with the preparation and fetching of instructions.

Simulation of Stretch programs on the 704 proved a performance of 100 × 704 speed in mesh-type calculations. Higher performance figures are achieved where double- or triple-precision calculations are required.

Circuits

Having reviewed the systems organization of Stretch, it is now of interest to discuss briefly the components, circuits, and packaging techniques used to implement the design.

The basic component used in Stretch is the high-speed drift transistor which exists in both an NPN and a PNP version. This transistor has a frequency cut-off of approximately 100 mc and for high-speed operation must be kept out of saturation at all times. This then explains why both the PNP and NPN version are used: mainly to avoid the problem of level translation, which would be required due to the potential difference of the base and the collector. This difference is 6 volts, an optimum point for this device.

Figure 11 shows the basic circuit configuration. It consists of a current source, represented by the -30 volt supply and resistor R. The functional operation of the circuits consists of two possible

	OPERATION	IBM 704	IBM 705	STRETCH
	1. FLOATING POINT	± 128		± 2048
	EXPONENT RANGE	± 2		± 2
	MANTISSA BITS	27		48
	FLOATING ADD	84 μSEC		1.0 μSEC
	FLOATING MPY	204 μSEC		1.8 μSEC
	FLOATING DIV	216 μSEC		7.0 μSEC
	LOAD/STORE	24 μSEC		0.6 μSEC
	2. BINARY VARIABLE			
	FIELD LENGTH ARITH			
	BIT RANGE			1 TO 64
16	ADD/LOAD/STORE			2.0 μSEC
BIT	MPY			10.0 μSEC
FIELD	DIVIDE			15.0 μSEC
	3. DECIMAL			
	ARITHMETIC			
	DIGIT RANGE		1 $\rightarrow$ MEM CAPACITY	1 TO 21
FOR	ADD		119 μSEC	3.5 μSEC
5	MPY		799 μSEC	40.0 μSEC
DIGITS	DIVIDE		4828 μSEC	65.0 μSEC
	LOAD/STORE		204 μSEC	3.2 μSEC
	4. MISCELLANEOUS			
	ERROR CORRECTION	NO	NO	YES
	CHECKING	NO	YES	YES
	WORD SIZE	36 BITS		64 BITS

Fig. 10. Comparison of Stretch and 705/704 operation times.

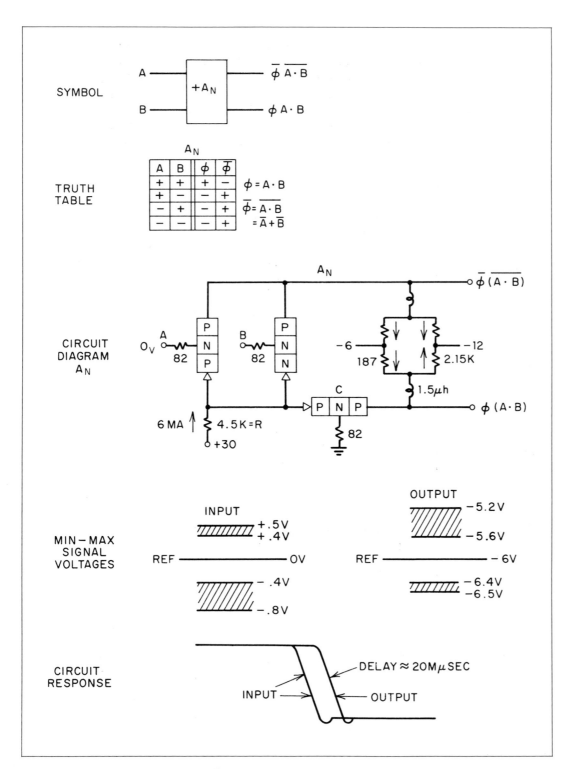

Fig. 11. Current switching circuits (+AND).

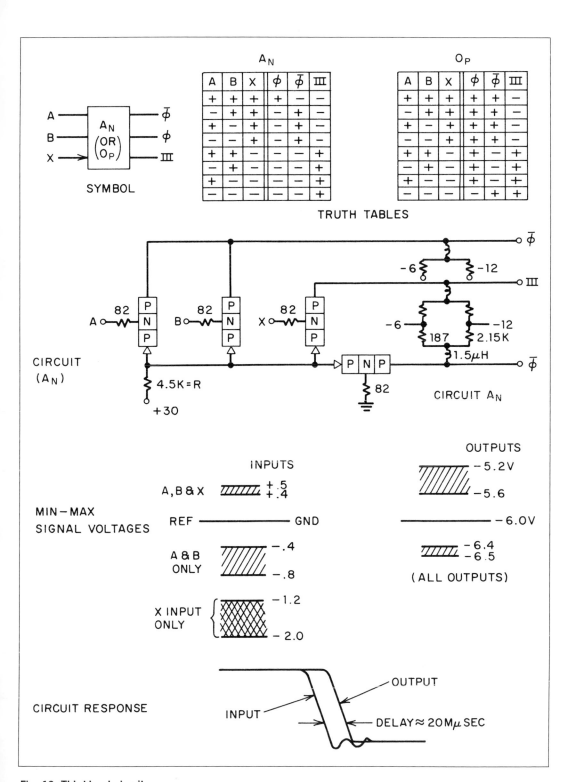

Fig. 12. Third-level circuit.

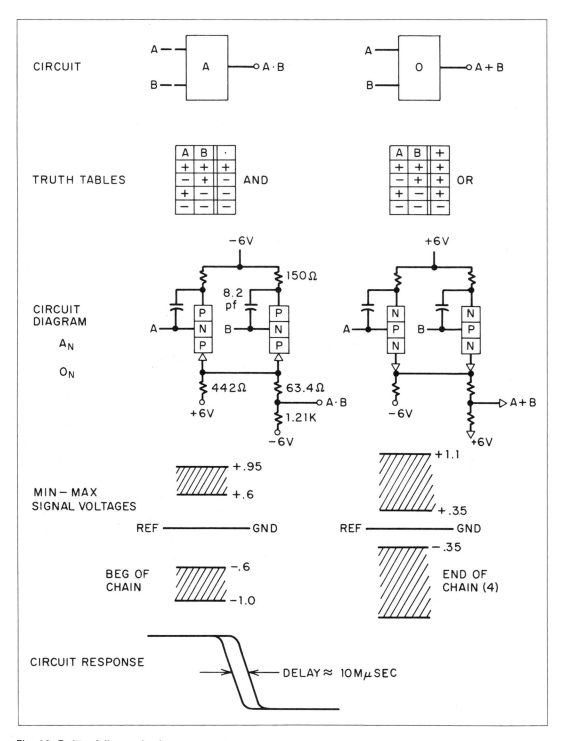

Fig. 13. Emitter-follower circuit.

paths represented by transistor A or C. Which path is chosen by the current depends on the condition existing on base A. If point A is positive with respect to ground by 0.4 volts, that particular transistor is cut off, making the emitter of transistor C positive with respect to the base and, therefore, making C conducting. The current supplied by the current source (6 ma) will then flow through transistor C to the load ϕ. Output ϕ, then, is positive by 0.4 volts with respect to the −6 volt reference. This indicates at ϕ the equivalent function impressed on A. At the same time, $\bar{\phi}$ is negative with respect to the −6 volt power supply by 0.4 volt, representing, therefore, the inverse of the function impressed on A. Conversely if A is negative with respect to the ground reference, transistor A is the conducting one, keeping emitter C negative with respect to its base. The current flows through transistor A, making $\bar{\phi}$ positive with respect to −6 and ϕ negative with respect to −6. Again, the output of ϕ reflects the function impressed on A, whereas $\bar{\phi}$ represents the inverse of the function.

If an additional transistor now is paralleled with A, it becomes obvious that only if both bases A and B are positive will output

ϕ be positive and $\bar{\phi}$ negative. If any or none of the bases A and B are positive, then ϕ will be negative and $\bar{\phi}$ will be positive. In other words, an AND function is obtained on output ϕ.

This principle, which is reflected in all the circuits, is essentially the principle of current switching or current steering.

Logical functions for the PNP circuits are, therefore, a +AND or −OR. Two outputs from each circuit block are available: the AND function and the inverse of the AND function.

A dual circuit exists for NPN transistors with input levels at −6 volts and output levels at ground. This circuit will give the +OR or −AND function.

A thorough investigation of the systems design showed that the circuits described so far are versatile enough to be used throughout the system. However, there are enough special cases (resulting from the many data buses and registers throughout the machine) that could use a distributor function or an overriding function. This caused the design of a circuit which permitted great savings in space and transistors by adding a third voltage level. Figure 12 shows the PNP version of the third-level circuit.

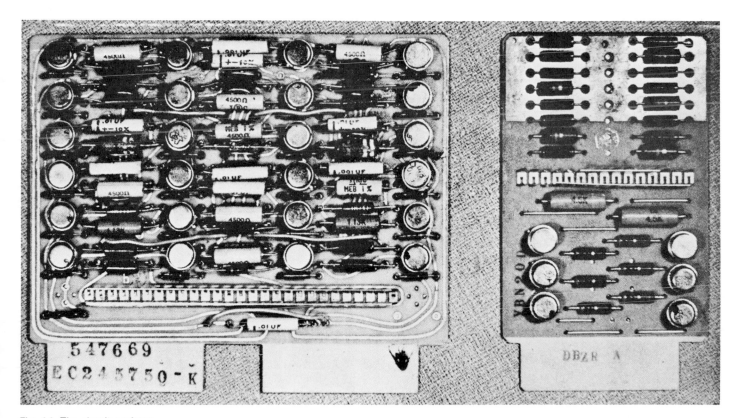

Fig. 14. The circuit package.

If transistor X were eliminated, then transistors A and B in conjunction with the reference transistor C would work normally as a current switching circuit, in this case a $+$AND circuit. If transistor X is added with the stipulation that the down level of X is more negative than the lowest possible level of A or B, it becomes apparent that when X is negative, the current will flow through that branch of the circuit in preference to branch ϕ or $\bar{\phi}$, regardless of inputs A and B. Therefore, the output of ϕ and $\bar{\phi}$ will be negative, provided input X is negative. Output III is the inverse of input X. If, however, X is positive, then the status of A and B will determine the function ϕ and $\bar{\phi}$ implicitly. This demonstrates the overriding function of input X.

Similarly, the NPN version (not shown) results in the OR function of ϕ if input X is negative and in a positive output at ϕ and $\bar{\phi}$, regardless of status A and B, if X is positive. Again minimum and maximum signal swings are shown in Fig. 12.

The speed of the circuits described so far depends on the number of inputs and the number of circuits driven from each load. The response of the circuit is anywhere between 12 and 25 mμsec per logical step with 18 to 20 mμsec average. The number of inputs allowable per circuit is eight. The number of driven circuits is three. Additional circuits are needed to drive more than three bases and where current switching circuits communicate over long lines, termination networks must be added to avoid reflections.

To improve the performance of the computer in certain critical places, emitter-follower logic is used as shown in Fig. 13. These circuits, having a gain less than one, after a number of stages require the use of current switching circuits as level setters and gain devices. Both AND and OR circuits are available for both a ground-level and a -6-level input. Change from a -6-level circuit to a ground-level circuit is obtained by applying the appropriate power supply levels. Due to the variations in inputs and driven loads, the circuits must be designed so that the load can vary over a wide range. This resulted in instability which had to be offset by the feedback capacitor C shown in the circuit.

All functions needed in the computer can be implemented by the use of the aforementioned circuits, including flip-flop operation, which is obtained by tying a PNP current switch block and an NPN current switch block together with proper feedback.

Packaging

The circuits described in the last paragraph are packaged in two ways:

A circuit package using the smaller of the two printed circuit boards shown in Fig. 14, called a single card, contains AND or OR circuits. It should be mentioned that the printed wiring is one-sided and that besides the components and transistors, a rail is added which permits the shorting or addition of certain loads depending on the use of the circuits. This rail then has the effect of reducing the different types of circuit boards in the machine. Twenty-four different boards are used and of these, two types reflect approximately 70% of the total single card population.

Due to the large number of registers, adders, and shifters used in the computer, it seems reasonable that functional packages could be employed economically, because of wide usage. This results in the high-density package also shown in Fig. 14, called

Fig. 15. The back panel.

a Double Card, which has 4 times the capacity of a single card and which has wiring on both sides of the board. Furthermore, components are double-stacked; and again, the rail is used to effect circuit variations due to different applications. Eighteen double card types are used in the system. Approximately 4,000 double cards are used, housing 60% of the transistors. The rest of the transistors are on approximately 18,000 single cards.

The cards, both single and double, are assembled in gates, and two gates are assembled into a frame. Figure 15 shows the gate back-panel wiring, using wire-wraps; and Figs. 16 and 17 the frame construction, both in a closed and open version.

To achieve high performance, special emphasis must be placed on keeping noise to a low level. This required the use of a plane

Fig. 17. The frame (extended).

Fig. 16. The frame (closed).

which overlies the whole back panel, against which the intercircuit wiring is laid. In addition, the power-supply distribution system must be of such a low impedance that extraneous noise cannot induce circuit malfunction. For this reason, a bus system, consisting of laminated copper sheets, is used to distribute the power to each row of card sockets. The wiring rules are such that single-conductor wire is used up to a maximum of 24″, twisted pair to a maximum of 36″, unterminated coax to a maximum of 60″, and terminated coax to a maximum of 100 feet. The whole back-panel construction and the application of single wire, twisted pair, or coax are calculated by a computer program to minimize the noise on each circuit node.

The two gates of a frame are a sliding pair with the power supply mounted on the sliding portion. All connecting wires between frames are coax and arrayed in layers which are formed into a drape.

References

BlaaG59; BrooF57a, 59; BuchW58; DunwS56; RobeJ58; BlosR60; BuchW57, 62; BrooF60; CockJ59; CoddE59, 62.

Chapter 35

PILOT, the NBS multicomputer system[1]

A. L. Leiner / W. A. Notz / J. L. Smith
A. Weinberger

Summary   PILOT, the new NBS system, possesses both powerful external control capabilities and versatile internal processing capabilities. It contains three independently operating computers. The primary and secondary computers each utilize only 16 basic types of instructions, thus providing a simple code structure; but because so many variations of the formats are possible, a wide variety of computing, data-processing, and information-retrieval operations can be performed with these instructions. The secondary computer is specially adapted for performing so-called "red-tape" operations, and both the secondary and the primary computers, acting co-operatively, can carry out special complex sorting or search operations. The third computer in the system, called the format controller, is specially adapted for performing editing, inspecting, and format modifying operations. The system is equipped to transfer information concurrently along several input-output trunks, though only two are planned for the near future. Using two such trunks, it is possible to maintain two continuous streams of data simultaneously flowing between any two external units and the internal memory, without interrupting the data-processing program. The system can operate with a wide variety of input-output devices, both digital and analog, either proximate or remotely located. The external control capabilities of the system enable the machine to supervise this wide family of external devices and, on an unscheduled basis, to interrupt or redirect its overall program automatically, in order to assist or manage them.

At the National Bureau of Standards (NBS) a new large-scale digital system has been designed for carrying out a wide range of experimental investigations that are of special importance to the Government. The system can be utilized for investigating new or stringent applications of these general types: (1) data-processing applications, in which the system can be used for performing accounting and information-retrieval operations for management purposes; (2) mathematical applications, in which the system can be used for performing mathematical calculations for scientific purposes, including scientific data-reduction; (3) control applications, in which the system can be used for performing real-time control and simulation operations, in conjunction with analog computer facilities or in conjunction with other instrument installations, remotely located if necessary; and (4) network applications,

in which the system can be used in conjunction with other digital computer facilities, forming an interconnected communication network in which all the machines can work together collaboratively on large-scale problems that are beyond the reach of any single machine.

Because the system was designed for such varied uses (ranging from automatic search and interpretation of Patent Office records to real-time scheduling and control of commercial aircraft traffic), the system is characterized by a variety of features not ordinarily associated with a single installation, namely: a high computation rate, highly flexible control facilities for communicating with the outside world, and a wide repertoire of internal processing formats. The system contains three independently programmed computers, each of which is specially adapted for performing certain classes of operations that frequently occur in large-scale data-processing applications. These computers intercommunicate in a way that permits all three of them to work together concurrently on a common problem. The system thus provides a working model of an integrated multicomputer network.

System organization

Exclusive of data-storage and peripheral equipment, the central processing and control units of the over-all system contain approximately 7,000 vacuum tubes and 165,000 solid-state diodes. The basic component for these units is a modified version of the one megacycle package used in the NBS DYSEAC, which in turn was evolved from the hardware used in NBS Electronic Automatic Computer (SEAC). As a result of a more effective logical design and faster memory, however, the new NBS system will run more than 100 times faster than SEAC on programs involving only fixed-point operations; for programs involving floating-point manipulations, the advantage exceeds 1,000. The arithmetic speed of the new system derives in a large part from connecting a novel type of parallel adder to a diode-capacitor memory capable of providing one random access per microsecond.

The system contains seven major blocks, which are indicated in Fig. 1, namely: (1) the primary computer, in the lower center

[1]*Proc. EJCC*, 71–75 (1958).

Table 1 Arithmetic operation times

(including 4 random access times to last memory)

Operation	Total time (microseconds)	
	Average	Minimum-maximum
Fixed-point Addition, Subtraction, Comparison	7.5	6–9
Fixed-point Multiplication	31	22–40
Fixed-point Division	73	72–74
Floating-point Addition, Subtraction†	20	19–21
Floating-point Multiplication	37	28–46

† For shift of 4 bits.

of the figure, (2) the primary storage, upper center; (3) the secondary computer and the secondary storage, right; (4) the input-output control, upper left; (5) the external storage units, upper far left; (6) the external input-output units such as readers, printers, and displays, lower far left; and (7) lower left, the external control containing the special features that facilitate communication with people and devices in the world outside the system which is remotely located if necessary. Interchanges of information between the system and the outside world can take place at any time, on a completely impromptu basis, at the instigation of either the system or the external world, or both acting jointly.

The primary computer, a high-speed general-purpose computer, contains both an arithmetic unit and a program control unit of considerable versatility. This computer can carry out a variety of high precision arithmetic and logical processing operations, in either binary or decimal code and in a wide variety of word lengths and formats. Its partner computer, the secondary computer, specializes in short-word operations, usually manipulations on address numbers or other "red-tape" information, which it supplies automatically as needed to the primary program. The third computer of the system, called the format controller (see input-output control in Fig. 1), is specially designed for carrying out editing, inspecting, and format-modifying operations on data that are flowing in or out of the internal memory via the peripheral external units of the system. All three computers, and all the external units of the system, share access privileges to the common high-speed internal memory, which is linked to the input-output and external storage units via independent trunks for effecting data-transfers. Transfers of data can take place between the external units, the memory units, and the computers concurrently without interrupting the progress of the computational program. Because of the flexibility of the format controller, incoming data can be accepted

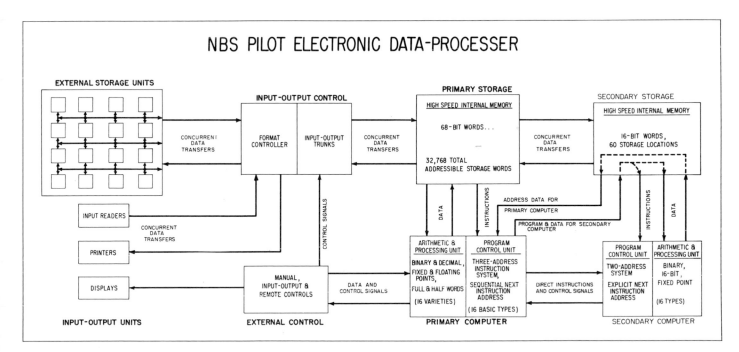

Fig. 1. Over-all block diagram for PILOT.

from a wide variety of external devices and in a wide variety of formats.

Functions of the major units

The specific functions of the major units can be described briefly as follows:

Primary computer

Arithmetic and processing unit. Using a 64-bit number word with algebraic sign, this unit carries out 7 different types of arithmetical operations, 5 types of choice (branch) operations, and 2 types of logical pattern-processing operations. See Table 2. Arithmetical operations can be performed in any of 16 possible formats. For example, arithmetic can be performed using either a pure binary or a binary-coded decimal number code, and in both fixed-point and floating-point notation. Fixed-point operations can also be carried out in a special half-word format in which two independently addressable half-words are stored in a single full-word storage location. These two half-words can be processed either separately, as independent words, or concurrently in duplex format. In duplex format, the respective lefthand and righthand halves of each double operand are processed simultaneously in a single instruction time, and the two independent half-word results are written back in the corresponding halves of the full-length result location.

Program control unit. The program control unit interprets and regulates the sequencing of instructions in the program. It operates with a 68-bit binary-coded 3-address instruction word. See Table 3. Each instruction word contains three 16-bit codes which specify the addresses of each of two operands, alpha and beta, and usually the address of the result of the operation, gamma, in the main memory. The memory location of the next instruction word is specified by a 16-bit address number contained in one of 16 possible base registers; a 4-bit code in the instruction word (*d*-digits) specifies which one of the base registers contains the desired word. Whenever a register is so used as a next-instruction address source, its contents are automatically increased by unity. Choice instructions, used for program branching, from time to time may cause a new alternative address number to be inserted in any one of the base registers. This register is then used as the source of the address number of the next instruction.

Table 2   Types of internal operations

Primary computer	
Name	Abbreviation
Arithmetic operations:	
Add	AD
Augment	AG
Subtract	SB
Multiply	MP
Divide	DV
Square-root	SQ
Shift	SH
Nonnumerical processing operations:	
Transplant Segment with Shift	TL
Generate Boolean Functions	GB
Choice operations:	
Compare, Algebraic	CA
Compare, Modulus	CM
Compare, Equality	CE
Check Scale	CS
Compare Boolean Functions	CB
Control operations:	
Transfer Between Storage Units	TS
Regulate Secondary Computer	RS

Secondary computer	
Name	Abbreviation
Clear add	ca
Hold add	ha
Store positive	sp
Transfer	tr
Increase	in
Decrease	de
Logical Multiply	lm
Compare, Zero	cz
Compare, Righthand Bit	cr
Compare, Lefthand Bit	cl
Compare, Negative	cn
Check Primary and Proceed	cp
Check Primary and Wait	cw
Regulate Primary Computer	rp
Replace Primary Instruction	ri
Secondary Take Input from Primary	si

Leiner, Notz, Smith, Weinberger—PILOT

Table 3 Contents of primary instruction word

						Digits numbered 1 through 68										
68–65	64–61	60–57	56–53	52–49	48–45	44–41	40–37	36–33	32–29	28–25	24–21	20–17	16–13	12–9	8–5	4–1
Tags	Address alpha				Address beta				Address gamma				Next Instn.	Code for Operation		Mon. Break Point
000±	a-Digits				b-Digits				c-Digits				d-Digits	Param-eter	Basic Type	e-Digits

Addresses alpha, beta, and gamma written in the instruction word are subject to automatic modification if desired by writing a 1-digit in a specified bit position. Such addresses are called relative addresses. Each of the three addresses (α, β, and γ) in each instruction word contains a 4-bit code group, called the *a-*, *b-*, and *c*-digits respectively, in which any base register identification number (0 through 15) may be written. When this is done, the address number to which the computer actually refers is equal to the sum (modulo 2^{16}) of the address number stored in the designated base register plus an address-modification constant, indicated in the remaining 12 bits of the 16-bit address segment of the instruction word.

Primary storage units

Fast access memory. Because of budget limitations, the initial installation of the system will contain only a relatively small section of internal memory of the diode-capacitor type. This diode-capacitor memory, originally developed at NBS in 1953, is very fast; i.e., capable of providing one random access per microsecond, but it has the disadvantage of relatively high cost per word of storage. This type of memory is available in modules of 256 words subdivided as follows:

Numerical information	64 bits
Algebraic signs and tags	4 bits
Parity check digits	4 bits
Total word length	72 bits

The over-all system is designed to accommodate up to 32,768 internally-accessible full-words, which may be held in storage units with access times ranging from 1 microsecond (μsec) to 32 μsec. Thus the minimum fast access memory can be backed up with a much larger and slower magnetic-core memory.

Inter-memory transfer trunk. Provision is made for transferring blocks of information between the various internal storage units in the system, concurrently with computation. The size of the block transferred may range from a single word to the entire contents of the memory, and the addresses between which the information is transferred are specified by a single programmed inter-memory transfer instruction. Automatic interlocks are provided to insure that all future references which the program may make to any memory positions involved in the inter-memory transfer operation are automatically made after the data have been shifted to the new locations.

Secondary computer

Arithmetic and processing unit. The secondary computer is a high-speed independently programmable general-purpose computer that operates in conjunction with the primary computer and can perform 16 distinct types of operations using 16-bit words. These operations include 6 arithmetic-processing operations, 4 choice operations, 1 nonnumerical processing operation, and 5 operations that transfer digital information or control-signals between the primary and the secondary computers. See Table 2. Operation times for the secondary computer average about 2 μsec.

Both computers operate concurrently and can transfer information back and forth between each other. One of the principal functions of the secondary computer is to carry out so-called "red-tape" operations, such as: (1) counting iterations, (2) systematically modifying the addresses of the operands and instructions referred to by the primary program, (3) monitoring the primary program, and (4) various special tasks. Through the use of special subroutines for the secondary computer, both computers acting co-operatively can be made to carry out a wide variety of complex operations without unduly complicating the writing of the primary computer programs. Examples of such operations are: (1) special types of sorting, (2) logarithmic search, (3) routines involving cross-referencing, or items selected according to an attached code, (4) error analyses, and (5) operations involving small numerical fields.

Secondary storage unit. Associated with the secondary computer is the secondary storage unit which consists of 60 storage locations containing 16-bit words. Sixteen of these locations can be used as base registers by the primary computer and may be selected by the primary computer according to the *a-*, *b-*, *c-*, and *d-*digits in the primary instruction word. The contents of the registers selected by the primary computer in this way are automatically added to the address numbers specified in the primary computer instruction word. The secondary storage unit is also capable of being addressed directly by the primary computer. The fifteen 4-word blocks of the secondary storage are identified by 15 special primary address numbers. Other addressable registers associated with the secondary storage hold the address numbers of current and next instruction words in the primary program.

Program control unit. The secondary computer program operates with a 2-address instruction system, the addresses referring to words in the secondary storage unit, including the base registers. See Table 4. From time to time the primary instruction program may order the insertion of a new instruction into the secondary instruction register or may order the transfer of data in either direction between the primary storage units and the secondary storage unit. The secondary computer program may also cause data to be transferred into the secondary storage unit from the primary instruction register and can also cause information to be transferred into the primary instruction register from a location in the main memory.

Using these facilities, the secondary computer can inspect each instruction word in the primary program as it is selected from the primary store and, acting upon specifications written into the secondary program, can cause the primary instruction either to be executed as written or to be replaced by a new instruction word from a memory location determined by the secondary. Other types of discrimination can be effected by the secondary that depend upon the result of a primary operation, such as an overflow, jump, etc. These features facilitate the use of interpretive programming methods.

Input-output control

Concurrent input-output trunks. The concurrent input-output trunks have the function of controlling the transfer of information in either direction between the internal memory and the external storage units. All input-output transfers are initiated by a single internally programmed instruction, and are carried out by the trunk units with the aid of automatic interlocks similar to those used in the inter-memory transfer trunk for preventing interference with the progress of the computing program. The size of the block of data that is transferred may range from a single word to the entire contents of the memory and may be directed to any addresses. Using two such trunks, it is possible to maintain two continuous streams of data simultaneously flowing between the internal memory and any two external storage units without interrupting the progress of the computations.

Format controller. Data that are passing in and out of the internal storage system via the input-output trunks are subject to further concurrent processing by the format controller. The format controller is an independent internally-programmed data-processing unit specially designed for carrying out general-purpose editing, inspecting, and format-modifying operations on incoming or outgoing data. Programs for the format controller are stored on removable plugboards, and the primary computer program is able to direct the format controller to select whichever particular format program may be appropriate from among the small library of format programs contained on the boards currently attached to the machine. Among the typical kinds of programs that the format controller can carry out are: (1) searching of magnetic tapes for words bearing identifying addresses or other coded labels specified by the internal program, with selective input or output of data at these selected tape locations, (2) insertion of incoming data for the internal storage units of the system into address locations specified by the incoming data itself, (3) conversion and rearrangement of data that are stored on external units in formats not compatible with the formats used in the internal units; e.g., binary-decimal character conversion, adjustment of word-length modules, etc.

External storage

External storage in the initial installation of the system will consist mainly of magnetic tape units. Because of the flexibility of the format controller, it will be possible to supplement these tape units later with a wide variety of other types of external units without making any significant changes in the existing equipment.

Table 4   Contents of secondary instruction word

Digits numbered 1 through 16		
16 13	12 7	6 1
Operation code (0–15)	Address "g"	Address "h"

Input-output units

The system is designed to operate with a wide variety of input-output devices, both digital and analog.

Input readers and printers. Flexowriter units and paper-tape readers and punches will be available in the initial installation. Punched card input readers and high-speed printers, along with their auxiliary controls, may be attached to the format controller in the manner indicated in the preceding paragraph.

Displays. Two types of displays are provided for: (1) pilot-light display of data and control information in the various registers and flip-flops throughout the system, in order to aid the rapid diagnosis of equipment malfunctions of programming faults, and (2) picture-tube display of real-time data stored in the internal memory of the system. This kinematic diagram type of display is very important when performing dynamic simulation operations which require visual presentation of the simulated data in real-time to the human operators.

External control

Manual-monitor control. The term "manual-monitor" was coined at NBS several years ago to describe certain types of control operations that are initiated either manually by the machine operator or by the machine itself under conditions which are specified by means of external switch settings. The former is referred to as a manual operation and the latter is called a monitor operation because the machine must monitor its internal program to determine precisely when the operation should be performed. The type of operation to be performed as well as the conditions under which it is to be performed are specified by means of external switch settings.

This feature provides for convenient communication between the data-processor and the operator, and allows the operator to monitor the progress of the program automatically, to insert new data and instructions, and to withdraw intermediate results conveniently, without need for advance preparation of special programs. This is particularly useful in debugging programs and in checking equipment malfunctions.

Monitor operations are performed by the machine whenever the conditions specified by the external switch settings occur in the course of the program; e.g., every time the program refers to a new instruction, any time the program refers to an instruction to which a special monitor breakpoint symbol (*e*-digits) is attached, any time an arithmetic overflow occurs, etc. By pairing a particular type of manual-monitor operation with a selected set of conditions, a variety of special composite operations can be performed.

Remote controls. Manual-monitor operations can be specified and initiated by external devices as well as by human operators. Since all of the external switch settings control only d-c voltages, the external devices can even be remote from the machine itself, and from a distance, via ordinary electrical transmission lines, they can exercise supervisory control over the internal program of the machine. This makes it possible to harness together two or more remotely located data-processing machines, and have them work together co-operatively on a common task. Each member of such an interconnected network of separate data processors is free at any time to initiate and dispatch special control orders to any of its partners in the system. As a consequence, the supervisory control over the common task may be shared among the various members of the system, and may be passed back and forth from one machine to the other as the need arises.

References

LeinA57, 59

Section 3

Computers for multiprocessing and parallel processing

The computers in this section are probably the most general in the book. Although the general PMS model for a computer in Chap. 3, page 65, characterizes these computers, the structure by Lehman (Chap. 37) most closely fits the model. The Burroughs computers that are presented have multiple Pc's;[1] however, K's are used for control of device K's, rather than Pio's—perhaps a wise choice.

D825—a multiple-computer system for command and control

The Burroughs D825 computer is discussed, together with other stack processors, in Part 3, Sec. 5, page 257. Chapter 36 emphasizes the PMS structure and operating system characteristics necessary in a multiprocessor system.

[1]As does the B 8500, a successor to the D825; however, its successor, the B 8501, is designed with Pio's.

Design of the B 5000 system

This computer (Chap. 22) is discussed, together with other stack processors, in Part 3, Sec. 5, page 257.

A survey of problems and preliminary results concerning parallel processing and parallel processors

Chapter 37, by M. Lehman, provides a very good introduction to the concepts of multiprogramming, multiprocessing, and parallel processing. A specific multiprocessor computer structure is postulated to provide parallel processing. The processing ability of the structure is analyzed at the instruction level. It is significant that the paper is by an IBM scientist. IBM has not been particularly advanced in the use of multiple arithmetic processor computers.

Chapter 36

D825—a multiple-computer system for command and control[1]

James P. Anderson / Samuel A. Hoffman
Joseph Shifman / Robert J. Williams

Introduction

The D825 Modular Data Processing System is the result of a Burroughs study, initiated several years ago, of the data processing requirements for command and control systems. The D825 has been developed for operation in the military environment. The initial system, constructed for the Naval Research Laboratory with the designation AN/GYK-3(V), has been completed and tested. This paper reviews the design criteria analysis and design rationale that led to the system structure of the D825. The implementation and operation of the system are also described. Of particular interest is the role that developed for an operating system program in coordinating the system components.

Functional requirements of command and control data processing

By "command and control system" is meant a system having the capacity to monitor and direct all aspects of the operation of a large man and machine complex. Until now, the term has been applied exclusively to certain military complexes, but could as well be applied to a fully integrated air traffic control system or even to the operation of a large industrial complex. Operation of command and control systems is characterized by an enormous quantity of diverse but interrelated tasks—generally arising in real time—which are best performed by automatic data-processing equipment, and are most effectively controlled in a fully integrated central data processing facility. The data processing functions alluded to are those typical of data processing, plus special functions associated with servicing displays, responding to manual insertion (through consoles) of data, and dealing with communications facilities. The design implications of these functions will be considered here.

Availability criteria. The primary requirement of the data-processing facility, above all else, is availability. This requirement, essentially a function of hardware reliability and maintainability,

is, to the user, simply the percentage of available, on-line, operation time during a given time period. Every system designer must trade off the costs of designing for reliability against those incurred by unavailability, but in no other application are the costs of unavailability so high as those presented in command and control. Not only is the requirement for hardware reliability greater than that of commercial systems, but downtime for the complete system for preventive maintenance cannot be permitted. Depending upon the application, some greater or lesser portion of the complete system must *always* be available for primary system functions, and *all* of the system must be available *most* of the time.

The data processing facility may also be called upon, except at the most critical times, to take part in exercising and evaluating the operation of some parts of the system, or, in fact, in actual simulation of system functions. During such exercises and simulations, the system must maintain some (although perhaps partially and temporarily degraded) real-life and real-time capability, and must be able to return quickly to full operation. An implication here, of profound significance in system design, is, again, the requirement that *most* of the system be *always* available; there must be no system elements (unsupported by alternates) performing functions so critical that failure at these points could compromise the primary system functions.

Adaptability criteria. Another requirement, equally difficult to achieve, is that the computer system must be able to analyze the demands being made upon it at any given time, and determine from this analysis the attention and emphasis that should be given to the individual tasks of the problem mix presented. The working configuration of the system must be completely adaptable so as to accommodate the diverse problem mixes, and, moreover, must respond quickly to important changes, such as might be indicated by external alarms or the results of internal computations (exceeding of certain thresholds, for example), or to changes in the hardware configuration resulting from the failure of a system component or from its intentional removal from the system. The system

[1]*AFIPS Proc. FJCC*, vol. 22, pp. 86–96, 1962.

must have the ability to be dynamically and automatically restructured to a working configuration that is responsive to the problem-mix environment.

Expansibility criteria. The requirement of expansibility is not unique to command and control, but is a desirable feature in any application of data processing equipment. However, the need for expansibility is more acute in command and control because of the dependence of much of the efficacy of the system upon an ability to meet the changing requirements brought on by the very rapidly changing technology of warfare. Further, it must be possible to incorporate new functions in such a way that little or no transitional downtime results in any hardware area.

Expansion should be possible without incurring the costs of providing more capability than is needed at the time. This ability of the system to grow to meet demands should apply not only to the conventionally expansible areas of memory and I/O but to computational devices, as well.

Programming criteria. Expansion of the data-processing facility should require no reprogramming of old functions, and programs for new functions should be easily incorporated into the overall system. To achieve this capability, programs must be written in a manner which is independent of system configuration or problem mix, and should even be interchangeable between sites performing like tasks in different geographic locales. Finally, because of the large volume of routines that must be written for a command and control system, it should be possible for many different people, in different locations and of different areas of responsibility, to write portions of programs, and for the programs to be subsequently linked together by a suitable operating system.

Concomitant with the latter requirement and with that of configuration-independent programs is the desirability of orienting system design and operation toward the use of a high-level procedure-oriented language. The language should have the features of the usual algorithmic languages for scientific computations, but should also include provisions for maintaining large files of data sets which may, in fact, be ill-structured. It is also desirable that the language reflect the special nature of the application; this is especially true when the language is used to direct the storage and retrieval of data.

Design rationale for the data-processing facility

The three requirements of availability, adaptability, and expansibility were the motivating considerations in developing the D825 design. In arriving at the final systems design, several existing and proposed schemes for the organization of data processing systems were evaluated in light of the requirements listed above. Many of the same conclusions regarding these and other schemes in the use of computers in command and control were reached independently in a more recent study conducted for the Department of Defense by the Institute for Defense Analysis [Kroger et al., 1961].

The single-computer system. The most obvious system scheme, and the least acceptable for command and control, is the single-computer system. This scheme fails to meet the availability requirement simply because the failure of any part—computer, memory, or I/O control—disables the entire system. Such a system was not given serious consideration.

Replicated single-computer systems. A system organization that had been well known at the time these considerations were active involves the duplication (or triplication, etc.) of single-computer systems to obtain availability and greater processing rates. This approach appears initially attractive, inasmuch as programs for the application may be split among two or more independent single-computer systems, using as many such systems as needed to perform all of the required computation. Even the availability requirement seems satisfied, since a redundant system may be kept in idle reserve as backup for the main function.

On closer examination, however, it was perceived that such a system had many disadvantages for command and control applications. Besides requiring considerable human effort to coordinate the operation of the systems, and considerable waste of available machine time, the replicated single computers were found to be ineffective because of the highly interrelated way in which data and programs are frequently used in command and control applications. Further, the steps necessary to have the redundant or backup system take over the main function, should the need arise, would prove too cumbersome, particularly in a time-critical application where constant monitoring of events is required.

Partially shared memory schemes. It was seen that if the replicated computer scheme were to be modified by the use of partially shared memory, some important new capabilities would arise. A partially shared memory can take several forms, but provides principally for some shared storage and some storage privately allotted to individual computers. The shared storage may be of any kind—tapes, discs, or core—but frequently is core. Such a system, by providing a direct path of communication between computers, goes a long way toward satisfying the requirements listed above.

The one advantage to be found in having some memory private to each computer is that of data protection. This advantage vanishes when it is necessary to exchange data between computers, for if a computer failure were to occur, the contents of the private memory of that computer would be lost to the system. Furthermore, many tasks in the command and control application require access to the same data. If, for example, it would be desirable to permit some privately stored data to be made available to the fully shared memory or to some other private memory, considerable time would be lost in transferring the data. It is also clear that a certain amount of utilization efficiency is lost, since some private memory may be unused, while another computer may require more memory than is directly available, and may be forced to transfer other blocks of data back to bulk storage to make way for the necessary storage. It might be added in passing that if private I/O complements are considered, the same questions of decreased overall availability and decreased efficiency arise.

Master/slave schemes. Another aspect of the partially shared memory system is that of control. A number of such systems employ a *master/slave* scheme to achieve control, a technique wherein one, designated the master computer, coordinates the work done by the others. The master computer might be of a different character than the others, as in the PILOT system, developed by the National Bureau of Standards [Leiner et al., 1957], or it may be of the same basic design, differing only in its prescribed role, as in the Thompson Ramo Wooldridge TRW400 (AN/FSQ-27) [Porter, 1960]. Such a scheme does recognize the importance, for multicomputer systems, of the problem of coordinating the processing effort; the master computer is an effective means of accomplishing the coordination. However, there are several difficulties in such a design. The loss of the master computer would down the whole system, and the command and control availability requirement could not, consequently, be met. If this weakness is countered by providing the ability for the master control function to be automatically switched to another processor, there still remains an inherent inefficiency. If, for example, the workload of the master computer becomes very large, the master becomes a system bottleneck resulting in inefficient use of all other system elements; and, on the other hand, if the workload fails to keep the master busy, a waste of computing power results. The conclusion is then reached that a master should be established only when needed; this is what has been done in the design of the D825.

The totally modular scheme. As a result of these analyses, certain implications became clear. The availability requirement dictated a decentralization of the computing function—that is, a multiplicity of computing units. However, the nature of the problem required that data be freely communicable among these several computers. It was decided, therefore, that the memory system would be completely shared by all processors. And, from the point of view of availability and efficiency, it was also seen to be undesirable to associate I/O with a particular computer; the I/O control was, therefore, also decoupled from the computers.

Furthermore, a system with several computers, totally shared memory, and decoupled I/O seemed a perfect structure for satisfying the adaptability requirements of command and control. Such a structure resulted in a flexibility of control which was a fine match for the dynamic, highly variable, processing requirements to be encountered.

The major problem remaining to realize the computational potential represented by such a system was, of course, that of coordinating the many system elements to behave, at any given time, like a system specifically designed to handle the set of tasks with which it was faced at that time. Because of the limitations of previously available equipment, an operating system program had always been identified with the equipment running the program. However, in the proposed design, the entire memory was to be directly accessible to all computer modules, and the operating system could, therefore, be decoupled from any specific computer. The operation of the system could be coordinated by having any processor in the complement run the operating system only as the need arose. It became clear that the master computer had actually become a program stored in totally shared memory, a transformation which was also seen to offer enhanced programming flexibility.

Up to this point, the need for identical computer modules had not been established. The equality of responsibility among computing units, which allowed each computer to perform as the master when running the operating system, led finally to the design specification of identical computer modules. These were freely interconnected to a set of identical memory modules and a set of identical I/O control modules, the latter, in turn, freely interconnected to a highly variable and diverse I/O device complement. It was clear that the complete modularity of system elements was an effective solution to the problem of expansibility, inasmuch as expansion could be accomplished simply by adding modules identical to those in the existing complement. It was also clear that important advantages and economies resulting from the manufacture, maintenance, and spare parts provisioning for identical modules also accrue to such a system. Perhaps the most important result of a totally modular organization is that redundancy of the required complement of any module type, for greater reliability, is easily achieved by incorporating as little as one additional module of that type in the system. Furthermore, the additional module of each type need not be idle; the system may be looked upon as operating with active spares.

Thus, a design structure based upon complete modularity was set. Two items remained to weld the various functional modules into a coordinated system—a device to electronically interconnect the modules, and an operating system program with the effect of a master computer, to coordinate the activities of the modules into fully integrated system operation.

In the D825, these two tasks are carried out by the *switching interlock* and the *Automatic Operating and Scheduling Program* (AOSP), respectively. Figure 1 shows how the various functional modules are interconnected via the interlock in a matrix-like fashion.

System implementation

Most important in the design implementation of the D825 were studies toward practical realization of the switching interlock and the AOSP. The computer, memory, and I/O control modules permitted more conventional solutions, but were each to incorporate some unusual features, while many of the I/O devices were selected from existing equipment. With the exception of the latter, all of theses elements are discussed here briefly. (A summary of D825 characteristics and specifications is included at the end of the paper.)

Switching interlock. Having determined that only a completely shared memory system would be adequate, it was necessary to find some way to permit access to any memory by any processor, and, in fact, to permit sharing of a memory module by two or more processors or I/O control modules.

A function distributed physically through all of the modules of a D825 system, but which has been designated in aggregate the switching interlock, effects electronically each of the many brief interconnections by which all information is transferred among computer, memory, and I/O control modules. In addition to the electronic switching function, the switching interlock has the ability to detect and resolve conflicts such as occur when two or more computer modules attempt access to the same memory module.

The switching interlock consists functionally of a crosspoint switch matrix which effects the actual switching of bus interconnections, and a bus allocator which resolves all time conflicts resulting from simultaneous requests for access to the same bus

or system module. Conflicting requests are queued up according to the priority assigned to the requestors. Priorities are preemptive in that the appearance of a higher priority request will cause service of that request before service of a lower priority request already in the queue. Analyses of queueing probabilities have shown that queues longer than one are extremely unlikely.

The priority scheduling function is performed by the bus allocator, essentially a set of logical matrices. The conflict matrix detects the presence of conflicts in requests for interconnection. The priority matrix resolves the priority of each request. The logical product of the states of the conflict and priority matrices determines the state of the queue matrix, which in turn governs the setting of the crosspoint switch, unless the requested module is busy.

The AOSP: an operating system program. The AOSP is an operating system program stored in totally shared memory and therefore available to any computer. The program is run only as needed to exert control over the system. The AOSP includes its own executive routine, an operating system for an operating system, as it were, calling out additional routines, as required. The configuration of the AOSP thus permits variation from application to application, both in sequence and quantity of available routines and in disposition of AOSP storage.

The AOSP operates effectively on two levels, one for system control, the other for task processing.

The system control function embodies all that is necessary to call system programs and associated data from some location in the I/O complement, and to ready the programs for execution by finding and allocating space in memory, and initiating the processing. Most of the system control function (as well as the task processing function) consists of elaborate bookkeeping for: programs being run, programs that are active (that is, occupy memory space), I/O commands being executed, other I/O commands waiting, external data blocks to be received and decoded, and activation of the appropriate programs to handle such external data. It would be inappropriate here to discuss the myriad details of the AOSP; some idea of its scope, however, can be obtained from the following list of some of its major functions:

1   Configuration determination

2   Memory allocation

3   Scheduling

4   Program readying and end-of-job cleanup

5   Reporting and logging

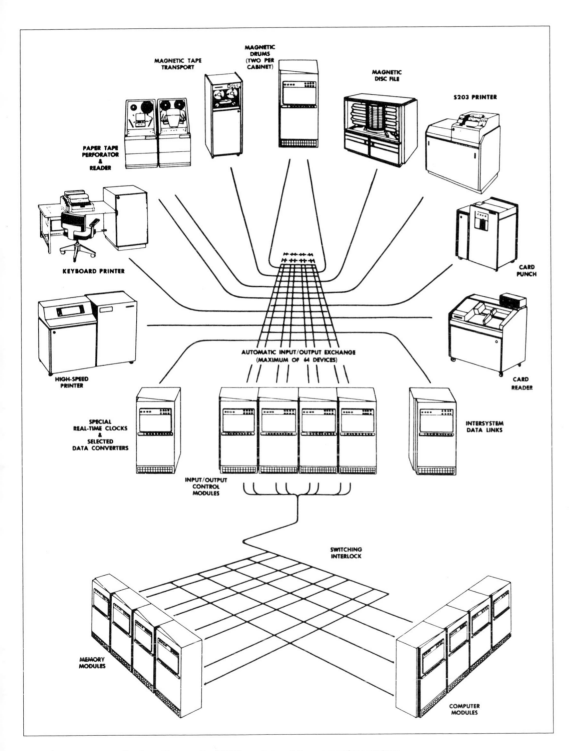

Fig. 1. System organization, Burroughs D825 modular data processing system.

6   Diagnostics and confidence checking

7   External interrupt processing

The task processing function of the AOSP is to execute all program I/O requests in order to centralize scheduling problems and to protect the system from the possibility of data destruction by ill-structured or conflicting programs.

AOSP response to interrupts. The AOSP function depends heavily upon the comprehensive set of interrupts incorporated in the D825. All interrupt conditions are transmitted to all computer modules in the system, and each computer module can respond to all interrupt conditions. However, to make it possible to distribute the responsibility for various interrupt conditions, both system and local, each computer module has an interrupt mask register that controls the setting of individual bits of the interrupt register. The occurrence of any interrupt causes one of the system computer modules to leave the program it has been running and branch to the suitable AOSP entry, entering a *control mode* as it branches. The control mode differs from the normal mode of operation in that it locks out the response to some low-priority interrupts (although recording them) and enables the execution of some additional instructions reserved for AOSP use (such as setting an interrupt mask register or memory protection registers, or transmitting an I/O instruction to an I/O control module).

In responding to an interrupt, the AOSP transfers control to the appropriate routine handling the condition designated by the interrupt. When the interrupt condition has been satisfied, control is returned to the original object program. Interrupts caused by normal operating conditions include:

1   16 different types of external requests

2   Completion of an I/O operation

3   Real-time clock overflow

4   Array data absent

5   Computer-to-computer interrupts

6   Control mode entry (normal mode halt)

Interrupts related to abnormalities of either program or equipment include:

1   Attempt by program to write out of bounds

2   Arithmetic overflow

3   Illegal instruction

4   Inability to access memory, or an internal parity error; parity error on an I/O operation causes termination of that operation with suitable indication to the AOSP

5   Primary power failure

6   Automatic restart after primary power failure

7   I/O termination other than normal completion

While the reasons for including most of the interrupts listed above are evident, a word of comment on some of them is in order.

The array-data-absent interrupt is initiated when a reference is made to data that is not present in the memory. Since all array references such as A[k] are made relative to the base (location of the first element) of the array, it is necessary to obtain this address and to index it by the value k. When the base of array A is fetched, hardware sensing of a presence bit either allows the operation to continue, or initiates the array-data-absent interrupt. In this way, keeping track of data in use by interacting programs can be simplified, as may the storage allocation problem.

The primary power failure interrupt is highest priority, and *always* pre-emptive. This interrupt causes all computer and I/O control modules to terminate operations, and to store all volatile information either in memory modules or in magnetic thin-film registers. (The latter are integral elements of computer modules.) This interrupt protects the system from transient power failure, and is initiated when the primary power source voltage drops below a predetermined limit.

The automatic restart after primary power failure interrupt is provided so that the previous state of the system can be reconstructed.

A description of how an external interrupt is handled might clarify the general interrupt procedure. Upon the presence of an external interrupt, the computer which has been assigned responsibility to handle such interrupts automatically stores the contents of those registers (such as the program counter) necessary to subsequently reconstitute its state, enters the control mode, and goes to a standard (hardware-determined) location where a branch to the external request routine is located. This routine has the responsibility of determining which external request line requires servicing, and, after consulting a table of external devices (teletype buffers, console keyboards, displays, etc.) associated with the interrupt lines, the computer constructs and transmits an input instruction to the requesting device for an initial message. The computer then makes an entry in the table of the I/O complete program (the program that handles I/O complete interrupts) to activate the appropriate responding routine when the message is read in. A check is then made for the occurrence of additional external requests. Finally, the computer restores the saved register contents and returns in normal mode to the interrupted program.

AOSP control of I/O activity. As mentioned above, control of all I/O activity is also within the province of the AOSP. Records are kept on the condition and availability of each I/O device. The locations of all files within the computer system, whether on magnetic tape, drum, disc file, card, or represented as external inputs, are also recorded. A request for input by file name is evaluated, and, if the device associated with this name is readily available, the action is initiated. If for any reason the request must be deferred, it is placed in a program queue to await conditions which permit its initiation. Typical conditions which would cause deferral of an I/O operation include:

1 No available I/O control module or channel.

2 The device in which the file is located is presently in use.

3 The file does not exist in the system.

In the latter case, typically, a message would be typed out on the supervisory printer, asking for the missing file.

The I/O complete interrupt signals the completion of each I/O operation. Along with this interrupt, an I/O result descriptor is deposited in an AOSP table. The status relayed in this descriptor indicates whether or not the operation was successful. If not successful, what went wrong (such as a parity error, or tape break, card jams, etc.) is indicated so that the AOSP may initiate the appropriate action. If the operation was successful, any waiting I/O operations which can now proceed are initiated.

AOSP control of program scheduling. Scheduling in the D825 relies upon a job table maintained by the AOSP. Each entry is identified with a name, priority, precedence requirements, and equipment requirements. Priority may be dynamic, depending upon time, external requests, other programs, or a function of many variable conditions. Each time the AOSP is called upon to select a program to be run, whether as a result of the completion of a program or of some other interrupt condition, the job table is evaluated. In a real-time system, situations occur wherein there is no system program to be run, and machine time is available for other uses. This time could be used for auxiliary functions, such as confidence routines.

The AOSP provides the capability for program segmentation at the discretion of the programmer. Control macros embedded in the program code inform the AOSP that parallel processing with two or more computers is possible at a given point. In addition, the programmer must specify where the branches indicated in this manner will join following the parallel processing.

Computer module. The computer modules of the D825 system are identical, general-purpose, arithmetic and control units. In determining the internal structure of the computer modules, two considerations were uppermost. First, all programs and data had to be arbitrarily relocatable to simplify the storage allocation function of the AOSP; secondly, programs would not be modified during execution. The latter consideration was necessary to minimize the amount of work required to pre-empt a program, since all that would have to be saved to reinstate the interrupted program at a later time would be the data for that program and the register contents of the computer module running the program at the time it was dumped.

The D825 computer modules employ a variable-length instruction format made up of quarter-word syllables. Zero-, one-, two-, or three-address syllables, as required, can be associated with each basic command syllable. An implicitly addressed accumulator stack is used in conjunction with the arithmetic unit. Indexing of all addresses in a command is provided, as well as arbitrarily deep indirect addressing for data.

Each computer module includes a 128-position thin-film memory used for the stack, and also for many of the registers of the machine, such as the program base register, data base register, the index registers, limit registers, and the like.

The instruction complement of the D825 includes the usual fixed-point, floating-point, logical, and partial-field commands found in any reasonably large scientific data processor.

Memory module. The memory modules consist of independent units storing 4096 words, each of 48 bits. Each unit has an individual power supply and all of the necessary electronics to control the reading, writing, and transmission of data. The size of the memory modules was established as a compromise between a module size small enough to minimize conflicts wherein two or more computer or I/O modules attempt access to the same memory module, and a size large enough to keep the cost of duplicated power supplies and addressing logic within bounds. It might be noted that for a larger modular processor system, these trade-offs might indicate that memory modules of 8192 words would be more suitable. Modules larger than this—of 16,384 or 32,768 words, for example—would make construction of relatively small equipment complements meeting the requirements set forth above quite difficult. The cost of smaller units of memory is offset by the lessening of catastrophe in the event of failure of a module.

I/O control module. The I/O control module executes I/O operations defined and initiated by computer module action. In keeping with the system objectives, I/O control modules are not assigned to any particular computer module, but rather are treated in much the same way as memory modules, with automatic resolution of conflicting attempted accesses via the switching interlock function. Once an I/O operation is initiated, it proceeds independently until completion.

I/O action is initiated by the execution of a transmit I/O instruction in one of the computer modules, which delivers an I/O descriptor word from the addressed memory location to an inactive I/O control module. The I/O descriptor is an instruction to the I/O control module that selects the device, determines the direction of data flow, the address of the first word, and the number of words to be transferred.

Interposed between the I/O control modules and the physical external devices is another crossbar switch designated the I/O exchange. This automatic exchange, similar in function to the switching interlock, permits two-way data flow between any I/O control module and any I/O device in the system. It further enhances the flexibility of the system by providing as many possible external data transfer paths as there are I/O control modules.

Equipment complements. A D825 system can be assembled (or expanded) by selection of appropriate modules in any combination of: one to four computer modules, one to 16 memory modules,

Fig. 2. Typical D825 equipment array.

Table 1 Specifications, D825 modular data processing system

Computer module:	4, maximum complement
Computer module, type:	Digital, binary, parallel, solid-state
Word length:	48 bits including sign (8 characters, 6 bits each) plus parity
Index registers: (in each computer module)	15
Magnetic thin-film registers: (in each computer module)	128 words, 16 bits per word, 0.33-μsec read/write cycle time
Real-time clock: (in each computer module)	10 msec resolution
Binary add:	1.67 μsec (average)
Binary multiply:	36.0 μsec (average)
Floating-point add:	7.0 μsec (average)
Floating-point multiply:	34.0 μsec (average)
Logical AND:	0.33 μsec
Memory type:	Homogeneous, modular, random-access, linear-select, ferrite-core
Memory capacity:	65,536 words (16 modules maximum, 4096 words each)
I/O exchanges per system:	1 or 2
I/O control modules:	10 per exchange, maximum
I/O devices:	64 per exchange, maximum
Access to I/O devices:	All I/O devices available to every I/O control module in exchange
Transfer rate per I/O exchange:	2,000,000 characters per second
I/O device complement:	All standard I/O types, including 67 kc magnetic tapes, magnetic drums and discs, card and paper tape punches and readers, character and line printers, communications and display equipment

one to ten I/O control modules, one or two I/O exchanges, and one to 64 I/O devices per I/O exchange in any combination selected from: operating (or system status) consoles, magnetic tape transports, magnetic drums, magnetic disc files, card punches and readers, paper tape perforators and readers, supervisory printers, high-speed line printers, selected data converters, special real-time clocks, and intersystem data links.

Figure 2 is a photograph of some of the hardware of a completed D825 system. The equipment complement of this system includes two computer modules, four memory modules (two per cabinet), two I/O control modules (two per cabinet), one status display console, two magnetic tape units, two magnetic drums, a card reader, a card punch, a supervisory printer, and an electrostatic line printer.

D825 characteristics are summarized in Table 1.

Summary and conclusion

It is the belief of the authors that modular systems (in the sense discussed above) are a natural solution to the problem of obtaining greater computational capacity—more natural than simply to build larger and faster machines. More specifically, the organizational structure of the D825 has been shown to be a suitable basis for the data processing facility for command and control. Although the investigation leading toward this structure proceeded as an attack upon a number of diverse problems, it has become evident that the requirements peculiar to this area of application are, in effect, aspects of a single characteristic, which might be called *structural freedom*. Furthermore, it is now clear that the most unique characteristic of the structure realized—integrated operation of freely intercommunicating, totally modular elements—provides the means for achieving structural freedom.

For example, one requirement is that some specified minimum of data processing capability be always available, or that, under any conditions of system degradation due to failure or maintenance, the equipment remaining on line be sufficient to perform primary system functions. In the D825, module failure results in a reduction of the on-line equipment configuration but permits normal operation to continue, perhaps at a reduced rate. The individual modules are designed to be highly reliable and maintainable, but system availability is not derived solely from this source, as is necessarily the case with more conventional systems. The modular configuration permits operation, in effect, with active spares, eliminating the need for total redundancy.

A second requirement is that the working configuration of the system at a given moment be instantly reconstructable to new forms more suited to a dynamically and unpredictably changing work load. In the D825, all communication routes are public, all modules are functionally decoupled, all assignments are scheduled dynamically, and assignment patterns are totally fluid. The system of interrupts and priorities controlled by the AOSP and the switching interlock permits instant adaptation to any work load, without destruction of interrupted programs.

The requirement for expansibility calls simply for adaptation on a greater time scale. Since all D825 modules are functionally decoupled, modules of any types may be added to the system simply by plugging into the switching interlock or the I/O exchange. Expansion in all functional areas may be pursued far beyond that possible with conventional systems.

It is clear, however, that the D825 system would have fallen far short of the goals set for it if only the hardware had been considered. The AOSP is as much a part of the D825 system structure as is the actual hardware. The concept of a "floating" AOSP as the force that molds the constituent modules of an equipment complement into a system is an important notion having an effect beyond the implementation of the D825. One interesting by-product of the design effort for the D825 has, in fact, been a change of perspective; it has become abundantly clear that computers do not run programs, but that programs control computers.

References

AndeJ62; KrogM61; LeinA57; PortR60; ThomR63

Chapter 37

A survey of problems and preliminary results concerning parallel processing and parallel processors[1]

M. Lehman

Summary   After an introduction which discusses the significance of a trend to the design of parallel processing systems, the paper describes some of the results obtained to date in a project which aims to develop and evaluate a unified hardware-software parallel processing computing system and the techniques for its use.

1.   Multiprogramming, multiprocessing, and parallel processing

A brief review of the literature, of which a partial listing is given in the bibliography, reveals an active and growing interest in multiprogramming, multiprocessing, and parallel processing. These three terms distinguish three modes of usage and also serve to indicate a certain historical development. We cannot here attempt to trace this history in detail and so must rely on the bibliography to credit the contributions from industrial, university, and other research and development organizations.

The·emergence of autonomous input-output devices first suggested [Gill, 1958] the time-sharing of the processing and peripheral units of a computing system among several jobs. Thus surplus capability that could not be applied to the processing of the leading job in a batch processing load, at any stage of the computation, could be usefully applied to successor jobs in the work load. In particular, while any computation was held up for some I/O activity, the single main processor could be used for other computation. The necessary decision-taking, scheduling, and allocation procedures were vested in a supervisor program, within which the user-jobs were embedded, and the resultant mode of operation was termed *Multiprogramming*.

The use of computers in on-line control situations and for other applications giving rise to ever-more stringent reliability and availability specifications, resulted in the construction of systems including two or more central processing units [Leiner et al., 1959; Bright, 1964; Desmonde, 1964; McCullough et al., 1965]. Under normal circumstances, with all units operational, each could be assigned a specific activity within an overall control program. As a result of the multiplicity of units in such *Multiprocessing Systems*, failure of any one would degrade, but not immobilize, the system, since a supervisor program could re-assign activities and configure the failed unit out of the system. Subsequently, it was recognized that such systems had advantages over a single processor system in a more general environment, with each processor in the system having a multiprogramming capability as well.

Finally, following from ideas first exploited in the Gamma 60 Computer [Dreyfus, 1958], there has come the realization that multi-instruction counter systems can speed up computation, particularly of large problems, when these may be partitioned into sections which are substantially independent of one another, and which may therefore be executed concurrently—that is, in parallel. When the several units of a multiprocessing system are utilized to process, in parallel, independent sections of a job, we exploit the macro-parallelism [Lehman, 1965] of the job, which is to be distinguished from micro-parallelism [Lehman, 1965], the relative independence of individual machine instructions, exploited in look-ahead machines. This mode of operation is termed *Parallel Processing* and, as in PL/I [IBM OS/360, PL/I Language Specification, Form C28-6571, p. 74], the execution of any program string is termed a *Task*. We note that parallel processing may, and normally will, include multiprocessing activity.

2.   The approach to parallel processing system design

In the previous section we indicated that the prime impetus for the development of parallel processing systems arose from their potential for high performance and reliability. These systems may operate as pools of resources organized in symmetrical classes and it is this property that promises *High Availability*. They also possess a great reserve of power which, when applied to a single problem with the appropriate degree of parallelism, can yield high

[1] *Proc. IEEE,* vol. 54, no. 12, pp. 1889–1901, December, 1966.

performance and fast turn around time. Surplus resources can be applied to other jobs, so that the system is potentially efficient, displaying a peak-load averaging effect and hence high utilization of hardware [Corbato and Vyssotsky, 1965]. The concept of sharing in parallel processing systems and its related cost reduction is not, however, limited to hardware. Perhaps even more significant is the common use of data-sets maintained in a system library or file, and even concurrent access during execution from a high-speed store. This may represent considerable economy in storage space and in processing time for I/O and internal memory-hierarchy transfers. But above all [Corbato and Vyssotsky, 1965] it facilitates the sharing of ideas, experience, and results and a cross fertilization among users, a prospect which from a long term point of view represents perhaps the most significant potential of large, library-oriented, multiprocessing systems. Finally, in this brief summary of the basic advantages of parallel processing systems, we refer to their intrinsic modularity, which may yield an expandable system in which the only effect of expansion on the user is improved performance.

Adequate performance of parallel processing systems is, however, predicated on an appropriately low level of overhead. Allocation, scheduling, and supervisory[1] strategies, in particular, must be simplified and the related procedures minimized to comprise a small proportion of the total activity in the system. The system design must be based on performance objectives that permit a user to specify a time period and a tolerance within which he requires and expects to receive results, and the cost for which these will be obtained. In general the entire system must yield minimum throughput time for the large job, adequate response time to the terminal requests in conversational mode, guaranteed throughput time for real-time tasks, and minimum cost processing for the batch-processed small job. These needs require the development of an executive and supervisory system integrated with the hardware into a single, unified computing system. Finally, the techniques and algorithms of classical computation, of problem analysis, and of programming, must be modified and new, intrinsically parallel procedures developed if full advantage is to be gained from exploitation of these parallel systems.

Our studies to date represent but a small fraction of the ground that will have to be covered if effective parallel processing systems are to come into their own. It is, however, abundantly clear that such systems will yield their potential only if the design is approached on a broad but unified front ranging from problem analysis and usage techniques, through executive strategies and operating systems, to logic design and technology. We therefore present concepts and results from each of these areas, as obtained during our preliminary investigation into the design and use of parallel processing systems.

3. Language

3.1 Parallelism in high level languages

The analysis of high level language requirements for parallel processing has received considerable attention in the literature. We may refer in particular to the paper by Conway [1963] which discussed the concepts of Fork, Join, and Quit, and the recent review by Dennis and Van Horn [1966].

Recognizing that programming languages should possess capabilities that express the structure of the computational algorithm, Schlaeppi [19??] has proposed augmentations to PL/I-like languages that portray the macro-parallelism in numerical algorithms. These in turn have been reflected in proposals for machine-language implementation. As examples we discuss *Split, Terminate, Assemble, Test and Set or Wait* (interlock), *Resume, Store-Test and Branch*, and *External Execute* instructions. We describe here only the basic functional elements, from which machine instructions for actual realization will be composed as suggested by practical programming experience.

3.2 Machine level instructions for tasking

Split provides the basic task-generating capability. It indicates that in addition to continuing the execution of the present instruction string in normal fashion a new task, or set of tasks, may be initiated, execution starting at a specified address or set of addresses. Such potential tasks will be queued to await pick-up by an appropriate processing unit.

Terminate causes cessation of activity on a task. The terminating unit will, of its own volition, access an appropriate queue to obtain its next task. Alternatively, it may execute an executive allocation-task to determine which of a number of task-queues is to be accessed next according to the current urgency status of work in the system.

Assemble permits the merging of several tasks. The first $(n - 1)$ tasks in an n-way parallel set belonging to a single job, reaching the assemble instruction terminate. The nth task, however, will proceed to execute the program string which constitutes the continuation of all n tasks.

---

[1] We differentiate intuitively between executive and supervisory activities. The former are those whose costs should be chargeable to the individual user directly, whereas the latter are absorbed in the system running costs.

Test and Set or Wait provides an interlock facility. Thus a number of tasks all operating on a common data set may be required to filter through certain sections of program or data, one at a time. This may be achieved by an instruction related to the S/360 test and set instruction [Falkoff et al., 1964], but causing the task finding the specified location to be already set to go into a wait state. System efficiency requires that processors do not idle, so that the waiting task will generally be returned to queue and the processor released for other work.

Resume directs a processor or processors waiting as a result of a test on a specified location, to proceed, or more generally, that specified waiting tasks that have been returned to queue be re-activated to await the spontaneous availability of an appropriate processor.

Test and Branch Storage Location permits communication between parallel tasks based on tests analogous to the register tests of uniprocessors, but associated with the contents of storage locations. This is desirable since processor registers are private to the processor and inaccessible from outside.

External Execute is a special case of the general interaction facility discussed in Section 4 that permits related tasks to influence one another. This can be achieved through the application of instructions already discussed. It is, however, more efficient to provide a new facility akin to the *Interrupt* concept. By applying this *Interaction* function, a task may cause other specified tasks to execute an instruction at a specified location, each on completion of its present instruction. Thus, for example, a number of processors searching for a particular item in a partitioned list can be caused to abandon the search when the item has been located by one, while processors searching for other items, or otherwise busy, will not be redirected.

4. Interaction

4.1   The interaction concept

An extension of the task interaction concept introduced in the preceding section is fundamental to efficient parallel processing. In the particular example cited, the interaction, in the form of an external execute instruction, forms part of the computational procedure. In fact, many other situations arise in which processing for inter-task communication may be detached from problem processing and be carried through concurrently in autonomous units, thereby increasing system utilization.

We therefore propose to associate with each active unit in the system an autonomous *Interaction Controller*. Groups of controllers

are linked by a special bus. This provides facilities whereby any one unit may, at a given time, act as a command or signal source with all other units potential recipients. By thus systemizing inter-unit communication and making it a concurrent activity, we both increase system utilization and remove a maze of interconnecting cables. Succeeding subsections describe some of the functions that the controllers fulfill and, briefly, one hardware proposal for their realization.

4.2   Interaction activities

In present-day systems there already exist activities of the type to be classified as interaction. Thus, for example, in System/360 we find a CPU to Channel *Halt I/O* facility, channel interruptions of processors, and timer interruptions. In extending the concept we differentiate among three classes of interaction.

PROBLEM INTERACTION. These relate to logical dependencies between tasks, and will generally require waits, forced branches, or terminations. Search termination, previously discussed, is an example of this type interaction, as are data and instruction-sequence interlocks.

EXECUTIVE INTERACTION. This activity is concerned primarily with the allocation of system resources. Consider, for example, the problem of processing interrupts in a parallel processing system. These will usually not need to interrupt a computing activity, but may await the spontaneous availability of a unit at a *Terminate*, a natural breakpoint.[1] If an interrupt does become critical it should not be applied to a specific physical unit. Instead the interruption should be steered to that unit which, by virtue of the work it is processing, may be classed as *Most Interruptable*. Selection of the latter may be obtained ahead of time and is maintained by the interaction system, on the basis of the relative urgency of tasks.

Another example of executive interaction concerns the constant provision of queue status information to all active units. Besides simplifying scheduling activity this may prevent units from accessing empty queues, reducing both storage and executive interference. Similarly, units can be caused to access a previously empty queue when an entry is made, obviating continuous testing of queue status.

[1] This is possible in a parallel processing system since tasks are smaller than jobs and since there are many processors. Furthermore, units operate anonymously. That is, on picking up a task, a unit records the task identity in an internal register and its own identity in a table associated with the work queue. Other processors do not, therefore, know how tasks and processors are matched at any time, since this is a matter of chance, and determination would require an extensive and wasteful table search.

The interaction system also supports other activities associated with accounting, recording, and general system supervision.

SYSTEM INTERACTION. System interaction provides controls and interlocks for operation and maintenance of the physical system. It includes, for example, interchange of information between active units about the validity of storage map entries, storage protection control, queue interlocks, checks and counts of unit availability, the initiation of routine and emergency diagnostic and maintenance activity, and the isolation of malfunctioning units.

SUMMARY. The preceding paragraphs have indicated some of the many applications of an interaction controller. The common property which, for practicality, has been used to identify potential interaction activities is that they should be autonomous relative to the main computational stream and that their execution should not require access to storage.

4.3 The interaction controller

4.3.1. *The basic system hardware architecture.* It is not intended to give a full description of an interaction controller in the present paper. We shall, however, outline its basic structure, indicate its mode of operation, and list some of the proposed interaction instructions, termed *Directives*.

As a first step we introduce, in Fig. 1, a diagrammatic description of an overall representative hardware system. This consists of central processors (Pi) with local storage (LSi), I/O processors (SCi), storage modules (Si), a requestor-storage queue (Qi), and a communication system functionally equivalent to a crossbar switch. An I/O area, including a bulk-store, files, channels (Ch), devices, device control units (Cu), and interconnection networks, is indicated in less detailed fashion.

4.3.2. *Interaction controllers.* Interaction controllers (IC) are associated with all central and I/O processors, and communicate with each other over a special bus. Similarly localized interaction systems may provide a facility for certain classes of I/O units or devices to interact amongst themselves.

To be economically feasible, the Interaction Controller must be simple. Figure 2 illustrates a structure which includes about two hundred and fifty bits of storage, of which about half are organized in registers. The remainder are used as status bits or appear in the controller-processor interface. Control is obtained from a read-only store, whose capacity depends on the size of the directive repertoire (an interaction directive being analogous to

a processor instruction) and the number of interaction functions it is required to implement.

Controller connection to the ten-bit wide interaction bus is by means of OR gates. When an interaction is occurring, one and only one controller will be in command of the bus. Figure 3 illustrates the sequence of events required to implement an interaction.

The controller required by its associated processor to initiate an activity will await availability of the bus, indicated by an ALL ZERO state, and will then attempt to seize control by transmitting a unique identifying four-out-of-eight code. Should more than one controller attempt to seize the bus at the same time, a conflict resolution procedure is initiated. This is based on the simultaneous transmission by all requesting controllers of a second, two byte, identifying code. Each byte consists of one or more ones followed by all zeros. A simple comparison by each controller of its trans-

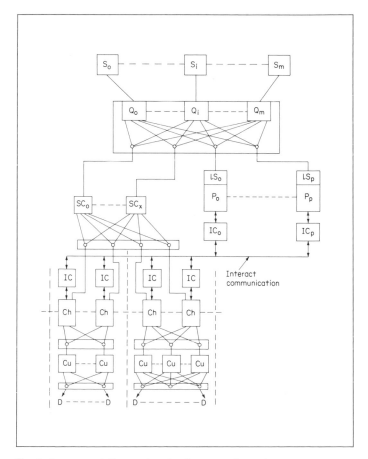

Fig. 1. A representative system hardware configuration.

Fig. 2. The interaction controller.

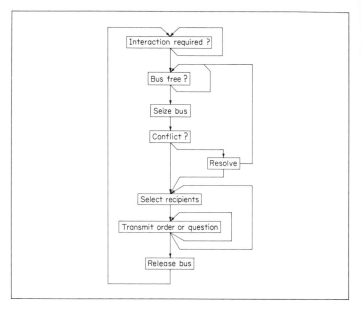

Fig. 3. The interaction sequence.

mitted signals with the state of the bus, identifies to itself that controller having the most ones in each byte, since it will have found a match on both comparisons. This enables it to seize the bus and to switch to the command state. All remaining controllers remain in the listening state.

The controller in command of the bus then transmits signals which select recipients for the directives which are to follow. Other controllers ignore all further communications until the next selection signal appears.

4.4 Interaction directives

A signal designating the interaction function required by a processor is transmitted across the processor/controller interface, as the result of the execution of some processor instruction. The processor will then generally continue its execution sequence unless or until it is required to pass on a second interaction function before a previously issued function has been completed. Upon receipt of the interaction command, and after successful seizure of the bus as described, the command controller may initiate

execution of the interaction by transmitting a sequence of one or more directives to the selected units. A basic set of directives is listed in Table 1.

The *Compare* directives are most frequently used to seize the bus and to select a subset of the controllers for the receipt of subsequent directives. The remaining units ignore further directives until alerted by an *Attention* signal or until *Free Bus* provides the release that permits waiting controllers to attempt to seize the bus. *Receive* provides for transmittal of data between controllers; for example, transmission of a machine instruction to a selected set of controllers, followed by the directive *Interact*. Thus this sequence could realize the basic interaction function. *External Execute* is, however, considered so fundamental to efficient exploitation of a parallel processing system that we include it as an

Table 1

Send and Compare
Compare
Received
Set Status Bits
Interact
External Execute
Attention
Free Bus

explicit directive. Status bits that may be set or reset by appropriate directives, provide data on the status of various systems queues, on the interruptability of given processors, on Wait status, and so on.

5. Storage communication

The fact that interest in large parallel processing systems is increasing rapidly as technology enters into the integrated or monolithic era is no coincidence. Such systems will not, in fact, be practical for general purpose application until miniaturization reaches the stage where the large amount of hardware required can be assembled in compact fashion. This need is most apparent when one considers communication between the high-speed store and the various classes of processors, which may collectively be termed *Requestors*. Already in presently available systems, the transmission delay between storage and requestors is of the same order of magnitude as the storage cycle time; and cycle times are still decreasing.

Formulation of a hardware model as in Fig. 1 led to the immediate conclusion that feasibility of the interconnection of large numbers of units had first to be established. Many possible systems were considered, and preliminary studies concluded that the crossbar switch was the most appropriate system for early study in view of its regular structure, simplicity, and basic modularity. More particularly, monolithic crossbar modules are visualized which it will be possible to interconnect to provide networks of any required dimensions. Alternatively, or additionally, other interconnections of these modules can provide highly available, multi-level trunking systems.

In addition to the switch proper, the crossbar network requires a selection and control mechanism. It is moreover appropriate to locate the queues, which store all but one of a group of conflicting requests, within the switching area. A switch complex, as in Fig. 4, has been designed for a system configuration including twenty-four requestors, thirty-two memory modules, thirty-two data plus four parity bit words, and sixteen plus two parity bit addresses.

The result of this design study shows that the size and complexity of such a switch is not excessive for a large scale system. In its simplest form and using standard high-performance logical devices, with a fan-in of four, a fan-out of ten and a four-way OR capability, its use leads to a worst case delay of some seven logical levels in the control and queue decision circuits and two levels in each direction of the switch. The switch uses between two and three times as many circuits as a central processor such as the model 75 of System/360. While this, in itself, represents a consid-

erable amount of hardware, it is still an order of magnitude less than the hardware found in the units that the switch is interconnecting. Moreover, its regular structure and simple, repetitive logic suggest ultimate economical realization using monolithic circuit techniques.

6. Usage

6.1 The executive system

The basic properties outlined in Sec. 2 give parallel processing systems the potential to overcome many of the ills and shortcomings that presently beset computer systems. For maximum effectiveness, the system must be library- or file-oriented. It can, however, be exploited efficiently only if the overhead resulting from executive control and supervisory activity does not strangle the system. More particularly, the gains from the sharing of resources and any peak averaging effect must exceed any additional overhead due to resource allocation procedures, conflict resolutions, and other processing activity arising from the concurrent operation of many units. Thus a unified and integrated design approach is required in which software and hardware, operating system and processing units, lose their separate identities and merge into one

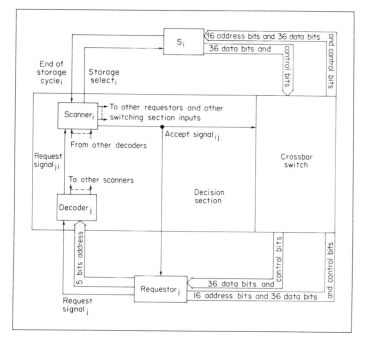

Fig. 4. The centralized crossbar switch.

overall complex, for which allocation and scheduling procedures, for example, are as basic and as critical as arithmetic operations.

Equally significant to the successful exploitation of parallel processing potential are the problems of data management, man-machine interactions; and, most generally, problem preparation and usage of the system. We restrict the present discussion to brief comments on programming techniques for task generation and on the development of algorithms possessing macro-parallelism. In particular we indicate that multi-instruction-counter systems can be profitably applied to the solution of the large problems whose computing requirements tax the speed capability and storage of the largest computer and the patience of their users. In the following section we evaluate these proposals by quoting some performance measurements obtained from an executing simulator.

6.2 Programmed task generation

Study of the usage of parallel processing systems for the rapid solution of large real-time problems involves two aspects. On the one hand we must consider the development of algorithms displaying an appropriate form of macro-parallelism. On the other hand programming techniques must be developed for efficient exploitation in terms of both problem- and machine-oriented instructions, such as those discussed in Sec. 4.

It is appropriate to discuss programmed task generation first. For simplicity we consider a job segment that requires n executions of a procedure I. The procedure will itself include modification of index registers or other changes that distinguish the individual tasks. We assume that on completion of all n tasks, a new procedure J should be initiated. Moreover, should processing power be available at a time when n executions of I have been initiated but not all n completed, we assume that an independent procedure K, belonging to the same job, may be initiated. In the simplest case K will be a terminate instruction which releases the processor, and makes it available to process other work as determined from the work-queue complex.

	$A = 0$	
	$B = 0$	
	$C = 0$	
ST	IF $N - B \leq 1$ THEN GO TO IN	Suppress split if nth task being initiated
	$A = A + 1$	
	IF $A \geq P$ THEN GO TO IN	Split if less than p processors allocated
	SPLIT TO ST	
	$B = B + 1$	
IN	IF $B > N$ THEN GO TO FIN	If all n I-tasks started, proceed with K

	CALL I PROCEDURE	
	$C = C + 1$	
	IF $C < N$ THEN GO TO IN	If all n I-tasks completed, proceed with J
	CALL J PROCEDURE	
FIN	CALL K PROCEDURE	

Execution of split and terminate instructions involves executive overheads, so that these instructions should not be used indiscriminately. Within a system in which a maximum of p processors are available to a job, it is pointless to partition a job, at any one time, into more than p tasks. It is, however, undesirable to guarantee a user that p processors, or even more than one processor, will execute his program. A simple task generation scheme that makes as many entries in the task queue as there are potentially concurrent parts of the algorithm (for example, from a loop containing a split instruction) is inefficient when that number is much larger than the number of processors that happen to be available. The technique also leads to very large queues. An alternative, termed *Onion Peeling* by us, puts the instruction sequence containing the split at the head of procedure I and ends each execution of the procedure with a terminate. This restricts the queue length for this job segment to one but it otherwise is as inefficient as the previous method.

A *Modified Onion Peeling* scheme (MOP) restricts the split and terminate overhead to at most one more[1] than the number of processors actually applied to the segment. It also ensures that processing is completed as quickly and as efficiently as possible with the number of processors that become available to the job segment. Thus if during execution no further processors are freed, the n tasks are executed sequentially with only one split and no terminate. If, on the other hand, some other number of processors is used for execution, the procedure is speeded up accordingly. The maximum number p of processors that may be applied to the job may be limited by the number of processors in the system and available, or by executive edict.

The basic scheme was illustrated by the above program, in which the first expressions following the ZEROing of counters ensures that no unnecessary splits are queued.

---

[1] This is not quite accurate. The simple MOP algorithm presented here does not explicitly interlock the split sequence. There is therefore a possibility that unnecessary task-calls may be queued during the execution of the split which is to generate the nth task. The probability of this is, however, small, while the degradation arising from an interlock could be significant, and the algorithm in the form given appears more economical.

6.3 Macro-parallelism

Commonly used numerical algorithms, data processing procedures, and computer programs are generally sequential in nature. The reason for this is largely historical, a consequence of the fact that the *Mechanisms*, human, mechanical, and electronic, used in developing and executing these procedures have been incapable of significant parallel activity, other perhaps than the simultaneous, coordinated use of many humans. The advent of parallel processing systems thus calls for the modification of accepted techniques to expose any inherent parallelism. The resultant procedures must then be further adapted to make parallel tasks of such a magnitude that the overhead involved in their generation becomes insignificant. But the ultimate benefit from parallel execution will be obtained only by going back to the problems themselves. These must be analyzed anew. Algorithms must be developed that make it possible to exploit the parallel executing capability, by introducing into the mathematical and program model parallelism that ultimately reflects the parallelism of the physical system or phenomena being studied. In this need to return to fundamentals, the situation is somewhat analogous to the early days of electronic computing, when attempts at commercial application were largely frustrated until it was realized that widespread application required the development of new techniques, rather than the adaptation and mechanization of existing procedures.

At the present time, however, our direct activity in problem analysis has concentrated mainly on the adaptation of existing numerical techniques for parallel processing, for problems in which the basic macro-parallelism was self-evident. These include, for example, linear algebra and the solution of elliptic partial differential equations. In these areas the extent and nature of the parallelism had previously led to proposals for vector processing systems such as Solomon [Slotnick et al., 1962; Gregory and McReynolds, 1963] and Vamp [Senzig and Smith, 1965]. Other areas in which the parallelism is self-evident but where vector processors prove less effective are those in which the algorithms model distinct physical activities such as in file processing and Monte Carlo techniques. For all significant problems investigated [Schlaeppi, 19??] it was possible to establish the existence of parallel tasks of such a length that tasking overheads could be expected to be negligible.

Other classes of problems have been studied, both in terms of the extension of existing algorithms and the development of new ones. In particular we refer to the extraction of polynomial roots [Shedler and Lehman, 1966], solution of equations [Shedler, 1966], and the solution of linear differential equations [Nievergelt, 1964], [Miranker and Liniger, 1967]. These various studies, not all directly related to the present project, were more mathematical in nature, and to the best of our knowledge, no attempt has yet been made to develop efficient parallel computer programs. Thus, while numerical methods are beginning to emerge which enable the exploitation of macro-parallelism in the solution of time-limited problems, and from which it appears that significant reductions may be obtained in throughput times, much work remains to be done on re-programming the problems themselves.

7. Simulation

7.1 Simulation as a design tool

It has been our experience with simulation that its principal function as a design tool is to focus attention on features that require investigation and explanation. Many results, qualitative and quantitative, that are obtained during simulation experiments may also be obtained analytically. It is, however, the insight and understanding gained from the design of simulation experiments and the analysis of their results that draws attention to specific details and difficulties. The undeniable value of simulation in development and design is therefore quite different from that in system evaluation, where meaningful performance figures may be obtained when the work load is well defined.

7.2 The executing simulator

In the present study simulation was seen as fulfilling a number of additional functions. In particular it made available a *usable* working model of a parallel processing system. This would give potential users the incentive to undertake actual programming and to gain limited operational experience. An executing simulator was also required for the investigation of what is commonly regarded as the most immediate question in parallel processing, the extent of performance degradation due to storage-access interference and executive (queue-access) interference. Such an executing simulator is now operational and its use is discussed in the next section. We note parenthetically that a limitation of this type simulator is its speed. For the evaluation of total system performance over any length of time, particularly when using a computer itself much slower than the simulated system, only gross, nonexecuting, simulation is reasonable [Katz, 1966].

The system presently modeled in the executing simulator includes the processors, switch, and *Storage Modules* of Fig. 1. The storage modules are accessed through a fully interleaved address structure, though it is clear that in any realization interleaving will be partial, both to sustain high availability and to decrease storage interference between independent jobs. The individual processors have a System/360-like structure [Blaauw and Brooks, 1964] and execute an augmented subset of S/360 machine language. The nonstandard instructions added to the repertoire include the functions discussed in Section 4. The local store LSi, to be used also as an instruction buffer, is however not included in the model for which the interference results are quoted in the next section. The simulator configuration is parameterized so that, for example, the numbers of storage modules and processors, instruction execution times (in storage cycles), and the nature of statistics gathered and printed may be selected for each run. The program itself is modular, and both system features and measurement facilities may be expanded or modified as required.

7.3   Simulator experiments

7.3.1   *Kernels.* Simulation experiments first concentrated on an investigation of storage interference arising in the execution of typical kernels from numerical analysis. The results indicated that under the limited condition of the experiments and for a storage module-to-processor ratio of two, interference would degrade performance by less than twenty percent, dropping to some five percent for storage module-to-processor ratio of eight. Addition of a local processor store and its use as an instruction buffer effectively eliminated interference, as expected, indicating that it had been substantially due to instruction-fetch interference.

These results were considered to have been generated under conditions too restrictive to permit generalization. In particular each set referred only to concurrent executions of a single loop. Thus more recent experiments have included many runs of a matrix-multiply subroutine and the solution of an electrical network problem using an appropriately modified version of the Jacobi variant of the Gauss-Seidel solution of a set of linear algebraic equations.

7.3.2   *The matrix multiplication.* The *Matrix Multiply* program was written in two versions. A classical sequential program excluding all the special instructions provided the standard on which measurement of the parallelism overhead and interference could be based. The second, parallel, program used the onion peeling rather than the MOP algorithm described in Sec. 7.2. The product matrix was partitioned by rows, with the computation of each comprising one task. The experiments were performed for square matrices of dimensions thirty-nine and forty with from one to sixteen processors and sixteen to sixty-four storage modules. Two

sizes of matrices were used to isolate the effect of commensurate periodicities of array mapping with the address structure of the store, which demonstratively had significant influence on the results.

Instruction execution times for the most frequently executed instructions used in the experiment are given in Table 2.

These times exclude the instruction fetch time (one instruction for each fetch), since these are overlapped unless storage conflict occurs, when a request must be queued. The arithmetic operations may also include a data fetch (RX instructions) in which case a further store access time is required.

In the absence of an internal instruction buffer, processors executing the same program string interfere with each other continuously during instruction fetches. To minimize this effect for loops that are short relative to the width of the interleaving, it is profitable to unwind such loops by repetition so that the resultant string stretches as far as possible across the interleaved store. The program was unwound in this way. We note, however, that it is in fact better [Rosenfeld, 1965] to repeat the loop, appropriately modified, several times across the interleaved store, directing successive processors to successive, but unconnected, loops. This can decrease interference by as much as twenty percent over the previous case.

Some results of the simulation are given in Table 3 and plotted in Figs. 5 and 6.

We note that running time (col. 4) is defined as the interval between the start of the first processor on its first task and the completion, by the last processor to finish, of its final task. Since an onion peel technique has been used for the splitting, there is an interval (of order 70 storage cycles) between the start of successive tasks. There is also an initial interval (87 memory cycles) in which the first processor initializes the program. Finally, the finish of processors is staggered and, in particular, for the sixteen-processor case, eight processors are assigned two tasks (rows) in succession, and eight, three tasks. The former processors will, of

Table 2

Instruction	Execution time in storage cycles
Fixed Point Addition	0.4
Floating Point Addition	0.5
Floating Point Multiplication	1.0
Floating Point Division	2.0
Split	25.0
Terminate	25.0
New Task Fetch (Part of Terminate)	25.0

Table 3   Results of the matrix multiply simulation

1	2	3	4	5	6	7	8	9	10	11
No. of proc.	No. of storage mods.	Matrix dim.	Run time	Total proc. time	Storage interference		Exec. interf. %	No. of storage accesses	Storage utilization %	Notes
					Time	%				
1	64	40	427	427	1.02	.2	NA	459K	1.69	Sequential program
1	64	40	429	429	0.21	0.05	NA	460K	1.68	Interference between
2	64	40	216	432	1.77	0.4	0.33	460K	3.3	instruction & data fetches
4	64	40	109	436	5.79	1.3	0.39	460K	6.6	
8	64	40	56	445	14.4	3.3	0.68	460K	13.0	
16	64	40	35	461	30.3	7.0	0.76	460K	25.0	
16	32	40	38	507	75.9	17.7	0.88	460K	45.4	
16	16	40	47	639	207.0	48.2	0.64	460K	72.1	
16	64	39	33	428	26.1	6.5	NV	427K	26.9	

Note: All times in thousands of storage cycles.
NA—Not Applicable
NV—Not Available

$$\% \text{ Storage Utilization} = \frac{\# \text{ Acc.} \times \# \text{ Proc.}}{\text{Proc. time} \times \# \text{ Mods.}}$$

$$= \frac{\text{Col. 9} \times \text{Col. 1}}{\text{Col. 5} \times \text{Col. 2}}$$

course, terminate considerably earlier than the latter. Thus, as indicated by the corresponding entry in column four, the particular mode of partitioning is not optimum if the shortest execution time is to be obtained. From a system efficiency point of view, however, and in actual operation with other jobs and tasks in the system, it is of no consequence since processor idling does not actually occur. New tasks, perhaps arising from quite different jobs, are initiated, according to some scheduling strategy, whenever a processor becomes spontaneously available.

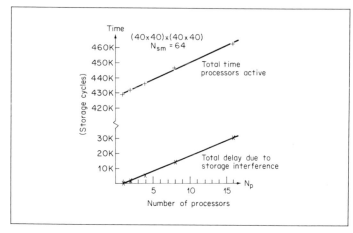

Fig. 6. Total processor time and interference in matrix multiply modules.

In addition to run time, we define a total processor time (col. 5). This represents the sum total of time that individual processors were active in the program and is therefore a reflection of total processor running cost. Storage interference (cols. 6, 7) measures the total time that processors were inactive due to attempts to initiate simultaneous accesses to the same storage module. It occurs also when only a single processor is applied, when it represents a conflict between a data fetch and an attempt by the overlap circuit to initiate an instruction fetch from the same module.

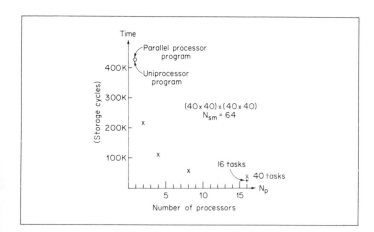

Fig. 5. Execution time for matrix multiply.

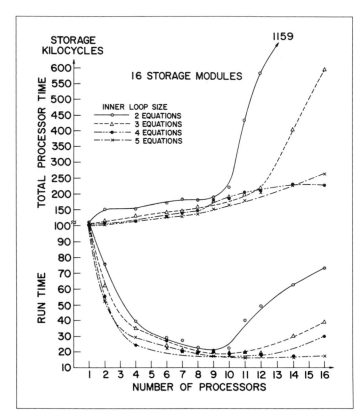

Fig. 7. Total processor and throughput times in electrical network analysis—16 storage modules.

to get program and data into the high-speed store and to output results. We include utilization figures for these executions in Table 3, to aid in analysis of the system behavior but not for evaluation purposes.

7.3.3 *The electrical network analysis problem.* This problem represents the solution of a set of simultaneous linear equations, described by a sparse coefficient matrix. The technique used for its solution on the executing simulator essentially comprises a relaxation procedure. Extensive runs have been made using a specific thirty-six node network, yielding twenty-six equations with up to four terms in each equation.

From the wealth of results obtained we present representative sets that indicate some general trends related to the characteristics and performance of the parallel processing system. Available space will not permit, however, detailed analysis in the present paper, nor does it permit a discussion of the equally interesting results obtained concerning speed of convergence, in particular, and other

Executive interference (col. 8) represents processor hold-ups due to the simultaneous attempts by two or more processors to access the system work-queues. These interferences are of course representative of a whole class of effects that can lead to performance degradation in parallel processing systems.

In Table 3 interference has been related to the number of interleaved storage modules and to the number of processors. In an actual system it is of course a complex function of the number of storage modules, of the degree of address interleaving, of the relationship between active jobs and the degree of program and data sharing, and of the total system utilization of storage. In optimizing a design, the numbers of processors and storage modules and the addressing scheme must be fixed subject to constraints related to cost, total storage capacity, the capacity of available storage modules, the degree of availability desired, and the expected nature of the work load. Processor utilization of storage alone is not very significant, since a critical factor is the I/O storage activity present, the degree of storage utilization required

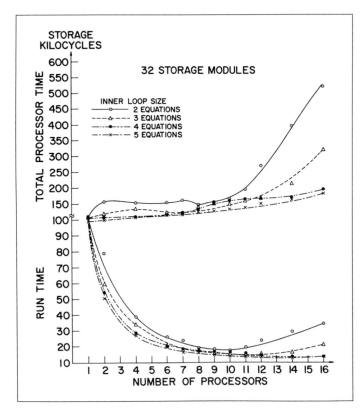

Fig. 8. Total processor and throughput times in electrical network analysis—32 storage modules.

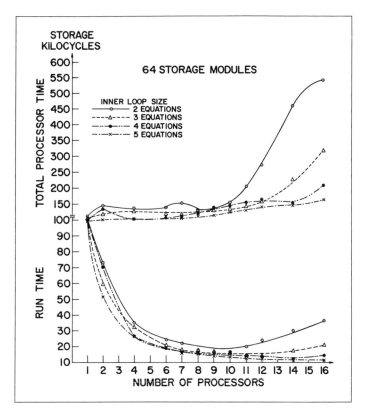

Fig. 9. Total processor and throughput times in electrical network analysis—64 storage modules.

effects which must be understood within the framework of a numerical analysis of the relaxation solutions.

Figures 7, 8, and 9 present the basic performance data, throughput time, and total processor time, for a total of one hundred and forty-four cases. The variables are the number of processors in the system (12 cases), the size of the inner loop as represented by the number of currents (from 2 to 5) evaluated in the loop, and the number of interleaved storage modules (16, 32, 64).

These curves clearly indicate the reduction in throughput time to be obtained from the use of parallel processing, the consequent increase in processor cost due to interferences of various sorts, the resultant effect of diminishing returns, and the actual increase in throughput time, when too many processors chase too few equations and generally get seriously "into each other's way."

For the smaller inner loops and when interference between processors is low, total processor times vary somewhat erratically. The causes for this are related to the relaxation pattern and the rate of convergence in each case. In fact there appears strong

circumstantial evidence that an ad hoc procedure, which does not guarantee sequential evaluation of the equations, improves performance. This point, however, requires further study.

Figure 10 reproduces some of the results of the previous three figures for the case of a five-equation inner loop. Table 4 lists these same results as a percentage of the time using one processor and compares them with the reciprocal of the number of processors.

Figure 11 indicates storage interference and parallel processing overheads as a function of the number of processors, with storage modularity again a parameter and an inner loop again comprising

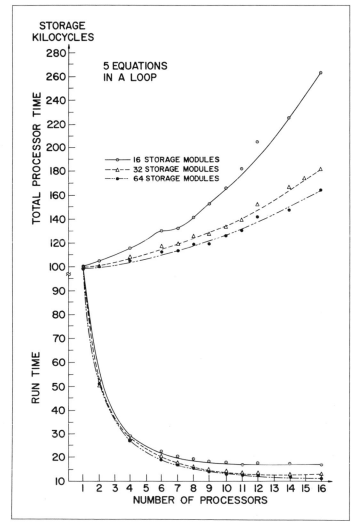

Fig. 10. Total processor and throughput times in electrical network analysis with number of storage modules as a parameter.

Table 4   Run time for resistor network system relative to the run time using one processor, with a five equation inner loop

| Number of processors | Relative time | | | 100 |
	16 Storage modules	32 Storage modules	64 Storage modules	No. of processors
1	100%	100%	100%	100%
2	52.8	51.2	51.2	50.0
4	29.5	27.9	27.1	25.0
6	22.4	20.3	19.5	16.7
7	20.9	17.9	17.1	14.3
8	19.2	16.8	15.8	12.5
9	17.8	15.2	14.2	11.1
10	17.6	14.5	13.7	10.0
11	16.8	13.9	12.9	9.1
12	17.5	13.9	13.0	8.3
14	17.3	13.2	11.7	7.2
16	17.7	13.7	11.7	6.3

the evaluation of five currents. Storage interference has previously been defined. The parallel processing overhead represents as a percentage the excess of total number of storage cycles required for execution, *excluding* storage interference cycles, when more than one processor is used, relative to the number of cycles required by a one-processor execution.

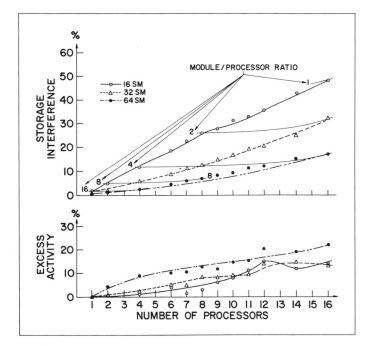

Fig. 11. Storage and executive interference.

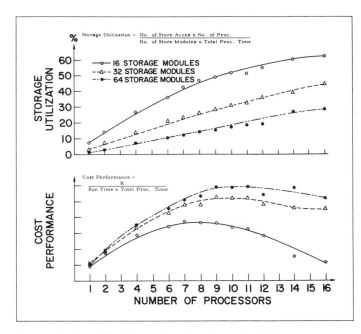

Fig. 12. Storage utilization and cost/performance factors.

Actual counts during execution show that in general some sixty-seven percent of store access are instruction fetches in this program and some thirty-three percent are data fetches. Thus incorporation of a substantial instruction buffer in each processor clearly reduces all interference by an order of magnitude, since of the four ways in which a storage interference can occur, only one—a data fetch conflicting with a data fetch—remains in the inner loop. Moreover, these measurements refer to a processor in which arithmetic speeds, as in Table 2, are of the order of magnitude of a memory cycle time, which implies a somewhat powerful processor. Thus in every sense the interference figures are worst case results which, with the performance curves to which they relate, support the view that storage interference is not a serious obstacle to parallel processing.

The four contours drawn on these curves represent lines of constant storage module-to-processor ratio. They slope slightly upward due to the statistical *Marbles and Boxes* [Rosenfeld, 1965] effect previously referred to.

Figure 12 presents two sets of data, based on the five-equation line loop. The upper family of curves relates to storage utilization. The reservations made at the end of Sec. 7.3.2, with reference to the significance of utilization figures, also apply. The second family of curves represents a first attempt at estimating the relative quality of processing, that is, some function of a cost/performance factor. Such a factor is intuitive and environment-sensitive, depending on the relative concern for speed and for costs of various sorts. For the present data we have chosen to display a function:

$$Q = \frac{K}{\text{throughput time} \times \text{total processor time}}$$

where K is a constant, throughput time a measure of the speed of computation, and total processor time a measure of the cost.

8. Conclusion

In this paper we have presented some thoughts on parallel processing. In particular we have chosen to survey the topic by including an extensive bibliography and some of the results of our work in this area. The discussion has had to be brief, but our intention has been to convey the picture of the potential that parallel processing systems offer for the future development of computing.

The key to successful exploitation lies in a new, unified, and scientific approach to the entire problem of the design and usage of computing systems. The development of large, integrated systems raises many problems, but there can be no doubt that economic solutions to these will be found. Their development should comprise a significant part of the computer system architectural design effort of the next few years.

Any ultimate evaluation of a parallel processing system within a working environment depends on actual operating experience. This in turn requires the existence of a system and the interest of users. Only when usable systems become available will the concept of parallel processing in integrated systems be accurately evaluated.

References

BlaaG64; BrigH64; ConwM63; CorbF65; DennJ66; DesmW64; DreyP58; FalkA64; GillS58; GregJ63; KatzJ66; LehmM65; LeinA59; McCuJ65; MiraW67; NievJ64; RoseJ65; SchlH??; ShedG66a, b; SlotD62; SmitR64; PL/I Language Specification, FormC28-6571

Bibliography

AlleM63; AmdaG62; AndeJ62, 65; ArdeB66; BaldF62; BlaaG64; BrigH64; BuchW62; BussB63; CoddE62; ComfW65; ConwM63; CorbF62, 65; CritA63; DaleR65; DennJ65, 66; DesmW64; DijkE65; DreyP58; ErnsH63; EstrG60, 63; EwinR64; FalkA64; ForgJ65; FranJ57; GillS58; GlasE65; GregJ63; HellH61, 66; KatzJ66; KinsH64; KnutD66; LehmM63a, 63b, 65; LeinA59; LourN59; MarcM63; McCaJ62; McCuJ65; MeadR63; MillW63; MiraW67; NievJ64; OssaJ65; PennJ62; RoseJ65; SchlH??; SeebR63; SenzD65; ShedG66a, 66b; SlotD62; SmitR64; SquiJ63; StraC59; VyssV65; WirtN66; IBM OS/360 PL/I Language Specification, Form C 28-6571; Proc. IFIP1962, "Symposium on Multi-Programming" 1963.

Section 4

Network computers and computer networks

The RW-400 and the CDC 6600 are actually computer networks by our definition of a computer (Chap. 2, page 17). Yet because of the restrictions on the quantity and location of the components in these structures, we still consider them to be computers. On the other hand, two or more computers which are separated physically, yet connected, constitute a computer network. Computer networks will appear in the future; it is important to understand the basis for them.

The RW-400—a new polymorphic data system

Chapter 38 presents the RW-400 (also called the AN/FSQ-27), a later version of the Ramo-Wooldridge RW-40 originally designed in 1959. The diagram (page 478) gives an indication of the relationship and names of the components. The PMS structure in Fig. 1 has more configuration details. At least six RW-400's were built for military command and control applications (although the number of computers of a type in existence has little to do with a machine's worth or ability).

The RW-40 ISP as given in Appendix 1 of Chap. 38 is a good example of a processor with a two-address instruction set. The ISP does not have index registers; it has a small state consisting of the accumulator (A), a limited extended accumulator (B), the program counter (P), and about 6 state bits. The Pc is limited by its ability to address directly only a 1,024-word Mp. The ISP is undoubtedly sufficient for solving the kinds of problems encountered by the computer and compares favorably with Whirlwind and the IBM 1800.

The RW-40 introduced multiple parts for reliability [Rothman, 1959]. Multiple C's (or Mp—Pc and Mp—Pio) are provided for redundancy and capacity. However, the S('Central Exchange) which provides communication among the C's may not have redundant parts. The multiple-computer concept can be viewed as the forerunner to our present computer networks, in which the central switching element is the Telephone Exchange. Over a longer time span, the RW-400 may be most significant as a pioneer. However, the whole system, with the exception of the small Mp's, is nicely designed. The problem of low speed T(typewriter, display)'s is handled well by transferring data from Mp—Pc to Ms(drum) for concurrent and independent T and P activity. Similar solutions are common for managing T activity by using an M, local to particular T's, and local C's.

The structure should be compared with the CDC 6600 (Chap. 39) and the network examples in Chap. 40.

The CDC 6400, 6500, 6600, 6416, and 7600

The CDC 6600 development began in 1960, using high-speed transistors and discrete components of the second generation. The first 6600 was delivered in September, 1964. Subsequent compatible successors included the 6400, in April, 1966, which was implemented as a conventional Pc(a single shared arithmetic function unit instead of the 10 D's); the 6500 in October, 1967, which uses two 6400 Pc's; and the 6416 in 1966, which has only peripheral and control processors. The first 7600, which is nearly compatible, was delivered in 1969. The dual processor 6700, consisting of two 6600 Pc's was introduced in October, 1969. Subsequent modifications to the series in 1969 included the extension to 20 peripheral and control processors with 24 channels. CDC also marketed a 6400 with a smaller number of peripheral and control processors (e.g., 6415-7 with 7). Reducing the maximum PCP number to 7 also reduced the overall purchase cost by approximately $56,000 per processor.

The computer organization, technology, and construction are described in Chap. 39. ISP descriptions for both the Pc and Pc ('Peripheral and Control Processors/PCP) are given in Appendices 1 and 2 of Chap. 39.

To obtain the very high logic speeds, the components are placed close together. The logic cards use a cordwood-type construction. The logic is direct-coupled transistor logic, with 5 nanoseconds propagation time and a clock of 25 nanoseconds. The fundamental minor cycle is 100 nanoseconds and the major cycle is 1,000 nanoseconds, also the memory cycle time. Since the component density is high (about 500,000 transistors in the 6600), the logic is cooled by conduction to a plate with Freon circulating through it.

This series is interesting from many aspects. It has remained the fastest operational computer for many years. Its large

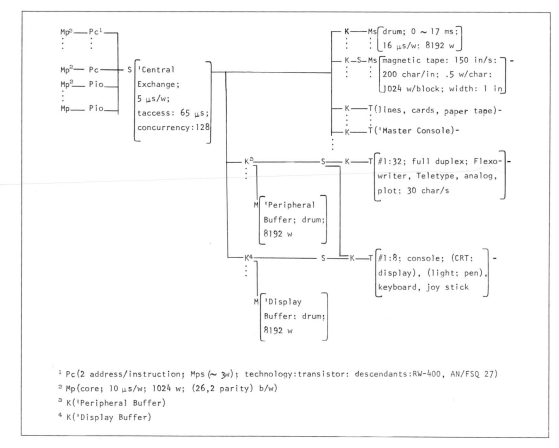

Fig. 1. RW-40 (Polymorphic) PMS diagram.

PMS structure

A simplified PMS structure of the C('6400, '6600) is given in Fig. 2. Here we see the C(io; #1:10) each of which can access the central computer (Cc) primary memory (Mp). Figure 2 shows

component count almost implies it cannot exist as an operational entity. Thus it is a tribute to an organization, and the project leader–designer Seymour Cray, that a large number exist. There are sufficiently high data bandwidths within the system so that it remains balanced for most job mixes (an uncommon feature in large C's). It has high performance Ms.disks and T.displays to avoid bottlenecks. The Pc's ISP is a nice variation of the general-registers processor and allows for very efficient encoding of programs. The Pc is nicely multi-programmed and can be switched from job to job more quickly than any other computer. Ten smaller C's control the main Pc and allow it to spend time on useful (billable) work rather than its own administration. The independent multiple data operators in the 6600 increase the speed by at least $2\frac{1}{2}$ times over a 6400 which has a shared D. Finally, it realizes the 10 C's in a unique, interesting, and efficient manner. Not many computer systems can claim half as many innovations.

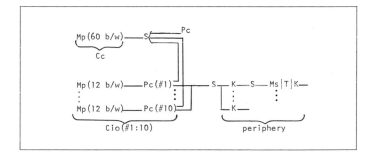

Fig. 2. CDC 6600 PMS diagram (simplified).

why we consider the 6600 to be fundamentally a network. Each Cio (actually a general-purpose, 12-bit C) can easily serve the specialized Pio function for Cc. The Mp of Cc is an Ms for a Cio, of course. By having a powerful Cio, more complex input-output tasks can be handled without Cc intervention. These tasks can include data-type conversion, error recovery, etc. The K's which are connected to a Cio can also be less complex. Figure 2 has about the same information as Thorton's Fig. 1 block diagram (Chap. 39).

A detailed PMS diagram for the C('6400, '6416, '6500, and '6600) is given in Fig. 3. The interesting structural aspects can be seen from this diagram. The four configurations, 6400 ~ 6600, are included just by considering the pertinent parts of the structure. That is, a 6416 has no large Pc; a 6400 has a single straightforward Pc; a 6500 has two Pc's; and the 6600 has a single powerful Pc. The 6600 Pc has 10 D's, so that several parts of a single instruction stream can be interpreted in parallel. A 6600 Pc also has considerable M.buffer to hold instructions so that Pc need not wait for Mp fetches.

The implementation of the 10 Cio's can be seen from the PMS diagram (Fig. 3). Here, only one physical processor is used on a time-shared basis. Each 0.1 μs a new logical P is processed by the physical P. The 10 Mp's are phased so that a new access occurs each 0.1 μs. The 10 Mp's are always busy. Thus the i.rate is 10×12 b/μs or 120 megabits/s. This process of shifting a new Pc state into position each 0.1 μs has been likened to a barrel by CDC. A diagram of the process is shown in Fig. 4.

The T's, K's, and M's are not given, although it should be mentioned that the following units are rather unique: a K for the management of 64 telegraph lines to be connected to a Cio; an Ms(disk) with four simultaneous access ports, each at 1.68 megachar/s data transfer rate, and a capacity of 168 megachar; an Ms(magnetic tape) with a K(#1:4) and S to allow simultaneous transfers to 4 Ms; the T (display) for monitoring the system's operation; K's to other C's and Ms's; and conventional T(card reader, punch, line printer, etc.).

ISP

The ISP description of the Pc is given in Appendix 1, Chap. 39. The Pc has a very clean, straightforward scientific-calculation-oriented ISP. We can consider it a variation on the general-register structure because the Pc state has three sets of general registers. Their use is explained both in Chap. 39 and its Appendix 1. This structure assumes that a program consists of several read accesses to a large array(s), a large number of operations on these accessed elements, followed by occasional write accesses to store results. We would agree that this is a valid assumption for scientific programs (e.g., look at a FORTRAN arithmetic statement), and it is probably valid for most other programs as well.

Cc has provisions for multiprogramming in the form of a protection and relocation address. The mapping is given in the ISP description for both Mp and Ms('Extended Core Storage-/ECS).

Appendix 2, Chap. 39, has an ISP description of the PCP. Appendix 2 includes a figure which shows the instruction decoding and execution as well. The 6600 PCP is about the same as the early CDC 160. The PCP has an 18-bit A register because it has to process addresses for the large Cc.

One interesting aspect of the 6600 which we question is the lack of communication among all components at the ISP (programming) level. When Pc stops, it has no way of explicitly informing any other components. There are no interprocessor interrupts. An io device cannot interrupt a Pio, nor can Pio's communicate with one another except by polling. The state switching for Pc is, however, elegant, since a Pio can request Pc to stop a job, store Mps, and resume a new task in one instruction. (The t.save + t.restore ~ 2 μs.)

The operating system

The Cio's functions are data transmission between a peripheral device and the large Cc via the Cio's Mp with some data transformation or conversions; complete task management, including initiation, termination, and error handling; and management of Pc. The Cio's perform in about the same manner as the C('Attached Support Processor) in the N('360 ASP) (Chap. 40, page 506). The operating-system software is managed by a single fixed Cio. The remaining nine Cio's are free, and as io tasks arise in the system, the Cio's assign themselves to particular tasks, carry out the tasks, and then free themselves to take on other tasks. The operating-system software resides in Mp(Pc) (that is, Cc) accessible to all Cio's and includes:

1 The variables which determine the state of a particular job, e.g., data pointers to Ms(disk, 'ECS), running time, a list of jobs to do, etc.

2 Programs for the Cio's
   a Parts of the operating system used by the Cio responsible for the system management
   b IO management programs (or programs to get the task management program from Ms) which the Cio's use

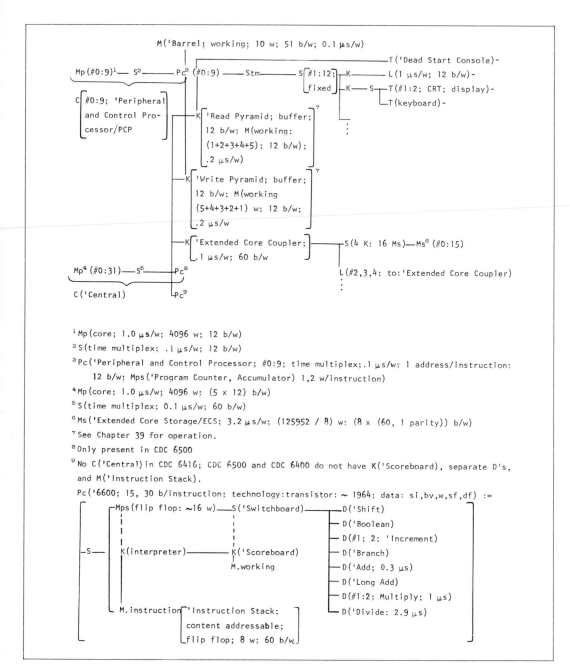

M('Barrel; working; 10 w; 51 b/w; 0.1 μs/w)

¹Mp(core; 1.0 μs/w; 4096 w; 12 b/w)

²S(time multiplex; .1 μs/w; 12 b/w)

³Pc('Peripheral and Control Processor; #0:9; time multiplex;.1 μs/w: 1 address/instruction:
    12 b/w; Mps('Program Counter, Accumulator) 1,2 w/instruction)

⁴Mp(core; 1.0 μs/w; 4096 w: (5 x 12) b/w)

⁵S(time multiplex; 0.1 μs/w; 60 b/w)

⁶Ms('Extended Core Storage/ECS; 3.2 μs/w; (125952 / 8) w: (8 x (60, 1 parity)) b/w)

⁷See Chapter 39 for operation.

⁸Only present in CDC 6500

⁹No C('Central) in CDC 6416; CDC 6500 and CDC 6400 do not have K('Scoreboard), separate D's,
    and M('Instruction Stack).

Fig. 3. CDC 6400, 6416, 6500, and 6600 PMS diagram.

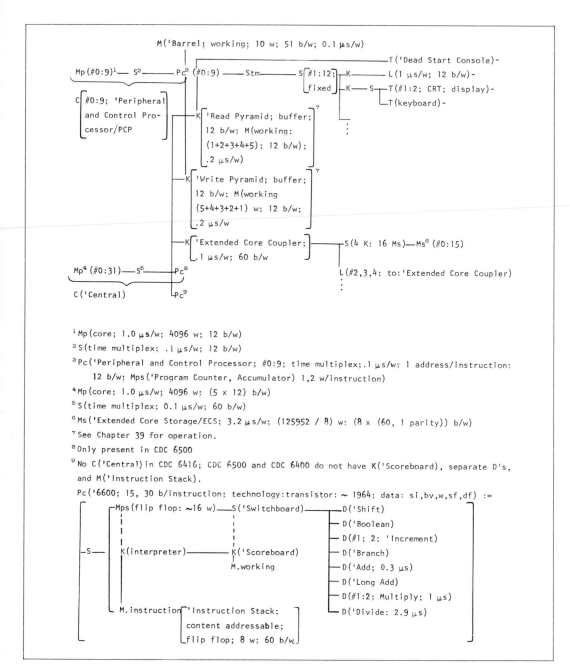

[1] Mp(core; 1.0 μs/w; 4096 w; 12 b/w)

[2] S(time multiplex; .1 μs/w; 12 b/w)

[3] Pc('Peripheral and Control Processor; #0:9; time multiplex;.1 μs/w: 1 address/instruction:
    12 b/w; Mps('Program Counter, Accumulator) 1,2 w/instruction)

[4] Mp(core; 1.0 μs/w; 4096 w: (5 x 12) b/w)

[5] S(time multiplex; 0.1 μs/w; 60 b/w)

[6] Ms('Extended Core Storage/ECS; 3.2 μs/w; (125952 / 8) w: (8 x (60, 1 parity)) b/w)

[7] See Chapter 39 for operation.

[8] Only present in CDC 6500

[9] No C('Central) in CDC 6416; CDC 6500 and CDC 6400 do not have K('Scoreboard), separate D's,
    and M('Instruction Stack).

Pc('6600; 15, 30 b/instruction; technology:transistor: ~ 1964: data: si,bv,w,sf,df) :=

Fig. 3. CDC 6400, 6416, 6500, and 6600 PMS diagram.

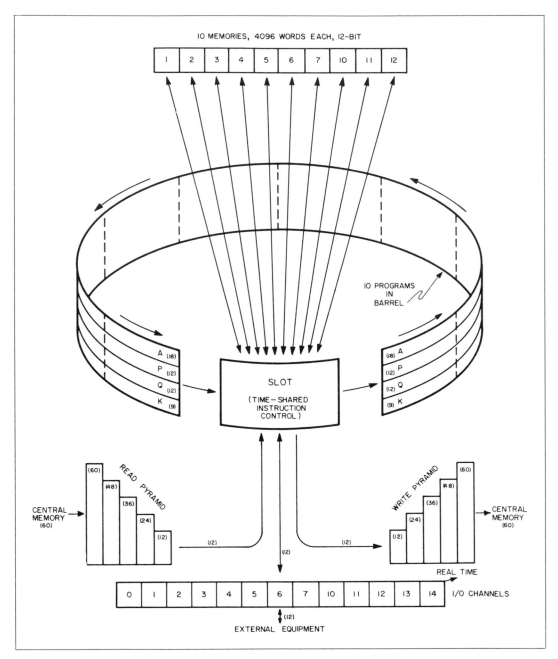

Fig. 4. CDC 6600 peripheral and control processors. (*Courtesy of Control Data Corporation.*)

In a typical system, one might expect to find the following assignment of PCP's to be:

1 Operating-system execution, including scheduling and management of Cc and all Cio's

2 Display of job status data on T(display)

3 Ms(disk) transfer management

4 T(printers, card reader, card punch)

5 L(# 1:3; to:C.satellite)

6 Ms(magnetic tape)

7 T(64 Teletypes)

8 Free to be used with Ms(disk) and Ms(magnetic tape)

9 Free

10 Free

CDC 7600

The CDC 7600 system is an upward compatible member of the CDC 6000 series. Although the main Pc in the 7600 is compatible with the main Pc of the 6600, instructions have been added for controlling the io section and for communicating between Large Core Memories/LCM and Small Core Memory/SCM. It is expected to compute at an average rate of four to six times a C('6600).

The PMS structure (Fig. 5) is substantially different from that of the 6600. The C('7600 Peripheral Processing Unit/PPU), unlike the C('6600 Peripheral and Control Processor)'s, has a loose coupling with the main C. The PPU's are under control of the main C when transferring words into SCM via K('Input-Output Section). The 15 C('PPU)'s have 8 input/output channels. These channels, which can run concurrently, provide the link between C('PPU) and peripheral Ms's and T's. Some of the PPU's are located in the same physical space as the Pc.

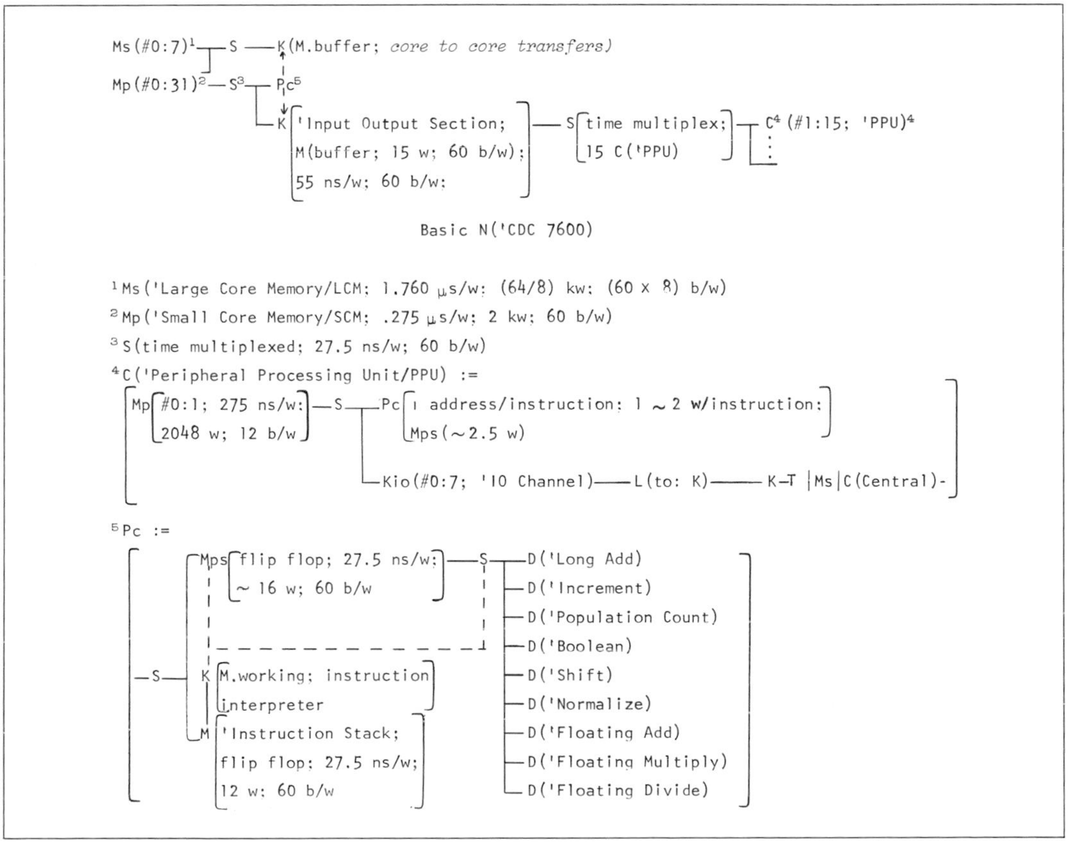

Fig. 5. CDC 7600 computer PMS diagram.

The 7600 Pc can be interrupted by a clock, the PPU's, and trap condition within the Pc. A breakpoint address, BPA, can be set up within Pc such that, on the program reaching BPA, a trap is initiated. This interruption scheme is in contrast to that of the 6600, which could not be interrupted or trapped. The 7600 interrupt may be a reaction to the lack of intercommunication in the 6600.

Conclusions

Although the 6600 was somewhat behind its announced delivery schedule and represented a significant drain on the financial resources of CDC, it is now clear that it is a successful product.

There have been instances of very large computers not being carried to completion either for financial or technical reasons. The 6600 seems to be the first large computer to achieve these marks of success. Here we are interested in the 6600 because it has held the ''world's largest computer'' title for so long.

Computer-network examples

In Chap. 40, we present examples of seven computer networks. There is a dearth of both computer networks and of papers on computer networks.

This chapter takes examples from papers and from knowledge of several existing or proposed networks.

Chapter 38

The RW-400—a new polymorphic data system[1]

R. E. Porter

Summary The RW-400 Data System, based upon modularly constructed, independently operating and flexibly connected components, is the logically evolved successor to conventional computer designs. It provides the means by which information processing requirements can be met with equipment capable of producing timely results at a cost commensurate with problem economic value. System obsolescence is minimized by the expandability in numbers and types of processing modules. Real time reliability is assured by component duplication at minimum cost and by the advanced design techniques employed in the system's manufacture. Man-machine communication facilities are program controlled for maximum flexibility. Parallel processing and parallel information handling modules increase the system's speed and adaptability when handling complex computing workloads. This polymorphic design truly represents an extension of man's intellect through electronics.

The RW-400 Data System is a new design concept. It was developed to meet the increasing demand for information processing equipment with adaptability, real-time reliability and power to cope with continuously-changing information handling requirements. It is a polymorphic system including a variety of functionally-independent modules. These are interconnectable through a program-controlled electronic switching center. Many pairs of modules may be independently connected, disconnected, and reconnected, in microseconds if need be, to meet continuously-varying processing requirements. The system can assume whatever configuration is needed to handle problems of the moment. Hence it is best characterized by the term "polymorphic"—having many shapes.

Rapid, program-controlled switching of many pairs of functionally-independent modules permits nondisruptive system expandability, operating reliability, simultaneous multi-problem processing capability, and man-machine intercommunication feasibility. These are only partially found in computers of conventional design.

Computer users have been forced heretofore to match problems to computer limitations. Problem changes posed serious reorientation and reprogramming difficulties. Changes from one computer

to another model, due to growth in applications, often resulted in large expenditures of time and money. During maintenance or malfunction of a conventional computer its entire processing capacity is shut down. Real time processing reliability cannot be maintained on an around-the-clock basis. The conventional machine must process its problems serially. This serious limitation is only partially alleviated by time-sharing or computing-element-doubling designs. The high cost-per-hour of conventional computer operation rules out direct man-machine intercommunication during other than emergency situations.

The radically-new polymorphic design concept of the RW-400 Data System was evolved by Ramo-Wooldridge engineers to provide a practical solution to those information processing problems now inadequately handled by conventional computer designs. The RW-400 is a powerful new tool in the field of intellectronics—the extension of man's intellect by electronics.

System description

The RW-400 Data System contains an optional number and variety of functionally-independent modules. These communicate via a central electronic switching exchange. Each module is designed, within practical economic and functional limits, to maximize system adaptability over a wide range of problem types and sizes. This new design embodies the latest proven electronic design techniques, assuring high processing speeds and high equipment reliability. The RW-400's modularity assures reliable, round-the-clock processing of information with controllable computing capacity degradation during module maintenance or malfunction. Practical man-machine intercommunication is achieved in the RW-400 system by use of program-controlled information display and interrogation consoles.

Figure 1 shows the over-all system design. Modules of various types communicate through a central exchange switching center. Computing and buffering modules provide control for the system. These modules are self-controlled and make possible completely independent processing of two or more problems. One of the computer modules may be designated the master computer and

[1]*Datamation*, vol. 6, no. 1, pp. 8–14, January/February, 1960.

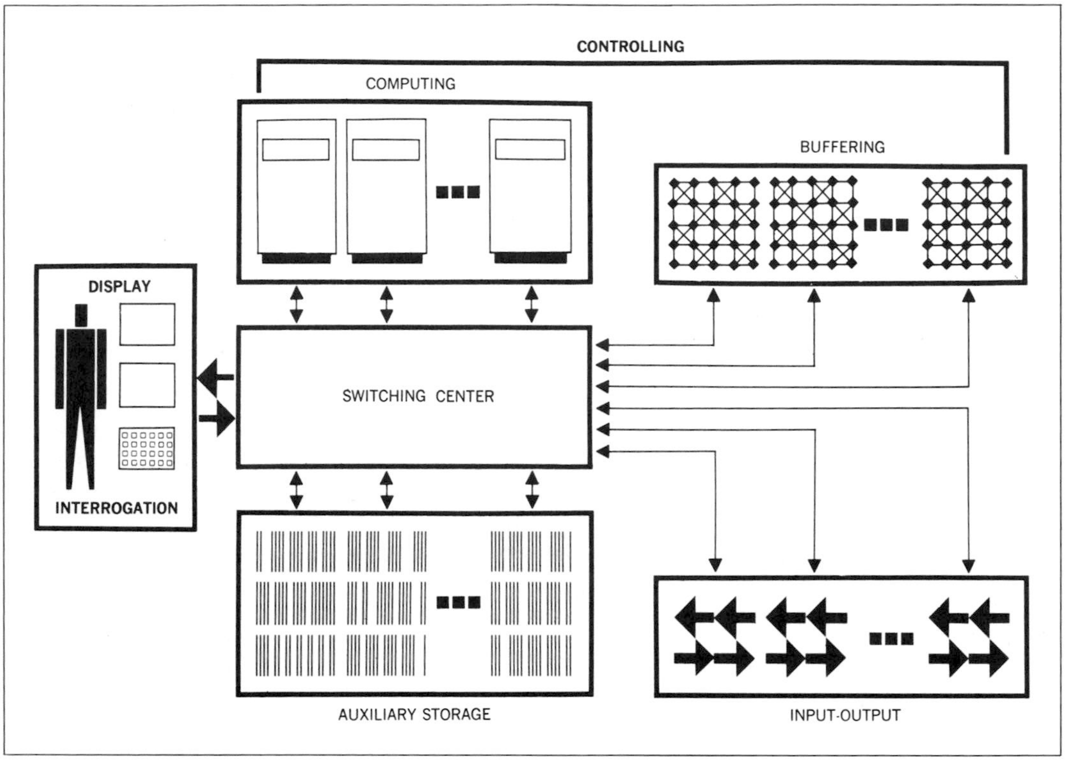

Fig. 1. The RW-400 data system.

in this role initiates and monitors actions of the entire system. An alert-interrupt network is provided to allow coordinated system action. Therefore, the system as applied to given information processing problems may change on a short range (microsecond) basis, thus providing, through programming, a self-organizing aspect to the system. In addition, the system may change through the years as the applications change. The most efficient and economical complement of equipment is applied to the problem at all times.

An RW-400 system is built around an expandable Central Exchange (CX) to which a number of primary modules may be attached. These are: Computer Modules (CM); self-instructed Buffer Modules (BM); Magnetic Tape Modules (TM); Magnetic Drum Modules (DM); Peripheral Buffer Modules (PB); and console communication Display Buffer Modules (DB). How many modules are put together in a system is entirely a function of system application. In addition to primary system modules, punched card, punched tape, high speed printing and control console devices are available. These handle nominal system in-

put/output requirements. Additional man-machine communication devices such as interrogation, display and control consoles, may be included in the system as problem requirements dictate. A Tape Adapter (TA) module is available to provide compatibility with magnetic tape of other computers. Information generated at Flexowriter inquiry and recording stations may be directly received by the system via the Peripheral Buffer Module. This latter module also buffers the receipt of TWX and punched tape information.

The way in which a particular RW-400 Data System functions depends on the number and type of each module included. It may initially be composed of the minimum number and variety of modules needed to do a small problem or the initial part of some large but yet-to-be-defined problem. Such a system would work much like a conventional computer. It would probably include a buffer module and thus have a parallel data handling capability not found in the conventional design at a comparable price. The initial system installation may then be augmented by the timely addition of modules.

A buffer module (BM) has the capability to control its acquisition and dissemination of information independently. The buffer provides a computer module with parallel data handling capability without complicating the problem processing program with the conventional intermixture of arithmetic and housekeeping instructions. Information previously generated by the processing program may be appropriately disposed of within the system while processing continues. Data needed at a subsequent time in the processing may be retrieved from system storage in advance of need while processing progresses. The simultaneity of these operations not only materially increases over-all processing speed but also increases the practical utility of the less costly types of internal system storage such as a magnetic tape.

The computer (CM) or buffer (BM) modules, when acting in a controlling capacity, may initiate connection to an information storage or handling module during that part of the processing program when the two can work profitably in unison. The pair of modules thus interconnected neither affect nor are affected by other modules. Logical interlocks prevent unwanted cross talk among modules. An intermodule communication system lets controlling modules signal status or alert other such modules of their need to communicate. The decision by a module receiving an alert signal to permit interruption or to proceed is optional with that module. The optional interrupt feature is that needed to make the often-discussed but seldom-used program interrupt capability both useful and practical. Programs may thus permit interruptions only at convenient points in the processing sequence.

Modules may be assigned, under program control, to work together on a problem in proportion to its needs. As soon as a module's function is complete for a given problem, that module may be released for reassignment to some other task. The system is thus self-controlled to match processing capacity to each problem for the time necessary to do the job. Full system capacity may be brought to bear upon a very large problem when needed. This capacity may be apportioned among a number of smaller problems for simultaneous processing, program compilation, program checkout, module maintenance etc., when it is not needed for maximum system effort.

From the preceding system description, it is apparent that such equipment can be expanded from a modest initial installation into a very powerful and comprehensive information processing center as requirements warrant. More specific descriptions of principal system modules follow to give the reader a better feel for how this system might perform his information processing work.

The functional modules

The key to appreciative understanding of the power of the RW-400 lies in knowledge of intermodule connection. It is appropriate to describe the Central Exchange (CX) unit first, then follow with descriptions of the various modules.

The central exchange

The Central Exchange performs the vital function of interconnecting a pair of modules whenever requested to do so by either a computer or a buffer module. Since internal programmed control is only possible within a computer or a buffer module, one of the interconnected pair of modules must be either a computer or a buffer. The time in which any connection may be made or broken is about 65 microseconds. An exchange has basic capacity to connect any of 16 computer or buffer modules to any of 64 auxiliary function modules. There is nothing sacred about the number 16 since it is possible to extend the CX module's interconnection matrix through design modification when need arises. The CX is an expandable, program-controlled, electronic switching center capable of connecting or disconnecting any available pair of modules in roughly the time of one computer instruction execution. Figure 2 illustrates the permissible module interconnections within the Central Exchange.

Every intersection on the illustration represents a possible connection between modules. The "x-ed" intersections indicate typical connections in force at any point in time. The control logic of the CX module's connection table prevents more than one interconnection on any horizontal (controlling) or vertical (controlled) data path representation on the diagram. When connection is requested of the Central Exchange while one of the required modules is already carrying out a previous assignment, the requesting module can be programmed to sense this condition and wait until connection can be made without interference. Should waiting be undesirable, the requesting module can go on about its business and check back later to see when the desired connection can be made. There is an implication here, of course, that knowing the kind of a system he is dealing with, a programmer requests connections in advance of need whenever possible.

Provision for master-slave control is included via an Assignment Matrix established within the CX module by a computer module previously assigned to master status. Such a provision is necessary to preclude inadvertent connection requests from unchecked programs or malfunctioning control modules from affecting sets of modules simultaneously processing another problem. Connection requests are therefore essentially filtered through both an assignment and an interconnection validity matrix prior to being acted

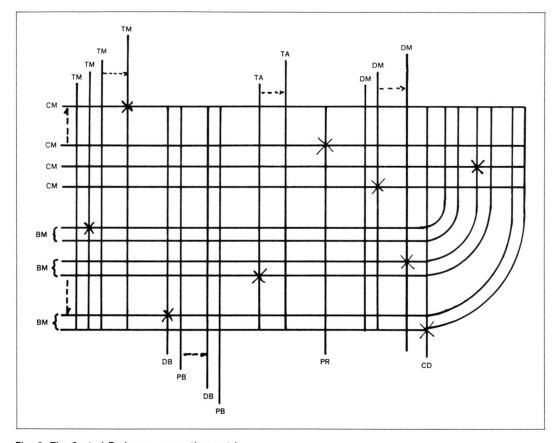

Fig. 2. The Central Exchange connection matrix.

upon by the Central Exchange. The computer module manually assigned to master status is the only one permitted to cause the interconnection of a pair of modules which does not include itself.

The computer module (See Fig. 3)

The Computer Module (CM) is a self-sufficient, general purpose, two-address, parallel word, fixed point, random access computer. Its internal magnetic core memory has a capacity of 1024 words. A computer word consists of 26 information bits and 2 parity bits. Each parity bit is associated with the 13-bit half word transferred in parallel via the Central Exchange to other system modules. The instruction repertoire of the CM consists of 38 primary instructions whose various modes effectively result in over 300 different operations. Of the 39 available CM-400 instructions, 24 may be classified as "arithmetic" and 10 as "program control" or "sequence determining" instructions. Five additional instructions may be

classified as "external" or "input/output" instructions. All but three of the 24 arithmetic instructions fit into a symmetric scheme of classification wherein there are seven basic operations, each having three distinct modes. The seven basic operations are—add, subtract, absolute subtract, multiply, divide, square root and insert. The three modes are—Replace, Hold and Store. If we let the capital letter "G" identify the first operand, "H" identify the second operand, an "∘" signify an arbitrary operation, the symbol "→" indicate replace, and "A" the word in the accumulator, then the three modes may be characterized as:

Replace: $H \circ G \rightarrow H, A$
  Hold: $H \circ G \rightarrow A$
  Store: $A \circ G \rightarrow H, A$

The three remaining arithmetic operations are Add Accumulate wherein the contents of H and G are added to the Accumulator;

Multiply Accumulate wherein the contents of H are multiplied by G and added to A; and Transmit where the contents of G are stored in H.

The ten program control instructions are Store, Store Double Length Accumulator, Load Accumulator, Insert Mask in the S Register, Stop, Link Jump, Compare Jump, Tally Jump, Test Jump and a Multi-purpose Shift.

The five external instructions are those which cause data to be transmitted to or received from a device external to the computer. Each command is multi-purpose in nature and hence equivalent to several conventional external instructions. The commands are—Command Output, Data Input, Conditional Data Input, Data Output and Character Transfer. A comprehensive discussion of the variation of each of these commands is not pertinent to this article.

Suffice it to say that commands are available for carrying out a wide variety of intermodule data communication.

The interrupt capability of a Computer Module is a logical generalization of the "trapping" feature found on several conventional computers. It permits the automatic interruption of a program, at the option of the program, when the computer module receives an "alert" that a condition requiring attention has arisen. It can be used to warn the program when an error of some type has occurred, minimize unproductive computer waiting time while another module completes its task, eliminate many programmed status test instructions and provide a convenient means of subjecting one computer module to the control of another. Program control of interruptions within a CM-400 is accomplished through the sense register S. This register may be filled with an interrupt

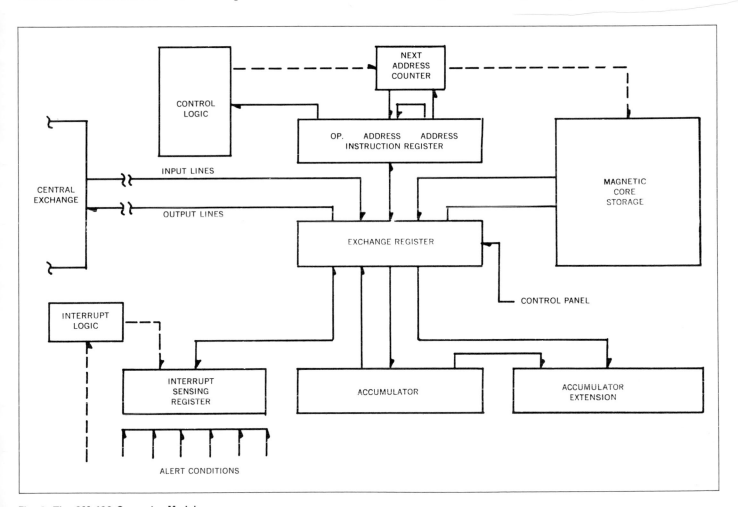

Fig. 3. The CM-400 Computer Module.

RW-400 analysis console.

mask by means of the Insert S instruction. A bit by bit correspondence exists between the S register and the interrupt register and the interrupt register I to which the alert lines are connected. A Test Jump instruction can be used to examine the coincidence between these registers of an alert signal in a bit position corresponding to a one in the S register mask. If an alert is received by the computer during the execution of an instruction, control will be transferred to memory location "O" at the end of the instruction if, and only if, (a) the sense bit corresponding to the alert is a "one," (b) the master sense bit is a "one," and (c) the instruction was not an "Insert S." The master sense bit in the S register may be programmed to permit the interrupt to take place according to the interrupt mask or to inhibit interrupt until the program can conveniently cope with it. All instructions being executed at the time an interrupt condition occurs are completed before the interruption is allowed to take place.

Figure 3 schematically illustrates the Computer Module's primary registers and the interconnecting information paths.

Typical two-address addition and subtraction times are approximately 35 microseconds including memory access time. Multiplication takes about 80 microseconds, and division and square root about 130 and 170 microseconds respectively.

Before attempting to draw a comparison between a CM and a deluxe conventional computer the reader should bear in mind

the trade offs in features versus cost; parallel processing versus sequential processing; independent information handling versus program complicating "housekeeping"; and real time system reliability versus periodic inoperability. The only valid comparison is that between the RW-400 Data System and a conventional computer applied to the same task. The contribution to the RW-400 system made by the Buffer Modules can be better assessed by the reader after the following description has been considered.

The buffer module

A Buffer Module consists of two independent logical buffer units, each having 1024 words of random access magnetic core storage and a number of internal registers used in performing its functions when in the self-controlling mode. A Buffer Module may be connected to a Computer Module so that the Buffer's core storage is accessible to the computer as an extension of the computer's own storage. A Buffer may also serve as an intermediary device between a computer and another module, such as a tape or drum, to minimize time conventionally lost in data transfers. The Buffer is capable of recognizing and executing certain instructions stored in its own memory. It can therefore be left to perform data handling functions on its own while computer modules are otherwise occupied.

A Buffer Module may be connected to a Computer Module and the buffer 1024 word storage used as an indirectly addressed extension of the computer's own working storage. When the address 1023 (all ones) appears in the operand field of a computer instruction to be executed, the computer is signalled that the operand refers to some cell in buffer storage. The computer then uses the number in the buffer read register R (or in the case of a few instructions, the buffer write register W) as the effective address designated by the operand field of the instruction. Extended addressing may be used in either the first or second operand field of the instruction or in both operand fields. If extended addressing is used in only one operand field, the effective address designated by that field is the number in register R. A "1" is automatically added to the contents of the R register after the instruction is executed. If extended addressing is used in both operand fields of an instruction, the effective address of the first operand is the number in register R and the effective address of the second operand is one more than the number in register R. A "2" is automatically added to the contents of register R after the execution of this type of instruction. The R (or W) register may be preset to any desired initial condition by means of the computer's Command Output instruction. All the commands being executed by the computer must be stored within the computer module's storage and may not be in buffer cells addressed by the computer at execution time. The extended addressing and buffer register indexing may be used to materially simplify repetitive data acquisition operations.

The primary function of a Buffer Module is not, however, that of an auxiliary computer storage unit. The drum and tape modules more aptly serve this function in the RW-400 system. A Buffer Module is capable of operating autonomously and of controlling other modules such as Tape Modules, Drum Modules, Peripheral Buffers, Display Buffers, Printers or Plotters. This capability enables the Buffer Modules in a system to perform routine tape searching and data transferral tasks thereby freeing the Computer Modules to do more computing. In its "self-instruction" mode, the buffer executes its own internally stored program in much the same fashion as a computer. The memory of a Buffer Module will therefore be occupied by its own control programs as well as blocks of data which it is holding for transmission to other units. The buffer is used to acquire information from the relatively slower auxiliary storage and communication modules while the computer proceeds at high speed. Blocks of information retrieved in advance of computer need by the buffer may then be rapidly transferred to the computer's own storage or operated upon as they stand in the buffer via the indirect addressing capability of the computer. Another feature of the buffer is its switching capability. Each Buffer Module is composed of two buffer units tied together. A unit function switching feature permits the employment of the two units together in an alternating mode of operation. Continuous information transfer from tape to computer, for example, may be accomplished without stopping the tape unit. A switching instruction executed simultaneously by both units of a Buffer Module causes whatever devices were connected to the first unit to be connected to the second and vice versa.

Now that the functional controlling modules and the module interconnection concept have been discussed, the more conventional auxiliary storage modules available with the system may be described to round out the processing capability of the system.

The tape modules

A Tape Module consists of an altered Ampex FR-300 tape transport plus the necessary power supplies and control circuitry to effect information reading, writing and control. One inch mylar tape is used. Information is written on 16 channels—two of which are clock channels. The remaining 14 channels consist of 13 information bits plus parity. The information reading or recording rate is 15,000 computer words per second. Data may be recorded on tape in variable blocks up to a maximum of 1024 words per block

(the size of the storage available to hold the data in a sending or receiving module). Each block is preceded by a block identification which permits selective tape information searching by a Buffer Module. Single blocks imbedded in a tape file of other blocks can be overwritten. A two-stack head permits automatic verification of each block as it is written. Readback parity errors are automatically detected during the writing process. Thus dropout areas may be determined while the data is still available in a computer or buffer for recording elsewhere.

A description of the RW-400's tape handling capability would not be complete without mentioning the Tape Adapter (TA) module. This is a self-contained unit capable of performing the reading and writing of magnetic tapes in a format acceptable to the IBM 704 and 709 systems. The TA consists of an Ampex FR-300 half-inch digital tape transport, including dual gap head and servo control system; reading, writing and control circuits; and a module housing with its own blower and power supply.

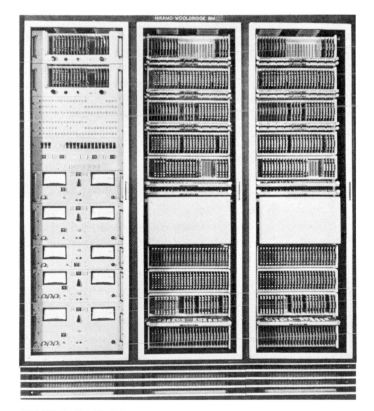

RW-400 Buffer Module.

The drum module

The Drum Module (DM) contains a magnetic drum with storage capacity of 8192 words. It may be connected to either a Computer or a Buffer Module through the Central Exchange. Average access time to the first word position on the drum is $8\frac{1}{2}$ milliseconds. Successive words are transmitted at the rate of 60,000 computer words per second. The Drum Module is conventionally used as an intermediate item storage device to minimize tape handling time.

Special system communication modules

The external data and man-machine communication of the RW-400 Data System are handled via drum buffer modules. A wide variety of asynchronously operated equipment is speed matched and program controlled through the features designed into these special system communication modules.

The Peripheral Buffer (PB) provides input/output buffers for communication between Computer or Buffer Modules and relatively slow speed external devices such as Flexowriters, Plotters, Punched Tape Handlers, Teletype Lines and Keyboard Operated Equipment. The Peripheral Buffer stores its information in four pairs of bands which operate alternately as circulating registers. Each band contains eight input and eight output buffers for a total of 32 input buffers and 32 output buffers in each Peripheral Buffer Module. Each buffer is a drum band sector 64 computer words long. Conventionally one input and one output buffer sector are connected to each external device (such as a Flexowriter) to permit two-way communication between the external device and the RW-400 system.

The display buffer

A Display Buffer (DB) acts as a recirculating storage for the cathode ray tube display units in a Display Console. Information to be displayed is sent to the DB band associated with a particular display tube via the Central Exchange. The Display Buffer sends only status information back to other system modules upon request. The information displayed on any tube is controlled by the bit pattern sent to the Display Buffer. The display pattern is regenerated 30 times per second to minimize image fading and flicker. The preceding explanation of the Display Buffer has little meaning to a reader unfamiliar with the features of the Display Console itself. This console is therefore described in more detail in the following paragraphs.

Display consoles

Display Consoles can give a problem "analyst" or "monitor" a visual picture of the status or results of any information being handled by the RW-400 system. In addition to the actual Cathode Ray Tube, numerical indicator, signal lamp and typewriter information outputs, several types of keyboard activated system control and parameter entry facilities are provided on the console. The total man-machine communication facility represented by each console is designed to be primarily a function of the computer control programs initiated by the analyst via his console.

A set of Display Control Keys generate messages which are recorded on a Peripheral Buffer sector for later interpretation and display generation by a computer program. A set of Process Step Keys are provided the analyst so that he can initiate preprogrammed system processing variations. Associated with the Process Step Keys is an overlay or "program card" which permits the assignment of a variety of meanings to the set of Process Step Keys. Insertion of the overlay by the analyst gives him a unique label for each Process Step Key and automatically cues the controlling computer to assign the corresponding set of programs to each key message. A Data Entry Keyboard is provided on the console so that the analyst can enter control parameters when asked to do so via the display devices.

A Joystick Lever affords the console operator a means of controlling the position of cross hair markers on the cathode ray display tubes. Associated with the joystick are control keys which may be used to send a message to the controlling computer specifying the coordinates of the cross hairs. Control programs may be written, for example, to act upon this information to reorient the display with respect to the area selected by the cross hair position.

A Light Gun is also provided as a means of selecting any point on the cathode ray tube displays. The gun emits a small beam of light. With the beam centered on a given point on the cathode ray display tube, pressing the trigger results in the automatic generation of a message to the Peripheral Buffer specifying the address in the Display Buffer containing the coordinates of the selected point.

A set of Status and Error lights are contained on the Display Console to provide the console operator with over-all knowledge of the system and thus minimize conflicting control requests and intermodule interference. For example, a Peripheral Buffer may not be ready to accept a console key message until after certain previously requested control actions have been completed. The Status Lights indicate this condition to the console operator so that he may act accordingly.

The printer module

The Printer Module (PR) is basically a 160 column, 900 line per minute Anelex type printer. It receives information from either a Computer or a Buffer module via the Central Exchange. Individual characters to be printed are represented by a 6-bit code and are transmitted four to a computer word. Zero suppression, line completion and information block end codes are included for format control. A plugboard is provided for flexibility in columnar data arrangement. Paper feed is controlled by means of a loop of 7-channel punched paper tape. Control of the printing operation has been arranged so that the connected control module may send line headings from one set of memory locations, stop sending information while going to a different part of the memory, and then proceed to send data from this new set of memory locations to complete a line of print.

The punched card modules

The RW-400 Data System may be equipped with a high speed punched card reading module (CR) and an IBM card punch. The CR communicates with Computer or Buffer modules via the Central Exchange. It is capable of reading 80 column punched cards at the rate of 2,500 cards per minute. The card punch is connected to the system through the Peripheral Buffer Module (PB) since it is a relatively low speed device. Emphasis has not been placed on directly connected punched card equipment since the sources of large volumes of punched cards usually convert this data into magnetic tape form which may be more rapidly handled using the Tape Adapter Module (TA).

References

RothS59; WestG60

APPENDIX 1   RW 40 ISP DESCRIPTION

```
                                        Appendix 1

                                  RW-40 ISP Description

    The description does not include Input-Output instructions, interrupts and communication with the other computers or processors.
    The description was taken from the Preliminary Manual of Information on the RW-40 and is no doubt changed in final machines.
    Pc State
       A<26:1>                                             Arithmetic register
       B<26;1>                                             extension to A
       AB[0:1]<26:1>:= A□B                                 Arithmetic register (double)
       P<10:1>                                             Program Counter
       Ov                                                  Overflow for arithmetic shifts, +, -, and /
       SR<20:1>                                            Sense Register
       Parity error                                        for Mp and transfer to other computers
       Program error                                       undefined command or incorrect sequence of IO commands
       Run

    Mp State
       M[1:1022]<26:1>                                     Mp registers 0 and 1023 are inaccessible

    Pc Console State
       CJS<8:1>                                            conditional jump switches
       Control␣panel␣test                                  communication indicator

    External State for IO and Other Computers
       Tape␣read                                           tape search flag
       External␣Address/EA<10:1>                           register associated Pc to address another module
       M[0:1023]<26:1>                                     extra memory being accessed by External Address register
       I␣cond<19:1>                                        interrupt conditions to Pc
       IO␣Select<3:1>                                      1 of 8 IO devices can be selected
       IO␣Data<13:1>                                       IO device Data

    Instruction Format
       instruction/i<26:1>
          f/op<6:1> := i<26:21>                            function or op code bits
          g<10:1>   := i<20:11>                            first address
          j<5:1>    := g<5:1>                              test selection parameter
          h<10:1>   := i<10:1>                             second address

    Operand Calculation Process
       G<26:1> := (G'; next                                first operand
          (g = 1777₈) →External␣Address ← External␣Address + 1)
          G'<26:1> := ((g = 0) → 0;
                      (0 < g < 1777) →M[g]<26:1>;
                      (g = 1777) → M[External␣Address]<26:1>)
       H<26:1> := (H'; next                                second operand
          (h = 1777) →External␣Address ←External␣Address + 1)
          H'<26:1> := ((h = 0) → 0;
                      (0<h<1777) → M[h]<26:1>
                      (g = 1777) → M[External␣Address]<26:1>)
```

Instruction Interpretation Process

Run → (instruction ← M[P]; P ← P + 1; next   *fetch*

    Instruction_execution)   *execute*

Instruction Set and Instruction Execution Process

    Instruction_execution := (

Transmit	(:= op = 27) → (H ← G);

Arithmetic (1's complement)

Replace Add	(:= op = 0) → (0v,A ← H + G; next H' ← A);
Hold Add	(:= op = 1) → (0v,A ← H + G);
Store Add	(:= op = 2) → (0v,A ← A + G; next H' ← A);
Replace Subtract	(:= op = 3) → (0v,A ← H - G; next H' ← A);
Hold Subtract	(:= op = 4) → (0v,A ← H - G);
Store Subtract	(:= op = 5) → (0v,A ← A - G; next H' ← A);
Replace Absolute Subtract	(:= op = 6) → (A ← abs(H) - abs(G); next H' ← A);
Hold Absolute Subtract	(:= op = 7) → (A ← abs(H) - abs(G));
Store Absolute Subtract	(:= op = 10) → (A ← abs(A) - abs(G); next H' ← A);
Replace Multiply	(:= op = 11) → (AB ← H x G; next H' ← A);
Hold Multiply	(:= op = 12) → (AB ← H x G);
Store Multiply	(:= op = 13) → (AB ← A x G; next H' ← A);
Replace Divide	(:= op = 14) → ((H ≥ G) → 0v ← 1;
	(H < G) → (
	A,B ← H/G; next H' ← A));
Hold Divide	(:= op = 15) → ((H ≥ G) → 0v ← 1;
	(H < G) → (A,B ← H/G));
Store Divide	(:= op = 16) → ((A ≥ G) → 0v ← 1;
	(A < G) → (
	A,B ← A/G; next H' ← A));
Replace Square Root	(:= op = 17) → (A ← sqrt(H+G); next H' ← A);
Hold Square Root	(:= op = 20) → (A ← sqrt(H+G));
Store Square Root	(:= op = 21) → (A ← sqrt(A+G); next H' ← A);
Accumulate Add	(:= op = 25) → (A ← 0v□A + H + G);
Accumulate Multiply	(:= op = 26) → (A ← 0v□A + H x G);

Shift, g⟨10:1⟩ is used to control the shift as follows:

  g⟨5:1⟩ *specifies number of shifts*

  g⟨6⟩ = 1 → *shift left*; g⟨6⟩ = 0 → *shift right*

  g⟨7⟩ = 1 → *indicate an overflow*

  g⟨8⟩ = 1 → *round the result*

  g⟨9⟩ = 1 → *signed arithmetic*; g⟨9⟩ = 0 → ⌊*logical*⌋

  g⟨10⟩= 0 → *A is the operand*; g⟨10⟩ = 1 → *AB is the operand*

  Shift := (op = 30) → (0v□A ← f(A x $2^{g\langle 5:1\rangle}$, B x $2^{g\langle 5:1\rangle}$, g⟨6:10⟩); next H ← A);

Logical or boolean vector data:

Replace Insert	(:= op = 22) → (A ← (H ∧ ¬ G) ∨ (A ∧ G); next H' ← A);
Hold Insert	(:= op = 23) → (A ← (H ∧ ¬ G) ∨ (A ∧ G));
Store Insert	(:= op = 24) → (A ← A ∧ G; next H' ← A);
Test Jump	(:= op = 31) → (

```
        (g<10:7> ≠ 0) → (
          A ← ((g<10> → I cond; ¬ g<10> → 177777777) ∧
               (g<9> → SR; ¬ g<9> → 177777777) ∧
               (g<8> → IO Select IO Data; ¬ g<8> → 177777777) ∧
               (g<7> → CJS; ¬ g<7> → 177777777)); next
        (g<6> ⊕ Test) → (P ← h));
The Test condition is a selected bit of A, or other Pc or IO bits.
        Test := ((j = 0) → 0;
                (i ≤ j ≤ 32) → A<j>;
                (j = 33) → (Ov; Ov ← 0);
                (j = 34) → (Parity error; Parity error ← 0);
                (j = 35) → (Control panel test; Control panel test ← 0);
                (j = 36) → (Tape read; Tape read ← 0);
                (j = 37) → (Program error; Program error ← 0))
Link Jump              (:= op = 32) → ((g ≠ 0) → (P ← h; G<10:1> ← P);
                                        (g = 0) → (P ← h));
Tally Jump             (:= op = 33) → ((G = ¬ 0) → (P ← h);
                                        (G = 0) → ;
                                        (G > 0) → (G' ← G - 1; P ← h);
                                        (G < 0) → (G ← G + 1));
Compare Jump           (:= op = 37) → (A ≤ G) → P ← h;
Load A                 (:= op = 34) → (A ← 0 g h);
Insert S               (:= op = 35) → (S ← (A ∧ (0 g h)) ∨ (S ∧ ¬ (0 g h)));
Store AB               (:= op = 36) → (G ← B; H ← A;
                                        (g = 0) ∧ (h = 0) → (A ← B; B ← A))
                                        )                    end Instruction execution
```

Chapter 39

Parallel operation in the Control Data 6600[1]

James E. Thornton

History

In the summer of 1960, Control Data began a project which culminated October, 1964 in the delivery of the first 6600 Computer. In 1960 it was apparent that brute force circuit performance and parallel operation were the two main approaches to any advanced computer.

This paper presents some of the considerations having to do with the parallel operations in the 6600. A most important and fortunate event coincided with the beginning of the 6600 project. This was the appearance of the high-speed silicon transistor, which survived early difficulties to become the basis for a nice jump in circuit performance.

System organization

The computing system envisioned in that project, and now called the 6600, paid special attention to two kinds of use, the very large scientific problem and the time sharing of smaller problems. For the large problem, a high-speed floating point central processor with access to a large central memory was obvious. Not so obvious, but important to the 6600 system idea, was the isolation of this central arithmetic from any peripheral activity.

It was from this general line of reasoning that the idea of a multiplicity of peripheral processors was formed (Fig. 1). Ten such peripheral processors have access to the central memory on one side and the peripheral channels on the other. The executive control of the system is always in one of these peripheral processors, with the others operating on assigned peripheral or control tasks. All ten processors have access to twelve input-output channels and may "change hands," monitor channel activity, and perform other related jobs. These processors have access to central memory, and may pursue independent transfers to and from this memory.

Each of the ten peripheral processors contains its own memory for program and buffer areas, thereby isolating and protecting the

[1]*AFIPS Proc. FJCC*, pt. 2 vol. 26, pp. 33–40, 1964.

more critical system control operations in the separate processors. The central processor operates from the central memory with relocating register and file protection for each program in central memory.

Peripheral and control processors

The peripheral and control processors are housed in one chassis of the main frame. Each processor contains 4096 memory words of 12 bits length. There are 12- and 24-bit instruction formats to provide for direct, indirect, and relative addressing. Instructions provide logical, addition, subtraction, shift, and conditional branching. Instructions also provide single word or block transfers to and from any of twelve peripheral channels, and single word or block transfers to and from central memory. Central memory words of 60 bits length are assembled from five consecutive peripheral words. Each processor has instructions to interrupt the central processor and to monitor the central program address.

To get this much processing power with reasonable economy and space, a time-sharing design was adopted (Fig. 2). This design contains a register "barrel" around which is moving the dynamic information for all ten processors. Such things as program address, accumulator contents, and other pieces of information totalling 52 bits are shifted around the barrel. Each complete trip around requires one major cycle or one thousand nanoseconds. A "slot" in the barrel contains adders, assembly networks, distribution network, and interconnections to perform one step of any peripheral instruction. The time to perform this step or, in other words, the time through the slot, is one minor cycle or one hundred nanoseconds. Each of the ten processors, therefore, is allowed one minor cycle of every ten to perform one of its steps. A peripheral instruction may require one or more of these steps, depending on the kind of instruction.

In effect, the single arithmetic and the single distribution and assembly network are made to appear as ten. Only the memories are kept truly independent. Incidentally, the memory read-write cycle time is equal to one complete trip around the barrel, or one thousand nanoseconds.

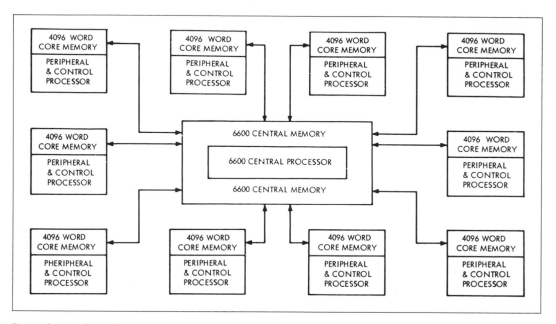

Fig. 1. Control Data 6600.

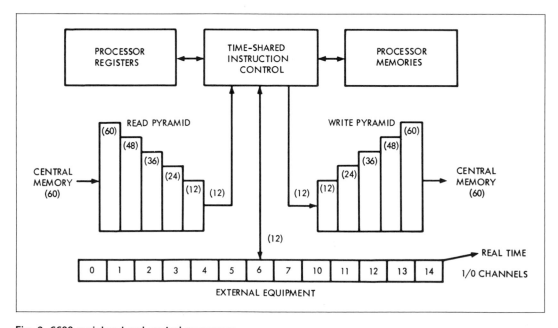

Fig. 2. 6600 peripheral and control processors.

Input-output channels are bi-directional, 12-bit paths. One 12-bit word may move in one direction every major cycle, or 1000 nanoseconds, on each channel. Therefore, a maximum burst rate of 120 million bits per second is possible using all ten peripheral processors. A sustained rate of about 50 million bits per second can be maintained in a practical operating system. Each channel may service several peripheral devices and may interface to other systems, such as satellite computers.

Peripheral and control processors access central memory through an assembly network and a dis-assembly network. Since five peripheral memory references are required to make up one central memory word, a natural assembly network of five levels is used. This allows five references to be "nested" in each network during any major cycle. The central memory is organized in independent banks with the ability to transfer central words every minor cycle. The peripheral processors, therefore, introduce at most about 2% interference at the central memory address control.

A single real time clock, continuously running, is available to all peripheral processors.

Central processor

The 6600 central processor may be considered the high-speed arithmetic unit of the system (Fig. 3). Its program, operands, and results are held in the central memory. It has no connection to the peripheral processors except through memory and except for two single controls. These are the exchange jump, which starts or interrupts the central processor from a peripheral processor, and the central program address which can be monitored by a peripheral processor.

A key description of the 6600 central processor, as you will see in later discussion, is "parallel by function." This means that a number of arithmetic functions may be performed concurrently. To this end, there are ten functional units within the central

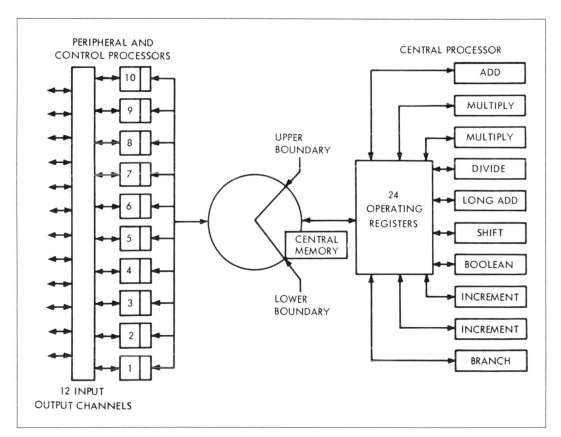

Fig. 3. Block diagram of 6600.

processor. These are the two increment units, floating add unit, fixed add unit, shift unit, two multiply units, divide unit, boolean unit, and branch unit. In a general way, each of these units is a three address unit. As an example, the floating add unit obtains two 60-bit operands from the central registers and produces a 60-bit result which is returned to a register. Information to and from these units is held in the central registers, of which there are twenty-four. Eight of these are considered index registers, are of 18 bits length, and one of which always contains zero. Eight are considered address registers, are of 18 bits length, and serve to address the five read central memory trunks and the two store central memory trunks. Eight are considered floating point registers, are of 60 bits length, and are the only central registers to access central memory during a central program.

In a sense, just as the whole central processor is hidden behind central memory from the peripheral processors, so, too, the ten functional units are hidden behind the central registers from central memory. As a consequence, a considerable instruction efficiency is obtained and an interesting form of concurrency is feasible and practical. The fact that a small number of bits can give meaningful definition to any function makes it possible to develop forms of operand and unit reservations needed for a general scheme of concurrent arithmetic.

Instructions are organized in two formats, a 15-bit format and a 30-bit format, and may be mixed in an instruction word (Fig. 4). As an example, a 15-bit instruction may call for an ADD,

designated by the f and m octal digits, from registers designated by the j and k octal digits, the result going to the register designated by the i octal digit. In this example, the addresses of the three-address, floating add unit are only three bits in length, each address referring to one of the eight floating point registers. The 30-bit format follows this same form but substitutes for the k octal digit an 18-bit constant K which serves as one of the input operands. These two formats provide a highly efficient control of concurrent operations.

As a background, consider the essential difference between a general purpose device and a special device in which high speeds are required. The designer of the special device can generally improve on the traditional general purpose device by introducing some form of concurrency. For example, some activities of a housekeeping nature may be performed separate from the main sequence of operations in separate hardware. The total time to complete a job is then optimized to the main sequence and excludes the housekeeping. The two categories operate concurrently.

It would be, of course, most attractive to provide in a general purpose device some generalized scheme to do the same kind of thing. The organization of the 6600 central processor provides just this kind of scheme. With a multiplicity of functional units, and of operand registers and with a simple and highly efficient addressing system, a generalized queue and reservation scheme is practical. This is called the *scoreboard*.

The scoreboard maintains a running file of each central register, of each functional unit, and of each of the three operand trunks to and from each unit. Typically, the scoreboard file is made up of two-, three-, and four-bit quantities identifying the nature of register and unit usage. As each new instruction is brought up, the conditions at the instant of issuance are set into the scoreboard. A snapshot is taken, so to speak, of the pertinent conditions. If no waiting is required, the execution of the instruction is begun immediately under control of the unit itself. If waiting is required (for example, an input operand may not yet be available in the central registers), the scoreboard controls the delay, and when released, allows the unit to begin its execution. Most important, this activity is accomplished in the scoreboard and the functional unit, and does not necessarily limit later instructions from being brought up and issued.

In this manner, it is possible to issue a series of instructions, some related, some not, until no functional units are left free or until a specific register is to be assigned more than one result. With just those two restrictions on issuing (unit free and no double result), several independent chains of instructions may proceed concurrently. Instructions may issue every minor cycle in the

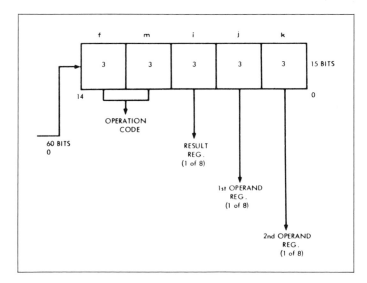

Fig. 4. Fifteen-bit instruction format.

absence of the two restraints. The instruction executions, in comparison, range from three minor cycles for fixed add, 10 minor cycles for floating multiply, to 29 minor cycles for floating divide.

To provide a relatively continuous source of instructions, one buffer register of 60 bits is located at the bottom of an instruction stack capable of holding 32 instructions (Fig. 5). Instruction words from memory enter the bottom register of the stack pushing up the old instruction words. In straight line programs, only the bottom two registers are in use, the bottom being refilled as quickly as memory conflicts allow. In programs which branch back to an instruction in the upper stack registers, no refills are allowed after the branch, thereby holding the program loop completely in the stack. As a result, memory access or memory conflicts are no longer involved, and a considerable speed increase can be had.

Five memory trunks are provided from memory into the central processor to five of the floating point registers (Fig. 6). One address register is assigned to each trunk (and therefore to the floating point register). Any instruction calling for address register result implicitly initiates a memory reference on that trunk. These instructions are handled through the scoreboard and therefore tend to overlap memory access with arithmetic. For example, a new memory word to be loaded in a floating point register can be brought in from memory but may not enter the register until all previous uses of that register are completed. The central registers, therefore, provide all of the data to the ten functional units, and receive all of the unit results. No storage is maintained in any unit.

Central memory is organized in 32 banks of 4096 words. Consecutive addresses call for a different bank; therefore, adjacent addresses in one bank are in reality separated by 32. Addresses may be issued every 100 nanoseconds. A typical central memory information transfer rate is about 250 million bits per second.

As mentioned before, the functional units are hidden behind the registers. Although the units might appear to increase hardware duplication, a pleasant fact emerges from this design. Each unit may be trimmed to perform its function without regard to others. Speed increases are had from this simplified design.

As an example of special functional unit design, the floating multiply accomplishes the coefficient multiplication in nine minor cycles plus one minor cycle to put away the result for a total of 10 minor cycles, or 1000 nanoseconds. The multiply uses layers of carry save adders grouped in two halves. Each half concurrently forms a partial product, and the two partial products finally merge while the long carries propagate. Although this is a fairly large complex of circuits, the resulting device was sufficiently smaller than originally planned to allow two multiply units to be included in the final design.

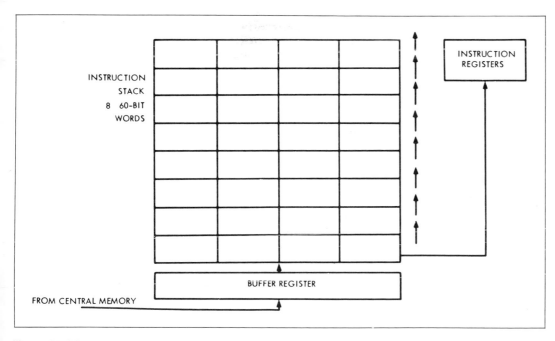

Fig. 5. 6600 instruction stack operation.

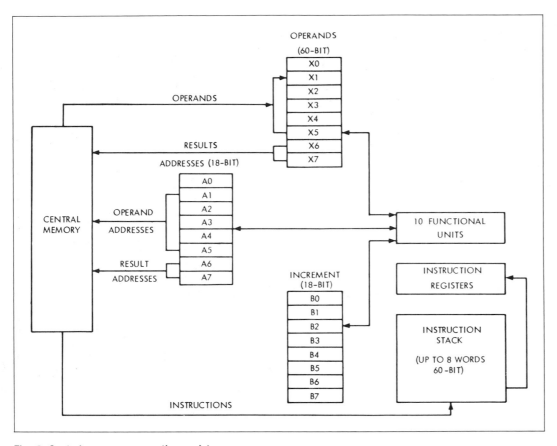

Fig. 6. Central processor operating registers.

To sum up the characteristics of the central processor, remember that the broadbrush description is "concurrent operation." In other words, any program operating within the central processor utilizes some of the available concurrency. The program need not be written in a particular way, although certainly some optimization can be done. The specific method of accomplishing this concurrency involves *issuing* as many instructions as possible while handling most of the conflicts during *execution*. Some of the essential requirements for such a scheme include:

1   Many functional units

2   Units with three address properties

3   Many transient registers with many trunks to and from the units

4   A simple and efficient instruction set

Construction

Circuits in the 6600 computing system use all-transistor logic (Fig. 7). The silicon transistor operates in saturation when switched "on" and averages about five nanoseconds of stage delay. Logic circuits are constructed in a cordwood plug-in module of about $2\frac{1}{2}$ inches by $2\frac{1}{2}$ inches by 0.8 inch. An average of about 50 transistors are contained in these modules.

Memory circuits are constructed in a plug-in module of about six inches by six inches by $2\frac{1}{2}$ inches (Fig. 8). Each memory module contains a coincident current memory of 4096 12-bit words. All read-write drive circuits and bit drive circuits plus address translation are contained in the module. One such module is used for each peripheral processor, and five modules make up one bank of central memory.

Logic modules and memory modules are held in upright hinged chassis in an X shaped cabinet (Fig. 9). Interconnections between modules on the chassis are made with twisted pair transmission

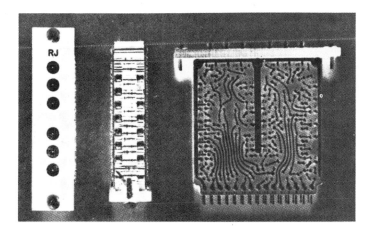

Fig. 7. 6600 printed circuit module.

lines. Interconnections between chassis are made with coaxial cables.

Both maintenance and operation are accomplished at a programmed display console (Fig. 10). More than one of these consoles may be included in a system if desired. Dead start facilities bring

Fig. 9. 6600 main frame section.

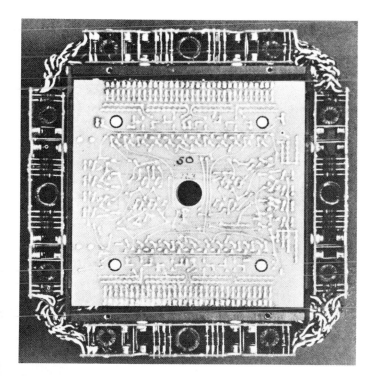

Fig. 8. 6600 memory module.

Fig. 10. 6600 display console.

the ten peripheral processors to a condition which allows information to enter from any chosen peripheral device. Such loads normally bring in an operating system which provides a highly sophisticated capability for multiple users, maintenance, and so on.

The 6600 Computer has taken advantage of certain technology advances, but more particularly, logic organization advances which now appear to be quite successful. Control Data is exploring advances in technology upward within the same compatible structure, and identical technology downward, also within the same compatible structure.

References

AllaR64; ClayB64

APPENDIX 1   CDC 6400, 6500, 6600
CENTRAL PROCESSOR ISP DESCRIPTION

Appendix I

CDC 6400, 6500, 6600 Central Processor ISP Description

Pc State

P<17:0>   *Program counter*

X[0:7]<59:0>   *Main arithmetic registers. X[1:5], are implicitly loaded from*
                                                      Mp when A[1:5] are loaded. X[6:7] are implicitly stored in
A[0:7]<17:0>   *Mp when A[6:7] are loaded.*

B[0]<17:0> := 0   *B registers are general arithmetic registers, and can be used*
B[1:7]<17:0>   *as index registers.*

Run   *1 if interpreting instructions, not under program control.*

EM<17:0>   *Exit mode bits*

    Address_out_of_range_mode := EM<12>

    Operand_out_of_range_mode := EM<13>

    Indefinite_operand_mode   := EM<14>

The above description is incomplete in that the above 3 mode's alarm allow conditions to trap Pc at Mp[RA]. Trapping occurs if
an alarm condition occurs "and" the mode is a one.

Mp State

Mp[0:777777$_8$]<59:0>   *main core memory of 2^{18} w, (256 kw)*

Ms[0:2015232]<59:0>   *ECS/Extended Core Storage Program can only transfer data between*
                                                     Mp and Ms. Program cannot be executed in Ms.

RA<17:0>   *reference (or relocation) address register to map a logical Mp'*
                                                     into physical Mp

FL<17:0>   *field length - the bounds register which limits a program's*
                                                     access to a range of Mp'

RAECS<59:36>   *reference or relocation register for Ms(Extended Core Storage)*

FLECS<59:36>   *field length for ECS*

Address_out_of_range   *a bit denoting a state when memory mapping is invalid*

Memory Mapping Process
  This process maps or relocates a logical program, at location Mp', and Ms', into physical Mp and Ms.

Mp'[X] := ((X < FL) →Mp[X + RA]);   *logical Mp'*

         (X ≥ FL) → (Run ←0; Address_out_of_range ←1))

Ms'[X] := ((X < FLECS) →Ms[X]+ RAECS]);   *logical Ms'*

         (X ≥ FLECS) → (Run ←0; Address_out_of_range ←1))

Exchange jump storage allocation map at location, n within Mp:

  The following Mp" array is reserved when Pc state is stored, and switched to another job. The exchange jump instruction in
a Peripheral and Control Processor enacts the operation: (Mp" ← Mp; Mp ← Mp").

Mp''[n]<53:0>   := P□A[0]□0000000$_8$

Mp''[n+1]<53:0>   := RA□A[1]□B[1]

Mp''[n+2]<53:0>   := FL□A[2]□B[2]

Mp''[n+3]<53:0>   := EM□A[3]□B[3]

Mp''[n+4]   := RAECS□A[4]□B[4]

Mp''[n+5]   := FLECS□A[5]□B[5]

Mp''[n+6]<35:0>   := A[6]□B[6]

Mp''[n+7]<35:0>   := A[7]□B[7]

Mp''[n+10$_8$:n+17$_8$]:= X[0:7]

Instruction Format

 instruction<29:0>

              although 30 bits, most instructions are 15 bits; see
               Instruction Interpretation Process

  fm<5:0>   := instruction<29:24>   *operation code or function*

  fmi<8:0>   := fm⬜i   *extended op code*

  i<2:0>   := instruction<23:21>   *specifies a register or an extension to op code*

  j<2:0>   := instruction<20:18>   *specifies a register*

  k<2:0>   := instruction<17:15>   *specifies a register*

  jk<5:0>   := j⬜k   *a shift constant (6 bits)*

  K<17:0>   := instruction<17:0>   *an 18 bit address size constant*

  long_instruction := ((fm < 10_8) ∨   *30 bit instruction*

         (50 ≤ fm < 53) ∨

         (60 ≤ fm < 63) ∨

         (70 ≤ fm < 73))

  short_instruction := ¬ long instruction   *15 bit instruction*

Instruction Interpretation Process
 *A 15 bit (short) or 30 bit (long) instruction is fetched from Mp'[P]<p × 15 + 15 - 1:p × 15> where p = 3, 2, 1, or 0. A 30
bit instruction cannot be stored across word boundaries (or in 2, Mp' locations).*

 p<1>$_4$ *a pointer to 15 bit quarter word which has instruction*

 Run → (instruction<29:15> ←Mp'[P]<(p × 15 + 14):(p × 15)>; next *fetch*

  p ←p - 1; next

  (p = 0) ∧ long_instruction →Run ←0;

  (p ≠ 0) ∧ long_instruction → (

   instruction<14:0> ←Mp'[P]<(p × 15 + 14):(p × 15)>;

    p ←p - 1); next

  Instruction_execution; next   *execute*

  (p = 0) → (p ←3; P ←P + 1))

Instruction Set and Instruction Execution Process
 *Operand fetches or stores between Mp' and X[i] occur by loading or storing registers A[i]. If (0 < i < 6) a fetch from
Mp'[A[i]] occurs. If (i ≥ 6) a store is made to Mp'[A[i]]. The description does not describe Address_out_of_range case,
which is treated like a null operation.*

 Instruction_execution := (

 Set A[i]/SA

  "SAi Aj + K" (fm = 50) → (A[i] ←A[j] + K; next Fetch_Store);

  "SAi Bj + K" (fm = 51) → (A[i] ←B[j] + K; next Fetch_Store);

  "SAi Xj + K" (fm = 52) → (A[i] ←X[j]<17:0> + K; next Fetch_Store);

  "SAi Xj + Bk" (fm = 53) → (A[i] ←X[j]<17:0> + B[k]; next Fetch_Store);

  "SAi Aj + Bk" (fm = 54) → (A[i] ←A[j] + B[k]; next Fetch_Store);

  "SAi Aj - Bk" (fm = 55) → (A[i] ←A[j] - B[k]; next Fetch_Store);

  "SAi Bj + Bk" (fm = 56) → (A[i] ←B[j] + B[k]; next Fetch_Store);

  "SAi Bj - Bk" (fm = 57) → (A[i] ←B[j] - B[k]; next Fetch_Store);

  Fetch_Store := (

   (0 < i < 6) → (X[i] ←Mp'[A[i]]);   *process to get operand in X or store operand from X when A*
                      is written

   (i ≥ 6) → (Mp'[A[i]] ← X[i]))

 Operations on B and X

 Set B[i]/SBi

  "SBi Aj + K" (fm = 60) → (B[i] ←A[j] + K);

```
    "SBi Bj + K" (fm = 61) → (B[i] ← B[j] + K);
    "SBi Xj + K" (fm = 62) → (B[i] ← X[j]<17:0> + K);
    "SBi Xj + Bk" (fm = 63) → (B[i] ← X[j]<17:0> + B[k]);
    "SBi Aj + Bk" (fm = 64) → (B[i] ← A[j] + B[k]);
    "SBi Aj - Bk" (fm = 65) → (B[i] ← A[j] - B[k]);
    "SBi Bj + Bk" (fm = 66) → (B[i] ← B[j] + B[k]);
    "SBi Bj - Bk" (fm = 67) → (B[i] ← B[j] - B[k]);
```

Set X[i]/SXi

```
    "SXi Aj + K" (fm = 70) → (X[i] ← sign_extend(A[j] + K));
    "SXi Bj + K" (fm = 71) → (X[i] ← sign_extend(B[j] + K));
    "SXi Xj + K" (fm = 72) → (X[i] ← sign_extend(X[j] + K));
    "SXi Xj + Bk" (fm = 73) → (X[i] ← sign_extend(X[j] + B[k]));
    "SXi Aj + Bk" (fm = 74) → (X[i] ← sign_extend(A[j] + B[k]));
    "SXi Aj - Bk" (fm = 75) → (X[i] ← sign_extend(A[j] - B[k]));
    "SXi Bj + Bk" (fm = 76) → (X[i] ← sign_extend(B[j] + B[k]));
    "SXi Bj - Bk" (fm = 77) → (X[i] ← sign_extend(B[j] - B[k]));
```

Miscellaneous program control

```
    "PS" (:= fm = 0) → (Run ← 0);                               program stop
    "NO" (:= fm = 46) → ;                                        no operation; pass
```

Jump unconditional

```
    "JP Bi + K" (:= fm = 02) → (P ← B[i] + K; p ← 3);            jump
```

Jump on X[j] conditions

```
    "ZR Xj K" (:= fmi = 030) → ((X[j] = 0) → (P ← K; p ← 3));   zero
    "NZ Xj K" (:= fmi = 031) → ((X[j] ≠ 0) → (P ← K; p ← 3));   non zero
    "PL Xj K" (:= fmi = 032) → ((X[j] ≥ 0) → (P ← K; p ← 3));   plus or position
    "NG Xj K" (:= fmi = 033) → ((X[j] < 0) → (P ← K; p ← 3));   negative
    "IR Xj K" (:= fmi = 034) → (                                out of range constant tests
       ¬((X[j]<59:48>= 3777)∨(X[j]<59:48> 4000)) →P ← K; p ← 3);
    "OR Xj K" (:= fmi = 035) → (
       (X[j]<59:48>=3777) ∨ (X[j]<59:48>=4000)→ (P ← K; p ← 3));
    "DF Xj K" (:= fmi = 036) → (                                indefinite form constant tests
       (X[j]<59:48>=1777) ∨ (X[j]<59:48>=6000)  → (P ← K; p ← 3));
    "ID Xj K" (:= fmi = 037) → (
       (X[j]<59:48>=1777) ∨ (X[j]<59:48>=6000) → (P ← K; p ← 3));
```

Jump on B[i], B[j] comparison

```
    "EQ Bi Bj K" (:= fm = 04) → ((B[i] = B[j]) → (P ← K; p ← 3));   equal
    "NE Bi Bj K" (:= fm = 05) → ((B[i] ≠ B[j]) → (P ← K; p ← 3));   not equal
    "GE Bi Bj K" (:= fm = 06) → ((B[i] ≥ B[j]) → (P ← K; p ← 3));   greater than or equal
    "LT Bi Bj K" (:= fm = 07) → ((B[i] < B[j]) → (P ← K; p ← 3));   less than
```

Subroutine call

```
    "RJ K" (:= fmi = 010) → (                                   return jump
       M[K]<59:30> ← 04₈□00₈□(P + 1)□000000₈; next
       (P ← K + 1; p ← 3));
```

Reading (REC) and writing (WEC) Mp with Extended Core Storage, subjected to bounds checks, and Ms', Mp' mapping

```
    "REC Bj + K" (:= fmi = 011) → (                             read extended core
```

```
        Mᴅ'[A[0]:A[0] + B[j] + K-1] ←Ms'[X[0]:X[0] + B[j] + K-1]);  ¬
    "VEC Bj + K" (:= fmi = 012) → (                                        write extended core
        Ms'[X[0]:X[0] + B[j] + K-1] ← Mp'[A[0]:A[0] + B[j] + K-1]);
```

Fixed Point Arithmetic and Logical operations using X

```
    "IXi Xj + Xk" (:= fm = 36) → (X[i] ←X[j] + X[k]);                      integer sum
    "IXi Xj - Xk" (:= fm = 37) → (X[i] ←X[j] - X[k]);                      integer difference
    "CXi Xk" (:= fm = 47) → (X[i] ←sum_modulo_2(X[k]);                     count the number of bits in X[k]
    "BXi Xj" (:= fm = 10₈) → (X[i] ←X[j]);                                 transmit
    "BXi Xj * Xk" (:= fm = 11₈) → (X[i] ←X[i] ←X[j] ∧ X[k]);               logical product
    "BXi Xj + Xk" (:= fm = 12) → (X[i] ←X[j] ∨ X[k]);                      logical sum
    "BXi Xj - Xk" (:= fm = 13) → (X[i] ←X[j] ⊕ X[k]);                      logical difference
    "BXi - Xk" (:= fm = 14) → (X[i] ←¬ X[k]);                              transmit complement
    "RXi - Xk * Xj" (:= fm = 15)→ (X[i] ←X[j] ∧ ¬ X[k]);                   logical product and complement
    "BXi - Xk + Xj" (:= fm = 16)→ (X[i] ←X[j] ∨ ¬ X[k]);                   logical sum and complement
    "BXi = Xk - Xj" (:= fm = 17)→ (X[i] ←X[j] ⊕ ¬ X[k]);                   logical difference and complement
    "LXi jk" (:= fm = 20) → (X[i] ←X[i] × 2ʲᵏ {rotate});
    "AXi jk" (:= fm = 21) → (X[i] ←X[i] / 2ʲᵏ);                            arithmetic right shift
    "IXi Bj Xk" (:= fm = 22) → (                                          left shift nominally
        ¬B[j]<17> → X[i] ←X[k] × 2^B[j]<5:0> {rotate};
        B[j]<17> → X[i] ←X[k] / 2^¬ B[j]<10:0>);
    "AXi Bj Xk" (:= fm = 23) → (                                          arithmetic right shift nominally
        ¬B[j]<17> → X[i] ←X[k] / 2^B[j]<10:0>;
        B[j]<17> → X[i] ←X[k] × 2^¬ B[j]<5:0> {rotate});
    "MXi jk" (:= fm = 43) → (                                             form mask
        X[i]<59:59-jk+1> ←2ʲᵏ - 1;
        (jk = 0) →X[i] ←0);
```

Floating Point Arithmetic using X
Only the least significant (lo) part of arithmetic is stored in Floating DP operations.

```
    "FXi Xj + Xk" (:= fm = 30) → (X[i] ←X[j] + X[k] {sf});                floating sum
    "FXi Xj - Xk" (:= fm = 31) → (X[i] ←X[j] - X[k] {sf});                floating difference
    "DXi Xj + Xk" (:= fm = 32) → (X[i] ←X[j] + X[k] {ls.df});             floating dp sum
    "DXi Xj - Xk" (:= fm = 33) → (X[i] ←X[j] - X[k] {ls.df});             floating dp difference
    "RXi Xj + Xk" (:= fm = 34) → (
        X[i] ← round(X[j]) + round(X[k]) {sf});
    "RXi Xj - Xk" (:= fm = 35) → (                                        round floating difference
        X[i] ← round(X[j]) - round(X[k]) {sf});
    "FXi Xj * Xk" (:= fm = 40) → (X[i] ←X[j] × X[k] {sf});                floating product
    "RXi Xj * Xk" (:= fm = 41) → (                                        round floating product
        X[i] ← X[j] × X[k] {sf}; next X[i] ← round(X[i]) {sf});
    "DXi Xj * Xk" (:= fm = 42) → (X[i] ←X[j] × X[k] {ls.df});             floating dp product
    "FXi Xj / Xk" (:= fm = 44) → (X[i] ←X[j] / X[k] {sf});                floating divide
    "RXi Xj / Xk" (:= fm = 45) → (X[i] ←round(X[j] / X[k]) {sf}); round floating divide
    "NXi Bj Xk" (:= fm = 24) → (                                          normalize
        X[i] ← normalize(X[k]) {sf};
        B[j] ← normalize_exponent(X[k]) {sf});
```

```
''ZXi Bj Xk'' (:= fm = 25) → (                              round and normalize
    X[i] ← round(X[k]) {sf}; next
    X[i] ← normalize(X[i]) {sf};
    B[j] ← normalize_exponent(X[i]) {sf});
''UXi Bj Xk' (:= fm = 26) → (B[j] ← X[k]<58:48> {si};       unpack
                             X[i] ← X[k]<59,47:0> {si});
''PXi Bj Xk'' (:= fm = 27) → (X[k]<58:48> ← B[j] {si};      pack
                             X[k]<59,47:0> ← X[i] {si})
                             )                              end Instruction_execution
```

APPENDIX 2 CDC 6400, 6500, 6600, AND 6416 PERIPHERAL AND CONTROL PROCESSORS, PCP, ISP DESCRIPTION

Appendix 2

CDC 6400, 6500, 6600, and 6416
Peripheral and Control Processors/PCP, ISP Description

Pc State

  A<17:0>   *accumulator*

  P<11:0>   *Program Address Counter*

Mp State

  M[0:4095]<11:0>   *Mp*

  M index[0:63]<11:0>:= M[0:63]<11:0>   *special array in Mp reserved for index register*

C('Central) State

  CP␣P<17:0>   *the main Pc instruction address counter*

  CPM[0:777777$_8$]<59:0>   *the Mp of main C*

IO Registers for C('PCP)

  C␣DATA[0:63]<11:0>   *data buffers at peripheral K's*

  C␣ACT[0:63]   *a bit to denote if 1 of the 64 K's is active*

  C␣FLG[0:63]   *denotes a full (or empty) buffer at the K*

  C␣FCN[0:63] <11:0>   *function or instruction register at a specific K*

Instruction Format

  Ins[0:1]<11:0>   *instruction*

    long␣instruction   *2 w instruction: defined in terms of op codes, see Table, page 503*

    short␣instruction := ¬ long␣instruction   *1 w instruction*

  F<5:0>   := Ins[0]<11:6>   *function or op code*

  d<5:0>   := Ins[0]<5:0>

  m<11:0>   := Ins[1]   *address part*

  dm<17:0> := d␣m

  i<11:0>   := Ins[1]<11:0>   *indirect bit*

  d␣sign<11:0> := (

    ¬d<5> → 0␣d;

    d<5> →¬ d)

  md<11:0> := (

    (d = 0) → m;

    (d ≠ 0) → m + M[d])

Effective Address Calculation Process

  z := ((F<5:3> = 3) → d;

    (F<5:3> = 4) → i;

    (F<5:3> = 5) → md)

Instruction Interpretation Process

  Run → (Ins[0] ←M[P]; P ← P + 1; next   *fetch*

      long␣instruction → (Ins[1] ← M[P]; P ← P + 1); next

      Instruction␣execution)   *execute*

Implementation
The 10 x 52 bits in the barrel for the 10 PCP ISP include:

A[0:9]<17:0>   *accumulators*

P[0:9]<11:0>   *instruction address counters*

Temporary Hardware registers (not in the ISP)

Q[0:9]<11:0>   *low order 6 bits of an instruction or address data*

K[0:9]<5:0>   *six bits hold the operation code. The 3 bits specify the*
                                                     trip count or state of an instruction's interpretation.
T[0:9]<2:0>

F = X_8Y_8 Instruction execution := (

X_8 \ Y_8	00	01	02	03	04	05	06	07
00	PSN →; *null*	LJM → (P← md);	RJM → (M[md] ←P; P← md+1);	UJN →((	ZJN → ((A=0) → (	NJN → ((A≠0) → (P← P+d⌣sign));	PJN → (¬ A<17> → (	MJN → (A<17> → (
10	SHN → (A←A×2^(d⌣sign)) *	LMN → (A←A⊕d) ; *	LPN → (A←A∧d) ; *	SCN → (A←A∧¬d) ; *	LDN → (A←d) ; *	LCN → (A ←-d) ; *	ADN→ (A←A+d) ; *	SBN → (A←A-d) ; *
20	LDC → (A← dm) :	ADC → (A← A+dm) ;	LPC → (A←A∧dm) ;	LMC → (A←A⊕dm) ;	PSN →; *null*	PSN →;	EXN → (CP⌣P←A) ;	RPN → (A←CP⌣P) ;
30	LDD → (	ADD → (	SBD → (	LMD → (	STD → (	RAD → (	AOD → (	SOD → (
40	LDI → (	ADI → (	SBI → (	LMI → (	STI → (	RAI → (	AOI → (	SOI → (
50	LDM → (A←M[z]) :	ADM → (A← A+M[z]) ;	SBM → (A←A-M[z]) ;	LMM → (A ←A⊕M[z]) ;	STM → (M[z]←A) ;	RAM →(A←A+M[z]; next M[z]←A) ;	AOM → (A←M[z]+1; next M[z]←A) ;	SOM (A←M[z]-1; next M[z]←A) ;
60	CRD → (M[d:d+5]← CPM[A]);	CRM → (M[m:m+ 5×M[d]-1]← CPM[A:A+ M[d]-1]);	CWD → (CPM[A]← M[d:d+5]);	CWM → (CPM[A:A+ M[d]-1]← M[m:m+ 5×M[d]-1]);	AJM → (C⌣ACT[d]→ (	IJM → (¬ C⌣ACT[d]→ (P←m));	FJM → (C⌣FLG[d]→ (	EJM → (¬ C⌣FLG[d]→ (
70	IAN → (A← C⌣DATA[d]):	IAM → (C⌣FLG[d]→ (M[m:m+A]← C⌣DATA[d]));	OAN → (C⌣DATA[d] ←A);	OAM → ((¬ C⌣FLG[d]→ C⌣DATA[d] ←M[m:m+A]));	ACN → (C⌣ACT[d] ←1);	DCN → (C⌣ACT[d] ←0);	FAN → (C⌣FCN[d] ←A);	FNC → (C⌣FCN[d] ←m);

) end Instruction⌣execution

*1 word or short⌣instruction

Chapter 40

Computer-network examples

We are just entering the era in which general-purpose networks of computers make technical and economic sense. The requisite hardware and software development of operating systems and multiprogramming capability is still maturing. Thus, unlike the other PMS structures discussed in this book, there is no supply of operational systems with published descriptions upon which we can draw. Consequently, we have assembled several brief examples of networks to provide at least some illustrations of what is sure to be an important aspect of computer systems in the near future. The more interesting of these examples are still in the planning stages; those that exist currently are still highly specialized.

Spatially distributed intercommunicating networks of digital devices have existed for a long time. But many of the ones that come most easily to mind are not computer networks. For example, the various airline reservation systems like American Airline's SABRE [Plugge and Perry, 1961] have spatially distributed terminals (T's) with a single Pc, possibly mediated by Pio's or Cio's. When there are several Pc's, they are functionally integrated so as to provide the total capacity and reliability needed. Some military networks, such as the SAGE Air Defense System [Everett et al., 1957] have multiple computers (SAGE actually has a very large number). But they transmit to each other highly specialized data streams (for example, aircraft positional information for control). The National Physics Laboratory of England has made a very comprehensive proposal for a general-purpose network [Davies et al., 1967], although we do not include it as a chapter. Again, it is just in the proposal stage. The Lawrence Radiation Laboratory (at Livermore) is no doubt the earliest and most impressive network.

In terms of our PMS descriptions, a computer network (N) requires at least two C's not connected through primary memory. Thus each C has a Pc and an Mp of its own and has to communicate with other C's through messages. Duplex computers are thus defined as networks, provided they do not share Mp. For networks, links (L's) are usually shown explicitly. In spatially distributed systems, both the time delays and the flow rates of the links are significant. The latter is so partly because the networks must make use of the telephone communication system, which exists independently of the networks, thus having parameters that do not correspond with any of the internal parameters of the individual computers. There may also be limitations of reliability, cost,

accessing characteristics, and the size of the information unit that derive wholly from the links. For instance, many computer networks would like to buy their transmissions from the telephone system for very short intervals (milliseconds), at very high data rates, and with short switching time (milliseconds), i.e., bursts. Switching time and pricing policies within the telephone system conspire to make this a difficult thing to do. Thus, with networks, links become important independent components.

One classification of networks (N's) is by fixed or variable interconnection structure. Fixed structure may mean that the links are fixed permanently over the life of the network. However, fixed structure may mean only that connections once made must be held for long periods of time relative to the message flows. An example is the telephone switching system mentioned above, which looks like a variable switching structure at the level of human conversations, but like a fixed switching structure at the level of computer conversations. Figures 1a and 1c show variable-structure systems; Fig. 1b shows a fixed-structure system. In the former, any C can talk directly to any other C. In the latter, each C talks directly to only a few C's; thus, to communicate with the other C's, it must transmit through them as links; that is, it must use another C as an L.

A second classification of N's is by the nature of the delays suffered by the messages as they travel from an initiating C to a target C. Communication can be direct, in which case the only delays are those through the switches (S) and links (L) between the two C's (Figs. 1a and 1b). Alternatively, communication can involve storing messages at intermediate nodes (called store-and-forward communication), thus introducing additional memory delays into the communication but decreasing the demands for coordination between the two C's. Although store-and-forward systems can be built with the intermediate nodes being K's with buffer memories, in the present context the natural form for such a system uses the other C's in the system as the intermediate nodes, as in Fig. 1c.

Several kinds of reasons can justify the existence of a particular network. The following list is adapted from Roberts [1967]:

Load sharing. A problem (program and data) initiated at one C that is temporarily overloaded is sent to another for processing. The cost of transshipment must clearly be less than the costs of

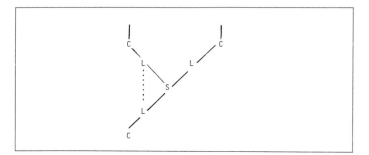

Fig. 1a. Variable-structure direct switching network PMS diagram.

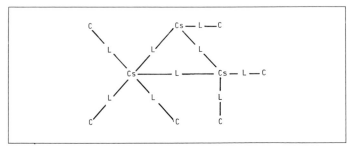

Fig. 1c. Store-and-forward network PMS diagram (using C switching).

delay in getting the problem processed. Load sharing implies highly similar facilities at the nodes of the network.

Data sharing. A program is run at a node that has access to a large, specialized data base, such as a specialized automated library. It is less costly to bring the program to the data than to bring the data to the program.

Program sharing. Data are sent to a C that has a specialized program. This might happen because of the size of the program (hence, fundamentally the same reason as data sharing), but it might also happen because the knowledge (i.e., initialization and error rituals) to run the program is available at one C but not at another.

Specialized facilities. Within the network there need exist only one of various rarely used facilities, such as large random-access memories, or special display devices, or special-purpose array processors.

Message switching. There may be a communication task of such magnitude that sophisticated switching and control are worthwhile.

Reliability. If some components fail, others can be used in their place, thus permitting the total system to degrade gracefully. (At the present state of the art, peripheral computers are needed to isolate the periphery from the unreliability of the network, and vice versa.)

Peak computing power. Large parts of the total system can be devoted for short periods to a single task, if there are important real-time constraints to be met. This depends on being able to fractionate the task into independent subtasks.

Communication multiplexing. Efficient use of communication facilities is obtained by multiplexing a number of low data-rate users, for example, T(typewriter; 150 b/s)'s. This may not be a reason for a network per se but may justify a larger network, provided that there is some reason for having one in the first place.

Better communication. A community of users (e.g., a scientific or engineering community) that could mutually use the same programs and data bases and converse about these directly (i.e., not by writing about them but in the context of mutual use) might become a much more productive community, with less duplication of work, faster communication of results, etc.

Better load distribution through preprocessing. Some tasks require very high-data-rate communication with a computer. By doing preprocessing in a smaller computer, a reduced information rate can be sent to the more general system.

With this general view of networks, let us consider several examples.

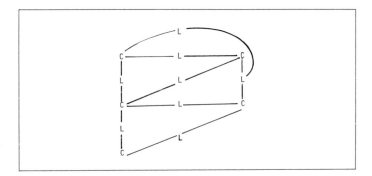

Fig. 1b. Fixed-network PMS diagram.

IBM ASP (*Attached Support Processor*)

This first example (Fig. 2) is the simplest of all computer networks, consisting of two computers tied together, with each functionally specialized (and in addition required to be physically close). The function of C.support is job setup and breakdown, that is, preprocessing and postprocessing. All T's for the network are handled by it (except for T.console on C.main). The function of C.main is to process data. Thus this is an escalated version of the Pc-n Pio organization, where the Pio's have been made into a C.support and thus can take on additional functions. It should be compared with the CDC 6600 organization, which is C.main-10 Cio, but where the Cio's are rather small Cio(4096 w; 12 b/w) compared with the C.support. The ASP organization is the 360 analog of a system consisting of an IBM 7090-IBM 7040 which emerged spontaneously in the early sixties at several IBM installations in order to deal with 7090 I/O bottlenecks. Thus this kind of simple computer network has been with us for some time.

In more detail, the advantages that are claimed for ASP are in reducing resource interference:[1]

[1] Adapted from IBM System/360 Attached Support Processor (ASP) System Description, H20-0223-0.

1 The addition of smaller modules of Mp in the form of a second processor. The processing of the application is divided between the main processor and the support processor, with each performing those functions for which it is best suited. The core requirements for the support processor are small in comparison with those for the main processor. With this division of responsibilities, the system can expand its capabilities with a minimum addition of storage.

2 The elimination of concurrent use of Pc time on the main processor for processing support functions (such as printing). Because the clerical functions are assigned to the support processor, the main processor no longer shares Pc time between the support functions and the application programs. Therefore, the application has the opportunity to use all the resources of the main processor to full capacity.

3 The addition of selector channels. The channel capacity of the system has been increased by one or more additional selector channels attached to the support processor.

4 An algorithm for efficient management of the direct-access storage devices for system input/output data sets. The algorithm was designed specifically to accommodate the data demands, the data set characteristics, and the available private devices. The input/output routines always know the position of the access mechanism, thereby ensuring minimum seek time when data are transferred to the devices.

IBM cites the above reasons for using the ASP system. These views differ from ours on its usefulness. Ideally, a multiprogrammed single-processor or multiprocessor structure would easily provide all the above advantages without the overhead of having large Mp's on two computers (both of which hold nearly the same operating system). Also, as we note in the introduction to the System/360 (page 584), the support-computer functions can be handled in the main computer with very little loss of large Pc power (3 to 10 percent). A multiprocessor structure should also cause less overhead, by not passing data sets between two C's. (Alternatively, in ASP this could be done by an S to common Ms from both C's.)

University of Texas network

The structure shown in Fig. 3 is similar to ASP in that a C.main is used, with some job setup and breakdown being done in several other C's. However, there are several of these C's, and they provide independent power for small tasks where the setup time for the large system is greater than the computation time. They are also physically remote from C.main and thus serve to make the power of the central facility available at local sites. The Teletypes are

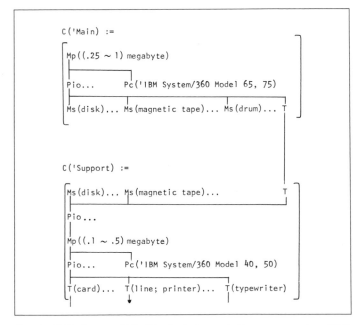

Fig. 2. IBM System/360 Attached Support Processor system/ASP PMS diagram.

```
T(Teletype)...
|
|
|
S('Telephone Exchange)
|
|
|
C('CDC 6600; Computation Center)
|...
S('Telephone Exchange )
|  |   |  L─ C('CDC 1700; Linguistic Research Laboratory)-
|  |   └── L ──── C('CDC 3100: College of Business Administration)-
|  └─── L ──── C('8231 Computer Terminal)→
|  :          |  |  T(card)-
|  :          |  T(line: printer)→
└─L(to:  other C's off campus)-
```

Fig. 3. The Computation Center, University of Texas, (Austin) Network PMS diagram.

used to enter jobs directly to the C.main, where they are run in a batch mode.

The network of Fig. 3 is that at the University of Texas, as derived from its internal planning memoranda. Similar systems are in existence or under construction at other universities.

M.I.T. proposed network

Figure 4 shows a network that is proposed for the M.I.T. campus [Bhushan, Stotz, and Ward, 1967]. It moves to a more complex switching system, partly because there are two C.main's. Here an S(direct) is used in a non-store-and-forward mode as each C communicates directly with another. The communication rate between C's is 40 ~ 230 kb/s. (Note that at higher data rates a fairly large computer is necessary just to handle the store-and-forward message switching information rates.) The purpose of the network is to allow users of the small or terminal C's to get access to C('IBM 360/67) and C('GE-645). These two C's can, of course, communicate with one another. A large number of users are connected to T(typewriters) via the S('Telephone Exchange).

The Lawrence Radiation Laboratory (at Livermore) network

The LRL network, started in 1964, appears to be the earliest general-purpose-computer network. It serves a user population of approximately 1,000, with several hundred simultaneous on line users. The network consists of five large computers (three CDC 6600s and two CDC 7600s), a switching computer (a DEC

PDP-6 with two Pc's and a 262 kword Mp and a 10^9-bit fixed-head disk for fast-access files), three terminal control computers (DEC PDP-8's), and a large central file (a 10^{12}-bit IBM Photostore controlled by an IBM 1800 computer). Hardwired 4 megabit per second links connect the large computers to the switching computer. The terminal computers and the large file are also connected to the switching computer.

The main purpose of the network is to gain access to the central filing, printing, and terminal facilities. Load sharing is not an important consideration because each of the large computers operates nearly autonomously. Thus little change was required in each system to be integrated to the network. Jobs enter the network in any of three ways—by the batch input terminals of a large computer; by the typewriter inputs of a large computer; or by the typewriter inputs of the terminal control computer which in turn connects to the central switch. Unlike most university computation centers, which provide service for many users with small jobs, the LRL network is oriented to users with (multiple) large jobs.

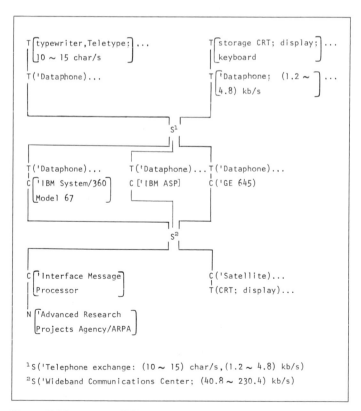

¹S('Telephone exchange: (10 ~ 15) char/s,(1.2 ~ 4.8) kb/s)
²S('Wideband Communications Center; (40.8 ~ 230.4) kb/s)

Fig. 4. M.I.T.-network PMS diagram (proposed).

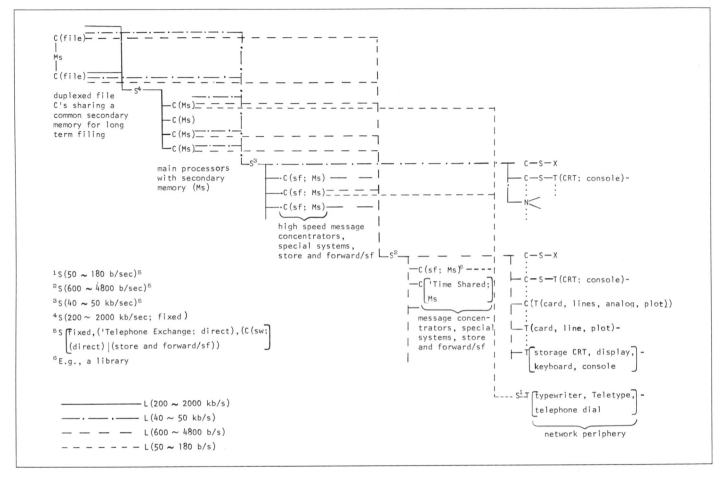

Fig. 5. Typical computer network PMS diagram.

Typical local network

We summarize in Fig. 5 the direction in which the last three networks are moving by presenting a hypothetical, local network, as it may mature on many large university campuses (and large industrial establishments). The network is conceived as a single computing facility, to serve a clientele with many heterogeneous but partially overlapping computing needs. An essential feature of the environment of the network is that the collection of computing resources it connects are not planned all at once but keep growing and changing in imperfectly controlled ways. This arises from the quasi-independent nature of the subparts of large universities and engineering establishments. In any event, the network is a mixture of functionally independent and functionally specialized C's. One probable feature is the duplexed C.files which handle all the Ms functions for all C's, except the C(library). A library's computer, though strongly coupled to the network, would have its own files and specialized terminals, including hard copy devices oriented to library needs. The C.file increases the requirements for the S.central but provides much more economic Ms, as well as easing the ability to connect new C's into the system, since they immediately have access to an organized Ms.

The reader should note that the four switches (S's) can be either fixed links, variable switches (e.g., Telephone Exchange), or a computer used as a direct switch or as a store-and-forward switch.

The most interesting aspect of this network is that it has a general hierarchical structure and is like other hierarchical organizations. Here, the levels of the organization are based on data rates. For example, there is a very low-level computer which deals with the basic communication to typewriters at ~ 150 b/s. This

C switch concentrates several typewriters into a time-multiplexed 2,400-b/s link. Several of the 2,400-b/s links can in turn be concentrated prior to transmitting via a 50-kb/s link. Thus the general organizing principle, like that of most large organizations, is to handle problems at the lowest (cheapest) possible level. Another organization principle of the hierarchy is that only relevant information be passed between the levels. For example, encoding would be used so that only some fraction of the bits flowing at the periphery would enter the highest-level computers. At each of the levels we assume that specialized, time-shared computers are employed to handle the very simpler tasks of editing, simple calculations, etc.

At the network periphery there are a number of terminal computers, i.e., C(terminal; CRT, card, lines, analog, plot, keyboard). Although they are computers, they behave as terminals. The DEC 338 (Chap. 25) is typical of this terminal class. Part of the periphery connects to other networks and part connects to specialized processes, e.g., a process control, or experimental apparatus on a dedicated basis. The peripheral computers are able to do local tasks independently of the larger, more unreliable computers.

Combat Logistics Network/ComLogNet

ComLogNet was developed for the U.S. Air Force in the early 1960s for the purpose of sending messages (or information) among T's [Segal and Guerber, 1961]. It is built to transmit both at low

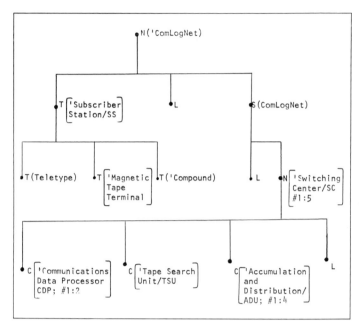

Fig. 6b. Combat Logistics Network/ComLogNet component relationships.

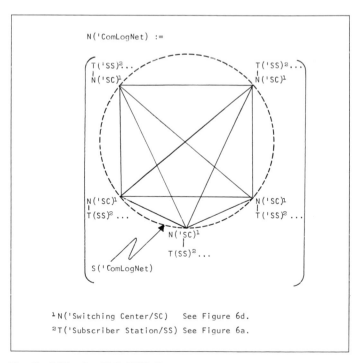

Fig. 6c. S('ComLogNet) PMS diagram.

Fig. 6a. Combat Logistics Network/ComLogNet PMS diagram.

(10 char/s) and medium (1,200 ~ 4,800 b/s) speed, as shown in Fig. 6a. In this regard the network is simply a message switch for the three terminal types. It employs C's for the switching elements and is fundamentally a store-and-forward system. Had it not been for security, reliability, response time, and other considerations, it would have been possible to construct an equivalent system using standard lease wire switches (or telephone exchanges). In Fig. 6b a tree is used to present the relationship of constituent members of ComLogNet. From it we see that at the first level ComLogNet has just a switch, links, and terminals (as shown in Fig. 6a). The network's switch employs five specialized N('Automatic Electronic Switching Centers/SC)'s which communicate among each other (Fig. 6c). Terminals connect to the individual N('SC)'s and messages are routed between two T's, either by a store-and-forward process within N('SC) or among two N('SC)'s.

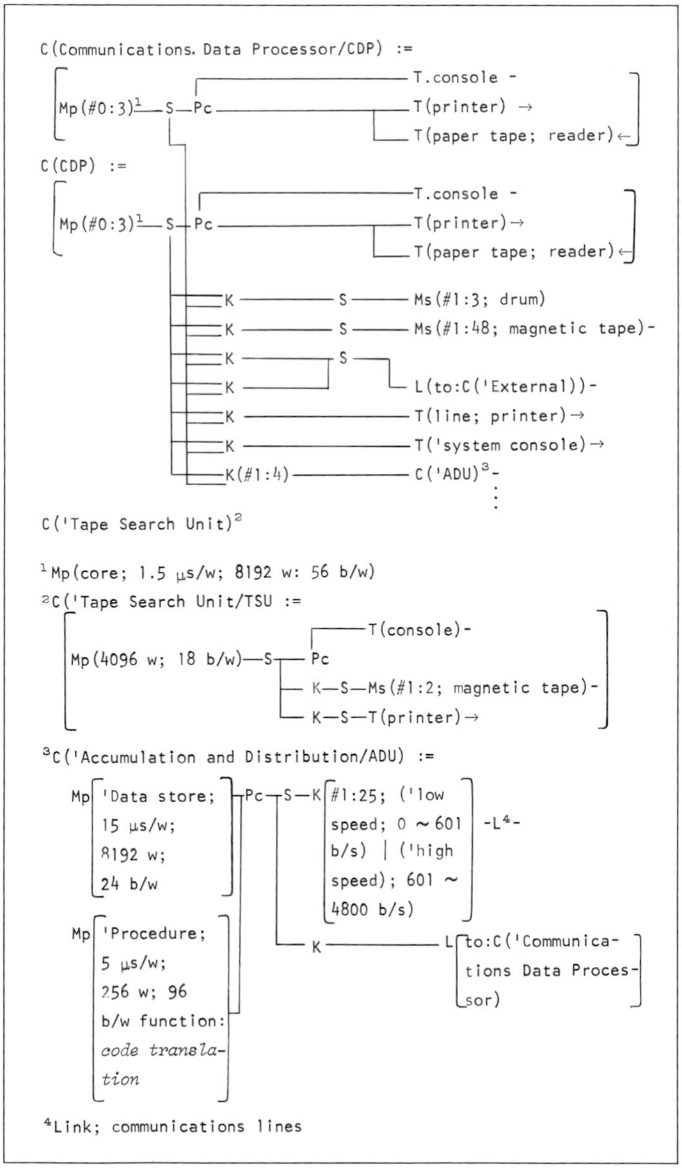

The individual N('SC)'s are located at five specific locations and consist of fixed computer configurations of five to seven C's. The structure of N('SC) (Fig. 6d) is formed basically by a duplex C structure which handles most processing. Attached to the two C('Communications Data Processor/CDP) are two to four C('Accumulation and Distribution Unit/ADU) which handle communication-link processing. A C('Tape Search Unit) is used off line to process data from Ms(magnetic tape). The structures of C('CDP), C('Tape Search Unit), and C('ADU) are defined within Fig. 6d.

ARPA network [1]

An experimental computer network (Fig. 7a) is operational and connects 19 computer facilities associated with the contractors of the Information Technology Branch of the Advanced Research Projects Agency (ARPA). These contractors, all of whom are engaged in advanced research in computer science and technology, form a community in which to attempt a general-purpose network. Since several of the nodes in this network (e.g., M.I.T.; see Fig. 4) will themselves be constructing networks at their own sites, the system has faced a good many of the design problems associated with such a network. Unlike many of the other networks discussed in this chapter, the ARPA network consists of sites that are physically remote, that are each developing as total systems under independent management, and that have no agreed-upon functional specialization vis-à-vis each other. Furthermore, the uses that each node will make of other nodes will be the fairly general ones cited at the beginning of this chapter, as generated by a general scientific community. Since many of the institutions that will be tied in are major academic institutions, diversity will be guaranteed. The motivation behind the experiment is to reveal and begin to solve the technical problems of such general networks, while also discovering which of the several advantages of using networks listed earlier (or others unmentioned) emerge as important.

Fig. 6d. ComLogNet N('Switching Center/SC) PMS diagram.

[1] The Specific links, sites, etc., change with time; thus the actual structures we present are, by the nature of the experiment, almost guaranteed to be in error.

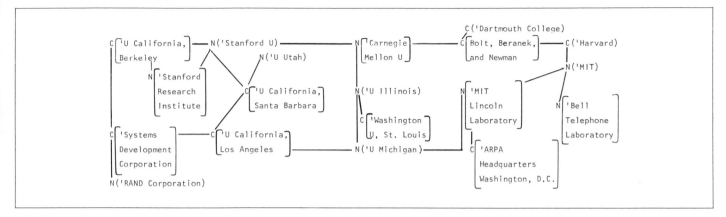

Fig. 7a. Advanced Research Projects Agency (ARPA) network PMS diagram (tentative).

```
C('Local) :=

┌                                                  ┐
│ C('Host)────C('Interface Message Processor/IMP)  │
│ │...        L(40.8 kb/s; to:N('ARPA))             │
│             │...                                  │
└                                                  ┘
```

Fig. 7b. Advanced Research Projects Agency (ARPA) local-computer PMS diagram.

design is to make only small perturbations to the larger host computers. The C('IMP) is responsible for network messages among other nodes (i.e., to their C('IMP)'s) and for the interface between the network and the C (or N) at the local site. The local computer C('Host)-C('IMP) interface is shown in Figs. 7b and 7c

```
N('Local) :=

┌                                                   ┐
│ C┌#1:(n-1);                                        │
│  │          ⟩ C('Interface Message Processor/IMP) │
│  └'Host                                            │
│ C(#n; 'Host)    L(40.8 kb/s; to:N('ARPA))          │
│                 │...                               │
└                                                   ┘
```

Fig. 7c. Advanced Research Projects Agency (ARPA) local-computer-network PMS diagram (tentative).

Technically, the goals of the network are (1) to make a user (T) at any site behave as though it were a T at another site and (2) to let a C at any site use a C at another site for load, program, and data sharing. To each site has been added a special C('Interface Message Processor/IMP). The C('IMP) has been designed by the creators of the network, and it provides the communality that will permit the network to function. One constraint in the network

```
        X(#1:3)²                    X(#1:6)²
        │                           │
        S('Lodi, California)¹       S('Littleton, Massachusetts)¹
        ‖                          ╱
        S('Mojave, California)¹───S('Williamstown, Kentucky)¹
        │                           │
        X(#1:3)²                    X(#1:3)²


  ¹ S(manual; 50 kb/s; 'Telephone Switching Centers)
  ² X(C('local)|N('local)) These N or C may communicate directly
    with one another or by using more L's can communicate via the S's.
```

Fig. 7d. Advanced Research Projects Agency (ARPA) fixed switching centers PMS diagrams (tentative).

for a local computer and local network cases, respectively. The C('IMP) is a C('Honeywell 516; 16 b/w; 12 ~ 16 kw; 1 μs/w) with capability to connect to four to six links at a 50-kb/s data rate.

The ARPA network leases a set of fixed links, L(50 kb/s). These emanate from four S.fixed, as shown in Fig. 7d. Thus the fixed links between the various sites, as shown in Fig. 7a, are composed of the links in Fig. 7d. For example, the L(Carnegie-Mellon University; Bolt Beranak and Newman) goes from Carnegie-Mellon University in Pittsburgh, Pa., to Williamstown, Ky., to Littleton, Mass. (on one of the two links) to Bolt Beranak and Newman in Boston, Mass. The other L(Littleton; Williamstown) is part of L(University of Michigan; Lincoln Laboratory). With such a fixed-link system the network must operate in a store-and-forward fashion, with C('IMP)'s at each site carrying out this function. Thus the C('IMP) is required at each site, since there is no uniformity in the other C's that are at a site and no control over their operation.

Conclusions

We feel the network is the most important computer structure in the book. Through understanding it, we will be able to organize more computing power than with any other structure and to achieve more reliability. The issues of switches and links are so vital that through understanding of them all computer structures will improve.

References

BhusA67; DaviD67; EverR57; PlugW61; RobeL67; SegaR61; IBM System/360 Attached Support Processor (ASP) System Description, H20-0223-0

Part 6

Computer families

The three groups or families of computers described in this part are each built around a single ISP and PMS structure. The IBM 701-7094 II sequence (Sec. 1) shows the evolution of a series. The reader can trace a number of incremental changes, or features, such as the addition of index registers, indirect addressing, I/O processors, and larger random-access memories. The SDS 900-9000 series and the IBM System/360 are both families in which successor models are within a planned framework; evolution occurs mainly in the implementations, not in the ISP.

Section 1

The IBM 701-7094 II sequence, a family by evolution

The IBM 701, 704, 709, 7090, 7040, 7044, 7094 I, and 7094 II sequence relationship is shown in Fig. 1. The group is not a compatible series. The IBM 701 [Astrahan and Rochester, 1952; Buchholz, 1953] is a forerunner of the series; all except the 701 are painfully compatible. The sequence is included because the 7090 is a reference or benchmark of scientific-computer power. All machines use 36-bit words. The 701 stores two instructions/word in the same manner as the IAS computer (Chap. 4), whereas all others in the sequence store only one instruction/word. The 701, 704, and 709 are first-generation, vacuum-tube technology; the rest are second-generation.

The IBM 7094 II description given in Chap. 41 is based directly on information in the Programming Reference Manual, but the Appendices of that chapter give the ISP of the Pc, a Pio, and a K as inferred by the authors of this book. The description of the Pc gives the instructions in the 704 and 7044

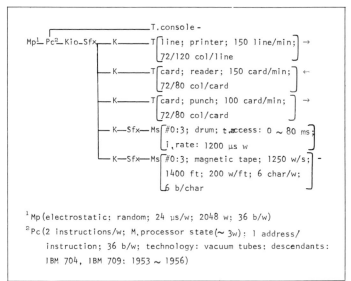

^{1}Mp(electrostatic; random; 24 μs/w; 2048 w; 36 b/w)

^{2}Pc(2 instructions/w; M.processor state($\sim$ 3w): 1 address/ instruction; 36 b/w; technology: vacuum tubes: descendants: IBM 704, IBM 709: 1953 $\sim$ 1956)

Fig. 2. IBM 701 PMS diagram.

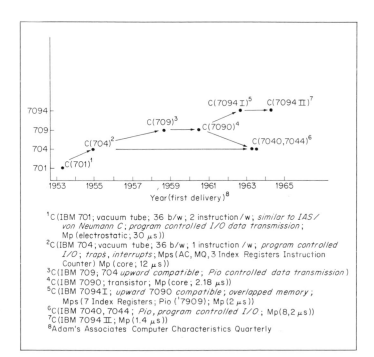

^{1}C(IBM 701; vacuum tube; 36 b/w; 2 instruction/w; *similar to IAS/ von Neumann C*; *program controlled I/O data transmission*; Mp(electrostatic; 30 μs))

^{2}C(IBM 704; vacuum tube; 36 b/w; 1 instruction/w; *program controlled I/O*; *traps, interrupts*; Mps(AC, MQ, 3 Index Registers Instruction Counter) Mp(core; 12 μs))

^{3}C(IBM 709; 704 *upward compatible*; *Pio controlled data transmission*)

^{4}C(IBM 7090; transistor; Mp(core; 2.18 μs))

^{5}C(IBM 7094 I; *upward 7090 compatible*; *overlapped memory*; Mps(7 Index Registers; Pio ('7909); Mp(2 μs))

^{6}C(IBM 7040, 7044; *Pio, program controlled I/O*; Mp(8,2 μs))

^{7}C(IBM 7094 II; Mp(1.4 μs))

^{8}Adam's Associates Computer Characteristics Quarterly

Fig. 1. Relationships among IBM 701, 704, 709, 7094 series.

to show an evolution. However, the major evolutionary change does not appear in Pc's ISP but in the PMS structure.

The 704 structure, like that of the 701 (Fig. 2), provides only for peripheral transfers to primary memory via Pc under programmed control with no interrupt system. As such, only one T or Ms could operate easily at a time. The 709 introduced the Pio('Data Channels) to improve the ability to transfer data between Mp and Ms without requiring Pc intervention. Concurrent operation of several I/O devices is carried out by multiple Pio's along the lines of the 7094 II PMS structure (Fig. 1, Chap. 41, page 518). However, the utilization of the data channels tends to be rather low, particularly when the data channel is controlling very slow devices (e.g., card equipment and line printers). When operating a high-speed tape unit at 90,000 × 6 bits/sec the utilization of the data channel is still only approximately 3 percent. A program interrupt method of data transfers would have been sufficient.

The incompatibility among the machines, especially the 7090-7040-7094, is disheartening, both from the point of view of a user and an engineer. The incremental hardware needed to achieve compatibility is inexpensive when the system price is considered. Also, the incremental changes in the ISP do little to increase the Pc performance. Compared with the 704, the extensive order code of the 7094 shows an evolution in which for marketing, emotional, or analytic reasons new instructions were added. The index registers and their instructions are a good example of this trend. The 7094 has a very general set of index-register transmission instructions; if implemented properly, they are probably easier to provide than the original 704 instructions.

In the implementation of the double-precision floating-point hardware, the sense-indicator register is needed for temporary storage. Thus a user has to preserve this register when double-precision floating-point instructions are given. The reason for this undoubtedly relates to field modifications and cost. In an original design this would be inexcusable; in this case double-precision floating point is undoubtedly worth the loss of sense indicators.

All in all, the designers of the 704-7094 II provided increased generality through evolution. They gradually ran out of patching time, technology, instruction encoding space, and memory addressing bits, while exceeding compatibility constraints. It was indeed time to create the IBM System/360.

Chapter 41

The IBM 7094 I, II

Introduction

The IBM 7094 I and 7094 II computers are the last of a series of computers beginning with the IBM 704 (Fig. 1, page 515). The series is an outgrowth of the IBM 701. Although the series is designed for scientific (arithmetic) calculations, its speed and structure allow it to be used for general-purpose computation. Business-type processing which uses string data is efficiently handled by conversion into fixed-length fields at input and output. From about 1956 to 1966 the family was the standard of large computers in the United States, there being approximately 20 701, 50 704, 20 709, 50 7090, 130 7094 I, 125 7094 II, 120 7040, and 120 7044 computers in existence.

The PMS structure is a single central processor (Pc) with multiple input/output processors (Pio's) (for all except the 701 and 704). The Pio's provide for multiple transfers to primary memory (Mp) at high information flow rates. The structure allows for duplex connection to terminal (T) or secondary-memory (Ms) control (K). This provision permits the system to be used in real-time applications requiring significant computation, high-data-rate transfers with other systems, and high availability. However, the system was not initially designed for time sharing and multiprogramming use, and the attempt to so use it required modification [Corbato et al., 1962].

The word length is 36 bits. There is one single-address instruction/word. In all but the 7094 the processor interprets instructions serially. In the 7094 one register instruction look-ahead is used. The Pc has index registers, the 704 being the first IBM computer to use them. Their number increased from three in the 704 $\sim$ 7090 to seven in the 7094, as their usefulness became apparent.

Structure

A simple tree-structured IBM 7094 I using PMS is shown in Fig. 1 and using a conventional block diagram in Fig. 2.

Primary memory (Mp) and P-Mp switch

The primary memory, Mp('7302 Core Storage), has a capacity of 32,768 36-bit words with a cycle time of 2 microseconds. The actual memory has a 72 + 1 parity bit word for even and odd addresses of 36-bit words. A request for two 36-bit words can be acknowledged in one 2-microsecond memory cycle. Thus Mp is Mp('7302 Core Storage; 2 μs/w; 16384 w; (72, 1 parity) b/w) for the 7094 I, and Mp(1.4 μs/w; 16384 w; (72, 1 parity) b/w) for the 7094 II.

The S('7606 Multiplexor; time multiplexed) provides access to Mp from any one of nine P's. Only Pc can request two 36-bit words at a time from Mp for instruction look-ahead and double-word operations. There can be only one Pc in the system.

Processors, P

Three processors are described: Pc('7109, 7110 Central Processing Unit/CPU), Pio('7607 Data Channel), and Pio('7909 Data Channel).

All P's behave similarly in that Pc instructions and Pio commands[1] are fetched (or requested) from Mp and then interpreted in P. An instruction location counter in P addresses the next instruction. A processor instruction may, in turn, require the processor to access Mp for data, to perform transfers, to modify its state, etc. Although structurally the P's are similar, organizationally the Pc is superior to the Pio('Data Channel)'s; Pc issues programs to Pio's and start and stops (controls) Pio's.

Two-way communication is required between Pc and the Pio's. Tasks (jobs or programs) for Pio's are first set up in Mp by Pc. Pc then demands that Pio execute the program independently under its own control. Initialization takes place when Pc sets the instruction counter of a Pio. Upon task completion in Pio, an interrupt request is sent to Pc from Pio.

Below we first give a description of the Pc. Then the Pio('7909) is presented in detail and the Pio('7607) is outlined. The reader should compare the two Pio's. The Pio('7909) is a later design than the Pio('7607). It interprets instructions for the block of data being transferred and issues instructions to the KMs or KT. The earlier Pio('7607) interprets the instructions for controlling the information being transferred; the Pc interprets and issues the instructions to KMs or KT. The 7909 is therefore able to control more closely a T or Ms using a single program without need for Pc intervention.

---

[1] IBM attempts to distinguish between Pc and Pio's terminologically by "instruction" and "command." We make no such distinction in the following discussion; P's interpret instructions.

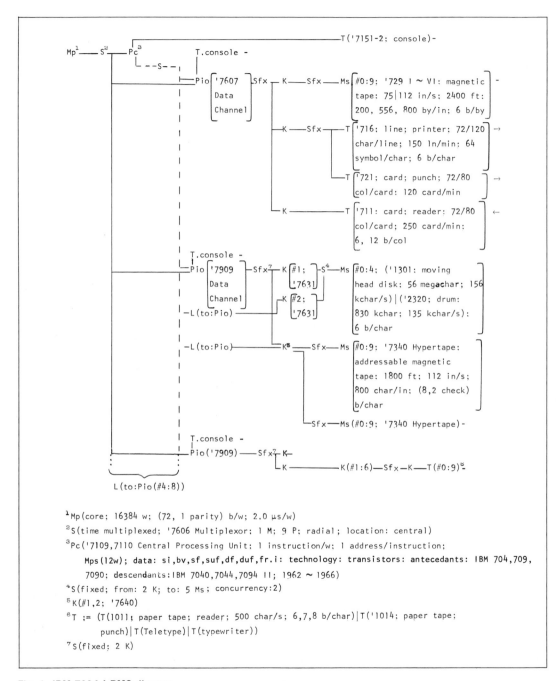

Fig. 1. IBM 7094 I PMS diagram.

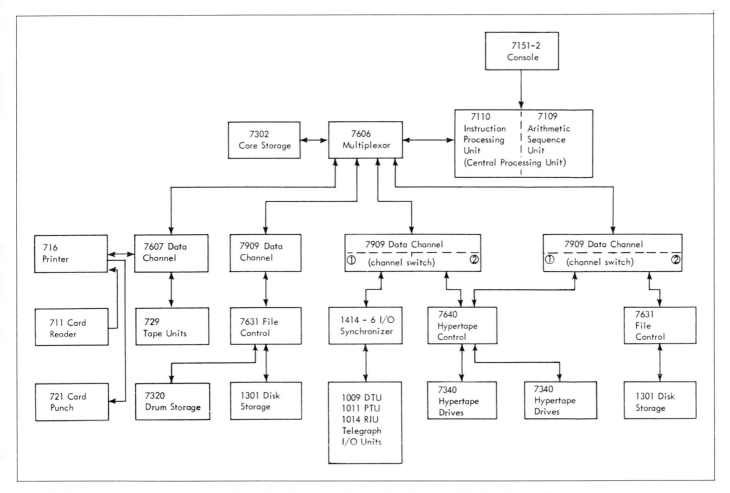

Fig. 2. IBM 7094 data-processing system configuration. (*Courtesy of International Business Machines Corporation.*)

Central processing unit, Pc

The Pc has three physical parts: the T('7151-2 Console), the D('7110 Instruction Processing Unit), and the K('7109 Arithmetic Sequence Unit). In terms of gross PMS parameters the 7094 I can be described similar to footnote three of Fig. 1 as

Pc['7109, 7110, 7151-2 Central Processing Unit/CPU;
  36 b/w; 1 address; 1 instruction/w;
  data: (si, bv, sf, suf, df, duf, fr.i);
  number representation: sign, magnitude;
  Mps('Accumulator, 'Multiplier_Quotient, 'Index_Registers
    [1:7], 'Sense_Indicators, 'Instruction_Counter, 'Trap_En-
    able⟨1:12⟩, 'miscellaneous_bits⟨1:7⟩)]

The Pc will be discussed in two parts: the Register-transfer level implementation and the Instruction-set Processor. These are partially redundant, but they offer another opportunity to compare the two types of descriptions. The Pc hardware will be described by first giving the registers and the interregister transfer paths. Then the process by which instructions are interpreted will be described. (Interpretation occurs in a distinct set of memory cycles, called instruction/I, execute/E, logic/L, and buffer/B, which are sometimes mentioned in describing registers and will be fully discussed later.)

Processor registers and mode bits registers

Figure 3 gives the Pc registers and the data transfer paths. Both the ISP registers (denoted by °) and the temporary registers are given. The ISP registers and modes are controlled by the program.

Instruction counter (IC)°. The Instruction Counter, IC, is 15 bits. It is used by the processor to locate the next instruction in Mp. Once the program is started, the IC can be set to an address specified by a transfer instruction. For most instructions, the IC is stepped sequentially by 1 with each new instruction. The IC is normally advanced at the end of each instruction (I cycle).

Instruction backup register (IBR). The Instruction Backup Register, IBR, is a 36-bit register, ⟨S, 1:35⟩, and is used to buffer the next instruction. Pc attempts to have the next instruction available in IBR, since the Mp permits 72-bit transfers, thus avoiding an unnecessary reference to Mp. When the instruction reference is to an even location, the IBR is loaded with the contents of the next higher odd address after the contents of the even address have been placed in the Storage Register. The IBR is also used for fetching operands in double-precision operations.

Address register (AR). The Address Register, AR, is 15 bits and receives information from the Storage Register, Instruction Backup Register (at the beginning of a storage reference I or E cycle), Index Register, and Index Adder. The contents of the AR are sent to the Multiplexor Address Switch to select the core memory location.

Instruction register (IR). The 18-bit Instruction Register, IR, is divided into two parts: bits ⟨S, 1:9⟩ always contain the operation part of the instruction, and bits ⟨10:17⟩ form the Shift-counter Register. The Shift Counter is used during shifting, multiplication, division, and floating-point instructions. Bits ⟨10:17⟩ may also contain a sense instruction address, operation codes for those instructions which require an address part, and the class and unit codes for input/output instructions.

Storage register (SR). The 36-bit Storage Register, SR, stores information that comes from or goes to core storage.

Adders (not a register). The Adders furnish a 36-bit path for data going from the storage register to other registers in the processor.

Accumulator register (AC)°. The Accumulator Register, AC, is 38 bits (a 35-bit word with a 1-bit sign, and 2 bits for overflow conditions, P and Q). The AC is used to hold one factor during arithmetic or logical operations and to receive results from the adders.

Information may be shifted into the accumulator from the MQ, 1 bit at a time.

Multiplier-quotient register (MQ)°. The MQ Register is 36 bits. During a multiply instruction, MQ contains the multiplier; during a divide instruction, MQ receives the quotient. It can be shifted right or left, independently, or combined with AC into a 72-bit register.

Sense indicator register (SI)°. The Sense Indicator Register, SI, is 36 bits. SI is normally used as a set of binary program switches which can be set and tested. However, it is also used as a temporary register in double-precision arithmetic operations.

Index registers (XR)°. Seven 15-bit Index Registers, XRs, in the 7094 system are used for address modification. They are specified by the tag bits of an instruction (bits ⟨18:20⟩) and modify an address by adding the two's complement of their contents to the address. In the earlier 7090 (and 7044) only XR[1, 2, 4] are available.

Multiple tag mode°. In Multiple Tag Mode only Index Registers 1, 2, and 4 can be specified. The indexing function specified is determined by the "logical-or" of each index register specified. When not in Multiple Tag Mode, each 3-bit number selects one of seven index registers. The 1-bit Multiple-Tag-Mode Register maintains the state of the mode. The requirement for the two modes comes entirely from the need to maintain compatibility between the 704, 709, 7090, 7040, and 7044 (which have three index registers addressed as in Multiple Tag Mode) and the 7094 I and 7094 II which have seven index registers.

Tag register (TR). This temporary register holds the tag field of the instruction being executed and is used to select the Index Register being addressed.

Index adders (XAD) (not a register). A separate 15-position Index Adder is used for the Index-register operations. All storing, loading, changing, and modifying of Index Registers is via the Index Adders.

Accumulator overflow°. The Accumulator Overflow Indicator is turned on whenever a 1 passes into or through position P from position 1 of the AC as a result of the execution of a fixed-point arithmetic or a shifting instruction.

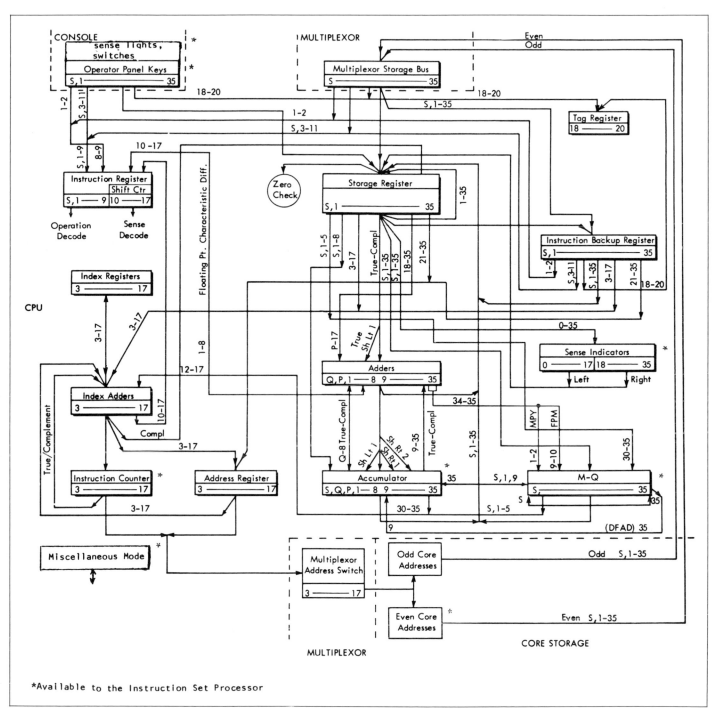

Fig. 3. IBM 7094 central-processing-unit information flow. (*Courtesy of International Business Machines Corporation.*)

Divide-check°. The Divide-Check Indicator is turned on, in fixed-point or floating-point division, if the magnitude of the number in the AC (dividend) is greater than or equal to the magnitude of the number in memory (divisor).

Input-output check°. The Input-Output Check Indicator (I-O check) is turned on by the attempted execution of an input/output instruction without first selecting an input/output unit.

Transfer trap mode°. The computer can be operated in a special Transfer Trap Mode. Operation in the Trap Mode permits the program to run at normal speed with interruptions of normal operation only at transfer points. At such points the location of the last sequential instruction is saved, and a transfer of control is made to a fixed location.

Sense switches°. Six Sense Switches are located on the console. They may be turned on or off manually, and there are instructions which sense them.

Sense lights°. Four Sense Lights are also on the console. Any one of these lights may be turned on, off, or the status tested by instructions.

Panel in-out switches°. These 36 switches on the console may be read by an instruction.

Instruction-set interpretation

The basic computer clock cycle is 2.0 μs in 7094 I and 1.4 μs in 7094 II, as dictated by Mp. Within the single 2- (or 1.4-) micro-second cycle, up to 10 sequential register transfers and/or data operations can take place, each of which transfers information among the Pc's registers; several operations may occur simultaneously. In Pc four different cycles are used: instruction/I, execute/E, logic/L, and buffer/B. The cyclic sequence of an instruction is fixed, always beginning with an I cycle and progressing to E, L, or B cycles, depending on the instruction. The number of cycles required for an instruction may vary from 1 (e.g., transfer) to 19 (e.g., double-precision floating-point divide).

Instruction cycle (I). The I cycle begins when IC furnishes the instruction location to Mp, via S('Multiplexor). The addressed instruction word taken from Mp goes to the Multiplexor Storage Bus (Fig. 3). From the Multiplexor Storage Bus the instruction is read into the Storage Register where it is separated into the operation portion and the address portion of the instruction word.

The operation portion of the Storage Register goes into the Instruction Register, where the operation code is decoded and the execute control circuitry is set up to perform the operation specified by the instruction. The address portion of the instruction word, now located in the Storage Register, may be used directly. Normally, however, it goes to the Address Register and then to the Multiplexor Address Switch to locate the appropriate data word in Mp. If the address is to be modified, it is routed from the Storage Register to the Index Adders for Index-register modification. The modified address is then brought to the Address Register and on to the Multiplexor Address Switch to locate the data word in core storage.

Concurrently, during the same instruction cycle, a second instruction, located at the immediately higher odd-numbered Mp address location, is brought to the Instruction Backup Register/ IBR. While in the IBR, the odd-numbered instruction is partially decoded to determine if it meets certain criteria for concurrent execution, thus saving a second Mp reference. If the instruction in the IBR cannot be executed with the current instruction, it is ignored in the current I cycle and is brought into the Storage Register on the next I cycle.

Execution cycle (E). The execution (E) cycle is used when a reference to core storage is needed. All instructions requiring an operand have an E cycle following the I cycle.

Indirect addressing of an instruction requires an extra E cycle. In other words, an instruction that normally goes from I to E to be executed will go to I, E, and again to E if it is indirectly addressed.

Logic cycle (L). The L cycle is an execute cycle that does not require a reference to Mp. Many instructions use both E and L cycles when information is required from storage and the instruction cannot be completed during an E cycle. Other instructions require no reference to storage and, therefore, use only I and L cycles for their completion.

Buffer cycle (B). A buffer (B) cycle is a null Pc cycle; it is used when the data channels get information from or put information into core storage. This information can be either data or data-channel commands. All demands for B cycles come from the channels themselves. Because of the nature of Ms's and T's, the demand for a B cycle takes precedence over an instruction being performed by Pc. If Pc is in its logic cycle, then both an L and B cycle occur simultaneously.

Instruction interpretation. Instruction flow diagrams for the CLA, CAL, and CLS instructions are given in Fig. 4. These diagrams show the sequential process of instruction execution. Although the flow diagrams for these instructions are trivial, the general process is still apparent. The more complex instructions, for example, double-precision floating-point divide, are carried out in a similar fashion, but with many more operations. The registers, transfer paths, and interregister data operations are the register-transfer-level primitives from which the ISP is implemented. The data flow diagram (Fig. 3) explicitly defines the main registers and register operations within Pc.

Pc ISP

The Pc Instruction-set Processor is given in Appendix 1 of this chapter. The instructions are arranged in groups according to the location of operands. These groups are:

Operations on Mp
Mp ← u Mp   (*unary operation/u on Mp*)
Mp ← u Mps   (*unary operation on Mprocessor state/Mps*)
Mp ← Mp b Mps   (*binary operation/b*)

Operations on AC and MQ
Mps ← u Mps
Mps ← u Mp
Mps ← Mps b Mp

Operations on the index registers

Operations on the sense indicators

Instruction for program control

Memory mapping for multiprogramming and Mp(65536 w)

A special option provides multiprogramming by allowing a program to run in a protected area of Mp. Two registers are used: The base register establishes the lower bound of the program, and the length register establishes the upper bound. Pc checks that all program references are within the protected area.

Two Mp(32678 w)'s can be used on the computer. Mp is then considered as A core and B core for addresses 0:32767 and 32768:65535. A 1-bit register is used to select whether A or B core is to be used for data; and one 1-bit register is used to select whether A or B core is to be used for the instruction. These modifications were used at M.I.T. in their Compatible Time Sharing System/CTSS [Corbato et al., 1962] which used a 7094 II.

Pio('7607 Data Channel)

The Pio('7607 Data Channel) executes programs which transfer data between Mp and Ms(magnetic tape) or T(card; reader, punch), (line; printer)). The paths and structure can be seen in Fig. 1.

Transferring blocks of data between Mp and an Ms or a T via the 7607 data channel takes places as follows:

1   Pc sets up the block transfer program in Mp for Pio.

2   Pc attaches a K for Ms(magnetic tape) or for T(card;reader) to Pio. (Faults in the connection may cause K to interrupt Pc.)

3   Pc starts the Pio by loading the Pio's instruction counter.

4   The data transmission takes place. On input, for example, T or Ms transmits a 6-bit character (or a 72-bit word) to K. The characters are buffered (collected) in K and sent on to Pio. Pio then requests a memory access from Mp via the S('7606 Multiplexor) and, finally, a data word is transmitted to Mp.

5   At the termination of a simple data block transfer, Pio fetches the next instruction from Mp. If the next instruction-task type is the same, Pio and K remain logically linked and continue to transmit data.

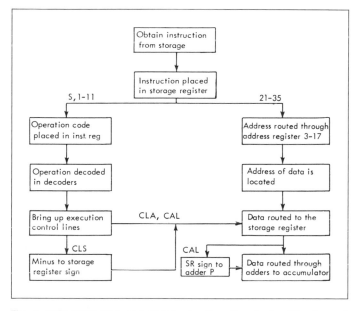

Fig. 4. IBM 7094 CLA and CLS instruction flowcharts. (*Courtesy of International Business Machines Corporation.*)

6 At the termination of the task, the completion signal from Pio causes Pc to interrupt and Pio may also halt.

Pio('IBM 7909 Data Channel)

Ms('1301 Disk Storage, '7340 Hypertape Drives) and the T('Tele-Processing equipment) communicate with Mp via the Pio('7909 Data Channel). Four 7909 Data Channels may be attached to a 7094 I or II system.

K('7631 File Control) is required for M(disks). Several K('7631) can be used with the 7094 system alone or shared with an IBM 1410 system or shared with another IBM 7000 series (not 7072 system).

When Ms('7340 Hypertape Drives) are attached to the 7094 system, K('7640 Hypertape Control) is used between the 7909 data channel and the drives. One K('7640) may be attached to a 7094 system; it has two paths, each of which can be used for data transmission.

The K('1416-6 Input-Output Synchronizer) is used with T('Tele-processing Equipment)'s. The structure for these T's is rather elaborate, yet only six T's can be active at a time.

Transferring data from Mp to a T or an Ms via the 7909 takes place as follows:

1 Pc sets up the data-transfer management program in Mp for a Pio.

2 Pc starts Pio by setting Pio's command (instruction) location counter at the origin of the task program in Mp. (Faults in the connection may cause Pio interrupts to Pc.)

3 Pio issues an instruction to be executed by K. This establishes a state in K which selects and initializes the particular Ms or T and attaches the peripheral device K to Pio. (Faults in this selection may cause interruption of Pio.)

4 The data-transmission instruction is read and initializes Pio.

5 The data transmission takes place under control of Pio-K. The K of the selected device assembles characters. Input characters are transferred to Pio which assembles them into words and in turn transfers them to Mp.

6 At the termination of a data block transfer instruction, another instruction is fetched from Mp by Pio. This instruction may be to another K.

7 At the termination of the Pio program, Pio signals completion by interrupting Pc.

This discussion is based on information taken from the IBM 7094 Reference Manual. The body of the description is contained

in ISP descriptions (Appendices 2, 3 and 4 of this chapter). The main registers of Pio are shown in Fig. 5. These registers are declared and their function is explained in the first section of the ISP description of Pio (Appendix 2). The remainder of the ISP description is concerned with defining the interpreter and the ISP instruction set.

There are about 50 bits in the K's (see Appendix 3). A knowledge of K's state and the K process is required for understanding the Pio. A description of the K and Pio data-transmission processes is given in Appendix 2.

The Pc instructions controlling Pio are presented in Appendix 4.

The level of detail in the appendices is slightly greater than that in normal ISP description. It is, however, not completely precise, as the behavior is extremely time- and Ms- or T-dependent. The sequence check conditions are incomplete; that is, the

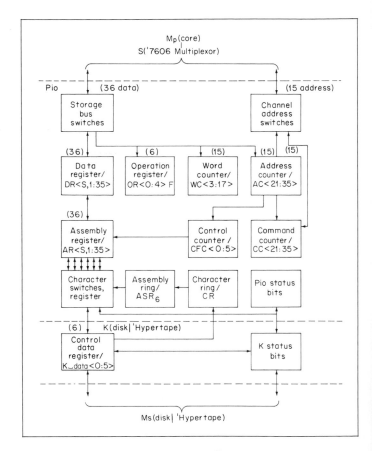

Fig. 5. IBM 7909 data-channel-registers diagram.

conditions for illegal instruction sequences are not given. Both ISP and text descriptions are given for parts which are particularly complex.

The ISP description should be observed in the following sequence: Pio State; K State (Appendix 3); Pio Instruction Format; Pio Interpreter; Pio Instruction—Control (or Initialization) instructions, Block Transfer (or Copy) instructions, Conventional Move and Transfer instructions, and Interrupt Control instructions; Instructions in Pc (Appendix 4); Interrupt Operation; and Processes defining data movements between K and Pio (Appendix 2). The Pio, K, and Ms or T processes are, in several ways, more complex than those of a Pc. First, Ms or T activity is not categorized as nicely as a Pc instruction set. The T or Ms events occur at times peculiar to the device—not a simple synchronous clock. Finally, the peripheral components have a large number of error states.

Conclusions

The series ending with the IBM 7094 II is a significant member of the computer population. It provides a good example of the evolution in computer systems that occurred from 1954 to 1965.

References

CorbF62; FrizC53; GreeJ57; GrumM58; RossH53; SaxoJ63; StevL52; A22–6703 IBM 7094 Principles of Operation

APPENDIX 1   IBM 7094 PC ISP

Appendix 1

IBM 7094 Pc ISP Description

Pc State

The description does not include the two protection and relocation schemes used for the 7040 and 7094. The Trap-Mode flip-flop is declared; its action is not described. Trap-Mode allows any change of the Instruction Counter to cause a trap. The Instruction Backup Register is not described, although it is used to save time in program execution. The description of the arithmetic functions is highly simplified.

AC<Q,P,S,1:35>	* Accumulator, 38 bits
ACs<S,1:35> := AC<S,1:35>	* signed AC word
ACl<P,1:35> := AC<P,1:35>	* logical AC word
P := AC<P>	* carry for AC<1:35>; AC overflow is also set
Q := AC<Q>	* carry for bits<P,1:35>
S := AC<S>	* sign bit of AC
MQ<S,1:35>	* Multiplier-Quotient
ACMQ<S,Q,P,1:71>:= AC□MQ<1:35>	* double word accumulator
SI<0:35>	Sense Indicators or program flags must be preserved if double precision floating point instructions are given.
XR'[1:7]<3:17>	Index Registers in 7094
XR''[A,B,C]<3:17> := XR[1,2,4]<3:17>	* Index Registers for 704, 7090
Multiple_Tag_Mode	program switch to force compatibility with 704, 7090; only 3 index registers XR[A,B,C] are in 704, 7090
IC<3:17>	* Instruction Location Counter
Run	* indicates whether machine is executing instructions
Divide_Check	*
AC_overflow	*
MQ_overflow	*
Input_Output_check	*
Trap_request<A:H>	Request to trap Pc from Pio #A...#H
Trap_Mode	* Allows trapping or not of transfer instructions (not described)

Pc Console State

Keys<0:35>	* console data
Sense_Switches<0:5>	
Sense_Lights<0:3>	

Mp State

M[0:32768-1]<S,1:35>	Primary Memory of 2^{15} w

Instruction Format

instruction<S,1:35>	corresponds to the physical Storage Register
Y<21:35> := instruction<21:35>	generally the address part; used to calculate the effective address; corresponds to the physical Address Register
T<18:20> := instruction<18:20>	the XR to use: 1,...7; 0 means no indexing; corresponds to a physical register
F<12:13> := instruction<12:13>	indirect address specification
indirect := (F<12:13> = 11)	
op<S,1:11> := instruction<S,1:11>	op code; corresponds to a physical register
hi_op<0:2> := instruction<S,1,2>	special op codes

* Denotes subset ISP, IBM 704, 7044 series

R<18;35> := instruction<18;35>   *right half of instruction used to select SI bits*

D<3:17>   := instruction<3:17>   *Decrement part of instruction, used to directly modify XR's*

C'<12:17> := instruction<12:17>   *specifies variable length part of operation*

C<10:17> := instruction<10:17>   *convert instruction parameter*

c<15:17> := instruction<15:17>   *specifies character position in 7040, 7044 or extends op code*

Effective Address Calculation Process

e<21:35> := (¬ indirect →e';   *effective address calculation*

   indirect →instruction<18:35> ←M[e']<18:35>; next e')   *1 level indirect addressing*

e'<21:35>:= ((T = 0) →Y;   *indexed effective*

            (T ≠ 0) →Y−XR[T])

e''<23:35>:= e'<23:35>

sc<28:35>:= e'<28:35>   *a truncation of e, used for specifying number of shifts: corresponds to a physical register*

XR[T]<3:17> := (*index registers are or'd together in multiple tag mode*

   ¬Multiple_Tag_Mode →XR'[T];

    Multiple_Tag_Mode → (

     (T<13>→ XR''[A]) ∨ (T<19> →XR''[B]) ∨ (T<20> →XR''[C])))

The description for Multiple Tag Mode is incomplete for the case of writing in several index registers at one time. The only way this could be accomplished in the description would be to define each load index register instruction as microprogrammed.

Data Formats

sl<S,1:35>   *logical data; unsigned integer/boolean vector*

sx<S,1:35>   *single precision fixed point (integer) data*

   sx sign := sx<S>

   sx magnitude<1:35> := sx<1:35>

sf<S,1:35>   *single precision floating point value of: sf_sign□sf_mantissa*
                                                   $\times 2^{sf\_exponent}$
   sf sign := sf<S>

   sf exponent<1:8> := 200₈ − sf<1:8>

   sf mantissa<0:26>:= sf<9:35>

df[0:1]<S,1:35>   *double precision floating point value of: df_sign□df_mantissa*
                                                   $\times 2^{df\_exponent}$
   df sign := df[0]<S>

   df exponent<1:8> := 200₈ − df[0]<1:8>

   df mantissa<0:53>:= df[0:1]<9:35>

Instruction Interpretation Process

Run → (instruction ←M[IC]; IC ← IC+1; next   *fetch*

        instruction_execution)   *execute*

Instruction Set and Instruction Execution Process

   Instruction_execution := (

Operations on M: M[e] ← f; or M[e] ← f(M[e]);

   STZ (:= op = 600) →M[e] ←0;   * *store zero*

   MSP (:= (op = −1623) ∧ (c = 7)) →M[e]<S> ←0);   *make sign positive; 704 series only*

   MSM (:= (op = −1623) ∧ (c = 6)) →M[e]<S> ←1);   *make sign minus; 704 series only*

Block transfer of data, M← M (704 series only)

   TMT (:= op = −1704) → (M[AC<21:35>: (AC<21:35> + e'<28:35>)] ←

                      M[AC<3:17>:(AC<3:17> + e'<28:35>)]);

Single word data transmission to M, M[e] ← Register

STQ (:= op = -600) → (M[e] ← MQ);   * *store MQ*

SLQ (:= op = -620) → (M[e]<S,1:17> ← MQ<S,1:17>);   * *store left half MQ*

STO (:= op = 601) → (M[e] ← ACs);   * *store*

SLW (:= op = 602) →(M[e] ← AC1);   * *store logical word*

STP (:= op = 630) →(M[e]<S,1,2> ← AC<P,1,2>);   * *store prefix*

STD (:= op = 622) →(M[e]<3:17> ← AC<3:17>);   * *store decrement*

STT (:= op = 625) → (M[e]<18:20> ← AC<18:20>);   * *store tag*

STA (:= op = 621) → (M[e]<21:35> ← AC<21:35>);   * *store address*

STL (:= op = -625) → (M[e]<21:35> ← IC);   *store instruction location counter*

STR (:= hi‿op = -1) → (M[0]<21:35> ← IC; IC ← 2);   *store instruction location counter and trap*

STI (:= op = 604) → (M[e] ← SI);   *store indicators*

Double length data transmission to M from A

DST (:= op = -603) → (M[e]□M[e+1] ← ACs□MQ);   *double store*

Binary operation with AC: M[e] ← AC b M[e];

ORS (:= op = -602) → (M[e] ←AC1 ∨ M[e]);   * *or to storage*

ANS (:= op = 320) → (M[e] ←AC1 ∧ M[e]);   * *and to storage*

6 bit character to M from AC, (7040 only);

SAC (:= op = -1623) → (M[e]<c × 6 :(c × 6+5)> ← AC<30:35>);

Operations to the AC,MQ, or AC□MQ with AC,MQ,ACMQ, Keys and M operands:

CLM (:= (op = 760) ∧ (e' = 0)) → (AC<Q,P,1:35> ←0);   *clear magnitude*

SSP (:= (op = 760) ∧ (e' = 3)] → (AC<S> ← 0);   * *set sign plus*

SSM (:= (op = -760) ∧ (e' = 3)) → (AC<S> ←1);   * *set sign minus*

CLA (:= op = 500) → (AC ←0; next ACs ←AC+M[e]);   *clear and add*

CAL (:= op = -500) → (AC ←0; next AC1 ←AC1+M[e]);   *clear and add logical*

CLS (:= op = 502) → (AC ←0; next AC ←AC-M[e]);   *clear and subtract*

LDQ (:= op = 560) → (MQ ←M[e]);   *load MQ*

ENK (:= (op = 760) ∧ (e' = 4)) → (MQ ← Keys);   *enter Keys*

PIA (:= op = -46) → (AC ← SI);   *place indicators in AC*

DLD (:= op = 443) → (ACs□MQ ← M[e]□M[e+1]);   *double load*

Operations with AC,AC ← f(AC)

CHS (:= (op = 760) ∧ (e' = 2)) → (AC<S> ←¬ AC<S>);   *change sign*

COM (:= (op = 760) ∧ (e' = 6)) → (AC<Q,P,1:35> ←¬ AC<Q,P,1:35>);   **complement magnitude*

RND (:= (op = 760) ∧ (e' = 10)) →MQ<1> →AC ← AC + 1;   * *round*

FRN (:= (op = 760) ∧ (e' = 11)) → (AC ← round(ACMQ){sf});   * *floating round*

ALS (:= op = 767) → (AC<Q,P,1:35> ←AC<Q,P,1:35> × 2^{sc});   * *AC left shift*

ARS (:= op = 771) → (AC<Q,P,1:35> ←AC<Q,P,1:35>/ 2^{sc});   * *AC right shift*

LLS (:= op = 763) → (ACMQ' ← ACMQ' × 2^{sc});   * *long left shift*

LRS (:= op = 765) → (ACMQ' ← ACMQ'/ 2^{sc});   * *long right shift*

     ACMQ'<0:71> := AC<Q,P,1:35> □ MQ<1:35>

LGL (:= op = -763) → (ACMQ'' ← ACMQ'' × 2^{sc} {logical});   * *logical left shift*

LGR (:= op = -765) → (ACMQ'' ← ACMQ'/ 2^{sc} {logical});   * *logical right shift*

     ACMQ''<0:72> := AC<Q,P,1:35> □ MQ<S,1:35>

RQL (:= op = -773) → (MQ ← MQ × 2^{sc} {rotate});   * *rotate MQ left*

Exchange of Data between registers, AC, and MQ

XCA (:= op = 131) → (AC ← MQ; MQ ← AC);   *exchange AC and MQ*

```
        XCL (:= op = -130) → (MQ ←AC1; AC1 ←MQ; AC<S,Q> ←0);        exchange logical AC and MQ
6 bit character to AC from M (704 only)
        PCS (:= op = -1505) → (AC<30:35> ←M[e]<(c x 6): (c x 6 + 5)>);place character from storage
Binary operations with M,AC← AC b M;
        ADD (:= op = 400) → (AC ←AC + M[e]);                        * add
        ADM (:= op = 401) → (AC ←AC + abs(M[e]));                   * add magnitude
        SUB (:= op = 402) → (AC ←AC - M[e]);                        * subtract
        SBM (:= op = -400) → (AC ←AC - abs(M[e]));                  * subtract magnitude
        MPY (:= op = 200) → (ACMQ ←MQ x M[e]; AC<Q,P> ←0);         * multiply
        MPR (:= op = -200) → (ACMQ ←MQ x M[e]; next                 * multiply and round
                            MQ<1> →AC ←AC + 1; AC<Q,P> ←0);
        DVH (:= op = 220) → (AC,MQ ←ACMQ / M[e]; next
                            Divide check →Run ←0);                  * divide or halt
        DVP (:= op = 221) → (AC,MQ ←ACMQ / M[e]);                   * divide or proceed; Divide check may be set
        ACL (:= op = 361) → (AC1 ←AC1 + M[e]);                      * add and carry logical word

    The following are variable length x and / operations.  C' specifies the length of divisor or multiplier.

        VLM (:= op = 204) → (ACMQ ←MQ x M[e] {vl});                 variable length multiply
        VDP (:= op = 225) → (AC,MQ ←ACMQ / M[e] {vl});             variable length divide or proceed
        VDH (:= op = 224) → (AC,MQ ←ACMQ / M[e] {vl}; next        variable length divide or halt
                            Divide check →Run ←0);
    Single precision floating point
        FAD (:= op = 300) → (AC,MQ ←AC + M[e] {sf});               * add
        FAM (:= op = 304) → (AC,MQ ←AC + abs(M[e]) {sf});          * add magnitude
        FSB (:= op = 302) → (AC,MQ ←AC - M[e] {sf});               * subtract
        FSM (:= op = 306) → (AC,MQ ←AC - abs(M[e]) {sf});          * subtract magnitude
        FMP (:= op = 260) → (AC,MQ ←MQ x M[e] {sf});               * multiply
        FDH (:= op = 240) → (AC,MQ ←AC / M[e] {sf}: next           * divide or halt
                            Divide check →Run ←0);
        FDP (:= op = 241) → (AC,MQ ←AC / M[e] {sf});               * divide or proceed
    Unnormalized single precision floating point
        UFA (:= op = -300) → (AC,MQ ←AC + M[e] {suf});             * add
        UAM (:= op = -304) → (AC,MQ ←AC + abs(M[e]) {suf});        * add magnitude
        UFS (:= op = -302) → (AC,MQ ←AC - M[e] {suf});             * subtract
        USM (:= op = -306) → (AC,MQ ←AC - abs(M[e]) {suf});        * subtract magnitude
        UFM (:= op = -260) → (AC,MQ ←MQ x M[e] {suf});             * multiply
    Double precision floating point
    In DF operations, the SI are used as temporary registers and will be changed.
        DFAD (:= op = 301) → (
            ACMQ ←ACMQ + M[e]□M[e+1] {df}; SI ←?);                 *add
        DFAM (:= op = 305) → (
            ACMQ ← ACMQ + abs(M[e]□M[e+1]) {df}; SI ← ?);          * add magnitude
        DFSB (:= op = 303) → (
            ACMQ ←ACMQ - M[e]□M[e+1] {df}; SI ←?);                 * subtract
        DFSM (:= op = 307) → (
            ACMQ ←ACMQ - abs(M[e]□M[e+1]) {df: SI ←?);             * subtract magnitude
```

```
    DFMP (:= op = 261) → (                                    * multiply
       ACMQ ← ACMQ x M[e]□M[e+1] {df}; SI ←?);
    DFDH (:= op = -240) → (                                   * divide or halt
       ACMQ ← ACMQ / M[e]□M[e+1] {df}; SI ←?; next
       Divide_check → Run ← 0);
    DFDP (:= op = -241) → (                                   * divide or proceed; Divide check may be set
       ACMQ ← ACMQ / M[e]□M[e+1] {df}; SI ←?);
Unnormalized double precision floating point
    DUFA (:= op = -301) → (                                   * add
       ACMQ ← ACMQ + M[e]□M[e+1] {duf}; SI ←?);
    DUAM (:= op = -305) → (                                   * add magnitude
       ACMQ ← ACMQ + abs(M[e]□M[e+1]){undf}; SI ←?);
    DUFS (:= op = -303) → (                                   * subtract
       ACMQ ← ACMQ - M[e]□M[e+1] {duf}; SI ←?);
    DUSM (:= op = -307) → (                                   * subtract magnitude
       ACMQ ← ACMQ - abs(M[e]□M[e+1]){duf}; SI ←?);
    DUFM (:= op = -261) → (                                   * multiply
       ACMQ ← ACMQ x M[e]□M[e+1] {duf}; SI ←?);
Logical
    ORA (:= op = -501) → (AC1 ←AC1 ∨ M[e]);                   * or to accumulator
    ANA (:= op = -320) → (AC1 ←AC1 ∧ M[e]);                   * and to accumulator
    ERA (:= op = 322) → (AC1 ← AC1 ⊕ M[e]);                   * exclusive or to accumulator
```

The convert instructions are not described in detail. These instructions take a table in memory, addressed by the 6, 6 bit characters in AC or MQ and form a sum of products in the AC or MQ for each character component of the word.

```
    CVR (:= op = 114) → (AC,MQ ← f(AC,C,XR[1],M[Y:Y+63]));       convert by replacement from the AC
    CRQ (:= op = -154) → (AC,MQ ← f(MQ,C,XR[1],M[Y:Y+63]));      convert by replacement from the MQ
    CAQ (:= op = -114) → (AC,MQ ← f(AC,MQ,C,XR[1],M[Y:Y+63]));   convert by addition from the MQ
```

Transmission between M,XR[T], and AC
If tag,T,=0, then a no operation occurs

```
    PDX (:= op = -734) → (XR[T] ←AC<3:17>);                   * place decrement in index
    PAX (:= op = 734) → (XR[T] ←AC<21:35>);                   * place address in index
    PDC (:= op = -737) → (XR[T] ←2^15 - AC<3:17>);            * place complement of decrement in index
    PAC (:= op = 737) → (XR[T] ←2^15 - AC<21:35>);            * place complement of address in index
    LXD (:= op = -534) → (XR[T] ←M[Y]<3:17>);                 * load index from decrement
    LXA (:= op = 534) → (XR[T] ←M[Y]<21:35>);                 * load index from address
    LDC (:= op = -535) → (XR[T] ←2^15 - M[Y]<3:17>);          * load complement of decrement in index
    LAC (:= op = 535) → (XR[T] ←2^15 - M[Y]<21:35>);          * load complement of address in index
    AXT (:= op = 774) → (XR[T] ←Y);                           * address to index true
    AXC (:= op = -774) → (XR[T] ←2^15 - Y);                   * address to index complement
    PXD (:= op = -754) → (AC ←0; next AC<3:17>←XR[T]);        * place index in decrement
    PXA (:= op = 754) → (AC ← 0; next AC<21:35> ← XR[T]);     * place index in address
    PCD (:= op = -756) → (AC ←0; next AC<3:17> ← 2^15 - XR[T]);    * place complement of index in decrement
    PCA (:= op = 756) →(AC ←0; next AC<21:35> ← 2^15 - XR[T]);     * place complement of index in address
    SXD (:= op = -634) → (M[Y]<3:17> ←XR[T]);                 * store index in decrement
    SXA (:= op = 634) → (M[Y]<21:35> ←XR[T]);                 * store index in address
```

SCD (:= op = -636) → (M[Y]<3:17> ←2¹⁵ - XR[T]);   * store complement of index in decrement

SCA (:= op = 636) → (M[Y]<21:35> ←2¹⁵ - XR[T]);   * store complement of index in address

Transmission to Sense Indicators

PAI (:= op = 44) → (SI ←AC1);   place accumulator in indicators

LDI (:= op = 441) → (SI ←M[e]);   load indicators

OAI (:= op = 43) → (SI ←SI ∨ AC1);   or accumulator to indicators

RIA (:= op = -42) → (SI ←SI ∧ ¬ AC1);   reset indicators from accumulator

IIA (:= op = 41) → (SI ←SI ⊕ AC1);   invert indicators from accumulator

OSI (:= op = 442) → (SI ←SI ∨ M[e]);   or storage to indicators

RIS (:= op = 445) → (SI ←SI ∧ ¬ M[e]);   reset indicators from storage

IIS (:= op = 440) → (SI ←SI ⊕ M[e]);   invert indicators from storage

SIL (:= op = -55) → (SI<0:17> ←SI<0:17> ∨ R);   set indicators of left half

RIL (:= op = -57) → (SI<0:17> ←SI<0:17> ∧ ¬ R);   reset indicators of left half

IIL (:= op = -51) → (SI<0:17> ←SI<0:17> ⊕ R);   invert indicators of left half

SIR (:= op = 55) → (SI<18:35> ←SI<18:35> ∨ R);   set indicators of right half

RIR (:= op = 57) → (SI<18:35> ←SI<18:35> ∧ ¬ R);   reset indicators of right half

IIR (:= op = 51) → (SI<18:35> ←SI<18:35> ⊕ R);   invert indicators of right half

Program flow control instructions

NOP (:= op = 761) → ;   no operation

HPR (:= op = 420) → (Run ←0);   * halt and proceed

HTR (:= op = 0) → (Run ←0; IC ←e);   * halt and transfer

TRA (:= op = 20) → (IC ←e);   * transfer

XEC (:= op = 522) → (instruction ←M[e]; next   execute
    Instruction␣execution);

Conditional transfers

TZE (:= op = 100) → ((AC<Q,P,1:35> = 0) →IC ←e);   * transfer on zero

TNZ (:= op = -100) → (¬ (AC<Q,P,1:35> = 0) →IC ←e);   * transfer on no zero

TPL (:= op = 120) → (¬ AC<S> →IC ←e);   * transfer on plus

TMI (:= op = -120) → (AC<S> →IC ←e);   * transfer on minus

TOV (:= op = 140) → (AC␣overflow →IC ←e;   * transfer on overflow
                     AC␣overflow ←0);

TNO (:= op = -140) → (¬ AC␣overflow →IC ←e;   * transfer on no overflow
                      AC␣overflow ←0);

TQP (:= op = 162) → (¬ MQ<S> →IC ←e);   * transfer on MQ plus

TQO (:= op = 161) → (MQ␣overflow →IC ←e;   * transfer on MQ overflow
                     MQ␣overflow ←0);

TLQ (:= op = 40) → ((AC > MQ) →IC ←e);   * transfer on low MQ

TIO (:= op = 42) → ((AC1 = (AC1 ∧ SI)) →IC ←e);   * transfer when indicators on

TIF (:= op = 46) → ((0 = (AC1 ∧ SI)) →IC ←e);   * transfer when indicators off

Index manipulation and control and subroutine calling

TSX (:= op = 74) → (XR[T] ←2¹⁵ - IC; IC ←Y);   * transfer and set index

TSL (:= op = -1627) → (M[e]<21:35> ← IC; IC ←e + 1);   * 704

Loop control

TXI (:= hi␣op = 1) → (XR[T] ← XR[T] + D; IC ← Y);   * transfer with index incremented

TXH (:= hi␣op = 3) → ((D < XR[T]) →IC ← Y);   * transfer on index high

TXL (:= hi⎵op = -3) → ((D ≥ XR[T]) → IC ← Y);   * transfer on index low or equal
TIX (:= hi⎵op = 2) → ((XR[T] > D) → (XR[T] ← XR[T] - D;   * transfer on index
              IC ← Y));
TNX (:= hi⎵op = -2) → ((XR[T] > D) → XR[T] ← XR[T] - D;   * transfer on no index
              (XR[T] ≤ D) → IC ← Y);

Skip tests

MIT (:= (op = -1341) ∧ (c = 7)) → (M[e]<S> → IC ← IC + 1);   storage minus test; 704 series only
PLT (:= (op = -1341) ∧ (c = 6)) → (¬ M[e]<S> → IC ← IC + 1);   storage plus test; 704 series only
CCS (:= ((op = -1341) ∧ (c < 6)) → (compare
    (AC<30:35> = M[e]<(c × 6):(c × 6 + 5)>) → IC ← IC + 1;   character with storage; 704 series only
    (AC<30:35> < M[e]<(c × 6):(c × 6 + 5)>) → IC ← IC + 2));
PBT (:= (op = -760) ∧ (e'' = 1)) → (AC<P> → IC ← IC + 1);   * P bit test
DCT (:= (op = +760) ∧ (e'' = 12)) → (Divide⎵check → IC ← IC + 1) * Divide⎵check test
LBT (:= (op = +760) ∧ (e'' = 1)) → (AC<35> → IC ← IC + 1);   * low bit test
ZET (:= op = +520) → ((M[e] = 0) → IC ← IC + 1);   * storage zero test
NZT (:= op = -520) → ((M[e] ≠ 0) → IC ← IC + 1);   * storage own zero test
CAS (:= op = +340) → (* compare AC with storage
    (ACs = M[e]) → IC ← IC + 1;
    (ACs < M[e]) → IC ← IC + 2);
LAS (:= op = -340) → (* logical compare AC with storage
    (AC<Q,P,1:35> = M[e]<S,1:35>) → (IC ← IC + 1);
    (AC<Q,P,1:35> < M[e]<S,1:35>) → (IC ← IC + 2));
SWT (:= (op = 760) ∧ (e'<9:14> = 16)) → (Sense⎵Switches test
    Sense⎵Switches<e'<15:17>> → IC ← IC + 1);
SLF (:= (op = 760) ∧ (e' = 140)) → (Sense⎵Lights<0:3> ← 0);   Sense⎵lights off
SLN (:= (op = 760) ∧ (e'<9:14> = 14) ∧ (e'<15:17> ≠ 0)) → (Sense⎵lights on
    Sense⎵Lights<e'<15:17>> ← 1);
SLT (:= (op = -760) ∧ (e'<9:14> = 14)) → (Sense⎵lights test
    Sense⎵Lights<e'<15:17>> → (IC ← IC + 1; Sense⎵Lights<e'<15:17>> ← 0));
ETM (:= (op = 760) ∧ (e' = 7)) → (Trap⎵Mode ← 1);   enter Trap⎵Mode
LTM (:= (op = -760) ∧ (e' = 7)) → (Trap⎵Mode ← 0);   leave Trap⎵Mode
EMTM (:= (op = -760) ∧ (e' = 16)) → (Multiple⎵Tag⎵Mode ← 1);   enter Multiple⎵Tag⎵Mode
LMTM (:= (op = 760) ∧ (e' = 16)) → (Multiple⎵Tag⎵Mode ← 0);   leave Multiple⎵Tag⎵Mode
)   end Instruction⎵execution

APPENDIX 2   IBM 7909 DATA CHANNEL ISP DESCRIPTION (A PIO)

Appendix 2

IBM 7909 Data Channel ISP Description (a Pio)

Although the following description is of a Pio, signals generated in Pc, M, and K are necessary. Appendices 1, 3, and 4 are also necessary for a complete description. The Ms attached to K controls the precise time information flows.

Pio State

CC<21:35>   *Command Counter; 15 bit command (or instruction) counter containing the location of the next command*

AC<21:35>   *Address Counter; during vector data transfers AC contains the address of the next data word to transfer. During a transfer command AC is set to the address of the next command*

AR<S,1:35>   *Assembly Register; a buffer for data flow between the data register and the device control registers*

ARc[0:5]<0:5> := AR<S,1:35>   *character array defined by AR; a character is normally selected ARc[ASR]*

CTC<0:5>   *Control Counter; a 6 bit register which can be loaded and stored by the ISP*

WC<3:17>   *Word Counter; a counter controlling the number of words left to transfer during a command*

Data transmission modes in Pio for the K-Pio dialogue;
These control the flow direction and data types between K and Pio. Although not described as such, each indicator is mutually exclusive of the others.

SNI   *Sense Indicator; K is transmitting sense data to Pio.*

WRI   *Write Indicator; K is receiving data from Pio.*

RDI   *Read Indicator; K is transmitting data to Pio.*

Wait   *bit denotes a halted condition in Pio; instructions are not executed*

IL := 42₈   *Interrupt Location for Pio #A to interrupt itself. Each of the 8 Pio's have special locations. two locations, IL, IL+1, are reserved*

Interrupt_Request := ((CKC<1:6> ∧ CKCI<30:35>) ≠ 0) *signifies a request to interrupt Pio from K or within Pio*

Pc_Trap_Request   *signifies a request to trap Pc from Pio*

Interrupt Mode   *bit to denote that an interrupt program is running in Pio*

CKC<1:6>   *Check Conditions in K that cause an interrupt of the Pio*

CKC<1>/Input_Output_Check/I_O_Check

CKC<2>/Sequence_Check

CKC<3>/K_Unusual_End

CKC<4:5>/Attention Conditions<1:2>

CKC<6>/K_Check

CKCI<30:35>   *a mask to inhibit Pio interrupts from CKC*

The CKC indicators are described as follows;

Input_Output_Check
This condition occurs when the channel fails to obtain a storage reference cycle in time to satisfy demands of the attached IO device. The condition is also monitored in the Pc. I_O_Check is turned off when an LIP or LIPT command is executed or when the Pc executes an RSC or RIC instruction.

When an I_O_Check occurs, the adapter is disconnected and an interrupt occurs when the K_End signal is received from the adapter (K). The command counter contains the location plus one of the present command. The address counter contains the location plus one or two of the last word transmitted if the operation was a write or control, or the location plus one of the last word transmitted if the operation was a read or sense.

If an I_O_Check occurs while the channel is in interrupt mode, the I_O_Check is not recognized and is not saved.

Sequence Check

  A Sequence Check indicates an invalid sequence of channel commands. If a Sequence Check occurs during data transmission, the adapter is logically disconnected and the interrupt occurs when the K End signal is received.
  The following instructions cause a Sequence Check and a channel interrupt. (The checks are not described in the ISP description.)
  1. *If a CTLW, CTLR, or SNS is followed by CTL, CTLW, WTR, TWT, or SNS.*
  2. *If an SNS or CPYP is followed by any command other than a CPYP, CPYD, TCH, or TDC.*
  3. *If a TCH or TDC following an SNS or CPYP transfers control to any command other than a CPYP, CPYD, TCH, or TDC.*
  4. *If a CPYP or CPYD has not been properly preceded by a CTLW, CTLR, or SNS.*

K Unusual End

  This signal indicates an error condition recognized by K. It causes an immediate interrupt to Pio. The signal may be determined by sensing the K error indication.

Attention Conditions

  This is a signal indicating a change in status of the attached input output device. For example, during disk operations, an attention signal is generated when an access mechanism has completed a seek operation. The particular access mechanism that generated this indication may be determined from sense data.

K Check

  Adapter check (K Check) indicates an error and is recognized by the 7909, but does not necessarily indicate a K malfunction. The conditions which cause an adapter check are:
  1. *Circuit failure occurs in the ASR or CR.*
  2. *The character rate of the attached IO device exceeds the capability of the channel.*
  3. *The adapter (K) is not operational. This indication occurs if power is off on the adapter and an attempt is made to read, write, control or sense.*

Hardware Switches

  These gates route information among the registers on a selected basis. They are not under control of the program and are not registers.

  Storage Bus Switches$<$S,1:35$>$ *These 36 switches (and/or gates) provide the data path to and from the 7606 Multiplexor for data or command entry into the Pio.*

  Channel Address Switches$<$21:35$>$ *These 15 switches provide the Mp with address information. Address information is selected from the Address Counter or the Command Counter.*

  Character Switches$<$0:5$>$ *These 6 bit switches enable the character to be read from or written into the Assembly Register.*

Pio State (not in ISP)
  Hardware registers not in ISP but used in the description and the Pio.

  OR$<$0:4$>$ *Operation Register. The register containing the operation part of the instruction. OR is made up from $i<$S,1:3,19$>$.*

  DR$<$S,1:35$>$ *Data Register. A buffer for data flow between M and the AR.*

  CR   *Character Ring. A register to control the timing or transmission into AR.*

  ASR$_6$ *Assembly Ring. The counter to control the gates to/from AR from/boK. Data are sent to or received from the control, K, one 6-bit character at a time via the Character Switches under control of ASR.*

Instruction Format

  i$<$S,1:35$>$ *instruction: normally IBM calls these commands because a Pio executes them*

  f   := i$<$18$>$ *indirect*
  op$<$0:4$>$:= i$<$S,1:3,19$>$ *operation code*
  y$<$0:14$>$:= i$<$21:35$>$ *address*
  c$<$0:14$>$:= i$<$3:17$>$ *count part*
  c'$<$0:2$>$:= i$<$3:5$>$
  m$<$0:5$>$:= i$<$12:17$>$ *mask*

```
        e<21;35>  := (¬ f →y; f →M[y]<21;35>);                  1 level of indirect addressing
```

Mp State
```
   M[0:32768-1]<S,1:35>                                         Computer's primary memory
```

Instruction Interpretation Process
```
   ¬ Interrupt_request ∧ ¬ Wait → (Instruction ←M[CC];          fetch, no interrupt
        CC ←CC+1; next
      Instruction_execution);                                   execute, no interrupt
    Interrupt_request ∧ ¬ Interrupt_mode → (                    interrupt process
      (M[IL]<21;35> ← CC;M[IL]<3:17> ←CC;
          Interrupt_mode ←1; next CC ←IL+1);
```

Pio Interrupts and Pc Traps

    The Pio is capable of having its stored program interrupted independently of other P's. This operation is separate and distinct from a data channel trap in which Pio interrupts the Pc. On recognition of an interrupt condition the Pio stores the contents of the command and address counters in a fixed memory location, IL, and then executes the command located in the next location.

    If the 7909 channel is to be diverted from normal command execution sequence, the command in the fixed location must be one that will change the contents of the command counter (TCH, LIPT, or successful TDC or TCM). If this command is other than a successful transfer, the channel executes it and resumes operation at the location immediately following the location where the interrupt occurred. If the command at the fixed location is a WTR or TWT, the channel suspends operation as described in the channel command section, but the command counter contains the location plus one of the command responsible for the interrupt.

    Interrupt conditions are stored in a six-position register in the data channel and may be examined with the TCM command. Any combination of interrupt conditions causes an interrupt; however, once interrupted the channel is placed in interrupt mode and further attempts to set the interrupt condition or to interrupt are inhibited. The channel remains in interrupt mode until an LIP or LIPT command is executed by the channel or an RIC instruction is executed by the CPU. If a channel is in interrupt mode and an RSC instruction is executed by the CPU before the channel executes a LIP or LIPT command, the interrupt condition register is reset but the channel remains in interrupt mode. An LIP or LIPT command or a RIC instruction is the only program means available to cause the channel to exit from interrupt mode and become receptive to further interrupt conditions.

    Interrupts are also inhibited if channel trap is in process on that channel. This inhibiting persists until either an RSC or STC instruction (depending on whether the channel was enabled) is executed by the Pc.

    This command, when decoded by a channel not prepared to read or write, causes a sequence check and, thus, a channel interrupt. If the channel is prepared to read or write, this command causes c words to be transmitted between the channel and Mp, starting with M[e]. Data transmission continues until c is reduced to zero or a K_End signal is received by the channel. In either case, the channel read or write indicator is reset. If, while a CPYD is being executed a K_End signal is received before the count is reduced to zero, the channel read or write indicator is reset, and the channel obtains a new command from the next sequential location.

    If the next command is other than a copy, the channel executes that command. If the next command is a copy, the channel interrupts on a program sequence check. The last word transmitted to storage under CPYD control remains in the assembly register if a K_End signal is received before the word count reaches zero.

    If the count for the CPYD goes to zero before the K_End signal is received, the channel initiates a disconnect but does not get the next sequential command until a K_End or K_Unusual_End signal is obtained. In general, when operating under CPYD control, the channel does not obtain the next sequential command until either a K_End or a K_Unusual_End signal causes an interrupt.

Instruction Set and Instruction Execution Process

    The following control commands transmit instructions (orders) or operation information to K. Information is sent to K from M[e] starting with the high order 6 bit character and continues until a K_End is received by Pio from K. If more than one control word is required, the next words come from M[e+1,e+2,...].

    For CTL, CTLR, and CTLW instructions, the control words are first transmitted. Next the Read or Write indicator is set in Pio.

```
   Instruction execution := (
   CTL   (:= op = 01000) → (AC ← e;                             control
                      Move_word_from_M; ASR ← 0: next
                      Move_control_char_to_K);
   CTLR (:= op = 01001) → (AC ← e;                              control and read
                      Move_word_from_M; ASR ← 0· next
                      Move_control_char_to_K; RDI ← 1);
   CTLW (:= op = 01010) → (AC ← e; next                         control and write
                      Move_word_from_M; ASR ← 0· next
                      Move_control_char_to_K; WRI ← 1);
```

```
CPYD  (:= op = 101$01) → (AC ← e;                              copy and disconnect
                    Copy_data_block; next
                    RDI□SNI□WRI ← 0; K_end_wait);
CPYP  (:= op = 100$0)    (AC ← e;                              copy and proceed
                    Copy_data_block);
SNS   (:= op = 01011) → (SNI ← 1);                            sense
```
Execution of this command must be followed by a copy command. The data in K's sense indicators are sent via the K_Data
register through AR and DR to M.

```
SMS   (:= (op = 11100) ∧ (c'=0)) → CKCI ← e<29:35>:          set mode and select
LCC   (:= op = 11011) → (AC ← e; next                        load control counter
                    CTC ← AC<30:35>):
TDC   (:= op = 11010) → (AC ← e; next                        transfer and decrement counter
    (CTC = 0) → ;
    (CTC ≠ 0) → (CTC ← CTC-1; CC ← AC));
ICC   (:= op = 111$1) → (                                     insert control counter
    (0 < c' < 7) → ARc[c']← CTC;
    (c' = 0) → ARc[5] ← CKCI;
    (c' = 7) → ;);
TCM   (:= op = 101$1) → (                                     transfer on conditions met
    ((c' = 0) ∧ ¬ i<11> ∧ (m = CKC)) → (CC ← e);
    ((c' = 0) ∧ i<11> ∧ ((m ∧ CKC) = m)) → (CC ← e);
    ((0 < c' < 7) ∧ ¬ i<11> ∧ (m = ARc[c])) → (CC ← e);
    ((0 < c' < 7) ∧ i<11> ∧ ((m ∧ ARc[C]) = m)) → (CC ← e);
    ((c' = 7) ∧ (m = 0)) → (CC ← e));
TCH   (:= op = 001$0) → (CC ← e);                             transfer in channel
LAR   (:= op = 01100) → (AC ← e; next AR ← M[AC]);            load assembly register
SAR   (:= op = 01101) → (AC ← e; next M[AC] ← AR);           store assembly register
XMT   (:= op = 00011) → (AC ← e; WC ← c; next                an instruction to move c words in M[CC:(CC + c)]  to M[e:(e + c]
                    M_block_move)
```
XMT is actually a vector move within Mp.

```
XMT   (:= op = 000$1) → ((c ≠ 0) → (                         vector move
                    M[e:(e + c - 1)] ← M[CC:(CC + c - 1)];
                    WC ← 0; AC ← AC + c;                      fix end conditions
                    CC ← CC + c)):
WTR   (:= op = 000$0) → (AC ← e; Wait ←1);                   wait and transfer
TWT   (:= op = 01110) → (AC ← e; Wait ←1;                    trap and wait
                    Pc_Trap_Request ←1):
LIP   (:= op = 11001) → (                                     leave interrupt program
    CC ← M[IL]<21:35>;
    CKC ← 0; Interrupt_Mode ←0);
LIPT  (:= op = 001$1) → (                                     leave interrupt program and transfer
    (CC ← e; CKC ← 0;
    Interrupt_Mode ←0)
                    )                                         end Instruction_execution
```

K, Pio, and M Data Movement Processes
The following processes define the movement of characters and words among the registers and Memory. The principle activity is
copy data block. On writing, a word is taken from M and placed in Pio, then transferred character by character to K. On
reading, a character is taken from K and assembled in Pio, then transferred as a word to M. The following processes move
either characters or words in a direction relative to Pio.

Move char to K	*writing into K*
Move control char to K	*setting up instruction in K*
Move char from K	*reading from K*
Move word to M	*writing into M*
Move word from M	*reading from M*
M block move	*read M, write M on a word by word basis*
K end wait	*process to wait for K end signals*

Copy data block := (
    RDI → (Move char from K; ASR ← 0);
    SNI → (Move char from K; ASR ← 0);
    WRI → (Move word from M; ASR ← 0; WC ← WC - 1; next
      Move char to K))
Move char to K := (*K ← Pio ← M data movement*
    K End ∨ (WC = 0) → ;   *stop at end*
    ¬ K End ∧ (WC ≠ 0) ∧ K Data Rq → (*transmit a char*
    (ASR = 0) → Move word from M; WC ← WC-1; next
    K Data ← ARc[ASR]; ASR ← ASR + 1; next
    Move char to K);
    ¬ K End ∧ (WC ≠ 0) ∧ ¬ K Data Rq → (*idle till char arrives*
    Move char to K))
Move control char to K := (*K ← Pio ← M*
    K End → ;   *stop at end*
  ¬ K End ∧ K Data Rq → (*transmit a char*
    (ASR = 0) → Move word from M; next
    K Data ← ARc[ASR]; ASR ← ASR + 1; next
    Move control char to K);
  ¬ K End ∧ ¬ K Data Rq → Move control char to K))   *idle, till char arrives*
Move char from K := (*M ← Pio ← K data movement*
    K End ∨ (WC = 0) → ;   *stop at end*
    ¬ K End ∧ (WC ≠ 0) ∧ K Data Rq → (*receive a char*
    ARc[ASR] ← K Data; ASR ← ASR + 1: next
    (ASR = 0) → (Move word to M; WC ← WC - 1); next
    Move char from K);
    ¬ K End ∧ (WC ≠ 0) ∧ ¬ K Data Rq → (*idle till char arrives*
    Move char from K))
Move word to M := (DR ← AR; next   *M ← Pio data movement*
          M[AC] ← DR: AC ← AC + 1)
Move word from M := (AR ← DR; next   *Pio ← M data movement*
          DR ← M[AC]; AC ← AC + 1)
M block move := (*M ← M block move process for moving WC words within M, i.e.,*

```
        (WC = 0) → ;                                       M[CC:(CC + WC)] ← M[AC:(AC + WC)]
        (WC ≠ 0) → (DR ← M[CC]; CC ← CC + 1; next
          M[AC]  ← DR; AC ← AC + 1; WC ← WC - 1; next
          M⌴block⌴move))
    K⌴end⌴wait := (                                        Process to idle until K transmits an end signal
       ¬ (K⌴End ∨ K⌴Unusual⌴End) → K⌴end⌴wait;
         (K⌴End ∨ K⌴Unusual⌴End) → ;)
```

APPENDIX 3   K('HYPERTAPE) AND 'KDISK ISP DESCRIPTIONS

Appendix 3

K('Hypertape) and K(disk) ISP Descriptions

These K depend on control and state definitions from Pio of Appendix 2.

K State

K‿op<0:1>$_{16}$ *the operation or instruction register in K*

K‿Data<0:5>   *data buffer in K; used for transmitting and receiving characters*

K‿Data‿Rq   *used to control data flow between ARc[ASR] and K‿Data: signal in K denoting K‿Data requires new data if writing, or has a full data buffer if reading*

K‿End   *set by K at the completion of reading or writing a block of data*

K‿Unusual‿End   *set by K when an error is detected during writing or reading and data flow must be terminated*

The following sense data bits for tape originate in Ms and K. These registers can be read by Pio using the Pio SNS instructions Some of the bits are set using the CTL, CTLR, or CTLW instructions from Pio as control words

```
SDT[0:1]<S,1:35>                                                sense data for K('Hypertape)
   SDT[0]<1>/Operator Required := (
      SDT[0]<13>/Selected Drive Not Ready V
      SDT[0]<15>/Selected Drive Not Loaded V
      SDT[0]<16>/Selected Drive File Protected V
      SDT[0]<17>/Operation Not Started)
   SDT[0]<3>/Program Check := (
      SDT[0]<19>/Invalid Order Code V
      SDT[0]<21>/Selected Drive Busy V
      SDT[0]<22>/Selected Drive at Beginning of Tape V
      SDT[0]<23>/Selected Drive at End of Tape)
   SDT[0]<4>/Data Check := (
      SDT[0]<25>/Correction Occurred V
      SDT[0]<27>/Channel Parity Check V
      SDT[0]<28>/Code Check V
      SDT[0]<29>/Envelope Check V
      SDT[0]<31>/Overrun or Character Lost Check V
      SDT[0]<33>/Excessive Skew Check V
      SDT[0]<34>/Track Start Check or Clock Lost Check)
   SDT[0]<5>/Exception Conditions := (
      SDT[1]<1>/Selected Drive Read a Tape Mark V
      SDT[1]<3>/Selected Drive in End of Tape Warning Area)
   SDT[0]<7,9:11>/Selected Tape Unit Address 0:3
   SDT[1]<7>/Read Section Busy
   SDT[1]<9>/Write Section Busy
   SDT[1]<11>/Backward Mode
   SDT[1]<13,15:17,19,21:23,25,27>/Drive Attention[0:9]
SDF[0:1]<S,1:35>                                                sense data for the K('Disk)
   SDF[0]<3>/Program Check := (
      SDF[0]<7>/Invalid Sequence V
      SDF[0]<9>/Invalid Code V
      SDF[0]<10>/Format Check V
      SDF[0]<11>/No Record Found V
      SDF[0]<13>/Invalid Address)
   SDF[0]<4>/Data Check := (
      SDF[0]<15>/Response Check V
      SDF[0]<16>/Data Compare Check V
      SDF[0]<17>/Parity or Cyclic Code)
   SDF[0]<5>/Exception Condition := (
      SDF[0]<19>/Access Inoperative V
      SDF[0]<21>/Access Not Ready V
      SDF[0]<22>/Disk Circuit Check V
      SDF[0]<23>/File Circuit Check)
   SDF[0]<7>/six Bit Mode/Status Bit
   SDF[0]<31,33:35>□SDF[1]<1,3:5,7,9>/Access 0, Module[0:9]
```

Control Orders, i.e
Instruction Names and Numbers for K(disk)
These instructions are set in the K op register by the CTL instructions from Pio. The instructions are then executed by the
K's. They will only be given as names, mnemonics, and operation codes.

DNOP (:= K␣op = AA) → *no operation*

DREL (:= K␣op = A4) → *release*

DEBM (:= K␣op = A8) → *eight bit mode*

DSBM (:= K␣op = A9) → *six bit mode*

DSEK (:= K␣op = 8A) → *seek*

DVSR (:= K␣op = 82) → *prepare to verify (single record)*

DWRF (:= K␣op = 83) → *prepare to write format*

DVTN (:= K␣op = 84) → *prepare to verify (track with no addresses)*

DVCY (:= K␣op = 85) → *prepare to verify (cylinder operation)*

DWRC (:= K␣op = 86) → *prepare to write check*

DSAI (:= K␣op = 87) → *set access inoperative*

DCTA (:= K␣op = 88) → *prepare to verify (track with addresses)*

DVHA (:= K␣op = 89) → *prepare to verify (home address)*

Control Orders,.i.e.
Instruction Names and Numbers for K('Hypertape)

HNOP (:= K␣op = AA) → *no operation*

HEOS (:= K␣op = A1) → *end of sequence*

HRLF (:= K␣op = A2) → *reserved light off*

HRLN (:= K␣op = A3) → *reserved light on*

HCLN (:= K␣op = A5) → *check light on*

HSEL (:= K␣op = A6) → *select*

HSBR (:= K␣op = A7) → *select for backward reading*

HCCR (:= K␣op = 28) → *change cartridge and rewind*

HRWD (:= K␣op = 3A) → *rewind*

HRUN (:= K␣op = 31) → *rewind and unload cartridge*

HERG (:= K␣op = 32) → *erase long gap*

HWTM (:= K␣op = 33) → *write tape mark*

HBSR (:= K␣op = 34) → *backspace*

HBSF (:= K␣op = 35) → *backspace file*

HSKR (:= K␣op = 36) → *space*

HSKF (:= K␣op = 37) → *space file*

HCHC (:= K␣op = 38) → *change cartridge*

HUNL (:= K␣op = 39) → *unload cartridge*

HFPN (:= K␣op = 42) → *file protect on*

APPENDIX 4   IBM 7094 PC INSTRUCTIONS TO PIO('7909)

Appendix 4

IBM 7094 Pc Instructions to Pio('7909)

Pc State

Pc␣trap␣enable<A,B,C,D,E,F,G,H>   An 8 bit register in Pc which is used to mask or allow trap
                                                        requests from Pio. (#A,B,...H)

Instruction Set
   The following instructions in Pc are used to operate on each Pio state; thus, each instruction is actually 8 instructions.

RSC → (Wait → (CC ← e; Wait ← 0);   reset and start channel
      ¬Wait → RSC);   initializes a Pio
STC → (Wait → (CC ← AC; Wait ← 0);   start the Pio program
      ¬Wait → STC);
SCH → (M[e]<21:35> ← CC; M[e]<3:17> ← AC);   store channel. Checks status of a Pio.
ENB → (Pc␣trap␣enable ← M[e]<28:35>);   enable from effective address
RIC → (CTC□AC□AR□CC□WC□Wait ← 0);   reset channel
TCO → (¬ Wait → IC ← e);   transfer on channel in operation
TCN → (Wait → IC ← e);   transfer on channel not in operation

Section 2

The SDS 910-9300 series, a planned family

The Scientific Data System 900-9000 series consists of the SDS 910, 920, 925, 930, 940, 945, and 9300 computers. The series includes capabilities and features found in most 24-bit machines. The design implementation is among the best for 24-bit machines, as measured by equipment utilization, the processor state, implementation technology, and ease of use.

The first delivery dates for the members of the series are 910 (August, 1962), 920 (September, 1962), 925 (February, 1965), 930 (June, 1964), 940 (April, 1966), 945 (~1968), and 9300 (December, 1964).

The 910 and 920 were designed at the same time as a planned series of compatible computers which spanned a range of performance. The 910 has instructions which facilitate defining 920 instructions by software. For example, these include the multiply and divide step[1] (see page 544) instructions in the 910 for programming the multiply and divide instruction in the 920.

The I/O facility evolved to a clean structure, with the potential for having a high degree of T and Ms data-transfer concurrency at a comparatively low cost. The IBM 7094 should be studied for a contrasting (more expensive) approach.

The instructions which help manipulate floating-point data are interesting and useful. The machine's ability to execute closed floating-point arithmetic subroutines is fairly good considering that the instructions are not hardwired.

The Programmed Operator (POP) instructions provide the ability to define an instruction set for efficient encoding. The idea appeared earlier in Atlas. However, the POP instruction calls subprograms in primary memory, instead of in fixed memory like Atlas.

A nice scheme[1] is described for increasing the memory address space from 16,384 to 32,768 words. Other schemes which switch memory banks, like those in the PDP-8 (Chap. 5)

[1] We believe this appeared originally in the DEC PDP-1 introduced in November, 1960.

and in the 65,384-word 7094 II (Chap. 41), tend to be less desirable and flexible.

The SDS 930 was used at the University of California (Berkeley) as the base machine for the design of the Berkeley Time Sharing System (Chap. 24). SDS later marketed the system as the SDS 940.

The 9300 was not a member of the original 910-930 series. There is almost symbolic language program compatibility. Several registers and extra memory transfer paths were added to form the 9300 from the 930. The power of the 9300 is only a factor of 2 times the 930 for simple instructions. However, the hardwired floating-point instructions in the 9300 increases the power over the 930 by a factor of almost 10 for arithmetic problems. It is hard to believe that the incompatible 9300 was a wise choice. (We suggest a more reasonable alternative could have been a two-processor 930'. The 930' processor would be a 930 but with hardwired floating-point arithmetic instructions.) The 9300 has interesting twin-mode instructions for simultaneously operating on 12-bit data pairs. The 24-bit fixed-point word is sufficient for the real-time applications for which the computer was designed.

A flaw in the series is the sharing of K's among peripheral T's and Ms's. This problem can be seen by looking at the PMS structure (Chap. 42, Fig. 2, page 546). The connection to the peripheral K from K('Channel) requires a continuous connection during the data-transfer dialogue to Mp. This structure is especially bad in the case of a slow T, for example, a typewriter. A single character transmission requires that K('W, 'Y) be assigned to the typewriter during the complete message transmission (at a connected time of 100 milliseconds/character). The problem can be avoided by placing a character memory in each slow KT. Multiple devices could then run concurrently without requiring the elaborate K('W, 'Y) to be attached to them. The structure does not preclude such an improvement.

A complete description of the input/output and interrupt system is given and should be read carefully.

Chapter 42

The SDS[1] 910-9300 series

Introduction

The SDS 910, 920, 925, and 930 form a compatible series of computers. The 9300, though not compatible with the series, was an outgrowth of it. The 9300 uses the Ms and T devices of the 930. The 940 was designed initially at the University of California, Berkeley (see Chap. 24) for time sharing, and the 945 is a successor to the 940. The word length is 24 bits, and one single address instruction is encoded per word. The state of the machine consists of Mp(2048 ~ 32768 w) and Mps('P/Program Counter, 'A/Accumulator, 'B/Extended Accumulator, 'X/Index register).

These computers have been designed to process data originating from physical processes in real time. This design goal leads to a priority interrupt system with many (1,024) levels. The multiple interrupts facilitate programming and decrease the interrupt response time. A 24-bit word or two 12-bit words are a reasonable size for the problem types encountered. A multiple of 6 bits was chosen because of the (then) standard 6-bit magnetic-tape character. The relatively efficient storage representation and processing of floating-point data allow these computers to be used for general-purpose computation. However, only the 9300 has built-in floating-point operations. The 9300 has extensive capability for more general-purpose use. It is also used for operations on half-length data.

The data types processed by the 910-930 include words, integers, addresses, and boolean vectors. Several special instructions aid processing of types floating-point and double-length integers. The 9300 processes the additional data-types single- and double-length floating point. The 9300 has twin-mode instructions which operate on two half-length data (12 b) simultaneously. The two's complement representation is used for negative numbers.

The multiply, divide, and several other instructions are not wired into the 910, and compatibility between the 910 and 920-930 cannot be completely obtained by programming, although the 910 is a subset of the 920-930. Likewise, a smaller minimum Mp is available on the 910 (2,048 word versus 4,096 word). The 920 and 930 have identical instruction sets and differ in memory and logic performance. The 930 has a t.cycle: 1.75 μs, and the 910-920 has t.cycle: 8 μs. The more elaborate PMS structure of the 930 allows for greater growth, (e.g., by having more access ports to Mp).

The 9300's instruction set is different from the 930's. There are three index registers. The PMS structure is similar (and nearly compatible) with the 930. There are more (and better) working registers in the 9300 Pc to increase performance. The 9300 has two memory-access links, and the Pc can fetch instructions and data simultaneously. The instructions in the various C's appear in Table 1 for comparison purposes.

The SDS 925, a 1.75-μs version of the SDS 910, was available only for a brief time and will not be discussed further.

The machines process instructions (operations to the accumulator) in the following times (microseconds):

Instruction	910	920	930	9300
Fixed-Point Add	16	16	3.5	1.75
Fixed-Point Multiply	248	32	7.0	7.0
Floating-Point Add	896	384	92	14.0
Floating-Point Multiply	1696	656	147	12.25

Structure

The structure of these computers is given with PMS and conventional diagrams in Figs. 1 to 4.

The SDS channel is a Kio('Channel) and not a Pio, since it has no program counter and uses Pc. However, it can be as effective as a Pio. Of course, the cost is lower since Pc is shared. If K('W, 'Y) requires memory accesses, they must wait until suitable times in the Pc instruction-interpretation process to communicate with memory (Fig. 1).

The PMS structural detail (Fig. 2) does not show the algorithm by which simultaneous Kio('W, 'Y, 'C, 'D) and Pc requests for Mp are resolved. K has the highest priority, and further resolution among K's is determined by the K with the fullest buffer memory. Thus the priority is variable.

There are three basic K types, or channels (Fig. 2), in the 930 and 9300:

1 K('Time Multiplexed Communications Channel/TMCC)

2 K('Direct Access Communications Channel/DACC)

3 K('Data Subchannel/DSC)

[1] Scientific Data Systems merged with Xerox Corporation in 1969. The divisional name became Xerox Data Systems (XDS).

Table 1   SDS 910, 920, 930 and 9300 instruction sets[1]

Mnemonic		Name	Mnemonic		Name
LOAD/STORE			**REGISTER CHANGE**		
+LDA	M, T	Load A	RCH	M, T	Register Change
+STA	M, T	Store A	AXB	M, T	Address to Index Base
+LDB	M, T	Load B	‡∘CLA		Clear A*
+STB	M, T	Store B	‡∘CLB		Clear B*
LDP	M, T	Load Double Precision	∘CLR		Clear AB
STD	M, T	Store Double Precision	‡∘CAB		Copy A into B*
LDS	M, T	Load Selective (Masked)	∘ABC		Copy A into B, Clear A
STS	M, T	Store Selective (Masked)	‡∘CBA		Copy B into A*
+LDX	M, T	Load Index X	∘BAC		Copy B into A, Clear B
+STX	M, T	Store Index X	∘XAB		Exchange A and B
+EAX	M, T	Copy Effective address into Index Register 1	‡∘CBX		Copy B into Index*
			‡∘CXB		Copy Index into B*
STZ	M	Store Zero	‡∘XXB		Exchange Index and B*
+XMA	M, T	Exchange M and A	‡∘STE		Store Exponent*
XMB	M, T	Exchange M and B	‡∘LDE		Load Exponent*
XMX	M, T	Exchange M and Index Register	‡∘XEE		Exchange Exponents*
			‡∘CXA		Copy Index into A*
			‡∘CAX		Copy A into Index*
ARITHMETIC			∘XXA		Exchange Index and A*
+ADD	M, T	Add M to A	‡∘CNA		Copy Negative into A*
DPA	M, T	Double Precision Add	∘CLX		Clear X
+SUB	M, T	Subtract M from A	COPY		Copy
DPS	M, T	Double Precision Subtract			
MPO	M, T	M Plus One	**BRANCH**		
∘MIN	M, T	M Increment (M + 1)	+BRU	M, T	Branch Unconditionally
MPT	M, T	M Plus Two	+BRX	M, T	Increase Index and Branch
‡+ADM	M, T	Add to Memory	+BRM	M, T	Mark Place and Branch
‡+MUL	M, T	Multiply	BRC	M, T	Branch and Clear Interrupt
‡+DIV	M, T	Divide	BMA	M, T	Branch and Mark Place or Argument Address
TMU	M, T	Twin Multiply			
DPN	M, T	Double Precision Negate	+BRR	M, T	Return Branch
×MUS	M, T	Multiply Step			
×DIS	M, T	Divide Step	**TEST/SKIP**		
‡∘SUC	M, T	Subtract with Carry	‡+SKE	M, T	Skip if A Equals M
‡∘ADC	M, T	Add with Carry*	+SKG	M, T	Skip if A Greater than M
×∘MDE	M, T	M Decrement	SKL	M, T	Skip if A Less than M
			+SKM	M, T	Skip if A equals M on B Mask
			SKU	M, T	Skip if A Unequal M
ARITHMETIC, FLOATING-POINT (OPTIONAL)			SKQ	M, T	Skip if Masked Quantity in A Greater than M
FLA	M, T	Floating Add			
FLS	M, T	Floating Subtract	SKF	M, T	Skip if Floating Exponent in B is Greater than or Equal
FLM	M, T	Floating Multiply			
FLD	M, T	Floating Divide	+SKA	M, T	Skip if A and M do not Compare Ones Anywhere
LOGICAL			‡+SKB	M, T	Skip if B and M do Compare Ones Anywhere
+ETR	M, T	Extract	+SKN	M, T	Skip if M is Negative
+MRG	M, T	Merge	‡+SKR	M, T	Reduce M, Skip if Negative

Table 1 SDS 910, 920, 930 and 9300 instruction sets (Continued)

Mnemonic		Name	Mnemonic		Name
+EOR	M, T	Exclusive OR	SKP	M, T	Skip if Bit Sum Even
			+SKS	M, T	Skip if Signal Not Set
REGISTER SHIFT			∘SKD	M, T	Difference Exponents; Skip*
SHIFT	M, T	Shift	**FLAG REGISTER**		
ARSA	N, T	Arithmetic Right Shift A	FRTS	M	Flag Indicator Reset Test/Set
ARSB	N, T	Arithmetic Right Shift B	FLAG	M	Flag
∘RSH,		Arithmetic Right Shift AB	FIRS	M	Flag Indicator Reset/Set
ARSD	N, T;	Arithmetic Right Shift Double	FSTR	M	Flag Indicator Set Test/Reset
ARST	N, T	Arithmetic Right Shift Twin	FRST	M	Flag Indicator Reset/Set Test
LRSA	N, T	Logical Right Shift A	SWT	M	SENSE Switch Test
LRSB	N, T	Logical Right Shift B			
∘LRSH	(930 only),	Logical Right Shift AB	**INTERRUPTS**		
LRSD	N, T;	Logical Right Shift Double	+EIR		Enable Interrupts
LRST	N, T	Logical Right Shift Twin	+DIR		Disable Interrupts
CRSA	N, T	Circular Right Shift A	+EIT		Interrupt Enabled Test
CRSB	N, T	Circular Right Shift B	+IDT		Interrupt Disabled Test
∘RCY,		Circular Right Shift AB	+AIR		Arm Interrupts
CRSD	N, T;	Circular Right Shift Double			
CRST	N, T	Circular Right Shift Twin	**MEMORY EXTENSION (930 ONLY)**		
∘LSH;		Arithmetic Left Shift AB	∘		Set Extension Register
ALSA	N, T	Arithmetic Left Shift A	∘		Extension Register Test
ALSB	N, T	Arithmetic Left Shift B			
ALSD	N, T	Arithmetic Left Shift Double	**BREAKPOINT TESTS (SENSE SWITCHES IN 9300)**		
ALST	N, T	Arithmetic Left Shift Twin	∘BPT	4	Breakpoint No. 4 Test
LLSA	N, T	Logical Left Shift A	∘BPT	3	Breakpoint No. 3 Test
LLSB	N, T	Logical Left Shift B	∘BPT	2	Breakpoint No. 2 Test
LLSD	N, T	Logical Left Shift Double	∘BPT	1	Breakpoint No. 1 Test
LLST	N, T	Logical Left Shift Twin			
CLSA	N, T	Circular Left Shift A	**OVERFLOW (FLAG IN 9300)**		
CLSB	N, T	Circular Left Shift B	∘ROV		Reset Overflow
∘LCY,		Circular Left Shift AB	∘REO		Record Exponent Overflow
CLSD	N, T;	Circular Left Shift Double	∘OVT		Overflow Test; Reset
CLST	N, T	Circular Left Shift Twin			
NORA	N, T	Normalize A	**PROGRAMMED OPERATORS**		
∘NOD	N, T	Normalize; Decrement X	∘POP	M, T	Programmed Operator (64 instructions)
NORD	N, T;	Normalize Double			
CONTROL					
+HLT		Halt			
+NOP	M, T	No Operation			
+EXU	M, T	Execute			
INR	M, T	Interpret			
REP	M, T	Repeat			

[1]M-Memory or Memory Address; N-number of shifts; T-tag field; +-also in the 910, 920 and 930; x-910 only; ∘-not in the 9300; ‡-not in the 910.

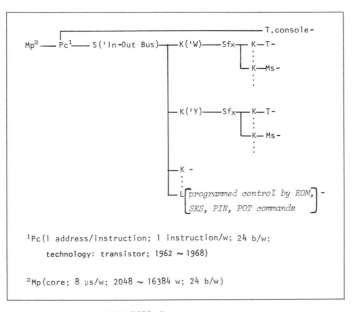

Fig. 1. SDS 910 and 920 PMS diagram.

The links between KT or KMs and any one of K('TMCC), K('DACC), and K('DSC) are identical. The KT or KMs assembles/ disassembles characters into/from words and transmits/receives them to/from the Kio('Channel). The channel communicates with Mp or Pc for data transmission and finally communicates with Pc at task completion (the block of data transferred). Task alarms may cause Kio to interrupt Pc. Each Kio('Channel) can assemble data on a 6-, 12-, or 24-bit basis for Mp accesses. A K('Channel) recognizes two types of information: data being transmitted between Mp and the peripheral K, and initialization or controlling information from Pc.

In the 930 or 9300 K's the principal distinction is that the actual data-path switching routes differ. From a program operation and control viewpoint the Time Multiplexed and the Direct Access Communication Channels (TMCC and DACC) and the Data Subchannels (DSC) behave almost identically. The TMCC and DSC differ from DACC in that the block control information (number of words and location in memory) for the channel may be either in primary memory or in local hardware memory associated with the channel hardware.

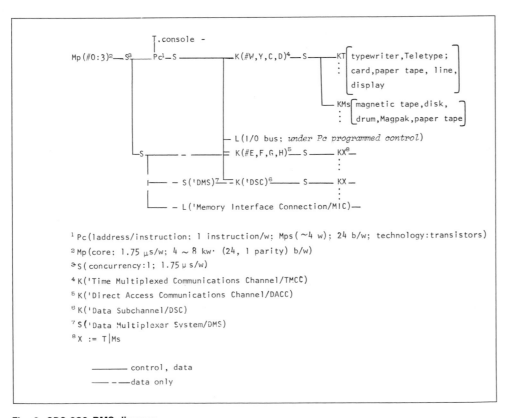

Fig. 2. SDS 930 PMS diagram.

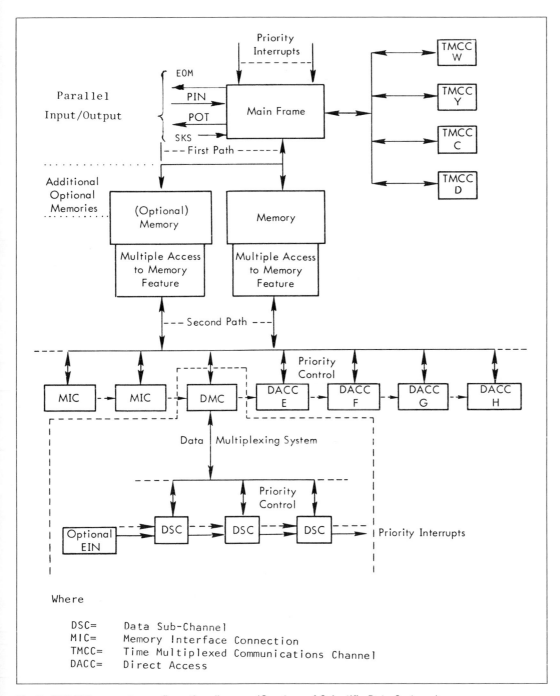

Fig. 3. SDS 390 computer-configuration diagram. (*Courtesy of Scientific Data Systems.*)

The 9300 structure, though not given in the PMS diagram, is essentially that of the 930 (Figs. 3 and 4). In the 9300, Mp has three access ports or a S('Memory-Processor; 8 Mp; 3 P,K). The Pc('9300) requires two of the access ports for independent access of instructions and data, leaving one for K transfer to Ms and T.

Instruction-set processor

The interesting parts of the ISP are discussed informally below. The formal ISP description given in Appendix 1 of this chapter should be read. The descriptions are partially taken from the SDS Programming Reference Manuals.

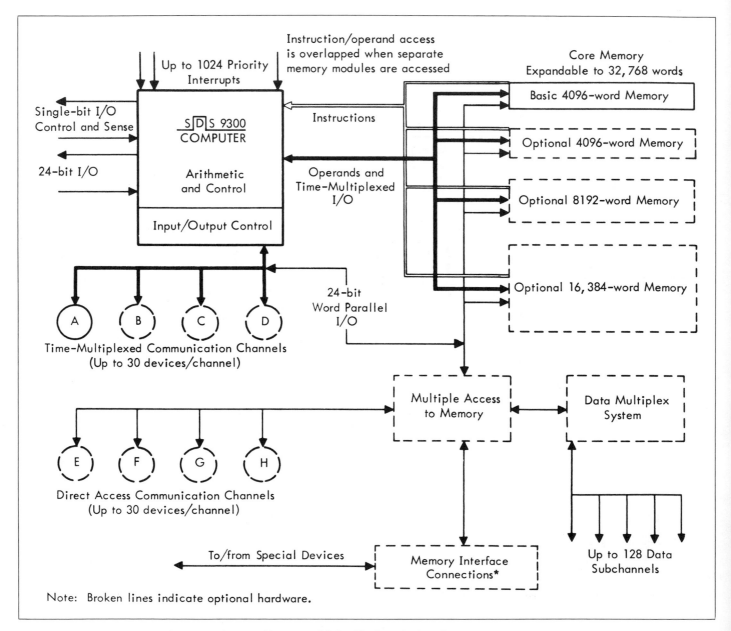

Fig. 4. SDS 9300 computer-configuration diagram. (*Courtesy of Scientific Data Systems.*)

Registers and memory (930)

The Pc state is declared in the ISP description. The ISP registers are A, B, X, P, M, and miscellaneous bits for overflow, carry, etc. Overflow can be turned on for arithmetic overflow in addition, subtraction, multiplication, division, and left-shift instructions.

Data formats

General. A computer word, W, is 24 binary digits (bits) or 8 octal digits. A word is numbered $W\langle 0:23\rangle$ from left to right or alternatively $W\langle 0:7\rangle_8$.

Fixed-point data format. Fixed-point numbers are represented in two's complement form with the sign at $W\langle 0\rangle$. A 23-bit fraction $W\langle 1:23\rangle$ can be assumed. The binary point is to the left of bit position 1 ($W\langle 1\rangle$). For integers, the binary point is to the right of $W\langle 23\rangle$.

Floating-point data format. Subroutines perform double- and single-precision floating-point arithmetic. A floating-point word is defined as $f\langle 0:47\rangle := W[n:(n + 1)]\langle 0:23\rangle$. Of course, single-precision floating point requires less processing time.

The fractional portion (mantissa), $f\langle 0:38\rangle$, of a double-precision floating-point number is a 39-bit proper fraction with the leading bit being the sign bit and the binary point located to the left of the most significant magnitude bit, $f\langle 1\rangle$.

The floating-point exponent is a 9-bit integer, $f\langle 39:47\rangle$, with the leading bit being the sign, $f\langle 39\rangle$. The standard routines operate on both fraction and exponent in two's complement form. If F represents the contents of the fractional field and E represents the contents of the exponent field, the number has the form $F \times 2^E$.

Standard subroutines assume that the more significant word is in the A register and that the less significant word is in the B register. Correspondingly for Mp, the more significant word is in Mp[x] and the least significant word in Mp[x + 1].

The single-precision floating-point representation is identical to that of double-precision floating point; i.e., it takes two words. However, the least significant bits of the mantissa, $f\langle 24:38\rangle$, are not processed; thus there is a saving in time but not in space for using single precision.

Instruction word format (930)

The computer instruction word format is given in Fig. 5.

$W\langle 0\rangle$ is the Relative Address bit, R. Standard software loading programs use this bit; central processor decoding logic does not use or sense this bit. A 1 in $W\langle 0\rangle$ causes some loading programs

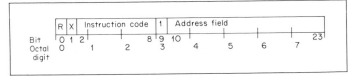

Fig. 5. SDS 930 instruction-format diagram.

to add the assigned location of the instruction to the address field contents prior to actual storage into the assigned location.

$W\langle 1\rangle$ is the Index Register bit, X. It determines whether or not the index register will be added to calculate the effective address.

$W\langle 2:8\rangle$ is the Instruction Code field and determines the operation to be performed. The Programmed Operator facility is selected by $W\langle 2\rangle$; it is part of the Tag field $W\langle 0:2\rangle$.

$W\langle 9\rangle$ is the Indirect Address bit, I. It determines whether or not e or M[e] is to be used as the effective address (see below).

$W\langle 10:23\rangle$ is the Address field and for most instructions represents the location of the operand called for by the instruction code.

Address modification. Index and indirect addressing, used singly or in combination, perform address modification after bringing the instruction from memory but before executing it. The instruction remains in memory in its original form. The results of indexing and/or indirect addressing form the "effective address," e.

INDEXING   If the content of the index bit in an instruction is a 1, prior to execution the computer adds the contents $X\langle 10:23\rangle$, of the index register to the contents of the address field of the instruction. This addition does not keep any overflow or carry beyond the fourteenth address bit. This addition occurs prior to any indirect action.

INDIRECT ADDRESSING   A 1 in the indirect address bit causes the computer to decode the contents of the effective address, accessed as described above, as if it were an instruction without an instruction code; that is, the address logic reinitiates address decoding, using the word in the effective location (the memory cell whose address is the effective address). This is an iterative process and provides multilevel indirect and indexed addressing. Each level of indirect addressing adds an additional cycle time to the instruction execution time.

930 memory extension control registers. Core memory in the 930 is expandable to 32,768 words. However, the address field in the instruction format is 14 bits long, allowing direct access of only up to 16,384 words. Memory extension in the 930 contains two 3-bit memory extension registers, EM2 and EM3, and allows addressing of memories of 32,768 words. The program loads either or both of the registers and activates them as desired. Each register can become the most significant digit (fifth octal) of any operand address.

The program uses the first extension register, EM3, by calling for an address with an 11_2 in the most and next most significant address bits, respectively (a 3 for the most significant octal digit). The program calls for EM2, the second extension register, by setting the same two address bits to 10_2 (a 2 for the most significant octal digit). In this way, normal addressing compatible with the 910 and 920 occurs by setting a 3 in EM3, and a 2 in EM2.

910-930 instructions

Programmed Operators (POP's) enable subroutines to be called with a single instruction. This provides definable instructions of the same form as built-in machine instructions. The computer decodes the operation codes $100_8 \sim 177_8$ as special instructions and transfers to a subroutine whose address is uniquely determined by the code. The computer records the address of the POP instruction at location 0 together with an indirect address bit so that the program continuity may be maintained. By indirect addressing which refers to location 0, which in turn refers to the POP instruction, the subroutine can gain access to the effective address of the operand associated with the POP instruction.

The instruction set for the computers in this series is listed in Table 1. The table should be used to compare the machines.

There are two instructions in the 910 which are not in the 920 or 930: Multiply Step and Divide Step. These instructions facilitate writing subroutines for multiplication and division. The Multiply Step (MUS) instruction is defined:

$$\text{MUS} \rightarrow (B\langle 23\rangle) \rightarrow A \leftarrow A + M[e]; \text{ next } AB \leftarrow AB/2);$$

9300 instructions

The instruction word format in the central processor is shown in Fig. 6.

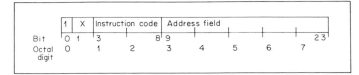

Fig. 6. SDS 9300 instruction-format diagram.

$W\langle 0\rangle$ contains the Indirect Address bit I.

$W\langle 1:2\rangle$ contains the Index Register bits $X\langle 0:1\rangle$.

$W\langle 0:2\rangle$ is called the Tag field.

$W\langle 3:8\rangle$ contains the Instruction code; the contents of this field determine the operation to be performed.

$W\langle 9:23\rangle$ contains the Address; for most instructions, the contents of this field represent the memory location of the operand called for by the instruction code.

Address modification. Each index register contains an unsigned base address of 15 magnitude bits and a signed increment of 9 bits. The increment contains 8 magnitude bits and a sign bit and is held in two's complement form.

Index registers are modified by adding the signed-increment value to the base address using two's complement arithmetic. Since the increment and base address fields are of unequal lengths, the sign bit (bit 0) of the increment field is extended six positions to the left prior to the addition. This 15-bit sum is then stored in the base address field of the index register. The index register may be incremented by any value from -256_{10} to 255_{10} using a single instruction. Incrementing and testing for a "terminal condition" is done by the instruction Increase Index And Branch (BRX), as follows:

If the index register has been negatively incremented, a terminal condition exists when the base address has been reduced below the zero value.

If the index register has been positively incremented, a terminal condition exists when the resultant base address has been increased beyond the maximum address value (077777_8).

If the terminal condition exists, the next instruction is taken in sequence. If the terminal condition does not exist, program control is transferred to the location specified.

The instruction set for the 9300 is given in Table 1.

Pc implementation

All the processors of the series have basically similar register configurations because of the common Instruction-set Processor. However, the increasing complexities of the machines can be seen by comparing the register structures of the 910-930 (Fig. 7) with the 9300 (Fig. 8). The figures show both the registers accessible to the program or defined by the ISP (denoted by °) and the temporary registers which are necessary for the implementation.

910, 920, 930 registers (Fig. 7)

ISP registers (°). The A register is the main accumulator of the computer. The B register is an extension of the A register. The

The 24-bit C register communicates with memory. Instructions are temporarily held in C before instruction decoding. It is used as an arithmetic and control register in multiply, divide, and other operations. Address modification and parity generation/detection use the C register.

The O register is a 6-bit register that contains the instruction or operation code of the instruction being executed.

The M' register is a 24-bit register that holds each word as it comes from memory. Recopying a word into memory takes place from the M' register.

9300 registers (*Fig. 8*)

ISP registers (°). The A and B registers of the 9300 are the same as in the 900 series computers; however, the P register is P⟨9:23⟩.

There are three 24-bit index registers, X[1:3]. Each index register is composed of a base address of 15 bits and a signed increment of 9 bits.

The Flag register, F, is a 6-bit register that may be set and/or sensed by the program. The first bit position of this register is the overflow indicator.

Hardware registers not in the ISP. The C register holds the 24-bit operand word as it is transmitted to, or received from, memory.

The D register holds the next 24-bit instruction word as it is received from memory.

The 15-bit S register contains the address of the memory location to be accessed for either instruction or operand.

The 6-bit O register contains the instruction code of the instruction being executed.

The A' register is an optional 15-bit register used for the floating-point option. It temporarily extends the A register during the execution of floating-point instructions.

The B' register is an optional 15-bit register which temporarily extends the B register during the execution of floating-point instructions.

Instruction interpretation in the 900 series

The instruction-interpretation process can be explained in terms of the processor's registers (Fig. 7). The ADD instruction execution (not including memory mapping) defined in ISP as A ← A + M[e] is interpreted as

$$S \leftarrow P; \; P \leftarrow P + 1; \; \text{next} \qquad \textit{fetch the instruction}$$

$$M' \leftarrow \text{Memory}[S]; \; \text{next}$$

$$C \leftarrow M'; \; \text{next}$$

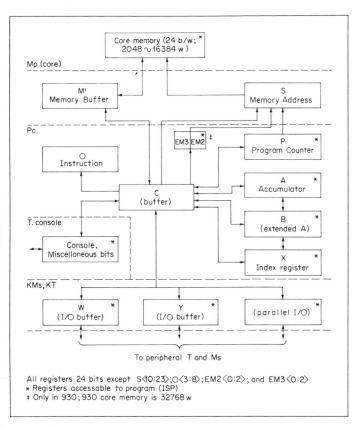

All registers 24 bits except S⟨10:23⟩; O⟨3:8⟩; EM2⟨0:2⟩; and EM3⟨0:2⟩
* Registers accessible to program (ISP)
‡ Only in 930; 930 core memory is 32768 w

Fig. 7. SDS 910, 920, and 930 registers diagram.

B register contains the less significant portion of double-length numbers. Overflow and carry bits are used with A and B operations.

The index register X, used in address modification, is a full-word register. Index-register operations use the least significant 14 bits.

The P register is a 14-bit register that contains the memory address of the current instruction. Unless modified by the program, the contents of P increase by 1 at the completion of each instruction.

The memory extension registers, EM3 and EM2, are 3-bit registers that specify the portion of extended memory being used. They exist only in the 930.

Hardware registers not in the ISP. The S register is a 14-bit register that contains the address of the memory location to be accessed for instructions or data. The 15-bit address is formed by S and one of the memory extension registers.

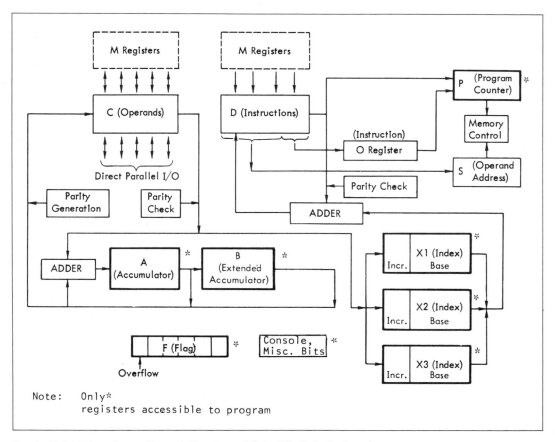

Fig. 8. SDS 9300 registers diagram. (*Courtesy of Scientific Data Systems.*)

$O \leftarrow C\langle 0:5 \rangle$; next

$(O = 05) \rightarrow ($ *ADD execution*

operand effective-address-calculation process (including indexing and indirect addressing)

$S \leftarrow C\langle 10:23 \rangle$; next   *final operand fetch*

$M' \leftarrow \text{Memory}[S]$; next

$C \leftarrow M'$; next

$A \leftarrow A + C)$ *add operation*

Input/output processing

Introduction

There are several methods of transferring data between Mp and the K's. These methods will be described independently, and in

order of increasing complexity. They are:

1a   Single bit sent to a selected K (EOM instruction).

1b   Single bit sense (or bit detection) from a K (SKS instruction).

2   Word parallel to/from a K (POT/PIN instruction).

3   Interrupt from one of 1,024 K's on a priority basis to Pc. K can signal Pc to execute a particular program.

4a   Time Multiplexed Communication Channel/TMCC (Internal Interlace[1] feature).

4b   Time Multiplexed Communication Channel (External Interlace[2]).

5   Direct Access Communication Channel/DACC (External Interlace).

---

[1] The control information for the location of the next word transferred and the number of words to transfer are kept in Mp.
[2] The control information is taken from registers within K.

6a   Data Subchannel/DSC (Internal Interlace).

6b   Data Subchannel (External Interlace).

7   Memory Interface Connection/MIC link. A component has a link to Mp.

Methods 1 to 3 above are completely under control of a program and are simple time-independent instructions (or methods) of transferring data to K's (and onto KT or KMs). The ISP description (Appendix 1 of this chapter) has a detailed description of the I/O devices and these I/O instructions.

Single-bit control and sense

Two instructions provide for single-bit ON/OFF control signals. The first, EOM, transmits a control signal and a 14-bit address to an external device or a function within the computer. The second, SKS, selects an external device or computer function and skips in response to a false (0) signal. Up to 16,384 control signals can be sent and 16,384 input signals tested theoretically. (A more reasonable number of physical destinations would be 50.) Execution of an EOM causes a signal of approximately 1.4 microseconds duration to be transmitted.

EOM instruction format. EOM is used to select a specific I/O device by placing a 1 in its select register. EOM requires one cycle.
  $W\langle 2 \rangle = 0$.
  $W\langle 0:1 \rangle$ is reserved for special system address bits.
  $W\langle 3:8 \rangle$ contains the EOM instructions code, 02.
  $W\langle 10:11 \rangle$ contains the system mode specifier.
  $W\langle 12:23 \rangle$ contains the 12-bit address field that specifies special system destinations.

SKS format. The SKS instruction format has each corresponding bit field identical to the system EOM format. Execution of an SKS causes a 14-bit address to be presented to all K's; the K being addressed responds and is tested. If the addressed external K supplies a "set" signal to the central processor, the computer executes the next instruction in sequence from the SKS. If no signal is set, the computer skips the next instruction in sequence and executes the following instruction. No registers are affected except the P register. SKS requires two or three Mp cycles if no skip or skip, respectively, is executed.

Word parallel instructions

Two instructions, Parallel Output (POT) and Parallel Input (PIN), permit any word in Mp to be presented in parallel on a physical connector to a K or, inversely, permit signals sent from a K to be stored in Mp. The execution of a POT or PIN instruction sends a signal to the external device involved in the input/output operation, which notifies the device to send its data word as soon as it is operational. When the device becomes operational during a Read or PIN operation, it transmits a Ready signal to the central processor while at the same time presenting a data word to Pc.

During the execution of a POT instruction, the central processor transmits a signal to the external device, alerting it to receive a data word. When the device becomes operational, it transmits a Ready signal to the central processor, which releases the data word to the external device.

Selective input/output with these devices is accomplished by preceding POT or PIN with an EOM to alert (select) the desired device by a specific address. By preceding the POT or PIN with an SKS, the Ready signal of the special device can be tested after the execution of the EOM but prior to execution of the parallel transfer instruction; a possible Pc "hangup" can thus be avoided. The Ready signal can also set one of the priority interrupts.

PIN stores the contents of 24 input lines in parallel in the effective-memory location. PIN or POT requires four cycles plus any waiting time for Ready.

Interrupt

The interrupt provides program control of input/output operations, aids in programming simultaneous input/output and compute operations, and allows immediate recognition of special external conditions by causing Pc to execute an instruction in a selected Mp location at the end of the execution cycle of the current instruction. Without disturbing the program register, the processor executes an instruction in one of a selected set of memory locations. A Mark Place and Branch (BRM) instruction in this location saves the contents of the program register, EM3, EM2, and overflow indicator and transfers to the particular interrupt servicing routine required. To exit from the interrupt service routine, a Branch Unconditionally (BRU) instruction using indirect addressing returns control to the next instruction in proper sequence in the main program; it also clears the interrupt. Processor state (that is, A, B, Overflow, and X) must be preserved and restored by the program if the registers are used by the program.

The priority interrupt system has up to 1,024 interrupts arranged in levels. The levels have priority according to a priority number; the higher priority levels have a smaller number. Interrupt channels are installed in Pc in groups of 16. The assignment of physical memory locations to interrupt levels is shown in Appendix 1 of this chapter; the assignment is in order of decreasing priority from location 200_8 (highest) to 1477_8 (lowest). Interrupt requests can also be programmed. The power fail-safe (for power supply off) interrupts and out-of-order interrupts have the highest priority.

Besides the interrupt mechanism just discussed, there is also a single instruction interrupt. This permits the execution of only one instruction before automatically being cleared and returning to the program that was interrupted. For example, if an external clock source is connected to the computer so that it pulses an interrupt line at set intervals, the program can maintain a programmed real-time clock. Each time the external pulse causes an interrupt, the program executes the single instruction, Memory Increment (MIN), to add 1 to the memory word selection for use as a programmed real-time clock. (The main program can examine this memory location whenever necessary to determine how many time increments have elapsed since the clock was started.)

Interrupts can be single or normal-instruction interrupts in any combination desired.

An interrupt has three operational states: inactive, waiting, and active states.

In the inactive state, no interrupt signal has been received into the level and none is currently being processed by its interrupt servicing subroutine.

In the waiting state, an interrupt has been received but is not being processed. This situation may arise when an interrupt of higher priority is being processed. When all higher waiting interrupts have been processed, this level goes to the active state.

In the active state, the interrupt has caused the main program to recognize its presence and has transferred to its assigned interrupt location where it is being processed.

Two program control features are Arm/Disarm and Enable/Disable. Arm/Disarm controls whether an interrupt can proceed from the inactive state to the waiting state. When armed, an interrupt signal sets the interrupt to the waiting state. Enable/

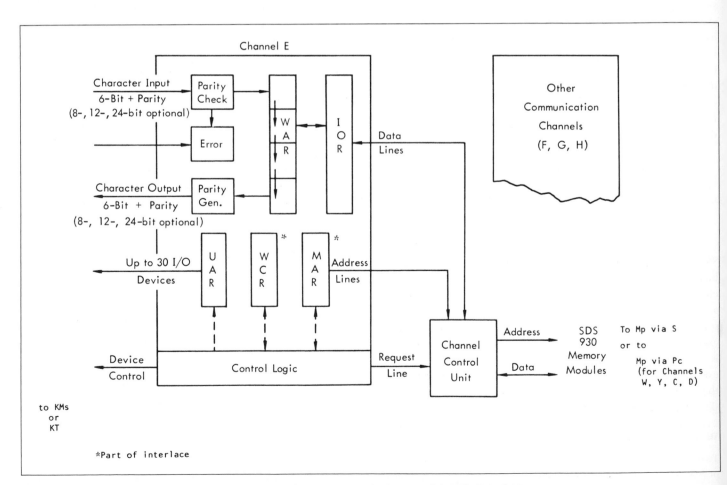

Fig. 9. SDS 930 direct-access communication-channel register diagram. (*Courtesy of Scientific Data Systems.*)

Disable operates on the entire interrupt system. (When the interrupt system is enabled, interrupts can occur.)

Communications channels—Kio('Channel)'s

Kio('Communication Channels) provide buffering, input/output control, and data transmission simultaneously with computation. There can be up to eight independent communication channels and a large number of subchannels in a single system. Figure 9 shows the registers in a K('Channel).

Each channel can control up to 30 KT's or KMs's. The channel handles character, word assembly and disassembly, input/output parity detection and generation, data transmission to and from memory, and end-of-transmission detection.

All channels are bidirectional and can communicate with 6-bit character devices or word devices in 6, 12, and 24 bits. The main program that initializes a K specifies the number of characters to be contained in each word during the transmission.

The channel interlace controls the transfer of the data words going through the associated channel buffer, supplies the memory address of data coming from or going to memory, and maintains the word count determining the number of words transferred. This interlace information can be either in K hardware (external interlace) or in Mp (internal interlace). The terminal interrupts, End of Record and Zero Word Count, come from the interlace and are under its control.

The time-multiplexed channels use the memory-access logic of Pc to transmit input and output of data words and require two memory cycles (see Fig. 2). Each direct-access channel has independent memory-access logic and requires one memory cycle (see Fig. 2).

Communication-channel description. Up to 30 peripheral devices (K's for T or Ms) may be connected to one K('Channel) (Fig. 9). Each device has a unique, 2-digit, octal address by which it is selected for an input/output operation. To select the peripheral device, the program loads the proper unit address into the 6-bit Unit Address Register (UAR) in the channel. This address selects both the device and, if appropriate, the function to be performed. Placing a nonzero unit address in the unit address register connects the peripheral unit addressed to the channel, and the unit becomes active. When the UAR contains a zero address, or any time that a terminal or initial condition clears the contents of UAR, the channel becomes inactive.

The 24-bit data Word Assembly Register (WAR) contains the data word actively being received or transmitted during an input or output operation. During input, 6-bit characters (plus parity) enter the Single-Character Register (SCR) where the channel buffer assembles them, one at a time, into the WAR.

The channel interlace contains two working registers: the Word Count Register (WCR) and the Memory Address Register (MAR). A channel may have these registers either in K or in Mp. In the setup sequence for an interlaced input/output operation, the POT instruction transmits to the interlace a data word made up of the word count (that is, length) and the starting address of the data block. The 15-bit Word Count Register (WCR) contains the data word count during a data transfer. The number of data words is decremented by 1, and the new count replaces the old one in the WCR for each word transmitted.

The Memory Address Register (MAR) contains the starting destination or source address in memory of the transmitted data. The memory locations to or from which data words are to be transmitted enter the MAR at the same time the word count does. During transmission of data, the interlace increments the MAR after each word as it decrements the contents of the WCR. These two registers provide the interlace control of block transmissions. Obviously, if the interlace control registers are in Mp, then two extra accesses are required for each word transferred.

Memory interface connection link

Once a computer is equipped with a multiple-access-to-memory feature, one or more Memory Interface Connections (MIC) can be attached. The MIC is a general interface to the computer that allows special devices to access Mp. It preserves the integrity of the memory by generating the parity of incoming data words and checking the parity of words read from memory to indicate memory failures. The device that is connected to the MIC must hold both the data and the address until the transmission to/from memory is completed (that is, MIC does not have registers).

Conclusions

The SDS computers appear to be the first attempt to design several computers at the same time with a common ISP. Over a longer time span other compatible computers were added to the original 910 and 920 as technology (and marketing) dictated. The series is characteristic of well-designed typical 24-bit computers. By increasing the arithmetic capability, the series could also be used more generally.

References

Scientific Data Systems Reference Manuals for the 930 and 9300 computers

APPENDIX 1   SDS 930 ISP DESCRIPTION

Appendix 1

SDS 930 ISP Description

The description defines the Instruction Set without exact assignment of operation codes to instruction names. Input-output instruction actions are given for the simple controls, but do not include the action of the channels or the devices.

Pc State

A<0:23>	*Accumulator; main arithmetic register*
B<0:23>	*secondary arithmetic register for multiplier, quotient, etc.*
AB<0:47> := A□B	*combined 48 bit arithmetic register*
X<0:23>	*Index Register*
P<10:23>	*Program or instruction location counter for 16 kw*
Overflow/Ov	*set on integer operations*
Carry := X<0>	*used in multiple precision operations to link words*
Run	

Mp State

Memory [0:77777$_8$]<0:23>   *32 kw primary memory*

Two 3 bit map (or extension) registers extend the address space of Mp to 32 kw. EM2 holds a 4 kw block number when addresses 20000$_8$-27777$_8$ are used. EM3 holds the 4 kw block number for addresses 30000-37777$_8$.

EM2<0:2>   *Extension Memory registers*

EM3<0:2>

Memory Mapping Process
This process maps the 16 kw address space into the 32 kw physical memory.

M<0:23>[a] := (

    (a < 20000$_8$) →Memory [a]<0:23>

    (20000$_8$ ≤ a ≤ 27777$_8$) →Memory [EM2<0:2>□a<12:23>]<0:23>

    (30000$_8$ ≤ a) →Memory [EM3<0:2>□a<12:23>]<0:23>

Pc Console State
Individual registers in Pc can be read and written from the console.

BPT<1:4>   *Breakpoint or sense switches*

Instruction Format

    instruction/i<0:23>

relative	:= i<0>	*unused by ISP; software relocation bit*
index bit/xb	:= i<1>	
op code/op<2:8>	:= i<2:8>	
pop code<0:5>	:= i<3:8>	*programmed operation code value*
indirect bit/ib	:= i<9>	
y<10:23>	:= i<10:23>	*address field for 16 kw*

    μ *microcoded instruction bits within an instruction*

Effective Address Calculation Process

    e<10:23>:= (¬ ib →(*iterative process of indefinite indirect addressing until*
                                                no indirect bit, ib, is found

    ¬ xb →y;

      xb →y + X);

             ib →(

    ¬ xb →(i <0□9:23> ←M [y]<0□9:23>

      xb →(i<0□9:23> ←M [y + X]<0□9:23>); next e))

    e1<18:23> := e<18:23>   *shift count*

Instruction Interpretation Process

```
¬ Interrupt⌴interpretation → (                                    normal interpretation
    instruction ←M[P]; P ←P + 1; next                            fetch
        Instruction⌴execution);                                  execute
    Interrupt⌴interpretation → (                                 interrupt interpretation
        instruction ←M[200₈ + 20₈ × K⌴address + I⌴address]: next
            Instruction execution)
```

Instruction Set and Instruction Execution Process

```
    Instruction⌴execution := (
```

Load and Store Group

```
    LDA → (A ←M[e]);                                            load A
    STA → (M[e] ←A);                                            store A
    LDB → (B - M[e]);                                           load B
    STB →M[e]←B);                                               store B
    LDX → (X ←M[e]);                                            load index
    STX → (M[e] ←X);                                            store index
    EAX → (X ←e);                                               load index from e
    XMA → (M[e] ←A; A ←M[e]);                                   exchange A and M
```

Arithmetic Group

```
    SUB → (Ov,Carry□A ←A - M[e]);                              subtract
    ADD → (Ov,Carry□A ←A + M[e]);                              add
    SUC → (Ov,Carry□A ←A - M[e] - Carry);                      subtract with Carry
    ADC → (Ov,Carry□A ←A + M[e] + Carry);                      add with Carry
    MIN → (Ov,M[e] ←M[e] + 1);                                 memory increment
    ADM → (Ov,M[e] ←M[e] + A);                                 add to memory
    MUL → (Ov,AB ←A × M[e]);                                   multiply
    DIV → (Ov,B ←AB/M[e]; A ←AB mod M[e]);                     divide
```

Logical Group

```
    ETR → (A ←A ∧ M[e]);                                       extract
    MRG → (A ←A ∨ M[e]);                                       merge
    EOR → (A ←A ⊕ M[e]);                                       exclusive or
```

Microcoded Register Exchange Instruction
Each instruction can be formed from a series of microprogrammed operations. Compound microcoded instructions are shown below without a μ).

```
    CLA → (A ←0);                                              μ, clear A
    CLB → (B ←0);                                              μ, clear B
    CLR → (AB ←0);                                             clear A and B
    CLX → (X ←0);                                              μ, clear X
    CAB → (B ←A);                                              μ, copy A into B
    CBA → (A ←B);                                              μ, copy B into A
    XAB → (A ← B; B ←A);                                       exchange A and B
    CXB → (B ←X);                                              μ, copy X into B
    CBX → (X ←B);                                              μ, copy B into X
    XXB → (X ←B; B ←X);                                        exchange X and B
    CAX → (X ←A);                                              μ, copy A into X
```

```
CXA → (A ← X);                                              μ, copy X into A
XXA → (A ← X; X ← A);                                       exchange X and A
CNA → (A ← ¬ A);                                            μ, not A
BAC → (A ← B; B ← 0);                                       copy B into A, clear B
ABC → (B ← A; A ← 0);                                       copy A into B, clear A
STE → (X<15:23> ← B<15:23>; X<0:14> ←sign_extend(B<15>));   μ, store exponent: exponent control bit
        B<15:23 > ← 0);
LDE → (B<15:23> ← X<15:23>);                                load exponent
XEE → (<B15:23> ← X<15:23>; X<15:23> ← B<15:23>;            exchange exponent
        X<0:14> ← sign_extend(B<15>));
    End of microcoded instruction group
Shift Group
    LRSH → (AB ← AB / 2^el {logical});                      logical right shift
    RSH → (AB ← AB / 2^el);                                 right shift
    RCY → (AB ← AB / 2^el {rotate});                        right cycle
    LSH → (Ov,AB ← AB x 2^el);                              left shift
    LCY → (AB ← AB x 2^el {rotate});                        left cycle
    NOD → (X ← X - normalize_exponent(AB)                   normalize, decrease X
            AB ← normalize(AB));
Skip Test Group
    SKE → ((A = M[e]) → (P ← P + 1));                       skip if A = M
    SKB → ((M[e] ∧ B) = 0) → (P ← P + 1);                   skip if B and M don't compare 1's
    SKN → (M[e]<0> → (P ← P + 1));                          skip if M negative
    SKR → (Ov,M[e] ← M[e] - 1; next M[e]<0> → (P ← P + 1)); reduce M, skip < 0
    SKM → ((M[e] ∧ B) = (A ∧ B)) → (P ← P + 1);             skip on masked M
    SKG → (A > M[e]) → (P ← P + 1);                         skip if greater than M
    SKD → (XR<0:23> ← abs(B<15:23> - M[e]<15:23>);          difference exponents and skip
            (M[e]<15:23> > B<15:23>) → (P ← P + 1)):
    SKA → ((M[e] ∧ A) = 0) → (P ← P + 1);                   skip if A and M don't compare 1's
Branch Group
    BRU → (P ← e);                                          branch unconditionally
    BRX → (X ← X + 1; X<9> → P ← e);                        increment Index, Branch
    BRM → (M[e]<0> ← Ov; M[e]<3:5> ← EM3; M[e]<1,2,9> ← 0;  mark place and branch
            M[e]<6:8> ← EM2; M[e]<10:23> ← P; next          used to call subroutines
            P ← e + 1);
    BRR → (P ← M[e] + 1; Ov ← Ov ∨ M[e]<0>);               branch return; used in terminating subroutines
Control Group
    HLT → (Run ← 0);                                        halt
    NOP → :                                                 no operation
    EXU → (instruction ← M[e];                              execute
            Instruction_execution);
Overflow Test Group
    OVT → (0v → (P ← P + 1); (0v ← 0));                     overflow test
    ROV → (0v ← 0);                                         reset overflow
    REO → (X<14> ⊕ X<15>) → (0v ← 1);                       record exponent
```

Breakpoint Test Group

$((BPT 1 \wedge BPT<1>) \vee (BPT 2 \wedge BPT<2>) \vee (BPT 3 \wedge BPT<3>) \vee (BPT 4 \wedge BPT<4>)) \rightarrow (P \leftarrow P + 1);$

Memory Extension Register Control Group

$SET \rightarrow (instruction<17> \rightarrow (EM2 \leftarrow instruction<21:23>);$

　　$instruction<16> \rightarrow (EM3 \leftarrow instruction<18:20>));$

$EXT \rightarrow condition \rightarrow (P \leftarrow P + 1);$

　　$condition := ((instruction<22> \wedge (EM2 = 2)) \wedge (instruction<23> \wedge (EM3 = 3)))$

$POP \rightarrow (M[0]<0,9:23> \leftarrow 0v\square l\square P; P \leftarrow 100_8 + pop\_code);$ *programmed operator; 64 user defined instructions called via subroutine link in M[0]*

$EOM \rightarrow IO\_instruction\_execution;$ *see the definition of the IO instruction set below*

$POT \rightarrow IO\_instruction\_execution;$

$PIN \rightarrow IO\_instruction\_execution;$

$SKS \rightarrow IO\_instruction\_execution;$

　　　　　$)$ *end Instruction_execution; not including Input Output instructions*

Input-Output Control from the Pc

KT and KMs State

 Devices consist of the following parts:

$IO\_Device[0:77777_8]$ *name (or address) of a specific IO device; the EOM command is first given to select the specific device; subsequent commands are implicitly to the selected device*

$IO\_Output[0:77777_8]<0:23>$ *Input and Output Data buffers associated with specific devices*

$IO\_input[0:77777_8]<0:23>$

$IO\_Ready[0:77777_8]$ *bit for each device to denote when device is ready to transmit data*

$IO\_Select[0:77777_8]$ *a bit within each device denoting it has been selected for an operation*

$io\_unit<0:14>$ *the particular io device selected by the EOM command;*

IO Instruction Set

$EOM \rightarrow (io\_unit \leftarrow e);$ *command to select or address the device: energize output M*

$POT \rightarrow (IO\_Select[io\_unit] \wedge IO\_Ready[io\_unit] \rightarrow ($ *output data command*

　　$IO\_Output[io\_unit] \leftarrow M[e]; io\_unit \leftarrow 0):$

　　$IO\_Select[io\_unit] \wedge \neg IO\_Ready[io\_unit] \rightarrow (POT)):$ *wait until ready*

$PIN \rightarrow (IO\_Select[io\_unit] \wedge IO\_Ready[io\_unit] \rightarrow ($ *input data command*

　　$M[e] \leftarrow IO\_Input[io\_unit]; io\_unit \leftarrow 0);$

　　$IO\_Select[io\_unit] \wedge \neg IO\_Ready[io\_unit] \rightarrow (PIN));$ *wait until ready*

$SKS \rightarrow (io\_unit \leftarrow e: next$ *skip if signal is not set*

　　$(IO\_select[io\_unit] \wedge IO\_Ready[io\_unit] \rightarrow ($

　　$P \leftarrow P + 1);$

　　$io\_unit \leftarrow 0);$

Interrupt System States

Interrupt   *controls whether interrupts will be processed*

$I\_RQ[0:63]<0:15>$ *array of 1024 interrupt requests*

$I\_ON[0:63]<0:15>$ *array of interrupt enable to enable or inhibit interrupt requests*

$I\_Signal[0:63]<0:15> := I\_RQ[0:63]<0:15> \wedge I\_ON[0:63]<0:15>$

$K\_address<0:5>$ *group number*

$I\_address<0:3>$ *level number within a group of the active interrupt*

The I address and K address combine ($200_8 + 20_8 \times$ K address + I address) to establish an interrupt address, 200_8 is the highest priority and $200_8 + 1477_8$ the lowest priority.

Interrupt level state$[0:63]<0:15>_3$

There are three states associated with each interrupt, *Inactive, Waiting*, and *Active*:

Inactive means no I signal is present.

Waiting means the I signal has been received but is waiting to be processed.

Active means the interrupt has caused the main program to recognize its presence.

The instruction in M[$200_8 + 20_8 \times$ K address + I address] is executed upon interrupt. There are two kinds of interrupts: single instruction allows one instruction to be executed and the interrupt level state is changed from active to inactive; and normal requires that a mark place and branch, BRM, instruction to be executed to save P. At the completion of the interrupt program, a branch unconditional (BRU) indirectly via the BRM instruction restores the interrupt level. (That is, the Interrupt level state is changed from Active to Inactive, and another I Signal can be processed.)

Interrupt interpretation

A state denoting that an interrupt is to be processed or the interrupt level state to be changed from Waiting to Active for normal interrupts and Waiting to Active to Inactive for single interrupts. The interrupt processed is the highest of those waiting provided there are no interrupts of highest level in the Active state.

Interrupt Control Instructions

EIR → (Interrupt ← 1);	*enable interrupt; turn on mode*
DIR → (Interrupt ← 0);	*disable interrupt; turn off*
IET → (Interrupt → P ← P + 1);	*interrupt test; skip if on*
IDT → (¬ Interrupt → P ← P + 1);	*interrupt disable test; skip if off*

POT instruction to control the Interrupt System. EOM[20020] is first given to select the Interrupt System.

(POT ∧ IO Ready[20020]) → (	*interrupt control instructions*
(c = 1) → I ON[a]<0:15> ← I ON[a]<0:15> ∨ B<0:15>;	*arm a channel level group*
(c = 2) → I ON[a]<0:15> ← I ON[a]<0:15> ∨ ¬ B<0;15>;	*disarm a channel level group*
(c = 3) → I ON[a]<0:15> ← b<0:15>);	*set a channel level group*
a<0:5> := M[e]<0:5>	*group select or K address*
b<0:15>:= M[e]<8:23>	*data for I address*
c<0:1> := M[e]<6:7>	*command control bits*

Section 3

The IBM System/360—
a series of planned machines which span
a wide performance range

In this introduction, besides making some general comments on the IBM System/360, we will attempt an analysis of the performance and costs of the series. Performance is notoriously difficult to measure, as we noted in Chap. 3, and costs are even more so. With respect to the latter, what is publicly available are price data, not manufacturing-cost data.

These prices reflect not only marketing policies but also accounting policies within the organization for the attribution of costs to product lines. For example, we have had to determine Pc and Mp prices on the basis of incremental Mp prices within a C. Nevertheless, the 360 series provides two things which make a comparative analysis worthwhile. First, the common ISP makes simple performance measures more comparable; second, the common manufacturer makes relative prices more a reflection of relative costs than would otherwise be the case. Neither of these aspects is perfect, as we will note at several points in the discussion. Nevertheless, the 360 series provides as good an opportunity to attempt cost/performance analysis as we know. Indeed, this opportunity has already been grasped in a paper by Solomon [1966], which we have found very valuable and use to provide a basis of Pc power.

Analyses of the type we attempt here produce only rather crude pictures and are subject to question if all the input data are not very carefully checked. We have not done the latter, depending instead on published sources. For the purpose of this book, illustration of the style of analysis seems sufficient. In addition, using a performance measure based only on Pc power measurements, as we do here, leaves many questions unanswered because it does not address the soft areas of analysis relating to throughput, task environment, and the operating system software.

Unlike the other introductions in this book, the reader may find it worthwhile to scan this one, read the chapters in the section, and then return to this introduction when the system has become somewhat familiar.

The IBM System/360 is the name given to a third-generation series of computers which constitute the current primary IBM product line. They all have a common ISP but differ in inter-preter speeds and PMS structure. Many PMS elements are used in common, particularly K's, Ms's, and T's.

The System/360 series is presented both because IBM's market dominance makes it the most prevalent current computer and because its implementations span the largest performance and price range of any series. The C('360) models should be compared with one another (Table 1) to be aware of their capabilities. Their introduction dates and their relationship are shown in Fig. 1. Chapters 43, 44, and 32 discuss the logical structure of the system, the implementations,[1] and the microprogrammed Model 30.

A succinct description of the design goals and innovations is given in the abstract of the paper Architecture of the IBM System 360 [Amdahl et al., 1964a]:

[1]Chapters 43 and 44 are from *IBM Systems Journal*, vol. 3, no. 2, 1964, which was devoted exclusively to the System/360. The other articles (listed in the bibliography) are recommended for additional details.

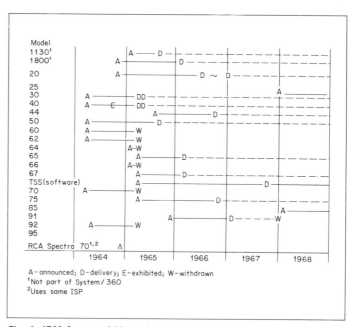

Fig. 1. **IBM System/360 models introduction dates.**

The architecture° of the newly announced IBM System/360 features four innovations:

1 An approach to storage which permits and exploits very large capacities, hierarchies of speeds, read-only storage for microprogram control, flexible storage protection, and simple program relocation.

2 An input/output system offering new degrees of concurrent operation, compatible channel operation, data rates approaching 5,000,000 characters/second, integrated design of hardware and software, a new low-cost, multiple-channel package sharing mainframe hardware, new provisions for device status information, and a standard channel interface between central processing unit and input/output devices.

3 A truly general-purpose machine organization offering new supervisory facilities, powerful logical processing operations, and a wide variety of data formats.

4 Strict upward and downward machine-language compatibility over a line of six models having a performance range factor of 50.

The above four featured innovations are all stated as IBM Corporation design results. It seems better to analyze them in terms of design constraints and implementation results. It appears that the design constraints, from marketing and management directions, were compatibility (item 4 above) and the use of common peripheral equipment (item 2 above). Thus we can measure the 360 design in terms of how well it meets these constraints. With some minor exceptions, all the peripheral components existed at the time of the design and had been used with other IBM computers; thus a goal was already realized. A measure of the design can also be based on a comparison with alternative designs. In the following sections we suggest that several forms of multiprocessing would yield higher performance at lower cost. A difficult and important constraint, though not mentioned above, is the necessity of program compatibility with almost all earlier IBM computers.

It should be noted that, at the outset of the IBM System/360 announcement, another company, RCA, adopted the 360 ISP as a design constraint for its own future computer development. Although some price-performance characteristics appear to be better in the RCA series, the implementation scheme is similar.

The lower RCA prices do not reflect entirely implementation and technology but include RCA marketing and profit strategy. In addition, of course, there should have been lower development costs.

An interesting aspect of the design is the method used to implement the individual computer models (of the range) and their associated costs. From the standpoint of innovation, the 360 was the first computer series to cover a wide range. The more basic P's (Models 20 ~ 65) were implemented via a microprogrammed processor. This is based on a computer program within an M(read only), i.e., a Read Only Storage/ROS, to interpret the common ISP. A payoff from this implementation strategy is a solution to the "compatibility design constraint," which is the ability to provide compatibility with the customer's previous (IBM) machine, which of course was not a member of the 360 series. This is undoubtedly the most difficult constraint to meet in the P designs, and probably the most significant real innovation. From the marketing viewpoint, it provided the user with a crutch to go from a former IBM computer to the System/360. This is accomplished through "emulation," which (as defined by IBM) means the ability of one C to interpret another's programs *at a reasonable performance level.* These emulations are realized by various microprogrammed P's being designed to interpret both the 360 ISP and one or more of IBM 704, 709, 1401, 1410, 1440, 1460, 1620, 7010, 7040, 7044, 7070, 7074, 7090, 7094.

Most of the above ISP's have a different structure from the 360 ISP. For example, the 1401 (Chap. 18) series instructions and data are variable-length character strings; the 1620 has variable-length data strings; the 704 series process fixed- and floating-point data with single-address instructions; and the 7070 is a fixed-word decimal computer. Thus the 360 C's represent the first machines to be two logical processors in the same physical implementation.

The emulated speeds are often better than that of the original hardwired computer. This is not surprising, considering the change in technology; it is a very attractive feature. The 360 Mp performance is often a factor of 5 to 10 times the "emulated" computers; and the M(ROS) data rates are a factor of 25 times the Mp's. For example, the Model 65 emulating a 7090 runs faster than a hardwired 7090 (Table 1). The use of an M(ROS) for defining an ISP is questionable if we ignore the emulation constraint. Note, by way of evidence, that the hardwired models 91 and 44 have the lowest cost-to-performance ratios in the series.

There are minor deviations in the particular models, but all

° The term *architecture* is used here to describe the attributes of a system as seen by the programmer, i.e., the conceptual structure and functional behavior, as distinct from the organization of the data flow and controls, the logical design, and the physical implementation.

Table 1† IBM System/360 Models, IBM 1130, and IBM 1800 computer characteristics

Parameter	1130[a]	1800[a]	20[b]	25	30	40	44	50	65|67	75	85	91	
Pc (technology: (hybrid/h|µ.ro|µ.rw)); Pio (technology)	h;h	h;h	µ.ro	µ.rw	µ.ro	µ.ro	h;h	µ.ro	µ.ro;h	h;h	h,µ.ro,µ.rw,h	h;h	
M (ro|rw; t.cycle: µs/w;	···	···	?	(Mp)	1.0	0.625	···	0.5	0.2	···	0.08	0.08	
size:w;	···	···	4096	···	4096	4096	···	2816	4000	···	2000,500	···	
b/w;	···	···	60	···	60	60	···	90	100	···	108	108	
technology: (ind|cap|core);	···	···	core	core	cap	ind	···	cap	cap	···	ro.rw	ro.rw	
ISP's implemented in P.microprogram	···	···	1401[c]	1401[c]	1401| 1620[d]	1401| 1410[e]	···	1410| 7070[f] 7090[g]	7070| 7090[g]	···	7090	7090	
S (concurrency: (Mp;Pc))	1;1	1;1	1;1	1;1	1;1	1;1	1;1	1;1	1;2|8;5	1,2,4;1	(4,1)|(1,1)	16;1	
Mp (i.width: (by); (8, 1 parity) b/by;	2	2	1	2	1	2	4	4	8	8	16	8	
t.cycle: µs/w;	3.6	2|4	7.2	0.9	1.5	2.5	1.0	2.0	0.75	0.75	(0.96,1.04)0.08[h]	0.75	
size: $\log_2$(by);	13~14;	13~16;	12~14;	14~15;	13~16;	14~18;	15~18	16~19,	17~(20|24);	18~20;	19~22;	20~22;	
i.avg: $\log_2$(by);	13.5	14.5	13	14.75	14.5	16	6.5	17.5	18.5	19	20.5	21	
i.rate: b/µs;	4.45	8|4	1.1	17.8	5.3	6.4	32	16	(85~170)|425	85,170,340	~512|1600	1370	
t.1-bit: µs;	0.22	0.125|0.25	0.9	0.056	0.19	0.16	0.031	0.063	0.017|0.0025	0.017	$2\times10^{-3}\mid6.3\times10^{-4}$	7.3×10^{-4}	
t.64-bit: µs;	14.	8|16	58.	3.6	12.1	10	2	4.0	0.75,0.375,0.147	0.75,0.375,0.18	0.125|0.04	0.047	
C(t.matrix.q: µs;	···	···	···	···	71	23.4	···	7.35	1.8	1.02	···	···	
t.sqrt.q: µs;	···	···	···	···	118	26.8	···	6.8	1.97	1.24	···	···	
t.field-scan.q: µs;	···	···	···	···	35	12.8	···	5.8	1.8	1.64	···	···	
t.scientific-mix.q.µs;	···	···	···	···	47	15	···	8.0	2.4	1.45	···	···	
t.all.q: µs;	···	···	148	110	88	25.4	5.7	8.5	2.3	1.55	*	*	
t.avg.q: µs;	42[o]	24|48[o]	120	80	60	20	4	8	1.9|2.2	1.3	0.5[k]	0.4[k]	
power/p_1(1./t.avg.q):	2.9	5.0|2.5	1.0	1.5	2.0	6.0	30	15	63|54	92,100[j]	252[j]	314[j]	range: 1~314
power/p_2(1./t.64-bit):	4.2	7.2|3.6	1.0	16.	4.8	5.8	29	14	77,155|394	77,155,310	465|1450	1230	range: 1~1450
power/p_3[Stevens, 1964][i]:	···	···	···	···	2.0	7.0	···	20	42~60	100	···	···	range: 2~100
power/P_4[Conti, 1968][j]:	···	···	···	···	···	···	···	···	1	1.58	3.9~4.3[m]	5	range: 1~5
Mp utilization efficiency/ (t.64-bit/t.avg.q))	···	···	0.49	0.045[n]	0.2	0.5	0.5	0.5	0.37~0.18	0.54~0.27	0.25/0.08	0.2	
Pc(cost:$/s):	0.00064	0.0019|0.0016	0.00049	0.00050	0.0013	0.0030	0.0041	0.012	0.022|0.029	0.037	0.087	0.091	range: 1~186
Mp(cost.avg:$/s):	0.00049	0.0014|0.0012	0.00065	0.0027	0.0023	0.0049	0.0050	0.0084	0.023|0.032	0.031	0.080	0.069	range: 1~123
C(cost.min: $/s):	0.00096	···	0.0019	···	0.0043	0.008	0.008	0.022	0.054|	0.075	0.18	0.20	range: 1~105
C(cost.avg: $/s):	0.0018	0.0077	0.0045	0.0085	0.0130	0.027	0.024	0.051	0.08|	0.128	0.18	0.30	range: 1~65
Pc(cost) + Mp(cost.avg)	0.00113	0.0033|0.0028	0.00114	0.0032	0.0036	0.0079	0.0091	0.020	0.045|0.061	0.068	0.167	0.160	range: 1~160
C(cost.min: $/s)/$p_1$	0.00069	···	0.0038	···	0.0043	0.003	0.00053	0.0029	0.0017	0.0016	0.0013	0.0013	
C(cost.avg: $/s)/$p_1$	0.013	0.0031	0.009	0.011	0.0130	0.009	0.0016	0.0068	0.0025	0.0028	0.00143	0.0019	
Pc(cost)/.5p_1	0.00046	0.0008|0.0013	0.00098	0.00067	0.0013	0.0010	0.00028	0.0016	0.0007|0.0011	0.0008	0.00069	0.00058	
C(cost.min)/C(cost.avg.)	0.5	···	0.42	···	0.32	0.3	0.33	0.43	0.68|	0.59	···	0.66	avg: .47
Pc(cost)/Mp(cost.avg.)	1.3	1.4|1.3	0.75	1.85	0.57	0.61	0.82	1.4	0.96|0.91	1.2	1.1	1.3	avg: 1.1
Pc(cost)/C(cost.avg.)	0.35	0.25	0.11	0.06	0.10	0.11	0.17	0.24	0.28|	0.29	0.47	0.3	avg: 0.23

† This table is presented as PMS expressions.
[a] Not IBM System/360 compatible, but made with hybrid technology.
[b] Similar, but not identical to System/360 ISP.
[c] C(IBM 1401, 1440, 1460).
[d] C(IBM 1620).
[e] C(IBM 1410, 7010).
[f] C(IBM 7070, 7074).
[g] C(IBM 709, 7040, 7044, 7090, 7094).
[h] Two M's; an M(content addressable) working with Mp.
[i] Estimated. see Chap. 44.
[j] See Conti [1968], based on running many programs.
[k] Models 85, and 91 are too difficult to predict because of instruction buffering based on Conti [1968].
[l] Cost derived from purchase cost;/45.
[m] Varies depending on buffering and multiply options.
[n] Meaningless per se; Mp is used by microprogram defining System/360 ISP.
[o] 1130 and 1800 are not program-compatible. The very high penalty factor of 3 is used to compare them to System/360 ISP.

implementations belong to a common ISP subset. The Model 20 and the Model 91, the extremes of the series, deviate most from the standard 360 ISP. The range of models (Table 1) shows the comparative effects of implementation on the actual processing times. For example, the designers of the various C's were constrained by memory bandwidths. Since the core memories have about the same cycle time (0.75 ~ 2.0 microseconds), variation in bandwidth is obtained by increasing the data path width from 8 to 64 bits and by increasing the number of independent Mp's. By looking at just Mp bandwidth, for models 30 ~ 65, we obtain a range of 5.3 to 85 megabits/s, corresponding to a performance range of about 1 to 16. By doubling the number of independent memories, this factor can be increased to 32. These models correspond to a Pc performance range of 1 to 32. Although we might expect a narrower range (based on Mp speed), the range can be increased by performance suppression (at the low end). Power range can be increased by lowering the absolute performance of Model 30. This is accomplished by making performance tradeoffs to lower cost.

Logic technology

The logic of the 360 series is realized in a hybrid technology, composed partly of integrated-circuit techniques and partly of the solid-state techniques standard in second-generation machines. It is a "thick-film" technology that deposits the circuitry on a ceramic substrate. This is called Solid Logic Technology (SLT) and is used solely by IBM. This production technique allows only for the fabrication of passive circuit elements on the substrate. The semiconductor elements (diodes and transistors) are produced independently, using standard semiconductor production techniques on a wafer. The semiconductors are then cut and bonded to the substrate, and the complete SLT logic unit is encapsulated. The substrates correspond roughly to logic elements (gates, inverters, flip-flops, etc.). The SLT units are placed on larger printed-circuit boards.

Although SLT differs fundamentally from integrated-circuit technology, the overall size of the final printed-circuit boards is about the same. At the time the decision was made to develop the technology, it was unclear that integrated-circuit technology would reach mass-production state. Thus the SLT program was an intermediate design prior to integrated-circuit technology. The two approaches are about the same from the standpoint of reliability, especially when one considers the soldered printed-circuit mounting. The number of connections to the printed-circuit board are about the same. The production technology of the 360 series is outstanding, perhaps surpassed only by the 360 marketing plan.

The Instruction-set processor

The following discussion covers only the Pc. The instruction set consists of two classes, Scientific ISP and Data Processing ISP, which operate on the different data-types. These data-types correspond roughly to the IBM 7090 (Chap. 41) and IBM 1401 (Chap. 18). For the scientific ISP they are half- and single-word integers, address integers, single, double, and quadruple (Model 85) floating point, and logical words (boolean vectors); for the data-processing ISP they are address or single-word integers, multiple byte strings, and multiple digit decimal strings. These many data-types give the 360 strength in the minds of its various types of users. The many data types may be of questionable utility and constrain the ISP design by having to perform few operations, rather than having a more complete operation set for a few basic data types. The viewpoint taken here is a biased one; we feel that, unless a particular data-type adds significant processing and storage capability, it should not be fundamental to the ISP. The decimal-string integers appear to cost in storage and processing time. Their redeeming virtues are that little or no conversion is required at input or output time, and their internal representation is easily recognized by people.

Advantages of general-registers organization

The ISP uses a general-register organization. The ISP power can be compared with several similar general-register ISP structures such as those of the UNIVAC 1107, 1108; the DEC PDP-6, PDP-10; the SDS Sigma 5, Sigma 7; and the early general-registers-organized machine Pegasus (Chap. 9). Of the above machines the 360 Scientific ISP appears to be the weakest in terms of instructions and the completeness of the instruction set.

For example, in Pegasus, PDP-6, and the UNIVAC 1107 symmetry is provided in the instruction set. For any binary operation b the following are possible:

GR ← GR b Mp
GR ← GR b GR
Mp ← GR b Mp
Mp ← Mp b Mp

The 360 ISP provides only the first two. Additional instructions (or modes) would increase the instruction length.

In the System/360 the only advantage taken of general registers is to make them suitable for use as index registers, base registers, and arithmetic accumulators (operand storage). Of course, the commitment to extend the general-purposeness of these general registers would require more operations. Chapter 3 (page 61) suggests advantages for general register organizations.

The 360 has a separate set of general registers for floating-point data. This provides more processor state and temporary storage but again detracts from the general-purpose ability of the existing registers. Special commands are required to manipulate the floating-point registers independent of the other general registers. Unfortunately the floating-point instruction set is not quite complete (e.g., fixed- to floating-point conversion), and several instructions are needed to move data between the fixed and floating registers.

When multiple data-types are available, it is desirable to have the ability to convert among them unless the operations are complete in themselves. The System/360 might use more data conversion instructions, for example, between the following:

1 Fixed precision integers and floating-point data

2 Address-size integers and any other data

3 Half-word integer and other data

4 Decimal and byte string and other data (decimal string to and from byte string conversion is provided)

Some of the facilities are redundant and might be handled by better but fewer instructions. For example, decimal strings are not completely variable-length (they are variable up to 31 digits, stored in 16 bytes), and so essentially the same arithmetic results could be obtained by using fixed multiple length binary integers. This would remove the special decimal arithmetic and still give the same result. If a large amount of fixed field decimal or byte data were processed, then the binary-decimal conversion instructions would be useful.

The communication instructions between Pc and Pio are minimal. The Pc must set up Pio program data, but there are inadequate facilities in Pc for quickly forming Pio instructions (which are actually yet another data-type). There are, in effect, a large number of Pio's as each device is independent of all others. However, signaling of all Pio's is via a single interrupt channel to Pc.

The Pc state consists of 26 words of 32 bits each:

1 Program state word, including the instruction counter (2 words)

2 Sixteen general registers (16 words)

3 Four 2-word floating-point general registers (8 words)

Many instructions must be executed (taking appreciable time) to preserve the Pc state and establish a new one. A single instruction would be preferable; even better would be an instruction to exchange processor states, as in the CDC 6600 (Chap. 39).

Addressing and multiprogramming

The methods used to address data in Mp have some disadvantages. It is impossible to fetch an arbitrary word in Mp in a single instruction. The address space is limited to a direct address of only 2^{12} bytes. Any Mp access outside the range requires an offset or base address to be placed in a general register. Accesses to several large arrays may take significant time if a base address has to be loaded each time. The reason for using a small direct address is to save space in the instruction. We know of no published attempt to analyze the tradeoffs, even of instruction efficiency alone, although undoubtedly such comparisons were made within IBM.

Another difficulty of the 360 addressing is the inhomogeneity of the address space. Addressing is to the nearest byte, but the system remains organized by words; thus, many addresses are forced to be on word (and even double-word) boundaries. For example, a double-precision data-type which requires two words of storage must be stored with the first word beginning at a multiple of an 8-byte address. (However, the Model 85, which is a late entry in the series, allows arbitrary alignment of data-types with word boundaries.) When a general register is used as a base or index register, the value in the index register must correspond to the length of the data-type accessed. That is, for the ith value of a half integer, single integer, single floating, double floating (long), and quadruple floating (extended), i must be multiplied by 2, 4, 4, 8, and 16, respectively, to access the proper element.

A single instruction to load or store any string of bits in Mp (as provided in the IBM Stretch) would provide a great deal of generality. Provided the length were up to 64 bits, such an instruction might eliminate the need for the more specialized data-types.

A basic scheme for dynamic multiprogramming is nonexistent (i.e., although static multiprogramming is done, relocation hardware is not present). Only a simple method of Mp protection is provided, using protection keys (see Chap. 43, page 597). This scheme associates a 4-bit number (key) and a 1-bit write protect with each 2 kby block, and each Pc access must have the correct number. Both protection of Mp and assignment of Mp to a particular task (greater than 2^4 tasks) are necessary in a dynamic multiprogramming environment. Although the architects of System/360 advocate its use for multiprogramming, the operating system does not enforce conventions to enable a program to be moved, once its execution is started. Indeed, the nature of the 360 addressing is based on absolute binary addresses within a program. The later experimental Model 67 does, however, have a very nice scheme for protection, relocation, and name assignment to program segments [Arden et al., 1966].

PMS structures and implementations of the computer

The PMS structures of the various models in System/360 are basically similar, except for the upper end of the series and for the Model 44 (complete compatibility can be purchased as an option). We take up the main group first and then discuss the others individually.

Models 30, 40, 50, and 65

The PMS of Models 30, 40, and 50 is the tree-structured Mp-Pc shown in Fig. 2.[1] They all use a P.microprogram, although with different ISP's. Some gross characteristics are given in Table 1. The Pc of Model 65 is also microprogrammed, but it has hardwired Pio's. A PMS diagram of Model 65 (and Model 75) is given in Fig. 3.

The C structures with M(ROS) use a single physical P.microprogram to realize the Pc, the Pio('Multiplexor Channel), and the Pio('Selector Channel). This technique of using a single shared physical P for multiple logical P's with fast changing of P.state is the same one that Pio('Multiplexor) uses. The

[1] The structure of the Mp's does not include the local M's used for access control, i.e., the storage protect key mechanism, which it is hoped the student will forget about (forever).

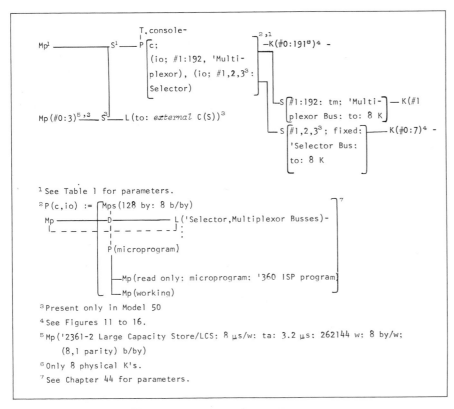

Fig. 2. IBM System/360 Models 30, 40, and 50 PMS diagram.

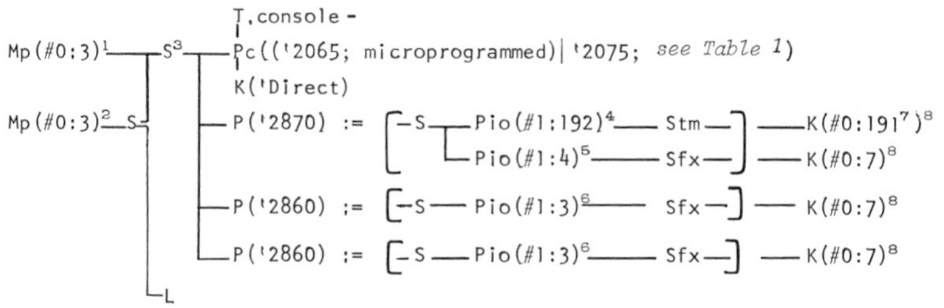

```
                     T.console -
Mp(#0:3)¹ ──S³──── Pc(('2065; microprogrammed)│'2075; see Table 1)
                     K('Direct)
Mp(#0:3)²─S─      ── P('2870) :=  ┌─S ─┬─ Pio(#1:192)⁴ ──── Stm ──┐ ──── K(#0:191⁷)⁸
                                  │    └─ Pio(#1:4)⁵ ──── Sfx ──┘ ──── K(#0:7)⁸
                  ── P('2860) := ┌─S ── Pio(#1:3)⁶ ──── Sfx ─┐ ──── K(#0:7)⁸
                  ── P('2860) := ┌─S ── Pio(#1:3)⁶ ──── Sfx ─┐ ──── K(#0:7)⁸
         └L
```

¹Mp('2365-3) := (Mp(#0,1; '2365-2; core: .75 μs/w; 8 by/w; 16 kw; (8,1 parity) b/by)-S-)

²Mp('2361-2 Large Capacity Store/LCS; 8 μs/w; t.access:3.2 μs; 262 kw; 8 by/w; (8,1 parity)
     b/by)

³S(8 M; 4 P; time multiplexed; concurrency:1; 'Bus Control Unit/BCU)

⁴Pio('2870 IO Multiplexor Channel)

⁵Pio('2870 IO Selector Subchannel)

⁶Pio('2860 Selector Subchannel)

⁷Only 8 physical K's

⁸See Figures 11 to 16.

Fig. 3. PMS structure for IBM System/360 Models 65 and 75 PMS diagram.

Pio('Multiplexor) is equivalent to multiple Pio's. Within the physical P both interrupts and polling are used to switch among the P's. Polling is used to service the several P's since the main program loop of the ISP interpreter returns to a common point each time the next instruction is fetched. That is, the interpretation cycle for the 360 ISP starts by fetching the instruction, proceeds to fetch the operands, executes the instruction, and then returns results to Mp. The instruction-interpretation process takes only a few Mp references for most instructions.

A few instructions require a long (or indefinite) interpretation time, e.g., character translate, edit, etc., since the operations are on character strings. Here, the iterative program loop which operates on each character of the string must test the attached K's to detect when the Pio interpreter is to be run for data transfers. The long instructions can take several hundred microseconds and cannot be interrupted; thus the response time for an interrupt can be very poor. Figure 4 gives a simplified picture of the registers organization of a Model 50, but it is also typical of Models 30, 40, and 65.

The actual System/360 ISP interpretation program in each of the models is different. In addition, each model has microprograms for interpreting other ISP's through emulation. Tucker [1967] discusses how the models were changed as the emulation constraint was added. Table 1 gives the computers which each of the models can emulate. A register structure of the C('30) and the operation for the P.microprogram ISP are given in Chap. 32, page 386. Tables 2 and 3 in Chap. 44 give the additional parameters which influence the instruction interpretation rate of the P.microprogram. The significant parameters for a P.microprogram are the M(ROS) hardware characteristics (speed, size, and information width); the number of fields in the M(ROS) instructions, which gives an indication of the number of control functions performed in parallel; the M(general register) rates and their location in the structure; the Mp data rate; and the characteristics of M(temporary) within P. The activity of transferring data from a K, via the Pio('Selector), is done concurrently with normal instruction interpretation in Models 30, 40, and 50. A program in M(ROS) sets up the data transmission with Mp, and transmission is controlled by an independent hardware control.

Model 20

This model is a subset of the System/360. It has eight 16-bit general registers. It is possible to write programs which will run on both the Model 20 and other models. Model 20 does not have Pio's, and Pc issues instructions to control the attached K's.

Model 25

The Model 25 is an interesting C. Perhaps some of the interest of the authors is caused by the mystery (to the authors) as to what its ISP is. Its ISP is no doubt described in maintenance

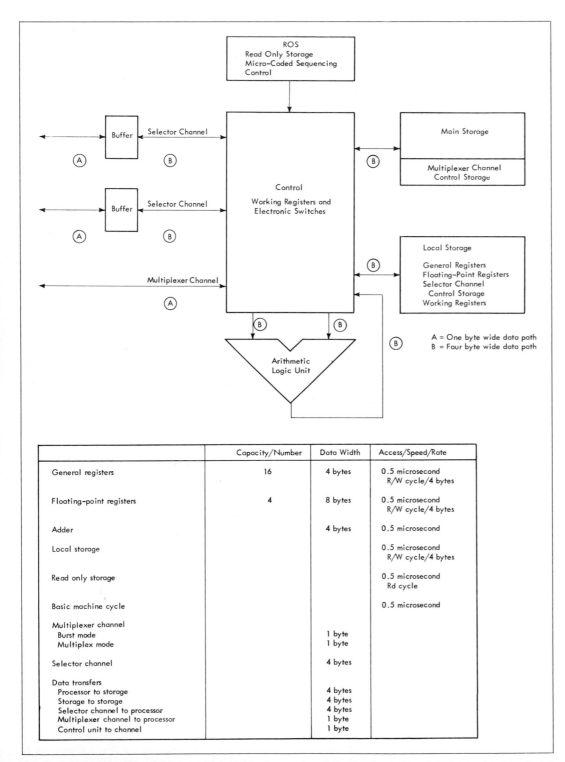

Fig. 4. IBM System/360 Model 50 data-flow diagram and system characteristics. (*Courtesy of International Business Machines Corporation.*)

manuals. We can make the following observations based on its characteristics taken from its manual of Functional Characteristics. These appear in Table 1. The observations are:

1 It has a very high-performance Mp, namely, Mp(core; .9 μs/w; 16|24|32|48 kby; 2 by/w); the Mp power is almost that of a Model 50.

2 There is a relatively straightforward Pc which is microprogrammed. The Pc uses Mp for its memory. The System/360 ISP is defined in conventional M(read,write). Of the Mp(48 kby) 16 kby is reserved for a microprogram.

3 Its performance is between that of Models 20 and 30, performing a 360 ISP instruction in about 80 μs.

4 The penalty paid (slowdown factor) to interpret the 360 ISP is therefore 80/1.8 $\simeq$ 45.

5 A small 180-nanosecond local store is used for operands.

6 The Pc cost appears to be about the lowest in the series.

We should ask ourselves:

1 Why do we want an intermediate-level P.microprogram with its own M.read-only, as in the other processors? These P's just seem to waste power.

2 Why should we bother to implement an intermediate-level 360 ISP? We know the final user will write programs in a much higher level language. Thus two levels of interpretation are required instead of one. It is assumed that to program a given task will take, say, x μs if using the 360 ISP. We assume the same task programmed directly in the Pc could take as short a time as x/45 μs if the Pc were used directly.

We assume that if the P.microprogram, which is used to define the System/360 ISP, were used to interpret a FORTRAN ISP, the speed for a Model 25 FORTRAN ISP might easily approach that of the Model 50.

Model 44

Model 44 does not use M(ROS), but its Pc and Pio are hardwired (Models 75 and 91 are also hardwired). The PMS structure of the Model 44 is given in Fig. 5. Model 44 (and 91) stand out as having better performance per unit of cost than their nearest neighbors, which are implemented with M(ROS), as can be seen from Table 1. It must be noted that Models 44 and 91 are not strictly compatible with the 360 ISP since they do not process variable-string and variable-decimal-data formats, although Model 44 options can make it completely compatible. (Subroutines will probably perform satisfactorily for most applications.)

The PMS structure of the Model 44 (Fig. 5) is a tree. The C('44) structure indicates 2-Pio('High Speed Multiplexor Channels/HSMPX) which are between a P('Selector) and P('Multiplexor) in power, since a single physical P('HSMPX) with four subchannels can behave as four independent Pio's. The organization of the Model 44 Pc registers is given in Fig. 6, which reveals a straightforward implementation. The heavy lines in Fig. 6 indicated an ORing of register outputs to form a single data bus (usually 16 or 32 bits wide). The 16-bit crossover function box allows the right and left halves (16 bits) of the input to be exchanged when output. Almost all the units are registers (except the adders, parity generators, and ORers). The A, Ax, B, and Bx registers are used as the M.working for performing instructions, where the x indicates an extension register used in the 64-bit floating-point operations. The C register

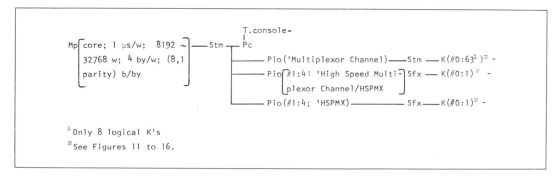

Fig. 5. IBM System/360 Model 44 PMS diagram.

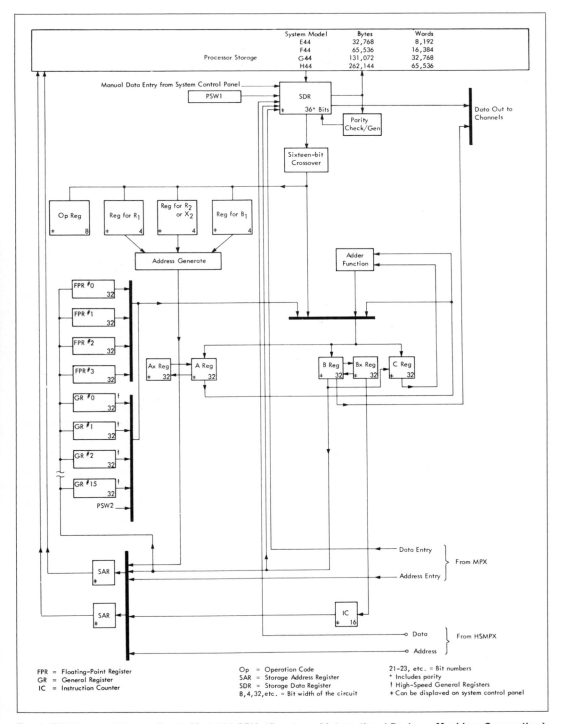

Fig. 6. IBM System/360 data flow in Model 44 CPU. (*Courtesy of International Business Machines Corporation.*)

is a second operand register used for arithmetic and logical operations.

Model 75

The PMS structure of Model 75 is given in Fig. 3. Models 65, 67, 75, and 91 all use the same basic Mp('2365; core). The S(n Mp; mP), which switches between the n Mp modules and the m Pc and Pio's, varies with model, however. C('65) and C('75) use a simple time-multiplexed S in Pc, called the S('Bus Control Unit/BCU). This S makes decisions about which P is to use which Mp, rather than having each Mp arbitrate the P requesting service locally. When the memories are all about the same speed, such an S is all right; however, it has severe limitations when slow speed (8 microseconds for the large core store) and high-speed memories (0.75 microsecond) are intermixed. The principal difference between Models 65 and 75 is that C('75) is hardwired and, depending on the size of the configuration, may have lower cost/performance.

The simplified functional unit diagram of C('75) (Fig. 7) is more abstract than the register interconnection diagram of a C('44) (Fig. 6). From this description (Fig. 7) of the logic design, one is able to conjecture what is necessarily within the instruction, execution, variable field length, and decimal functional units. The diagram is presented at a nonuniform level at both the PMS and register-transfer levels. There is somewhat more detail than in the PMS structure (Fig. 3). The Model 75 is possibly the first System/360 to require an intermediate-level diagram between a PMS structure and a register-transfer diagram. The instruction unit contains the instruction location counter (part of the ISP) and is responsible for obtaining the next instruction and the operands. Since there can be overlap in the instruction fetching process, this unit is responsible for holding a number of instructions and stores up to 128 bits (2 double words) of instructions at a time. The execution unit and the variable field and decimal units carry out operations on data. The execution unit processes floating-point and fixed-point data.

Model 67

The Model 67 was introduced in April, 1965, for the purpose of time sharing. The entry was prompted by M.I.T.'s project MULTICS. M.I.T. had ordered a GE 645 for experimental research in time sharing. IBM formed a group for the development of a time-shared computer and responded with the Model 67. The Model 67 is essentially a Pc('65) with adequate S's for multiprocessing and a K between Mp and Pc for multiprogram-

ming and memory mapping. Because of software uncertainties, the Model 67 ran as a Model 65 in most installations (in 1968). The University of Michigan and M.I.T.'s Lincoln Laboratory, the first two customers having considered the MULTICS proposal, were instrumental in outlining the specifications [Arden, et al 1966]. Several 67's have been delivered, and the software continues to evolve and be scheduled for completion (see Fig. 1). Questions of costs per console must wait until the system is stable enough to test and evaluate, although in April, 1969 IBM considered the system attractive (operational) enough to market. The most significant outcome of the experiment to date is:

1 The hardware seems capable of supporting a straightforward time-sharing system [Corbato et al., 1962]. Had IBM first developed a simple system based on proved concepts, they would be capable of undertaking research into more complex systems like the version to which they originally committed themselves. (Vendors should have some basis of actual operating experience before committing a product to market.)

2 The problems of building really large-scale software systems are not fully understood yet.

3 The idea of a virtual memory with a large address space (2^{32}w) is excellent. Many storage allocation problems are simplified by this concept. Unfortunately, the system software builders seem well on their way to filling such a memory. Thus the new freedom allows relaxation in this level of programming.

4 There is a problem of getting users into Mp.core so that Pc can be kept busy. Thus a swapping system is often found waiting for Ms.drum or Ms.disk information. Work at Carnegie-Mellon University using a Mp('LCS; core; $.5 \sim 1$ mw; 8 by/w; 8 μs/w) seems to indicate that a large number of users can have adequate response from the Model 67 if the users reside in core and are not subjected to swapping [Lauer, 1967; Fikes et al., 1968].

The above items relate to the software. The hardware (Fig. 8) is interesting from several aspects. First, there are adequate facilities for memory mapping and program segmentation. This general scheme is outlined in Fig. 9. In the Model 67 a user's segment and page maps are in Mp, and these maps point to physical Mp blocks of the program. Each time a reference is made, the map is checked for the actual reference. In order to avoid the accesses to Mp for each Mp reference, a K, with an M(content address), is located between Pc and Mp to trans-

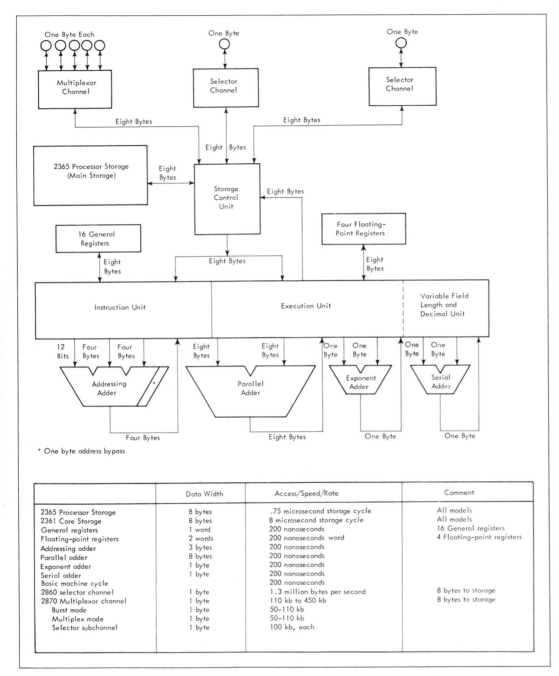

Fig. 7. IBM System/360 Model 75 data-flow diagram and system statistics. (*Courtesy of International Business Machines Corporation.*)

	Data Width	Access/Speed/Rate	Comment
2365 Processor Storage	8 bytes	.75 microsecond storage cycle	All models
2361 Core Storage	8 bytes	8 microsecond storage cycle	All models
General registers	1 word	200 nanoseconds	16 General registers
Floating-point registers	2 words	200 nanoseconds word	4 Floating-point registers
Addressing adder	3 bytes	200 nanoseconds	
Parallel adder	8 bytes	200 nanoseconds	
Exponent adder	1 byte	200 nanoseconds	
Serial adder	1 byte	200 nanoseconds	
Basic machine cycle		200 nanoseconds	
2860 selector channel	1 byte	1.3 million bytes per second	8 bytes to storage
2870 Multiplexor channel	1 byte	110 kb to 450 kb	8 bytes to storage
Burst mode	1 byte	50–110 kb	
Multiplex mode	1 byte	50–110 kb	
Selector subchannel	1 byte	100 kb, each	

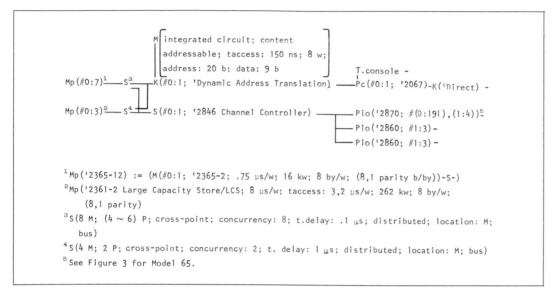

^{1}Mp('2365-12) := (M(#0:1; '2365-2; .75 μs/w; 16 kw; 8 by/w; (8,1 parity b/by))-S-)
^{2}Mp('2361-2 Large Capacity Store/LCS; 8 μs/w; taccess: 3,2 μs/w; 262 kw; 8 by/w;
      (8,1 parity)
^{3}S(8 M; (4 ~ 6) P; cross-point; concurrency: 8; t.delay: .1 μs; distributed; location: M;
      bus)
^{4}S(4 M; 2 P; cross-point; concurrency: 2; t. delay: 1 μs; distributed; location: M; bus)
^{5}See Figure 3 for Model 65.

Fig. 8. IBM System/360 Model 67 PMS diagram.

form a 24- or 32-bit virtual address in Pc into an actual 19- to 22-bit physical address in Mp. This K is not shown in Fig. 9 because it is not logically necessary. The scheme suggested in Fig. 9 uses control bits in the map to determine legal Mp accesses. In the Model 67 the storage key mechanism holds whether a given page can be accessed by a given numbered user (instead of associating the control with the mapping as shown in Fig. 9).

Second, the Model 67 is the first acknowledgment by IBM of multiprocessor computers, since it provides adequate switching to allow multiple Pc's. The C('65) multiprocessing configuration has been introduced based on Model 67 structure. Multiprocessors are necessary for reliability, not solely for performance reasons.

The PMS structure of C('67) in Fig. 8 does not have to use the S('Bus Control Unit/BCU),[1] as in the C('65). The C('67) can have an S in each Mp, so that four P's can communicate with an Mp, as shown in Fig. 8. Each Mp makes the decision about the P request to be honored next. Thus the problem of having an "all knowing" S('BCU) is solved by allowing each Mp to do local scheduling, rather than having a dialogue with another component (with time delays). The S('BCU) in a duplex C('67) is still present, but with less power, in the form of the S('2846

Channel Controller). It is used to arbitrate the Pio accesses to Mp.

Without multiprocessing, the Pc seems very badly mismatched with respect to Mp. Consider, for instance, the data rates on the C('67). From Fig. 8 its maximum possible Mp data rates are:

For 1 Mp('2365-12):

$$\frac{2 \times 64 \text{ bits}}{0.75 \text{ μs}} = 171 \text{ megabits/sec}$$

and for 1 Mp('2361 Large Core Store):

$$\frac{64 \text{ bits}}{8 \text{ μs}} = 8 \text{ megabits/sec}$$

Thus the total data rate is

$$171 \times 8 + 8 \times 4 = 1,368 + 32 \text{ megabits/sec}$$
$$= {\sim}1,400 \text{ megabits/sec}$$

The processing rate is approximately

$$\frac{64 \text{ bits}}{2.2 \text{ μs}} = 29 \text{ megabits/sec}$$

An Ms.drum rate is approximately

$$\frac{8b \times 1.2}{\text{μs}} = 10 \text{ megabits/sec}$$

[1]A system with only one port at Mp, controlled by BCU, is called a simplex. A system with multiport Mp is called a duplex.

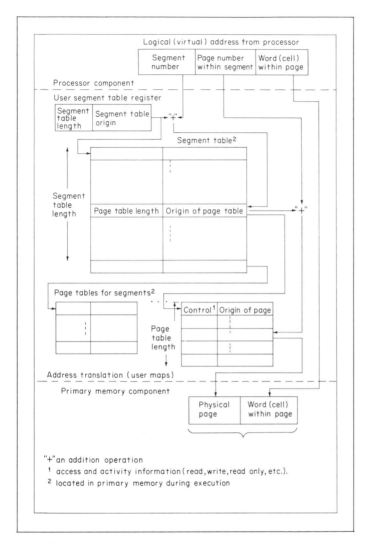

Fig. 9. Memory allocation using pages and segments.

Thus, for the several P's, an effective Mp request rate of 100 megabits/sec might be needed. The data-flow mismatch (between Mp and the P's) occurs because of the P's, the S (the L's connecting P and Mp), the lack of P's, and the fact that t.access = $\sim \frac{1}{2}$ t.cycle.

The Pio('2870), used in Model 65 and above, is described at two structural levels in Fig. 3. The Pio includes a large M.working to store the state of each of the logical Pio's. This Pio state includes the instruction location counter, the control state bits (active, running, interpreting an instruction, process-

ing data, etc.), and buffering (one 8-byte word). By having an M.buffer, the demands on Mp from the Pio's are reduced by a factor of 8. Although the expected data rate from many K's does not require the extra M, there are possible times when the uncertainty of the access times for Mp might cause data loss. Since the M.working is necessary to store the Pio state, the additional space for buffering is not expensive. An alternative design might use Mp for this buffering.

The four Pio('2860 Selector Channel)'s are implemented as independent Pio's, using conventional hardwired logic and buffering. However, they are packaged as one unit.

Model 85

The Model 85 was announced in February, 1968, with the goal of being the highest-performance Model 360 in production. The performance is $\sim(3 \sim 5)$ times the Model 65 and in some cases outperforms a Model 91 [Conti et al., 1968].

The PMS diagram of the Model 85 is shown in Fig. 10. The Pio, T, Ms structure is identical to that of Models 65 and 75 (Fig. 3). The two interesting aspects of the structure in Fig. 10 are the M(content addressable; 'Buffer Storage; 16|32 page; 1024 by/page) and the Pc. The pages are filled in groups of 64 bytes, as references to a particular physical block in Mp.core are made. Conti [1968] gives running times for various programs as a function of buffer memory size. Multiprogramming may degrade the performance more than any other case. This process, which has been referred to as "look aside," or a "slave memory," was suggested by Wilkes [1965]. It is completely analogous to the Model 67 M(content_addressable; 8 w) which is used to hold the segment-page map for a multiprogrammed time-sharing system. It is also analogous to a one-level storage system (Atlas; see Chap. 23) which is formed from two physical M's whose performance differs significantly. Here, the effect is to try to approximate a computer with a large Mp(80 ns/w) by using a large Mp(1 μs/w) and a small Mp(80 ns/w). The CDC 7600 (page 475) has a similar structure, but the Mp-Ms migration is under programmed control.

The P.microprogram used for controlling the Pc(K('Execution Unit)) allows for great flexibility in the definition of ISP's. An Mp(500 w) is available for the user; this may be loaded by a program, and it specifies an ISP. One standard option is to emulate the 704-7094 series.

The Model 85 removes the restriction of aligning words at particular boundaries. Thus any logical word, independent of its length, can be located at any physical location addressed in bytes.

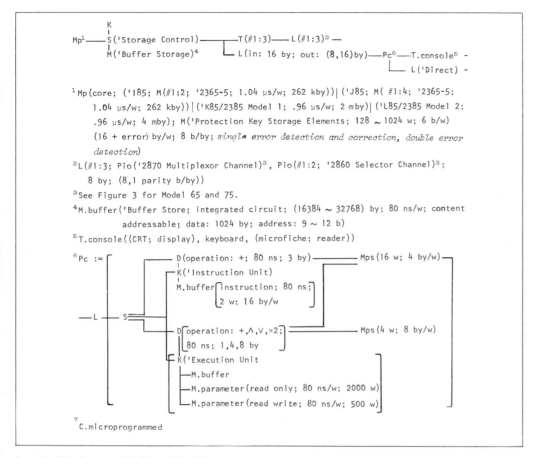

Fig. 10. IBM System/360 Model 85 PMS diagram.

The Pc's data operation performance is impressive. A fixed-point multiply is done in 0.4 μs, and a floating-point multiply takes 0.56 μs (not including accesses).

The data-type, extended floating-point number, is used in Model 85. Thus a 24-, 56-, or 112-bit fraction part can be used.

Model 91

This model has a very low cost/performance ratio (see Table 1). Only about 20 Model 91's were produced before it was withdrawn from the market. It has the highest performance of the series. The Mp is 0.75 μs, but 16 are overlapped to provide a theoretically maximum bandwidth of $16 \times 64/0.75 = 1,370$ megabits/s. About 2.5 mega-instructions/s are executed; thus, a total of 160 megabits/s of Mp are absorbed by Pc.

There are other interesting models in the '90 series; the Model 92 was a paper machine,[1] and the Model 95 was unannounced but produced, a version of the Model 91 with an Mp(integrated circuit; 60 ns/w; 8 by/w). The Model 91 is not covered in any detail here because of space limitations. It is similar to other very large computers in that many techniques are employed to obtain parallelism. The January, 1967, *IBM Journal of Research*[1] is devoted to design issues of the Model 91.

Models 1130 and 1800

These computers are presented as reference points and have nothing to do with the C('360). They are implemented outside the System/360 framework but use its technology, and so cost comparisons are still somewhat meaningful. These computers

---

[1]See bibliography at the end of this chapter.

are straightforward, and for a given task which does not use floating-point arithmetic, they should perform as well as any System/360 model. The arguments we use for the intermediate Pc for the Model 25 apply equally well here, too. Namely, why have such a complex ISP when simple ones will do just as well?

The programmed floating-point arithmetic times for a 4-μs 1800 and the "hardwired" (microprogrammed) System/360 Model 30 are compared in Table 2. We would expect the 2-μs 1800 to be better by a factor of 2. Note that the times are about the same for Model 30 and the slower 1800. The cost/performance is especially low with the 1130 (Table 1). In Chap. 33 we discuss the 1800. It is interesting to speculate why the 1130 and 1800 cannot be implemented within the System/360 framework. Are they "loss leaders"? Are they in response to more sophisticated, performance-oriented users?

The PMS structure of the controls, terminals, secondary memories, and special processors

There are many common components which attach to the C's (Figs. 11 to 17). Most of the components which attach to a Pio are not especially interesting, but they give an idea of the behavior and parameters. For example, the expression T('1403 Model 3; line; printer; 1100 line/min; 132 char/line; 8 bits/ character; 64 ~ 240 character set) pretty well describes a typical line printer. From the above description one can deduce the data rate of a T(line printer). It is 132 char/line × 1100 line/min × $\frac{1}{60}$ min/s × 8 b/char = 19.4 kb/s.

The channel-to-channel adapter control. The most interesting group of components (outside the C structures) are the special components shown in Fig. 11. The K('Channel to Channel Adapter) allows two P's, either on the same or a different C, to communicate with one another. This K is used in the con-

Table 2   IBM 1800 (4 μs) and IBM System/360 Model 30 floating-point arithmetic timing

	Operation times (μs)	
Operation	*1800 (4 μs)*	*System/360 Model 30*
+{sf}; +{df}	460; 440	75; 115
×{sf}; {df}	560; 790	320; 1060
÷{sf}	766	600
√ {f}	4500	2965
sin {f}	3000	3876
exponential {f}	2000	4173

```
—L(C(Pio))
 |
 K('Channel to Channel Adapter·
 |  used to transfer data among 2 C's)
—L(C(Pio))

a.  Interconnection of 2 computers (or within a computer)
    for transmission of Information

—L(S('Selector Channel:
 |  used in place of regular channel))
 P(block transfer; 'Storage to Storage Channel)

b.  Processor for the transmission of information (vectors)
    within Mp

—L(Pio)—┬—S—K(#A; '2903 Special Control Unit/SCU)-X¹
         |    |
         |    S
         |    |
         └—S—K(#B; 'SCU)-X¹

c.  Interconnection to other controls and computers

—L(S('Selector Channel, Models 44, 65, 75;
 |  used in place of regular Channel)
 P (array: '2938; microprogrammed: Mps(~ 64 w; 32 b/w):
     operations: (vector move, vector multiplication,
     vector inner product, sum of vector elements, sum of
     squares, convolution, difference equation, fixed float-
     ing conversion); data lengths; scalar, vector, matrix;
     data-types: fixed, floating)

d.  Array Processor


¹X := (C|K|T|Ms)
```

Fig. 11. IBM System/360 special P's and K's PMS diagrams.

struction of a dual C system or the N('Attached Support Processor/ASP) in Chap. 40, page 506. A C('40|'50) is attached to a C('65|'75). The C('40|50) is used as a Cio with file processing capabilities. The K has M.buffer. Data can flow in only one direction at a time.

The special control unit. The K('2903 Special Control Unit/SCU) consists of two independent K's which are physically packaged together and allow users to interface with the Pio's. Although it has not been discussed, the actual interconnection with a Pio, via the S(Pio; K), is via a physical I/O bus which is arranged in a bus (or chained) fashion. Such a single interface to handle a wide range of needs (high and low response and data rates) via a single set of electrical conductors requires a great deal of control information to be passed along the link. Therefore a K must have a great deal of knowledge of the dialogue in order to communicate. The hardware to attach to the I/O bus at a K is costly and must be designed carefully. The K('SCU) provides a rather simplified interface to the Pio. All I/O bus synchronization control, communication protocol control, buffering, and electrical isolation are within K('SCU). The K('SCU) is fairly flexible, in that devices connected to it can communicate with one another without Pio (see Fig. 11).

Storage-to-storage-channel processor. The P('Storage to Storage Channel) is a special processor which performs the sole function of transferring data blocks (a word vector) between one location in Mp to another in Mp. It qualifies as a P, since it takes an instruction from Mp containing the location and length, and once the instruction is executed, another is fetched and executed (if it exists). Thus the component has a well-defined interpretation cycle and set of operations. This P is useful in

a multiprogrammed environment requiring programs to be moved.

The 2938 array processor. The P.array('2938) is an extremely interesting special P (Fig. 11). It can be connected to Models 44, 65, or 75. It has a limited instruction repertoire, but the instructions it interprets are more complex than those in the ISP of the Pc. The instructions are algorithms for operating on an array (a vector or a matrix). These instructions include:

1 Vector move, similar to the P('Storage to Storage) described above, with conversion either way between fixed and floating point

2 An element-by-element vector sum

3 An element-by-element vector multiplication

4 A row-by-column vector inner product

5 A convolution multiply

6 The solution to a step in a difference equation

The P.array is microprogrammed, using an M(ROS), which

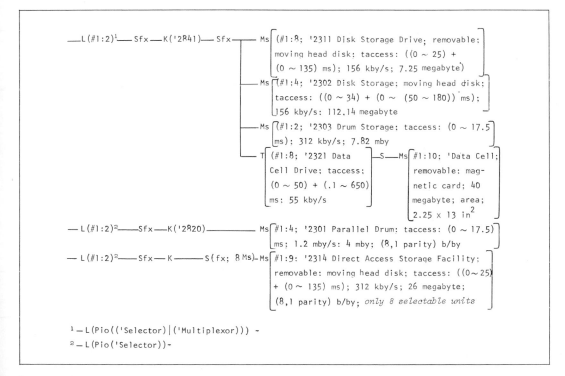

Fig. 12. IBM System/360 Ms(drum, disk, data cell) PMS diagrams.

```
  - L¹──K('2415)── Sfx ── Ms ⎡('2415; magnetic tape:                      ⎤ -
                              ⎢ 18.75 in/s; area: (.5 in x 1800 ft):      ⎥
                              ⎢(model: #; by/in; b/by): (                 ⎥
                              ⎢ (1; 1:2; 200,556,800; (6+1),(8+1)) |      ⎥
                              ⎢ (2; 1:4; 200,556,800; (6+1),(8+1)) |      ⎥
                              ⎢ (3; 1:6; 200,556,800; (6+1),(8+1)) |      ⎥
                              ⎢ (4; 1:2; 200,556,800,1600; (8+1)) |       ⎥
                              ⎢ (5; 1:4; 200,556,800,1600; (8+1)) |       ⎥
                              ⎣ (6; 1:6; 200,556,800,1600; (8+1)))        ⎦

  - L ── K('2802)── Sfx⁴── Ms ⎡#1:8; '7340-3 Hypertape;                   ⎤ -
                              ⎢addressable magnetic tape;                 ⎥
                              ⎢170|340 kby/s; 1511|3022 by/in;            ⎥
                              ⎢112.5 in/s; 1800 ft; (8,2 parity) b/by     ⎥
                              ⎣error correction                           ⎦
K('2403) := (
  - L ── K('2803)── Sfx⁴── Ms(#1; '2401² | '2402³ ) -
                          └─ Ms(#2:8; '2401² | '2402³ ) - )

K('2404) := (
  - L ── K('2804)── Sfx⁴── Ms(#1; '2401² | '2402³ ) -
                          └─ Ms(#2:8; '2401² | '2402³ ) - )

  - L(#1:2)── Sfx ── K('2803)── Sfx⁴── Ms⎡#1:8; '2401² | '2402³ ;⎤ -
                                          ⎣magnetic tape           ⎦

  - L(#1:2) ───── K⎡#1:2; ⎤─S⎡fx; ⎤Ms⎡#1:8; '2401² | '2402³ ;⎤ -
                    ⎣'2804⎦   ⎢in:1⎥  ⎣magnetic tape          ⎦
                              ⎣out:⎦


¹ - L(to:Pio('Selector|'Multiplexor)) -
² - Ms⎡'2401; magnetic tape; area:(.5 in x 1800 ft);⎤ -
     ⎢(model; in/s; by/in; b/by):(                  ⎥
     ⎢ (1; 37.5; 200,556,800; (6+1),(8+1)) |        ⎥
     ⎢ (2; 75; 200,556,800; (6+1),(8+1)) |          ⎥
     ⎢ (3; 112.5; 200,556,800; (6+1),(8+1)) |       ⎥
     ⎢ (4; 37.5; 200,556,800,1600:(8+1)) |          ⎥
     ⎢ (5; 75; 200,556,800,1600; (8+1)) |           ⎥
     ⎣ (6; 112.5; 200,556,800,1600; (8+1)))         ⎦
³ - Ms( 2402) := (Ms(#1:2; '2401; magnetic tape unit))
⁴Sfx := (S(fx; 1 K; 8 Ms)|S(fx; 2 K; 8 Ms; concurrency: 2)|
             S(fx; 4 K; 16 Ms; concurrency: 4))
```

Fig. 13. IBM System/360 Ms(magnetic tape) PMS diagrams.

makes it possible to construct complex algorithms in a flexible manner. The hardware logic is capable of doing a combined floating-point multiplication and addition in 200 nanoseconds. The impressive results this P achieves in the interpretation of the algorithms are principally because the time to access the algorithm has gone to zero. A measure we might apply to a P is the ratio of the time it spends fetching the algorithm's data to the total time it spends executing the algorithm. In a conventional computer Pc we suggest that a ratio of nearly ½ is very good. Two fetches are usually required—one for data, one for the instruction. This P has a ratio near one, as it is always accessing data (and rarely instructions).

Secondary-memory structure. Figures 12 and 13 present the Ms PMS structures. All the K's have an optional S, which can be placed between the K and the S(P;K) to allow two Pio's to access a common K (from either of two C's or two Pio's of the same C). The K('2841 Storage Control) is interesting only in being able to control a series of quite disparate devices, on a one-at-a-time basis.

Figure 13 presents all the M(s; magnetic tape)'s. The switch is interesting as it can be used for up to four K's to access simultaneously any of 16 M.tapes. (The vast array of very similar devices is due undoubtedly to marketing rather than production or engineering reasons.) It should be noted that there are two distinct M.tapes: conventional magnetic tape and Hypertape. Hypertape is explicitly addressed and has built-in error-correction coding.

Terminal structure. Figure 14 shows the T(cathode ray tube; display) and T(audio; output). There are terminals for writing and reading from photographic film (35 mm). The two approaches used for audio (vocal) output are noteworthy. One uses an M.drum to record a fixed vocabulary of words; the other uses an encoding mechanism to allow digital information stored in Mp to be transferred via the K('7772 Audio Response) to transforming a coded voice back to an audio output form. The S at the output of the T(audio) provides for audio signals to be switched on a word-by-word basis to any of several output telephone lines.

The structure of the vast array of printing devices that can attach to the C('360) is shown in Fig. 15. Some of the devices are interesting, such as the one that reads pencil-marked or typewritten paper. The main parameters of significance to PMS are the rate the device reads paper together with the kind of paper.

The T and K's which connect to external processes are given in Fig. 16. The K('1827) is used to connect with analog processes and is actually part of the IBM 1800 computer system (Chap. 33). The other K's are important, though not especially interesting, since they provide the K to T(Teletypes), K(telephone lines), and T(typewriters). The K('2701) and K('2702) are built to transform unsynchronized parallel data from the C into the synchronized serial form required by the telephone line. The K('2701) controls a small number of lines of high data rates; the K('2702) controls a large number of lines at low data

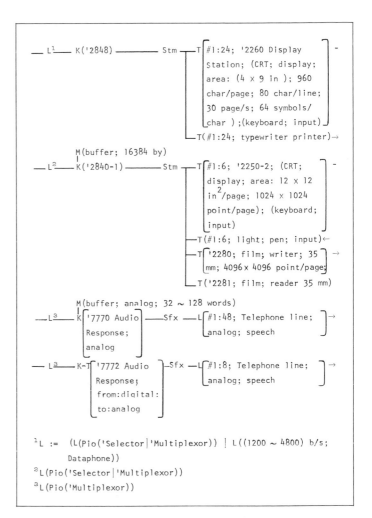

Fig. 14. IBM System/360 T(audio, display) PMS diagrams.

rates. The K('2702) is actually an array of up to 31 K's that are time-multiplexed, using an M.core to hold the state of each K.

Peripheral switching. For performance, communications, and reliability reasons it is necessary to provide access to K's, M's, or T's from several C's or Pio's. A sample structure of a possible configuration, using the above components, is given in Fig. 17. The PMS diagram also shows the physical structure of S(from:Pc; to:K).

Performance and costs

The System/360 series is perhaps the only group of computers for which a valid comparison of performance and cost can be

— L^1 ——KT('1442-N2; card; punch; 160 col/s)→

— L ——KT('1442-N1; card; (reader; 400 card/min), (punch; 160 col/s); half duplex)-

— L ——KT(card; reader, ('2501-B1; 600 card/min)|('2501-B2; 1000 card/min))←

— L ——KT('2520-B1; card; reader, punch; 500 card/min; half duplex)-

— L ——KT('2520; card; punch; ('model B2; 500 card/min)|('model B3; 300 card/min))→

— L ——K('2821)————T['2671-1; paper tape; reader; 1 kchar/s;]←
                        [5,6,7,8 b/char; area: ~1 x .1 in^2/char]

—L ——K('2821)—S(3T)—T[#1:3; '1403; line printer; *chain*:   →
                         ('Model; line/min; col/line): (2;
                         600; 132)|(3; 1100; 132)|(7; 600;
                         120)|(N1; 1100; 132; 48,96,144,192,
                         240 symbol/char;
                      —T['1404 Bill Feed; Printer Model 2;   →
                         600 lines/min; 132 col/line
                      —T['2540 card (reader; 1000 card/min),   ←
                         (punch; 300 card/min); full duplex

— L ————————————KT('1053; character; printer; 14.8 char/s)→

— L —KT['1231-N1; optical; pencil mark page; reader; area: (8.5 x 11) in^2/page;]←
         [1.8 s/page

— L —KT['1285; optical; printed character roll paper; reader; width: (.9375 ~ 3.5)]←
         [in; 22 char/col; 300 char/s

— L —KT['1287 Models 1 and 2; optical; reader; handprinted; roll, document:]←
         [area: (2.25 x 3 in^2)|(5.91 x 9 in^2)

— L —KT['1418, 1428 'Models 1,2,3; optical; typewritten character; reader; area:]←
         [(2.75 x 3.66 in^2)|(5.875 x 8.75 in^2)|(2.33 x 4.18 in^2)|(3 x 8.75 in^2);
         [288 ~ 420 documents/min

— L —KT('1445 Printer-N1; magnetic character line; printer; 190,240,525 lin/min)→

— L —KT[magnetic; character; reader; bank checks; ('1412; 950 document/min)|('1419;]←
         [1600 document/min)

^{1}L(Pio('Selector|'Multiplexor))

Fig. 15. IBM System/360 T(printer, reader, punch) PMS diagrams.

made. The models use essentially the same technology, implement the same ISP, and are probably constrained by a common corporate profit goal. Even here, as we noted earlier, comparisons are difficult to make.

In Table 3 we present the costs for various PMS component primitives. From this table, costs (relative to other components) can be obtained. These costs are expressed as dollars per second ($/s) to rent the equipment. They have been derived from the IBM monthly rental prices. The computer prices are based on estimates of minimum, average, and maximum configurations in the *Adams Computer Characteristics Quarterly* [Adams Associates]. The conversion factors are

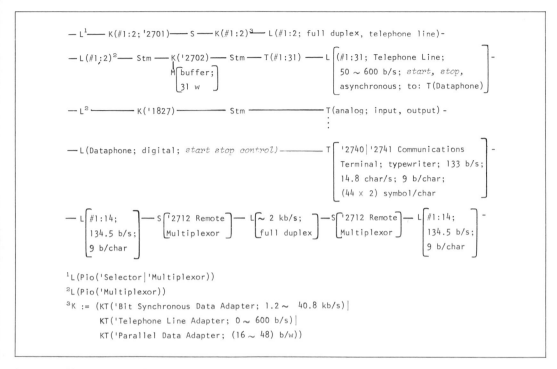

Fig. 16. IBM System/360 T(telephone line, analog, typewriter) PMS diagrams.

$/s = 1/[(173.3 \text{ hour/month}) \times 3,600 \text{ s/hour}]$
$= 1.6 \times 10^{-6}$ \$/month

$/month = 0.625 \times 10^6$ \$/s

The cost to buy, in dollars, is approximately

$ = 45 \times ($/month)

$ = 45 \times 0.625 \times 10^6 ($/s) = 2.82 \times 10^7 \times ($/s)

Table 1 is written as a single, large PMS expression, thus, the attributes are:

Pc(cost: ($/s|$)): = c.Pc : = cost of Pc alone

Mp(cost.avg) : = c.Mp.avg : = cost of average-size Mp for a model

C(cost.min:) : = c.C.min : = cost of minimum-size computer configuration

C(cost.avg:) : = c.C.avg : = cost of average-size computer configuration

Primary memory

The graph of Fig. 18 gives the Mp costs, c, (in \$/s) versus memory size (information/i). The line $i = 1.43 \times 10^7 \times c$ is

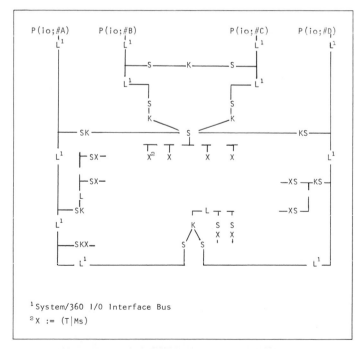

Fig. 17. IBM System/360 peripheral-switching PMS diagram.

Table 3   IBM System/360 component costs

Component	Cost ($/s)
Mp (core; cost: $/(kby × s))	
Mp ('Large Capacity Storage/LCS; cost: $/(kby × s))	
Pc ('20│25│30│40│44│50│65│67│ 75│85│91)	
P.array ('2938)	
Pio ('2860)	
Pio ('2870)	
Ms ('2415; magnetic tape) K ('2415)	
Ms ('2401; magnetic tape) K ('2803│2804)	
Ms ('7340 Hypertape) K ('2802)	
Ms ('2311; removable disk) K ('2814; #1:8)	
KMs ('2314; #1:9, removable disk)	
Ms ('2321 Data Cell) K ('2814; #1:8)	
Ms ('2303; drum) K ('2814; #1:8)	
Ms ('2301; drum) K ('2820)	
S ('2816; Ms.magnetic_tape; K)	
T ('2741; typewriter)	
T ('2260; display) K ('2848; #1:8, 16, 24)	
KT ('2250; display)	
T ('2761; paper tape; reader) K ('2822)	
KT ('7772/7770; audio)	
T ('1403/1404 line; printer) K ('2821; #1:3)	
KT ('1443│1445; line; printer)	
T ('2540; card; reader│punch) K ('2821; #1:3)	
KT ('1442│2501│2520; card; reader│punch)	
K ('2701 Data Adapter)	
K ('2702; typewriter; Teletype)	

Axis (Cost, $/s): 4   8   0.0001   2   4   8   0.001   2   4   8   0.01   2   4   8   0.1

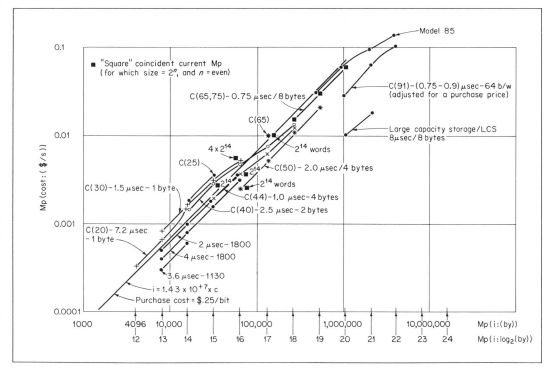

Fig. 18. Graph of IBM System/360 core-memory cost versus core-memory size.

plotted in terms of $/(by/s) and allows us to compute the purchase cost of a bit. The purchase cost of most Mp.core is $0.25/bit, according to the line. The 8-μs Large Capacity Storage/LCS cost is $0.032/bit. There appear to be slight cost savings for large Mp's and a significant saving for lower performance in the case of LCS, a factor of 8. A reasonable formula for Mp cost is: $c = (7 \times 10^6 \times i)/[t.cycle: (\mu s)]$. This formula would account for Model 50 Mp and LCS costs, but not Model 25 and 30 Mp costs. We really need an $i^{1/2}$ term in the formula to make a good fit (and also a constant). The value $i^{1/2}$ should be present, if purchase prices are related to manufacturing costs, because coincident current selection cost is inherently proportional to $i^{1/2}$.

An odd pricing point is the Model 44; it was developed after the other models and is either implemented better or priced differently. The anomalies in Mp('65; 2^{14} words), Mp('30; 2^{14} words), Mp('40; 2^{17} bytes), and Mp('44) are undoubtedly due to pricing-strategy differences. In the case of the Model 30 the incremental cost to increase the Mp size from 2^{13} to 2^{16} bytes is the addition of only a different core array (with no change

in electronics), at a small incremental manufacturing cost of goods.

The Mp size range within a model varies by a factor of 8 for Models 30, 40, 44, 50, 65, and 75, although by only a factor of 4 at the ends of the line (Models 20 and 91). The Mp implementation is usually a single common set of electronics to drive 2^{14} (16,384) words in a square or coincident-current-selection system of 2^7 by 2^7. These square points are indicated on the graph, and they should be the most economical memories. Smaller Mp's are implemented simply by using smaller core-memory arrays, but with the same basic electronic configuration, e.g., the Model 30 above. Larger Mp's are obtained by replicating the whole Mp system including the core array and the electronics.

An Mp size range of 8 for a given model presupposes a certain structuring of problems. That is, the models assume a fixed relationship between Pc capacity and Mp size requirements. An ideal system might let Pc power, Pc quantity, Mp power, and Mp size be completely variable. These parameters would all be selected independently to match the work load.

Central processors

The relative Pc powers (in 360 instructions/s) and costs are given in the graph of Fig. 19 and in Table 1. The most significant fact from the graph is that the cost/power ratio is roughly constant for each of the Pc's (especially if we ignore Model 44 and Model 50). Figure 19 gives the relative computing power versus cost for various configurations. Table 1 also shows a number of relationships. One interesting relationship (Table 1) is the ratio of actual Pc power to maximum possible Pc power for a model. This can be based on Mp utilization:

$$\frac{\text{Actual Pc power}}{\text{Maximum Pc power}} = \sim \frac{\text{Mp cycles utilized by Pc}}{\text{Mp cycles available}}$$

This ratio must be less than 1 unless there are many Pc's or a single Pc has more power than Mp. In every case, the Pc is far from fully utilizing the Mp. The technique of buffering instructions in a local Pc memory can increase this ratio to be >1 (although no computers ever do so). In the higher model numbers the utilization is low because a large number of cycles have to be available in order to avoid conflicts when a given cycle is requested—using an Mp with a long t.cycle. In the case of Model 25, the cycles are lost because the microprogram is being executed from Mp. (A ratio of 0.045 indicates 21 cycles are used for microprograms to every 1 of program.)

In the case of the Model 30 the power is limited by holding the general registers in Mp. For example, by using an additional fast M to hold the general registers and working data, the Pc power could increase. Unfortunately, such a change might cause the cost of other parts of the system to be increased, so that it would not be just a simple incremental addition. The C('30) performs well for the field-scan problem [Solomon, 1966] (see Table 1). The data structure for the field-scan problem coincides with the 1-byte Mp organization. C('65) and C('75) perform the worst for field scan because of the mismatch between Mp organization (8 bytes) and program data (1 byte).

C('65) and C('75) have the same Mp structure and hence have the same potential power available from Mp. In the case

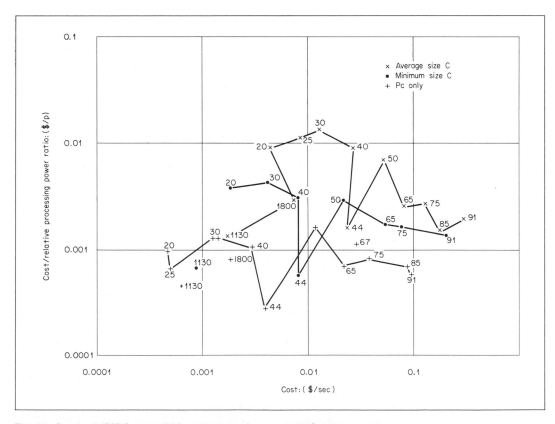

Fig. 19. Graph of IBM System/360 cost/processing power ratio versus cost.

of C('75) the power of the Mp is more nearly utilized. Unfortunately for the more complex Mp structures, which have more potential Mp cycles, the Pc is not able to utilize them. The C('65) and C('75) have several registers concerned with obtaining the next instruction and holding it for execution while other instructions are obtained (look-ahead). The hardwired Model 75 Pc may account for the improvement over the Model 65 P.microprogrammed.

The performance of C('20) is inaccurately high since it is a limited subset of the 360 ISP. (C('20) does not have floating-point or fixed-point multiply and divide instructions, and it has only eight 16-bit general registers.) The hardwired Model 44 has a better cost/power characteristic than any of the other C's, by any measured criteria (see Fig. 19). In the case of the Model 44, the Pc price also includes Ms.disk. Perhaps the Model 44, designed initially for real-time scientific problem solving, is priced more competitively with similar machines (DEC PDP-10 and SDS Sigma 5, 7), whereas the other models compete in a performance-insensitive, competition-free market for general-purpose business data processing. Thus its anomalous position may be due to external market pressures and not manufacturing cost.

The design of the IBM System/360 models is undoubtedly predicated on the basis that performance or computing power is proportional to the cost raised to some power, g, greater than 1: power = $k \times$ cost^g; where $g > 1$.[1] Almost all models follow the above relationship with $g > 1$. When $g > 1$ there is an advantage to have large configurations since the cost/computation will decrease. If $g \leq 1$, then an alternative implementation for the 360 C's would simply use multiple C's or Pc's to obtain the same power. Unfortunately, such an approach does not provide for the interconnection of the components to function as a single unit. In many cases a single task cannot be broken into a number of parallel and independent subtasks. If the performance for the system varied by a factor of 100, then 100 Pc's or C's would be placed together. From Table 1 we see a power range of about 314 corresponds to a cost range of 65 to 114 (which tells us $g < 2$).

The following discussion takes computing power to be measured by instructions per second and Mp (size; t.cycle). Costs are measured in dollars per second of rental time. The graph (Fig. 20) shows the relationship to computing power p and costs. The power (actually p.Pc) is taken from the measures of instruction times for certain fixed work. Solomon ob-

served Grosch's law to hold for Models 30, 40, 50, 65, and 75. This line is drawn in Fig. 20 for C(cost.average). Considering Models 20, 25, 44, 85, and 91, a line with a less steep slope might fit the points better. If we consider C(cost.minimum), $g < 2$; considering only Pc, a $g = 1$ might be appropriate (see Fig. 20) in which the power/cost is essentially constant with cost.

Pc(cost)/Mp(cost.avg) : = c.Pc/c.avg.Mp = $\sim$ 1.1, the ratio of processor to memory cost

C(cost.min)/C(cost.avg) : = c.min.C/c.avg.C = $\sim$ 0.47, the ratio of the smallest computer configuration to an average configuration

Pc(cost)/C(cost.avg) : = c.Pc/c.avg.C = $\sim$ 0.23, the ratio of processor to computer cost

These are averages over all the series and can be rather misleading. For example, in higher-numbered models the C(cost.min)/C(cost.avg) : = c.min.C/c.avg.C is about 0.6. whereas in lower-numbered models the ratio is 0.3. We might have expected this, since it indicates that a higher proportion of system cost is in Ms and T on lower-number models.

An alternative computer series based on multiprocessing

In this section we suggest an alternative design providing a wide range of computing power but using multiprocessing. That is, rather than building a higher-performance model, we would have multiple lower-performance models. On the surface, this appears feasible only if the cost of the processor is a relatively small part of the computer, and if for a particular configuration there are memory cycles available in the system (so that a more costly memory system is not required). It is also desirable that the proposed multiprocessor configurations have rather large Mp's so that it can be assumed there will be several jobs in Mp waiting to run; i.e., we should be able to multiprogram rather than do parallel processing. These conditions are satisfied with the System/360 models. Although we do not address the question of development cost, it is clear that a multiprocessor system would have a lower development cost because fewer processors would be required. Within IBM we can assume that the development cost tends to go to zero because of the large production; unfortunately, even for IBM, the training cost for servicemen and salesmen does not go to zero but is proportional to the number of products. Thus, we would anticipate savings by having a smaller line.

The multiprocessor view is presented in Table 4; namely, we suggest dropping Models 20, 30, 40, 50, 65, 75, 85, and 91.

[1] Herb Grosch [Grosch, 1953] first noted this relationship and estimated g to be 2; thus we use g for this exponent. Adams suggested $g = \frac{1}{2}$ [Adams, 1962]. See also The Economics of Computers [Sharpe, 1969].

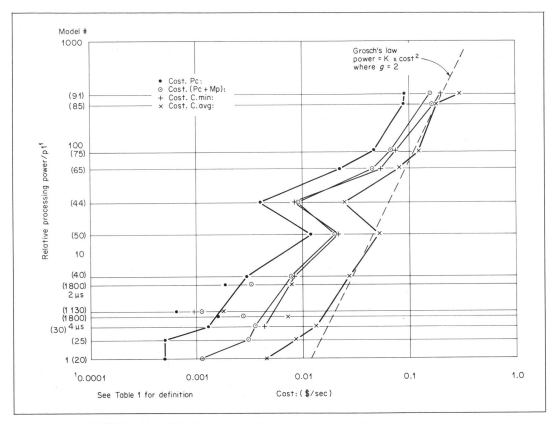

Fig. 20. Graph of IBM System/360 relative processing power versus cost.

These would be replaced with only Models 25 and 44. Note there are Pc's in Table 4 (other than 25 and 44) which when multiprocessed can perform better for lower cost, e.g., 2 Model 65's are >1 Model 75, for about the same cost. Admittedly there are major problems in multiprocessing with 11 Pc's, but other existence proofs [Anderson, 1961] have shown that two to four Pc's can be effective (Chap. 36). If we ignore Models 85 and 91, the worst case is for a maximum of four Pc's needed to obtain the power of model 40. Note that in the above cases the processor cost is about one-half the cost of a single Pc. This factor of 2 might be used to answer critics of the scheme. The reasons against the scheme are: There have to be good switches between Mp and Pc's; there has to be communication among the Pc's (which is about the same as what the Pc-Pio communication should be); and there has to be knowledge of the program environment to split tasks apart to run in parallel.

A less radical suggestion is also presented in Table 4: namely, examining the number of processor models which can be used to provide processing power for the next highest model.

Actually, if we carry this view further and were forced to build such a system, the view that the ideal machines are the Model 25 and 44 would undoubtedly change. Model 25 and 44 exist and can be used for the argument. The reader should note that there is a major flaw in our argument using a Model 25. The microprogrammed Model 25 Pc cost should include a 16-kby memory for the microprogram (actually one Mp should be included for each Pc to avoid memory-request conflict). Alternatively, if we use the Model 25 directly without a microprogram, we would lose performance range. With our present knowledge of multiprocessors, a responsible engineer would hardly suggest building a multiprocessor system with 11 processors as a sure-fire money-making venture. A more reasonable alternative would be to use the multiprocessor Model 75 as an alternative to Models 85 and 91. A reasonably safe alternative would be three basic processors and a four-processor multiprocessor structure. For a power range of 320:1, then the processors could be 1, 20, 80, giving powers of 1, 2, 3, 4, 20, 40, 60, 80, 160, 240, 320. This structure would leave a gap of a factor of

Table 4   IBM System/360 Pc (power: cost) and an alternative design based on multiprocessors

Given			Proposed multiprocessor alternatives			
Pc.model	Pc.power	Pc.cost	Quantity.Pc	Pc.model	Pc.power	Pc.cost
20	1	0.00049	1	25	1.5	0.0005
25	1.5	0.00050	1	25	1.5	0.0005
30	2	0.0013	2	25	3	0.001
			2	20	2	0.00098
40	6	0.003	4	25	6	0.002
			6	20	6	0.00294
44	30	0.0041	1	44	30	0.0041
50	15	0.012	1	44	30	0.0041
65	63	0.022	2	44	60	0.0082
75	92	0.037	3	44	90	0.012
			2	65	126	0.044
85	252	0.087	8	44	240	0.033
91	314	0.091	11	44	330	0.045

5 between a 4 × 1 power processor and 20 power processor. The largest gap in the System/360 is a factor of 3 between Models 30 and 40.

Conclusions

The IBM System/360, by achieving a production record, has fulfilled its principal design objective. The technical goals, however, are of interest to us here. The most interesting aspect of the design is achieving a performance range of 314 to 1 over a series of models, with a primary-memory size range of 2,048 to 1 for various computer configurations. Thus a user is given a very large set of configuration alternatives. The SLT technology, though not integrated-circuit, is certainly of the third generation. Using SLT the fabrication of the models is superb.

There is a vast array of secondary-memory and terminal devices to couple with almost any other system. The System/360 is the first computer to make extensive use of microprogramming. Microprogramming is used for the definition of the System/360 instruction-set processor, but, more important, microprograms define previous IBM computers so that a user can operate satisfactorily during the interim period when older programs are being updated to use the System/360. There are provisions for multicomputer structures. Within a single computer structure there is adequate means of peripheral switching so that reliable and high-performance structures can be assembled. Early structures do not provide multiprocessing; we have suggested multiprocessing as a technique to achieve the same performance-range objectives. The io processor, though rather elaborate, provides a certain commonality.

The instruction-set processor for the System/360, based on a general-registers structure, appears to be overly complex, yet incomplete, because there are so many data types. The addressing mechanism and lack of multiprogramming ability make the System/360 a hard machine to appreciate fully. Although we praise microprogramming as a means of accomplishing compatibility with the past, it appears to stand in the way of getting the most performance from the hardware. Perhaps of most significance, the System/360 may have a greater lifetime than any past computer.

Selected Bibliography

Architecture and logical structure: AmdaG64a (TeagH65)[1], BlaaG64a[2], BlaaG64b[2]; General implementations: AmdaG64b[2], CartW64, PadeA64[2], StevW64[2]; Microprogramming: GreeJ64, TuckS67, WebeH67; Formal description of Pc[5]; FalkA64[2]; Performance and reviews: HillJ66, SoloM66; Model 40 modifications for multiprogramming: LindA66; Model 67: ArdeB66, FikeR68, GibsC66, LaueH67; Model 85: ContC68[3], LiptJ68[3], PadeA68[3]; Model 91 architecture and technology: AndeD67[4], AndeS67[4], BolaL67[4], FlynM67[4]a, LangJ67[3], LloyR67[4], SechR67[4], TomaR67[4]; Model 92 (proposed): ContC64 (GrimR65a), AmdaG64c (GrimR65b), ChenT64 (GrimR65c); Serviceability: CartW64; Other references: AdamC62, CorbF62, GrosH53, SharW69, WilkM65; IBM reference manuals: IBM System/360 Functional characteristics manuals for each model, IBM System/360 Configurator (diagram) for each model, A22-6821-4 IBM System/360 Principles of Operation, A22-6810-8 IBM System/360 System Summary

[1]( ) denotes the review of previous article.
[2] *IBM Systems Journal*, vol. 3, nos. 2 and 3, 1964.
[3] *IBM Systems Journal*, vol. 7, no. 1, 1968.
[4] *IBM Journal of Research and Development*, vol. 11, no. 1, January, 1967.
[5] Given in A Programming Language/APL [Iverson, 1962].

Chapter 43

The structure of SYSTEM/360[1]

Part I—Outline of the logical structure

G. A. Blaauw / F. P. Brooks, Jr.

Summary A general introductory description of the logical structure of SYSTEM/360 is given. In addition, the functional units, the principal registers and formats, and the basic addressing and sequencing principles of the system are indicated.

In the SYSTEM/360 logical structure, processing efficiency and versatility are served by multiple accumulators, binary addressing, bit-manipulation operations, automatic indexing, fixed and variable field lengths, decimal and hexadecimal radices, and floating-point as well as fixed-point arithmetic. The provisions for program interruption, storage protection, and flexible CPU states contribute to effective operation. Base-register addressing, the standard interface between channels and input/output control units, and the machine-language compatibility among models contribute to flexible configurations and to orderly system expansion.

SYSTEM 360 is distinguished by a design orientation toward very large memories and a hierarchy of memory speeds, a broad spectrum of manipulative functions, and a uniform treatment of input/output functions that facilitates communication with a diversity of input/output devices. The overall structure lends itself to program-compatible embodiments over a wide range of performance levels.

The system, designed for operation with a supervisory program, has comprehensive facilities for storage protection, program relocation, nonstop operation, and program interruption. Privileged instructions associated with a supervisory operating state are included. The supervisory program schedules and governs the execution of multiple programs, handles exceptional conditions, and coordinates and issues input/output (I/O) instructions. Reliability is heightened by supplementing solid-state components with built-in checking and diagnostic aids. Interconnection facilities permit a wide variety of possibilities for multisystem operation.

The purpose of this discussion is to introduce the functional units of the system, as well as formats, codes, and conventions essential to characterization of the system.

[1] *IBM Sys. J*, vol. 3, no. 2, pp. 119–135, 1964.

Functional structure

The SYSTEM/360 structure schematically outlined in Fig. 1 has seven announced embodiments. Six of these, namely, Models 30, 40, 50, 60, 62, and 70, will be treated here.[1] Where requisite I/O devices, optional features, and storage capacity are present, these six models are logically identical for valid programs that contain explicit time dependencies only. Hence, even though the allowable channels or storage capacity may vary from model to model (as discussed in Chap. 44), the logical structure can be discussed without reference to specific models.

Input/output

Direct communication with a large number of low-speed terminals and other I/O devices is provided through a special *multiplexor* channel unit. Communication with high-speed I/O devices is accommodated by the *selector* channel units. Conceptually, the input/output system acts as a set of subchannels that operate concurrently with one another and the processing unit. Each subchannel, instructed by its own control-word sequence, can govern a data transfer operation between storage and a selected I/O device. A multiplexor channel can function either as one or as many subchannels; a selector channel always functions as a single subchannel. The control unit of each I/O device attaches to the channels via a standard mechanical-electrical-programming *interface*.

Processing

The processing unit has sixteen general purpose 32-bit registers used for addressing, indexing, and accumulating. Four 64-bit floating-point accumulators are optionally available. The inclusion of multiple registers permits effective use to be made of small high-speed memories. Four distinct types of processing are pro-

[1] A seventh embodiment, the Model 92, is not discussed in this paper. This model does not provide decimal data handling and has a few minor differences arising from its highly concurrent, speed-oriented organization. A paper on Model 92 is planned for future publication in the *IBM Systems Journal*.

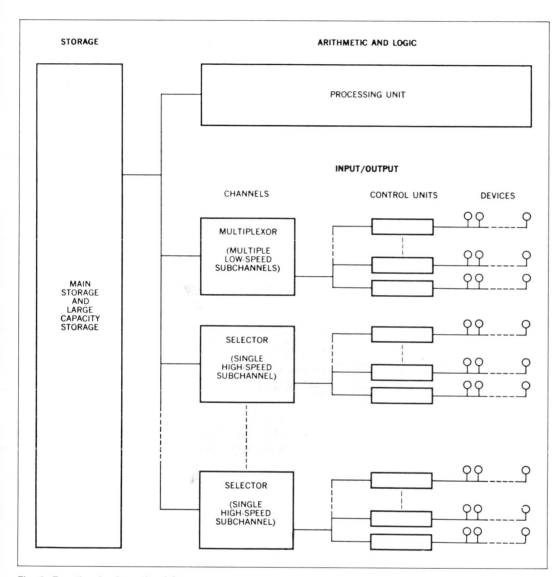

Fig. 1. Functional schematic of System/360.

vided: logical manipulation of individual bits, character strings and fixed words; decimal arithmetic on digit strings; fixed-point binary arithmetic; and floating-point arithmetic. The processing unit, together with the central control function, will be referred to as the central processing unit (CPU). The basic registers and data paths of the CPU are shown in Fig. 2.

The CPU's of the various models yield a substantial range in performance. Relative to the smallest model (Model 30), the internal performance of the largest (Model 70) is approximately 50:1 for scientific computation and 15:1 for commercial data processing.

Control

Because of the extensive instruction set, SYSTEM/360 control is more elaborate than in conventional computers. Control functions include internal sequencing of each operation; sequencing from instruction to instruction (with branching and interruption); governing of many I/O transfers; and the monitoring, signaling, timing, and storage protection essential to total system operation. The control equipment is combined with a programmed supervisor, which coordinates and issues all I/O instructions, handles excep-

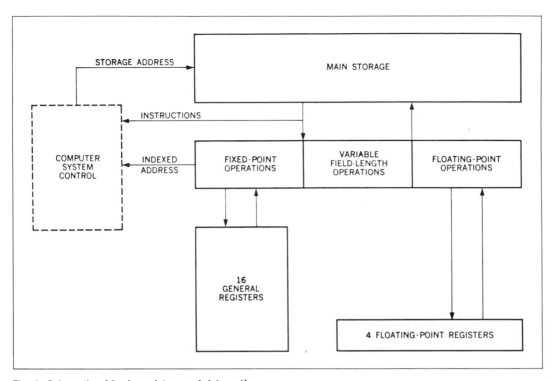

Fig. 2. Schematic of basic registers and data paths.

tional conditions, loads and relocates programs and data, manages storage, and supervises scheduling and execution of multiple programs. To a problem programmer, the supervisory program and the control equipment are indistinguishable.

The functional structure of SYSTEM/360, like that of most computers, is most concisely described by considering the data formats, the types of manipulations performed on them, and the instruction formats by which these manipulations are specified.

Information formats

The several SYSTEM/360 data formats are shown in Fig. 3. An 8-bit unit of information is fundamental to most of the formats. A consecutive group of *n* such units constitutes a *field of length n*. Fixed-length fields of length one, two, four, and eight are termed *bytes, halfwords, words,* and *double words,* respectively. In many instructions, the operation code implies one of these four fields as the length of the operands. On the other hand, the length is explicit in an instruction that refers to operands of variable length.

The location of a stored field is specified by the address of the leftmost byte of the field. Variable-length fields may start on any byte location, but a fixed-length field of two, four, or eight bytes

must have an address that is a multiple of 2, 4, or 8, respectively. Some of the various alignment possibilities are apparent from Fig. 3.

Storage addresses are represented by binary integers in the system. Storage capacities are always expressed as numbers of bytes.

Processing operations

The SYSTEM/360 operations fall into four classes: fixed-point arithmetic, floating-point arithmetic, logical operations, and decimal arithmetic. These classes differ in the data formats used, the registers involved, the operations provided, and the way the field length is stated.

Fixed-point arithmetic

The basic arithmetic operand is the 32-bit fixed-point binary word. Halfword operands may be specified in most operations for the sake of improved speed or storage utilization. Some products and all dividends are 64 bits long, using an even-odd register pair.

Because the 32-bit words accommodate the 24-bit address, the entire fixed-point instruction set, including multiplication, division,

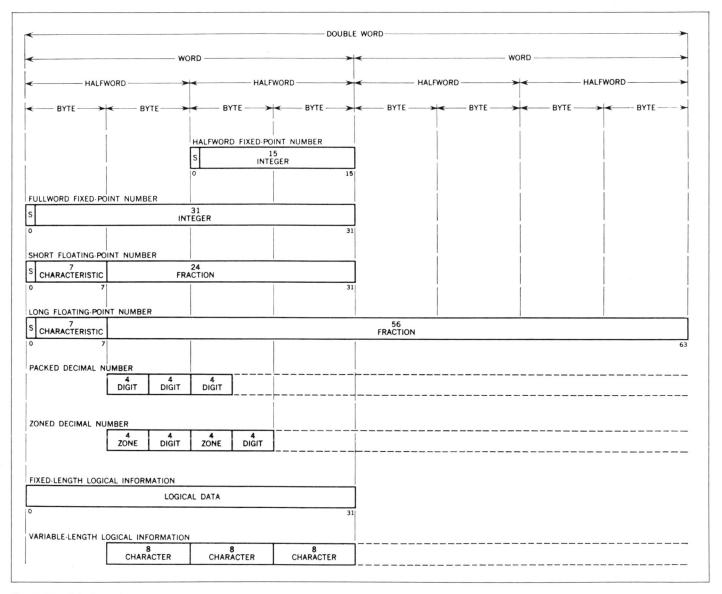

Fig. 3. The data formats.

shifting, and several logical operations, can be used in address computation. A two's complement notation is used for fixed-point operands.

Additions, subtractions, multiplications, divisions, and comparisons take one operand from a register and another from either a register or storage. Multiple-precision arithmetic is made convenient by the two's complement notation and by recognition of the carry from one word to another. A pair of conversion instruc-

tions, CONVERT TO BINARY and CONVERT TO DECIMAL, provide transition between decimal and binary radices without the use of tables. Multiple-register loading and storing instructions facilitate subroutine switching.

Floating-point arithmetic

Floating-point numbers may occur in either of two fixed-length formats—short or long. These formats differ only in the length of the fractions, as indicated in Fig. 3. The fraction of a floating-point number is expressed in 4-bit hexadecimal (base 16) digits. In the short format, the fraction has six hexadecimal digits; in the long format, the fraction has 14 hexadecimal digits. The short length is equivalent to seven decimal places of precision. The long length gives up to 17 decimal places of precision, thus eliminating most requirements for double-precision arithmetic.

The radix point of the fraction is assumed to be immediately to the left of the high-order fraction digit. To provide the proper magnitude for the floating-point number, the fraction is considered to be multiplied by a power of 16. The characteristic portion, bits 1 through 7 of both formats, is used to indicate this power. The characteristic is treated as an excess 64 number with a range from -64 through $+63$, and permits representation of decimal numbers with magnitudes in the range of 10^{-78} to 10^{75}.

Bit position 0 in either format is the fraction sign, S. The fraction of negative numbers is carried in true form.

Floating-point operations are performed with one operand from a register and another from either a register or storage. The result, placed in a register, is generally of the same length as the operands.

Logical operations

Operations for comparison, translation, editing, bit testing, and bit setting are provided for processing logical fields of fixed and variable lengths. Fixed-length logical operands, which consist of one, four, or eight bytes, are processed from the general registers.

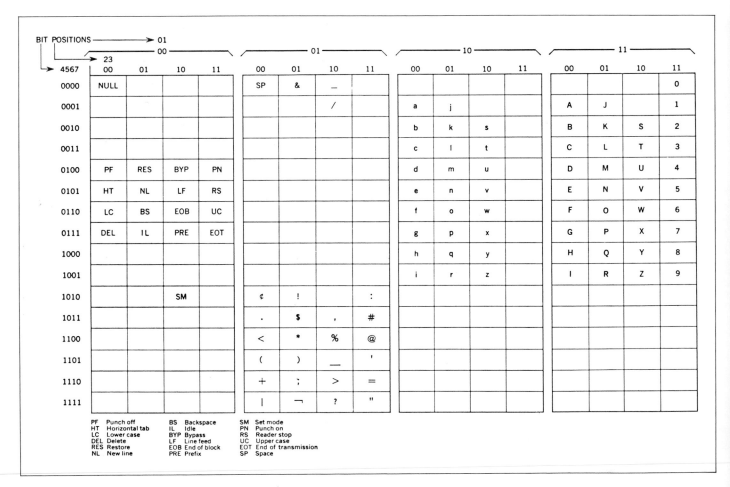

4567	00·00	00·01	00·10	00·11	01·00	01·01	01·10	01·11	10·00	10·01	10·10	10·11	11·00	11·01	11·10	11·11
0000	NULL				SP	&	_									0
0001							/		a	j			A	J		1
0010									b	k	s		B	K	S	2
0011									c	l	t		C	L	T	3
0100	PF	RES	BYP	PN					d	m	u		D	M	U	4
0101	HT	NL	LF	RS					e	n	v		E	N	V	5
0110	LC	BS	EOB	UC					f	o	w		F	O	W	6
0111	DEL	IL	PRE	EOT					g	p	x		G	P	X	7
1000									h	q	y		H	Q	Y	8
1001									i	r	z		I	R	Z	9
1010			SM		¢	!		:								
1011					.	$	,	#								
1100					<	*	%	@								
1101					(	)	_	'								
1110					+	;	>	=								
1111					|	¬	?	"								

PF	Punch off	BS	Backspace	SM	Set mode
HT	Horizontal tab	IL	Idle	PN	Punch on
LC	Lower case	BYP	Bypass	RS	Reader stop
DEL	Delete	LF	Line feed	UC	Upper case
RES	Restore	EOB	End of block	EOT	End of transmission
NL	New line	PRE	Prefix	SP	Space

Fig. 4. Extended binary-coded-decimal interchange code.

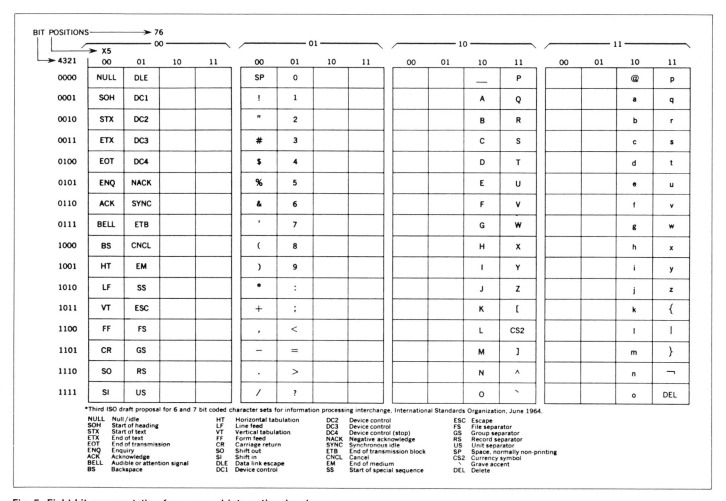

Fig. 5. Eight-bit representation for proposed international code.

Logical operations can also be performed on fields of up to 256 bytes, in which case the fields are processed from left to right, one byte at a time. Moreover, two powerful scanning instructions permit byte-by-byte translation and testing via tables. An important special case of variable-length logical operations is the one-byte field, whose individual bits can be tested, set, reset, and inverted as specified by an 8-bit mask in the instruction.

Character codes

Any 8-bit character set can be processed, although certain restrictions are assumed in the decimal arithmetic and editing operations. However, all character-set-sensitive I/O equipment assumes either the Extended Binary-Coded-Decimal Interchange Code (EBCDIC) of Fig. 4 or the code of Fig. 5, which is an eight-bit extension of a seven-bit code proposed by the International Standards Organization.

Decimal arithmetic

Decimal arithmetic can improve performance for processes requiring few computational steps per datum between the source input and the output. In these cases, where radix conversion from decimal to binary and back to decimal is not justified, the use of registers for intermediate results usually yields no advantage over storage-to-storage processing. Hence, decimal arithmetic is provided in SYSTEM/360 with operands as well as results located in storage, as in the IBM 1400 series. Decimal arithmetic includes addition, subtraction, multiplication, division, and comparison.

The decimal digits 0 through 9 are represented in the 4-bit binary-coded-decimal form by 0000 through 1001, respectively. The patterns 1010 through 1111 are not valid as digits and are interpreted as sign codes: 1011 and 1101 represent a minus, the other four a plus. The sign patterns generated in decimal arithmetic depend upon the character set preferred. For EBCDIC, the patterns are 1100 and 1101; for the code of Fig. 5, they are 1010 and 1011. The choice between the two codes is determined by a mode bit.

Decimal digits, packed two to a byte, appear in fields of variable length (from 1 to 16 bytes) and are accompanied by a sign in the rightmost four bits of the low-order byte. Operand fields can be located on any byte boundary, and can have lengths up to 31 digits and sign. Operands participating in an operation have independent lengths. Negative numbers are carried in true form. Instructions are provided for packing and unpacking decimal numbers. Packing of digits leads to efficient use of storage, increased arithmetic performance, and improved rates of data transmission. For purely decimal fields, for example, a 90,000-byte/second tape drive reads and writes 180,000 digits/second.

Instruction formats

Instruction formats contain one, two, or three halfwords, depending upon the number of storage addresses necessary for the operation. If no storage address is required of an instruction, one halfword suffices. A two-halfword instruction specifies one address; a three-halfword instruction specifies two addresses. All instructions must be aligned on halfword boundaries.

The five basic instruction formats, denoted by the format mnemonics RR, RX, RS, SI, and SS are shown in Fig. 6. RR denotes a register-to-register operation, RX a register and indexed-storage operation, RS a register and storage operation, SI a storage and immediate-operand operation, and SS a storage-to-storage operation.

In each format, the first instruction halfword consists of two parts. The first byte contains the operation code. The length and format of an instruction are indicated by the first two bits of the operation code.

The second byte is used either as two 4-bit fields or as a single 8-bit field. This byte is specified from among the following:

Four-bit operand register designator (R)

Four-bit index register designator (X)

Four-bit mask (M)

Four-bit field length specification (L)

Eight-bit field length specification

Eight-bit byte of immediate data (I)

The second and third halfwords each specify a 4-bit base register designator (B), followed by a 12-bit displacement (D).

Addressing

An effective storage address E is a 24-bit binary integer given, in the typical case, by

$$E = B + X + D$$

where B and X are 24-bit integers from general registers identified by fields B and X, respectively, and the displacement D is a 12-bit integer contained in every instruction that references storage.

The base B can be used for static relocation of programs and data. In record processing, the base can identify a record; in array calculations, it can specify the location of an array. The index X can provide the relative address of an element within an array. Together, B and X permit double indexing in array processing.

The displacement provides for relative addressing of up to 4095 bytes beyond the element or base address. In array calculations, the displacement can identify one of many items associated with an element. Thus, multiple arrays whose indices move together are best stored in an interleaved manner. In the processing of records, the displacement can identify items within a record.

In forming an effective address, the base and index are treated as unsigned 24-bit positive binary integers and the displacement as a 12-bit positive binary integer. The three are added as 24-bit binary numbers, ignoring overflow. Since every address is formed with the aid of a base, programs can be readily and generally relocated by changing the contents of base registers.

A zero base or index designator implies that a zero quantity must be used in forming the address, regardless of the contents of general register 0. A displacement of zero has no special significance. Initialization, modification, and testing of bases and indices can be carried out by fixed-point instructions, or by BRANCH AND LINK, BRANCH ON COUNT, or BRANCH ON INDEX instructions. LOAD EFFECTIVE ADDRESS provides not only a convenient housekeeping operation, but also, when the same register is specified for result and operand, an immediate register-incrementing operation.

Sequencing

Normally, the CPU takes instructions in sequence. After an instruction is fetched from a location specified by the instruction

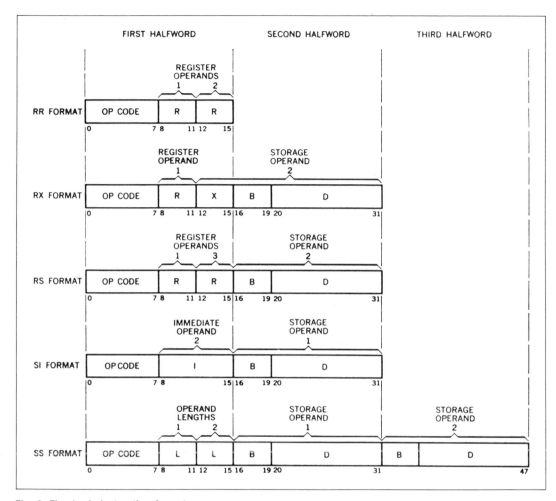

Fig. 6. Five basic instruction formats.

counter, the instruction counter is increased by the number of bytes in the instruction.

Conceptually, all halfwords of an instruction are fetched from storage after the preceding operation is completed and before execution of the current operation, even though physical storage word size and overlap of instruction execution with storage access may cause the actual instruction fetching to be different. Thus, an instruction can be modified by the instruction that immediately precedes it in the instruction stream, and cannot effectively modify itself during execution.

Branching

Most branching is accomplished by a single BRANCH ON CONDITION operation that inspects a 2-bit *condition register.* Many

of the arithmetic, logical, and I/O operations indicate an outcome by setting the condition register to one of its four possible states. Subsequently a conditional branch can select one of the states as a criterion for branching. For example, the condition code reflects such conditions as non-zero result, first operand high, operands equal, overflow, channel busy, zero, etc. Once set, the condition register remains unchanged until modified by an instruction execution that reflects a different condition code.

The outcome of address arithmetic and counting operations can be tested by a conditional branch to effect loop control. Two instructions, BRANCH ON COUNT and BRANCH ON INDEX, provide for one-instruction execution of the most common arithmetic-test combinations.

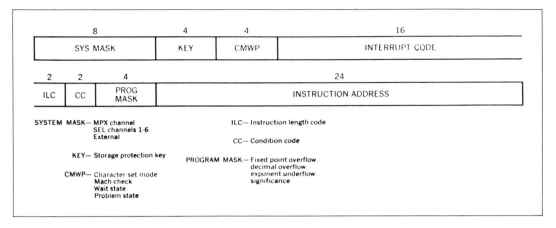

Fig. 7. Program status word format.

Program status word

A program status word (PSW), a double word having the format shown in Fig. 7, contains information required for proper execution of a given program. A PSW includes an instruction address, condition code, and several mask and mode fields. The active or controlling PSW is called the *current* PSW. By storing the current PSW during an interruption, the status of the interrupted program is preserved.

Interruption

Five classes of interruption conditions are distinguished: input/output, program, supervisor call, external, and machine check.

For each class, two PSW's, called *old* and *new*, are maintained in the main-storage locations shown in Table 1. An interruption in a given class stores the current PSW as an old PSW and then takes the corresponding new PSW as the current PSW. If, at the conclusion of the interruption routine, old and current PSW's are interchanged, the system can be restored to its prior state and the interrupted routine can be continued.

The system mask, program mask, and machine-check mask bits in the PSW may be used to control certain interruptions. When masked off, some interruptions remain pending while others are merely ignored. The system mask can keep I/O and external interruptions pending, the program mask can cause four of the 15 program interruptions to be ignored, and the machine-check mask can cause machine-check interruptions to be ignored. Other interruptions cannot be masked off.

Appropriate CPU response to a special condition in the channels and I/O units is facilitated by an I/O *interruption*. The addresses of the channel and I/O unit involved are recorded in the old PSW. Related information is preserved in a channel status word that is stored as a result of the interruption.

Unusual conditions encountered in a program create *program interruptions*. Eight of the fifteen possible conditions involve overflows, improper divides, lost significance, and exponent underflow.

Table 1   Permanent storage assignments

Address	Byte length	Purpose
0	8	Initial program loading PSW
8	8	Initial program loading CCW 1
16	8	Initial program loading CCW 2
24	8	External old PSW
32	8	Supervisor call old PSW
40	8	Program old PSW
48	8	Machine check old PSW
56	8	Input/output old PSW
64	8	Channel status word
72	4	Channel address word
76	4	Unused
80	4	Timer
84	4	Unused
88	8	External new PSW
96	8	Supervisor call new PSW
104	8	Program new PSW
112	8	Machine check new PSW
120	8	Input/output new PSW
128		Diagnostic scan-out area†

† The size of the diagnostic scan-out area is configuration dependent.

The remaining seven deal with improper addresses, attempted execution of privileged instructions, and similar conditions.

A *supervisor-call interruption* results from execution of the instruction SUPERVISOR CALL. Eight bits from the instruction format are placed in the interruption code of the old PSW, permitting a message to be associated with the interruption. SUPERVISOR CALL permits a problem program to switch CPU control back to the supervisor.

Through an *external interruption,* a CPU can respond to signals from the interruption key on the system control panel, the timer, other CPU's, or special devices. The source of the interruption is identified by an interruption code in bits 24 through 31 of the PSW.

The occurrence of a machine check (if not masked off) terminates the current instruction, initiates a diagnostic procedure, and subsequently effects a *machine-check interruption.* A machine check is occasioned only by a hardware malfunction; it cannot be caused by invalid data or instructions.

Interrupt priority

Interruption requests are honored between instruction executions. When several requests occur during execution of an instruction, they are honored in the following order: (1) machine check, (2) program or supervisor call, (3) external, and (4) input/output. Because the program and supervisor-call interruptions are mutually exclusive, they cannot occur at the same time.

If a machine-check interruption occurs, no other interruptions can be taken until this interruption is fully processed. Otherwise, the execution of the CPU program is delayed while PSW's are appropriately stored and fetched for each interruption. When the last interruption request has been honored, instruction execution is resumed with the PSW last fetched. An interruption subroutine is then serviced for each interruption in the order (1) input/output, (2) external, and (3) program or supervisor call.

Program status

Overall CPU status is determined by four alternatives: (1) *stopped* versus *operating* state, (2) *running* versus *waiting* state, (3) *masked* versus *interruptable* state, and (4) *supervisor* versus *problem* state.

In the stopped state, which is entered and left by manual procedure, instructions are not executed, interruptions are not accepted, and the timer is not updated. In the operating state, the CPU is capable of executing instructions and of being interrupted.

In the running state, instruction fetching and execution proceeds in the normal manner. The wait state is typically entered by the program to await an interruption, for example, an I/O interruption or operator intervention from the console. In the wait state, no instructions are processed, the timer is updated, and I/O and external interruptions are accepted unless masked. Running versus waiting is determined by the setting of a bit in the current PSW.

The CPU may be interruptable or masked for the system, program, and machine interruptions. When the CPU is interruptable for a class of interruptions, these interruptions are accepted. When the CPU is masked, the system interruptions remain pending, but the program and machine-check interruptions are ignored. The interruptable states of the CPU are changed by altering mask bits in the current PSW.

In the problem state, processing instructions are valid, but all I/O instructions and a group of control instructions are invalid. In the supervisor state, all instructions are valid. The choice of problem or supervisor state is determined by a bit in the PSW.

Supervisory facilities

Timer

A timer word in main storage location 80 is counted down at a rate of 50 or 60 cycles per second, depending on power line frequency. The word is treated as a signed integer according to the rules of fixed-point arithmetic. An external interrupt occurs when the value of the timer word goes from positive to negative. The full cycle time of the timer is 15.5 hours.

As an interval timer, the timer may be used to measure elapsed time over relatively short intervals. The timer can be set by a supervisory-mode program to any value at any time.

Direct control

Two instructions, READ DIRECT and WRITE DIRECT, provide for the transfer of a single byte of information between an external device and the main storage of the system. These instructions are intended for use in synchronizing CPU's and special external devices.

Storage protection

For protection purposes, main storage is divided into blocks of 2,048 bytes each. A four-bit *storage key* is associated with each block. When a store operation is attempted by an instruction, the *protection key* of the current PSW is compared with the storage key of the affected block. When storing is specified by a channel operation, a protection key supplied by the channel is used as the comparand. The keys are said to *match* if equal or if either is zero. A storage key is not part of addressable storage, and can be changed only by privileged instructions. The protection key of the CPU program is held in the current PSW. The protection key of a channel is recorded in a status word that is associated with the channel operation.

When a CPU operation causes a protection mismatch, its execution is suppressed or terminated, and the program execution is altered by an interruption. The protected storage location always remains unchanged. Similarly, protection mismatch due to an I/O operation terminates data transmission in such a way that the protected storage location remains unchanged.

Multisystem operation

Communication between CPU's is made possible by shared control units, interconnected channels, or shared storage. Multisystem operation is supported by provisions for automatic relocation, indication of malfunctions, and CPU initialization.

Automatic relocation applies to the first 4,096 bytes of storage, an area that contains all permanent storage assignments and usually has special significance for supervisory programs. The relocation is accomplished by inserting a 12-bit prefix in each address whose high-order 12 bits are zero. Two manually set prefixes permit the use of an alternate area when storage malfunction occurs; the choice between prefixes is preserved in a trigger that is set during initial program loading.

To alert one CPU to the possible malfunction of another, a machine-check signal from a given CPU can serve as an external interruption to another CPU. By another special provision, initial program loading of a given CPU can be initiated by a signal from another CPU.

Input/output

Devices and control units

Input/output devices include card equipment, magnetic tape units, disk storage, drum storage, typewriter-keyboard devices, printers, teleprocessing devices, and process control equipment. The I/O devices are regulated by control units, which provide the electrical, logical, and buffering capabilities necessary for I/O device operation. From the programming point of view, most control-unit and I/O device functions are indistinguishable. Sometimes the control unit is housed with an I/O device, as in the case of the printer.

A control unit functions only with those I/O devices for which it is designed, but all control units respond to a standard set of signals from the channel. This control-unit-to-channel connection, called the I/O *interface*, enables the CPU to handle all I/O operations with only four instructions.

I/O instructions

Input/output instructions can be executed only while the CPU is in the supervisor state. The four I/O instructions are START I/O, HALT I/O, TEST CHANNEL, and TEST I/O.

START I/O initiates an I/O operation; its address field specifies a channel and an I/O device. If the channel facilities are free, the instruction is accepted and the CPU continues its program. The channel independently selects the specified I/O device. HALT I/O terminates a channel operation. TEST CHANNEL sets the condition code in the PSW to indicate the state of the channel addressed by the instruction. The code then indicates one of the following conditions: channel available, interruption condition in channel, channel working, or channel not operational. TEST I/O sets the PSW condition code to indicate the state of the addressed channel, subchannel, and I/O device.

Channels

Channels provide the data path and control for I/O devices as they communicate with main storage. In the multiplexor channel, the single data path can be time-shared by several low-speed devices (card readers, punches, printers, terminals, etc.) and the channel has the functional character of many subchannels, each of which services one I/O device at a time. On the other hand, the selector channel, which is designed for high-speed devices, has the functional character of a single subchannel. All subchannels respond to the same I/O instructions. Each can fetch its own control word sequence, govern the transfer of data and control signals, count record lengths, and interrupt the CPU on exceptions.

Two modes of operation, *burst* and *multiplex*, are provided for multiplexor channels. In burst mode, the channel facilities are monopolized for the duration of data transfer to or from a particular I/O device. The selector channel functions only in the burst mode. In multiplex mode, the multiplexor channel sustains several simultaneous I/O operations: bytes of data are interleaved and then routed between selected I/O devices and desired locations in main storage.

At the conclusion of an operation launched by START I/O or TEST I/O, an I/O interruption occurs. At this time a channel status word (CSW) is stored in location 64. Figure 8 shows the CSW format. The CSW provides information about the termination of the I/O operation.

Successful execution of START I/O causes the channel to

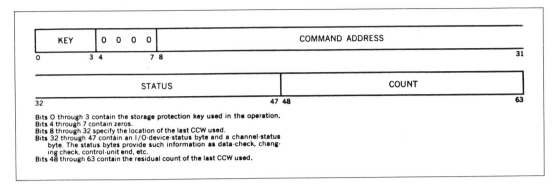

Bits 0 through 3 contain the storage protection key used in the operation.
Bits 4 through 7 contain zeros.
Bits 8 through 32 specify the location of the last CCW used.
Bits 32 through 47 contain an I/O-device-status byte and a channel-status byte. The status bytes provide such information as data-check, changing check, control-unit end, etc.
Bits 48 through 63 contain the residual count of the last CCW used.

Fig. 8. Channel status word format.

fetch a channel address word from main-storage location 72. This word specifies the storage-protection key that governs the I/O operation, as well as the location of the first eight bytes of information that the channel fetches from main storage. These 64 bits comprise a channel command word (CCW). Figure 9 shows the CCW format.

Channel program

One or more CCW's make up the channel program that directs channel operations. Each CCW points to the next one to be fetched, except for the last in the chain which so identifies itself.

Six channel commands are provided: read, write, read backward, sense, transfer in channel, and control. The read command defines an area in main storage and causes a read operation from the selected I/O device. The write command causes data to be written by the selected device. The read-backward command is akin to the read command, but the external medium is moved in the opposite direction and bytes read backward are placed in descending main storage locations.

The control command contains information, called an *order*, that is used to control the selected I/O device. Orders, peculiar to the particular I/O device in use, can specify such functions as rewinding a tape unit, searching for a particular track in disk storage, or line skipping on a printer. In a functional sense, the CPU executes I/O instructions, the channels execute commands, and the control units and devices execute orders.

The sense command specifies a main storage location and transfers one or more bytes of status information from the selected control unit. It provides details concerning the selected I/O device, such as a stacker-full condition of a card reader or a file-protected condition of a magnetic-tape reel.

A channel program normally obtains CCW's from a consecutive string of storage locations. The string can be broken by a transfer-in-channel command that specifies the location of the next CCW to be used by the channel. External documents, such as punched cards or magnetic tape, may carry CCW's that can be used by the channel to govern the reading of the documents.

The input/output interruptions caused by termination of an

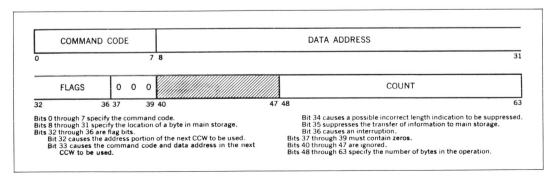

Bits 0 through 7 specify the command code.
Bits 8 through 31 specify the location of a byte in main storage.
Bits 32 through 36 are flag bits.
  Bit 32 causes the address portion of the next CCW to be used.
  Bit 33 causes the command code and data address in the next CCW to be used.
Bit 34 causes a possible incorrect length indication to be suppressed.
Bit 35 suppresses the transfer of information to main storage.
Bit 36 causes an interruption.
Bits 37 through 39 must contain zeros.
Bits 40 through 47 are ignored.
Bits 48 through 63 specify the number of bytes in the operation.

Fig. 9. Channel command word format.

Table 2 System/360 instructions

RR Format

xxxx	Branching and status switching 0000xxxx		Fixed-point fullword and logical -0001xxxx		Floating-point long 0010xxxx		Floating-point short 0011xxxx	
0000			LPR	LOAD POSITIVE	LPDR	LOAD POSITIVE	LPER	LOAD POSITIVE
0001			LNR	LOAD NEGATIVE	LNDR	LOAD NEGATIVE	LNER	LOAD NEGATIVE
0010			LTR	LOAD AND TEST	LTDR	LOAD AND TEST	LTER	LOAD AND TEST
0011			LCR	LOAD COMPLEMENT	LCDR	LOAD COMPLEMENT	LCER	LOAD COMPLEMENT
0100	SPM	SET PROGRAM MASK	NR	AND	HDR	HALVE	HER	HALVE
0101	BALR	BRANCH AND LINK	CLR	COMPARE LOGICAL				
0110	BCTR	BRANCH ON COUNT	OR	OR				
0111	BCR	BRANCH/CONDITION	XR	EXCLUSIVE OR				
1000	SSK	SET KEY	LR	LOAD	LDR	LOAD	LER	LOAD
1001	ISK	INSERT KEY	CR	COMPARE	CDR	COMPARE	CER	COMPARE
1010	SVC	SUPERVISOR CALL	AR	ADD	ADR	ADD N	ALR	ADD N
1011			SR	SUBTRACT	SDR	SUBTRACT N	SER	SUBTRACT N
1100			MR	MULTIPLY	MDR	MULTIPLY	MER	MULTIPLY
1101			DR	DIVIDE	DDR	DIVIDE	DER	DIVIDE
1110			ALR	ADD LOGICAL	AWR	ADD U	AUR	ADD U
1111			SLR	SUBTRACT LOGICAL	SWR	SUBTRACT U	SUR	SUBTRACT U

RX Format

xxxx	Fixed-point halfword and branching 0100xxxx		Fixed-point fullword and logical 0101xxxx		Floating-point long 0110xxxx		Floating-point short 0111xxxx	
0000	STH	STORE	ST	STORE	STD	STORE	STE	STORE
0001	LA	LOAD ADDRESS						
0010	STC	STORE CHARACTER						
0011	IC	INSERT CHARACTER						
0100	EX	EXECUTE	N	AND				
0101	BAL	BRANCH AND LINK	CL	COMPARE LOGICAL				
0110	BCT	BRANCH ON COUNT	O	OR				
0111	BC	BRANCH/CONDITION	X	EXCLUSIVE OR				
1000	LH	LOAD	L	LOAD	LD	LOAD	LE	LOAD
1001	CH	COMPARE	C	COMPARE	CD	COMPARE	CE	COMPARE
1010	AH	ADD	A	ADD	AD	ADD N	AE	ADD N
1011	SH	SUBTRACT	S	SUBTRACT	SD	SUBTRACT N	SE	SUBTRACT N
1100	MH	MULTIPLY	M	MULTIPLY	MD	MULTIPLY	ME	MULTIPLY
1101			D	DIVIDE	DD	DIVIDE	DE	DIVIDE
1110	CVD	CONVERT-DECIMAL	AL	ADD LOGICAL	AW	ADD U	AU	ADD U
1111	CVB	CONVERT-BINARY	SL	SUBTRACT LOGICAL	SW	SUBTRACT U	SU	SUBTRACT U

RS, SI Format

xxxx	Branching status switching and shifting 1000xxxx		Fixed-point logical and input/output 1001xxxx		1010xxxx		1011xxxx	
0000	SSM	SET SYSTEM MASK	STM	STORE MULTIPLE				
0001			TM	TEST UNDER MASK				
0010	LPSW	LOAD PSW	MVI	MOVE				
0011		DIAGNOSE	TS	TEST AND SET				
0100	WRD	WRITE DIRECT	NI	AND				
0101	RDD	READ DIRECT	CLI	COMPARE LOGICAL				
0110	BXH	BRANCH/HIGH	OI	OR				
0111	BXLE	BRANCH/LOW-EQUAL	XI	EXCLUSIVE OR				
1000	SRL	SHIFT RIGHT SL	LM	LOAD MULTIPLE				
1001	SLL	SHIFT LEFT SL						
1010	SRA	SHIFT RIGHT S						
1011	SLA	SHIFT LEFT S						
1100	SRDL	SHIFT RIGHT DL	SIO	START I/O				
1101	SLDL	SHIFT LEFT DL	TIO	TEST I/O				
1110	SRDA	SHIFT RIGHT D	HIO	HALT I/O				
1111	SLDA	SHIFT LEFT D	TCH	TEST CHANNEL				

SS Format

xxxx	1100xxxx		Logical 1101xxxx		1110xxxx		Decimal 1111xxxx	
0000								
0001			MVN	MOVE NUMERIC			MVO	MOVE WITH OFFSET
0010			MVC	MOVE			PACK	PACK
0011			MVZ	MOVE ZONE			UNPK	UNPACK
0100			NC	AND				
0101			CLC	COMPARE LOGICAL				
0110			OC	OR				
0111			XC	EXCLUSIVE OR				
1000							ZAP	ZERO AND ADD
1001							CP	COMPARE
1010							AP	ADD
1011							SP	SUBTRACT
1100			TR	TRANSLATE			MP	MULTIPLY
1101			TRT	TRANSLATE AND TEST			DP	DIVIDE
1110			ED	EDIT				
1111			EDMK	EDIT AND MARK				

NOTE: N = NORMALIZED   DL = DOUBLE LOGICAL   S = SINGLE
SL = SINGLE LOGICAL   U = UNNORMALIZED   D = DOUBLE

I/O operation, or by operator intervention at the I/O device, enable the CPU to provide appropriate programmed response to conditions as they occur in I/O devices or channels. Conditions responsible for I/O interruption requests are preserved in the I/O devices or channels until recognized by the CPU.

During execution of START I/O, a command can be rejected by a busy condition, program check, etc. Rejection is indicated in the condition code of the PSW, and additional detail on the conditions that precluded initiation of the I/O operation is provided in a CSW.

Manual control

The need for manual control is minimal because of the design of the system and supervisory program. A control panel provides the ability to reset the system; store and display information in main storage, in registers, and in the PSW; and load initial program information. After an input device is selected with the load unit switches, depressing a load key causes a read from the selected input device. The six words of information that are read into main storage provide the PSW and the CCW's required for subsequent operation.

Instruction set

The SYSTEM/360 instructions, classified by format and function, are displayed in Table 2. Operation codes and mnemonic abbreviations are also shown. With the previously described formats in mind, much of the generality provided by the system is apparent in this listing.

Chapter 44

The structure of SYSTEM/360[1]

Part II—System implementations

W. Y. Stevens

Summary   The performance range desired of SYSTEM/360 is obtained by variations in the storage, processing, control, and channel functions of the several models. The systematic variations in speed, size, and degree of simultaneity that characterize the functional components and elements of each model are discussed.

A primary goal in the SYSTEM/360 design effort was a wide range of processing unit performances coupled with complete program compatibility. In keeping with this goal, the logical structure of the resultant system lends itself to a wide choice of components and techniques in the engineering of models for desired performance levels.

This paper discusses basic choices made in implementing six SYSTEM/360 models spanning a performance range of fifty to one. It should be emphasized that the problems of model implementation were studied throughout the design period, and many of the decisions concerning logical structure were influenced by difficulties anticipated or encountered in implementation.

Performance adjustment

The choices made in arriving at the desired performances fall into four areas:

  Main storage

  Central processing unit (CPU) registers and data paths

  Sequence control

  Input/output (I/O) channels

Each of the adjustable parameters of these areas can be subordinated, for present purposes, to one of three general factors: basic speed, size, and degree of simultaneity.

[1] *IBM Sys. J*, vol. 3, no. 2, 136–143, 1964.

Main storage

Storage speed and size

The interaction of the general factors is most obvious in the area of main storage. Here the basic speeds vary over a relatively small range: from a 2.5-μsec cycle for the Model 40 to a 1.0-μsec cycle for Models 62 and 70. However, in combination with the other two factors, a 32:1 range in overall storage data rate is obtained, as shown in Table 1.

Most important of the three factors is size. The width of main storage, i.e., the amount of data obtained with one storage access, ranges from one byte for the Model 30, two bytes for the Model 40, and four bytes for the Model 50, to 8 bytes for Models 60, 62, and 70.

Another size factor, less direct in its effect, is the total number of bytes in main storage, which can make a large difference in system throughput by reducing the number of references to external storage media. This number ranges from a minimum of 8192 bytes on Model 30 to a maximum of 524,288 bytes on Models 60, 62, and 70. An option of up to eight million more bytes of slower-speed, large-capacity core storage can further increase the throughput in some applications.

Interleaved storage

Simultaneity in the core storage of Models 60 and 70 is obtained by overlapping the cycles of two storage units. Addresses are staggered in the two units, and a series of requests for successive words activates the two units alternately, thus doubling the maximum rate. For increased system performance, this technique is less effective than doubling the basic speed of a single unit, since the access time to a single word is not improved, and successive references frequently occur to the same unit. This is illustrated by comparing the performances of Models 60 and 62, whose only difference is the choice between two overlapped 2.0-μsec storage units and one single 1.0-μsec storage unit, respectively. The performance of Model 62 is approximately 1.5 times that of Model 60.

Table 1   System/360 main storage characteristics

	Model 30	Model 40	Model 50	Model 60	Model 62	Model 70
Cycle time (μsec)	2.0	2.5	2.0	2.0	1.0	1.0
Width (bytes)	1	2	4	8	8	8
Interleaved access	no	no	no	yes	no	yes
Maximum data rate (bytes/μsec)	0.5	0.8	2.0	8.0	8.0	16.0
Minimum storage size (bytes)	8,192	16,384	65,536	131,072	262,144	262,144
Maximum storage size (bytes)	65,536	262,144	262,144	524,288	524,288	524,288
Large capacity storage attachable	no	no	yes	yes	yes	yes

CPU registers and data paths

Circuit speed

SYSTEM/360 has three families of logic circuits, as shown in Table 2, each using the same solid-logic technology. One family, having a nominal delay of 30 nsec per logical stage or level, is used in the data paths of Models 30, 40, and 50. A second and faster family with a nominal delay of 10 nsec per level is used in Models 60 and 62. The fastest family, with a delay of 6 nsec, is used in Model 70.

The fundamental determinant of CPU speed is the time required to take data from the internal registers, process the data through the adder or other logical unit, and return the result to a register. This cycle time is determined by the delay per logical circuit level and the number of levels in the register-to-adder path, the adder, and the adder-to-register return path. The number of levels varies because of the trade-off that can usually be made between the number of circuit modules and the number of logical levels. Thus, the cycle time of the system varies from 1.0 μsec for Model 30 (with 30-nsec circuits, a relatively small number of modules, and more logic levels) and 0.5 μsec for Model 50 (also with 30-nsec circuits, but with more modules and fewer levels) to 0.2 μsec for Model 70 (with 6-nsec circuits).

Local storage

The speed of the CPU depends also on the speed of the general and floating-point registers. In Model 30, these registers are located in an extension to the main core storage and have a read-write

Table 2   System/360 CPU characteristics

	Model 30	Model 40	Model 50	Model 60/62	Model 70
Circuit family: nominal delay per logic level (nsec)	30	30	30	10	6
Cycle time (μsec)	1.0	0.625	0.5	0.25	0.2
Location of general and floating registers	main core storage	local core storage	local core storage	local transistor storage	transistor registers
Width of general and floating register storage (bytes)	1	2	4	4	4 or 8
Speed of general and floating register storage (μsec)	2.0	1.25	0.5	0.25	
Width of main adder path (bits)	8	8	32	56	64
Width of auxiliary transfer path (bits)		16	8		
Widths of auxiliary adder paths (bits)				8	8, 8, and 24
Approximate number of bytes of register storage	12	15	30	50	100
Approximate number of bytes of working locations in local storage	45 (main storage)	48	60	4	
Relative computing speed	1	3.5	10	21/30	50

time of 2.0 μsec. In Model 40, the registers are located in a small core-storage unit, called *local storage*, with a read-write time of 1.25 μsec. Here, the operation of the local storage may be overlapped with main storage. In Model 50, the registers are in a local storage with a read-write time of only 0.5 μsec. In Model 60/62, the local storage has the logical characteristics of a core storage with nondestructive read-out; however, it is actually constructed as an array of registers using the 30-nsec family of logic circuits, and has a read-write time of 0.25 μsec. In Model 70, the general and floating-point registers are implemented with 6-nsec logic circuits and communicate directly with the adder and other data paths.

The two principal measures of size in the CPU are the width of the data paths and the number of bytes of high-speed working registers.

Data path organization

Model 30 has an 8-bit wide (plus parity) adder path, through which all data transfers are made, and approximately 12 bytes of working registers.

Model 40 also has an 8-bit wide adder path, but has an additional 16-bit wide data transfer path. Approximately 15 bytes of working registers are used, plus about 48 bytes of working locations in the local storage, exclusive of the general and floating-point registers.

Model 50 has a 32-bit wide adder path, an 8-bit wide data path used for handling individual bytes, approximately 30 bytes of working registers, plus about 60 bytes of working locations in the local storage.

Model 60/62 has a 56-bit wide main adder path, an 8-bit wide serial adder path, and approximately 50 bytes of working registers.

Model 70 has a 64-bit wide main adder, an 8-bit wide exponent adder, an 8-bit wide decimal adder, a 24-bit wide addressing adder, and several other data transfer paths, some of which have incrementing ability. The model has about 100 bytes of working registers plus the 96 bytes of floating point and general registers which, in Model 70, are directly associated with the data paths.

The models of SYSTEM/360 differ considerably in the number of relatively independent operations that can occur simultaneously in the CPU. Model 30, for example, operates serially: virtually all data transfers must pass through the adder, one byte at a time. Model 70, however, can have many operations taking place at the same time. The CPU of this model is divided into three units that operate somewhat independently. The instruction preparation unit fetches instructions from storage, prepares them by computing their effective addresses, and initiates the fetching of the required data. The execution unit performs the execution of the instruction

prepared by the instruction unit. The third unit is a storage bus control which coordinates the various requests by the other units and by the channels for core-storage cycles. All three units normally operate simultaneously, and together provide a large degree of instruction overlap. Since each of the units contains a number of different data paths, several data transfers may be occurring on the same cycle in a single unit.

The operations of other SYSTEM/360 models fall between those mentioned. Model 50, for example, can have simultaneous data transfers through the main adder, through an auxiliary byte transfer path, and to or from local storage.

Sequence control

Complex instruction sequences

Since the SYSTEM/360 has an extensive instruction set, the CPU's must be capable of executing a large number of different sequences of basic operations. Furthermore, many instructions require sequences that are dependent on the data or addresses used. As shown in Table 3, these sequences of operations can be controlled by two methods; either by a conventional sequential logic circuit that uses the same types of circuit modules as used in the data paths or by a read-only storage device that contains a microprogram specifying the sequences to be performed for the different instructions.

Model 70 makes use of conventional sequential logic control mainly because of the high degree of simultaneity required. Also, a sufficiently fast read-only storage unit was not available at the time of development. The sequences to be performed in each of the Model 70 data paths have a considerable degree of independence. The read-only storage method of control does not easily lend itself to controlling these independent sequences, but is well adapted where the actions in each of the data paths are highly coordinated.

Read-only storage control

The read-only storage method of control is described elsewhere [Peacock, 19??]. This microprogram control, used in all but the fastest model of SYSTEM/360, is the only method known by which an extensive instruction set may be economically realized in a small system. This was demonstrated during the design of Model 60/62. Conventional logic control was originally planned for this model, but it became evident during the design period that too many circuit modules were required to implement the instruction set, even for this rather large system. Because a sufficiently fast read-only storage became available, it was adopted for sequence control at a substantial cost reduction.

Table 3 System/360 sequence control characteristics

	Model 30	Model 40	Model 50	Model 60/62	Model 70
Type	read-only storage	read-only storage	read-only storage	read-only storage	sequential logic
Cycle time (μsec)	1.0	0.625	0.5	0.25	0.2
Width of read-only storage word (available bits)	60	60	90	100	
Number of read-only storage words available	4096	4096	2816	2816	
Number of gate-control fields in read-only storage word	9	10	15	16	

The three factors of speed, size, and simultaneity are applicable to the read-only storage controls of the various SYSTEM/360 models. The speed of the read-only storage units corresponds to the cycle time of the CPU, and hence varies from 1.0 μsec per access for Model 30 down to 0.25 μsec for Models 60 and 62.

The size of read-only storage can vary in two ways—in width (number of bits per word) and in number of words. Since the bits of a word are used to control gates in the data paths, the width of storage is indirectly related to the complexity of the data paths. The widths of the read-only storages in SYSTEM/360 range from 60 bits for Models 30 and 40 to 100 bits for Models 60 and 62. The number of words is affected by several factors. First, of course, is the number and complexity of the control sequences to be executed. This is the same for all models except that Model 60/62 read-only storage contains no sequences for channel functions. The number of words tends to be greater for the smaller models, since these models require more cycles to accomplish the same function. Partially offsetting this is the fact that the greater degree of simultaneity in the larger systems often prevents the sharing of microprogram sequences between similar functions.

SYSTEM/360 employs no read-only storage simultaneity in the sense that more than one access is in progress at a given time. However, a single read-only storage word simultaneously controls several independent actions. The number of different gate control fields in a word provides some measure of this simultaneity. Model 30 has 9 such fields. Model 60/62 has 16.

Input/output channels

Channel design

The SYSTEM/360 input/output channels may be considered from two viewpoints: the design of a channel itself, or the relationship of a channel to the whole system.

From the viewpoint of channel design, the raw speed of the components does not vary, since all channels use the 30-nsec family of circuits. However, the different channels do have access to

different speeds of main storage and, in the three smaller models, different speeds of local storage.

The channels differ markedly in the amount of hardware devoted exclusively to channel use, as shown in Table 4. In the Model 30 multiplexor channel, this hardware amounts only to three 1-byte wide data paths, 11 latch bits for control, and a simple interface polling circuit. The channel used in Models 60, 62, and 70 contains about 300 bits of register storage, a 24-bit wide adder, and a complete set of sequential control circuits. The amount of hardware provided for other channels is somewhere in between these extremes.

The disparity in the amount of channel hardware reflects the extent to which the channels share CPU hardware in accomplishing their functions. Such sharing is done at the expense of increased interference with the CPU, of course. This interference ranges from complete lock-out of CPU operations at high data rates on some of the smaller models, to interference only in essential references to main storage by the channel in the large models.

Channel/system relationship

When the channels are viewed in their relationship to the whole system, the three factors of speed, size, and simultaneity take on a different aspect. The channel is viewed as a system component, and its effect on system throughput and other system capabilities is of concern. The speeds of the channels vary from a maximum rate of about 16 thousand bytes per second (byte interleaved mode) on the multiplexor channel of Model 30 to a maximum rate of about 1250 thousand bytes per second on the channels of Models 60, 62, and 70. The size of each of the channels is the same, in the sense that each handles an 8-bit byte at a time and each can connect to eight different control units. A slight size difference exists among multiplexor channels in terms of the maximum number of subchannels.

The degree of channel simultaneity differs considerably among the various models of SYSTEM/360. For example, operation of the Model 30 or 40 multiplexor channels in burst mode inhibits all

Table 4   System/360 channel characteristics

	Model 30	Model 40	Model 50	Model 60/62	Model 70
Selector channels					
Maximum number attachable	2	2	3	6	6
Approximate maximum data rate on one channel in Kbyps†	250	400	800 (1250 on high speed)	1250	1250
Uses CPU data paths for:					
initiation and termination	yes	yes	yes	yes	yes
byte transfers	no	no	no	no	nc
storage word transfers	no	low speed only	yes	no	no
chaining	yes	yes	yes	no	no
CPU and I/O overlap possible	yes	yes	regular—yes high speed—no	yes	yes
Multiplexor channels					
Maximum number attachable	1	1	1	0	0
Minimum number of subchannels	32	16	64		
Maximum number of subchannels	96	128	256		
Maximum data rate in byte interleaved mode (Kbyps)	16	30	40		
Maximum data rate in burst mode (Kbyps)	200	200	200		
Uses CPU data paths for all functions	yes	yes	yes		
CPU and I/O overlap possible in byte mode	yes	yes	yes		
CPU and I/O overlap possible in burst mode	no	no	yes		

† Thousand bytes per second.

other activity on the system, as does operation of the special high-speed channel on Model 50. At the other extreme, as many as six selector channels can be operating concurrently with the CPU on Models 60, 62, or 70. A second type of simultaneity is present in the multiplexor channels available on Models 30, 40, and 50. When operating in byte interleaved mode, one of these channels can control a number of concurrently operating input/output devices, and the CPU can also continue operation.

Differences in application emphasis

The models of system/360 differ not only in throughput but also in the relative speeds of the various operations. Some of these relative differences are simply a result of the design choices described in this paper, made to achieve the desired overall performance. The more basic differences in relative performance of the various operations, however, were intentional. These differences in emphasis suit each model to those applications expected to comprise its largest usage.

Thus the smallest system is particularly aimed at traditional commercial data processing applications. These are characterized by extensive input/output operations in relation to the internal processing, and by more character handling than arithmetic. The

fast selector channels and character-oriented data paths of Model 30 result from this emphasis. But despite this emphasis, the general-purpose instruction set of system/360 results in much better scientific application performance for Model 30 than for its comparable predecessors.

On the other hand, the large systems are expected to find particularly heavy use in scientific computation, where the emphasis is on rapid floating-point arithmetic. Thus Models 60, 62, and 70 contain registers and adders that can handle the full length of a long format floating-point operand, yet do character operations one byte at a time.

No particular emphasis on either commercial or scientific applications characterizes the intermediate models. However, Models 40 and 50 are intended to be particularly suitable for communication-oriented and real-time applications. For example, Model 50 includes a multiplexor channel, storage protection, and a timer as standard features, and also provides the ability to share main storages between two CPU's in a multiprocessing arrangement.

References

PeacA??

Appendix

PMS and ISP notations

This appendix provides complete definitions of the notations used for the PMS and ISP descriptions. It is intended to supplement Chap. 2, which provides an informal description of the notations along with some comments on motivation and underlying rationale.

The two descriptive systems are consistent with each other in two senses. First, certain general conventions that have to do with forming expressions and abbreviating apply to both systems. Second, the values of certain PMS attributes are describable in ISP but not in PMS. A complete "top down" development would thus embed ISP within PMS. Nevertheless, it appears appropriate to present them as two distinct notations: it makes reference easier and permits each to be organized around its own most important notions.

The style of presentation is moderately formal. Within a section, the syntax is presented, followed by remarks on the interpretation to be given to these syntactic forms (the semantics). Examples that help to pin down the notations are furnished throughout. Although not a computer language, we present it as if it were; thus, a number of elementary things are provided for in the definitions. (Part of the motivation for this is to introduce abbreviations.)

A language can be realized in many media. In this book we have taken some advantage of printing orthography insofar as it enhances communication. However, it may also be necessary to map the notations into various restrictive character sets—e.g., those of the typewriter and the computer. For the sake of brevity, we do not discuss this coding problem here.

The appendix is in three parts. The first part gives the general conventions common to both PMS and ISP. The second and third parts give PMS (page 615) and ISP (page 628), as discussed in Chap. 2.

General conventions

The conventions given in this section define the general nature of the syntax and semantics of both PMS and ISP.

These general conventions parallel closely natural usage by technically trained people familiar with programming languages, such as ALGOL. There is no need to consult these sections if the brief statements and illustrations following each subsection title are clearly understood.

1   Basic semantics
    The language can refer to any entities that are given by attributes and values.

2   Metanotation
    (There is no need for metanotation unless general conventions are to be read in detail.)

3   Basic syntax
    Expressions are built up from subexpressions and ultimately from names. Parentheses are used to avoid ambiguity.

4   Commands: assignments, abbreviations, variables, forms
    x := y assigns the name x to mean the same as the expression y.
    x / y establishes the name y as an abbreviation or alternative name (alias) for x.
    $x \div y := \min(x - y, 0)$ defines a new binary operation $(\div)$ by means of a form in the variables x and y.

5   Indefinite expressions
    a|b|c means one of a or b or c.
    x ∼ y means the interval from x up to and including y.
    ∼x means an interval around x of undetermined scope.

6   Lists and sets
    (3, 5, 1, 5) is a list of digits, which also could have been written (3; 5; 1; 5). Digit-list refers to all possible lists of digits. Digit-set refers to all possible sets digits, unordered and without repetition.

7   Definite expressions
    X := (size: integer; function: (primary|secondary); control: (yes| no)) defines X to be an entity with an attribute, size, taking any integer as value; with an attribute, function, taking primary or secondary as value; and with an attribute, control, taking yes or no as value.
    Y := X(size: 12 ∼ 20; primary; ¬control) defines Y as an entity of type X which is further specified by having size between 12 and 20, having the value of function be primary and the value of control be no.

8   Attributes
    3:Z is the third item on the list Z; −1:Z is the last item. (add-time, store-time) can be an attribute and then has values such as (10 μs, 6 μs).

9   Null symbol and optional expressions
    ∅ is the null symbol so that (x, ∅, y) is the same as (x, y). *x means that x is optional; defined as (x|∅)

10 Names
    Simple-names are strings of letters and digits, permitting concatenation with the space (␣) and the hyphen (-). 'The␣big␣instruction-set' is a simple-name.
    Memory.primary is a compound name, which is an abbreviation for Memory(primary).
    Classes of names can be constructed and assigned to be used for various entities—if for an entity, X, then called X-names.

11   Numbers
Numbers and arithmetic expressions are defined in the standard fashion.

12   Quantities, dimensions, and units
A quantity is just a dimensionalized number—a number of units along a given dimension.

13   Booleans and relations
Logical expressions involving and ($\wedge$), or ($\vee$), not ($\neg$), implies ($\supset$), equivalence ($\equiv$), and exclusive-or ($\oplus$) are defined in standard fashion, as are expressions involving the six basic relations ($=, \neq, <, >, \leq, \geq$).

1.   Basic semantics

1.1   We will use the term "entity" to refer to all things designatable by expressions in the language.

1.2   An entity is assumed to be fully characterizable by a set of attributes and associated values, which are themselves entities.

COMMENT   There will necessarily be entities with no further specification within the system—that, in effect, have only a name.

The semantics of the language consists in showing how expressions in the language determine the various attributes and values.

1.3   There are three types of expressions.

1   A definite expression designates an entity.

2   An indefinite expression defines a class of definite expressions; it designates one of the entities designated by members of this class.

3   A command designates the establishment of some purely linguistic convention.

EXAMPLES   'IBM 7090   is a definite expression.

Mp   is an indefinite expression (any primary memory).

SAM := Mp   is a command to give the name SAM to an Mp.

1.4   There are also English language comments, which are connected with the language only in being associated with particular occurrences of expressions (on which they comment) and in having a punctuation convention that allows them to be unambiguously distinguished from expressions in the language.

1   In the book we use italics.

EXAMPLE   *This is an example of a comment; it may appear anywhere.*

2.   Metanotation

2.1   The language itself is described by giving various classes of expressions and assigning meanings to the members of these classes (i.e., telling what they designate). We will generally do this in English but with a few special notations.

2.2   Expression-variables

1   Let **a**, **b**, . . . , **A**, **B**, . . . be variables whose domain is a set of expressions.

2   Let **class**(**a**) be the set of definite expressions defined by the indefinite expression **a**. This is extended to definite expressions, **x**, by defining **class**(**x**) = **x**.

COMMENT   Normally lowercase variables (e.g., **a**) stand for any legal expression, whereas uppercase variables (e.g., **A**) stand for any indefinite expression.

2.3   We will define the language by giving forms of expressions, that is, by writing down sequences of expressions and expression-variables. These forms are to be interpreted as permitting any expression that results from replacing the expression-variables with expressions from their respective domains.

EXAMPLE   If the form **x**|**y** is legal, where **x** and **y** range over components, then the expression M|P is legal.

2.4   The one special notation is the expression form

x o x . . .

which is to be taken as permitting an indefinite sequence of **x**'s separated by **o**'s, terminating with an **x**, where each occurrence is to be viewed as an independent variable. That is, **x o x** . . . is equivalent to

x

or

x o x

or

x o x o x

or

x o x o x o x

etc.

EXAMPLE   **d o d** . . . , where **d** ranges over digits and **o** over arithmetic operations, could have as instances: 5, 6 + 6, 7 − 2 + 3, etc.

COMMENT   Note that we have used the same variable several times, even though independently selected values are meant at each occurrence. It will always be clear from the context when this is being done.

3.   Basic syntax

3.1   An expression is either a name or a sequence of expressions.

3.2   A name is a sequence of characters written without spaces.

3.3   A character is a member of one of the following alphabets:

1	Capital letters	A B . . . Z
2	Small letters	a b . . . z
3	Digits	0 1 . . . 9
4	Marks	$\mid$; , : ← → ≡ ≢ ⊕ ⊃ ∨ ∧ ¬ = ≠ < > ≤ ≥ ? + − × / ∼ ↑ ↓ □ ⌞ . - \$ # * ' " ∅ Φ μ ( ) [ ] { } ⟨ ⟩

The characters of each alphabet are ordered as shown, from left (low) to right (high).

3.4   One or more spaces (freely determined) occur between names. The only exceptions are names that are single marks (alphabet 4, above) and can be disambiguated. For these, spaces can be omitted.

EXAMPLES   A, B instead of A , b

$\qquad$ −3 instead of − 3

$\qquad$ (A + B) instead of (A + B)

3.5   Parentheses are used around any expression that would otherwise be ambiguously interpreted. Conversely, parentheses can be dropped whenever there is no possibility of ambiguity.

3.6   To avoid excess parentheses, an order of precedence exists for names used as separators. The higher in the order, the greater the binding power, i.e., the greater precedence in being interpreted first. The following order is consistent with the alphabetical order:

$\qquad$:= | | | ; | , | : | → | ← | ≡ ≢ ⊕ | ⊃ |
$\qquad$ ∨ | ∧ | ¬ | = ≠ | < > ≤ ≥ | + − | × /
$\qquad$ | ∼ | ↑ | ↓ | □ | /(abbreviation)⌞ . - (hyphen)

3.7   Spacing on the page is freely determined (e.g., for legibility). An expression may run freely on several consecutive lines (with no explicit continuation mark).

EXAMPLE $z'\langle 0{:}11\rangle := (\neg$ ib → z"; $\qquad$ *This ISP expression and also*
$\qquad\qquad\qquad$ ib → M[z"]) $\qquad$ *this comment are on two lines.*

3.8   Subscripting and superscripting may be used interchangeably with the marks ↓ and ↑ respectively.

EXAMPLE $10 \downarrow 2$ is the same as 10_2

$\qquad$ $x \uparrow 2$ is the same as x^2

4.   Commands: assignment, abbreviation, variables, forms

4.1   If **x** is a free name [as defined in General Conventions section 10 (GC 10)] and **y** is any expression, then the command

$\qquad$ **x** := **y**

assigns the name **x** to the corresponding expression **y**. In particular,

$\qquad$ **class(x)** = **class(y)**

EXAMPLE   BILL := C(operation-rate: $10 \uparrow 6$ o/s) assigns a name to a particular (partially specified) computer.

4.2   If there are several assignment expressions for a single name **x**:

$\qquad$ **x** := **a**
$\qquad$ **x** := **b**

etc.; then **x** is assigned to be the name of the union of all the expressions:

$\qquad$ **class(x)** = union(**class(i)**)
$\qquad\qquad$ i = **a,b**, . . .

EXAMPLE   M.1 := M(size: 1000 w) and M.1 := M(size: 2000 w) would define M.1 to be memories of either 1,000 or 2,000 words.

4.3   If **x** is any name and **y** is any name, then the command

$\qquad$ **x** / **y**

assigns **y** to be an abbreviation (a synonym) for **x**. Abbreviation may occur on any occasion and not just when **x** is first defined. It may occur as a separate expression or it may occur in an expression in which x occurs, thus establishing the abbreviation in passing. A sequence of abbreviations may be defined in the same expression.

COMMENT   The abbreviation may not be a shorter phrase at all, but simply an alternative phrasing (say, one commonly known).

EXAMPLE   Memory / M, bit / b, second / sec / s
$\qquad$ multiplex / many channeled

COMMENT   / is also used for division, but no difficulties arise.

4.4   If **x** is any name and **D** is any indefinite expression, then the command

$\qquad$ **x** := **D**-variable

assigns **x** to be a variable with the set of entities of **class(D)** as the domain. If there are no restrictions on the domain of the variable, then the **D** may be dropped.

EXAMPLES   x := number-variable

y := component-variable

z := variable   *no restricted domain*

COMMENT   Note that these variables are over entities, not over expressions (as are the expression-variables **x, y, z**).

4.5   A form is any expression containing variables. If **f** is a form containing a single free name **x** (in addition to variables and defined subexpressions) and **g** is a form, then we extend the assignment command to include

f := **g**

which is taken as defining the name **x**. The variables occurring in **f** are called the operands of **x**. An occurrence of the form **f** with variables replaced by expressions designating in the domain of the variables is equivalent to the expression **g** with these same variables replaced by their values from the occurrence of **f**. This permits the definitions of functions and operations in which the operands (the variables in **f**) can be identified by the form of their occurrence.

EXAMPLES   x := number-variable   y := number-variable

x $\dot{-}$ y *is a form*

abs(x) *is a form*

abs(x) := (x $\geq$ 0 $\rightarrow$ x; x $<$ 0 $\rightarrow$ $-$x)   *defines abs(x)*

x $\dot{-}$ y := max(x $-$ y, 0)   *defines x $\dot{-}$ y*

5. Indefinite expressions

5.1   An indefinite expression is characterized completely by giving the **class** associated with the expression.

5.2   The basic evaluation rule is the following:

If **A** contains an occurrence of another indefinite expression **B**, then **class(A)** is the union of the **classes** of all the expressions formed by replacing the occurrence of **B** by each member of **class(B)**. In symbols,

$$\text{class}(\, A(\ldots B \ldots)\,) = \text{union}(\, \text{class}(\, A(\ldots b \ldots)\,)\,)$$
$$b \text{ in } \text{class}(B)$$

EXAMPLE   X := M(size: 1000 w)

Y := C(Mp: X)

class(Y) contains   C(Mp: M(size: 1000 w; width: 12 b))

C(Mp: M(size: 1000 w; width: 16 b))

C(Mp: M(size: 1000 w; speed: 1000 o/s))

etc.

5.3   Indefinite expressions can be formed in five ways:

1   Postulation: an expression is given in the initial definition in this appendix.
EXAMPLE   Entity is so defined in GC 7.

2   Specialization: If **A** contains an occurrence of another indefinite expression, **B** and **x** is any expression for a subset of **class(B)**; then the expression formed by replacing the occurrence of **B** in **A** by **x** yields a legitimate expression. In symbols, if **A**(... **B** ...) is legal and **x** is legal and **class(x)** $\subset$ **class(B)**, then **A**(... **x** ...) is legal.

EXAMPLE   In the example of GC 5.2, the expressions of the members of **class**(Y) are legal expressions.

3   Alternation: If **x, y**, ... are any expressions, then **x**|**y** ... is the indefinite expression "either **x** or alternatively **y** or alternatively. . . ." In symbols,

$$\text{class}(\mathbf{x}\,|\,\mathbf{y} \ldots) = \text{union}(\,\text{class}(\mathbf{i})\,)$$
$$\mathbf{i} = \mathbf{x,y}, \ldots$$

COMMENT   Note that **x** := **a** and **x** := **b** is equivalent to **x** := **a**|**b**.

EXAMPLE   number-name := integer|decimal

4   Range: If **x** and **y** designate members of an ordering, such that **x** $\leq$ **y**, then

x $\sim$ **y**

is the indefinite expression containing all members of the ordering starting with **x**, up to and including **y**.

EXAMPLE   7 $\sim$ 11 is equivalent to 7|8|9|10|11

5   Approximation: If **x** designates a member of an ordering, then $\sim$**x** is an indefinite expression containing **x** plus members of the order on both sides of **x**, without specification of the exact limits.

EXAMPLE $\sim$10 is a set of numbers around 10, possibly 8|9|10|11.

COMMENT   In the above five ways of defining indefinite expressions, specialization and alternation correspond to the usual definition of a simple-phrase structure grammar (Backus Normal Form, BNF); BNF is often used to define programming languages.

6.   Lists and sets

6.1   If **x** is any expression, then

   x-list

is an abbreviation either for

   x, x . . .

or for

   x; x . . .

x-list designates an ordered set of entities designated by **x**, with repetition permitted. The choice of a comma or a semicolon for the separator is semantically irrelevant. The two choices permit the nesting of comma lists within semicolon lists without parentheses. (Recall the order of precedence of comma over semicolon.)

EXAMPLE   4, 6, 3, 6, 9 is an instance of digit-list

      (3; 2, 5; 6; 4, 3, 8; 7) = (3, (2, 5), 6, (4, 3, 8), 7)

6.2   If **x** is any expression, then

   x-set

is an abbreviation either for

   x, x . . .

or for

   x; x . . .

except that no repetition is permitted. x-set designates an unordered set of entities designated by **x.** The choice of comma or semicolon is semantically irrelevant, as above.

EXAMPLE   (3, 6, 2) and (2, 3, 6) are the same entity, as instances of digit-set.

      (3, 3) is not an instance of digit-set.

7.   Definite expressions

7.1   All definite expressions can be defined by specialization of the indefinite expression entity. In the following, all names are legitimate, as defined in GC 10. Also, any expression that occurs without expression-variables in it is a legal expression of the language as it stands.

7.2   entity := (parameter-set)

   parameter := attribute: value
          := value
               if attribute can be inferred from value

          := attribute | ¬attribute
               if value is binary-value
          := quantity / entity
               if attribute can be inferred from entity.

  value := entity | ?
  binary-value := boolean | (1 | 0) | (on | off) | (high | low) |
               (exist | not-exist) | (+ | −) | (positive | negative)

An entity may be defined (or described) by listing its attributes and values explicitly. There is no natural ordering on the attributes, so they form a parameter-set rather than a parameter-list. The value may be any entity, but for each attribute there will be a domain of possible entities. This domain can always be given as an indefinite expression. The question mark can be used when the value is uncertain. A parameter always defines both an attribute and a value but may be abbreviated in several ways if the context makes clear what the attributes and values are.

1   Both the attribute and value may be given explicitly

   EXAMPLE   M(size: 100 w)

2   The attribute may be dropped, if the value uniquely determines the attribute.

   EXAMPLE   M(1000 w) is legal because the only attribute of a memory that has a number of words as value is size.

   COMMENT   What is inferable is somewhat ill-defined, because it depends on the information available to the reader of the expression (whether man or machine). The simplest case is when the value is a quantity whose unit is uniquely associated with the attribute, as in the example above. Another is when the value is a member of a class (or a subset of that class) and the attribute is the class name (see GC 8.5).

3   Binary-valued attributes may drop the value and use the occurrence of the attribute to symbolize the negative sense and the negated attribute to symbolize the negative sense.

   EXAMPLE   M(destructive_read) for M(destructive_read: yes)
             M(¬destructive_read) for M(destructive_read: no)

4   If the parameter gives some kind of unit quantity, then it is often natural to state the parameter in the form of quantity per entity (quantity / entity), where the attribute either is the attribute itself (the unit to be defined) or permits inference of the attribute.

   EXAMPLE   Memory(word: 32 bits) = Memory(32 bits/word)
             Control(number_devices_controlled: 3) = Control(3 devices / control)

COMMENT   The remark made in point 2 above on "inferable" holds here as well.

7.3   entity := attribute(entity)
An entity can be designated as the value of an attribute of some other entity.

COMMENT   This is simply standard functional notation.

EXAMPLE   Pc(speed: speed(Mp))

7.4   entity := A(parameter-set)
An entity can be defined as having all the parameters of the indefinite expression **A,** further specialized, modified, or augmented by the given parameter-set.

COMMENT   This permits one entity to be defined as an instance or further specification of another "general" entity, allowing the equivalent of sub-routining in building up a system of definitions. It also permits one entity to be defined as like another except in certain specified respects.

EXAMPLE   Let M := Component(size: +integer word; color: blue)

M(size: 100 ~ 1000 word)   *further specification*

M(size: 100; o-rate: 10 s/word)   *further specification, if Component defines o-rate*

M(color: red)   *definition by exception*

M(size: 100; weight: 300 lb)   *definition by augmentation*

7.5   entity := entity-set | entity-list | labeled-entity-set | labeled-entity-list
labeled-entity := label: entity
label := simple-name

An entity can be a set or a list of entities. It is possible to affix labels to the entities of a set or list to make referencing easier.

EXAMPLE   C(M: Mp, Ms, M.ps)   *declares the memory of C*
T(co-components: to: L.1, from: L.2)   *to and from are labels*

7.6   entity := +integer entity
An abbreviation for a list of a specified number (the +integer) of entities, as specified in the entity following the +integer. If the specifying entity is an indefinite expression, then each of the entities is independent.

EXAMPLE   12 M(tape) where each M(tape) may have different further specifications.

7.7   entity := number | quantity | predicate | entity-name
Each of these possibilities is taken up in later sections.

8.   Attributes

8.1   The following gives the possibilities for attributes. It also provides for the automatic definition of certain attributes. Throughout, let **x** be the entity whose attribute is being defined and let **V** be the domain of values of the attribute.

8.2   attribute := simple-name
Simple-names provide freely definable attributes, without restriction on use.

EXAMPLE   C(user_efficiency: fraction) an attribute called user_efficiency can simply be defined and given any domain desired.

8.3   attribute := label
                if **x** is a labeled-list or labeled-set

The labels of a labeled-list or labeled-set automatically become attributes.

8.4   attribute := **V**
Often there exists no separate name for an attribute other than the set of values it can take on (**V**), which already has an appropriate expression in the language.

EXAMPLE   C(Mp: M(1000 w; 32 b/w)) where Mp serves as the attribute, being also the domain.

8.5   attribute := attribute: attribute . . .
A sequence of attributes, interpreted as making an iterated sequence of selections, can serve as a single attribute. The first (leftmost) attribute determines a value of x; the next attribute determines a value in the parameter set of this value, and so on through the sequence.
In symbols:

$$\mathbf{a: b:} \ldots \mathbf{q(x)} = \mathbf{q(p(\ldots b(a(x)) \ldots))}$$

EXAMPLE   X := C(Mp(size: 1000 w))
size: Mp(X) = 1000 w

8.6   attribute := **a**
                if **q: p:** . . . **b: a** is an attribute of **x** and there is only one value of **x** to any depth with attribute **a**

The front end of an attribute sequence can be dropped if the remainder uniquely identifies the value; that is, if there is only one occurrence of **a** within **x** and its values.

EXAMPLE   X := C(Pc, Mp, Ms)
add-time(X) is defined, since only Pc has an add-time.
size(X) is not defined, since both Mp and Ms have size as an attribute.

8.7   attribute := attribute-list
The value is a value-list that corresponds one-to-one with the attributes of

the attribute-list. This is an abbreviation technique that permits writing the attribute names only once for a list of values, each of which has several subattributes.

EXAMPLE operation-times := (add-time, store-time) has values
(10 μs, 6 μs), (20 μs, 20 μs), etc.

8.8 attribute := x-name
This is a single special attribute, defined for each entity x. See GC 10.10 for definition.

8.9 attribute := index / #
where value(index) := + integer | − integer
if x is a list (more generally, of form z o z . . .)

The elements of a list (or other sequence) are automatically indexed by their number from the front (+ integer) or the end (− integer) of the list. This index can be used as an attribute.

EXAMPLE x := (Ma, Mb, Mc, Md)
x(index: 3) = x(#: 3) = x(3) = Mc
x.4 = x.−1 = Md

9. Null symbol and optional expression

9.1 Let ∅ be the null expression

class(∅) = the null class

∅ may occur as the defining expression in an assignment or as a member of an alternation:

x := ∅
x | ∅ | y

∅ may occur as a member of a set or list, in which case it may be deleted from the set or list.

x, ∅, y is equivalent to x, y

9.2 If x is any expression, define the optional expression

*x to be (x | ∅)

Thus, if *x occurs in any expression, it means that either x can occur there or ∅, that is, x has an optional occurrence.

EXAMPLE (1, *2, 3, *4) = (1, 2, 3, 4) | (1, 3, 4) | (1, 2, 3) | (1, 3)

10. Names

10.1 Names are expressions distinguished by two things:

1 They are composed of strings of characters, which are not themselves expressions.

2 They are written without spaces between the characters.

10.2 There is a special class of expressions called name-expressions, which are used to define names.

1 Name-expressions all have names that are of the form x-name, where x is a name.

2 Name-expressions are written with spaces, which are to be removed in generating strings of characters from them.

3 Name-expressions occur only in conjunction with name-expression names, either as an assignment:

x-name := name-expression

or as an attribute-value:

x-name: name-expression

Thus, it can always be determined when a name-expression occurs.

EXAMPLE Q-name := A B (1 | 2)   defines Q-name
AB1 and AB2 are the two possible Q-names.

10.3 Alphabets are defined as the alternates of their characters, e.g.,

digit := 0 | 1 | 2 | 3 | 4 | 5 | 6 | 7 | 8 | 9

Capital letters, small letters, marks, and characters, as laid out in GC 3.3, are defined similarly.

10.4 If x is any set of characters, then

x-string

is a string of such characters of indefinite length (at least one) with no spaces between.

EXAMPLE digit-string contains 1, 1354, 65487, etc.

COMMENT Note that expression-variables are being extended to cover sets of characters and character strings, even though these are not always expressions.

10.5 name := simple-name | compound-name | number-name | x-name

10.6 simple-name := primitive-name | phrase-name | hyphen-name
primitive-name := (capital-letter | small-letter | digit)-string
phrase-name := primitive-name_primitive-name . . .
hyphen-name := phrase-name-phrase-name . . .

Single-names are strings of letters and digits or phrases made up of such strings with space concatenation marks (_) (phrase-names) or with hyphens

(-) (hyphen-names). All simple-names function identically: they obtain their designations through assignment (:=) or abbreviation (/). They may thus be definite or indefinite, corresponding to the expressions they name. Any simple-name may be used if it has not already been used for a different expression or is not excluded by number-name or by a previously defined **x**-name (see below).

EXAMPLES   AB3   SAM   Baker   Instruction_set   input-register   13-B

ABBREVIATION   If there is no chance for ambiguity, phrase-names may be written with a space instead of the space-concatenation mark (_).

EXAMPLE   skip condition = skip_condition

ABBREVIATION   If the hyphen-name **x-a** is used within the scope of the definition of the entity **x**, then the name may be abbreviated to just **a**.

COMMENT   This permits the use of the same name in local contexts, where the name of the context (the expression being defined) serves to disambiguate the name where needed.

EXAMPLE   data-type := (. . . data-type-component: data-type . . .)
           data-type := (. . . component: data-type . . .) *alternative form*

10.7   compound-name := **S** . **v** . **v** . . .
                    where **S** is an indefinite simple-name and the **v** are simple-names.
The compound-name has the same designation as

   S(**v**; **v** . . .)

where each of the **v**'s defines a parameter whose attribute may be dropped because the **v** is self-identifying. Thus a compound-name is an abbreviation technique that constructs a name for an entity by conjoining a series of modifying attribute values to the type of the entity.

EXAMPLE   Memory.primary   is an abbreviation for
           Memory(function: primary)

ABBREVIATION   An intervening period may be dropped if no ambiguity results.

EXAMPLE   Mp is the same as M.p
           Mprimary is the same as M.primary   *though poor taste*

COMMENT   Compound names have the desirable feature that the leading symbol (leftmost) gives the kind of entity being designated, e.g., M.primary is a kind of memory.

10.8   number-name. Defined in GC 11.

10.9   x-name. The names to be used in defining an immediate instance of the entity x. If **x** is any entity and **y** is any name-expression, such that

   x := (x-name: **y**; . . .)

then any **z** which is an instance of **x**,

   z := **x**(. . . . .)

must be chosen from the name-expressions defined by **y**. This holds only for a single level. If **w** := **z**(. . .), then **w** is not constrained as to the name used.

EXAMPLE   component := (component-name: capital-letter)
           M := component (. . .)   is legal;
           SAM := component(. . .)   is not legal;
           SAM := M(. . .)   is legal.

11.   Numbers

11.1   number := number-name | number-variable | number ↓ base | arithmetic-expression | count-expression

number-name := integer | decimal

integer-name / integer := *sign digit-string
recall * *means optional*

sign := + | −

+ integer-name / + integer := digit-string   *includes 0*

− integer-name / − integer := − + integer

decimal-name / decimal := integer . digit-string

base := + integer

arithmetic-expression := unary-arithmetic-operation number |
   number binary-arithmetic-operation number |
   number n-ary-arithmetic-operation number . . . |
   arithmetic-function(number-list)

unary-arithmetic-operation := − | +

binary-arithmetic-operation := − | / | exponentiation / exp / ↑ | modulo / mod

n-ary-arithmetic-operation := + | ×

arithmetic-function-operation := log ↓ 2 | absolute-value / abs | entier | maximum / max | minimum / min | average / avg | sum | product / prod

count-expression := number(**x**-set) | number(**x**-list)

Numbers are defined in the standard way, starting with number-names for integers (1324 or − 14) and decimals (13.23). If the base of the number system is different from 10, it may be given explicitly (for example, $10 \downarrow 2 = 10_2 = 2$). Arithmetic expressions are formed from various arithmetic operations with numbers as operands. Operations are classified by their syntactic form: unary operations ($-(3)$ or $+(7)$); binary operations ($7 - 6$, $3/8$ or $3 \uparrow 2 = 3^2$); and n-ary operations ($3 + 8 + 6$ or $5 \times 6 \times 2 \times 3$). Functions are defined as taking a list of numbers as operands (abs(3) or max(5, 7, -12)). There is a counting function that takes any set or list of entities as inputs and produces their number (if X := (Ma, Mb, Mc) then number(X) = 3). Abbreviations are introduced for many of the operations and functions.

11.2   number-set-name := (digit | Φ)-string

A special subset of (alternative) numbers may be defined by substituting a Φ for a digit. The Φ stands for any digit (of the base of the number).

EXAMPLE $01\Phi = 010|011$ 01Φ *binary*
          $7\Phi = 70|71|...|77$ 7Φ *octal*

12.   Quantities, dimensions, and units

quantity := number unit

unit := (dimension; conversion-list) | unit-name := multiplier unit | simple-name

conversion := number-name unit | number-name / unit | arithmetic-expression(unit)

multiplier := pico / p := 10^{12} | nano / n := 10^9 | micro / μ / u := 10^{-6} | milli / m := 10^3 | centi / c := 10^2 | kilo / k := $(10^3 | 2^{10})$ | mega := 10^6 | giga / g := 10^9

dimension := (base-unit: unit) | [dimension-expression]

dimension-expression := dimension | dimension $\times$ dimension | dimension / dimension

A quantity is a number of units of a given dimension. A unit is defined by the dimension and the conversion between the given unit and other units of the same dimension. Conversions can be expressed either as the amount of the other unit for each of the given units (e.g., 1 minute is 60 seconds) or as the amount of the given unit per each of the other units (e.g., 1 minute is 1/60 per second = .0167 / second). When conversions are not linear, it is necessary to use functions of the other unit. Thus, for bits the conversion to states is $\log_2$(states) (e.g., 128 states is equivalent to $\log_2(128) = 7$ bits).

Each dimension has a base unit (e.g., seconds for the dimension of time). A dimension may also be given as a product of two other dimensions (e.g., [energy] is [force $\times$ distance]) or the ration of two other dimensions (e.g., [velocity] is [length / time]). We use the standard bracket notation to indicate dimension, (e.g., [l/t] for the dimension of velocity).

13.   Boolean and relations

boolean := true / t/1 | false / f/0 | boolean-variable | boolean-expression | relational-expression

boolean-expression := unary-boolean-operation boolean | boolean binary-boolean-operation boolean | boolean n-ary-boolean-operation boolean . . .

unary-boolean-operation := $\neg$

binary-boolean-operation := $\supset$ | $\equiv$

n-ary-boolean-operation := $\vee$ | $\wedge$ | $\oplus$

relational-expression := number relational-operator number

relational-operation := $=$ | $\neq$ | $<$ | $>$ | $\leq$ | $\geq$ | $\equiv$ | $\not\equiv$

There are two primary boolean values, true and false. Boolean-variables, boolean-expressions, and relational-expressions are expressions that evaluate (potentially) to true or false. Boolean expressions are made up from the standard operations on truth values: negation ($\neg$), implication ($\supset$), equivalence ($\equiv$), conjunction ($\wedge$), disjunction ($\vee$), and exclusive-or ($\oplus$). Relations are defined on numbers.

COMMENT   More general definitions for entities (for $=$ and $\neq$) and for ordered sets (for $<$, $>$, $\leq$, and $\geq$) are not needed.

PMS conventions

Making use of the prior general conventions, PMS is developed systematically through the definitions of the various components: P, M, S, etc. Much of the development repeats common abbreviations and conventions, simply to provide a self-contained notational system.

1   Dimensions

2   General units

3   Information units

4   Component

5   Link (L)

6   Memory (M)

7   Switch (S)

8   Control (K)

9   Transducer (T)

10   Data (D)

11   Processor (P)

12   Computer (C)

1. Dimensions

1.1   Definition of dimension, repeated from GC 12.

dimension := (base-unit: unit) | [dimension-expression]

1.2   Basic dimensions

time / [t] := dimension(base-unit: second)

length / [l] := dimension(base-unit: meter)

cost / [$] := dimension(base-unit: dollar)

weight := dimension(base-unit: kilogram)

power := dimension(base-unit: watt)

temperature := dimension(base-unit: degree-centigrade)

voltage := dimension(base-unit: volt)

current := dimension(base-unit: ampere)

component / [c] := dimension

operation / [o] := dimension

information / [i] := dimension(base-unit: bit)

state := dimension(base-unit: state)

2. General units

2.1   Definition of unit, repeated from GC 12.

unit := (dimension; conversion-list) | unit-name := multiplier unit | simple-name

conversion := number-name unit | number-name / unit | arithmetic-expression (unit)

2.2   We give the basic units, but no variations with multipliers.

second / sec / s := unit(dimension: time)

minute / min := unit(dimension: time; conversion: 60 s)

meter / m := unit(dimension: length)

foot / ft := unit(dimension: length; conversion: 3.28 / meter, 12 in)

inch / in := unit(dimension: length; conversion: 39.37 / meter, 12 / ft)

dollar / $:= unit(dimension: cost)

operation / o := unit(dimension: operation)

watt / w := unit(dimension: power)

volt / v := unit(dimension: voltage)

ampere / amp / a := unit(dimension: current)

kilogram / kg := unit(dimension: weight; conversion: 2.2 / lb)

pound / lb := unit(dimension: weight; conversion: 2.2 kg)

3. Information units

3.1   *Units*

state := unit(dimension: state; conversion: 2^x bits)

binary-digit / bit / b := unit(dimension: [i]; conversion: $\log_2(x)$ states)

octal-digit / od := unit(dimension: [i]; conversion: 3 bits)

decimal-digit / digit / d / dit *rare* := unit(dimension: [i]; conversion: $\log_2(10)$ bits, $\log_{10}(x)$ states)

hexa-decimal-digit / hex := unit(dimension: [i], conversion: 4 bits)

character / char / ch := unit(dimension: [i]; conversion: $4 \sim 8$ bits)

byte / by := unit(dimension: [i]; conversion: 8 bits)

COMMENT   The byte is almost standardized at 8 bits; occasional use otherwise, although not in this book.

3.2   *I-units*

i-unit := base-unit | length × i-unit | i-unit-name | (base-unit; length-list; content: product(length-list) base-unit; level:number(length-list))

i-unit-name := i-unit-prefix i-unit-name | simple-name

i-unit-prefix := +integer | multiple/m | quadruple/q | triple/t | double/d | *single/s | half/h | fractional/fr

base-unit := unit(dimension: [i])

length := +integer

The i-unit is a hierarchically organized information structure, in which each level consists of a number of subunits, all identically organized. The number of subunits in a level is called its length. Units eventually occur that cannot be decomposed further. These are called base-units and are some unit of information—e.g., the bit or the character. Thus, if the lengths are $L_1, L_2, \ldots, L_n$ and the base unit is the bit, then the total amount of information (the content of the i-unit) is $L_1 \times L_2 \times \ldots \times L_n$ bits and the number of levels is n. The i-unit may be likened to an n-dimensional rectangular volume of information (except that the "dimensions" —the lengths—occur in a fixed order).

COMMENT   Almost all information in computer systems is organized in terms of i-units—e.g., a memory consists of a number of words, each of a number of characters, each of a number of bits. More exotic data structures are invariably encoded into i-units and are not reflected in the hardware.

word := length $\times$ bits | length $\times$ character | length $\times$ base-unit

word-bit-length := 12 $\sim$ 64

word-character-length := 2 $\sim$ 8

block := length $\times$ word | length $\times$ character

record := length $\times$ word | length $\times$ character

file := + integer $\times$ block | + integer $\times$ record

IBM-card / card := column $\times$ row $\times$ card-hole

card-column / col := 80

card-row / row := 12

card-hold := 1 bit

print-line / line := print-column $\times$ character

print-column / col := 64 $\sim$ 132 | 72 | 80 | 120 | 132   *rarely* <64

4. Component

4.1   component := (

component-name: capital-letter;

manufacturer-name / ' : *manufacturer catalog-number;

operation-set;

operation-rate-set;

*subcomponents: (function-attribute: component)-set;

*cocomponents: (function-attribute: component)-set;
port-set;

function: (subcomponent-attribute | cocomponent-attribute);

logic-technology;

*technology;

reliability: (mean-operations-between-failure / MOBF, mean-time-
      between-failure / MTBF);

error-rate: (erroneous-operations / error-free operations);

cost: purchase, rental;

lineage;

history;

weight;

power;

volume;

area;

temperature)

This single definition of a computer component contains all of the attributes common to all components. All components can thus be given as further specifications of this definition. (Such definitions can add attributes not in the higher entity.) Examples are given in succeeding sections. We comment on some of the attribute domains below and provide an extensive listing of values for some.

4.2   Component-name. All components that are immediate instances of this definition are to have single-letter names—for example, P, M, S, etc. Names of instances of P, M, S, etc., are arbitrary.

4.3   Manufacturer-names | Proper-name. We provide a very short abbreviation (') to indicate that a string of characters is a manufacturer's name, since these names are arbitrary and need to be distinguished from other values. A proper name can also be given to a component.

EXAMPLES   'IBM System/360 Model 50.   'I/O_Bus

4.4   Operation-set and operation-rate-set. A component is defined fundamentally by the set of operations it can perform. In PMS such operations are defined informally and given names (e.g., read, transmit). Significant performance parameters may be defined, but complete definitions are given only in ISP. Each operation has a rate (number per unit time), which need not be constant.

EXAMPLE   A link might have an operation-set consisting of two transmission operations (one in each direction) of a single i-unit. The operation-rate might be $10 \uparrow 3$ o/s for each operation. If the i-unit were 10 b, it would be given an information-rate of $10 \uparrow 4$ b/s.

4.5   Subcomponents, cocomponents, function. In general, components consist of PMS structures of other components, which are called its subcomponents. Also, in general, a component participates in a PMS structure. The components to which it is connected are called its cocomponents. The connecting interface of a component and a cocomponent is called a port. Conventional names exist that describe the roles the components play in a PMS structure (e.g., central processor, buffer memory, address switch). These terms are called functions and can be used to label both subcomponents and cocomponents.

*4.6   port := (

operations: (output | input);

operation-rate / o-rate;

i-unit:[i];

information-rate / i-rate: ((i-unit / operation) $\times$ o-rate [i/t]);

concurrency: + integer;

concurrency-type: (simplex | half-duplex | full-duplex | time-multiplex |
      multiplex);

direction: (from / out / output / X →)|(to / in / into / input / X ←);

turn-around-time / t.turn: [t] *only for half-duplex carrier;*

carrier)

carrier := (

writability: (human / h|machine / mechanical process / m|
both machine and human / b);

readability: (human / h|machine / mechanical process / m|
both machine and human / b);

medium;

encoding)

medium := (electrical conduction := voltage|current)|

magnetic|electrostatic|radiowave|microwave|optical light|

(mechanical movement := tactile|linear position|angular position|

spatial position)|temperature / heat|

(acoustical / airpressure := high frequency audio)|memory technology
see PMS 6.2

encoding / modulation := continuous-modulation / analog|

digital / discrete-modulation

continuous-modulation := direct / null|amplitude / am|

pulse amplitude modulation / pam|pulse duration modulation / pdm|

time duration modulation|frequency modulation / fm

discrete-modulation := direct / pulse code modulation / pcm|

frequency shift keying / fsk|digital pulse|digital level|contact

The ports are the connection points (nodes or terminals) of a component at which cocomponents connect. A port is not a component but simply an interface with a characteristic i-unit that crosses it in one direction or the other. One can thus associate two operations with a port, namely, the transmission operations of its component and the cocomponent. The port introduces directionality: input is from the cocomponent into the port's component; output is from the port's component to the cocomponent.

The i-unit subcomponents usually correspond to physical subparts of the port. For conventional information-carrying structures, the base-unit is the encoding of information on a single wire of the port, i.e., a bit. The width is the number of wires available per unit time. The length is the number of (width × base-unit)'s which are necessary to transmit the i-unit. As such, the i-unit can be thought of as a message normally with length × width × base-unit. More complex messages can have multiple dimensional lengths (e.g., consider a record which is transmitted serially, where the base-unit is a bit, the width is 1, the length is an 8-bit byte, and the record length is 1,000 bytes).

The information rate as measured at the port is the flow of i-units per unit of time. An equivalent measure is the time for the i-unit to pass through the port. Concurrency is a measure of the number of simultaneous i-units the port can pass. Concurrency-type denotes both the number of simultaneous messages and the message direction. The simplex port allows only one message to enter or leave the port, not both. The half-duplex port allows a message to either enter or leave the port, but only on a time-multiplexed basis; that is, the port is simplex for one direction at a time. In the case of the half-duplex port, the turnaround time is a significant attribute that denotes the time taken to go from receiving to transmitting or vice versa. A full-duplex port allows information to flow in both directions at once (i.e., enter and leave the port simultaneously). Finally, the multiplex port denotes multiple ports that can be decomposed into the more elementary structures discussed above.

Direction is usually indicated on each port of a component to denote the direction of information flow. Direction must be specified for simplex ports (using arrowheads ←, →). Half- and full-duplex ports are shown with no arrowheads.

Carrier characterizes the form of information at a port. The two major attributes, writability and readability, define whether human beings, machines, or both human beings and machines are able to use (interpret) the carrier directly. Media denotes the technology of the carrier. Information can be carried by any of the media listed. It should be noted that memory technology is also listed as a media to carry information. Unlike the media that are instantaneous carriers, memory holds information over a long period of time. For each media, it is appropriate to encode information in particular ways. The two basic methods are continuous and discrete encoding (or modulation).

4.7   Logic-technology and technology. All devices have a logic technology and almost always only a single one (though exceptions exist, especially in compound components). They may also have other technology specific to the type of component (e.g., disk-memory technology). The logic technology is given here; other technologies are given with the specific component.

logic-technology := magnetic-core|cryogenic|

electro-mechanical|fluidic|hybrid-circuit|

monolithic integrated / integrated / ic|large scale integrated / LSI|

mechanical|integrated metal oxide silicon / MOS|

medium scale integrated / MSI|optical|

transistor|vacuum-tube

4.8   Reliability. Although of extreme importance, we list only two values for reliability, the mean number of operations between failures, and the mean time between failures. In essence, one can be derived from the other if the operation rate is known.

4.9 Error rate. Usually a ratio of the number of erroneous operations per error-free operations. Approximately 1/(probability of an error).

4.10 Cost. Only the two simplest cost numbers, purchase price and (monthly) rental are listed as attributes. Conventionally, purchase price is taken as 45 times monthly rental. In addition, one could list manufacturing costs, broken down into materials, labor, etc., and more elaborate sales costs, such as lease-purchase options. Most of these quantities are not relevant from an engineering viewpoint. Some that are important are unobtainable in general.

4.11 lineage := (

manufacturer: Burroughs|

Control Data Corporation / CDC|

Digital Equipment Corporation / DEC|

English Electric|

Ferranti|

General Electric / GE|

Honeywell|

International Business Machines / IBM|

International Computers and Tabulators / ICT|

Hewlett-Packard / HP|

Olivetti|

Radio Corporation of America / RCA|

Remington-Rand / UNIVAC|

Scientific Data Systems / SDS / Xerox Data Systems / XDS|

Westinghouse;

manufacturer-type: government / g|industrial / i|
research-laboratory / r|university / u;

country: Australia / A|Great Britain / B|Canada / C|Denmark / D|
France / F|Germany / G|Israel / H|Italy / I|Japan / J|
Netherlands / N|Russia / R|Sweden / S|United States / *U;

*descendants: component-set;

*antecedent: component-set)

The attributes are mostly self-descriptive. We have not attempted to list manufacturers other than the principle industrial ones. Descendants and antecedents are necessarily vague, since no precise notion of parenthood can be defined. It is not limited to computers built as a series (as in the IBM 704 being a descendant of the IBM 701) but includes any machine where the design bond is strong (e.g., IBM 709 and 7090).

4.12 history := (

t.conception / t.start: date;

*t.announcement / t.paper: date

*t.birth / t.prototype / t.operational: date;

*t.scheduled: date;

*t.exhibited: date;

*t.delivery / t.production: date-list;

*t.first-delivery / t.first: date;

*t.last-delivery / t.last / t.withdrawal: date;

*t.death / t.last-use: date;

*production: number(t.delivery))

date := year|month year|day month year|quarter year

quarter / q := winter / 1|spring / 2|summer / 3|fall / 4

The history of the component is viewed as a series of event dates, only the more important being given above. Often the same essential function is served by a variety of events (e.g., the announcement of a computer to the public can be made either by formal announcement, as happens with commercial systems, or by a technical paper). Delivery or production refers to the actual placing of systems and consists of a series of dates, one for each instance produced. This series is normally abbreviated to the first and last delivery, plus the number produced. None of the attributes beyond t.start need exist, as a computer system can be aborted at any time. For all attributes, the dates may be known only approximately.

4.13 Weight, power, volume, area, temperature. Since we concentrate on the informational aspects of components, other attributes are mentioned only briefly (and others, such as decor, are left out entirely). The values of these parameters are especially important in aerospace applications. They also show the effects of technology on packaging and computing power per unit volume.

5. Link

5.1 Link / L := simple-link|compound-link

5.2 simple-link := component (

cocomponents: (input: component, output: component, initiators: input|output|both);

subcomponents: (*control; *input-buffer: M.i-unit; *output-buffer: M.i-unit);

concurrency: 1;

concurrency-type: simplex;

information-rate / i-rate: (i-unit/operation) $\times$ o-rate [i/t];

i-unit: i-unit(input) equals i-unit(output);

delay / t.delay / td: [t];

carrier)

A simple-link has the capability of moving an i-unit from the input cocomponent to the output cocomponent. The simple-link has two simplex ports that connect to the ports of the two cocomponents and are separated by a delay. In essence, as the delay goes to zero, the input port and output ports become one. Initiation of the transmission may be fixed at one end or the other or be from either end, depending on the design of the link. The base-unit is usually a bit (i.e., two states), but it may be more. The width of the i-unit is the number of base-units transmitted in parallel; and the length is the number of widths serially transmitted in one operation. A simple-link permits transmission in one direction only (from input to output cocomponent); this is normally called a simplex link. The port-to-port delay is the time from the initiation of the transmit operation at one port to the arrival of the i-unit at the second port. (Occasionally, the arrival time between widths can be relevant operationally, and then a more precise characterization of the time structure would be required.) The rate of transmission (the information rate) may be calculated by taking the operation rate times the information transmitted per operation (i.e., the content of the i-unit). Links may—but need not—contain buffering at either end for a single i-unit. There may be a distinct control involved, especially if initiation and termination rituals must be accomplished; but it is possible to have links that are simple wires and simply present at the output terminal what was presented at the input.

EXAMPLE   L $\begin{bmatrix} \text{input: register A; output: register B; width: 36 b;} \\ \text{1 megawords/s} \end{bmatrix}$

5.3   compound-link := (

simple-link(concurrency: 1; concurrency-type: half-duplex)|

simple-link(concurrency: 2; concurrency-type: full-duplex)|

simple-link(concurrency: +integer; concurrency-type: broadcast;
    output: component-set)|

simple-link(concurrency: +integer; concurrency-type: network broadcast; input: component-set; output: component-set)|

simple-link(concurrency: +integer; concurrency-type: star)|

(simple-link)-set)

A compound-link is made up of several links, but such that no switching occurs. A half-duplex link permits information to flow from either terminal to the other, but transmission is possible in only one direction at

a time—which thus leads to a turnaround delay time. A full-duplex link permits simultaneous transmission in both directions. Broadcast links permit transmission to many receivers; thus the output components can be set. Network broadcast permits more than one terminal to be a source, though only one at a time. The star denotes all n components of a set to simultaneously communicate with one another via $(n/2) \times (n-1)$ full-duplex links.

Finally, a set of disjoint links (that is, inputs disjoint and outputs disjoint) can be considered to be a single link. This latter is essentially a convenience for naming a multiplex link.

EXAMPLES   L $\begin{bmatrix} \text{Dataphone; 1800 b/s; half-duplex; i-unit: (length: 8,} \\ \text{width: 1 b)} \end{bmatrix}$

L(Telephone; i-rate: 110 b/s; direction: full-duplex)

Telephone := L(110 b/s; full-duplex)   *alternative form*

I/O Bus := L $\begin{bmatrix} \text{half-duplex; i-unit: 1 w; 12 b/w;} \\ \text{operation-rate: 500 ko/s} \end{bmatrix}$

L $\begin{bmatrix} \text{'I/O Bus; half-duplex; i-unit: 1 w;} \\ \text{12 b/w; 500 kw/s} \end{bmatrix}$ *alternative form*

L $\begin{bmatrix} \text{'I/O Bus; half-duplex; i-unit (length:} \\ \text{12 b; width: 1 b); 6 megabits/s} \end{bmatrix}$ *alternative form*

6.   Memory

6.1   Memory / M := simple-memory|compound-memory

6.2   simple-memory := component (

cocomponents: read: component, write: component;

functions: *see Table 1;*

subcomponent: control;

word / w: i-unit [i];

size: 1 word [i];

operations: (read|write|read, write);

information-rate / i-rate: [i] / word $\times$ operation-rate [i/t];

access-time / ta: constant|~constant [t];

cycle-time / tc: time(read; next write) [t];

permanency: (decay|fast-read-slow-write / frsw|permanent / read-only / ro / ros / ROS / read-only-memory / rom / ROM| read-destruct|read-regenerate / rr|read-write / rw|write-only) [t];

portability: (portable / p|not portable / fixed / f);

technology: *see Table 2)*

Table 1   Memory functions

Within C

primary / p	Primary memory; holds directly executable programs; instructions and data for instructions are taken from Mp and it must be directly accessible by P
secondary / s	Secondary memory, in which data accessible to the ISP is stored; programs are not executed from secondary; normally Ms is much larger than Mp (and much slower); Ms holds files, programs (waiting to be executed), data, etc.

Within P, K

address	Holds operands
buffer / synchronizer	Holds data while synchronizing with another component
control	Used during instruction's interpretation; state of a K
data operands	Holds information that are operands or eventual operands
fixed	Used to define permanently the nature of a processor or a control
error detection	Holds detected error information, normally hardware errors
error accounting	Holds counts of errors; normally part of Mps; two major types or errors, machine (or hardware) errors and process (or program) errors, are accounted
instruction	Holds parts of instruction as it is being interpreted
processor state / ps	Includes all registers, state bits, and instruction counter associated with ISP; includes the following subcomponents:
program state word	Holds the state of the program flow, overflow bits, i.e., the instruction or program counter, and any state bits accessible to a program
process map	Used to locate programs within Mp (and Ms)
process registers	Specific arithmetic and indexing registers (e.g., AC, MQ, general registers, stack)
program address / instruction address / instruction location counter / program	Holds pointer to either the current or the next instruction the processor is to interpret
working / temporary	Holds intermediate results

Within T, L

buffer / synchronizer	Used for synchronizing purposes
control	The K part of T or L
working / temporary	Temporary results

Within D

control	K part of D
data operand	D may stack operands and results, synchronizing with some other process
instruction	Current operation D is performing
working / temporary	Temporary results of intermediate data

Within S

address	Position of switch, i.e., the information that holds gate-switches open or closed
buffer / synchronizer	Any synchronizing storage needed within S for links
control	The K part of S

Table 2   Memory technology

Machine readable; machine writable

	Access†	Portability†	Permanency†
capacitor	r	f	decay
core / magnetic core	r	f	rr
bulk core / large core storage / lcs /	r	f	rr
extended core storage / ecs	r	f	frsw
delay line / magnetostrictive delay line	c	f	rr
mercury delay line	c	f	rr
optical delay line	c	f	rr
disk / diskpak	1, c	p	rw
fixed head disk	1, c	f	rw
moving head disk	c	f	
drum / fixed head drum	c	f	rw
moving head drum	1, c	f	rw
electrostatic storage tube	r	f	decay
integrated circuit array	r / content	f	rw
logic / technology *See PMS 4.6 for logic used to make active bit, register and array memories*	r	f	rw
magnetic card *e.g., Datacell*	1, c	p	rw
magnetic tape / tape	1	p	rw
addressable magnetic tape	b	p	rw
carousel magnetic tape	c, 1	p	rw
magnetic wire	1	p	rw
photographic store *e.g., photostore*	1, r	p	wlro
film (write once)	1, r	p	wlro
plasma display *readability: both*	r	f	rw
thin film	r	f	rw

Machine readable; read-only; nonportable; random access

	Access†	Portability†	Permanency†
capacitor array	r	f	ro
diode array	r	f	ro
inductor array	r	f	ro
rope / transformer coupled braided	r	f	ro
rope resistor	r	f	ro

Memories which cannot be both read and written by a machine

	Writa-bility	Reada-bility	Access	Perma-nency
badge	b	b	1	ro
card / punched card	m|b	m|b	1	wlro
credit card	b	b	1	ro
cathode ray tube / CRT	m	h	r	decay
storage CRT	m	h	r	wo
garment tag	m	b	1	wlro
joystick	h	b	r	rw
keys / pushbuttons keyboard	h	b	r	rw
knobs	h	b	r	rw
page / impact printed page / paper	m	b	1	wlro
braille page	m	h	1	wlro
handprinted page	h	b	1	wlro
handwritten page	h	h	1	wlro
magnetic ink page	m	b	1	wlro
thermal page	m	b	1	wlro
typewritten page	b	b	1	wlro
xerographed page	m	b	1	wlro
paper tape / punched paper tape	m|b	m|b	1	wlro
plot / incremental point plot	m	h	1	wlro
analog plot *continuous*	m	h	r, 1	wlro
patchboard	h	b	r	rw
switches / toggle switches	h	b	r	rw

†See PMS 6.2 for abbreviations, also c/cylic, l/linear, r/random.

A simple-memory stores a single word of information by means of a read operation and delivers that word on subsequent write operations. There is no addressing, and the access time is a constant (or approximately so). The memory is connected to the larger system via one component for its read operation and one for its write operation. These are usually links and need not be distinct. The only subcomponent that need be distinguished in a simple-M is the control (though of course the word may be built up from a set of bit memories). The information rate is the amount of information in a word times the operation-rate. The cycle time is the time it takes to read the memory and then write new information into it; the ISP expression (read; next write) implies a sequential operation. The permanency describes what happens to information left in the memory as a function of time. This concept is often partially covered by other notions, such as reliability, volatility, destructive-nondestructive, etc. We give the main values that arise in practice: a rate of decay with time (which expands to an actual decay function); write-once-read-only (e.g., cards and photographs); read-write; fact-read-slow-write (a special case of read-write); destruction of the information upon reading; and permanent or read-only (as long as the system remains viable). Write-only refers to the characteristic of the memory from the point of view of the system under discussion; always there is some other system (usually a human being) who can read the memory. Whether the memory can be only read or only written

(readability, writability) or both read and written, and by whom (human or machine), is derived from the port characteristics. Portability denotes whether information can be carried away from the system or is non-portable (fixed). Two of the parameters, function and technology, are extensive enough to give by tables.

6.3   compound-memory := component (

cocomponents: read: component, write: component, address: component;

function: *see Table 1;*

subcomponents: control; address; switch; memory: M-set, *read-buffer: memory, *write-buffer: memory;

word: word(M.memory);

size: sum(word(M.memory));

operations: read-set, write-set;

information-rate: [i] / word × operation-rate [i/t];

access-time: access-time(S.address) [t] *random, cyclic, etc. see PMS 7.3;*

cycle-time: cycle-time(simple-M);

permanency: permanency(simple-M);

portability: portability(simple-M);

technology: *see Table 2)*

A compound-memory is a system of simple-memories, organized by an addressing switch. Thus memory is fundamentally defined recursively as a switch to other memories. At each switch stage the dimensionality of the overall i-unit is reduced by one. The addressing may be provided by a different cocomponent than those for the read and write data. All the submemories have the same word, and the size of the compound-memory is the sum of all these words. There may be additional subcomponent memory within a memory, such as buffer memories and a memory connected with the address switch and the control. However, none of these are available for storage purposes and are not counted in the size. The access time of the memory is defined by the access time of the address switch. A classification of these can be found under the definition of switch and is often used to classify memories (e.g., linear, random, cyclic, etc.). Some parameters are the same as those given for a simple-memory, and these are simply cross-referenced.

COMMENT   Not all conceivable memories come under the definitions just given (e.g., we have assumed constant word size); but in fact all memories used in existing digital computers do.

EXAMPLES   Mp(core; t.access: 2 us/w; 4096 w; 16 b/w)
M(fixed head disk; t.access: 0 ∼ 17 ms; i-rate: 300 kchar/s; size: 1 megaword)

7. Switch

7.1 Switch / S := gate-switch | simple-switch | compound switch

7.2 gate-switch := component (

cocomponents: (input: component, output: component: initiators: component);

subcomponents: (*control; *input-buffer: M.i-unit; *output-buffer: M.i-unit);

operation: (open | close);

concurrency: (1 | 2);

concurrency-type: (simplex | half-duplex | full-duplex / duplex);

i-rate: i-rate(link);

delay: delay(link);

hang-up-delay: [t];

access-time / ta: constant [t])

A gate-switch acts as a simple-link or as no connection. It is used to transmit information conditionally between the ports of two components. It can be used as a basic primitive to express the structure of other switches, including the simple-switch. The parameters will be discussed under the simple-switch.

7.3 simple-switch := component (

cocomponents: (input/from: component-set, output/to: component-set, initiator: component-set);

subcomponents: control, links: link-set, *address: memory;

operation: access;

size: size(output(cocomponents));

concurrency: + integer;

concurrency-type:(simplex | half-duplex | full-duplex/duplex | dual-simplex | dual half-duplex | dual full-duplex / dual-duplex | time-multiplexed-cross-point / 1 trunk | cross-point | dual-cross-point | k-trunk);

hierarchy: (hierarchical | nonhierarchical / anarchical);

location: (central | distributed (cocomponent set));

distribution: (radial | bussed / bus / chain / daisy chain);

access-time / ta: switch-type(address / a, prior-address / p)

switch-type := (

bilinear: constant + constant $\times$ abs$(a - p)$ |

cyclic: constant + constant $\times$ $(a - p)$ mod (size) |

interleave: (a interleave-relation $p \rightarrow$ random)-list |

linear: ($a \geq p \rightarrow$ constant + constant $\times$ $(a - p)$;

   $a < p \rightarrow$ reset-time + constant $\times$ a) |

first-in-first-out / fifo / queue: (constant | ~constant) |

   last-in-first-out / lifo / stack: (constant | ~constant) |

   dequeue: (constant | ~constant));

permanency: (decay | transmit-destruct | time-multiplexed / tmx / tm | moving | cyclic | permanent | irreversible | fixed until broken / fixed | manual);

hang-up-delay: [t];

delay: delay(links);

L-initiator: initiator(links);

technology)

A simple-switch consists of a set of potential links between a set of input and output components, with an operation (access) that can actualize some subset of the links. This is done according to an instruction called the address (which may or may not be held in a memory). For a switch, the cocomponent input and output ports are sometimes listed to specify the size of the switch.

An important parameter is the concurrency-type, which describes the various subsets that can be simultaneously realized. The values given correspond to practical alternatives—simplex, in which only a single simplex link may be established at a time; duplex, in which a single full-duplex link may be established; cross-point (also dual-cross-point), which permits true simultaneity; time-multiplexed-cross-point, in which functional simultaneity is established for many links by means of rapid switching within the course of transmission of an i-unit (in essence the time multiplexed-cross-point has 1-trunk, which permits 1 conversation); and finally k-trunks for k-simultaneous conversations. We often use a duplex switch instead of simplex or half duplex switch in PMS diagrams, even though the latter would be more accurate.

Hierarchy is a redundant attribute derived from the cocomponent set. As a rule, if there are n identical cocomponents each of which communicates with one another, there is no hierarchy. A telephone system is a typical nonhierarchical structure. Usually the switches internal to a computer are hierarchical in that there are n components of type a which communicate with m components of type b. The a's only communicate with the b's and vice versa: hierarchy does not determine the component initiating the dialogue.

The location of a switch refers to whether the hardware is localized within one of the components using the switch, whether it is separate (called central), or whether it is distributed through all the cocomponents.

An attribute that is not completely independent is distribution, which denotes whether the physical structure is a continuous bus or chain or is fed radially from a centralized component. See Fig. 13, Chap. 3, page 67 for common alternative physical structures.

A major way of classifying simple-switches is by their access time—cyclic, linear, random, etc. With each is given the type of formula that determines the actual access time. The two critical parameters in most switches are the address being sought (a) and the prior address (p), which represents the existing state of the switch. Thus, in a bilinear switch the access time consists of a start-up time plus a time proportional to the magnitude of the difference between the prior address and the desired address. This differs from a linear switch, which only permits movement in one direction and must reset to an initial state if an address lower than the existing address (p) is sought. An interleave memory is one that consists of a collection of random-access memories, depending on the relationship between a and p (usually a modular one, such as $(a = p \mod 4) \rightarrow$ long access; $a \neq p \mod 4 \rightarrow$ short access). Random access means that the access time is independent of both a and p. This constancy may be only approximate (as in using a drum with its cyclic character ignored). Queues and stacks differ from the other switches in having a degenerate addressing system such that the next link selected is determined by the state of the switch itself. Dequeues allows either of the two ends of a queue to be accessed.

Permanency refers to how long the switch maintains a link (or set of them) after establishing the link by an access operation. The three common values are (1) the destruction of the connection with the transmission of the i-unit across the link, (2) the maintenance of the connection permanently, and (3) the autonomous movement of the connection (as in disks and drums). The latter two give rise to the p used in the access formulas. Rarer is a decay function, in which the link remains established for some period of time, or an irreversible connection, which can be set just once and from then on operates like a simple-link.

Hang-up delay is the time taken to break a connection after the appropriate i-unit has been transmitted. Hang-up delay is given only for certain permanencies of fixed-until-broken and manual switches.

A number of parameters derive directly from the properties of the set of ports or links—the size of the i-unit, the information-rate, the link delay, the direction of data flow, and the component that can initiate data transmission (as opposed to initiating accessing). Finally, there is technology, which is not given in detail, since much of it is identical to memory technology.

EXAMPLES   S('I/O BUS; location: K; from:P; to:K; half-duplex; initiators: P, K; switch-type: random; ta: 5μs; concurrency: 1)

S(cross-point; 16 M; 6 (P + K); concurrency; 6; location: M)

7.4   compound-switch := simple-switch (

    subcomponents: control, links: link-set, subswitches: switch-set, *address: memory;

    access-time: (cascade: sum(access-time(subswitches)) |
                 parallel: max(access-time(subswitches))))

A compound-switch is an array of switches whose links are connected so that the outputs of some are inputs to others and thus effects a total set of links, which go from output to input component-sets. It can be defined as an extension of a simple-switch, since most parameters are defined identically for both. Many combinations of accessing arrangements are possible. The two most common are given above. A cascade-switch is one in which each accessing of the next subswitch must take place after the prior one so that the access times add. A parallel-switch makes all the accesses simultaneously, so that the total access time is simply the access time of the subswitch that takes longest. (In both cases, there can be additional overhead time, but this can usually be allotted to the subswitches and does not require separate terms in the expressions for access time.)

8. Control

8.1   Control / K := simple-control | compound-control

8.2   simple-control := component (

    cocomponents: controlled / object: component-set, *instruction: component-set, *data: component-set;

    subcomponents: *instruction: memory, working / w: memory, operations: data-operation;

    operations: evoke / $\rightarrow$, next-evoke / next, condition-operations;

    controlled-operations: (controlled-component: operation)-list;

    instruction-source: (none | data | instruction);

    instruction-set)

A simple-control is a logical circuit (usually sequential) that evokes operations in other components (the controlled, or object, components). Thus, its main operations are those of evoking and evoking-next (symbolized as $\rightarrow$ and *next* in ISP). However, it must also detect conditions on which such evoking depends, so that it has available additional operations, that are combined in an instruction-set (see ISP 2.1). These vary greatly in complexity, from boolean operations to arithmetic operations (such as counting the number of i-units processed).

A major distinction is the source of the external instructions that can be given the control. At one extreme there may be none, as in a clock whose function is to interrupt the system every millisecond. The common case is that in which all the external instruction comes via the data itself. More complex controls have a separate set of external instructions (often called control characters or commands). A control does not obtain its own next instruction, being dependent on an external component to set it into action. This is the primary characteristic that distinguishes it from a processor. It does have an instruction-set, which is the ISP expression that shows what conditions evoke what actions.

No technology is given, since controls are all realized in a logic technology, as given in the definition of component. Likewise, no function parameter is given, since there exists no special vocabulary to designate the different subspecies of control tasks.

EXAMPLES   K(Mp; input: Pc; output: Mp)
    K(D(multiply))

8.3 compound-control := simple-control (

 subcomponents: alternatives: simple-K-set, *instruction: memory,
  working: memory;

 instruction-source: mode-instructions)

A compound-control consists of a collection of alternative simple-controls and can be given as an extension of the simple-control. At any time, the control is one of these simple-controls. Determination of what simple-control is operative (often called the mode the control is in) is by a mode-instruction from some external component. This additional freedom requires a subcomponent, the control-state, to hold the current specification. (Thus it is possible, though rare, that the actual simple-K is determined by a sequence of mode-instructions, each determining some part of the control state.)

EXAMPLE   K(Instruction set processor/ISP; input:M.processor_state; output: D, K(Mp), K(L('I/O Bus)); M(read-write; 40 b; working); M(read only; 100 w; 36 b/w 1 μs/w))

9. Transducer

9.1 Transducer / T := simple-transducer | compound-transducer

9.2 simple-transducer := component (

 cocomponents: input: component, output: component, initiator:
  (input | output | both);

 subcomponents: input: L, output: L, *control;

 functional-name: (input: reader / sensor / pen / receiver; output:
  writer / punch / perforator / display / printer / transmitter;
  synchronizer isolator; transducer);

 operation: transduce (plus transmit) / ←;

 carrier *See port of component;*

 transduction: port(output) ← port(input);

 divergence: i-unit(output) − i-unit(input) [i];

 divergence-rate / divergence × o-rate [i/t];

 portability: (portable | not portable / fixed);

 concurrency-type: simplex;

 concurrency: 1;

 transduction-technology := (amplification | analog-digital | angular-
  linear | attentuation | electroluminescence | electromagnetic |
  electromechanical | electromechanical-acoustic | electro-optical |
  mechanical-indentation | photochemical | xerographic)

transducer-technology := (analog-digital converter | bell | buzzer | TV camera / vidicon | card reader | card punch | CRT display | storage CRT display | plasma display | 3 D display | printed document reader / document reader | document printer | magnetic character document reader | film reader | film | writer | gong | joystick | keys | keyboard | light gun | light pen | continuous line plotter | line printer / printer | linear actuator | SRI mouse | paper tape reader | paper tape punch | incremental point plotter | pressure transducer | speech synthesizer | Rand tablet | Sylvania tablet | telephone dial | push button telephone dial | thermocouple | Lincoln Laboratory Wand))

A simple-transducer is a pair of connected links that have different i-units and/or underlying carriers. As defined above, transduction is a digital operation, taking in an i-unit of the input link and producing an i-unit of the output link. Meaning is preserved; that is, only the encoding has changed. Preservation of meaning distinguishes transduction from data operation. The amount of information need not be preserved, so that information divergence is an additional characteristic of a transducer. It may be positive or negative, as the net number of bits is either increased or decreased.

 A simple-transducer is called a simplex, in that information flow is in one fixed direction only (as in a simple-link).

 Knowing the function of the transducer permits an inference of whether one interface of the transducer involves a human being. This inference can be derived from the port characteristics.

EXAMPLE   T(line printer; 1000 lines/m; 132 char/line; 8 bit/char)
    T(paper tape; reader; 300 char/s; 8 b/char; width: 1 in.)
    T(sense amplifier; i-rate: .5 w/s; 24 b/w; input: M(memory
     stack))

9.3 compound-transducer := (

 simple-transducer-set;
 concurrency-type: (half-duplex | full duplex);
 compound-transducer-technology;

 concurrency: +integer)

 compound-transducer-technology := card reader-punch | computer
  console / processor console / console | Dataphone | keyboard-CRT
  display | diskpak drive | film write-reader | magnetic card transport |
  magnetic tape transport | typewriter | Teletype | special purpose
  console := (airlines reservations | stock quotation | data collection)

A compound-transducer consists of a set of simple-transducers. The two simplest kinds are the half-duplex and the full-duplex, which are extensions of the simple-transducer, wherein the direction of information flow can be either way but only one way at a time (half-duplex) or can be both ways simultaneously (full-duplex). The more general case is simply a set of transducers with independent inputs and outputs (so that overall there is no switching function). It is common to call this a multiplexed transducer in which concurrency is specified by an integer.

EXAMPLES   T.half-duplex(typewriter; 15 char/sec; output: paper, video, audio; input: keyboard; 88 char; 8 b/char)

T.multiplex(console; keyboard, display, printer)

10.   Data-operations

10.1   Data operations/D := simple-data-operation | compound-data-operation

10.2   simple-data-operations := component (

cocomponents: inputs: components, output: component, initiator; input;

subcomponents: working: M-set, control: K-set;

operations: *see ISP data-operations, ISP 3.1;*

operation time: [t];

concurrency-type: simplex;

data-types: data-type(operations)   *see ISP data-types, ISP 1.3)*

A data-operation creates information (i.e., new instances of data-types) that has new meaning. It usually does this as a function of input information (e.g., a floating point multiply which creates a floating point number that represents the product of the two input numbers). It may or may not destroy some existing information (e.g., a tally operation, which modifies the existing number in creating the new one). A data operation differs from a transducer (T), since its output differs in meaning from its input. The T preserves meaning, while changing representation.

The data-operation takes the data-type i-units at the input ports, operates on the data, and presents the result at the output port. The simple-data-operation can perform only one operation at a time. The simplest D is just a set of transfer paths between registers for performing some operation on a boolean vector (that is, $A \wedge B$, $A \oplus B$, $\neg A$) or a combinational network (that is, $X = 0$). Slightly more complex D's are the additive operations on integers $(+, -)$. Operations like $\times, /$ are usually constructed from more primitive D's, $+$, $-$, and $(/2)$, with a subcontrol (K) to step through the various substeps of the arithmetic algorithm. Finally, a floating point multiply would be formed as a sequence of simple-data-operations controlled by one or more common subcontrols.

EXAMPLE $D \begin{bmatrix} \text{operation: } +; \text{ data-type: fixed; i-unit: 32 b;} \\ \text{operation-time: .2 } \mu s \end{bmatrix}$

$D \begin{bmatrix} \text{floating point multiplier; data-type: f; i-unit: 36 b;} \\ \text{operation-time: 2.0 } \mu s; \text{ M.working } (3 \times 36 + 10)b \end{bmatrix}$

10.3   compound-data-operation := simple-data-operation(

subcomponents: alternatives; simple-data-operation-set;

instruction: memory;

concurrency: + integer;

instruction-source: data, instructions, operator instruction)

A compound-data-operation consists of a collection of alternative simple-data-operations. Thus, a compound-data-operation is compound either in time, by having many varied operations which can be selected sequentially, or in space, by having many separate operations which can perform in parallel.

EXAMPLE $D \begin{bmatrix} \text{arithmetic unit; data-types; integer, floating, boolean vector} \\ \text{operations: } +, -, \times, / \wedge, \vee, \oplus, \neg, \text{ normalize; operation-} \\ \text{time: } 1 \sim 2.0 \ \mu s; \text{ input: } 2 \times 36 \text{ b; output: 36 b; M.working:} \\ \sim 4 \times 36 \text{ b} \end{bmatrix}$

11.   Processor

11.1   Processor / P := simple-processor | complex-processor

11.2   simple-processor := component (

cocomponents: primary: M-set, *secondary: M-set, controlled: component-set;

function: (microprogram | central / general purpose / c | input-output / io | display | array | vector move | special algorithm | language)

subcomponents: (interpreter: K; data-operations: D-set; M.processor-state / ps: *see PMS Table 1;* M.non-processor-state: *see PMS Table 1;*

operations: operations(data-operations), operations(cocomponents) *see ISP;*

data-types: data-type(operations)   *see ISP;*

cycle-time / tc: cycle-time(Mp);

i-rate: i-rate(Mp);

concurrency: (o-rate / cycle-time) [o];

program-switching-time: [t];

interrupt-response-time: [t];

instruction-set   *see ISP 2.1;*

instruction-efficiency: (operations / instruction) / instruction-size [o/i];

algorithm-encoding-efficiency: (sum(data i-units/[t])/ sum(data i-units + instructions)/[t]));

instruction-size: [i];

operation-code-size: [i];

address-size: [i];

addresses-per-instruction: (0 address / stack | 1 address / 1 | 1 + index / (1 + x) | 1 + general register address / (1 + g) | 2 address | 3 address | n + 1 address | compound))

A simple processor is always associated with a memory (its primary memory), which holds the program (and usually the data) for the processor. In addition, there may be secondary memories and also other components that are controlled by the processor.

The processor often functions as the main component of an essentially isolated system (often called stand-alone); it is then a central processor, Pc. Processors also occur as more specialized components in larger systems; e.g., to manage input/output (Pio) or display (P.display) or to do a subset of data-operations efficiently (P.data, P.vector_move, P.array, or P.special_-algorithm). Processors are sometimes built in hierarchy, using one processor to perform the interpretation and operations of another. Such processors have become known as microprogram processors.

The distinguishing feature of a processor is that it determines its own next instruction. The control that does this is called the interpreter. The repertoire of operations of the processor is partly a set of data-operations performed by its own subcomponents and partly the set of operations proper to a set of transducers, memories, links, and switches external to the processor but incorporated into its operation code. The operations are largely determined by the set of data-types (see the ISP section).

A processor may have considerable internal memory (called the processor state, Mps). Besides the instruction and instruction-address registers, which are necessary for interpretation, there may be various amounts of status information, accumulators, index registers, general registers, and accumulator stacks. No one system has all of these memories, since they often provide alternatives to each other (e.g., index registers and general registers).

Each of the operations has its own operation time and its own possibilities for being overlapped with other operations. Several parameters are given that summarize this array of information: the cycle-time of Mp, which in the long run limits the rate at which instructions and data can be accessed (and also determines the maximum throughput); the concurrency, which tells how many operations can be performed per cycle time (this requires an averaging of the various possibilities as given in the instruction set); and the program-switching time, which is the time required to change context from one program to another. In simple operating regimes (standard batch processing) program-switching time is not an important parameter; it becomes so when interrupts are permitted. For interrupts, the response time is critical. It is the time between when a request is made and when the request is acknowledged by P. The instruction set is really an entry point to the ISP description of the processor. One might give here simply the number of instructions, but this can be a very misleading number, since many variations of a basic instruction can be counted thus giving highly erroneous results. The algorithm-encoding-efficiency is the ratio of i-units used for data per unit time to the number of accesses for data + instructions per unit time. This efficiency is strongly affected by the address size, which is usually the address size of the Mp but need

not be if a processor uses an incremental or relative addressing system. The ratio can be measured at many levels of the ISP: instruction-by-instruction, on a subroutine, or for a whole program. In a simple computer, this ratio is near $\frac{1}{2}$. Vector operations can allow a ratio much closer to 1.

Common measures for the instructions give the size of the operation code, the address, and the instruction. The addresses per instruction is one of the best parameters to indicate the overall structure of the instruction set and is called the instruction-type. It ranges from 0 addresses (systems which execute a sequence of operations) through 1, 2, and 3 addresses per instruction to variable number of addresses. Between 1 and 2 addresses lie index register (1 + x) and general register (1 + g) machines. In a special class is the (n + 1) organization, which involves an additional address to obtain the next instruction; it can be added to any other organization.

EXAMPLES   Pc('DEC PDP-8; 1 address / instruction; ~2 w/ instruction; 12 b/w; 1.5, 3.0, 4.5 μs / instruction)
Pio('IBM 7909; 500 kw/s; data-types: words; integer; 1 address / instruction: 36 b/w)

11.3   complex-processor := simple-processor (

Mp-concurrency: (1 P | 1 P with interrupt | 1 program with multiple concurrent subprograms | 1 Pc - n Pio | monitor + 1 user program | monitor + 1 swapped program | fixed multiprogramming | multiprogramming | segmented-programming);
multiprogramming := (no relocation | protect only | 1 segment | 2 segment / pure | impure segments | > 1 segments | paging)
segmented-programming := (fixed length page segments | multiple length page segments | variable length page segments | named segments);

P-concurrency: (serial / serial by bit | parallel / parallel by word | multiple instruction streams | multiple data streams (arrays) | pipeline processing | instruction-memory);

instruction-memory := (none | 1 instruction look ahead | n instruction look ahead | cache / look aside / slave memory))

A complex processor is often an extension of a simple processor along the dimension of memory mapping, since a processor is already a highly structured and "complex" component.

Note that a collection of processors does not constitute a compound processor in a way similar to other PMS components; hence, we denote a general collection of processors as a computer. Thus, a complex processor can be written in terms of a simple-P with new values. The central processor using a microprogrammed processor contains a specialized processor as a subcomponent (P.microprogram).

Three attributes separate a simple processor from a complex processor: Mp-concurrency, P-concurrency, and instruction-memory. In essence, the simple processor has no Mp concurrency (interpreting a single program) and serial or parallel P concurrency, with no instruction-memory (buffering for multiple instructions). These attributes are independent of one another and are discussed in Chap. 3.

12. Computer

12.1   Computer / C := simple-computer | compound-computer | network

12.2   simple-computer := component (

structure: 1Pc | 1 Pc.interrupt;

subcomponents: Pc, Mp-set, *controlled: component-set(Pc);

cocomponents: none;

function: (scientific | business data processing | general purpose | process control / control | communication := (switching | store and forward) | terminal control / input-output / io | display | file processing / file control | time-sharing);

access-time: access-time(Mp);

cycle-time: min(cycle-time(Mp));

access-type: access-type(Mp.min);

instruction-type: instruction-type(Pc))

A simple computer consists of a single Pc (possibly with interrupt capability) with an Mp (possibly a set of them) plus some set of transducers, Ms's, switches, and controls. It is a complete system that can stand alone and accomplish processing for a wide variety of functions.

Almost all of its significant parameters are derived from those of the Pc or the Mp (using the Mp with the minimum cycle time if there are several Mp's).

EXAMPLES   C('Whirlwind I: Mp(core; 8μs/w; 2048w; 16 b/w);
Pc(M.processor_state: ~2w; 1 instruction/w; 1 address/instruction); 1948 ~ 1966)
C('LGP-30; technology: vacuum tubes; power: 1500 watts;
Mp(drum, 4096 w; 31 b/w; t.access: .260 ~ 16.6ms);
Pc(1 address/instruction; 1 instruction/word; Mps: ~2w))

12.3   compound-computer := simple-computer(

structure: ((1 Pc, n Pio) | (1 Pc, n Pio, P.display) | (2 Pc) | (n Pc multi-processor) | (n Pc, P(array) | (n Pc, special algorithm) | (n Pc parallel processor));

subcomponents: Pc-set, Mp-set, *controlled: component-set(Pc-set))

The essential feature of compound computers is to have more than one processor. This is indicated primarily by the structure parameter but re-

quires augmenting the subcomponents to include a set of Pc's. Other than this, compound-C's are the same as simple-C's, although some parameters (such as instruction-type) may not have simple values if several Pc's differ radically.

The simpler compound-C's retain a single Pc, but add input/output processors (Pio's and then P.display's). The next step is to limited multi-processing, with 2 Pc's, and on to n Pc's operating on many programs, and finally to parallel processing operation on many tasks of a single program. A parallel processor is distinguished from a network; namely, there is no way to decompose a parallel processor into disjoint C's (with Pc's and Mp's). In both multiprocessing and parallel processing there may or may not be Pio's, P.display's, and other special-function processors.

EXAMPLES   C(1 Pc-8 Pio; 'IBM 7094 II; Mp (32768 w; 1.4μs/w; 36 b/w);
Pc(1 address; 1 instruction / w; Mprocessor state: 12 w; data-types:(integer, word, bv, sf, suf, df, duf, fr.i); 1962 ~ 1966)
C(multiprocessor; 'Burroughs D-825; Mp(65 kw; 4.8μs/w; 48 b/w); 16 (Pc, Kio); Pc(stack; 12 b/syllable; 1 ~ 7 syllable / instruction; data-types: integer, floating, single character, boolean vector))

12.4   network/N := dual-C | network-C | C-set.

A network is any collection of two (dual-C) or more computers not interconnected through primary memory. The network-C is a special case of a single physical structure which is usually called a single C but by its structure is a network (for example, CDC 6600). Finally, a set of interconnected computers that are physically separate are the most general case of networks.

ISP conventions

Making use of the prior general conventions and the PMS definitions, ISP is developed systematically. We do this only for the processor and not for controls (though the system might be adapted to that end). Several notations are added to make ISP conform with currently existing notations.

The top-level entities of ISP—data-types, operations, the interpreter, and the instruction-set—are values of corresponding attributes in the PMS definition of a processor. An image of all the PMS structure for a computer system exists in the instruction set of the processors that control the PMS components. PMS notation is assumed for this. In ISP the primary memory (Mp) is usually named M; all other memories must be specifically declared and named.

1   Data-types

2   Instructions

3   Operations

4   Processors

1. Data-types

1.1 We give first a general definition of data-types (1.2), and then two shorter notations, which are the ones commonly used—i-units (1.3) and data-type-names (1.4).

1.2   data-type := (

referent: entity;

referent-expression;

*component-list:;

component: data-type;

carrier: i-unit;

format: (component: memory-expression)-list;

information-content: [i])

A data-type specifies the encoding of a meaning into an information medium. The meaning of the data-type (that which it designates or refers to) is called its referent (or value). The referent may be an entity, ranging from highly abstract (the uninterpreted bit) to highly concrete (the payroll account for a specific type of employee). The encoding of this referent either is directly understood (as when a bit encodes a bit) or must be given by the referent expression in terms of the component data-types.

EXAMPLE   binary-floating-point-number := data-type(

referent: number;

component-list: mantissa, exponent;

referent-expression: mantissa $\times 2 \uparrow$ exponent)

COMMENT   Note that in the referent expression the component data-types are taken to designate their values, i.e., a signed fraction and an exponent is an integer. This avoids a clumsier notation in which one could write:

referent(mantissa) $\times 2\uparrow$ referent(exponent).

Associated with every data-type is an i-unit, called its carrier, into which all its component data-types can be mapped. The carrier is used in storing the data-type in memories and in transmitting it over links. It must be extensive enough to hold all the component data-types, but it may be larger (having error-checking and -correcting bits, or even unused bits). It need not hold disjointly all the carriers of the component data-types, since packing may occur. However, the component data-types must all have their relative structures preserved (or they cannot be processed). The mapping of the component data-types into the carrier is called the format. It is given as a list that associates to each component a memory expression involving the carrier (see ISP 2 for definition of memory-expression).

EXAMPLE   floating-point-number := data-type (

component-list: mantissa, exponent;

mantissa := 23 b; exponent := 9 b;

carrier: word, 32 b/w;

format:(mantissa: word⟨0:22⟩, exponent: word⟨23:31⟩))

The five parameters—referent, referent-expression, component-list, carrier, and format—determine a data-type. The information content is simply a useful redundant parameter, which gives the amount of variety of the data-type. An upper bound, of course, is the amount of information in the carrier. A better estimate is the sum of the contents of the component data-types. A true value must take into account the dependencies between components. The efficiency of encoding (under the constraint that the encoding must be into the carrier and that all possible values must be represented, no matter how low their probability of occurrence) is the ratio of the information content to the carrier content.

1.3   data-type := i-unit

The simplest data-types are i-units. An i-unit as a data-type implicitly determines the five defining parameters given in ISP 1.2. The referent is the uninterpreted i-unit itself (i.e., a word is to be handled only as an uninterpreted unit of information). There is no need for a referent expression. The carrier is the i-unit itself, if it is an i-unit capable of independent storage and transmission in the system. If not, then the carrier is the smallest such i-unit that contains the given i-unit. The component data-types are the first sublevel of structures of the i-unit. There are no components if the i-unit is a base-unit (bit or undecomposable character). If the i-unit is the carrier, no format is needed. If a larger carrier is required, then a mapping is usually implicit (e.g., 1 bit in a word goes into the low-order position; 1 word in a block goes into the first word, etc.). If not, a format must then be given in the regular way.

1.4   data-type := data-type-name

data-type-name := i-unit-name | simple-name |
  component-name . length-type | precision . data-type-name |
  component . component . . .

length-type := array / a | string / st | vector / v

precision := + integer | multiple / m | quadruple / q | triple / t |
  double / d | *single / s | half / h | fractional / fr

A naming scheme is provided for data-types, which can be used as a basis for abbreviations. Some data-types have arbitrary simple names (e.g., character, floating point numbers); others are named by their value (e.g., integer). Data-types that are iterations of a basic component can be named by the component suffixed by a length-type. The length-type can be array/a, implying a multidimensional array of fixed but unspecified dimensions; a string/st, implying a single sequence of variable length (on each occurrence) or a vector/v, implying a one-dimensional array of a fixed but unspecified number of components. The length-type need not exist, and then this form of the name is not applicable.

Data-types are often of a given precision, especially when referring to numbers; it has become customary to measure this in terms of the number of components that are used, e.g., triple-precision integers. Names can be formed from the basic data-type-name by prefixing the precision. Note that a double-precision integer, while taking two words, is not the same thing as a two-integer vector; so that the precision and the length-type, although both implying something about the size of the carrier, do not express the same thing. Finally, it is possible to name a data-type by simply listing its components.

The main use of the data-type-name is to permit the short abbreviations which arise by replacing every part with its abbreviation and dropping the periods. Thus, double-precision integers have the data-type-name of double.integer, which can be replaced by d.i and then by di. Similarly, a vector of bits is bit.vector / b.v / bv. [The definition of data-type-name is consistent in its use of period with the definition of compound name (see GC 10)].

If a data-type is defined by giving just its name, conventions are required to define the five parameters of the data-type. The carrier is always taken to be the smallest i-unit that can contain the data-type with the following mapping. The format is taken to imply that the components are laid out in order (with no packing) into the subcomponents of the carrier i-unit. The referent of the data-type is given by context, e.g., if the data-type is simply an iteration of some kind of a data-type whose value is already understood, (e.g., in a vector of integers). Thus, there is no need for a referent expression.

1.5 We give below a number of basic data-types that need to be defined explicitly. Table 3 summarizes a large number of data-types and gives their standard abbreviations, as above. Figure 3 of Chap. 2 shows the lattice of data-types in which one data-type is connected to a higher one if it can be obtained by a further specification of the higher one. This is significant, since operations on higher data-types also apply to the lower ones. In the definitions below, which are the standard general data-types, we omit the referent expressions, carriers, and formats except those that are simple. (The fully general definition of radix-complement number representation, for example, is too extensive to be worthwhile here.)

base-data-type / radix := data-type(referent: (binary / 2 | octal / 8 | decimal / 10 | hexidecimal / 16); component: i-unit: (b | o | d | hex))

+ integer-data-type / ui / unsigned-integer / magnitude := data-type (referent: +integer; component: radix)

integer-data-type / i := sign-magnitude | radix-complement | (radix − 1)-complement

number-data-type := data-type(referent: number; normalization: (*normalized / n | unnormalized / u); name: normalization . number-data-type-name)

Table 3   Examples of commonly used data-types (organized by basic i-units)

bit / boolean / b
  bit.array / ba
  bit.vector / bv

byte / by
  byte.string / by.st
  10 byte.vector / 10 by.v

character / char / ch
  char.string / char.st
  10 char / 10 ch
  4 char.vector / 4 ch.v

complex / cx

digit / d

  10 digits / 10 d
  digit vector / d.v
  10 digit, array / 10 d.a

floating point / f / single floating point / sf
  unnormalized floating point / uf
  double floating point / df
  double unnormalized floating point / duf
  floating point vector / s.f.v / f.v

field

fraction / fr

integer / i
  integer vector / iv
  double integer / di

mixed / mx

word / w
  half word / hw
  double word / dw
  triple word / tw
  multiple word / mw
  word vector / wv
  word string / w.string
  half word vector / hw.v
  7 word / 7 w
  8 word vector / 8 w.v

COMMENT   The general data-type for number introduces a new parameter (normalization) to prefix the name of all numbers.

mixed / mx / fixed-point := number-data-type (components: integer-part, fractional-part)

floating-point / f := number-data-type(components: mantissa, exponent; value-expression: mantissa $\times$ radix $\uparrow$ exponent) complex := data-type(components: real, imaginary; *usually floating complex*)

field   := data-type(carrier: word; components: i-unit-list; format: ⟨element-range⟩)

COMMENT   A field is a subset of bits, or characters, or bytes in a word. It is usually, though not always, an interval. See ISP.2 for element range.

EXAMPLES

12, 101, 5; +125, −126;	*unsigned; and signed integers*
+72, −999;	*sign-magnitude*
101_2, 77_8, $A9_{16}$;	*binary, octal and hexidecimal*
+6.257; 6.257×10^0;	*mixed, and floating point*
(1, 2, 2.7);	*complex*
$1\Phi_2$; $7\Phi_8$	*digit set specification: stands for $10_2 \mid 11_2$; and $70_8 \mid 71_8 \mid \ldots \mid 77_8$ respectively*
?	*questionable value*

2. Instruction

2.1   instruction := data-type(referent: instruction-expression; operation-code: field; operand-list; operand: data-type)

instruction-expression := condition → action-sequence

action-sequence := (step | next step)-list

step := action | condition → action-sequence

action := memory-expression ← data-expression

memory-expression := (
   memory *[address-range]-list *⟨element-range⟩ character-base |
   memory-expression□memory-expression | memory-expression-list)

address-range := address | address: address | address-expression | address-range-list

address-expression := operation-expression(address-operations)

element-range := field | field-list

character-base := +integer   *base i-unit*

condition := boolean | memory-expression

data-expression := data-type | memory-expression | operation-expression | data-expression{data-type}

operation-expression := (nonary-operation |
   unary-operation data-expression |
   data-expression binary-operation data-expression |

data-expression n-ary operation data-expression . . . |
   function(data-expression-list) / f(data-expression-list) |
   operation-expression *{operation-modifier}

operation-modifier := data-type | name   *See GC 10*

2.2   The instruction is a data-type and thus has both a representation in memory and a referent, which is called the instruction-expression. The only fixed part of the instruction format is the operation-code. All the rest are operands to be used by the instruction-expression.

2.3   The instruction-expression, when interpreted, takes the processor through a sequence of steps which result (possibly) in some change of state of the computer system that holds past the period of interpretation, thus constituting a new initial condition for the next instruction. The action sequence has two structural features. First, steps (and subsequences of steps) may be conditional on a boolean value, developed according to a condition. Second, steps may be accomplished in parallel or in series. Any set of steps between two occurrences of the term "next," are to have all their data expressions developed prior to any transmission of data. Thus, all their data is a function of the existing state at the start of the sequence. At the occurrence of the term "next," all pending transmissions are made, so that the state for the following sequence of steps is now different (if there were in fact transmissions to be made).

2.4   All permanent changes in state are accomplished by means of actions, which take data developed according to a data expression and transmit it for storage in a memory, as designated by a memory expression.

EXAMPLES

A ← B; B ← D − G; B ← B + g

x1 ← ¬ x2;   x2 ← −x1;   a ← abs(a); a ← normalize(b)

AB ← a□b

x1 {float} ← x2 {fixed}	*fixed to floating data-type*
x1 ← x1 + x2 {floating}	*floating data*
$a \leftarrow a \times 2^n$ {logical}	*usually called logical shift, actually a boolean vector operation*

AC, MQ ← AC□MQ / M[z]

AC□MG ← AC × M[z]

A	← 6777	*nonary operation*
G	← f(A, B, C)	*general function*
A	← u B	*general unary operation*
A	← B b C	*general binary operation*
A	← max(a, B, XYZ, E, 4)	*n-ary operation*

2.5 The memory expression specifies the contents of a memory (an instance of a data-type) by giving the memory switch (possibly compound), as seen from PMS. However, all that is represented in ISP is the address that is used to control the switch. The address is a data-type, usually represented as a positive integer. The element-range is a field. In both cases it is possible to specify an arbitrary list of contents (addresses and fields), although in most processors this can never arise. The address-range x:y means from address x to address y inclusive.

EXAMPLES OF REGISTERS

A_2 *or* A;	*boolean-memories; scalar bits*
sign bit/sign_bit/sb	
1b; b2; 2C1; 2C2′; C″; C′″; "+"; "A"	
end_around_shift *or* end around shift	*identical names*
G_3	*ternary memory*
i⟨2⟩; Z⟨a⟩	*scalar bits of an array*
bc⟨12:8⟩ *or* bc⟨12, 11, 10, 9, 8⟩	*identical registers*
AC⟨P,Q,S,1:35⟩	*38 bit register*
X⟨0:7⟩$_8$ *or* X⟨0:23⟩$_2$ *or* X⟨0:23⟩	*identical registers*
M[0:7777$_8$]⟨0:11⟩ *or* M[0:4095]⟨0:3⟩$_8$	*identical vectors*
X[0:15][0:15]⟨31:0⟩	*16x16 matrix*
M[0:7][0:31][0:127]⟨0:11⟩	*3 dimensional array*

EXAMPLES OF RESTRUCTURING AND RENAMING

A⟨17⟩ := B⟨4⟩; A⟨0:1⟩ := B⟨0, 4⟩

op⟨0:2⟩ := i[1]⟨9:11⟩

A[0:3]⟨0:7⟩ := A′⟨0:31⟩

indicator[1100001$_2$] := sense_switch⟨A⟩

XR[1:2][1:3] ⟨ B, A, 8, 4, 2, 1⟩ := M[87:89,   *vectors formed from*
92:94]⟨B, A, 8, 4, 2, 1⟩   *single bit vector*

EXAMPLES OF REGISTERS FORMED BY CONCATENATION

LAC⟨L, 0:11⟩ := L□AC⟨0:11⟩

AB⟨0:47⟩ := A⟨0:23⟩□B⟨0:23⟩

EXAMPLES OF REGISTERS FORMED BY A LIST OF REGISTERS

C, D⟨0:4⟩ := B⟨7⟩, A⟨1:4⟩□Z⟨8⟩

2.6 An address-expression is an operation-expression on addresses, i.e., using only the address-operations available in the processor. An address-

expression may imply the use of memory if it involves nested parentheses; such memory is assumed to be temporary with no permanent effect on the memory state.

2.7 A condition is given as a boolean, that is, as either true or false (equivalently, 1 or 0), or the result of a boolean expression involving the logical connectives or relations among data-expressions (see Table 4, ISP 3, and also GC 13). A condition can also be given as a memory-expression, in which case the memory contents are normally evaluated as a boolean vector with all 0s being false, and not all 0s being true.

2.8 Data-expressions are either instances of data-types; the contents of a memory, as given by a memory-expression; or the results of operation-expressions, which is to say, the results of operating on data-types by the data-operations available in the processor. Data-expressions may imply the use of memory if they involve nested parentheses. Such memory is assumed to be temporary, with no permanent effects on the memory state of the processor or memory. The data-type name may sometimes follow the data-expression, {data-type}, in order to carry more information and avoid more complex names for memory-expressions, etc. (see Chap. 2, page 30, and ISP 3.1).

2.9 Operation-expressions are the form used by the operations (see ISP 3). Note that the operation-expression as a whole can be modified by an operation modifier enclosed in braces.

EXAMPLES OF INSTRUCTIONS

add (:= op = 101) → (L □ AC ← L □ AC + M[z])	*integer add*
jms (:= op = 100) → (M[z] ← PC; next PC ← z + 1)	*jump to subroutine*
FAD (:= op = +767) → (FAC ← FAC + M[z] {s.f})	*single precision floating point add*
add → (A ← A + M[z] {two's complement})	*the operation code need not be given*
skip (:= op = 67) → ((A > 0) → P ← P + 2; (A = 0) → P ← P + 1)	

add / "A" (:= op = 110001) →
   (0v, M[B] ← M[B] + M[A] {string})
"B" (:= op = 1) → (A ← M[t][s])

((A ∧ B) ∨ (C > F)) → (G ← G + H)

3. Operations

3.1 Operations are defined to produce results of specific data-types from operands of specific data-types. The data-types themselves determine by and large the possible operations that apply to them. No attempt will be made to define the various operations here, as they are all familiar. Table 4 gives the notation for the operation-types, organized by data-types. In

Table 4 Data-operations

operation-types := access-i-unit-operations | transmission-operation | control-operations | unary-arithmetic-operations | binary-arithmetic-operations | n-ary-arithmetic-operations | conversion-arithmetic-operations | unary-vector-operations | relational-i-unit-operations | relational-arithmetic-operations | boolean-operations

nonary-operation := memory-expression

unary-operation / u := unary-arithmetic-operations | unary-boolean-operation *see GC 13*

binary-operation / b := binary-arithmetic-operations | binary-boolean-operations *see GC 13*

n-ary-operation := n-ary-arithmetic-operations | n-ary-boolean-operations *see GC 13*

Operation	Abbreviation	Result[1]	Operation Form[1]	Comments
access-i-unit-operations				basic operation is to access an i-unit in a memory (e.g., word vector)
read		t_2	t_1	access t_1 for reading
write		t_2	t_1	access t_2 for writing
vector element write		t_3	$t.v_1[i_2]$	the i_2th element of vector$_1$ is read
vector element		$t.v_1[i_2]$	t_3	the i_2th element of vector$_1$ is written
concatenation	□	t_3	$t_1 \ \Box \ t_2$	t_1 and t_2 are combined to form t_3
extraction		t_2	t_1 ⟨element-range⟩	some part of t_1 forms t_2
transmission-operation				
transmit	←	t_2	t_1	t_2 receives 1-unit of t_1; involves read transmit and write
control-operations				
evoke	→		$b_1 \rightarrow$ action-sequence	if b_1 is *true* then action-sequence is applied; else the action-sequence is ignored
next				the occurrence of "next" implies operations following occur later
unary-arithmetic-operations				
absolute value or magnitude	abs	n_2	$abs(n_1)$	n_2 may be unsigned data-type
negate	−	n_2	$-n_1$	
reciprocal	1 /	n_2	$1 / n_1$	
integer part		n_2	integer_part(n_1)	n_2 is an integer data-type
fraction part		n_2	$frp(n_1)$	n_1 may be mixed	f	unf
sign		b_2	$sgn(n_1)$	n_1 may not be ui	ufr
round		n_3	round(n_1,n_2)	used with multiply, divide
normalize, mantissa part		n_2	normalize(n_1)	used with f arithmetic
normalize exponent, exponent part		n_2	normalize_exponent(n_1)	to fix numbers into a standard form
square root	sqrt	n_2	$sqrt(n_1)$	$(n_1 \geq 0)$
square	$(\ )^2$	n_2	$(n_1)^2$	
logarithms	log, 1n	n_2	$\log_{e,10}(n_1)$	$\log_{10}(n_1)$
exponential	e	n_2	e^{n_1}	$\log_e(n_1)$
trigonometric	trigfcn	n_2	trigfcn(n_1)	also sin, sin^{-1}, sinh, etc. for the separate trigonometric function (both radians and degrees)
random (parameter for particular distributions)		n_2	random(n_1)	n_1 may be previous pseudo-random number (seed)
arithmetic shift of radix, r	xr / r	n_3 n_3	$n_1 \times r^{i2}$ n_1 / r^{i2}	if i_2 is signed, then either form can be used for both x and /

[1] Results and operations forms given in terms of data-types to which they apply: b—booleans; i—integers; f—floating; n—any numeric data-type (e.g., floating, integer, mixed); t—all data-types; v—vectors.

Table 4 Data-operations (Continued)

Operation	Abbreviation	Result	Operation Form	Comments
binary-arithmetic-operations				
add	$+$	n_3	$n_1 + n_2$	
subtract	$-$	n_3	$n_1 - n_2$	
inverse subtract		n_3	$n_2 - n_1$	
multiply	$\times$	n_3 or n_3, n_4	$n_1 \times n_2$	
divide	$/$	n_4, n_5	$n_1, n_2 / n_3$	where only n_1 or n_2 may be used to give n_4 or n_5
inverse divide				see divide -similar to inverse subtract
modulo	mod	i_3	i_1 mod i_2	$i_1 - (i_1 / i_2) \times i_2$ remainder
conversion-arithmetic-operations				
fix-to-float		f_2	$float(i_1)$	integer or fixed to floating
float-to-fix		i_2	$fix(f_1)$	floating number to integer
unary-vector-operations				radix r; note if $r = 2$, the character is a bit
end-around-shift (rotate)		v_3 x r {rotate}	v_1 x r^{i_2}{rotate}	
		v_3 / r {rotate}	v_1 / r^{i_2}{rotate}	
logical-shift		v_3 x r {logical}	v_1 x r^{i_2}{logical}	the most or least significant digits receive 0's in the shift
		v_3 / r {logical}	v_1 x r^{i_2}{logical}	
tally/count		i_2	$tally(b.v)$	count 1's in a vector
sign extend		n_2	$sign\_extend(b.v_1)$	copy sign of $b.v$ to fill vector in n_2
n-ary-arithmetic-operations				
minimum	min	n_j	$min(n_1, n_2, \ldots, n_m)$	smallest of $n_1 \ldots n_m$
maximum	max	n_j	$max(n_1, n_2, \ldots, n_m)$	largest of $n_1 \ldots n_m$
summation	sum	n_{m+1}	$sum(n_1, n_2, \ldots, n_m)$	$n_1 + n_2 \ldots + n_m$
average	avg	n_{m+1}	$avg(n_1, n_2, \ldots, n_m)$	$n_1 + n_2 \ldots n_m) / m$
product	prod	n_{m+1}	$prod(n_1, n_2, \ldots, n_m)$	$n_1 \times n_2 \ldots \times n_m$
relational-i-unit-operations				comparison of two i-units
identical	$\equiv$	b_3	$d_1 \equiv d_2$	
not identical	$\not\equiv$	b_3	$d_1 \not\equiv d_2$	
relational-arithmetic-operations				comparison of two numbers
equality	$=$	b_3	$n_1 = n_2$	
inequality	$\neq$	b_3	$n_1 \neq n_2$	
less than	$<$	b_3	$n_1 < n_2$	
greater than	$>$	b_3	$n_1 > n_2$	
less than or equal to	$\leq$	b_3	$n_1 \leq n_2$	
greater than or equal to	$\geq$	b_3	$n_1 \geq n_2$	
boolean-operations				
false (0)	0	b_3	0	all 16 possibilities are listed
and	$\wedge$	b_3	$b_1 \wedge b_2$	
		b_3	$b_1 \wedge \neg b_2$	
null		b_3	b_1	
		b_3	$\neg b_1 \wedge b_2$	
null		b_3	b_2	
exclusive or;	$\oplus$	b_3	$\neg(b_1 \equiv b_2) = ((b_1 \wedge \neg b_2) \vee (\neg b_1 \wedge b_2)) = b_1 \oplus b_2$	
inclusive or	$\vee$	b_3	$b_1 \vee b_2$	
nor/Pierce stroke	$\downarrow$	b_3	$\neg b_1 \wedge \neg b_2$ or $\neg(b_1 \vee b_2)$	
coincidence or	$\equiv$	b_3	$b_1 \equiv b_2$ or $\neg(b_1 \oplus b_2)$ or $(b_1 \wedge b_2) \vee (\neg b_1 \wedge \neg b_2))$	

Table 4 Data-operations (Continued)

Operation	Abbreviation	Result	Operation Form	Comments
not	¬	b_3	$\neg b_2$	
implication-inverse		b_3	$b_1 \vee \neg b_2$	
not	¬	b_3	$\neg b_1$	
implication	⊃	b_3	$\neg b_1 \vee b_2 \text{ or } b_1 \supset b_2$	
nand/Sheffer stroke	↑	b_3	$\neg b_1 \vee \neg b_2 \text{ or } \neg(b_1 \wedge b_2)$	
true (1)	1	b_3	1	
boolean-operations (common set)				
not	¬	b_3	$\neg b_1$	
and	∧	b_3	$b_1 \wedge b_2$	
or	∨	b_3	$b_1 \vee b_2$	
exclusive or	⊕	b_3	$b_1 \oplus b_2$	
boolean-operations (sufficient sets)				
nand	↑	b_3	$\neg(b_1 \wedge b_2)$	
nor	↓	b_3	$\neg(b_1 \vee b_2)$	
ʃ not	¬	b_3	$\neg b_1$ ⎱ this pair of operations are required	
⎩ and	∧	b_3	$b_1 \wedge b_2$ ⎰ for sufficient set	

order to have an open-ended scheme for operating on many data-types and defining new operators, the operation modifier is used. The operation modifier enclosed in braces is used to distinguish operations from one another. The operation modifier is usually the name of a data-type, but it can also be a descriptive name applying to the operation (e.g., rotate). For example, the various add operations on differing data-types are specified by writing {data-type} after the operation (see Chap. 2, page 30).

3.2 Operations can be defined for the most inclusive data-types for which they will work and can then be applied to more specific data-types. The most general instance of this is the transmit operations which works on i-units, and is therefore used for all specific data-types, such as numbers (because it works on their carriers). Another example is the relational operations of equality and inequality.

3.3 New operations can be defined by means of forms (see GC 4.5). We simply give some examples.

EXAMPLES

$$X1 + X2 := (X1 + X2;$$
$$(X1 + X2 \geq 2^{12}) \rightarrow (Ov \leftarrow 1))$$

two's complement add side effect, set Ov

$$X1\langle 11{:}0 \rangle := X2 \times 2 \ \{rotate\} := ($$
$$X1\langle 11{:}1 \rangle := X2\langle 10{:}0 \rangle;$$
$$X1\langle 0 \rangle := X2\langle 11 \rangle)$$

rotate operation; end bits, X⟨11⟩ and X⟨0⟩, are connected

4. Processors

4.1 The ISP definition of a processor consists of a set of instructions, which involve a set of operations, data-types, memories, and other PMS components, plus an interpreter that finds the next instruction and executes it. These sets are all values of corresponding attributes of the PMS description of a processor. All these aspects of an ISP processor have to be declared in giving the description. In practice, some of them are given by having the PMS description available (e.g., word size, T's, Ms's, etc.); others declare themselves simply by occurring in the ISP expressions (e.g., most of the operations and data-types). We list below the common form of the machine ISP descriptions as a reader will find them in the chapter appendices of this book.

4.2 Memory (Mps, Mp and M(T.console)). The processor state memory is declared first. It holds the information necessary to restart the processor, if it is stopped between instructions. Table 1 (page 621) names the functions of the memory (e.g., program counter, accumulators, etc.). The state also includes the interrupt status, machine fault bits, etc. Any memory-mapping hardware registers are considered part of this state.

The primary memory, the largest state, is used to hold the program that the processor interprets. It also holds data.

The console state is accessible from the operator's console. Only the bits that are part of the ISP are relevant, i.e., bits that can be used to change the state of the primary memory or processor state. The switches that are used to start and stop the machine should also be given in a complete definition.

4.3   Instruction Format. The instruction formats are usually declared in the same fashion as memory and are not distinguishable as special non-memory entities. Normally, the instructions are carried in registers; it is thus natural to give declarations in this fashion. Usually only a single declaration is made, the instruction/i, followed by the declarations of the parts of the instruction—the operation code, the address fields, indirect bit, etc.

EXAMPLE

i/instruction[0:4]⟨0:7⟩	*five 8 bit byte instruction*
op⟨0:4⟩ := i[0]⟨0:4⟩	*opcode*
r⟨0:2⟩ := i[0]⟨5:7⟩	*register address*
d⟨0:15⟩ := i[1:2]⟨0:7⟩	*16 bit address*

4.4   Effective Address Calculation Process. This process is declared using the assignment command (:=) and is evoked each time an instruction makes reference to a variable that is taken to be an effective address or an operand. In the book operands have two forms. Most of the time they are expressed as memories and address expressions using the effect address calculation process; otherwise the operands are defined by a process.

EXAMPLES

Conditional register definition
$$z⟨0:11⟩ := (\neg i \rightarrow z';$$
$$i \rightarrow (M[z'] + 1;$$
$$M[z'] \leftarrow M[z'] + 1))$$ *effective address with side effects*

$$G := M[g]$$ *operand definition process*

shift_count / SC⟨0:25⟩ :=
$$(\neg F \leftarrow e'; F \rightarrow z)$$

$$E'⟨21:35⟩ :=$$
$$((T = 0) \rightarrow (T \neq 0) \rightarrow XR[T] + y)$$ *index convention*

Declarations in terms of a variable parameter
$$Mp[z] := ((z > FL) \rightarrow Mp[z + RA];$$
$$(z \geq FL) \rightarrow (Run \leftarrow 0;$$
$$violation \leftarrow 1))$$ *only side effects, no value*

Evaluated expressions
add_instruction := (op = 5)   *boolean*

$$z⟨0:6⟩ := (a⟨0:5,7⟩ + b⟨1:7⟩)$$ *7 bit value*

skip_condition := ($\neg Q \wedge d⟨15⟩ \vee z⟨6⟩$)

4.5   Data-type Format and Special Data-Operation Definitions. The component parts of the data-types are named, and their element ranges are

first defined, so that the data-operation definitions can use them. For example, a precise definition of an ISP would include the data-type formats (for example, floating-point), followed by a definition of each data operation (for example, $+$, $-$, $\times$, $/$). Normally, we do not give enough information about the data-type and its appropriate operation implementation in our description of machines, since the information for these descriptions is obtained from the programming manuals. If we were actually to use the ISP descriptions, as an interpreter using a compiled or interpreted language, then only a few well-defined primitives would exist in the language and all other operations would have to be defined in terms of these primitives for each ISP. ISP 2 and ISP 3 describe how the various data-types and operations are declared.

4.6   Instruction Interpretation Process. In the definition of processors, the only part that is executed is the instruction interpreter. All the other parts are memory data declarations and processes to be carried out as an indirect consequence of the interpretation process. The format for most interpreters is the familiar fetch-the-instruction then execute-the-instruction pair of states, and consists of only one ISP statement.

EXAMPLE

Run → (instruction ← M[PC];   *fetch (PC/program counter)*
PC ← PC + 1; next

Instruction_execution)   *execute*

In more complex processors the conditions for trapping and interrupting must be described. Also, in the interpretation process it is often more descriptive to carry out part of effective address calculation prior to Instruction_execution. See below.

EXAMPLE

¬ interrupt ∧ Run →

(op[0] ← M[PC]; PC ← PC + 1; next   *fetch*

long instruction →

(op[1] ← M[PC]; (op[1] ← M[PC];   *fetch more instruction*
PC ← PC + 1); next   *if a long instruction*

Instruction_execution)   *execute*

interrupt ∧ Run → (M[0] ← PC; PC ← 1;   *interrupt, save*

interrupt ← 0)   *PC and go to M[1]*

The IBM 1401 interpreter (Chap. 18) requires a separate process to fetch the operands addresses prior to execution in a variable-length instruction. The fetch is based on the specific instruction to be executed next.

Run → (op ← M[PC]; PC ← PC + 1; next   *fetch*

    Fetch_operands_addresses; next   *fetch operands*

    Instruction_execution)   *execute*

4.7   Instruction-Set and Instruction Execution Process. The instruction-set and the process by which each instruction is executed are usually given together in a single definition. This process is called Instruction_execution in all the ISP descriptions in this book. It usually includes the definition of the conditions for execution, the instruction (i.e., its operation code), the name of the instruction, its mnemonic name, and the process for execution.

Instruction_execution := (

    add → (A ← A + M[z];

    $\vdots$

    opr → (qqq);

    and → (A ← A M[q]))   *end Instruction_execution*

where

    qqq := (cb →　(A ← 0); next   *secondary definition*

           cmb → (A ← ¬A);

       $\vdots$

        pl →　(A ← A + 1))   *end qqq definition*

Bibliography

Abbreviations

Journals

ACM	Association for Computing Machinery
ADC	Automatic Digital Computation
AFIPS	American Federation of Information Processing Societies
AIEE-IRE Conf.	American Institute of Electrical Engineers—Institute of Radio Engineers Conference
Appl. Sci. Res.	Applied Scientific Research
EJCC	Eastern Joint Computer Conference
FJCC	Fall Joint Computer Conference
SJCC	Spring Joint Computer Conference
WJCC	Western Joint Computer Conference
IBM J. of Res. and Dev.	IBM Journal of Research and Development
IBM Sys. J.	IBM Systems Journal
ICIP	International Conference on Information Processing
IEE	Institution of Electrical Engineers, London
IEEE	Institute of Electrical and Electronics Engineers
IFIP	International Federation for Information Processing
IRE	Institute of Radio Engineers
Psychology Rev.	Psychology Review

General

Bull.	Bulletin
Comm.	Communications
Conf.	Conference
Cong.	Congress
J.	Journal
Proc.	Proceedings
Pt.	Part
Res. Rept.	Research Report
Supp.	Supplement
Symp.	Symposium
Trans.	Transactions

Reports, manuals, and miscellaneous

"Study of a Computer Directly Implementing an Algebraic Language," AD633-727, Air Force Office of Scientific Research Contract AF19(628)-2798.

Control Data 6600 Computer System Reference Manual, 1st ed. Publ. 450, Copyright © 1963, Control Data Corporation, Minneapolis 20, Minn.

"Digital Small Computer Handbook," 1967 Edition, Copyright © 1967, all rights reserved, Digital Equipment Corporation, Maynard, Mass.

Programmed Buffered Display 338 Programming Manual—PDP-8, DEC-08-G61C-D, Copyright © 1967, all rights reserved, Digital Equipment Corporation, Maynard, Mass.

A22-6703, IBM 7094 Principles of Operation, Data Processing System, Copyright © 1959, 1960, 1961, 1962, International Business Machines Corporation.

A22-6821-4 IBM System/360 Principles of Operation.

A22-6810-8 IBM System/360 System Summary.

IBM System/360 Functional Characteristics Manuals for each Model

IBM System/360 Configurator (diagram) for each Model.

IBM OS/360: PL/I Language Specification, Form C28-6571, p. 74.

H20-0223-0, IBM System/360 Attached Support Processor System (ASP) System Description, Copyright © 1966, International Business Machines Corporation.

A24-1403-5, IBM 1401 Reference Manual, Data Processing System, Copyright © 1960, 1961, 1962, International Business Machines Corporation.

225-6487-3, IBM 1401 Customer Engineering Reference Manual, Copyright © 1960, 1961, 1962, 1963, International Business Machines Corporation.

A26-5919-4, IBM 1800 Data Acquisition and Control System Configurator.

A26-5918-5, IBM 1800 Functional Characteristics, Copyright © 1966, International Business Machines Corporation.

IBM 1620 FORTRAN: Preliminary Specifications, Form J29-4200-2, April, 1960.

FORTRAN Specifications and Operating Procedures, IBM 1401, IBM Systems Ref. Lib. C24-1455-2.

International Business Machines Corporation, General Information Manual FORTRAN, Form F28-807401, December, 1961.

Type 650 Magnetic Drum Data-processing Machine (Manual of Operations), Form 22-60 60-1, International Business Machines Corporation, New York, 1955.

Librascope LGP-30, Manual, Librascope, Inc., 80 Western Ave., Glendale, Calif.

Olivetti Underwood Programma 101 General Reference Manual, Olivetti Underwood Corporation, One Park Avenue, New York, 10016.

Pegasus Maintenance Manuals, Ferranti Ltd., London.

Pegasus Programming Manual, Ferranti Ltd., London.

Proceedings Conference on Spaceborne Computer Engineering, Anaheim, Calif., Oct. 30–31, 1962.

Scientific Data Systems Reference Manual, SDS 930 Computer, Copyright © 1965, 1966, 1967, Scientific Data Systems, Inc., 1649 Seventeenth Street, Santa Monica, Calif.

Scientific Data Systems Reference Manual, SDS 9300 Computer, Copyright © 1963, 1964, 1965, 1966, 1967, Scientific Data Systems, Inc., 1649 Seventeenth Street, Santa Monica, Calif.

Symposium on Multi-programming (Concurrent Programs), Information Processing, 1962 Proc. IFIP Congress, pp. 570–575, North-Holland Publishing Company, Amsterdam, 1963.

Univac Scientific Electronic Computing System Model 1103A, Form EL338, Remington-Rand Corporation, 1902 West Minnehaha Ave., St. Paul W4, Minn.

"Comprehensive System Manual, A System of Automatic Coding for the Whirlwind Computer," Digital Computer Laboratory, Massachusetts Institute of Technology, Cambridge 39, Mass., August, 1955; revised, December, 1955.

Books and periodicals

AdamA60   Adams Associates: *Computer Characteristics Quarterly*, summary of the characteristics of computers being currently manufactured, Cambridge, Mass. Specific quarterlies used: January, 1966, vol. 6; no. 1; 1st and 2nd quarters, 1967, vol. 7, nos. 1, 2; 4th quarter, 1967, and 1st quarter, 1968, vol. 7, no. 4, vol. 8, no. 1, (first published in 1960).

AdamC60   Adams, C. W.: A Chart for EDP Experts, *Datamation*, vol. 6, pp. 13–17, November-December, 1960. See AdamA60.

AdamC62   Adams, Charles W.: Grosch's Law Repealed, *Datamation*, vol. 8, no. 7, pp. 38–39, July, 1962.

AinsE52   Ainsworth, Ernest: SEAC Input-Output Operating Experience, *AIEE-IRE-ACM Conf.*, pp. 44–47, December, 1952.

AlexS51   Alexander, S. N.: The National Bureau of Standards Eastern Automatic Computer (SEAC), *AIEE-IRE Conf.*, pp. 84–89, December, 1951.

AllaR64   Allard, R. W., K. A. Wolf, and R. A. Zemlin: Some Effects of the 6600 Computer on Language Structures, *Comm. ACM*, vol. 7, no. 2, pp. 112–119, February, 1964.

AlleM63   Allen, M. W., T. Pearcey, J. P. Penny, G. A. Rose, and J. G. Sanderson: CIRRUS, An Economical Multiprogram Computer with Microprogram Control, *IEEE Trans.*, vol. EC-12, no. 6, pp. 663–671, December, 1963.

AllmR62   Allmark, R. H., and J. R. Lucking: Design of an Arithmetic Unit Incorporating a Nesting Store, *Proc. IFIP Cong. 1962*, pp. 694–698, 1962.

AlonR60   Alonso, R. L., and J. H. Laning, Jr.: Design Principles for a General Control Computer, Institute of Aeronautical Sciences, New York, S. M. Fairchild Publ. Fund Paper FF-29, April, 1960.

AlonR61   Alonso, R. L., J. H. Laning, Jr., and H. Blair-Smith: Preliminary MOD 3C Programmers Manual, *M.I.T. Instrumentation Lab.*, *Rept.* E-1077, 1961.

AlonR62   Alonso, R. L., A. Green, H. Maurer, and R. Oleksiak: A Digital Control Computer; Development Model 1B, *M.I.T. Instrumentation Lab.*, *Rept.* R-358 (confidential), April, 1962.

AlonR63   Alonso, R. L., H. Blair-Smith, and A. L. Hopkins: Some Aspects of the Logical Design of a Control Computer, A Case Study, *IEEE Trans.*, vol. EC-12, no. 6, pp. 687–697, December, 1963.

AmdaG62   Amdahl, Gene M.: New Concepts in Computing

System Design, *Proc. IRE*, vol. 50, no. 5, pp. 1073–1077, May, 1962.

AmdaG64a   Amdahl, G. M., G. A. Blaauw, and F. P. Brooks, Jr.: Architecture of the IBM System/360, *IBM J. Res. and Dev.*, vol. 8, no. 2, pp. 87–101, April, 1964. Review TeagH65

AmdaG64b   Amdahl, G. M.: Processing Unit Design Considerations, *IBM Sys. J.*, vol. 3, no. 2, pp. 144–164, 1964.

AmdaG64c   Amdahl, G. M.: The Model 92 as a Member of the System 360 Family, *AFIPS Proc. FJCC, Pt. II*, vol. 26, pp. 69–72, 1964. Review GrimR65b

AndeD67   Anderson, D. W., F. J. Sparacio, and R. M. Tomasulo: The IBM System/360 Model 91: Machine Philosophy and Instruction Handling, *IBM J. of Res. and Dev.*, vol. 11, no. 1, pp. 8–24, January, 1967.

AndeJ61   Anderson, James P.: A Computer for Direct Execution of Algorithmic Languages, *AFIPS Proc. EJCC*, vol. 20, pp. 184–193, 1961.

AndeJ62   Anderson, James P., Samuel A. Hoffman, Joseph Shifman, and Robert J. Williams: D825—A Multiple Computer System for Command and Control, *AFIPS Proc. FJCC*, vol. 22, pp. 86–96, 1962.

AndeJ65   Anderson, James P.: Program Structures for Parallel Processing, *Comm. ACM*, vol. 8, no. 12, pp. 786–788, December, 1965.

AndeS67   Anderson, S. F., J. G. Earle, R. E. Goldschmidt, and D. M. Powers: The IBM System/360 Model 91: Floating-point Execution Unit, *IBM J. of Res. and Dev.*, vol. 11, no. 1, pp. 34–53, January, 1967.

ArbuR66   Arbuckle, R. A.: Computer Analysis and Thruput Evaluation, *Computers and Automation*, p. 13, January, 1966.

ArdeB66   Arden, B. W., B. A. Galler, T. C. O'Brien, and F. H. Westervelt: Program and Addressing Structure in a Time-sharing Environment, *J. ACM*, vol. 13, no. 1, pp. 1–16, January, 1966.

AstrM52   Astrahan, M. M., and N. Rochester: The Logical Organization of the New IBM Scientific Calculator, *Proc. ACM, Pittsburgh Conf.*, pp. 79–83, May, 1952.

BaldF62   Baldwin, F. R., W. B. Gibson, and C. B. Poland: A Multiprocessing Approach to a Large Com-

puter System, *IBM Sys. J.*, vol. 1, pp. 64–76, September, 1962.

BarnG68   Barnes, George H., Richard M. Brown, Maso Kato, David J. Kuck, Daniel L. Slotnick, and Richard A. Stokes: The ILLIAC IV Computer, *IEEE Trans.*, vol. C-17, no. 8, pp. 746–757, August, 1968.

BartR61   Barton, R. S.: A New Approach to the Functional Design of a Digital Computer, *Proc. WJCC*, pp. 393–396, 1961.

BashT64   Bashkow, T. R.: A Sequential Circuit for Algebraic Statement Translation, *IEEE Trans.*, vol. EC-13, no. 2, pp. 102–105, April, 1964.

BashT67   Bashkow, Theodore, Azra Sasson, and Arnold Kronfeld: System Design of a FORTRAN Machine, *IEEE Trans.*, vol. EC-16, no. 4, pp. 485–499, August, 1967.

Basil57   Basilewskii, Iu. Ia.: The Universal Electronic Digital Machine (URAL) for Engineering Research, *J. ACM*, vol. 4, no. 2, pp. 511–519, 1957.

BeckF61   Beckman, F. S., F. P. Brooks, Jr., and W. J. Lawless, Jr.: Developments in the Logical Organization of Computer Arithmetic and Control Units, *Proc. IRE*, vol. 49, no. 1, pp. 53–66, January, 1961.

BernA58   Bernstein, A., M. De V. Roberts, T. Arbuckle, and M. A. Belsky: A Chess Playing Program for the IBM 704, *Proc. WJCC*, pp. 157–159, 1958.

BhusA67   Bhushan, A., R. H. Stotz, and J. E. Ward: Recommendations for an Intercomputer Communications Network for M.I.T. *Memorandum MAC-M-355*, July, 1967.

BlaaG59   Blaauw, G. A.: Indexing and Control-word Techniques, *IBM J. of Res. and Dev.*, vol. 3, no. 2, pp. 288–301, July, 1959.

BlaaG64a   Blaauw, G. A., and F. P. Brooks, Jr.: The Structure of System/360, Part I—Outline of the Logical Structure, *IBM Sys. J.*, vol. 3, no. 2, pp. 119–135, 1964.

BlaaG64b   Blaauw, G. A.: Multisystem Organization, *IBM Sys. J.*, vol. 3, no. 2, pp. 181–195, 1964.

BlocE59   Bloch, Erich: The Engineering Design of the Stretch Computer, *Proc. EJCC*, pp. 48–58, 1959.

BlosR60   Blosk, R. T.: The Instruction Unit of the STRETCH Computer, *Proc. EJCC*, pp. 299–324, 1960.

BockR63   Bock, R. V.: An Interrupt Control for the B 5000 Data Processor System, *AFIPS Proc. FJCC*, vol. 24, pp. 229–241, 1963.

BolaL67   Boland, L. J., G. D. Granito, A. U. Marcotte, B. U. Messina, and J. W. Smith: The IBM System/360 Model 91: Storage System, *IBM J. of Res. and Dev.*, vol. 11, no. 1, pp. 54–68, January, 1967.

BoutE63   Boutwell, E., Jr., and E. A. Hoskinson: The Logical Organization of the PB 440 Microprogrammable Computer, *AFIPS Proc. FJCC*, vol. 24, pp. 201–213, 1963.

BowdB53   Bowden, B. V., editor: "Faster than Thought," Sir Isaac Pitman and Sons, Ltd., London, 1953.

BrigH64   Bright, H. S.: A Philco Multiprocessing System, *AFIPS Proc. FJCC*, pt. II, vol. 26, pp. 97–141, 1964.

BrooF57a   Brooks, F. P., Jr.: A Program-controlled Program Interruption System, *Proc. EJCC*, pp. 128–132, 1957.

BrooF57b   Brooks, F. P., Jr., A. L. Hopkins, Jr., P. G. Neumann, and M. V. Wright: An Experiment in Musical Composition, *IRE Trans.*, vol. EC-6, no. 3, pp. 175–182, September, 1957.

BrooF59   Brooks, F. P., Jr., G. A. Blaauw, and W. Buchholz: Processing Data in Bits and Pieces, *IRE Trans.*, vol. EC-8, no. 2, pp. 118–124, June, 1959.

BrooF60   Brooks, F. P.: The Execute Operations, A Fourth Mode of Instruction Sequencing, *Comm. ACM*, vol. 3, no. 3, pp. 168–170, March, 1960.

BrooR60   Brooker, R. A.: Some Techniques for Dealing with Two-level Storage, *Computer J.*, vol. 2, pp. 189–194, 1960.

BuchW53   Buchholz, Werner: The System Design of the IBM Type 701 Computer, *Proc. IRE*, vol. 41, no. 10, pp. 1262–1275, October, 1953.

BuchW57   Buchholz, W.: Design Objectives for the IBM STRETCH Computer, *New Computers, Rept. from the Manufacturers ACM Conf.*, pp. 99–104, 1957.

BuchW58   Buchholz, W.: The Selection of an Instruction Language, *Proc. WJCC*, pp. 128–130, 1958.

BuchW62   Buchholz, Werner, (ed.): "Planning a Computer System," McGraw-Hill Book Company, New York, 1962.

BurdE53   Burdette, E. W.: Characteristics of the Oracle, *Argonne Natl. Lab., Proc. Symp. on Large Scale Digital Computing Machines*, pp. 194–201, August, 1953.

BurkA62a   Burks, Arthur W., Herman H. Goldstine, and John von Neumann: Preliminary Discussion of the Logical Design of an Electronic Computing Instrument, Part I, *Datamation*, vol. 8, no. 9, pp. 24–31, September, 1962.

BurkA62b   Burks, Arthur W., Herman H. Goldstine, and John von Neumann: Preliminary Discussion of the Logical Design of an Electronic Computing Instrument, Part II, *Datamation*, vol. 8, no. 10, pp. 36–41, October, 1962.

BurkA63   Burks, Arthur W., Herman H. Goldstine, and John von Neumann: Preliminary Discussion of the Logical Design of an Electronic Computing Instrument (Pt. I, vol. 1), Rept. prepared for U.S. Army Ordnance Dept., 1946, in A. H. Taub (ed.), "Collected Works of John von Neumann," vol. 5, pp. 34–79, The Macmillan Company, New York, 1963.

BussB63   Bussell, B., and G. Estrin: An Evaluation of the Effectiveness of Parallel Processing, *IEEE Pacific Computer Conf.*, pp. 201–220, 1963.

CampR52   Campbell, Robert V. D.: Evolution of Automatic Computing, *Proc. ACM, Pittsburgh Conf.*, pp. 29–32, May, 1952.

CarlC63   Carlson, C. B.: The Mechanization of a Pushdown Stack, *AFIPS Proc. FJCC*, vol. 24, pp. 243–250, 1963.

CarrJ56   Carr, J. W., III, and N. R. Scott (eds.): "Notes on the Special Summer Conference on Digital Computers," Special Summer Conferences on Digital Computers, University of Michigan, Ann Arbor, Mich., 1956.

CarrJ59   Carr, John W., III: Programming and Coding, in Eugene M. Grabbe, Simon Ramo, and Dean E. Wooldridge (eds.), "Handbook of Automation, Computation, and Control," vol. 2, chap. 2, pp. 77–83, 93–98, 111–115, 115–121, John Wiley & Sons, Inc., New York, 1959.

CartW64   Carter, W. C., H. C. Montgomery, R. J. Preiss, and H. J. Reinheimer: Design of Serviceability Features for the IBM System/360, *IBM J. of Res. and Dev.*, vol. 8, no. 2, pp. 115–125, April, 1964.

CasaC62   Casale, Charles T.: Planning the CDC 3600, *AFIPS Proc. FJCC*, vol. 22, pp. 73–85, 1962.

ChasG52   Chase, George C.: History of Mechanical Computing Machinery, *Proc. ACM, Pittsburgh, Conf.*, pp. 1–28, May, 1952.

ChenT64   Chen, T. C.: The Overlap of the IBM System/360 Model 92 Central Processing Unit, *AFIPS Proc. FJCC*, Pt. II, vol. 26, pp. 73–80, 1964. Review GrimR65*c*

ChuC52   Chu, J. C.: The Oak Ridge Automatic Computer, *Proc. ACM, Toronto Conf.*, pp. 142–148, September, 1952.

ClarW57   Clark, Wesley A.: The Lincoln TX-2 Computer Development, *Proc. WJCC*, pp. 143–145, 1957.

ClayB64   Clayton, B. B., E. K. Dorff, and R. E. Fagen: An Operating System and Programming Systems for the 6600, *AFIPS Proc. FJCC*, Pt. II, vol. 26, pp. 41–57, 1964.

CochD68   Cochran, David S.: Internal Programming of the 9100A Calculator, *Hewlett-Packard J.*, vol. 20, no. 1, pp. 14–16, September, 1968.

CoddE59   Codd, E. F., E. S. Lowry, E. McDonough, and C. A. Scalzi: Multiprogramming STRETCH Feasibility Considerations, *Comm. ACM*, vol. 2, no. 11, pp. 13–17, November, 1959.

CoddE62   Codd, E. F.: Multiprogramming, "Advances in Computers," vol. 3, pp. 78–153, Academic Press, Inc., New York, 1962.

ComfW65   Comfort, W. T.: A Computing System Design for User Service, *AFIPS Proc. FJCC*, Pt. I, vol. 27, pp. 619–626, 1965.

ContC64   Conti, Carl: System Aspect: System/360 Model 92, *AFIPS Proc. FJCC*, Pt. II, vol. 26, pp. 81–95, 1964. Review GrimR65*a*.

ContC68   Conti, C. J., D. H. Gibson, and S. H. Pitkowsky: Structural Aspects of the System/360 Model 85, I. General Organization, *IBM Sys. J.*, vol. 7, no. 1, pp. 2–14, 1968.

ConwM58   Conway, Melvin E.: Proposal for an UNCOL, *Comm. ACM*, vol. 1, no. 10, pp. 5–8, October, 1958.

ConwM63   Conway, M. E.: A Multiprocessor System Design, *AFIPS Proc. FJCC*, vol. 24, pp. 139–146, 1963.

CorbF62   Corbato, Fernando J., Marjorie Merwin-Daggett, and Robert C. Daley: An Experimental Time-sharing System, *AFIPS Proc. SJCC*, vol. 21, pp. 335–344, 1962.

CorbF65   Corbato, F. J., and V. A. Vyssotsky: Introduction and Overview of the MULTICS System, *AFIPS Proc. FJCC*, Pt. I, vol. 27, pp. 185–196, 1965.

CoxJ68   Cox, Jerome R., Jr.: Economy of Scale and Specialization in Large Computing Systems, *Computer Design*, vol. 7, no. 11, pp. 77–80, November, 1968.

CrawP??   Crawford, P.: Thesis for Master's Degree, Massachusetts Institute of Technology, Cambridge, Mass.

CritA63   Critchlow, A. J.: Generalized Multiprocessing and Multiprogramming Systems, *AFIPS Proc. FJCC*, vol. 24, pp. 107–126, 1963.

DaleR65   Daley, R. C., and P. G. Neumann: A General-purpose File System for Secondary Storage, *AFIPS Proc. FJCC*, Pt. I, vol. 27, pp. 213–229, 1965.

DaleR68   Daley, Robert C., and Jack B. Dennis: Virtual Memory, Processes, and Sharing in MULTICS, *Comm. ACM*, vol. 11, no. 5, pp. 306–312, May, 1968.

DarrJ69   Darringer, John A.: The Description, Simulation, and Automatic Implementation of Digital Computer Processors, Thesis for Ph.D. degree, Carnegie-Mellon University, College of Engineering and Science, Department of Electrical Engineering, Pittsburgh, Pa., May, 1969.

DaviD67   Davies, D. W., K. A. Bartlett, R. A. Scantlebury, and P. T. Wilkinson: A Digital Communication Network for Computers Giving Rapid Response at Remote Terminals, *ACM Symp. on Operating System Principles, Gatlinburg, Tenn.*, Oct. 1–4, 1967.

DaviG60   Davis, G. M.: The English Electric KDF9 Computer System, *Computer Bull.*, pp. 119–120, December, 1960.

DennJ65   Dennis, J. B.: Segmentation and the Design of Multiprogrammed Computer Systems, *J. ACM*, vol. 12, no. 4, pp. 589–602, October, 1965.

DennJ66   Dennis, J., and E. C. Van Horn: Programming Semantics for Multiprogrammed Computations, *Comm. ACM*, vol. 9, no. 3, pp. 143–155, March, 1966.

DesmW64   Desmonde, W. H.: "Real Time Data Processing Systems," Prentice-Hall, Inc., Englewood Cliffs, N.J., 1964.

DijkE65   Dijkstra, E. W.: Solution of a Problem in Concurrent Programming Control, *Comm. ACM*, vol. 8, no. 9, p. 569, September, 1965.

DreyP58   Dreyfus, P.: System Design of the Gamma 60, *Proc. WJCC*, pp. 130–133, May, 1958.

DunwS56   Dunwell, S. W.: Design Objectives for the IBM STRETCH Computer, *Proc. EJCC*, pp. 20–22, 1956.

EcclW19   Eccles, W. H., and F. W. Jordan: A Trigger Relay, *Radio Rev.*, pp. 143–146, October, 1919.

EckeJ51   Eckert, J. Presper, Jr., James R. Weiner, H. Frazer Welsh, and Herbert F. Mitchell: The UNIVAC System, *AIEE-IRE Conf.*, pp. 6–16, December, 1951.

EckeJ59   Eckert, J. P., J. C. Chu, A. B. Tonik, and W. J. Schmitt: Design of Univac-LARC System, Part I, *Proc. EJCC*, pp. 59–65, 1959.

EdwaD60   Edwards, D. B. G., M. J. Lanigan, and T. Kilburn: Ferrite-core Memory Systems with Rapid Cycle Times, *Proc. IEE*, pt. B, vol. 107, pp. 585–598, November, 1960.

ElboR53   Elbourne, R. D., and R. P. Witt: Dynamic Circuit Techniques Used in SEAC and DYSEAC, *IRE Trans.*, vol. EC-2, no. 1, pp. 2–9, 1953.

ElliW51   Elliott, W. S.: Circuit Standardization in Series Working, High-speed Digital Computers, *Elliott J.*, vol. 1, no. 2, p. 49, September, 1951; also in *Proc. ACM*, March, 1950.

ElliW52   Elliott, W. S., H. G. Carpenter, and C. E. Owen: Development of Computer Components and Systems, *Proc. ACM, Toronto Conf.*, September, 1952.

ElliW53   Elliott, W. S., H. G. Carpenter, and A. St. Johnston: The Elliott-NRDC Computer 401, A Demonstration of Computer Engineering by Packaged Unit Construction, *Symp. ADC*, pp. 273–276, 1953.

ElliW56a   Elliott, W. S., C. E. Owen, C. H. Devonald, and B. G. Maudsley: The Design Philosophy of Pegasus, A Quantity-production Computer, *Proc. IEE*, Pt. B, vol. 103, Supp. 2, pp. 188–196, 1956.

ElliW56b   Elliott, W. S., R. C. Robbins, and D. S. Evans: Remote Position Control and Indication by Digital Means, *Proc. IEE*, Pt. B, vol. 103, Supp. 3, pp. 437–446, 1956.

EnglW62   England, W. A.: Subminiature Computer Designed for Space Environments, *Proc. Conf. on Spaceborne Computer Engineering, Anaheim, Calif.*, pp. 95–101, October, 1962.

ErnsH63   Ernst, H. A.: TCS, An Experimental Multiprogramming System for the IBM 7090, *IBM Res. Rept.* RJ248, 41 pp., Yorktown Hts., N.Y., June, 1963.

EstrG52   Estrin, G.: A Description of the Electronic Computer at the Institute for Advanced Studies, *Proc. ACM, Toronto Conf.*, pp. 95–109, September, 1952.

EstrG60   Estrin, Gerald: Organization of Computer Systems, the Fixed Plus Variable Structure Computer, *Proc. WJCC*, pp. 33–40, 1960.

EstrG63   Estrin, G., B. Bussell, R. Turn, and J. Bibb: Parallel Processing in a Restructurable Computer System, *IEEE Trans.*, vol. EC-12, no. 6, pp. 747–755, December, 1963. Article reviewed by E. G. Newman in *IEEE Trans.*, vol. EC-13, no. 5, p. 649, October, 1964.

EverR51   Everett, R. R.: The Whirlwind I Computer, *AIEE-IRE Conf.*, pp. 70–74, 1951.

EverR57   Everett, R. R., C. A. Zraket, and H. D. Benington: SAGE—A Data-processing System for Air Defense, *Proc. EJCC*, pp. 148–155, 1957.

EwinR64   Ewing, R. G., and P. M. Davies: An Associative Processor, *AFIPS Proc. FJCC*, Pt. I, vol. 26, pp. 147–158, 1964.

FaggP64   Fagg, P., J. L. Brown, J. A. Hipp, D. T. Doody, J. W. Fairclough, and J. Greene: IBM System/360 Engineering, *AFIPS Proc. FJCC*, Pt. I, vol. 26, pp. 205–231, 1964.

FairJ56   Fairclough, J. W.: A Sonic Delay-line Storage Unit for a Digital Computer, *Proc. IEE*, Pt. B, vol. 103, Supp. 3, pp. 491–496, 1956.

FalkA64   Falkoff, A. D., K. E. Iverson, and E. H. Sussenguth: A Formal Description of System/360, *IBM Sys. J.*, vol. 3, no. 3, pp. 198–261, 1964.

FikeR68   Fikes, Richard E., Hugh C. Lauer, and Albin L. Vareha, Jr.: Steps toward a General-purpose Time-sharing System Using Large Capacity Core Storage and TSS/360, *Proc. 23rd Natl. Conf. of*

ACM, Las Vegas, Nevada, pp. 7–18, August, 1968.

FlynM66   Flynn, Michael J.: Very High-speed Computing Systems, *Proc. IEEE,* vol. 54, no. 12, pp. 1901–1909, December, 1966.

FlynM67*a*   Flynn, M. J., and P. R. Low: The IBM System/360 Model 91: Some Remarks on System Development, *IBM J. of Res. and Dev.,* vol. 11, no. 1, pp. 2–7, January, 1967.

FlynM67*b*   Flynn, Michael J., and M. Donald MacLaren: Microprogramming Revisited, *Argonne Natl. Lab., Appl. Math. Div., Tech. Mem.* 134, pp. 1–17, Argonne, Ill., 1967.

ForgJ65   Forgie, James W.: A Time- and Memory-sharing Executive Program for Quick Response, On-line Applications, *AFIPS Proc. FJCC,* Pt. II, vol. 27, pp. 127–139, 1965.

ForrJ51   Forrester, J. W.: Digital Information Storage in Three Dimensions Using Magnetic Cores, *J. Appl. Phys.,* vol. 22, pp. 44–48, January, 1951.

FothJ61   Fotheringham, John: Dynamic Storage Allocation in the Atlas Computer, Including an Automatic Use of a Backing Store, *Comm. ACM,* vol. 4, no. 10, pp. 435–436, October, 1961.

FranJ57   Frankovich, J. M., and H. P. Peterson: A Functional Description of the Lincoln TX-2 Computer, *Proc. WJCC,* vol. 19, pp. 146–155, February, 1957.

FrizC53   Frizzell, Clarence E.: Engineering Description of the IBM Type 701 Computer, *Proc. IRE,* vol. 41, no. 10, pp. 1275–1287, October, 1953.

GibsC66   Gibson, C. T.: Time-sharing in the IBM System/360: Model 67, *AFIPS Proc. SJCC,* vol. 28, pp. 61–78, 1966.

GillS58   Gill, S.: Parallel Programming, *Computer J.,* vol. 1, no. 1, pp. 2–10, April, 1958.

GlasE65   Glaser, E. L., J. Couleur, and G. Oliver: System Design of a Computer for Time Sharing Applications, *AFIPS Proc. FJCC,* Pt. I, vol. 27, pp. 197–202, 1965.

GoldH63*a*   Goldstine, H. H., and John von Neumann: On the Principles of Large Scale Computing Machines, unpublished, 1946; in A. H. Taub (ed.), "Collected Works of John von Neumann," vol. 5, pp. 1–32, The Macmillan Company, New York, 1963.

GoldH63*b*   Goldstine, H. H., and John von Neumann: Planning and Coding Problems for an Electronic Computing Instrument (Pt. II, vol. 1), Rept. prepared for U.S. Army Ordnance Dept., 1947, in A. H. Taub (ed.), "Collected Works of John von Neumann," vol. 5, pp. 80–151, The Macmillan Company, New York, 1963.

GoldH63*c*   Goldstine, H. H., and John von Neumann: Planning and Coding of Problems for an Electronic Computing Instrument (Pt. II, vol. 2), Rept. prepared for U.S. Army Ordnance Dept., 1948, in A. H. Taub (ed.), "Collected Works of John von Neumann," vol. 5, pp. 152–214, The Macmillan Company, New York, 1963.

GoldH63*d*   Goldstine, H. H., and John von Neumann: Planning and Coding of Problems for an Electronic Computing Instrument (Pt. II, vol. 3), Rept. prepared for U.S. Army Ordnance Dept., 1948, in A. H. Taub (ed.), "Collected Works of John von Neumann," vol. 5, pp. 215–235, The Macmillan Company, New York, 1963.

GreeJ57   Greenstadt, J. L.: The IBM 709 Computer, *New Computers, Rept. from the Manufacturers ACM Conf.,* pp. 92–98, 1957.

GreeJ64   Greene, J. E., R. F. Dean, and B. M. Updike: Micro-programmed Implementation of the IBM System/360 Machine Organization, *IBM General Products Div., Development Lab., Engineering Publ., Dept. PTP* 792, Endicott, N.Y., April, 1964.

GreeJ66   Green, J.: Microprogramming, Emulators and Programming Languages, *Comm. ACM,* vol. 9, no. 3, pp. 230–231, March, 1966.

GreeS52   Greenwald, Sidney: SEAC Input-Output System, *AIEE-IRE-ACM Conf.,* pp. 31–36, December, 1952.

GreeS53   Greenwald, Sidney, R. C. Haueter, and S. N. Alexander: SEAC, *Proc. IRE,* vol. 41, no. 10, pp. 1300–1313, October, 1953.

GregJ63   Gregory, J., and R. McReynolds: The SOLOMON Computer, *IEEE Trans.,* vol. EC-12, no. 6, pp. 774–781, December, 1963.

GrimR65*a*   Grimsdale, R. L.: A Review of ContC64, *Computing Rev.,* vol. 6, no. 6, p. 430, November, December, 1965.

GrimR65*b*   Grimsdale, R. L.: A Review of AmdaG64c, *Computing Rev.,* vol. 6, no. 6, p. 429, November, December, 1965.

GrimR65c   Grimsdale, R. L.: A Review of ChenT64, *Computing Rev.*, vol. 6, no. 6, pp. 429–430, November, December, 1965.

GrosH53   Grosch, H. R. J.: High Speed Arithmetic: The Digital Computer as a Research Tool, *J. Optical Society of America*, vol. 4, no. 4, pp. 306–310, April, 1953.

GrueF68   Gruenberger, F. J.: The History of the JOHNNIAC, *Mem. RM-5654-PR*, prepared for United States Air Force Project Rand, The Rand Corporation, Santa Monica, Calif., October, 1968.

GrumM58   Grumette, Murray: IBM 704—Code Nundrums, *Comm. ACM*, vol. 1, no. 3, pp. 3–13, March, 1958.

HainL65   Haines, L. H.: Serial Compilation and the 1401 FORTRAN Compiler, *IBM Sys. J.*, vol. 4, no. 1, pp. 73–80, January, 1965.

HaleA62   Haley, A. C. D.: The KDF9 Computer System, *AFIPS Proc. FJCC*, vol. 22, pp. 108–120, 1962.

HambC62   Hamblin, C. L.: Translation to and from Polish Notation, *Computer J.*, vol. 5, pp. 210–213, October, 1962.

HaneF68   Haney, Frederick M.: Using a Computer to Design Computer Instruction Sets, Thesis for Ph.D. degree, Carnegie-Mellon University, College of Engineering and Science, Department of Computer Science, Pittsburgh, Pa., May, 1968.

HartD68   Hartley, D. F., B. Landy, and R. M. Needham: The Structure of a Multiprogramming Supervisor, *Computer J.*, vol. 11, no. 3, pp. 247–255, November, 1968.

HaucE68   Hauck, E. A., and B. A. Dent: Burroughs B 6500/B 7500 Stack Mechanism, *AFIPS Proc. SJCC*, vol. 32, pp. 245–251, 1968.

HaueR52   Haueter, R. C.: Auxiliary Equipment to SEAC Input-Output, *AIEE-IRE-ACM Conference*, pp. 39–44, December, 1952.

HellH61   Hellerman, H.: On the Organization of a Multiprogramming—Multiprocessing System, *IBM Res. Rept. RC-522*, 52 pp., Yorktown Hts., N.Y., September, 1961.

HellH66   Hellerman, H.: Parallel Processing of Algebraic Expressions, *IEEE Trans.*, vol. EC-15, no. 1, pp. 82–91, February, 1966.

HerwP60   Herwitz, Paul S., and James H. Pomerene: The Harvest System, *Proc. WJCC*, pp. 23–32, 1960.

HillJ66   Hillegass, John R.: Auerbach on Equipment IBM System 360—The First Two Years, *Data Processing Mag.*, vol. 8, no. 5, pp. 44–51, May, 1966.

HodgD64   Hodges, Donald: IPL-VC, A Proposal for a Computer System Having the IPL-V Instruction Set, *Argonne Natl. Lab., Appl. Math. Div., Tech. Mem. 66*, 22 pp., January, 1964.

HollJ59   Holland, John: A Universal Computer Capable of Executing an Arbitrary Number of Subprograms Simultaneously, *Proc. EJCC*, pp. 108–113, 1959.

HopkA63   Hopkins, A. L., R. L. Alonso, and H. Blair-Smith: Logical Description of the Apollo Guidance Computer (AGC 4), *M.I.T. Instrumentation Lab., Rept. R-393* (confidential), Cambridge, Mass., March, 1963.

HowaD61   Howarth, D. J., R. B. Payne, and F. H. Sumner: The Manchester University Atlas Operating System, Part II: User's Description, *Computer J.*, vol. 4, no. 3, pp. 226–229, October, 1961.

HowaD62   Howarth, D. J., P. D. Jones, and M. T. Wyld: The ATLAS Scheduling System, *Computer J.*, vol. 5, no. 3, pp. 238–244, October, 1962.

HowaD63   Howarth, D. J.: Experience with the Atlas Scheduling System, *AFIPS Proc. SJCC*, vol. 23, pp. 59–67, 1963.

HughE54   Hughes, E. S., Jr.: The IBM Magnetic Drum Calculator Type 650, Engineering and Design Considerations, *Proc. WJCC*, pp. 140–154, 1954.

IverK62   Iverson, Kenneth E.: A Common Language for Hardware, Software, and Applications, *AFIPS Proc. FJCC*, vol. 22, pp. 121–129, 1962.

JohnD52   Johnston, D. L.: Standardized Printed Circuit Units for Digital Computers, *Proc. ACM, Pittsburgh Conf.*, pp. 135–141, May, 1952.

KampT60   Kampe, Thomas W.: The Design of a General-purpose Microprogram-controlled Computer with Elementary Structure, *IRE Trans.*, vol. EC-9, no. 2, pp. 208–213, June, 1960.

KatzJ66   Katz, J. H.: Simulation of a Multiprocessor Computing System, *AFIPS Proc. SJCC*, vol. 28, pp. 127–139, 1966.

KilbT56   Kilburn, T., D. B. G. Edwards, and C. E. Thomas: The Manchester University Mark II Digital Computing Machine, *Proc. IEE*, Pt. B, vol. 103, Supp. 2, pp. 247–268, 1956.

KilbT60*a*   Kilburn, T., and R. L. Grimsdale: A Digital Computer Store with a Very Short Read Time, *Proc. IEE*, Pt. B, vol. 107, pp. 567–572, November, 1960.

KilbT60*b*   Kilburn, T., D. B. G. Edwards, and D. Aspinall: A Parallel Arithmetic Unit Using a Saturated Transistor Fast-Carry Circuit, *Proc. IEE*, Pt. B, vol. 107, pp. 573–584, November, 1960.

KilbT61*a*   Kilburn, T., D. J. Howarth, R. B. Payne, and F. H. Sumner: The Manchester University Atlas Operating System, Part I: Internal Organization, *Computer J.*, vol. 4, pp. 222–225, October, 1961.

KilbT61*b*   Kilburn, T., R. B. Payne, and D. J. Howarth: The Atlas Supervisor, *AFIPS Proc. EJCC*, vol. 20, pp. 279–294, 1961.

KilbT62   Kilburn, T., D. B. G. Edwards, M. J. Lanigan, and F. H. Sumner: One-level Storage System, *IRE Trans.*, vol. EC-11, no. 2, pp. 223–235, April, 1962.

KinsH64   Kinslow, H. A.: The Time-sharing Monitor System, *AFIPS Proc. FJCC*, Pt. I, vol. 26, pp. 443–454, 1964.

KistJ57   Kister, J., P. Stein, S. Ulam, W. Walden, and M. Wells: Experiments in Chess, *J. ACM*, vol. 4, no. 2, pp. 174–177, April, 1957.

KitoA56   Kitov, A. I.: Elektronnie Tsifrovie Mashiny (Electronic Digital Machines), Izdatelstvo Sovetskoe Radio, Moscow, partial translation available, 1956.

KleiR53   Klein, R. J., Jr.: The Oracle Memory System, *Argonne Natl. Lab., Proc. Symp. on Large Scale Digital Computing Machines*, pp. 47–58, August, 1953.

KnigK66   Knight, Kenneth E.: Changes in Computer Performance, *Datamation*, vol. 12, no. 9, pp. 40–54, September, 1966.

KnigK68   Knight, Kenneth E.: Evolving Computer Performance 1963–1967, *Datamation*, vol. 14, no. 1, pp. 31–35, January, 1968.

KnutD66   Knuth, D. E.: Additional Comments on a Problem in Concurrent Programming Control, *Comm. ACM*, vol. 9, no. 5, pp. 321–322, 1966.

KrogM61   Kroger, Marlin G., et al.: Computers in Command and Control, TR61-12, prepared for DOD:ARPA by Digital Computer Application Study, Institute for Defense Analyses, Research and Engineering Support Division, November, 1961.

KuckD68   Kuck, D. J.: ILLIAC IV Software and Application Programming, *IEEE Trans.*, vol. C-17, no. 8, pp. 758–770, August, 1968.

LampB65   Lampson, B. W.: Interactive Machine Language Programming, *AFIPS Proc. FJCC*, Pt. I, vol. 27, pp. 473–481, 1965.

LampB66   Lampson, B. W., W. W. Lichtenberger, and M. W. Pirtle: A User Machine in a Time-sharing System, *Proc. IEEE*, vol. 54, no. 12, pp. 1766–1774, December, 1966.

LangJ67   Langdon, J. L., and E. J. Van Derveer: Design of a High-speed Transistor for the ASLT Current Switch, *IBM J. of Res. and Dev.*, vol. 11, no. 1, pp. 69–73, January, 1967.

LaueH67   Lauer, Hugh C.: Bulk Core in a 360/67 Time-sharing System, *AFIPS Proc. FJCC*, vol. 31, pp. 601–609, 1967.

LebeS56   Lebedev, S. A.: The High-speed Calculating Machine of the Academy of Sciences of the USSR, *J. ACM*, vol. 3, pp. 129–133, 1956.

LehmM63*a*   Lehman, M., R. Eshed, and Z. Netter: SABRAC, A Time-sharing Low-cost Computer, *Comm. ACM*, vol. 6, no. 8, pp. 427–429, August, 1963.

LehmM63*b*   Lehman, M., R. Eshed, and Z. Netter: SABRAC —A New Generation Serial Computer, *IEEE Trans.*, vol. EC-12, no. 6, pp. 618–628, December, 1963.

LehmM65   Lehman, M.: Serial Mode Operation and High-speed Parallel Processing, *Proc. IFIP Cong. 1965*, Pt. 2, pp. 631–633, 1965.

LehmM66   Lehman, M.: A Survey of Problems and Preliminary Results Concerning Parallel Processing and Parallel Processors, *Proc. IEEE*, vol. 54, no. 12, pp. 1889–1901, December, 1966.

LeinA54   Leiner, A. L., and S. N. Alexander: System Organization of the DYSEAC, *Professional Group on Electronic Computers, Institute of Radio Engineers*, vol. EC-3, no. 1, pp. 1–10, March, 1954.

LeinA57   Leiner, A. L., W. A. Notz, J. L. Smith, and A. Weinberger: Organizing a Network of Computers to Meet Deadlines, *Proc. EJCC*, pp. 115–128, 1957.

LeinA58   Leiner, A. L., W. A. Notz, J. L. Smith, and A.

Weinberger: PILOT, The NBS Multicomputer System, *Proc. EJCC*, pp. 71–75, 1958.

LeinA59   Leiner, A. L., W. A. Notz, J. L. Smith, and A. Weinberger: PILOT, A New Multiple Computer System, *J. ACM*, vol. 6, no. 3, pp. 313–335, 1959.

LichW65   Lichtenberger, W., and M. W. Pirtle: A Facility for Experimentation in Man-Machine Interaction, *AFIPS Proc. FJCC*, Pt. I, vol. 27, pp. 589–598, 1965.

LindA66   Lindquist, A. B., R. R. Seeber, and L. W. Comeau: A Time-sharing System Using an Associative Memory, *Proc. IEEE*, vol. 54, no. 12, pp. 1774–1779, December, 1966.

LiptJ68   Liptay, J. S.: Structural Aspects of the System/360 Model 85, II. The Cache, *IBM Sys. J.*, vol. 7, no. 1, pp. 15–21, 1968.

LloyR67   Lloyd, R. H. F.: ASLT: An Extension of Hybrid Miniaturization Techniques, *IBM J. of Res. and Dev.*, vol. 11, no. 1, pp. 86–92, January, 1967.

LoneW61   Lonergan, William, and Paul King: Design of the B 5000 System, *Datamation*, vol. 7, no. 5, pp. 28–32, May, 1961.

LonsK56   Lonsdale, K., and E. T. Warburton: Mercury: A High Speed Digital Computer, *Proc. IEE*, Pt. B, vol. 103, Supp. 2, pp. 174–183, 1956.

LourN59   Lourie, N., H. Schrimpf, R. Reach, and W. Kahn: Arithmetic and Control Techniques in a Multiprogram Computer, *Proc. EJCC*, pp. 75–81, 1959.

McCaJ62   McCarthy, J.: "Time Sharing Computer Systems in Management and the Computer of the Future," The M.I.T. Press, Cambridge, Mass., 1962.

McCaJ63   McCarthy, J., S. Boilen, E. Fredkin, and J. C. R. Licklider: A Time-sharing Debugging System for a Small Computer, *AFIPS Proc. SJCC*, vol. 23, pp. 51–57, 1963.

McCoB63   McCormick, Bruce H.: The Illinois Pattern Recognition Computer—ILLIAC III, *IEEE Trans.*, vol. EC-12, no. 5, pp. 791–813, December, 1963.

McCuJ65   McCullough, J. D., K. H. Speierman, and F. W. Zurcher: Design for a Multiple User Multiprocessing System, *AFIPS Proc. FJCC*, Pt. I, vol. 27, pp. 611–617, 1965.

McPhJ51   McPherson, J. L., and S. N. Alexander: Performance of the Census Univac System, *AIEE-IRE Conf.*, pp. 16–22, December, 1951.

MaheR61   Maher, R. J.: Problems of Storage Allocation in a Multiprocessor Multiprogrammed System, *Comm. ACM*, vol. 4, no. 10, pp. 421–422, October, 1961.

MarcM63   Marcotty, M. J., F. M. Longstaff, and A. P. M. Williams: Time-sharing on the Ferranti-Packard FP6000 Computer System, *AFIPS Proc. SJCC*, vol. 23, pp. 29–40, 1963.

MeadR63   Meade, R. M.: 604 Machine Description, *IBM Internal Mem.*, 38 pp., December, 1963.

MeagR51   Meagher, R. E., and J. P. Nash: The Ordvac, *AIEE-IRE Conf.*, pp. 37–43, December, 1951.

MelbA65   Melbourne, A. J., and J. M. Pugmire: A Small Computer for the Direct Processing of FORTRAN Statements, *Computer J.*, vol. 8, pp. 24–27, April, 1965.

MendM66   Mendelson, M. J., and A. W. England: The SDS SIGMA 7: A Real-time, Time-sharing Computer, *AFIPS Proc. FJCC*, vol. 29, pp. 51–64, 1966.

MercR57   Mercer, Robert J.: Micro-programming, *J. ACM*, vol. 4, no. 2, pp. 157–171, 1957.

Merrl56   Merry, I. W., and B. G. Maudsley: The Magnetic-drum Store of the Computer Pegasus, *Proc. IEE*, Pt. B, vol. 103, Supp. 2, pp. 197–202, 1956.

MetrN52   Metropolis, N., E. F. Klein, W. Orvedahl, J. R. Richardson, H. B. Demuth, and J. B. Jackson: MANIAC, *Proc. ACM, Toronto Conf.*, pp. 13–17, September, 1952.

MillW63   Miller, W. F., and R. A. Aschenbrenner. The GUS Multicomputer System, *IEEE Trans.*, vol. EC-12, no. 6, pp. 671–676, December, 1963.

MiraW67   Miranker, W. L., and W. M. Liniger: Parallel Methods for the Numerical Integration of Ordinary Differential Equations, *Math. of Computation*, vol. 21, no. 99, pp. 303–320, July, 1967.

MolnC67   Molnar, Charles E., Severo M. Ornstein, and Antharvedi Anné: The CHASM: A Macromodular Computer for Analyzing Neuron Models, *AFIPS Proc. SJCC*, vol. 30, pp. 393–401, 1967.

MonnR68   Monnier, Richard E.: A New Electronic Calculator with Computerlike Capabilities, *Hewlett-Packard J.*, vol. 20, no. 1, pp. 3–9, September, 1968.

MorrD67   Morris, Derrick, Frank H. Sumner, and Michael T. Wyld: An Appraisal of the Atlas Supervisor, *Proc. ACM Natl. Meeting*, pp. 67–75, 1967.

MuntC62   Muntz, C. A.: A List Processing Interpreter for AGC4, *M.I.T., Instrumentation Lab., AGC Mem.* 2, Cambridge, Mass., January, 1962.

MurtJ66   Murtha, J. C.: Highly Parallel Information Processing Systems, in "Advances in Computers," vol. 7, pp. 2–116, Academic Press, Inc., New York, 1966.

MyerT68   Myer, T. H., and I. E. Sutherland: On the Design of Display Processors, *Comm. ACM*, vol. 11, no. 6, pp. 410–414, June, 1968.

NeweA56   Newell, A., and H. A. Simon: The Logic Theory Machine, *IRE Trans.*, vol. IT-2, no. 3, pp. 61–79, September, 1956.

NeweA57a   Newell, A., and J. C. Shaw: Programming the Logic Theory Machine, *Proc. WJCC*, pp. 230–240, February, 1957.

NeweA57b   Newell, A., J. C. Shaw, and H. A. Simon: Empirical Explorations of the Logic Theory Machine, *Proc. WJCC*, pp. 218–230, February, 1957.

NeweA58   Newell, A., J. C. Shaw, and H. A. Simon: The Elements of a Theory of Human Problem Solving, *Psychology Rev.*, vol. 65, pp. 151–166, March, 1958.

NievJ64   Nievergelt, J.: Parallel Methods for Integrating Ordinary Differential Equations, *Comm. ACM*, vol. 7, no. 12, pp. 731–733, December, 1964.

NiseN66   Nisenoff, N.: Hardware for Information Processing Systems: Today and in the Future, *Proc. IEEE*, vol. 54, no. 12, pp. 1820–1835, December, 1966.

OsboT68   Osborne, Thomas E.: Hardware Design of the Model 9100A Calculator, *Hewlett-Packard J.*, vol. 20, no. 1, pp. 10–13, September, 1968.

OssaJ65   Ossanna, J. F., L. E. Mikus, and S. D. Dunten: Communications and Input-Output Switching in a Multiplex Computing System, *AFIPS Proc. FJCC*, Pt. I, vol. 27, pp. 231–241, 1965.

PadeA64   Padegs, A.: Channel Design Considerations, *IBM Sys. J.*, vol. 3, no. 2, pp. 165–180, 1964.

PadeA68   Padegs, A.: Structural Aspects of the System/360 Model 85, III. Extensions to Floating-point Architecture, *IBM Sys. J.*, vol. 7, no. 1, pp. 22–29, 1968.

PapiW57   Papian, W. N.: High-speed Computer Stores 2.5 Megabits, *Electronics*, vol. 30, no. 10, pp. 162–167, October, 1957.

PatzW67   Patzer, William J., and Gilbert C. Vandling: Systems Implications of Microprogramming, *Computer Design*, vol. 6, no. 12, pp. 62–66, December, 1967.

PeacA??   Peacock, A.: Read-only Memory and Computer Control, to be published.[1]

PennJ62   Penny, J. P., and T. Pearcey: Use of Multiprogramming in the Design of a Low Cost Digital Computer, *Comm. ACM*, vol. 5, no. 9, pp. 473–476, September, 1962.

PikeJ52   Pike, James L.: Input-Output Devices Used with SEAC, *AIEE-IRE-ACM Conf.*, pp. 36–38, December, 1952.

PlugW61   Plugge, W. R., and M. N. Perry: American Airlines' "SABRE" Electronic Reservations System, *Proc. WJCC*, pp. 593–602, May, 1961.

PortR60   Porter, R. E.: The RW-400—A New Polymorphic Data System, *Datamation*, vol. 6, no. 1, pp. 8–14, January/February, 1960.

RajcJ43   Rajchman, J., Snyder, and Rudnick: RCA Laboratories Report, under terms of OSRD contract OEM-sr-591.

RandB68   Randell, B., and C. J. Kuehner: Dynamic Storage Allocation Systems, *Comm. ACM*, vol. 11, no. 5, pp. 297–306, May, 1968.

RichR55   Richards, R. K.: "Arithmetic Operations in Digital Computers" D. Van Nostrand Company, Inc., Princeton, N.J., 1955.

RobeJ58   Robertson, J. E.: A New Class of Digital Division Methods, *IRE Trans.*, vol. EC-7, no. 3, pp. 218–222, September, 1958.

RobeL67   Roberts, Lawrence G.: Multiple Computer Networks and Intercomputer Communication, *ACM Symp. on Operating System Principles, Gatlinburg, Tenn.*, Oct. 1–4, 1967.

RoseG67   Rose, Gordon A.: "Intergraphic," A Microprogrammed Graphical-Interface Computer, *IEEE Trans.*, vol. EC-16, no. 6, pp. 773–784, December, 1967.

---

[1] According to E. F. Codd, this article has not been published as of Jan. 23, 1968. However, "Microprogram Control for System/360" by S. G. Tucker, *IBM Sys. J.*, vol. 6, no. 4, 1967, has and covers the material that we think was intended to be in PeacA??.

RoseJ65   Rosenfeld, J.: Marbles and Boxes, *IBM Res. Project Rept.*, Yorktown Hts., N.Y., November, 1965.

RoseS67   Rosen, Saul: "Programming Systems and Languages," McGraw-Hill Book Company, New York, 1967.

RoseS69   Rosen, Saul: Electronic Computers: A Historical Survey, *Computing Surveys*, vol. 1, no. 1, pp. 7–36, March, 1969.

RosiR69   Rosin, Robert F.: Contemporary Concepts of Microprogramming and Emulation, *Computing Surveys*, vol. 1, no. 4, pp. 197–212, December, 1969.

RossH53   Ross, Harold D., Jr.: The Arithmetic Element of the IBM Type 701 Computer, *Proc. IRE*, vol. 41, no. 10, pp. 1287–1294, October, 1953.

RothS59   Rothman, S.: R/W 40 Data Processing System, *Intern. Conf. on Information Processing and Auto-math 1959*, Ramo-Wooldridge, Div. of Thompson Ramo Wooldridge, Inc., Los Angeles, Calif., June, 1959.

SaltJ66   Saltzer, J. H.: Traffic Control in a Multiplexed Computer System, *M.I.T. Tech. Rept.* MAC-TR-30, July, 1966.

SamuA57   Samuel, Arthur L.: Computers with European Accents, *Proc. WJCC*, pp. 14–17, 1957.

SaxoJ63   Saxon, J. A.: "Programming the IBM 7090," Prentice-Hall, Inc., Englewood Cliffs, N.J., 1963.

SchlH??   Schlaeppi, H. P.: Extensions of PL/I-like Languages for Parallel Processing, with Programming Examples, in preparation.

SchwJ64   Schwartz, J. I.: A General-purpose Time-sharing System, *AFIPS Proc. SJCC*, vol. 25, pp. 397–411, 1964.

SechR67   Sechler, R. F., A. R. Strube, and J. R. Turnbull: ASLT Circuit Design, *IBM J. of Res. and Dev.*, vol. 11, no. 1, pp. 74–85, January, 1967.

SeebR63   Seeber, R. R., and A. B. Lindquist: Associative Logic for Highly Parallel Systems, *AFIPS Proc. FJCC*, vol. 24, pp. 489–493, 1963.

SegaR61   Segal, R. J., and H. P. Guerber: Four Advanced Computers—Key to Air Force Digital Data Communication System, *AFIPS Proc. EJCC*, vol. 20, pp. 264–278, 1961.

SenzD65   Senzig, D. N., and R. V. Smith: Computer Organization for Array Processing, *AFIPS Proc. FJCC*, Pt. I, vol. 27, pp. 117–128, 1965.

SerrR62   Serrell, R., M. M. Astrahan, G. W. Patterson, and I. B. Pyne: The Evolution of Computing Machines and Systems, *Proc. IRE*, vol. 50, no. 5, pp. 1039–1058, May, 1962.

ShanC38   Shannon, E. C.: A Symbolic Analysis of Relay and Switching Circuits, *Trans. AIEE*, vol. 57, pp. 713–723, 1938.

SharW69   Sharpe, William F.: "The Economics of Computers," Columbia University Press, New York, 1969.

ShawJ58   Shaw, J. C., A. Newell, H. A. Simon, and T. O. Ellis: A Command Structure for Complex Information Processing, *Proc. WJCC*, pp. 119–128, 1958.

ShedG66a   Shedler, G. S., and M. Lehman: Parallel Computation and the Solution of Polynomial Equations, *IBM Res. Rept.* 1550, Yorktown Hts., N.Y., February, 1966.

ShedG66b   Shedler, G. S.: Parallel Numerical Methods for the Solution of Equations, *IBM Res. Rept. RC* 1619, Yorktown Hts., N.Y., June, 1966.

ShupP53   Shupe, P. D., and R. A. Kirsch: SEAC, Review of Three Years of Operation, *Proc. EJCC*, pp. 83–90, 1953.

SlotD62   Slotnick, Daniel L., W. Carl Borck, and Robert C. McReynolds: The SOLOMON Computer, *AFIPS Proc. FJCC*, vol. 22, pp. 97–107, 1962.

SlutR51   Slutz, Ralph J.: Engineering Experience with the SEAC, *AIEE-IRE Conf.*, pp. 90–94, December, 1951.

SmitR64   Smith, R. V., and D. N. Senzig: Computer Organization for Array Processing, *IBM Res. Rept. RC* 1330, Yorktown Hts., N.Y., December, 1964.

SoloM66   Solomon, Martin B., Jr.: Economies of Scale and the IBM System/360, *Comm. ACM*, vol. 9, no. 6, pp. 435–440, June, 1966.

SquiJ63   Squire, J. S., and S. M. Polais: Programming and Design Considerations of a Highly Parallel Computer, *AFIPS Proc. SJCC*, vol. 23, pp. 395–400, 1963.

SteeT61   Steel, T. B., Jr.: A First Version of UNCOL, *Proc. WJCC*, pp. 371–377, 1961.

StevL52   Stevens, L. D.: Engineering Organization of Input and Output for the IBM 701 Electronic

Data-processing Machine, *AIEE-IRE-ACM Conf.*, pp. 81–85, December, 1952.

StevW64   Stevens, W. Y.: The Structure of System/360, Part II—System Implementations, *IBM Sys. J.*, vol. 3, no. 2, pp. 136–143, 1964.

StraC59   Strachey, C.: Time Sharing in Large Fast Computers, *Proc. ICIP, UNESCO*, pp. 336–341, June, 1959.

SumnF62   Sumner, F. H., G. Haley, and E. C. Y. Chen: The Central Control Unit of the "Atlas" Computer, *Proc. IFIP Cong. 1962*, pp. 657–662, 1962.

TaylN51   Taylor, Norman H.: Evaluation of the Engineering Aspects of Whirlwind I, *AIEE-IRE Conf.*, pp. 75–78, December, 1951.

TeagH65   Teager, Herbert M.: A Review of AmdaG64a; *Computing Rev.*, vol. 6, no. 5, pp. 355–356, September-October, 1965.

ThomR63   Thompson, R. N., and J. A. Wilkinson: The D825 Automatic Operating and Scheduling Program, *AFIPS Proc. SJCC*, vol. 23, pp. 41–49, 1963.

ThorJ64   Thornton, James E.: Parallel Operation in the Control Data 6600, *AFIPS Proc. FJCC*, Pt. II, vol. 26, pp. 33–40, 1964.

TomaR67   Tomasulo, R. M.: An Efficient Algorithm for Exploiting Multiple Arithmetic Units, *IBM J. of Res. and Dev.*, vol. 11, no. 1, pp. 25–33, January, 1967.

TuckS67   Tucker, S. G.: Microprogram Control for System/360, *IBM Sys. J.*, vol. 6, no. 4, pp. 222–241, 1967.

TuriS59   Turing, Sara: "Alan M. Turing," W. Heffer and Sons, Ltd., Cambridge, England, 1959.

UngeS58   Unger, S. H.: A Computer Oriented toward Spatial Problems, *Proc. IRE*, vol. 46, no. 10, pp. 1744–1750, October, 1958.

VandW52   Van der Poel, W. L.: A Simple Electronic Digital Computer, *Appl. Sci. Res.*, *Sec. B*, vol. 2, pp. 367–400, 1952.

VandW56   Van der Poel, W. L.: The Logical Principles of Some Simple Computers, Thesis, Amsterdam, 1956.

VandW59   Van der Poel, W. L.: ZEBRA, A Simple Binary Computer, *Proc. ICIP, UNESCO*, pp. 361–365, June, 1959.

VyssV65   Vyssotsky, V. A., F. J. Corbato, and R. M. Graham: Structure of the Multics Supervisor, *AFIPS Proc. FJCC*, Pt. I, vol. 27, pp. 203–212, 1965.

WaleE62   Walendziewicz, E. T.: The D210 Magnetic Computer, *Proc. Conf. on Spaceborne Computer Engineering, Anaheim, Calif.*, pp. 117–127, Oct. 30–31, 1962.

WareW63a   Ware, W. H.: "Digital Computer Technology and Design," vol. 1, "Mathematical Topics, Principles of Operation, and Programming," John Wiley & Sons, Inc., New York, 1963.

WareW63b   Ware, W. H.: "Digital Computer Technology and Design," vol. 2, "Circuits and Machine Design," John Wiley & Sons, Inc., New York, 1963.

WebeH67   Weber, Helmut: A Microprogrammed Implementation of EULER on IBM System/360 Model 30, *Comm. ACM*, vol. 10, no. 9, pp. 549–558, September, 1967.

WeikM55   Weik, M. H.: A Survey of Domestic Electronic Digital Computing Systems, *Ballistic Research Laboratories*, Aberdeen, Md., *Rept.* 971, December, 1955.

WeikM61   Weik, Martin H.: A Third Survey of Domestic Electronic Digital Computing Systems, *Ballistic Research Laboratories*, Aberdeen, Md.; report supersedes *BRL Rept.* 1010, Department of the Army Project No. 5B03-06-002 (1961).

WeikM64   Weik, Martin H., Jr.: A Fourth Survey of Domestic Electronic Digital Computer Systems, *Ballistic Research Laboratories*, Aberdeen, Md., *Rept.* 1227; processed by Defense Documentation Agency, Defense Supply Agency No. 42900, January, 1964.

WestG60   West, George P., and Ralph J. Koerner: Communications within a Polymorphic Intellectronic System, *Proc. WJCC*, pp. 225–230, 1960.

WilkJ53   Wilkinson, J. H.: "The Pilot ACE," pp. 5–14, Automatic Digital Computation, National Physical Laboratory, Teddington, England, March 25–28, 1953.

WilkM51*a*   Wilkes, M. V.: The Best Way to Design An Automatic Calculating Machine, *Manchester University Computer Inaugural Conf.*, July, 1951. Published by Ferranti Ltd., London.

WilkM51*b*   Wilkes, M. V.: The Edsac Computer, *AIEE-IRE Conf.*, pp. 79–83, December, 1951.

WilkM52   Wilkes, M. V., D. J. Wheeler, and S. Gill: "The Preparation of Programs for a Digital Computer," Addison-Wesley Publishing Company, Inc., Reading, Mass., 1952.

WilkM53   Wilkes, M. V., and J. B. Stringer: Microprogramming and the Design of the Control Circuits in an Electronic Digital Computer, *Proc. Cambridge Phil. Soc.*, Pt. 2, vol. 49, pp. 230–238, April, 1953.

WilkM58*a*   Wilkes, M. V., W. Renwick, and D. J. Wheeler: The Design of the Control Unit of an Electronic Digital Computer, *Proc. IEE*, Pt. B, vol. 105, pp. 121–128, March, 1958.

WilkM58*b*   Wilkes, M. V.: Microprogramming, *Proc. EJCC*, pp. 18–20, 1958.

WilkM65   Wilkes, M. V.: Slave Memories and Dynamic Storage Allocation, *IEEE Trans.*, vol. EC-14, no. 2, pp. 270–271, 1965.

WilkM69   Wilkes, M. V.: The Growth of Interest in Microprogramming: A Literature Survey, *Computing Surveys*, vol. 1, no. 3, pp. 139–145, September, 1969.

WillC53   Williams, Charles R.: A Review of ORDVAC Operating Experience, *Proc. EJCC*, pp. 91–95, 1953.

WillF49   Williams, F. C., and T. Kilburn: A Storage System for Use with Binary-Digital Computing Machines, *Proc. IEE*, Pt. 3, vol. 96, pp. 81–100, March, 1949. Same paper in Pt. 2, vol. 96, pp. 183–202, April, 1949.

WirsJ66   Wirsching, Joseph E.: NOVA: A List-oriented Computer, *Datamation*, vol. 12, no. 12, pp. 41–43, December, 1966.

WirtN66*a*   Wirth, N., and H. Weber: EULER: A Generalization of ALGOL, and Its Formal Definition: Part I, *Comm. ACM*, vol. 9, no. 1, pp. 13–25, January, 1966.

WirtN66*b*   Wirth, N., and H. Weber: EULER: A Generalization of ALGOL, and Its Formal Definition: Part II, *Comm. ACM*, vol. 9, no. 2, pp. 89–99, February, 1966.

WirtN66*c*   Wirth, N.: A Note on "Program Structures" for Parallel Processing, *Comm. ACM*, vol. 9, no. 5, pp. 320–321, May, 1966.

ZadeL63   Zadeh, Lotfi A., and Charles A. Desoer: "Linear System Theory," McGraw-Hill Book Company, New York, 1963.

Machine and Organization Index

Page references in **boldface** refer to the Appendix, ISP descriptions, and PMS diagrams.

Page references in **boldface** refer to the Appendix, ISP descriptions, and PMS diagrams.